MATERIALS SCIENCE AND ENGINEERING

MATERIALS SCIENCE AND ENGINEERING

G. S. Upadhyaya
Former Professor

Anish Upadhyaya
Associate Professor

Department of Materials and Metallurgical Engineering
Indian Institute of Technology, Kanpur

Foreword
Professor Guenter Petzow
Max Planck Institute for Metals Research,
Stuttgart, Germany

Anshan

First published in the UK by

Anshan Limited
In 2007

6 Newlands Road,
Tunbridge Wells,
Kent.
TN4 9AT. UK

Tel: +44 (0) 1892 557767
Fax: +44 (0) 1892 530358

E-mail: info@anshan.co.uk
Web Site: www.anshan.co.uk

ISBN-10: 1 905740 51 4
ISBN-13: 978 1 905740 51 2

British Library Cataloguing in Publishing Data
A catalogue record for this book is available from the British Library

Printed and bound in India by Raj Press, R-3 Inderpuri, New Delhi 110 012.

Dedicated to all the students
who inspired us to write this book

CONTENTS

PREFACE

The aim of this book is to cover the engineering section of Materials Science and Engineering spectrum. The book describes structure-properties-processing-performance relationships of various classes of engineering materials: metallic, ceramic, polymeric and composites based on them. Unlike many authors, who have divided their textbooks into four distinct parts metals, ceramics, polymers and composites, we have adopted a different style. Each chapter discusses about all these materials, so that students are reminded of bonding and structure and its role on properties, processing and performance of materials at each stage of their learning. Compartmentalisation of materials acts as a hindrance in getting a holistic view of all the solid materials. Pedagogically our present approach is sound.

Chapter 1 introduces various classes of materials highlighting their distinctive features. Chapter 2 *Bonding and Structure of Materials* describes various types of bondings in solid materials followed by structures. Defect structures including micro and macrostructures are also highlighted. Chapter 3 *Phase Equilibria and Transformation* covers various types of binary phase diagrams with a brief introduction of ternary diagrams. It also highlights diffusion in solids. Phase transformation in general and of steel in particular has been included. Precipitation, recrystallization and phase transitions in glasses are also included. Chapter 4 *Physical Properties of Materials* covers properties like electrical, magnetic, thermal and optical. A brief introduction of electron energy levels has been given, so that the students may appreciate their role on various physical properties. Chapter 5 descibes *Mechanical Properties of Materials*. Role of dislocation in plasticity has been merely touched upon. We strongly feel that an elaborate treatment is uncalled for materials engineers. After covering the behaviour of solids under different types of stresses, the chapter includes topics like hardness, impact toughness, fatigue, creep and fracture toughness. Strengthening and toughening of metallic and ceramic materials have been included and here the role of defects and microstructure on the overall mechanical properties has been touched upon. Chapter 6 *Electrochemical Properties, Corrosion and Degradation* pertains to an equally important class of properties. It is logical that we place this chapter right in the vicinity of chapters related to properties. The chapter also includes high temperature oxidation and sulphidation of metals and alloys.

Chapter 7 *Engineering Materials* is a summarization of various types of materials. The treatment is rather descriptive, as the requisite 'why' of the properties is already covered in earlier chapters. A liberal inclusion of Tables makes this chapter useful as a reference. After covering the general aspects of engineering materials, the book ventures on *Materials Forming* (in Chapter 8), where solidification, deformation, powder metallurgy formings, and solid freeform fabrication are covered. Attempt has been made to treat these aspects quantitatively. The details of equipments used for forming are purposely excluded in order to limit the size of the book. Moreover the design of equipments is not a domain of materials engineering. Chapter 9 is related to *Materials Removal*. Emphasis has been more on structure-property relations and not on solid mechanics, which is in the realm of mechanical engineering. Friction and wear aspects of materials are also included. Next chapter (Chapter 10) is on *Materials Joining*, which covers welding processes, weldability and weld design apart from brazing, soldering and adhesive bonding. Modern developments in joining of ceramic-metal and ceramic-ceramic are also included. *Surface Modifications* of materials (Chapter 11) is an essential part of materials engineering, wherin too varied types of methods are used. The nomenclature surface modification is more wide and appropriate than mere 'coatings'. Chapter 12 is the most important and distinctive part of the book and covers *Materials Design*. This includes materials and process selections and describes through case studies (44 in number) how properties and their manipulation play a crucial role in materials design. The student should seek the synergy and see how far a particular material design fits under the existing economic constraint.

The book can suitably serve more than one modular course: Chapters 1 to 6 for Introductory Materials Science, Chapter 7 for Engineering Materials, Chapters 8 to 11 for Materials Processing and the final Chapter 12 for Materials Design. Practically all chapters contain solved examples. In addition at the end of each chapter a wide variety of questions and problems are included. Some of them are simple, while others do require more depth of thinking on the part of the students. Instructors of different institutes may chalk out the materials from the book as per the requirment of their syllabus and the course duration. We do not prefer to dictate in this aspect as different institutes have different rigours in their courses-some are more quantitative, while others are more narrative. Our aim is, however, clearly evident: the book will not disappoint any student or instructor of materials science and engineering. Further, the book is also suited as an introduction to design engineers in manufacturing industries engaged in selection of engineering materials.

As the book concerns with 'Engineering Materials', other classes of materials like biomaterials and sport materials are not included. The reason is not that the latter types of materials are not important, but a too wide coverage often makes the impact of a textbook less focussed. We were conscious of this fact.

We would like to express our gratitude to our students today and those of the past, who posed numerous challenging problems, which needed to be answered to their satisfaction. We have tried to include similar problems and their solutions in this book for the benefit of the students. In spite of our best efforts to keep away any mistakes from the book, it is possible that some might have crept in. We shall be obliged if the readers point out those to us for corrections in the next edition. We are extremely grateful to Professor G. Petzow who kindly accepted our request to write the foreword for this book.

Authors

FOREWORD

Materials are the key to realization of all technological concepts. Therefore, they play a very important, often even a decisive role. As a consequence engineers are frequently faced with the problem to select for a distinct application the most suitable material out of thousands of others with great diversity in properties, availability and costs. To overcome such a challenging task sound information on the internal structure of materials is extremely valuable and helpful.

The internal structure of materials is set by the atomic species to more or less tight, stable, and regular three-dimensional arrangements including crystal structure and configurations in the nano-micro- and macro-range, all of which contribute to the broad spectrum of materials properties. In other words: the atomic architecture of these structures determines the gross properties of each kind of materials. Such view has united the scientific knowledge and the empirical rules surrounding the traditionally distinct classes of materials (metals, ceramics, polymers) into a single science and technology. The basics of this modern understanding of materials have been established throughout the last half century by intensive research work mainly in the young scientific discipline of materials science. The practical outcome of materials science can be seen as a "road map" for materials selection for engineers involved in processing, production, treatment, application and recycling of materials. And exactly, these are the ideas and aim behind the textbook of G. S. and Anish Upadhyaya "MATERIALS SCIENCE AND ENGINEERING".

This is a very special book mainly for three reasons. First, it is bridging materials science with substantial engineering affairs by using basic results of materials science for a pragmatic solution of daily as well as "exotic" problems in engineering. Second, it avoids sophisticated explanations of scientific details relating to atomic or molecular design in full accordance with the fact that an engineer necessarily takes a greater interest in the gross properties than in atomistic kinetics of the material he chooses to meet the requirements of his problems in forming, joining, corrosion, surfacing, application, design and/or others. Third, with refreshing pedagogical skilfulness it clearly demonstrates by numerous examples and case studies how the optimum adjustment of the structure fits a specific property profile.

This book is mainly written for engineering students who seek an easy understandable and enjoyable introduction to the properties of materials and the structure relations which govern them. I am sure, the students will find the rather new way outlined in this handy textbook, appealing and helpful in their further professional life.

Guenter Petzow
Max Planck Institute for Metals Research
Stuttgart, Germany

1

INTRODUCTION

The field of *Materials Science and Engineering* deals with the synthesis and use of knowledge about structure, properties, processing and performance in order to develop, prepare, modify and apply materials to specific needs. Materials Engineering can be classed as one of the subdivisions of Manufacturing Engineering, which is more broad based interdisciplinary in approach. The word 'production' is often used interchangeably with the word 'manufacturing'. Materials Engineering has a more quantitative approach as compared to Materials Technology. In many books, design is dealt with separately from engineering. However, the present authors differ from this approach. For example, a mechanical engineer must be conversant with mechanical design, a structural engineer with structural design, and a chemical engineer with chemical processes design, so on and so forth. With this in mind, a materials engineer is ought to know about materials design.

The selection of materials and the most appropriate manufacturing process is dependent on a large number of factors, but the most important considerations are shape complexity and materials properties. The materials properties ultimately are linked with the microstructure and processing. For example, the mechanical properties of cast steel are different from that of the worked steels, single crystals are different than polycrystalline bulk materials, amorphous polymers are different from semicrystalline polymers etc. The later chapters in this book shall cover these aspects in detail.

Improper selection of materials and process leads to component failures, when it either stops functioning, or does not function properly, or becomes unreliable or unsafe for further use.

Materials Engineering covers materials aspects of both discrete products and continuous products. The examples of the former are engineering components like piston, gear, bearings, screws etc., while for the latter are rolled sheet coils of metals or plastics, spool of wire, lengths of tubing, hose and pipes, from which the discrete parts may be manufactured. A materials engineer has to be conversant with the 'Design for Manufacture' (DFM), which systematically maximizes the use of manufacturing processes. However, he is not expected to be expert in the 'Design for Assembly' (DFA), an area which strictly falls in the domain of manufacturing engineering. However, the current trends for efficient product design is to combine both these design aspects (DFMA). But here too preponderance of DFM is more.

1.1 ENGINEERING MATERIALS CLASSIFICATION

Engineering materials can be broadly classified into following categories:

- Metals
- Ceramics
- Polymers
- Composites

1.1.1 Metals

Metals are characterized by their properties, such as ductility, toughness, malleability, electrical and thermal conductivity, and thermal expansion. Metals compose about three fourth of the elements but few find service in their pure form. The desired properties sought in engineering are often found in alloys. Metallic materials are again subclassified as ferrous, nonferrous and intermetallics. They can be used either in bulk or in powder form. Metallic materials are always crystalline in nature, but amorphous (noncrystalline) alloys have been developed by very rapid cooling of a melt or by very high-energy mechanical milling.

Another material obtained through rapid solidification is *quasicrystals*, which are neither crystalline nor amorphous, but form an ordered structure somewhere between the two recognized structures. Such materials are expected to exhibit far reaching electrical properties.

1.1.2 Ceramics

Ceramics are crystalline compounds containing metallic and nonmetallic elements. Pure nonmetallic element e.g., graphite is also categorized in ceramics. Glass is grouped with ceramics because it has similar properties but most glasses are amorphous. Porcelain, abrasives, refractories and structural clay are grouped in wider definition of ceramics. Ceramics, including glass, are hard, brittle, stiff and have high melting points. Ceramics are widely categorized into oxide and nonoxide ceramics.

1.1.3 Polymers

Polymers are better known to the general public as plastics, but it is misnomer to term all polymers as plastics. Polymers are generally classified into three categories: thermoplastic polymers, thermosetting polymers and elastomers, better known as rubbers. Polymers are noted for their low density and their use as insulators, both thermal and electrical. They are poor reflectors of light, tending to be transparent or translucent often in thin sections. Some of them are flexible and subject to deformation. Polymers contain nonmetallic elements sharing electrons to build up large molecules, called macromolecules. These large molecules contain many repeating units, or mers, from which we get the term polymers. By using the term polymer, one often means organic polymers, although there are inorganic polymers too. Organic polymers can be further subdivided into natural polymers e.g., leather, shellac, silk, wood; elastomer-rubber e.g., butyl and into synthetic polymers e.g., synthetic rubber (e.g., styrene butadiene, aramide fiber, varnish, plastics (e.g., acrylics) and adhesives (e.g., cynoacrylates). Basic ingredients of polymers such as ethylene and naphtha are but two of the many products obtained in the cracking of crude oil.

Polymerization is the linking together of smaller units (monomers) into long chains. The repeating units of some polymer chains are identical, as in polyethylene (Figure 1.1a), Polystyrene, and polyvinyl chloride are termed *homopolars* (Figure 1.1b). Copolymers contain two different types of monomers such as polyvinyl chloride mixed with vinyl acetate to produce polyvinyl acetate (Figure 1.1c) and terpolymers such as ABS (Acrylonitrile – Butadiene – Styrene) contain three types of monomers. *Isomers* (Figure 1.1d) are variations in the molecular structure of the same composition. Isomers are found not only in hydrocarbons, but also in polymeric molecules. One class of isomers is known as *sterioisomers*, which in turn, is divided into three classes. The first category is called *isotactic* sterioisomerism. Using a single carbon chain of atoms with some hydrogen atoms being replaced by another atom or group of atoms, denoted by the symbol R, one can demonstrate that the R group are situated in the same side of the carbon chain. In *syndiotactic* stereoisomeric arrangement, the R group of atoms are located on opposite sides of the carbon chain in an alternative fashion. The third category is called *atactic* stereoisomerism, in which the R groups are in a random position. These three categories are graphically shown in Figure 1.1(e–f). It is noteworthy that copolymers and terpolymers consist of units from each contributing mer and are not an alloy of mers.

Polymerization processes are of two types:

- Addition (chain reaction) polymerization
- Condensation (step reaction) polymerization

The addition process is the simpler of the two. By use of heat and pressure in an autoclave, double bonds of unsaturated monomers break down and then link up into a chain. These addition reactions are atoms or group of atoms that attach themselves to the carbon atom at the sites of multiple bonds. Saturated hydrocarbons undergo substantial reactions in which hydrogen atoms are replaced by other atoms or groups of atoms. Some examples of addition polymerization products are polypropylene (PP), polyethylene (PE), polyvinyl acetate (PVA), polyvinyl chloride (PVC), acrylonitrile-butadine-styrene (ABS) and polythene fluoroethylene (PTFE).

Thermosetting polymers emerge from condensation polymerization. The condensate is often water, but it may be a gas also . Phenolic (PF), polyester (PET), silicone (Si) and Urethanes (PUR) are typical thermosets for the step reaction synthesis, while nylon (PA) and polycarbonates (PC) are thermoplastic resin synthesized through step polymerization.

The term *resin* covers both solid and semisolid organic polymers. Previous one often considered as the uncompounded ingredients or monomers that are mixed but not yet

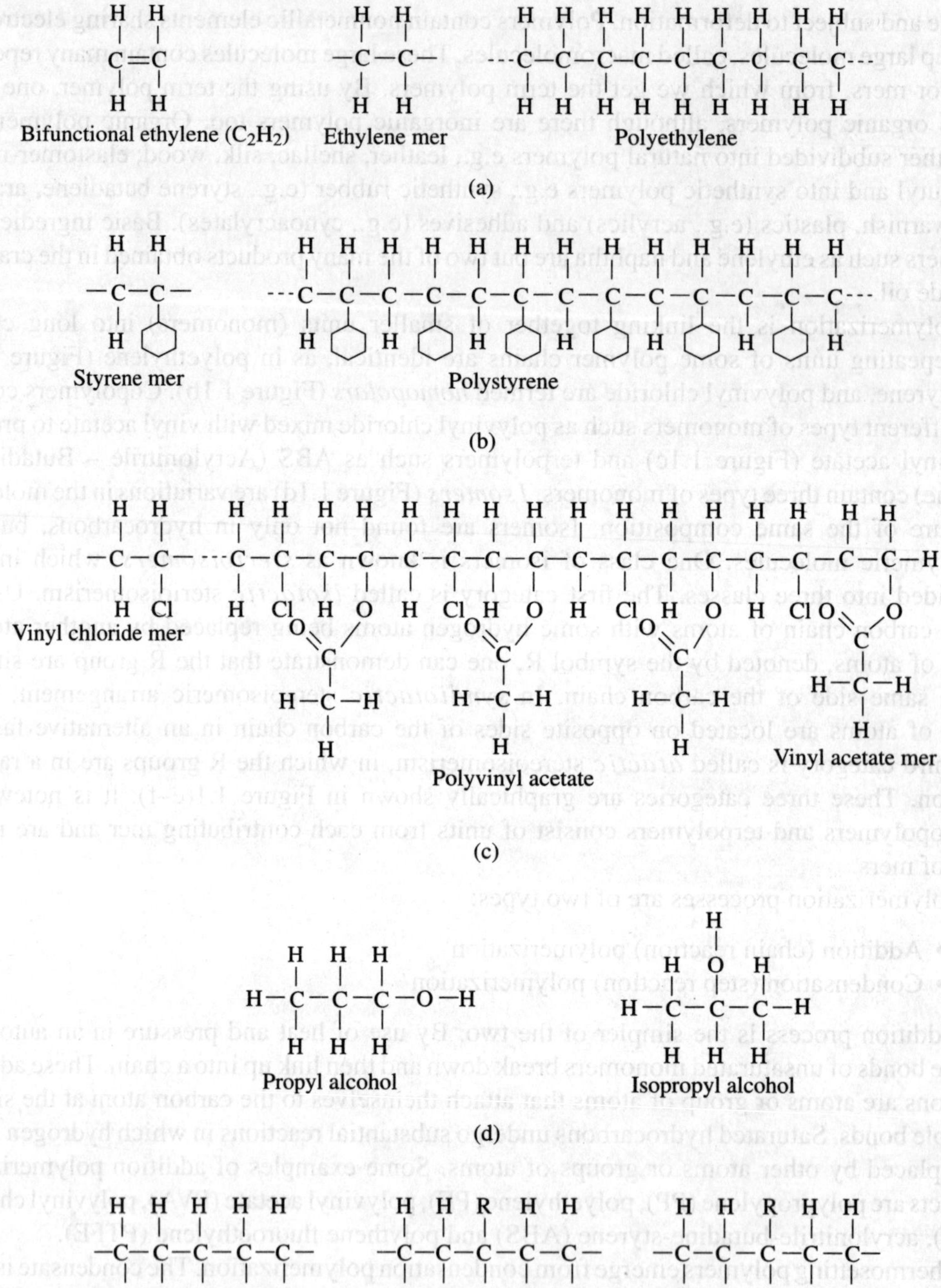

Fig. 1.1. Types of polymers (a) simple polymer, (b) homopolymers, (c) copolymers, (d) isomers, (e) isotactic stereoisomer, (f) syndiotactic stereoisomer, (g) atactic stereoisomer.

polymerized. For example, thermosetting resins as pellets are moulded into thermosetting plastic or elestomeric parts.

1.1.4 Composites

Composites are made from mixtures of the materials in the first three categories i.e., metals, ceramics and polymers. Figure 1.2 summarizes the relationship giving examples of the materials in these categories. The composite exhibits the best properties or qualities of its constituents or some properties possessed by neither. The combining of the two or more existing materials is done by physical means, as opposed in the chemical bonding that takes place in the alloys of monolithic solid materials. At the same time the significance of interfacial interaction with the matrix and the reinforcing component must not be under estimated. A true composite structure must show matrix material completely surrounding its reinforcing material in which the two phases act together to produce desired characteristics.

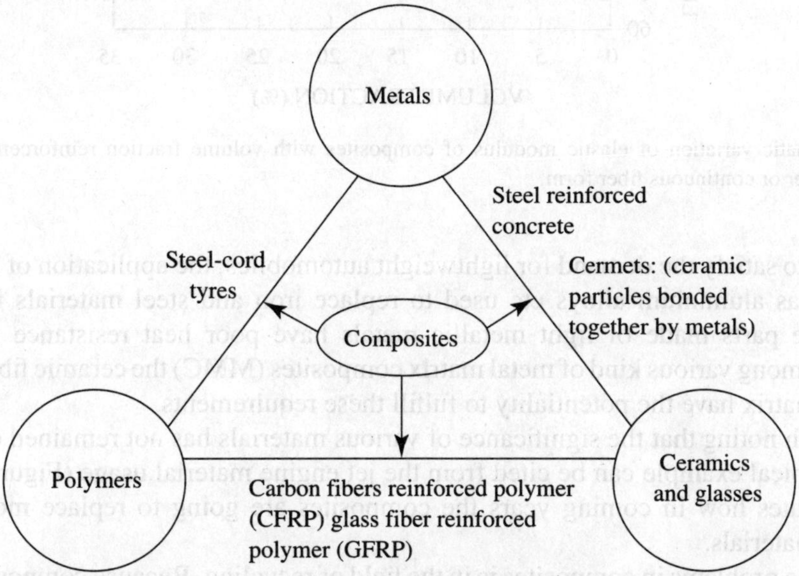

Fig. 1.2. Classification of engineering materials, showing evolution of various types of composites.

There exist many natural composites. Wood and many other plant tissues contain the cellulose chain polymer in a matrix of lignin, a phenol-propane polymer, and other organic compounds in a type of cellular structure.

One method of classifying engineered composite is according to the geometry of the reinforcing phase, which can be in fiber, whisker, particulates or laminar form, the latter often being made up of filaments or tows of fibers in matrices that are shaped in a laminar arrangement. Particulate composites have no distinct arrangement of the phases and the properties tend to be more isotropic than the fiber or laminar composites. The fibers may be continuous or discontinuous ones.

A schematic of the increase in elastic modulus is composite with the volume fraction for the same reinforcement, but a different form of reinforcement-continuous fiber, whisker or particulate is shown in Figure 1.3. Whiskers are very thin, single crystal of high perfection that has an extremely large length-diameter ratio. This illustrates the loss of reinforcement efficiency as one goes from continuous fiber to particle.

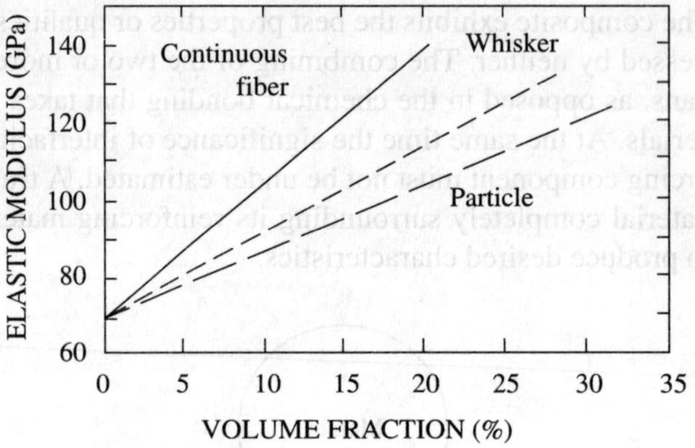

Fig. 1.3. Schematic variation of elastic modulus of composites with volume fraction reinforcement in particle, whisker or continuous fiber form.

In order to satisfy the demand for lightweight automobiles, the application of light metallic metals such as aluminium alloys etc used to replace iron and steel materials is increasing. However, the parts made of light metallic metals have poor heat resistance and abrasion resistance. Among various kind of metal matrix composites (MMC) the ceramic fiber-reinforced aluminium matrix have the potentiality to fulfill these requirements.

It is worth noting that the significance of various materials has not remained constant with history. A typical example can be cited from the jet engine material usage (Figure 1.4), which clearly indicates how in coming years the composites are going to replace more and more monolithic materials.

One of the problems in composites is in the field of recycling. Because composites are made using two dissimilar materials, they cannot be easily recycled or reused. Most polymer matrix composites currently available in the market are designed with long term durability in mind and are made using nondegradable polymeric resins, such as epoxies and polyurethane and high strength fibers, such as graphites, aramides and glass. Many of these polymers and fibers are derived from petroleum, a nonreplenishable commodity. The availability of inexpensive plant based fibers in every part of the world has, in part, promoted their use recently. These fibers offer several other advantages as well. They are nonabrasive to processing equipment, can be incinerated, are CO_2 neutral, when burned and because of their hollow and cellular nature, perform well in acoustic and thermal insulators.

Significant research efforts are currently being spent in developing a new class of fully biodegradable 'green' composites by combining the natural fibers with biodegradable resins.

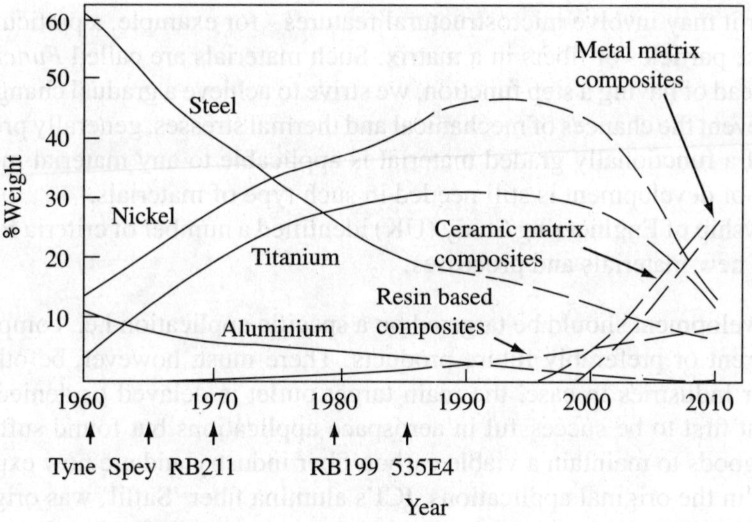

Fig. 1.4. Trends in jet engine material usage.

Most of these resins degrade through enzymatic reactions when exposed to a compost environment.

1.2 ADVANCED MATERIALS

During the 1980s, much attention has got focused on "new materials" (otherwise called "advanced materials," or "engineered materials"). These new materials exhibit high strength; great hardness; and superior thermal, electrical, optical, and chemical properties. They have dramatically altered communication technologies, reshaped data analysis, advanced space travel, restructured medical devices and transformed industrial production process. These materials often have following characteristics:

- they are created for specific purposes,
- they are highly processed and have a high value-to-weight ratio,
- they are developed and replaced with high frequency and
- they are frequently combined into new composites.

These materials are often synthesized from the by-products of conventional commodity materials. Gallium, indium, selenium, tellurium, and thallium-all by-products of aluminium, copper, iron, lead, and zinc processing are important starting materials or catalysts in the development of advanced materials. Thus, the supply of these new materials is closely linked to conventional mineral extraction and processing. The advanced polymers and many of the new composites are also largely developed from conventional organic chemicals. While the market for advanced materials have grown rapidly, the consumption of these materials in the production of finished goods is confined largely to the industrialized countries.

There is considerable interest in making materials that are graded in some respect. The gradient may be of the chemical composition, density or coefficient of thermal expansion of the material, or it may involve microstructural features - for example, a particular arrangement of second-phase particles or fibers in a matrix. Such materials are called *Functionally Graded Materials*. Instead of having a step function, we strive to achieve a gradual change. Such gradual change will prevent the chances of mechanical and thermal stresses, generally present otherwise. The concept of a functionally graded material is applicable to any material metal, ceramic or polymer. A lot of development is still needed in such type of materials.

The Fellowship of Engineering Study (UK) identified a number of criteria for the successful exploitation of new materials and processes:

- The development should be targeted on a specific application i.e., components required for current or preferably future products. There must, however, be other applications in other industries in case, the main target outlet is delayed or denied. Carbon fibers failed at first to be successful in aerospace applications but found sufficient outlets in sports goods to maintain a viable carbon fiber industry and are now experiencing rapid growth in the original applications. ICI's alumina fiber 'Saffil' was originally intended as an asbestos substitute, but is now being used as a reinforcing fiber in metal matrix composites.
- New process development is very risky. Many processes cannot be simulated adequately in the laboratory which means a commitment in expensive facilities in advance of detailed knowledge of performance and economics. Optimum process operation may depend on raw materials developments or modifications, which are outside the control of the process operator. Large production of identical components are often needed to justify expensive tooling.

In summary, it can be said that an ideal material is one where a core process technology, enjoying economics of scale, could be established and the needs of particular components accommodated by down stream processing, which tailors the material to the particular applications.

REFERENCES

The Science of Engineering Materials, C.R. Tottle, Chemical Publishing Co., N.Y., 1966.

Materials Engineering Science, R.W. Hank, Harcourt, Brace & World Inc., NewYork, 1970.

An Introduction to Engineering Materials. G.T. Murray, Marcel Dekker, Inc., N.Y., 1993.

Engineering Materials Technology, 3rd ed., J.A. Jacobs and T.F. Kilduff, Prentice Hall Int., N.J., 1997.

Engineering Materials Technology, 3rd ed., W. Bolton, Butterworth Heinemann, Oxford, 1998.

The Nature and Properties of Engineering Materials, 3rd ed., Z.D. Jastrzebski, John Wiley & Sons, N.Y., 1987.

Materials, J. Taylor, Cambridge University Press, Cambridge, 2002.

An Introduction to Materials Science, A.L. Ruoff, Prentice Hall, N.J., 1972.

The Principles of Engineering Materials, C. Barrett, W. Nix and A. Tetelman, Prentice Hall, N.J., 1973.

Materials Science, 4th ed., J.C. Anderson, K.D. Leaver, R.D. Rawlings and J.M. Alexander, Chapman and Hall, London, 1990.

Technology of Engineering Materials, M. Philips and W. Bolton, Butterworth-Heinemann, Oxford, 2002.

Materials Science and Metallurgy, P.W. Pollock, Reston Publishing Co., Reston, Virginia, 1977.

Engineering Materials Science, M. Ohring, Academic Press, San Diego, U.S.A., 1995.

The Science and Design of Engineering Materials, 2nd ed., J.P. Schaffer, A. Saxena, D. Antolowich, H.H. Sanders Jr. and S.B. Warner, McGraw Hill, Burr Ridge, IL, U.S.A., 2000.

The Science of Engineering Materials, 2nd ed., D.R. Askeland and P. Webster, Chapman and Hall, London, 1989.
Introduction to Materials Science and Engineering, K.M. Roll, T.H. Courtney and J. Wulff, John Wiley, N.Y., 1976.
Introduction to Materials Science for Engineers, 5th ed., J.F. Schackelford, Prentice Hall, N.J., 2000.
Principles of Materials Science and Engineering, 3rd ed., W.F. Smith, McGraw Hill, New York, 1995.
Elements of Materials Science and Engineering, 7th ed., L.H. Van Vlack, Addison-Wesley Longman, Reading, Mass., U.S.A., 1996.
Introduction to Properties of Materials, 2nd ed., D. Rosenthal, R.H. Asimow, Van Nostrand Reinhold Co., New York, 1971.
The Technology and Application of Engineering Materials, M.S. Ray, Prentice Hall Int., N.J., 1987.
Materials for Engineering, L.H. Van Vlack, Addison-Wesley Pub. Co., Reading, Mass., 1982.
Materials Science and Engineering, 6th ed., W.D. Callister Jr., John Wiley & Sons, Inc., N.Y., 2003.
Materials for Engineering, 2nd ed., J. Martin, Maney Publishing, Leeds, 2002.

2

BONDING AND STRUCTURE OF MATERIALS

The atom consists of nucleus containing neutrons and protons around which electrons orbit, the orbits being more or less confined to certain radii. The smaller the diameter of the orbit, the greater the attractive force between the electron and the nucleus and the greater the absolute binding energy. The electrons in the outer orbits are bound less tightly and are involved in the binding together of the atoms. The bonding strongly influences all the material properties. The full details of electronic configuration of various elements are well discussed in physics books and hence are not repeated here.

The bindings in materials can be classified as:

- Metallic Bond
- Ionic Bond
- Covalent Bond
- van der Waals Bond
- Mixed Bonds

The properties such as modulus of elasticity, heats of vaporization and fusion, and melting points are all measure of bond strengths. The higher these values, the stronger are the bonds. The heats of vaporization are the energies to dissociate the solids into the gaseous state and are approximately equal to the binding energies.

2.1 BONDING IN SOLIDS

2.1.1 Metallic Bond

In bulk metals each atom contributes its valence electrons to form a cloud of negative charge in which, the positive ions are arranged in a repetitive order. The metallic bond does not have directionality; however the properties do vary somewhat with crystallographic direction. The free valence electrons account for the high electrical and thermal conductivity of metals and to some extent their plasticity.

2.1.2 Ionic Bond

Ionic bonds are formed by a transfer of electrons from one atom to another. A typical example is sodium chloride, where sodium atom (atomic number 11 and electronic configuration $1s^2 2s^2 2p^6 3s^1$) donates one outer electron to chlorine atom (atomic number 17 and electronic configuration $1s^2 2s^2 2p^6 3s^2 3p^5$), thus completing its third shell of eight electrons.

2.1.3 Covalent Bond

Covalent bonds arise when two atoms share electrons in an attempt to fulfill their needs to completely fill the s, p, d and f suborbitals. From the knowledge of basic physical chemistry, we know that most stable states are those when the s orbitals contain two electrons, the p orbitals six electrons, the d orbitals 10 electrons, and the f orbitals 14 electrons. The simplest example for such bond is hydrogen molecule. Such bond is very common in organic molecules and occurs in many cases by the 'hybridization' of the s and p orbitals. The s shell is more or less spherical in shape, whereas the p shell is more like a dumbbell in shape. In pure carbon one has four equivalent sp^3 hybrid orbitals (Figure 2.1). Covalent bond is an extremely strong bond. An

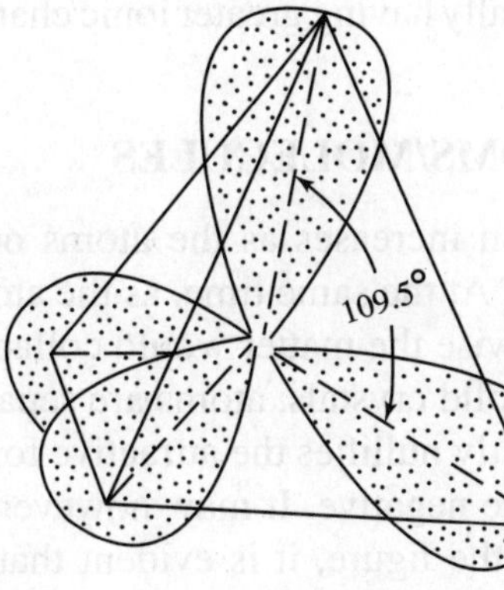

Fig. 2.1. sp^3 hybridization in carbon.

example is diamond, which is extremely hard and possesses high modulus of elasticity. Covalent bonds are directional and hence the properties are not isotropic. Now a question arises, why two forms of solid carbon between diamond and graphite differ in properties? The reason for this lies in different type of hybridization. In case of graphite the three electrons hybridize in bonds at 120° to each other, leaving one electron only weakly bonded to the other. This electron accounts for the high electrical and thermal conductivity of graphite compared to diamond. The layers are bonded together by weak secondary bonds, which allows them to glide past each other easily. This introduces lubricating properties in graphite.

2.1.4 Van der Waals Bond

This is a rather weak bond, which accounts for the weak attractive forces between gas molecules. These weak forces are due to both permanent and fluctuating dipoles. A dipole moment exists when there is a nonsymmetrical charge distribution. On approaching each other, even the inert gases experience this type of attractive force. Water is another example where the bond is due to permanent nonsymmetrical charge distribution in the molecules. In this case the hydrogen atoms are bonded to the oxygen atom at an angle of 105°.

2.1.5 Mixed Bond

In many substances, particularly in the inorganic compounds e.g., ceramics, the bond is neither pure covalent nor pure ionic. Here this is represented as the degree of covalency or ionicity. Pauling, using the electronegativity system expressed the degree of ionicity as:

$$\% \text{ ionic character} = [1 - \exp(-1/4)(X_A - X_B)^2] \times 100$$

where X_A and X_B are electronegativity of elements A and B respectively. Some of the ionicity values in ceramics such as SiC and MgO are 12% and 70% respectively.

Intermetallic compounds, such as nickel and titanium aluminides do also possess mixed bonding with some degree of ionicity and covalency mixed with the metallic character. The greater the difference in the electronegativity of the elements, the larger the ionic character. The titanium aluminides would be naturally having greater ionic character than the nickel aluminides.

2.2 FORCES BETWEEN ATOMS/MOLECULES

In all solids, the force of attraction increases as the atoms or molecules approach. At large separations the forces drop to zero. At the same time, as the atoms approach together repulsive forces also come into effect, otherwise the matter would collapse altogether. This is caused by the overlap of electron clouds. In solid crystals, atoms are separated at an equilibrium distance (r_o), where the repulsive force exactly nullifies the attractive force. This is shown in Figure 2.2, where attractive force is taken to be negative. It may, however, be pointed that in some books this convention is reversed. From the figure, it is evident that below r_o the resultant force is repulsive, which increases vary rapidly as r decreases. Above r_o the resultant force is attractive. As the separation increases beyond r_o the attractive force gets larger, but eventually it reaches a point when it begins to fall. If one sees Figure 2.2 carefully, one will notice that for values of r very close to r_o, the resultant graph is a straight line. The slope of the force-separation (k) could be taken as the measure of the modulus of elasticity (E). According to the sign convention followed, the 'k' value shall be negative on the graph. Therefore one can write

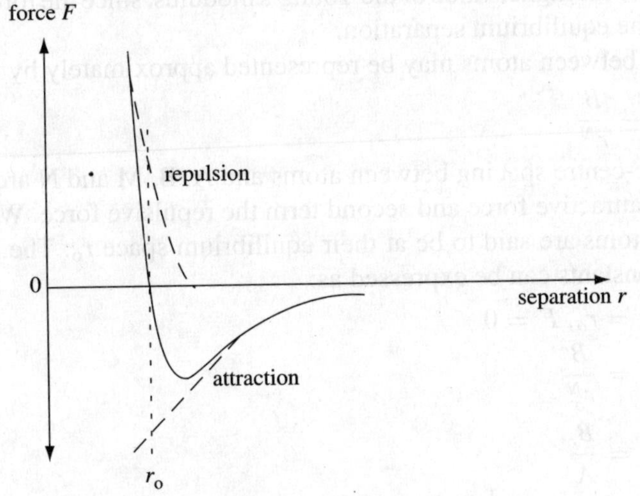

Fig. 2.2. Variation of repulsive and attractive forces with respect to interatomic separation. The solid curve shows the resultant force between pair of atoms. The equilibrium separation is r_o.

$$E = k/r_o$$

Example 2.1. **Force separation curves for two crystalline materials are shown in Figure Ex 2.1 one is weakly bonded, while the other is strongly bonded. Which of these materials will have the higher values of Young's Modulus?**

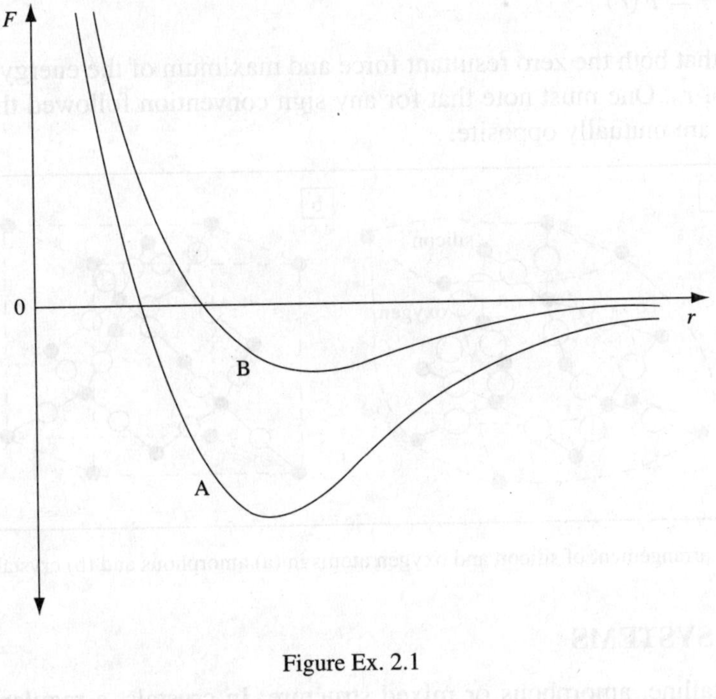

Figure Ex. 2.1

Solution. A will have the higher value of the Young's modulus, since the force-atomic separation curve is steeper at the equilibrium separation.

The body force between atoms may be represented approximately by the general equation

$$F = \frac{A}{r^M} - \frac{B}{r^N} \tag{2.1}$$

where r is the centre-centre spacing between atoms and A, B, M and N are constants. The first term represents the attractive force and second term the repulsive force. Where these are equal $F(r) = 0$, and the atoms are said to be at their equilibrium space r_o. The equilibrium spacing r_o in terms of the constants can be expressed as:

$$\text{at } r = r_o, F = 0$$

$$\text{i.e., } \frac{A}{r_o^M} = \frac{B}{r_o^N}$$

$$r_o^{N-M} = \frac{B}{A}$$

$$r_o = \left[\frac{B}{A}\right]^{\frac{1}{N-M}} \tag{2.2}$$

In most cases it is convenient to deal with energies than with forces. If U is the potential energy then

$$U(r) = \int F(r)dr$$

$$\frac{d.U(r)}{dr} = F(r)$$

This means that both the zero resultant force and maximum of the energy curve in the plot lie at the value of r_o. One must note that for any sign convention followed the signs for force and energy shell are mutually opposite.

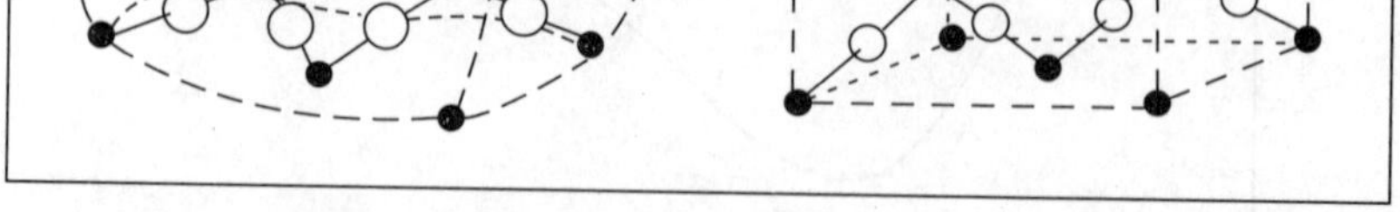

Fig. 2.3. Atomic arrangement of silicon and oxygen atoms in (a) amorphous and (b) crystalline silica SiO_2.

2.3 CRYSTAL SYSTEMS

Solids have crystalline, amorphous or mixed structure. In crystals, a regular arrangement of atoms extends over distances, which are very large compared with the diameter of the atoms.

In amorphous solids the atoms occupy random positions. For example Figure 2.3 shows typical atomic arrangement in both types of structures in silica. Amorphous or glassy structure looks as if the liquid structure has been frozen. In principle, all crystalline metals can also exist in an amorphous state if the molten material is cooled quickly enough. Amorphous metals are very uncommon but they do exist.

August Bravais, a French crystallographer, showed that there could be only 14 different and distinct ways in which point could be arranged geometrically. These are called Bravais lattices and shown in Figure 2.4. The Bravais lattices can be broken down in to seven crystals systems as shown in Table 2.1. Each system can be described in term of the relative spacing between the points and the angular relationship of the coordinate axes used to outline the points.

The different crystal systems can be characterized exclusively by their symmetries. The full discussion on these is beyond the scope of the book; however, it is sufficient to say that the cubic system can be described fully by four three-fold notations. Table 2.2 presents the different symmetry operations defining the seven crystal systems. For example, a threefold rotation is a rotation of 120° i.e., after 120°, the crystal system comes to a position identical to the initial one. The hexagonal system exhibits a six-fold rotation around the c axis, after each 60 °, the structure superimposes upon itself. It is worth mentioning here that quasicrystals have any rotational symmetry axes which are prohibited in crystals.

Example 2.2. Show that the face centred tetragonal Bravais lattice is equivalent to the body centred tetragonal and is not a true Bravais lattice. How do the lattice parameters a and c of the FCT compare with the lattice parameter a' and c' of the BCT?

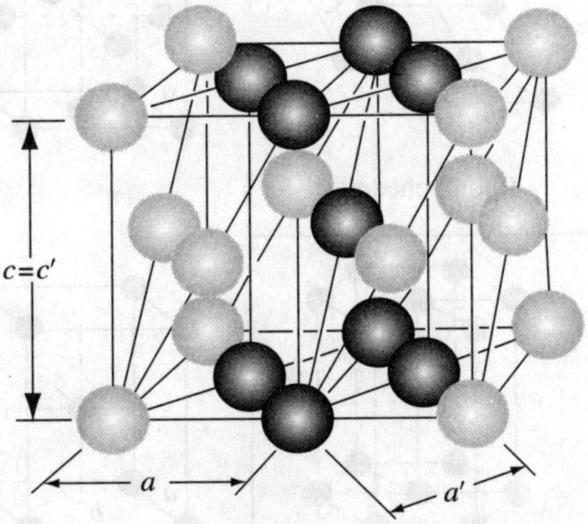

Figure Ex. 2.2

Solution. If two face centred tetragonal unit cells (Figure Ex. 2.2) are drawn side by side, a BCT cell may be generated along the plane separating the two unit cells.

From the figure, it is clear that the height of FCT and BCT cells are the same i.e., $c' = c$.

The lattice parameter a' is half of the face diagonal of the FCT cell, i.e., $a' = \dfrac{a\sqrt{2}}{2}$.

Example 2.3. TiAl intermetallic has an ordered Face Centred Tetragonal structure with $a = 4.005$Å, see Figure 2.13. Indicate the positions of the effective atoms of Al and Ti in the unit cell.

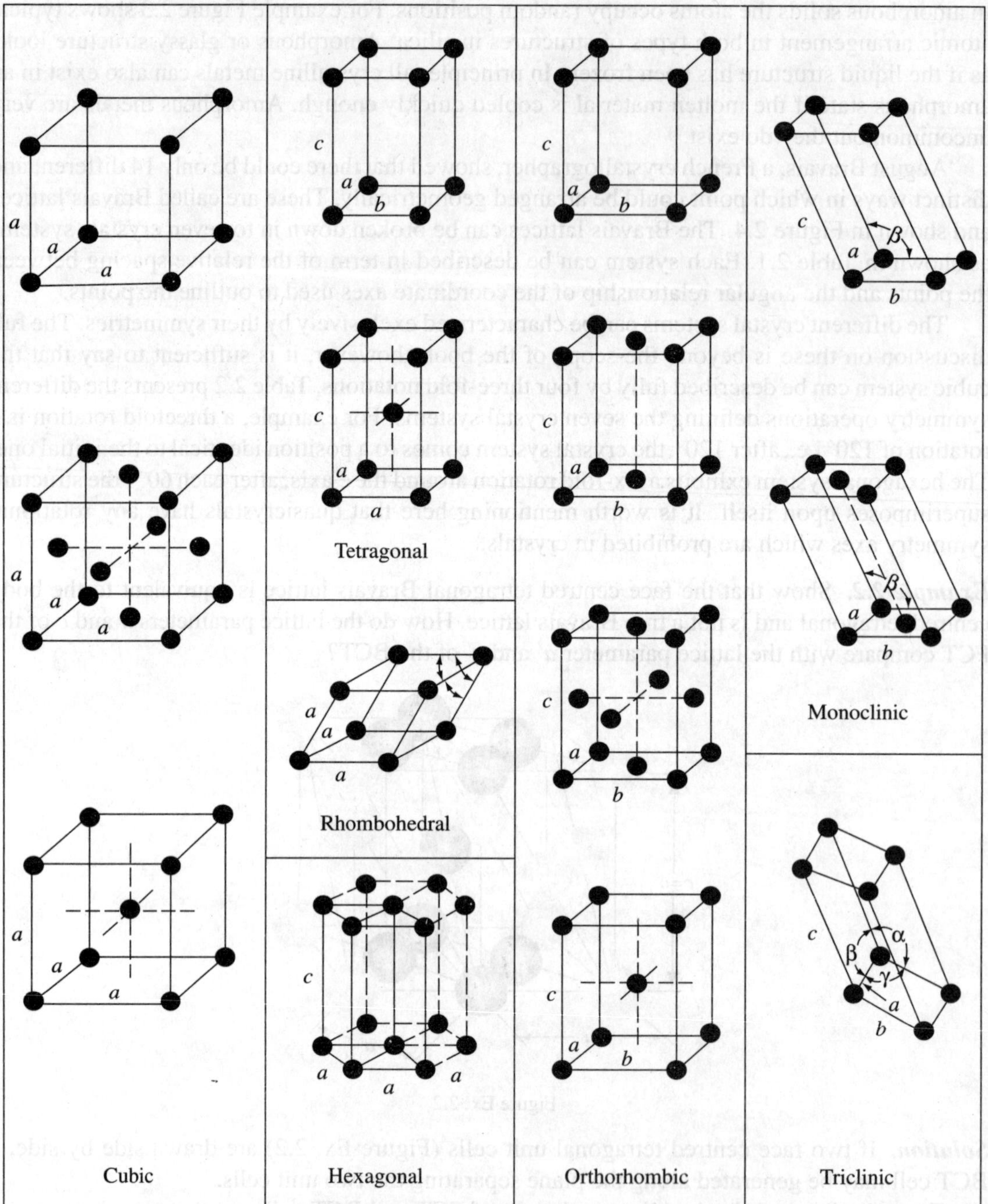

Fig. 2.4. The 14 Bravais lattices and the seven crystal systems (after W.G. Moffatt, G.W. Pearsell and J. Wulff, the Structure and Properties of Materials, Wiley, New York, 1964, p. 47.)

Table 2.1. Angular and Distance Relationships in the Seven Crystal System

System	*Axes and interaxial angles*	
Triclinic	Three axes not at right angles, of any lengths $a \neq b \neq c$	$\alpha \neq \beta \neq \gamma \neq 90°$
Monoclinic	Three axes, one pair not at right angles, of any lengths $a \neq b \neq c$	$\alpha = \gamma = 90° \neq \beta$
Orthorhombic (rhombic)	Three axes, at right angles, all unequal $a \neq b \neq c$	$\alpha = \beta = \gamma = 90°$
Tetragonal	Three axes at right angles,two equal $a = b \neq c$	$\alpha = \beta = \gamma = 90°$
Cubic	Three axes at right angles, all equal $a = b = c$	$\alpha = \beta = \gamma = 90°$
Hexagonal	Three axes coplanar at 120° equal Fourth axis at right angles to these $a_1 = b_2 = a_2 \neq c$ (or $a_1 = b \neq c$)	$\alpha = \beta = 90°, \gamma = 120°$
Rhombohedral (trigonal)	Three axes equally inclined, not at right angles; all equal $a = b = c$	$\alpha = \beta = \gamma \neq 90°$

Table 2.2. Minimum Number of Symmetry Operations in Various Systems

Systam	*Rotatation*
Triclinic	None (or center of symmetry)
Monoclinic	1 twofold rotation
Orthorhombic	2 perpendicular twofold rotations
Tertagonal	1 fourfold rotation around [001]
Rhombohedral	1 threefold rotation around [111]
Hexagonal	1 sixfold rotation around [0001]
Cubic	4 threefold rotations around <111>

Solution. Al: (0, 0, 0), (1/2, 1/2, 0)
Ti: (1/2, 0, 1/2), (0, 1/2, 1/2)

Example 2.4. Identify the position of all atoms in the face centred orthorhombic cell. How are these position coordinates related to those for the atoms in the FCC unit cell?

Solution. Referring the sketch of the Face Centred orthorhombic, the atomic position as expressed in terms of the fractions of the three lattice parameters a, b and c are

0, 0, 0; 1, 0, 0; 0, 1, 0; 0, 0, 1; 1, 1, 0;
1, 0, 1; 0, 1, 1; 1, 1, 1; 1/2, 1/2, 0; 1/2, 0, 1/2;
0, 1/2, 1/2; 1/2, 1/2, 1; 1/2, 1, 1/2; 1, 1/2, 1/2.

These are the same as the coordinates for the atoms of the FCC unit cell. Since the atomic positions are expressed in fraction of the lattice parameters, the positions coordinates for all face centred Bravais lattices are the same.

Example 2.5. Alpha ion transforms to gamma iron with little strain such that $(110)_{BCC} = (111)_{FCC}$, and $[111]_{BCC} = [110]_{FCC}$. Is this change slow or fast? Illustrate the feature schematically.

Solution. Figure Ex. 2.5 shows the plane and direction relations.

As the respective displacements of atoms are small, the transformation occurs rapidly.

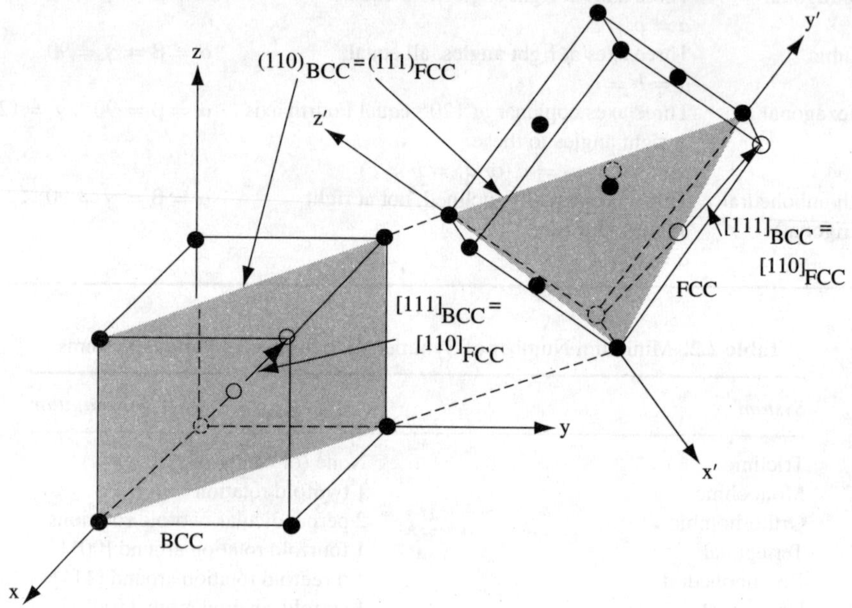

Figure Ex. 2.5

The determination of crystal system in a solid is carried out after x-ray and neutron diffraction studies. This comes strictly in the area of materials science and is not covered in this book.

Unit Cell is defined as the smallest repeating unit in the crystalline lattice pattern, which extends through the entire crystal. The distance that a unit cell must be translated in order to repeat itself is called the *lattice parameter*. Next question arises identification of crystal directions and planes. This was done by H. Miller and hence known as Miller indices. Once the origin and Cartesian xyz axes are specified, the crystallographic direction of a line from the origin of the unit cell to any point, whose vector coordinates are u, v, w is denoted as $[uvw]$. The coordinates of the end points of the direction vector are the same as the components of the vector resolved along the three coordinates axes. To avoid fractions, the components are reduced to the smallest integers. Negative point coordinates are indicated with a bar above the number. By convention, specific direction indices are bracketed by the [ ] configuration. The

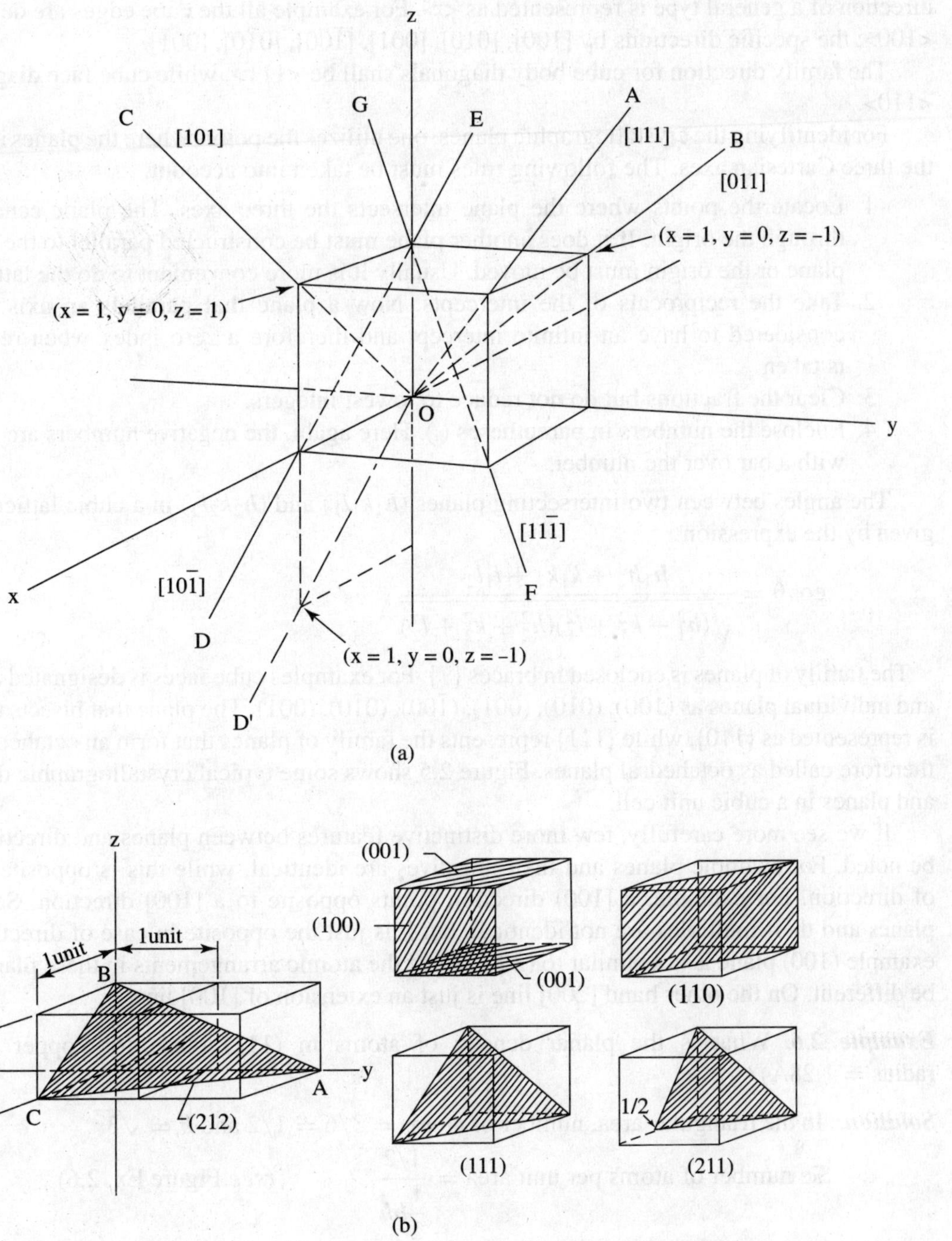

Fig. 2.5. Miller indices of some typical (a) directions and (b) planes in cubic lattice.

direction of a general type is represented as < >. For example all the cube edges are denoted as <100>; the specific directions by [100], [010], [001], [$\bar{1}$00], [0$\bar{1}$0], [00$\bar{1}$].

The family direction for cube body diagonals shall be <111>, while cube face diagonals is <110>.

For identifying the crystallographic planes, one utilizes the points where the planes intercept the three Cartesian axes. The following rules must be taken into account:

1. Locate the points where the plane intersects the three axes. The plane cannot pass through the origin. If it does another plane must be constructed parallel to the original plane or the origin must be moved. Usually it is more convenient to do the latter.
2. Take the reciprocals of the intercepts. Now a plane that parallels an axis may be considered to have an infinite intercept and therefore a zero index when reciprocal is taken.
3. Clear the fractions but do not reduce to lowest integers.
4. Enclose the numbers in parentheses ( ). Here again, the negative numbers are denoted with a bar over the number.

The angles between two intersecting planes $(h_1k_1l_1)$ and $(h_2k_2l_2)$ in a cubic lattice can be given by the expression

$$\cos\theta = \frac{h_1h_2 + k_1k_2 + l_1l_2}{\sqrt{(h_1^2 + k_1^2 + l_1^2)(h_2^2 + k_2^2 + l_2^2)}} \tag{2.3}$$

The family of planes is enclosed in braces { }. For examples cube faces is designated as {100} and individual planes as (100), (010), (001), ($\bar{1}$00), (0$\bar{1}$0), (00$\bar{1}$). The plane that bisects the cube is represented as {110}, while {111} represents the family of planes that form an octahedron and therefore called as octahedral planes. Figure 2.5 shows some typical crystallographic direction and planes in a cubic unit cell.

If we see more carefully, few more distinctive features between planes and directions can be noted. For example planes and their negatives are identical, while this is opposite in case of direction. For example, a [100] direction points opposite to a [$\bar{1}$00] direction. Secondly, planes and their multiples are not identical, but it is just the opposite in case of direction. For example (100) plane is not similar to (200), since the atomic arrangements in these planes may be different. On the other hand [200] line is just an extension of [100] line.

Example 2.6. What is the planar density of atoms in (111) plane for copper (atomic radius = 1.28Å)?

Solution. In the triangular area, number of atoms = 3/6 = 1/2 and $h = \sqrt{3}r$.

So number of atoms per unit area $= \dfrac{1/2}{\frac{1}{2}bh}$ (*see* Figure Ex. 2.6)

$$= \frac{1/2}{\frac{1}{2}(2 \times 1.28) \times \sqrt{3}(1.28)}$$

$$= 1.77 \times 10^{19}\ \text{m}^{-2}$$

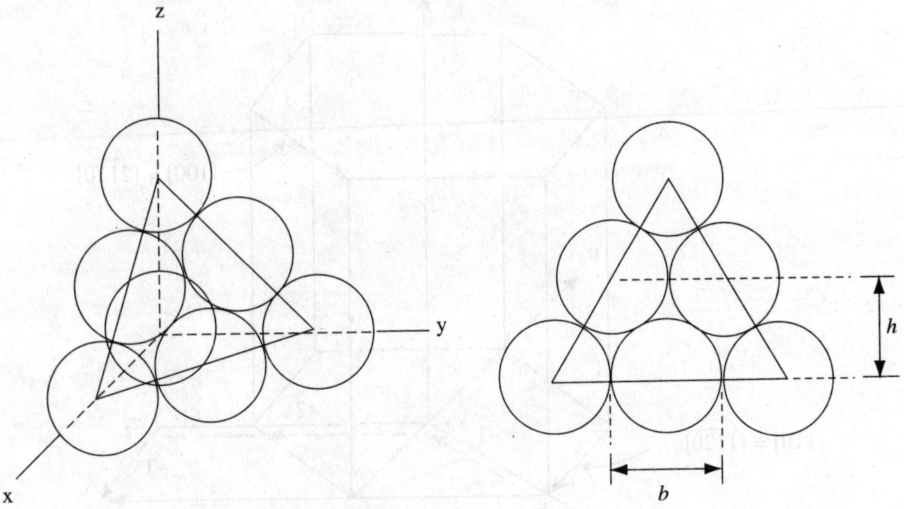

Figure Ex. 2.6

Example 2.7. Calculate the packing factor of any crystal having the BCC structure. Will the packing factor of one BCC metal will differ from another BCC metal?

Solution. The effective number of atoms per unit

$$\text{BCC Lattice} = \text{Centre atom} + 1/8\ (8\ \text{corner atoms}) = 2$$

From geometric consideration

$$a\sqrt{3} = 4R$$

where a is the lattice parameter and R is the atomic radius.

$$\text{Volume of two atoms} = 2(4\pi/3)R^3$$

$$\text{Volume of a unit cell} = a^3 = (4R/\sqrt{3})^3$$

$$\text{Therefore Packing factor} = \text{Volume of two atoms/Volume of a unit cell} = \frac{2(4\pi/3)R^3}{(4\pi/\sqrt{3})^3} = 0.68$$

Since all the atoms of a metal are identical, the packing factor shall the same for either BCC metal.

When we proceed to hexagonal unit cells, it becomes evident that the rules followed for cubic unit cells are no more tenable. In hexagonal cell, we use four axes rather than three (Figure 2.6). Three of the axes lie in what is called basal planes and are positioned 120 °C apart. The fourth axis, called the c axis, is perpendicular to the basal plane. In this unit cell the length of the a_1, a_2 and a_3 axes are all the same, including both positive and negative directions. There are

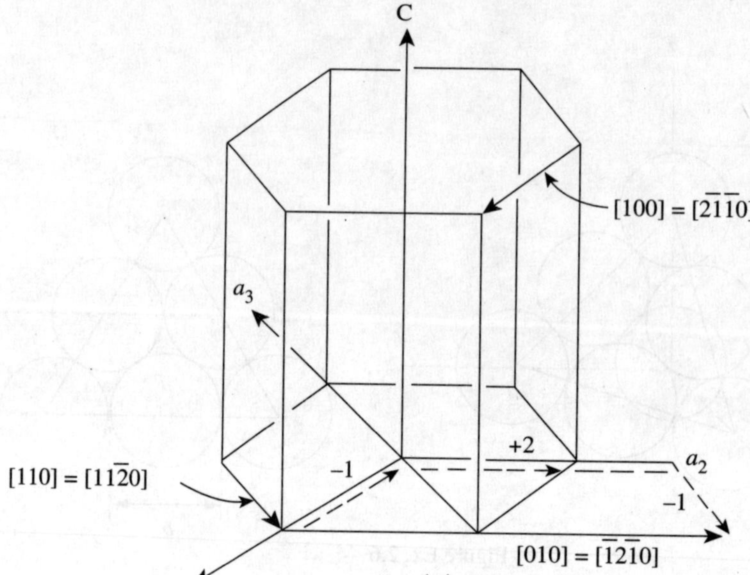

(a)

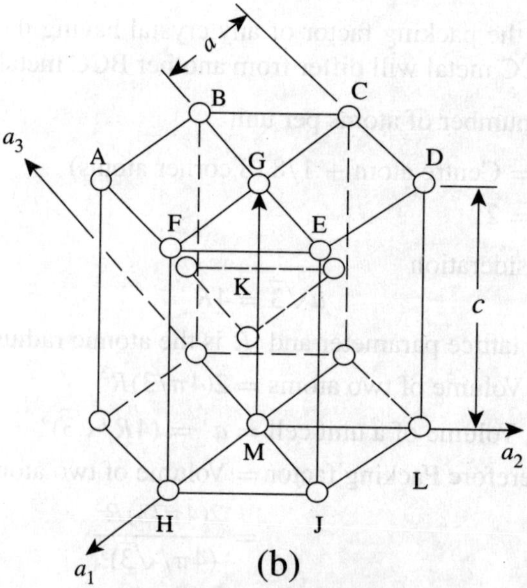

(b)

Basal plane (0001) – ABCDEF
Prism plane (10$\bar{1}$0) – FEJH
Pyramid planes: (1011) – GHJ
(10$\bar{1}$2) – KJH
(11$\bar{2}$1) – GHL
(11$\bar{2}$2) – KHL

Fig. 2.6. Representation of some crystallographic (a) directions and (b) planes in hexagonal lattice.

thus two lattice parameters, namely a and c with $c > a$. The four number system designation for hexagonal crystals is known as *Miller-Bravais Indices*. In hexagonal systems, for designating planes, we may have four intercepts, which are denoted as (h, k, i, l) indices. Because of the redundancy of the a_3 axes and the geometry of the lattice, the first three integers, corresponding to the a_1, a_2 and a_3 intercepts, are related by the equation $h + k = -i$. In other words, we may reduce the four digit indices to the three-digit notation. Figure 2.6 shows some examples of directions and planes in a hexagonal unit cell.

2.3.1 Packing in Crystals

While considering the packing of atoms in crystals, one must take a realistic picture i.e., hard sphere model, where atoms touch in some directions and nearly touch in all directions when they are at their equilibrium positions. The word 'hard' is used in the sense that the spheres are incompressible, due to the high repulsive forces exerted between atoms by the filled shells of electrons. To begin with, one should take simple crystal structures, when all atoms are alike, e.g., metals. A plane of spherical balls may be packed closely or nonclosely. In the former case, each sphere is surrounded by six others. If a second layer is packed on the top of the first, it will sit in the depression between the spheres in the first layer. Let us call this as B layer. When a third layer is added there are two possibilities. In case the spheres in the third layer fit into the depression lying directly above the spheres in layer 'A', the packing is known as hexagonal close packing and the stacking sequence is ABABAB… In other words, out of six depressions in A layer only three alternate ones shall be occupied by the sphere of identical size in next 'B' layer. In case of face centred cubic close packing the third layer atoms are just above the unoccupied depressions in the first layer. The stacking sequence is ABC ABC ABC… (Figure 2.7). It is evident that the stacking on any layer is to be done over a close packed layer. From this point of view the visualization in HCP is much simpler as its basal plane is closest packed. As the closest packed plane is (111) in FCC, one should try to build up on a three dimensional rather than visualizing an unit cell. However in real crystals the stacking order may not be ideal. For example in FCC metals the order may be interuppted like ABCABCCA…. Such interfacial defects are known as *stacking faults*. This shall be described more in detail in the section, defects in crystals.

Two other important characteristics of a crystal structure are the *coordination number* and the *atomic packing factor*. The number of nearest neighbour around an atom is known as coordination number. For example in body centred cubic structure it is 8, while for close packed hexagonal structures it is 12. The atomic packing factor is the fraction of solid sphere volume in a unit cell, assuming the atomic hard sphere model.

i.e., APF = Volume of atoms in a unit cell/total volume of unit cell

For the FCC structure, the atomic packing factor is 0.74.

In case of nonclose packed structures, body centred cubic is a typical example, where the coordination number (8) is lower than that for FCC or HCP (12). Another term used in crystal structure is the number of effective atoms in a unit cell. In BCC, this number is 2, the equivalent of one atom through the eight corners, each of which is shared among eight unit cells and the single center atom, which is wholly contained within its cell. For the FCC crystal structure,

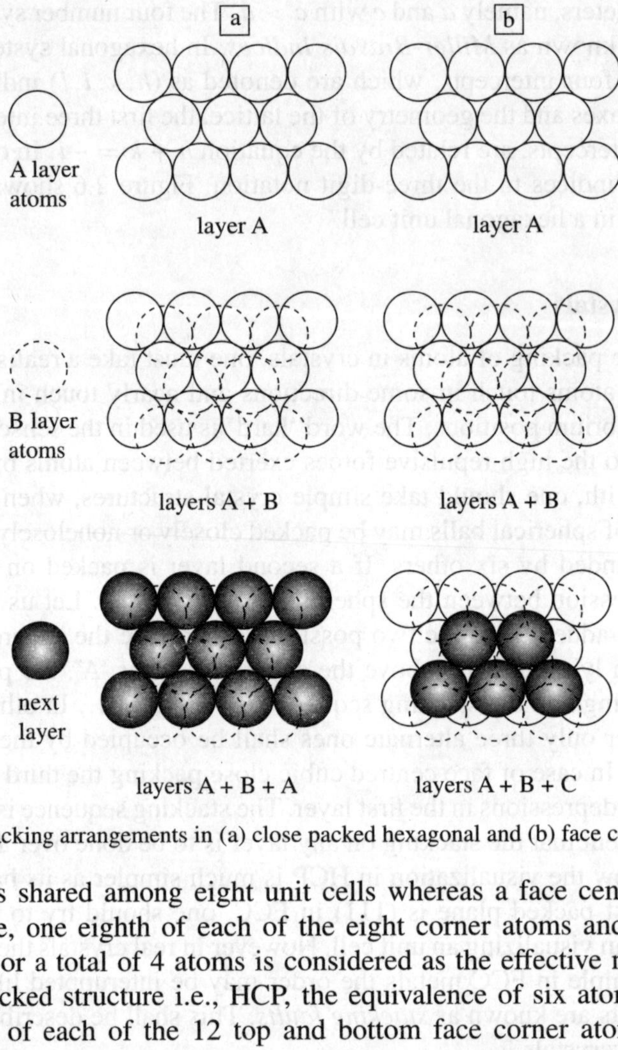

Fig. 2.7. Atomic stacking arrangements in (a) close packed hexagonal and (b) face centered cubic crystals.

each corner atom is shared among eight unit cells whereas a face centred atom belongs to only two. Therefore, one eighth of each of the eight corner atoms and one half of each of the six face atoms, or a total of 4 atoms is considered as the effective number of a unit cell. In another close packed structure i.e., HCP, the equivalence of six atoms is centred in each unit cell, one-sixth of each of the 12 top and bottom face corner atoms, one half of each of the 2 centre face atoms and all the three mid plane interior atoms. If a and c represent, respectively, the short and long unit cell dimensions, then c/a for ideal close packed HCP is 1.633 (*see* example).

Example 2.8. Show that in a close packed hexagonal system c/a ratio is 1.633.

Solution. Let us construct a tetrahedron along the c-axis of HCP. The height of this tetrahedron is $c/2$. If R = atomic radius

Then $a = 2R$. By simple geometry

$$\left(\frac{c}{2}\right)^2 = (2R)^2 - (\mathrm{AE})^2$$

Since the line AE bisects the angle at the corner of the tetrahedron (Figure Ex. 2.8)

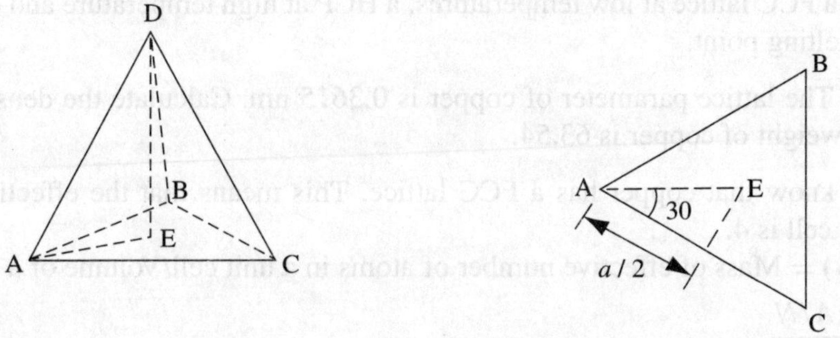

Figure Ex. 2.8

$$\cos 30 = \frac{a/2}{AE} = \frac{\sqrt{3}}{2}$$

$$\text{therefore, } AE = \frac{a}{\sqrt{3}}$$

Substituting this value in above expression, we get

$$c = 4R\sqrt{\frac{2}{3}}$$

$$\frac{c}{a} = 2\sqrt{\frac{2}{3}} = 1.633$$

By knowing the crystal structure of a solid, one can calculate the theoretical density ρ through the relationship

$$\rho = \frac{nA}{V_c N_A} \tag{2.4}$$

where, n = Effective number of atoms in each unit cell

A = Atomic weight

V_c = Volume of unit cell

N_A = Avogadro's number (6.023×10^{23} atom/mol)

Crystalline solids whether pertaining to a metal or nonmetal may have more than one crystal structure. This phenomenon is known a *polymorphism*. The crystal structure depends on both the temperature and the external pressure. For example, graphite is the stable polymorph of carbon at ambient conditions, whereas diamond is formed at extremely high pressure. Pure iron (Ferrite) has a BCC crystal structure at room temperature, which changes to FCC iron (Austenite) at 910 °C. On further heating the structure again changes to BCC at 1400 °C (δ iron), but with a different lattice parameter than ferrite.

The relationship between FCC and HCP structure is so close that transition from one to the other takes place in certain metals when thermal effects may allow some rearrangement. Cobalt, for example, exists as HCP Lattice at low temperatures and as a FCC one at high temperatures.

Strontium has a FCC lattice at low temperatures, a HCP at high temperature and changes again nearer to its melting point.

Example 2.9. The lattice parameter of copper is 0.3615 nm. Calculate the density of copper. Given atomic weight of copper is 63.54.

Solution. We know that copper has a FCC lattice. This means that the effective number of atoms per unit cell is 4.

Density (ρ) = Mass of effective number of atoms in a unit cell/Volume of a unit cell

$$= \frac{nA/N}{a^3}$$

where n = Effective number of atoms in an unit cell

A = Atomic mass

N = Avgadro's number

a = Lattice parameter

$$\rho = \frac{4 \times 63.54/6.02 \times 10^{23}}{(0.361 \times 10^{-9})^3}$$

$$= 8.94 \text{ g/cm}^3$$

2.3.2 Voids in Unit Cells

Voids invariably occur in the crystal structure and other smaller atoms may fit in to these spaces. This is important in inorganic crystals, as well as in certain metals. If we refer to the close packed layers described in the previous section, a *tetrahedral void* occurs when one atom is situated over the triangular space in the lower layer between three atoms. Figure 2.8(a) shows that four atoms exist around this void, hence the tetrahedral shape. The arrangement in Figure 2.8 (b) is one in which a triangular space in the first layer connects with a similar triangular space in the upper layer, so that six atoms are involved, arranged on the corners of an octahedron. These void are known as *octahedral voids*. Octahedral voids can just contain a sphere of radius 0.414 R, where R is the radius of the primary spheres of the lattice, and tetrahedral voids can just contain a sphere of radius 0.225 R. Figure 2.9 shows the voids in the cubic and hexagonal structures.

In HCP structures, the octahedral voids are situated between and below two atoms in the basal plane (along an a axis) and between and above the two nearest atoms in the plane below.

There are three such positions above layer B and three below in each unit (Figure 2.9). Similarly, the tetrahedral voids are directly above an atom in the B layer, and equivalent for three atoms in the basal plane, i.e., not along on an a-axis. For a close packed structure, the sizes of the voids are always in the same ratio to the atomic radius, whether FCC or HCP. If we examine carefully it can be established that in FCC structure there are 4 octahedral and 8 tetrahedral voids per unit cell. On the other hand in hexagonal closed packed structure there are 6 octahedral and 12 tetrahedral voids per unit cell.

In BCC structures the understanding of the voids can be more clear by considering two adjacent unit cells. Octahedral voids occur at center of cube faces and the center of cube edges

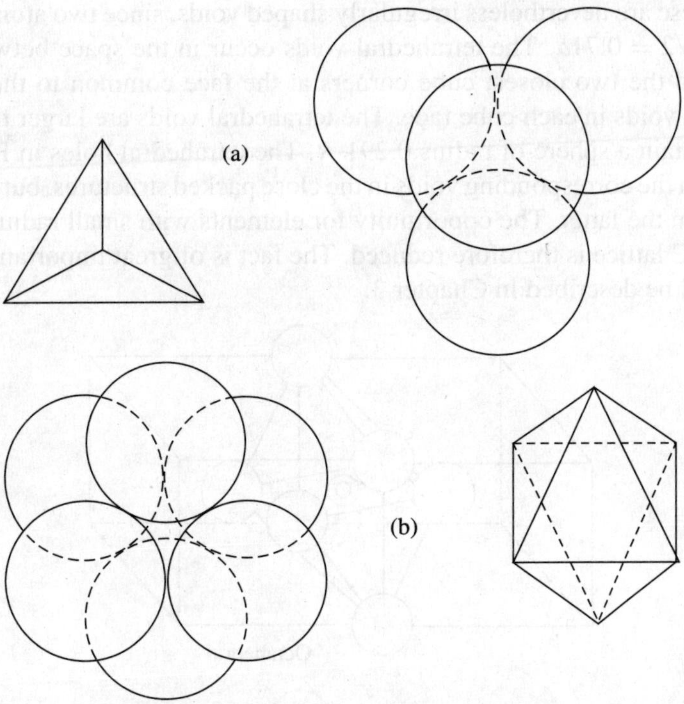

Fig. 2.8. (a) Tetrahedral voids and (b) octahedral voids.

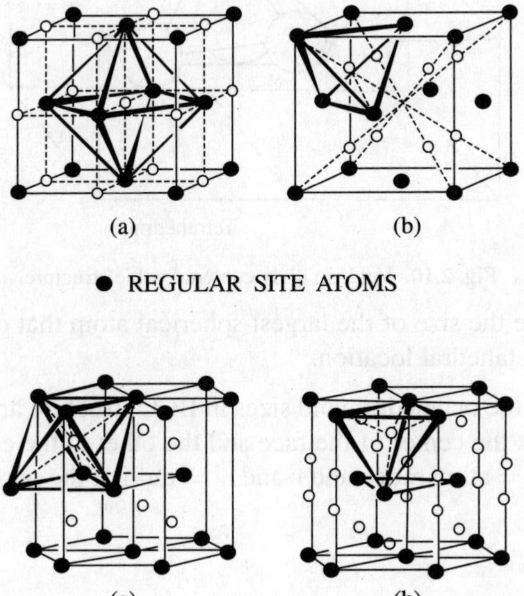

Fig. 2.9. Voids in close packed structures: Top: FCC; bottom: HCP; (a) octahedral void (b) tetrahedral void.

(Figure 2.10). These are nevertheless irregularly shaped voids, since two atoms are at a distance $1/2a$ and at $a/2\sqrt{2} = 0.71a$. The tetrahedral voids occur in the space between the two body centred atoms and the two closest cube corners at the face common to the two cells. There are thus four such voids in each cube face. The tetrahedral voids are larger than the octahedral voids and can contain a sphere of radius 0.291 R. The tetrahedral holes in BCC structures are slightly larger than the corresponding voids in the close packed structures, but much smaller than octahedral voids in the latter. The opportunity for elements with small radius atoms to occupy the voids in a BCC lattice is therefore reduced. The fact is of great importance in hardening of steels, which shall be described in Chapter 3.

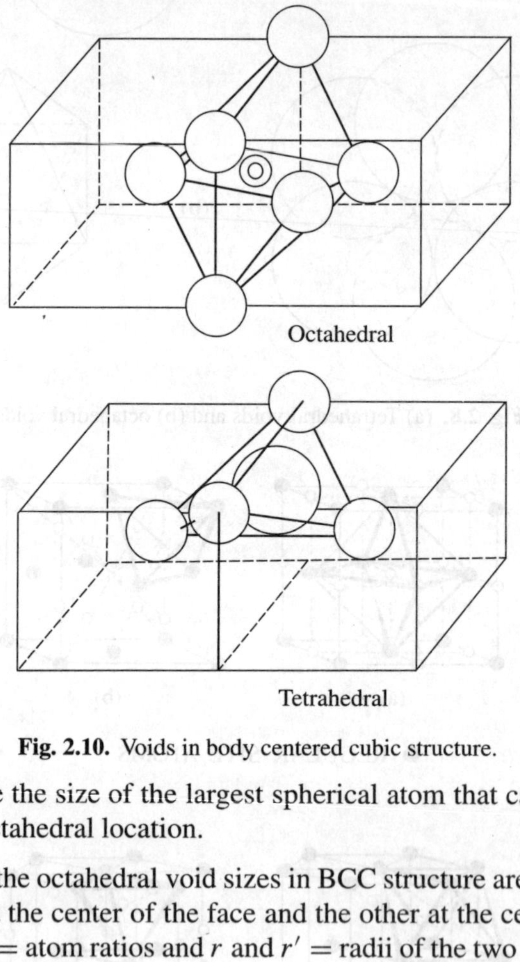

Fig. 2.10. Voids in body centered cubic structure.

Example 2.10. Calculate the size of the largest spherical atom that can just fit interstitially in a BCC structure at the octahedral location.

Solution. We know that the octahedral void sizes in BCC structure are not equal (Figure 2.10). One octahedral void is at the center of the face and the other at the center of the cube edge. If a = lattice parameter, R = atom ratios and r and r' = radii of the two types of octahedral void, then by simple geometry

$$4R = \sqrt{3}a$$

Maximum void size of (001) plane

$$2r = \sqrt{2}a - 2R$$

$$= \left(\sqrt{2} - \frac{2\sqrt{3}}{4}\right) a$$

$$= \left(\frac{2\sqrt{2} - \sqrt{3}}{2}\right) a$$

Maximum void size along [001] direction

$$2r' = 1 - 2R$$

$$= \left[\frac{2 - \sqrt{3}}{2}\right] a$$

It is evident that $r > r'$. After solving we get

$$r = 0.63R; r' = 0.154\ R$$

2.3.3 Interplaner Spacings

Interplaner spacing can be calculated for perfect crystals from the indices of a given plane by the methods of geometry. Since the Miller indices (hkl) are perpendicular to the reciprocals of the intercepts of the plane with the reference axes, one can calculate for the cubic system interplaner spacing d as

$$d = \left[\frac{a^2}{h^2 + k^2 + l^2}\right]^{1/2} \tag{2.5}$$

where a is the lattice parameter and (hkl) are the Miller indices of the plane. Table 2.3 gives the expressions for other crystal systems. Experimental value of interplaner spacing for real crystals are determined either by x-ray or by electron diffraction techniques. These techniques are very sensitive to small changes in interplaner spacings. The detailed theory behind such studies is not meant to be covered in this book.

2.4 CRYSTAL STRUCTURE OF CARBON

In diamond, a polymorph of carbon, the crystal structure consists of atoms at eight corners and in all six faces, plus atoms in interior tetrahedral positions of a cube. If we examine a single carbon atom bonded to its four nearest neighbours, a tetrahedron results (Figure 2.11 a). Four of these tetrahedral groups combine to form the large cube (Fig. 2.11 b), which is a variation of an FCC structure. There is space in the large cube for eight tetrahedra, but only four are used.

In graphite, another form of carbon, the carbon atoms form approximately hexagonal planes. Each carbon atom therefore, bonds with three other carbon atoms – the fourth valence electron from each atom behaves more like the delocalized electrons in a metal. It is these electrons which form the bonds between the planes of atoms. The presence of these unbound electrons means that graphite is a fairly good electrical conductor and like a metal it reflects light. However, graphite

Table 2.3. Interplaner Spacing of Simple Lattices*

Lattice	*Spacing, d*
Cubic	$\left[\dfrac{a^2}{h^2+k^2+l^2}\right]^{0.5}$
Tetragonal	$\left[\dfrac{a^2}{h^2+k^2}+\dfrac{c^2}{l^2}\right]^{0.5}$
Orthorhombic	$\left[\dfrac{a^2}{h^2}+\dfrac{b^2}{k^2}+\dfrac{c^2}{l^2}\right]^{0.5}$
Monoclinic	$\left[\dfrac{a^2\sin^2\beta}{h^2}+\dfrac{b^2}{k^2}+\dfrac{c^2\sin^2\beta}{l^2}-\dfrac{ac\sin^2\beta}{2hl\cos\beta}\right]^{0.5}$
Rhombohedral	$\left[\dfrac{a^2(1-3\cos^2\alpha+2\cos^3\alpha)}{(h^2+k^2+l^2)\sin^2\alpha+2(hk+kl+hl)(\cos^2\alpha-\cos\alpha)}\right]^{0.5}$
Hexagonal	$\left[\dfrac{3}{4}\cdot\dfrac{a^2}{h^2+hk+k^2}+\dfrac{c^2}{l^2}\right]^{0.5}$

**From The Solid State for Engineers,* by M.J. Sinnott.

has poor mechanical strength because of the weak bonds between the planes. In contrast, the diamond is a very poor electrical conductor, it is transparent, and is one of the hardest materials known.

Another form of carbon is fullerene which has a complex molecule with soccer ball shape, composed of 20 hexagons and 12 pentagon (Figure 2.11c). It is a sooty substance as helixes much like DNA, the basic building block of all living cells. In the solid state the C_{60} units form in a crystalline structure and pack together in a FCC array with a unit cell edge length of 1.41 nm. This is a fairly new material and is going to draw the attention of materials engineers in near future.

2.5 CRYSTAL STRUCTURES OF INORGANIC COMPOUNDS

We have already discussed the types of bonding in various crystalline materials in earlier section. The inorganic compounds' structure is related to the electronic structure and hence the nature of binding. We may here categorize some types of structure viz.

- Layer structure (e.g., Talc, Graphite)
- Ionic Lattice (e.g., NaCl or CaF_2)
- Complex ionic-covalent lattice (e.g., Silicates)
- Covalent lattice (e.g., Diamond, SiO_2)

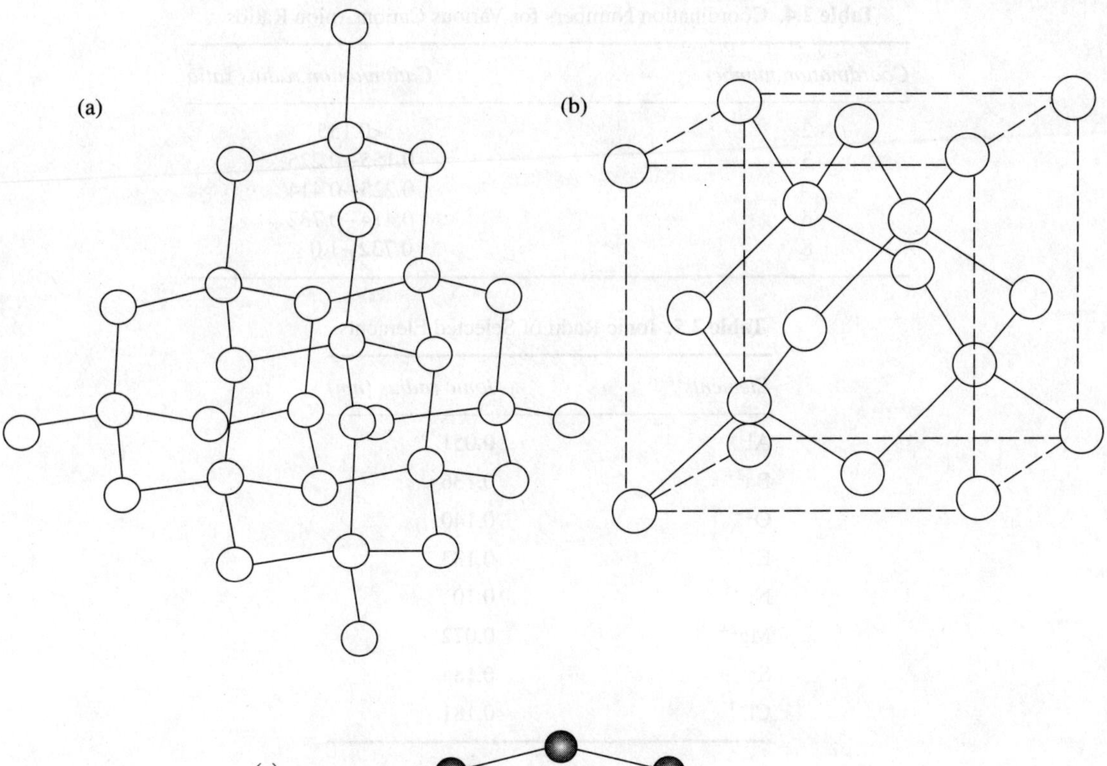

Fig. 2.11. (a) Atomic arrangement of atoms in diamond crystal (b) unit cell of diamond crystal. Similar structures are for silicon and germanium, (c) a C_{60} molecule.

Table 2.4. Coordination Numbers for Various Cation/Anion Ratios

Coordination number	*Cation/anion radius ratio*
2	<0.155
3	0.155–0.225
4	0.225–0.414
6	0.414–0.732
8	0.732–1.0

Table 2.5. Ionic Radii of Selected Elements

Elements	*Ionic radius (nm)*
Al^{3+}	0.051
Ba^{2+}	0.136
O^{2-}	0.140
F^{-1}	0.133
Na^{+1}	0.10
Mg^{2+}	0.072
S^{2-}	0.184
Cl^{-1}	0.181

2.5.1 AX Type Compounds

The crystal lattice of an ionic compound depends on the size of the ions and on their valency. For example sodium chloride (NaCl) is a simple compound of AB type, but the unit cell size is affected by the large chlorine atoms. Since they are so large, only six can surround a sodium ion, hence the adoption of the simple cube structure (Figure 2.12b). Compounds of AB type are not confined to this structure, however, and when the ions are of similar size a higher coordination number is possible as exists in caesium chloride CsCl (Figure 2.12a), in which the number is 8. When the discrepancy between ionic sizes is even wider than in NaCl, the coordination number may drop to 4, as in zinc blende (ZnS) (Figure 2.12 c). This structure is not unlike that of the diamond. The coordination number for the ranges of cation-anion ratios (r_c/r_a) are listed in Table 2.4 and the ionic radii for several ions in Table 2.5. In general, the larger the cation/anion ratio, the larger the coordination number.

Silicon carbides also has the diamond cubic structure, where each silicon atom is bonded to four carbon atoms and vice versa.

2.5.2 A_mX_p Type Compounds

CaF_2 is the typical example in this class of inorganic compound. Here one positive ion of Ca^{2+} is combined with two negative fluoride ions, resulting in charge neutrality. The r_{Ca}/r_f radius

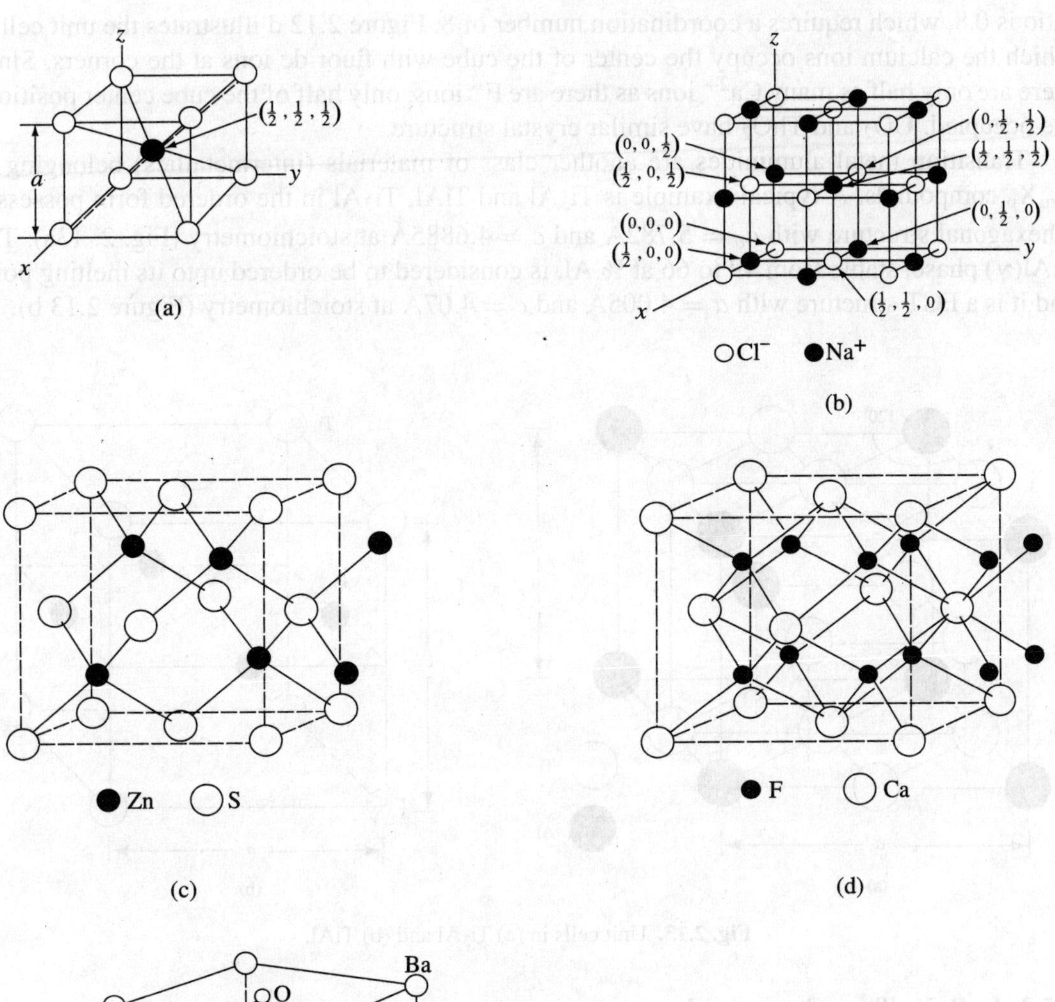

Fig. 2.12. Common cubic structures in inorganic compounds (a) CsCl; (b) NaCl; (c) ZnS; (d) CaF_2; (e) $BaTiO_3$.

ratio is 0.8, which requires a coordination number of 8. Figure 2.12 d illustrates the unit cell in which the calcium ions occupy the center of the cube with fluoride ions at the corners. Since there are only half as many Ca^{2+} ions as there are F^{-} ions, only half of the cube center positions are occupied. UO_2 and ThO_2 have similar crystal structure.

Transition metal aluminides are another class of materials (intermetallics) belonging to A_mX_p compounds. A typical example is Ti_3Al and TiAl. Ti_3Al in the ordered form possesses a hexagonal structure with $a_o = 5.782$Å and $c = 4.6885$Å at stoichiometry (Fig. 2. 13a). The TiAl(γ) phase, stable from 49 to 66 at % Al, is considered to be ordered upto its melting point and it is a FCT structure with $a = 4.005$Å and $c = 4.07$Å at stoichiometry (Figure 2.13 b).

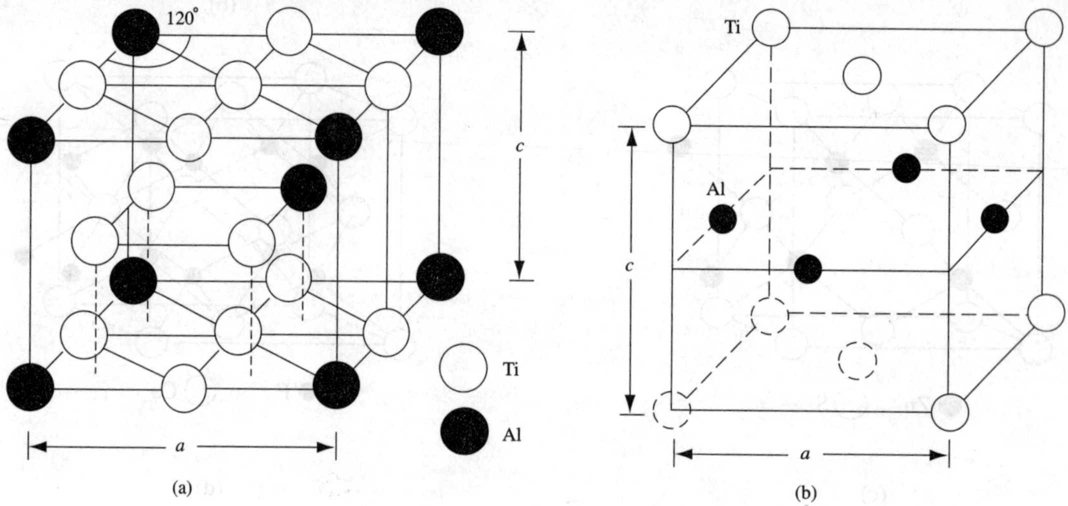

Fig. 2.13. Unit cells in (a) Ti_3Al and (b) TiAl.

2.5.3 $A_mB_nX_p$ Type Compounds

Perovskite structure represents such type of compounds. Barium titanate ($BaTiO_3$) is one of the typical examples of the perovskite. As shown in Figure 2.12e, Ba^{2+} ions are positioned at the corners of a cube, a single Ti^{4+} at the cube center and six O^{2-} ions in the center of the cube faces. $BaTiO_3$ is a very slightly distorted cubic structure at room temperature. The Ti^{4+} and O^{2-} atoms are shifted with respect to each other by almost 0.012 nm, resulting in a permanent dipole. The material is an excellent piezoelectric material and is used as electromechanical transducers. Other commercial perovskite structures include solid solutions of lead zirconate ($PbZrO_3$) with lead titanate ($PbTiO_3$).

Spinels are another examples in this group of compounds with a general formula AB_2O_4, where the metal ions A^{2+} and B^{3+} combine with O^{2-}. $MgAl_2O_4$ is typical example. The oxygen ions form a FCC lattice, with the A and B ions occupying tetrahedral and octahedral sites. All interstitial sites are not filled, nor do the B and A atoms always seek the same interstitial site. Figure 2.14 shows a typical unit cell of spinel.

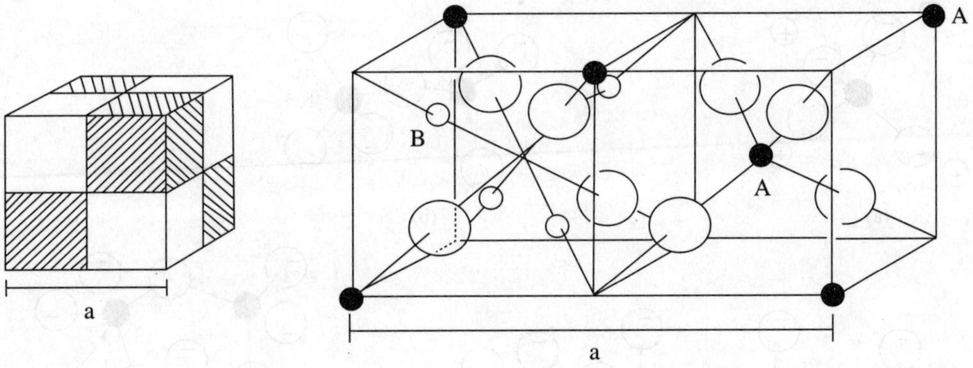

Fig. 2.14. The spinel structure. The unit cell can be divided into octants-tetrahedrally coordinated cations A, octahedrally coordinated cations B and oxygen atoms (large circles) shown in two octants only.

The combination of trivalent atoms (Fe) in tetrahedral voids and divalent and trivalent atoms mixed (Mg and Fe) in the octahedral voids is often called an *inverse spinel*. The formula is best described in general terms as $B(AB)O_4$. Many compounds of this type exist and their electrical and magnetic properties are of great importance. Magnetite Fe_3O_4 is of this type, the iron having divalent and trivalent ions arranged in the inverse spinel structure as $Fe^{3+}(Fe^{+2}.Fe^{3+})O_4$.

2.5.4 Silicates

The silicate structures are based on SiO_4^{4-} tetrahedral arrangement of silicon and oxygen. These tetrahedra can bond to metals and themselves and on many forms, such as chains, sheets, three dimensional networks etc. These materials can be explained better in terms of these structural forms rather than unit cells. For the various silicate minerals, one, two, or three of the corner oxygen atoms of the SiO_4^{4-} tetrahedra are shared by other tetrahedra to form some rather complex structures. Positive charged cations such as Ca^{2+}, Mg^{2+} and Al^{3+} serve two roles. Firstly, they compensate the negative charges from the SiO^{4+} units so that charge neutrality is achieved; and secondly, these cations ionically bond the SiO_4^{4-} tetrahedra together. Figure 2.15 summarizes these structures. When two Mg^{2+} ions are available to combine with one tetrahedron, a compound Mg_2SiO_4 or forsterite is produced. Similarly Fe^{2+} ions can combine with silica tetrahedra to produce $Fe_2SiO_4.Mg_2SiO_4$ and Fe_2SiO_4 form a series of solid solutions known as olivine or *Orthosilicates*. Two silicate tetrahedra can combine by sharing one corner to produce a double tetrahedra or a $Si_2O_7^{6-}$ ions. This ionic group can then combine with other ions to produce *pyrosilicate* or double tetrahedron compounds. When two corners of the tetrahedron are shared with other tetrahedral groups ring and chains with the formula $(SiO_3)_n^{2n-}$ form, where n gives the number of SiO_3^{2-} groups in the ring or chain. A large number of ceramic materials have this *metasilicate* structure. Wallastonite ($CaSiO_3$) is built from Si_3O_9 rings; beryl ($Be_3Al_2Si_6O_{18}$) contains larger Si_6O_{18} rings, and enstatite ($MgSiO_3$) has a chain structure. When the O:Si ratio gives the formula Si_2O_5, tetrahedra combine to give sheet structures, including clay and mica.

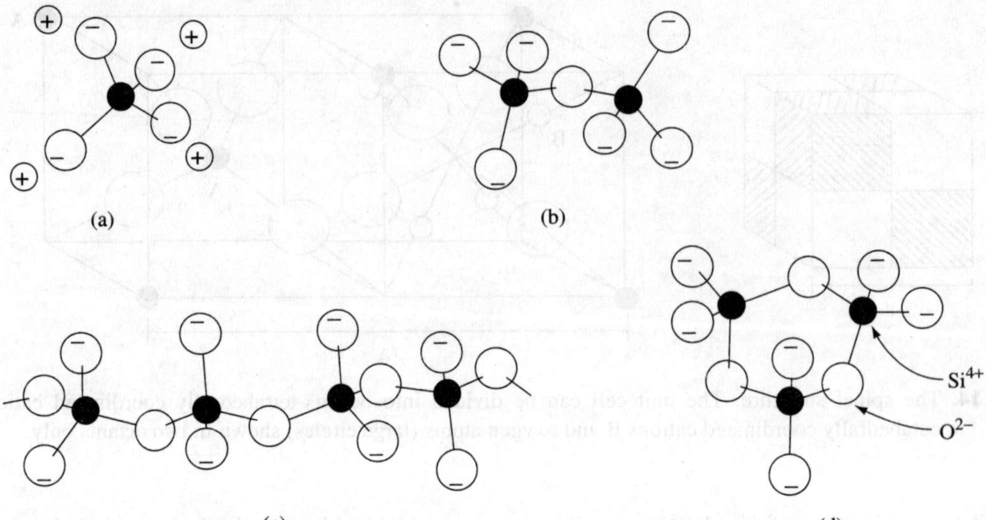

Fig. 2.15. Arrangement of silica tetrahedra (a) orthosilicate island; (b) pyrosilicate island; (c) chains; and (d) ring.

The size of the metal ions determines the composition of a silicate and affects the properties. Beryllium ions are small enough to replace silicon easily; aluminium ions can do so with slight lattice distortion. Aluminium ions fit more readily into octahedral spaces in the lattice of close packed oxygen ions, so do iron, magnesium and titanium. These form silicate lattices with very little distortion and hence fill up the spaces to make a more closely packed lattice. The larger metal ions such as sodium, potassium and calcium cannot fit into the lattice without some distortion and can only occupy octahedral spaces. These silicates are therefore less dense in the packing of their individual atoms than either pure silica or the group containing aluminium, iron, titanium, or magnesium.

Weathering of rocks produces some decomposition of the basic silicate minerals of the igneous rocks and a new set of silicates result. For example, weathering of feldspars produces the clay minerals, of which kaoline (China clay) is pure hydrated aluminium silicate. The clay minerals own their plasticity to the water of hydration and the hydrogen-hydroxyl bonding that results. Once the clay mineral losses its water, as it does when fired at high temperature, the bonding is changed, and becomes amorphous (vitrified). The water cannot be replaced easily and hence fired pottery remains reasonably strong but porous. Natural form of amorphous silica occurs in rocks that have been subjected to high temperature and rapid cooling owing to igneous activity or earth movements. Another example of silicate is Zeolites, which also occur as natural minerals and are referred as *Frame Work Silicates*. Zeolites are hydrated silicates [such as $NaCa_2(Al_5Si_5)O_{20}.6H_2O$] with an 'open' framework intersected by wide channels in which the cations and water molecules are located. The water can be expelled from these channels by heating and can be reversibly replaced or substituted by neutral molecules. In this way they can act as 'molecular sieves', permeable to only certain smaller molecules. Cations can also be reversibly exchanged within these channels in the crystal structure. This is the basis for ion exchange in the permutite system of water softening.

2.6 STRUCTURE OF GLASSES

Majority of glasses are based on silicates, but we have borate, phosphate and aluminate glasses as well. Some metallic oxides can be produced at least partially in the glassy form by rapid cooling from the melt. The replacement of silicon ion in the silica tetrahedra by sodium, potassium, calcium or lead can readily lead to production of glasses. Although there is room to spare in the silica tetrahedral network, nevertheless the specific volume is increased, partly because of the larger metallic ions, but also due to the difference in binding energy between Si^{4+} and the electrically equivalent $4K^+$ or $2Ca^{2+}$. These effects are responsible for the more open structure and for the relative ease with which crystallization can be prevented. The size of the metallic ions will thus affect the properties of the glass. Sodium and potassium ions can move through the lattice, even though amorphous, and a small degree of electrical conductively results in glasses containing these. Sodium ions are smaller than potassium, and lead ions smaller than calcium, so that the softer glasses tend to be those incorporating sodium and lead, the intermediate ones contain some calcium and the hardest potassium.

2.7 STRUCTURE OF POLYMERS

Polymers may be in semicrystalline, amorphous or in liquid crystal states. Figure 2.16(a) shows the semicrystalline configuration of polyethylene nylon or PVC containing both amorphous and crystalline regions. Such polymers are flexible and tough. In this type of polymer chains line up with each other. The three-dimensional network polymers are never crystalline. The long chain polymers may be examined under x-rays diffraction. It must be borne in mind that the crystals in crystalline polymer is referred to as molecular crystals. Figure 2.17 represents the crystalline arrangement in polyethylene, which belongs to long chain group. The cell is orthorhombic with 90 ° angles between the axes. But the unit cell points consist of a cluster of atoms, i.e., one carbon and two hydrogen, not the molecular C_2H_4. Polyethylene exists also in another crystal structure of monoclinic symmetry that can be formed as a result of deformation of the orthorhombic modification.

Figure 2.16(b) shows the disposition of chains as random coils in amorphous or glassy polymers.

There is another class of polymers known as *Liquid Crystal Polymers*, which are chemically complex and structurally distinctive materials. In the molten state, whereas other polymer molecules are randomly oriented, liquid crystal polymer molecules can become aligned in highly ordered patterns. In solid state, the molecules alignment remains (Figure 2.16 c). In addition these molecules form domain structure. There are three types of liquid crystal polymers, the full details of which are beyond the scope of discussion here. One of the biggest applications of such polymers are in liquid crystal displays e.g., digital watches, laptop computers and other digital displays. It should not be understood that they are in a liquid state, but it is entirely new state i.e., liquid crystalline state. These molecular crystals are also called *spherulites* which are

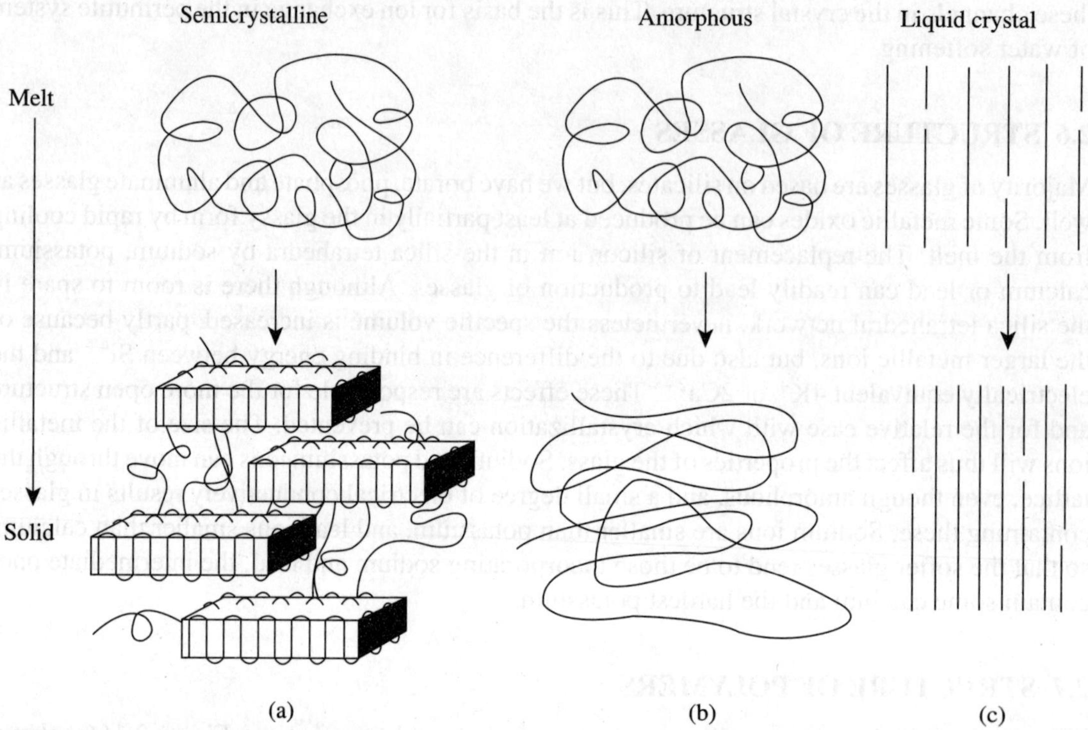

Fig. 2.16. Schematic representation of the molecular structures in both molten and solid states for (a) semicrystalline; (b) amorphous and (c) liquid crystal polymers. (Reproduced with permission from W.D. Callister, Jr, Materials Science & Engineering 5th ed John Wiley & Sons,Inc., Hoboken, NJ, 2001, p. 510).

polymer analogue of grains in polycrystalline metals and ceramics. Each spherulite is composed of many different lamellar crystals and in addition, some amorphous material. A characteristic Maltese cross pattern appears within each spherulite.

2.7.1 Structure of Polymer Chain

In a polymer chain the atoms of the backbone chain are connected by covalent bonds forming definite angles that may vary depending on the type of substituent atoms or group of atoms along the backbone chain (Figure 2.18). Such an arrangement is called the configuration of the polymer chain, which can be altered only by breaking or reforming covalent bonds. However, there is a continuous rotation or vibration of the atoms or groups of atoms about the main chain that causes a number of conformations of the polymer chain constantly changing with time.

If a polymer chain is fully extended, the carbon atoms of the backbone chain will assume a zigzag planar arrangement. The length of such an extended chain, L, can be determined as

$$L = Nd \sin \frac{\theta}{2} \tag{2.6}$$

where N is the number of links or bonds in the chain, d is the bond length and θ is the angle between the backbone chain atoms forming the bond.

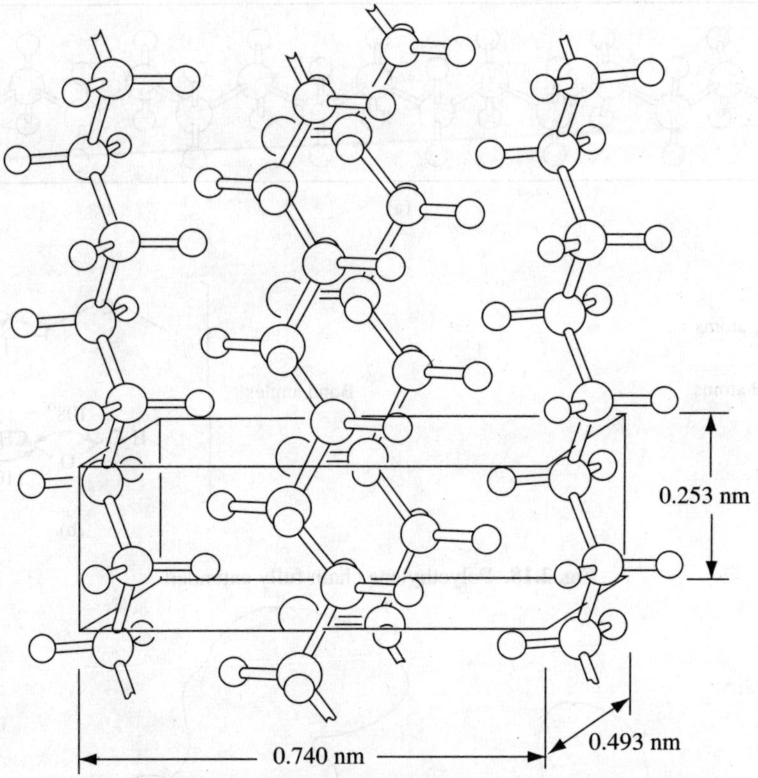

Fig. 2.17. Molecular crystal state of polyethylene (Reproduced with permission from C.W. Bunn, Chemical Crystallography, University Press, Oxford, 1945).

The size of the freely rotating polymer chain is usually determined by the mean root square distance between the first and the last link of the polymer chain (Figure 2.19). From statistical analysis the mean end-to-end distance is defined as

$$L^2 = cNd^2 \tag{2.7}$$

where c is a constant that may vary from 1 to 2. Ideally for a freely rotating chain with tetrahedral valency angles, $c = 2$.

Example 2.11. Calculate the length of an extended chain and the end-to-end chain distance of freely rotating chain for polyethylene of molecular weight 280,000. The angle between the backbone chain atoms is equal to 109.5 ° and bond length = 1.54Å.

Solution. The ethylene monomer is ($-CH_2-CH_2-$) and its molecular weight = 28

$$\text{Degree of polymerization} = \frac{\text{Molecular weight of the polymer}}{\text{Molecular weight of the monomer}}$$
$$= 280000/28$$
$$= 10,000$$

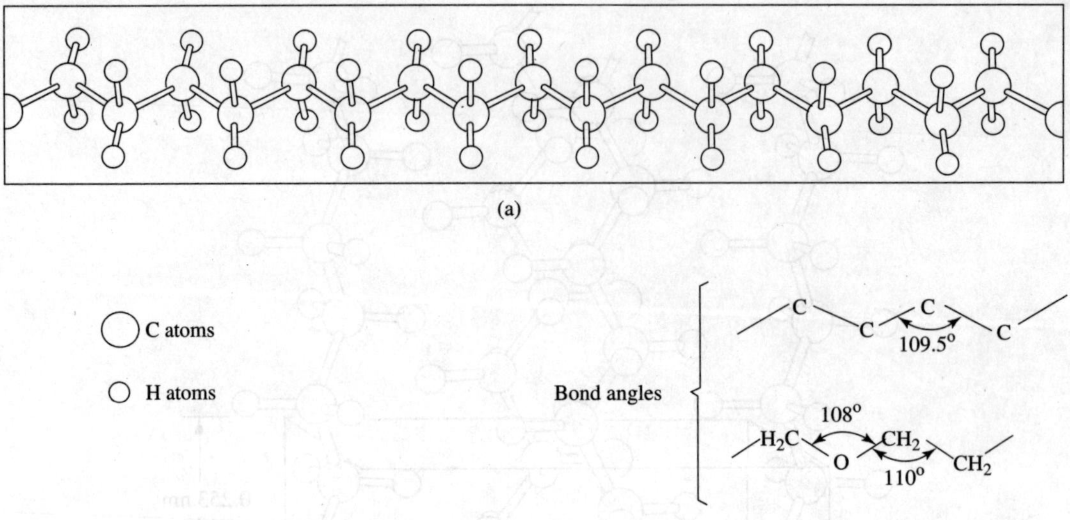

Fig. 2.18. Polyethylene chain fully extended.

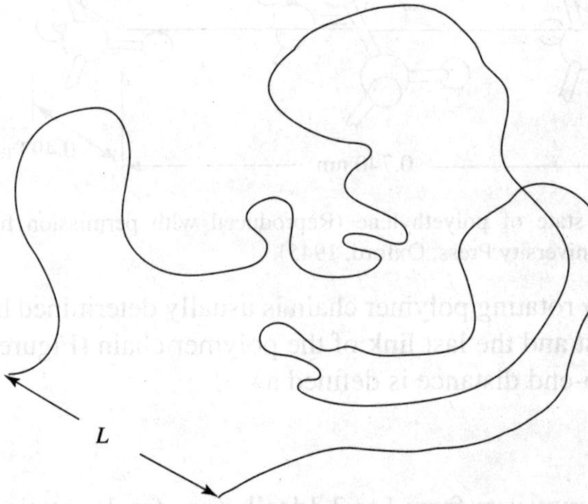

Fig. 2.19. End-to-end distortion in a freely rotating chain.

Therefore $N = 2n = 20000$

We know from equation 2.6

$$L = Nd \sin \theta$$
$$= 20000 \times 1.54 \times 10^{-2} \times \sin 53.75^\circ$$
$$= 2.54 \mu\text{m}$$

Further assuming c = 2, we get the length of a freely rotating chain (equation 2.7).

$\bar{L} = d\sqrt{Nc}$ .Substituting the values $\bar{L} = 308\text{Å}$.

The results show that the extended length is many times greater than the end-to-end distance.

2.8 DEFECT STRUCTURES IN CRYSTALS

There are three types of defects or imperfections in crystals. They are

- Point defects
- Line defects
- Surface or area defects
- Bulk or volume defects

2.8.1 Point Defects

These are vacancy and interstitials. Vacancy is nothing, but absence of an atom at a lattice site in the otherwise regular crystal. As a result, the electronic bonding of the adjacent atoms is disrupted, which changes the effective radii of these atoms. If sufficient vacancies are produced by heating a crystalline solid, the crystal structure would lose its large range symmetry and order, resulting ultimately in porosity. Such local imperfections in the crystal structure produce a disequilibrium that has great effect on the properties of crystal such as density, mechanical properties, diffusion and electrical conductivity. By themselves point defects do not affect strength as much as they affect diffusion – the migration of atoms.

Figure 2.20 shows various types of point defects. The atom which formerly occupied the site of the vacancy may be moved to the external surface of the crystal. Such type of defect is called *Schottky* defect. Another point defect is called *interstitials* which is produced by the presence of an extra atom in a void, the space between normal lattice positions (Figure 2.20b).

Another method of forming a vacancy consists in the removal of atom from a lattice site– the formation of the vacancy itself – and the transfer of this atom to an interstitial site. This method involves the formation of two point defects – a vacancy and an interstitial (Figure 2.20c). The interstitial need not be immediately adjacent to the vacancy. This type of defect is known as *Frankel defect* and differs from the Schottky defect in that the former involves two point defects while the latter involves only one.

The number of point defects at equilibrium at a certain temperature is given by

$$n_v = Ne^{-Ev/kT}$$

where n_v = The number of vacancies

N = The total number of atomic sites per cubic meter or per mole

E_v = The activation energy to form the vacancy

T = The absolute temperature

k = The Boltzmann constant

$= 1.38 \times 10^{-23}$ J/K

or 8.63×10^{-5} eV/K

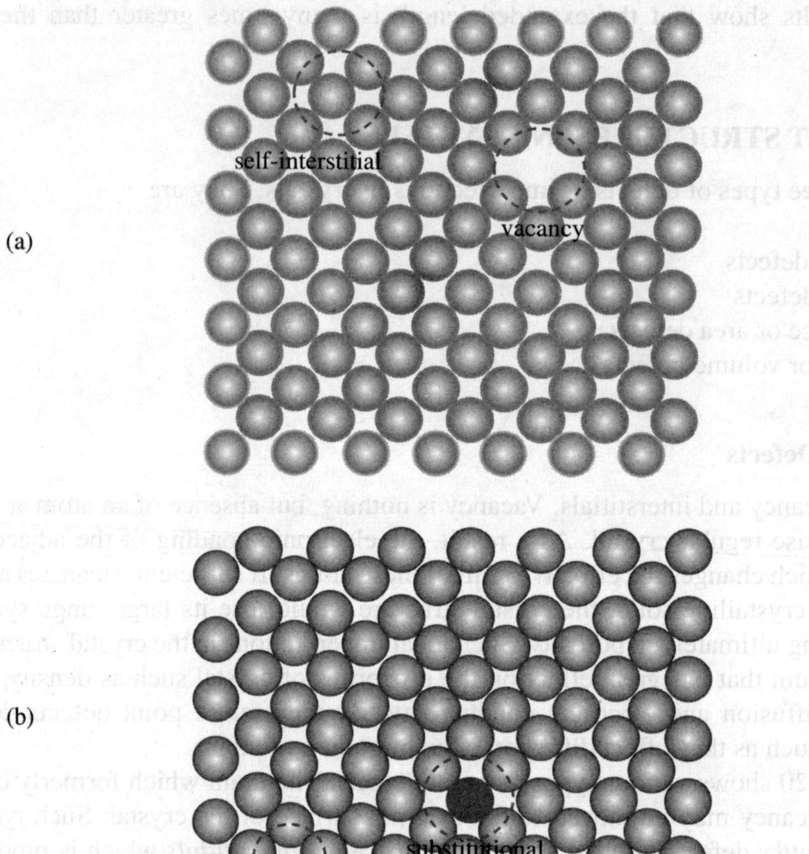

Fig. 2.20. (a) Vacancy and self interstitial (b) substitutional and interstitial impurity atoms (c) Frenkel and Schottky defects in ionic solids (Reproduced with permission from W.D. Callister, Jr, Materials Science & Engineering 5th ed, John Wiley & Sons , Inc., Hoboken, NJ, 2001, p. 68, 69 & 397).

Example 2.12. The number of vacancies present in Ag at 900 °C is $10^{24}/m^3$. Calculate the number of vacancies at 500 °C. Given that the activation energy for vacancy formation is 1.1 eV/atom.

Solution.

$$n_v(1173\text{ K}) = 10^{24}\text{ m}^{-3}$$

$$n_v(773\text{ K}) = ?$$

$$N = 10^{24}\exp\left[\frac{1.1eV}{8.62\times10^{-5}\,eV/K\times1173}\right]$$

$$= 5.3\times10^{28}\text{ m}^{-3}$$

$$n_v\,(773\;K) = 5.3\times10^{28}\exp\left[\frac{1.1}{8.62\times10^{-5}\times773}\right]$$

$$= 3.59\times10^{21}\text{ m}^{-3}$$

It is evident that the vacancy concentration shall be low at reduced temperature.

2.8.2 Line defects

These are also known as dislocations. It is a linear arrangement of atoms along which, there is some imperfections in the bonding of the atoms. Most substances deform by the gliding of atom planes; the plane and direction depending on the type of crystal structure. Calculations on the force required to cause such movements give results far above the experimentally determined values. This prompted Taylor in early 1930s to explain this discrepancy by advancing the hypothesis of an extra half plane of atoms, inserted between to parallel planes of a normal crystal. Slip would take place by movement of this extra half plane to left or right, with much less force than would normally be necessary. He called this imperfection as *edge dislocation* (Figure 2.21). In such dislocation, slip takes place along a plane perpendicular to the dislocation itself and the slip plane and slip direction remain in those of the particular lattice type involved. Such deformation requires a considerable area of the slip plane to move and this is not necessarily true in practice. Burgers proposed a second type of dislocation, in which the displacement of atoms is not confined to a half plane parallel to the normal planes of the lattice, but is arranged on two separate planes perpendicular to each other, in a helical fashion. This was called as

screw dislocation (Figure 2.22). In the screw dislocation the glide direction is parallel to the dislocation, in contrast to the perpendicular direction of the edge type. When a crystal contains both edge and screw dislocations, a well-defined area of slip results often in form of a regular figure which is a curved line called a *dislocation loop* (Figure 2.23). A dislocation line is a closed curve with a constant Burgers vector. The dislocation line shown in Figure 2.23 has a Burger Vector parallel to the line at point 2 and 4 and perpendicular to the line at points 1 and 3. The dislocation is pure screw at points 2 and 4 and pure edge at 1 and 3. Elsewhere it is mixed i.e., part edge and part screw. The motion of a screw dislocation is not confined to a single slip plane. When it moves into another slip plane having a slip direction common to the original one, the process is known as *cross slip*. Usually the cross slip plane is also a close packed plane, e.g., {111} in FCC crystals. The displacement of the half crystals meeting at the slip plane where

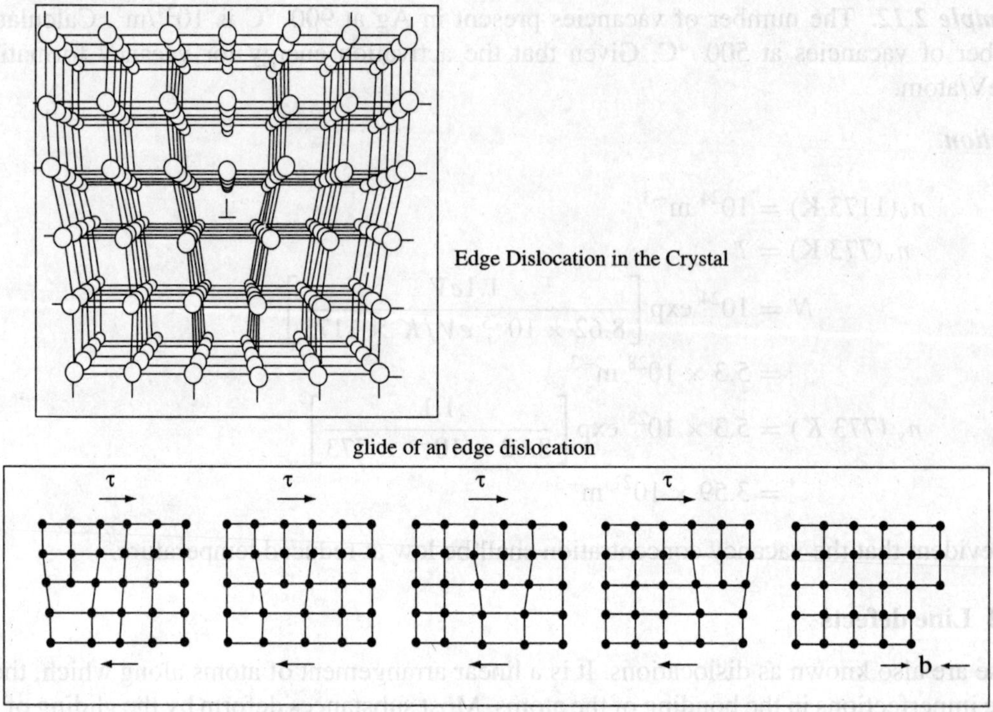

Fig. 2.21. Edge dislocation and its glide in a crystal.

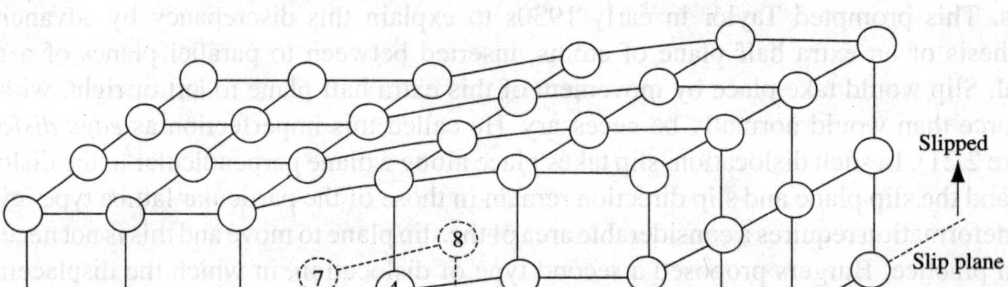

Fig. 2.22. A screw dislocation in a cubic lattice (the numbered atoms indicate the helical part of the screw).

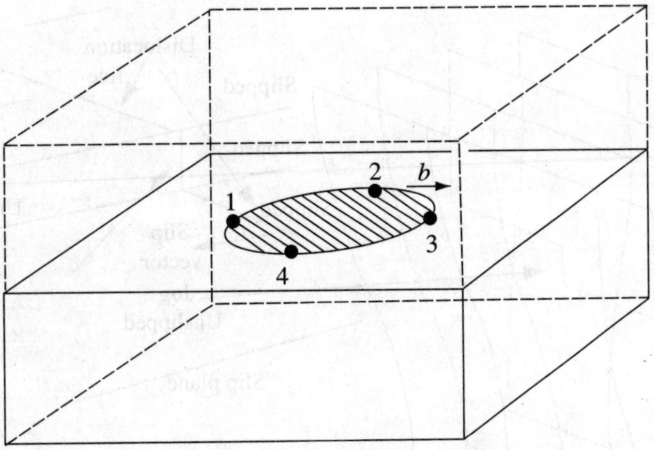

Fig. 2.23. A dislocation loop.

a dislocation moves is given the symbol, **b**, and called the *Burgers Vector*, a characteristic of the dislocation. *Dislocation Climb* is a phenomenon of the movement in the direction of the extra half plane of atoms of an edge dislocation. Dislocations can surmount obstacles in their path, by climbing into another slip plane and proceeding thence by normal slip. The addition or substraction of atoms from an edge dislocation need not be uniform across the half plane. In this case, the dislocation may operate on more than one slip plane when viewed three-dimensionally, one or more steps occurring between the various operative slip planes. These steps are known as *jogs*. Figure 2.24 illustrates a jog in an edge dislocation. In these cases, the jog can move with the dislocation by the normal process of slip. Screw dislocations may also have jogs, but since the slip direction is that of the dislocation line, the jog resembles small piece of edge dislocation, and cannot move so easily. If the jog climbs, by moving into another slip plane, point defects are created, but the screw dislocation containing the jog can continue to slip. Otherwise the jog must move sideways along the screw dislocation and attach itself to an edge component of the dislocation line. These movements are more difficult and are thought to be involved in the work hardening or strain hardening of metals and alloys.

Semicrystalline polymers exhibit line defects similar to that of crystalline materials. Screw dislocations with Burgers vector parallel to the chains in semicrystalline polyethylene have been observed by electron microscopy. These investigations also gave good evidence that the dislocations were activated by stress to generate slip steps.

Example 2.13. Show that the Burgers Vector in a crystal having a single mixed dislocation loop is constant.

Solution. Figure Ex. 2.13 shows a dislocation loop PQRS. Assume that a portion of the loop has Burgers Vector b_1, and the other b_2. Because dislocation lines must be closed curves, the two regions with different Burgers Vectors in PQRS must themselves be separated by a dislocation, say RTP, with Burgers Vector b_3. Therefore $b_3 = \pm(b_1 - b_2)$. However, if PQRS is to be single dislocation as proposed, then $b_3 = 0$. Therefore b_1 must be equal to b_2.

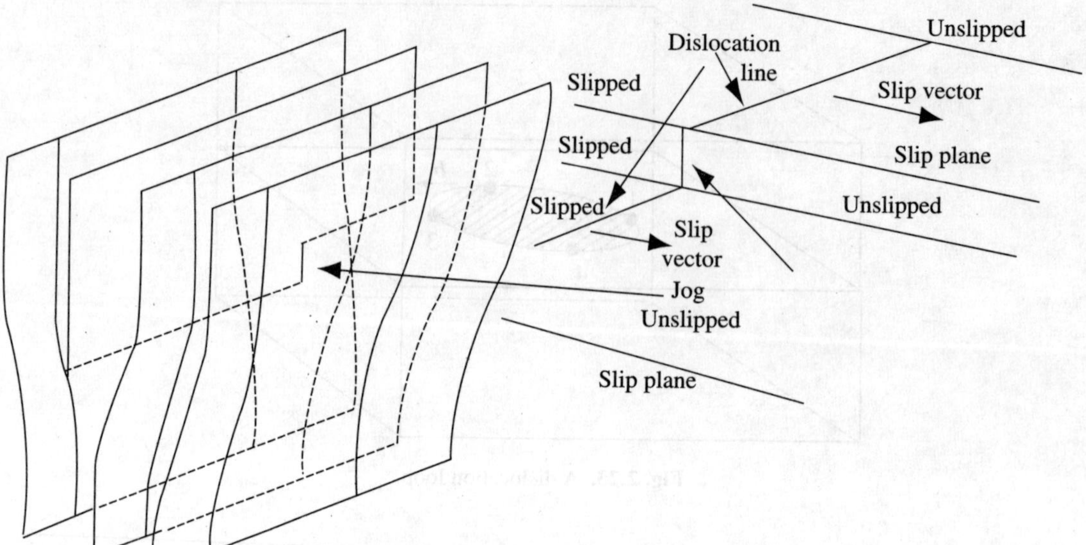

Fig. 2.24. A dislocation jog.

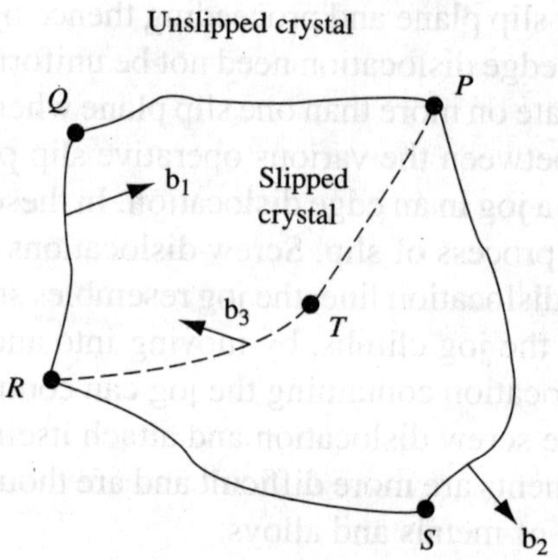

Figure Ex. 2.13

2.8.3 Surface Defects

Surface defects in crystals include external surfaces, grain boundaries, twin boundaries, stacking faults and phase boundaries. Since atoms in the free surface are not bonded to the maximum number of nearest neighbours, such defects emerge. The importance of such defects is in the sintering process, which shall be described in the Chapter 8 related to forming. Grain boundaries separate two grains or crystals having different crystallographic orientation in polycrystalline materials. Figure 2.25 shows the schematic sequence of formation of grain boundaries during solidification of a pure melt. Small angle grain boundary forms when the orientation mismatch

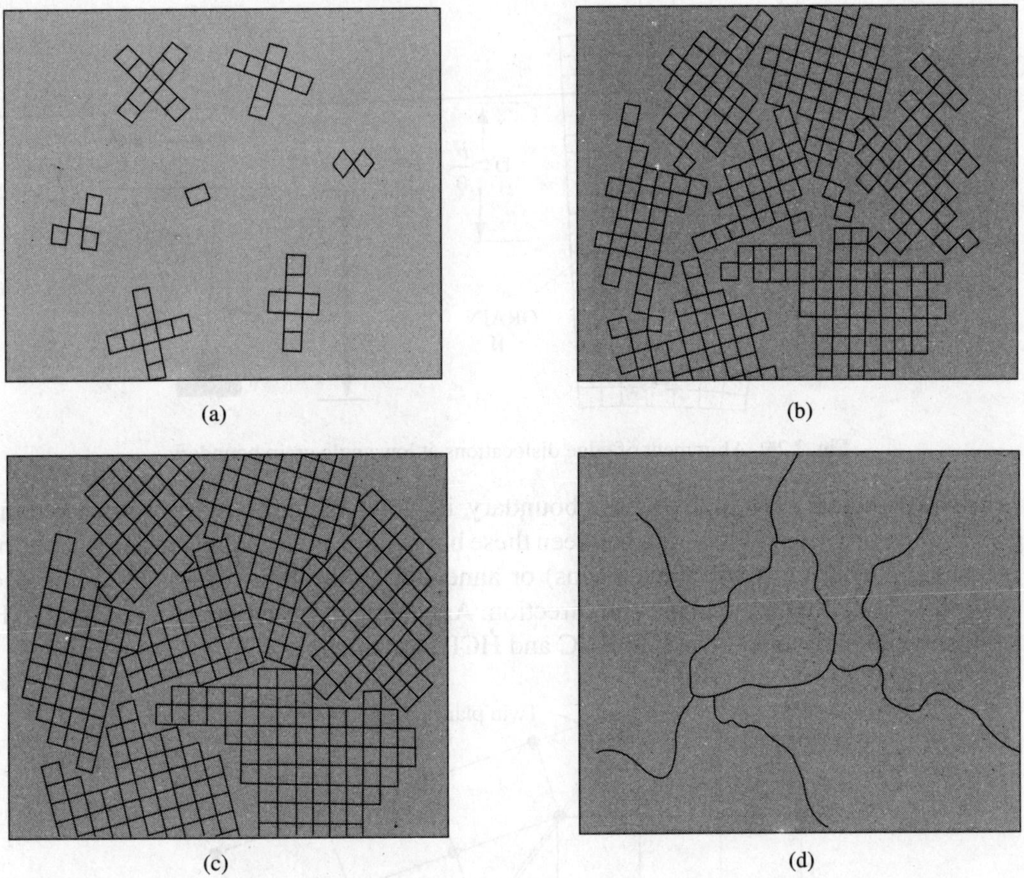

Fig. 2.25. Solidification of crystalline substances through the process of nucleation and growth. (adapted from W. Rosenhain, An Introduction to the Study of Physical Metallurgy, 2nd ed, Constable and Co. ltd., London, 1915).

is very low of the order of few degrees. Such a grain boundary is formed when edge dislocations are aligned as shown in Figure 2.26. The regions of crystals separated by small angle grain boundaries are called *subgrains*. When the grain boundary angles are high say greater than 10°, the energy values are also high.

The misorientation at the grain boundary is related to the spacing between dislocations D by the relation (Figure 2.26).

$$D = \frac{b/2}{\sin(\theta/2)} = \frac{b}{\theta} \text{ (For very small } \theta) \tag{2.8}$$

where b is the Burgers Vector. Boundaries consisting entirely of edge dislocations are called *tilt boundaries*, while the boundary consisting entirely of screw dislocation is called a *twist boundary*.

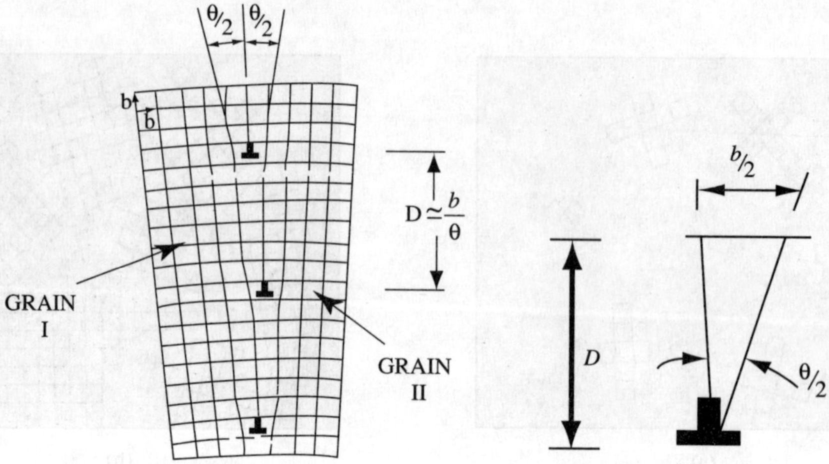

Fig. 2.26. Alignment of edge dislocations at low angle grain boundary.

The twin boundary is a type of grain boundary, in which there is a mirror lattice symmetry (Figure 2.27). The region of material between these boundaries is termed *twin*. These are formed by mechanical working (mechanical twins) or annealing (annealing twins). Twinning occurs on a specific crystallographic plane and direction. Annealing twins are generally found in FCC structures, while mechanical twins in BCC and HCP structures.

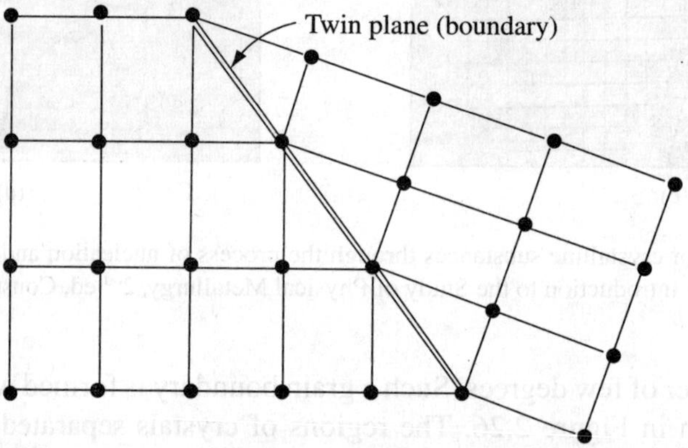

Fig. 2.27. Schematic diagram showing a twin plane or boundary.

Deformation twinning occurs when the applied stress is high due to work hardening, or low temperature or in the case of HCP metals when the resolved shear stress on the basal plane is low. This detail on the resolved shear stress shall be given in chapter 5. Copper and other FCC metals can be made to deform by twinning at very low temperatures or at very high strain rates. In HCP metals, the common twinning elements are the $(11\bar{1}2)$ plane and $[10\bar{1}\bar{1}]$ directions. Twinning results in a compression or elongation along the c-axis, depending on the ratio c/a.

Another form of surface defect i.e., stacking fault is already described earlier in section 2.3.1. Phase boundaries occur in multiphase materials e.g., $\alpha - \beta$ brass and steels.

Grain boundary triple junctions are sites where four grains or three grains boundaries meet. Such boundaries are commonly observed in crystalline solids. The number of triple junctions can have a great influence on the mechanical properties of the material. The number of triple junction in a polycrystalline solid depends on the grain size and crystal geometry. In nanocrystalline materials the volume fraction of grain boundaries is relatively very high.

Example 2.14. A low angle boundary is on array of dislocations. Calculate their spacings if the angle between the two grains is 5′. Assume all atomic spacing in the calculations as equal.

Solution. Convert angle into radians

$$5' = \left(\frac{5}{60}\right)^{\circ} = \left(\frac{5}{60} \times \frac{\pi}{180}\right) \text{ radians}$$

$$\theta = 0.00145 \text{ radians}$$

$$= 1.45 \times 10^{-3} \text{ radians}$$

$$d = b/\theta = b/1.45 \times 10^{-3} = 690b$$

$d =$ Distance between dislocations

$b =$ Burgers Vector

$\theta =$ Angle of misorientation

2.8.4 Bulk or Volume Defects

These defects are of larger dimensions than those described so far. These include pores, cracks, inclusions and other phases. These may be present in microlevel or even at macrolavels. These defects originate mainly due to poor chemistry of the material e.g., steels or during processing.

In steel making inclusion may be originated as:

- Precipitation due to reaction from the melt or during freezing.
- For nonmetallic materials which are refractories and slags that come in contact with molten metal by mechanical erosion with or without chemical corrosion, or by simple physical entrapment.
- Oxygen pick up by teeming stream and consequent oxide formation.

2.9 MICROSTRUCTURES

We have already described about grain and grain boundaries in polycrystalline materials. By studying microstructures one knows about the details of grain shape, size and their distribution. These are studied on a microscope either optical or electron. In electron microscope the resolution is very high and we can study very small size phases and other details. In optical microscopy careful surface preparations are necessary to reveal the major details of the microstructure. The specimen surface must be first ground and polished to a smoother and mirror like finish. Subsequently, they are etched using suitable chemical reagent, which attacks differentially on the grain and grain boundaries. As chemical reactivity is a function of grain orientation, in a single-phase material, the etching characteristics vary from grain to grain. Since grain boundaries possess higher energy than the grain itself, there shall be grooving along the grain boundary, which is evident under optical microscope. In two or multiphase alloys, each phase shall etch to varying degree.

In electron microscope both transmission and reflection beam mode of operation are possible. For transmission electron microscopy (TEM) the sample must be a thin foil, as crystals are highly absorptive to electron beams. The transmitted beam is projected on to a fluorescent screen or a photographic film for viewing the image. Magnification as high as 1,000,000 is possible under TEM. Fine observations of dislocation loops can be carried out under TEM. In case of *Scanning Electron Microscopy* (SEM), the specimen surface is scanned with an electron beam and the reflected or back-scattered beam of electrons is collected and displayed on a TV screen. In this case the surface may or may not be in polished condition, depending on the purpose of the study. The essential point is that the surface must be electrically conductive. In case the material is nonconductive, a very thin metallic surface coating is applied. The biggest advantage in this study is the great depth of field. Further attachment in the microscope may permit qualitative and semiqualitative analysis of the elemental composition of a very small localized surface area. Magnifications of varying level right from 10 to 50,000 times are possible with SEM. Application of SEM during failure analysis of components is very common.

Coming to the quantitative aspect of microscopy, the most important parameter is grain size. There are a number of techniques available in terms of grain volume, diameter or area. The most common one is the intercept method. Technologically the most common method is based on ASTM (American Society for Testing and Materials), which has prepared 10 standard comparison charts, all having different average grain sizes. To each is assigned a number range from 1 to 10, which is termed the *grain size number*. The sample's microstructure is photographed at 100 magnification and is matched with the standard chart. This method is very common in designating steel grain sizes. The general formula for such derivation is

$$N = 2^{n-1} \tag{2.9}$$

where $N =$ Average number of grains per square inch at magnificent of 100X.

$n =$ grain size number.

Since the grains are actually three dimensional, a two dimensional plane is not truly representative. In addition, the varying shape of the grains means that only a most probable grain size is used. If many grains are measured and the frequency of each grain diameter is plotted, a typical curve shown in Figure 2.28 is usually obtained. The value chosen to represent the grain size is the value associated with the peak of the frequency plot, d_1.

2.10 MACROSTRUCTURES

Macrostructures in polycrystalline materials are examined under the naked eyes or a low magnification using magnifying lens. This is useful in examining bulk or volume defects or nonuniformities. The examination may be carried out on unetched or etched samples. In case of inclusion impurities in cast alloys, there is fibering of inclusions during further hot working. Macrostructure of cast and welded part are often examined. Some examples are shown in Figure 2.29.

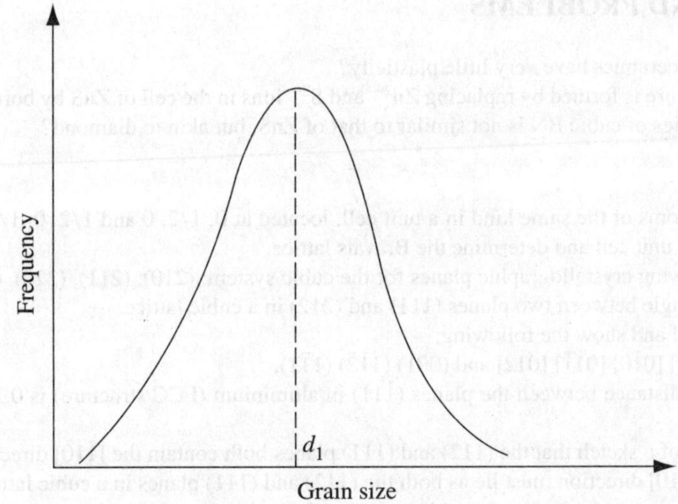

Fig. 2.28. Grain size distribution in a crystalline material.

Fig. 2.29. Macrostructures of etched polished sections of (a) aluminium alloy ingot; (b) weld deposit in steel.

QUESTIONS AND PROBLEMS

2.1 Why crystalline ceramics have very little plasticity?

2.2 Cubic B N structure is formed by replacing Zn^{2+} and S^{2-} ions in the cell of ZnS by boron and nitrogen atom. Why the properties of cubic BN is not similar to that of ZnS, but akin to diamond?

Crystal Structure

2.3 There are two atoms of the same kind in a unit cell, located at 0, 1/2, 0 and 1/2, 0, 1/2 in an orthorhombic lattice. Draw the unit cell and determine the Bravais lattice.

2.4 Sketch the following crystallographic planes for the cubic system: (210), (211), (220), (310), (311), (321).

2.5 Determine the angle between two planes (111) and (212) in a cubic lattice.

2.6 Sketch a unit cell and show the following:
[100] [110] [120] $[0\bar{1}0]$ $[01\bar{1}]$ [012] and (001) $(1\bar{1}\bar{2})$ $(1\bar{1}1)$,

2.7 The interplanar distance between the planes $(\bar{1}11)$ in aluminium (FCC structure) is 0.2338 nm. What is the lattice constant?

2.8 Show by means of a sketch that the (112) and (111) planes both contain the $[\bar{1}10]$ direction.

2.9 Show that the $[\bar{1}10]$ direction must lie as both the (112) and (111) planes in a cubic lattice.

2.10 Are planes farthest apart in the simple cubic crystal and a body centred cubic crystal identical? If not, explain why?

2.11 Determine the angle between the (100) and (101) planes in a body centred tetragonal crystal with lattice parameters $a = 5.60$ Å and $c = 3.13$ Å.

2.12 Draw a unit cell of closed packed hexagonal system and indicate the slip plane and direction as $\{11\bar{2}2\} < 11\bar{2}\bar{3} >$. Is this direction closest packed direction?

2.13 Indicate the following planes and direction in hexagonal unit cell $(1\bar{2}10)$, $(10\bar{1}2)$, $(\bar{1}011)$, [110] $[1\bar{1}1]$, [021].

2.14 HCP metals generally twins during deformation in $(10\bar{1}2)$ plane and $[10\bar{1}\bar{1}]$ direction. Show both the features in a unit cell.

2.15 Prove that the Miller indices of direction of two parallel vectors passing through different points are identical.

2.16 Prove that parallel planes which are crystallographically equivalent have identical Miller indices.

2.17 Show that the octahedral site in the center of the face of the BCC unit cell is equivalent to the center of an edge of the unit cell.

2.18 Determine the percentage volume change that occurs when titanium changes from a BCC structure to HCP structure. With the BCC structure the lattice parameter $a = 0.332$ nm and with HCP structure $a = 0.296$ nm and $c = 0.468$ nm.

2.19 An intermetallic compound of Ni-Al system has a cubic structure with nickel atoms at the cube corners and aluminium atoms at face centers. What is the compound?

2.20 The lattice parameter of the cubic cell of NaCl after x-ray measurement is 5.64 Å. How does it compare with the lattice parameter expected from the values of ionic radii $Na^+ = 0.98$Å and $Cl^- = 1.81$Å respectively?

2.21 Calculate the packing factor and density of NaCl if the ionic radii of Na^+ and Cl^{-1} are 0.97Å and 1.81 Å respectively. The atomic masses of Na and Cl are 22.99 g/g.mole and 35.45 g/g mole respectively.

2.22 NaCl has a cubic lattice with 4 Na^+ and 4 Cl^- ions per unit cell located as follows:

Na^+	000	1/2 1/2 0	1/2 0 1/2	0 1/2 1/2
Cl^-	1/2 1/2 1/2	0 0 1/2	0 1/2 0	1/2 0 0

Draw the unit cell of NaCl indicating the positions of Na^+ and Cl^- ions.

2.23 From the knowledge of the crystal structure, calculate the density of diamond (radius of carbon atom = 0.071 nm).

2.24 Establish that the packing factor for the diamond crystal is 46% that in the FCC structure.

2.25 If Ti atoms assume an FCC array and all octahedral interstices are occupied by C atoms, we have the structure of TiC. Sketch the unit cell of TiC and compare this with the NaCl structure unit cell. TiC is largely metallic with some covalent bonding character, whereas NaCl is ionic. In terms of bonding and atomic or ionic sizes, explain why identical structures result for TiC and NaCl.

2.26 GaAs is a typical semiconductor compound that has the Zinc blende structure
(a) Draw the cubic unit cell of ZnS structure showing the positions of Ga and As atoms.
(b) Make a drawing showing the crystallographic directions and the positions of the atom for the (111) plane.

2.27 Boron nitride can be easily machined, silicon nitride cannot. What structural differences account for this?

2.28 Describe the physical and structural differences of an amorphous metal above and below its glass transition temperature.

2.29 The C–C single bond distance in polyethylene is 1.54 Å. Assuming that the molecules are perfectly extended, determine the length of a polyethylene molecule with a degree of polymerization 40,000. In reality the bonds are zigzag. Does it increases or decrease the true distance?

2.30 The density of a polyethylene crystal is 1.014 mg m^{-3} at 20 °C. The density of amorphous polyethylene at 20 °C is 0.84 mg m^{-3}. Estimate the percentage crystallinity in a low density polyethylene with a density of 0.92 mg m^{-3} at 20 °C.

2.31 Nylon drape alternatively stretches and shrinks in climates in which humidity varies greatly. Explain why it is so sensitive to water.

Defect Structures

2.32 Calculate the ratio of the equilibrium concentration of vacancies in pure copper (Q_f = 120 kJ/mol) at 300 K to that produced at 1000 K.

2.33 A crystal has an enthalpy of formation of vacancies of 1.5 eV, establishing a certain equilibrium concentration of vacancies at the temperature 1200 K. By how much does the temperature has to be raised, in order to increase the vacancy concentration by a factor of 10?

2.34 Describe two methods of vacancy formation. Illustrate with help of sketches.

2.35 Explain the difference between screw and edge dislocations.

2.36 What is the effect of two edge dislocations of same and opposite signs on the same glide plane in a crystal?

2.37 What is dislocation annihilation? Explain how it can occur by (a) glide and (b) climb.

2.38 A single crystal of copper is subjected to shear stress. Indicate where the positive edge dislocation moves (a) towards the surface or (b) towards the axis. What about the screw dislocations?

2.39 Draw a closed dislocation loop consisting of (a) positive edge, (b) right hard screw, (c) negative edge and (d) left hard screw.

2.40 Draw schematically surface energy of a tilt boundary vs. tilt angle curves for two cases (a) small angle and (b) large angle.

2.41 Deformation twinning is observed in hexagonal metals. Indicate other crystal systems which undergo such type of phenomenon. Give example of metals after referring Metals Hand Book.

2.42 Explain how surface, grain boundaries and dislocations (edge and screw) can act as sources or sinks for vacancies.

2.43 Why electron microscopic technique is not often applied in studying the defect structures in polymers?

Microstructure

2.44 How the microstructure of a ceramic is different from that of a recrystallized metal?

2.45 Show with help of neat microstructures (schematic) in what possible manners two phases of varying sizes and shapes can be distributed.

2.46 Explain how by mere macrostructural examination of an engineering component you can assess its quality.

REFERENCES

Structure and Properties of Engineering Materials, B. Harris and A.R. Bunsell, Longman, London, 1977.
Modern Physical Metallurgy and Materials Engineering, 6th ed., R.E. Smallman and R.J. Bishop, Butterworth-Heinemann, Oxford, 1999.
An Introduction to Ceramic Science, D.W. Budworth, Pergamon Press, Oxford, 1970.
The Structure of Materials, S.M. Allen and E.L. Thomas, MIT Press, Boston, U.S.A., 1999.
Structure and Properties of Engineering Materials, 5th ed., D.P. Henkel and D.W. Pense, McGraw Hill, N.Y., 2000.

3

PHASE EQUILIBRIA AND TRANSFORMATION

Phase equilibria study of a single component (e.g., carbon) or a multicomponet system (e.g., steels) is important in order to know the phase stability under various conditions of temperature and pressures. As materials processings are mostly carried out at normal atmospheric pressure, one generally studies phase diagrams depicting stability of various phases only with respect to temperature. Before we go in details, it is worth while to define few terms.

Alloy: Alloy is a macroscopically homogeneous mixture of two or more substances. Steel is an iron alloy made by combining iron, carbon and some other elements. Brass is an alloy of copper and zinc, while duralumin is an Al-Cu alloy. Like metallic alloys, polymer alloys are blends of polymers or copolymers with other polymers or elastomers. In strict sense all composites are also alloys, including intermetallics.

Solid Solution: A solid solution is a solution in solid state that consists of two or more kind of atoms combined in one type of space lattice. There are two kinds of solid solutions. *Substitutional Solid Solutions* is one in which the solute atom occupies, in the solvent crystal a position belonging to one of the atoms of the solvent metal. *Interstitial Solid Solution* is one, in which the solute atom enters the vacant spaces between atoms in the lattice of the solvent crystal, without displacing a solvent atom.

Isomorphous Solid Solutions are always of substitutional kind. Cu-Ni, Cu-Sn are examples of substitutional type, while the ferrite and austenite phases in steels pertain to interstitial type of solid solution.

A cooling curve of a pure metal or alloy is a convenient tool to construct a phase diagram. However it is important that the cooling be carried out under equilibrium condition, i.e., very very slowly. Figure 3.1(a) shows the cooling curve of a pure metal. At melting point there is an

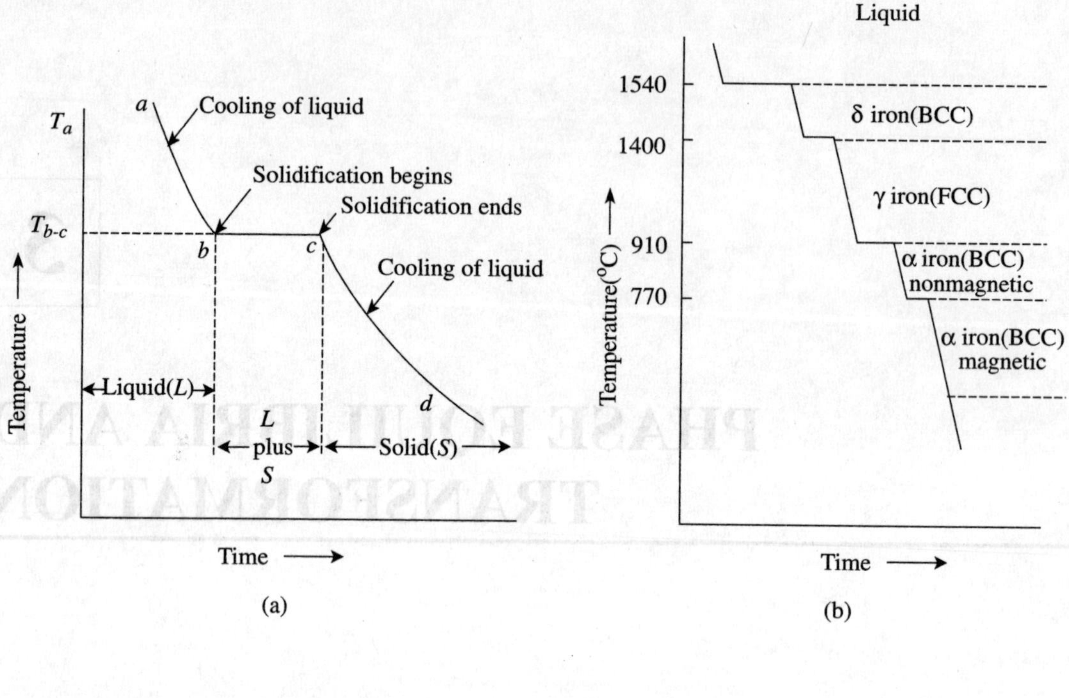

Fig. 3.1. Cooling curves of (a) pure metal without undergoing any polymorphism, (b) pure iron undergoing polymorphism and (c) solid solutions.

arrest in the cooling curve, when the latent heat of fusion is liberated. In case of a solid solution e.g., Cu-Ni alloy, solidification takes place at a range of temperature (Figure 3.1c). At point b (liquidus) solidification begins and at 'c' (solidus) completes. Here there is no horizontal arrests in the cooling plot. In case a pure metal undergoing polymorphism is cooled from the melt, at different critical transformation temperatures called isotherms, there are different arrests. At these temperatures one form of crystalline arrangement transforms into another undergoing solid to solid transformation (Fig. 3.1b). An intermetallic component of a fixed stoichiometry A_xB_y or a pure ceramic material as Al_2O_3 will have a cooling curve similar to that of a pure metal, i.e., they will have a fixed melting point. In binary phase diagrams they are shown as vertical lines.

3.1 THE PHASE RULE

This states that the maximum number of phases *P* which can coexist in a chemical system, or alloy, plus the number of degree of freedom *F* is equal to the sum of the components *C* of the system plus 2.

$$P + F = C + 2 \tag{3.1}$$

Phases are the homogeneous parts of a systems which having definite bounding surfaces, are separable by mechanical means alone, for example gas, liquid and solid.
Degree of Freedom are those externally controllable conditions of temperature, pressure and composition, which are independently variable and which must be specified in order to define completely the state of the system at equilibrium. One should mark the word independently variable.
Components are the smallest number of substances of independently variable composition making up the system. In case of metallic alloys, it is usually sufficient to count the number of elements present. In case of mixture of stable compounds, such as sugar and water, the number of component may be taken as two, unless carried out to a degree of temperature and pressure where one or both of the compounds decomposes.
Phase diagram or equilibrium diagram of an alloy system is a graph that identifies alloy phases occurring at various temperatures and percentages of alloying elements. It can be binary, ternary or quaternary depending on whether the number of components are two, three or four. In comparison to copious information on the phase diagrams of crystalline alloys, very little information is available for polymeric systems. Only recently the polyethylene/diphenyl ether system has been examined and a phase diagram established. The classical phase rule does not apply in such systems.

3.2 UNARY PHASE DIAGRAM

Figure 3.2 shows a typical phase diagram of a single component i.e., pure metal. Trasitions such as melting, sublimation and vaporization occur as boundaries between the three single phase fields are crossed. At triple point 0, all the three phases coexist simultaneously. Applying the phase rule, here the degree of freedom is zero, i.e., the system is invariant. A point within one

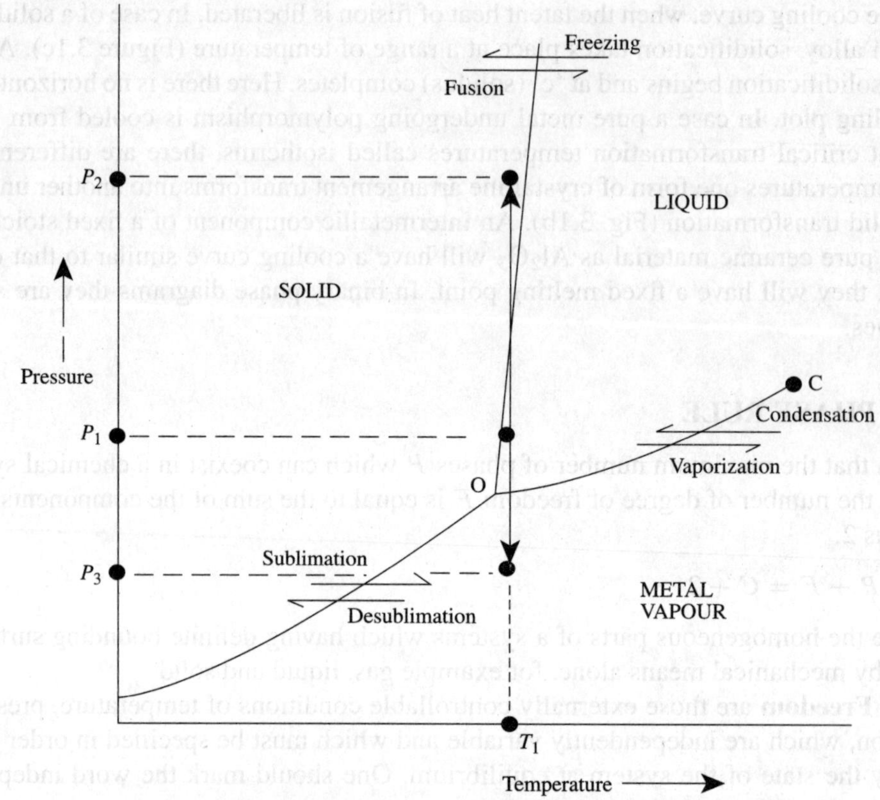

Fig. 3.2. Phase diagram (P, T plot) for a pure metal.

of the single phase fields represents a system, where degree of freedom is two. Both temperature and pressure must be specified if its state is to be fully defined. The line for solid/liquid transition is interesting, as here unlike in the system water the slope is towards the right. Bismuth and gallium behave like water, as they expand on freezing. Increase in pressure from P_1 to P_2 leads to freezing. The pressure at the critical point C, beyond which liquid and vapor are indistinguishable, is usually extremely high and considerably greater than one atmosphere. In case pressure is reduced from P_1 to P_3, vaporization is favoured. For most metals, the pressure value at the triple point is far below one atmosphere. For copper and lead, it is in the order of 10^{-6} atm. The pressures associated with the solid/vapor line are even much lower. In metals where the vapor pressure is high i.e., mercury, zinc, cadmium and magnesium, the full knowledge of pressure-temperature diagrams is a must.

3.3 BINARY PHASE DIAGRAMS

3.3.1 Isomorphous System

A binary isomorphous system is one in which two metals or compounds are mutually soluble in all proportions and in all states, the solid, liquid and gas. In order to produce a complete phase diagram of a binary alloy system in which the three externally variables i.e., pressure, temperature and composition are all represented, it is necessary to use a three dimensional space model, which is known as PTX (pressure-temperature-composition) diagram.

As in practice, engineers are mostly confronted with processing at 1 atmosphere of pressure, the use of TX diagram (temperature-composition diagram) would suffice. In other words TX diagrams are all horizontal sections taken arbitrarily through the PTX diagram at a constant pressure of 1 atmosphere. If we take help of cooling curves of a series of solid solutions of varying composition, we can construct a binary phase diagram (A-B) as shown in Figure 3.3.

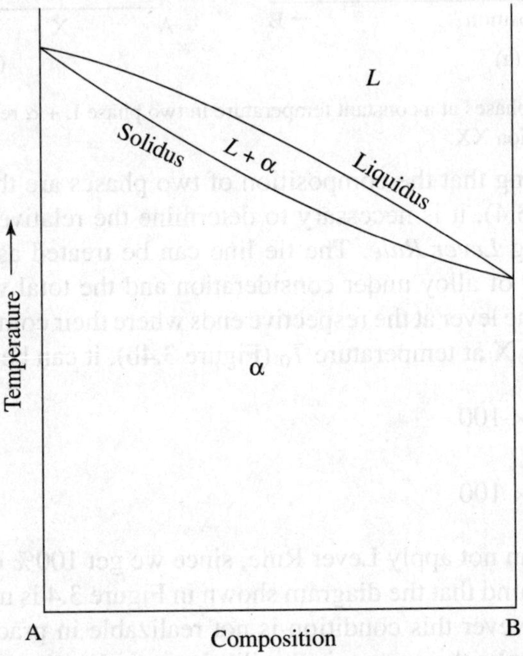

Fig. 3.3. A phase diagram of isomorphous solid solution system.

In this case at temperature T_1, except the ends all points lie within the $L + \alpha$ field and must represent the coexistence of liquid and solid solution (α) (Figure 3.4a). The point α_1 denotes the composition of the solid solution where the quantity of liquid reaches zero. Similarly point L_1 denotes the composition of the liquid solution, where the amount of solid reaches zero. In between α_1 and L_1, the phases are L and α; but the relative ratios of these are different. The line $\alpha_1 - L_1$ is called *tie line*. Figure 3.4b shows, where an alloy of fixed composition (X%B) is selected. Here the vertical line intersects the phase diagram at two temperatures liquidus and solidus. At an intermediate temperature T_0, again within the tie line both α and liquid phases coexist, but their compositions are different i.e., α_0 and L_0 respectively. By applying phase rule ($P + F = \mathrm{C} + 1$), it is evident that within the two phase area, the degree of freedom is one. In other words one can vary either temperature or composition independently, without disturbing the chemical equilibrium.

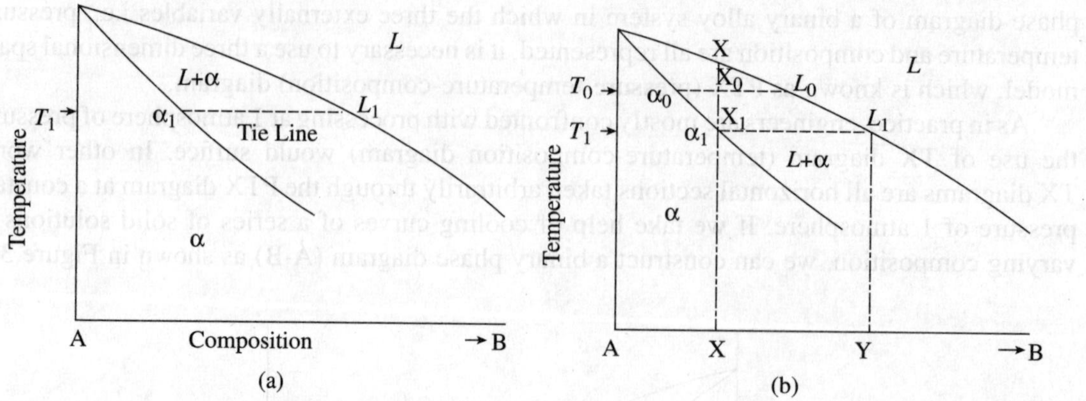

Fig. 3.4. (a) Composition of phases at a constant temperature in two phase L + α region (b) composition of phases for a fixed composition XX.

Now after ascertaining that the composition of two phases are the same in all mixtures at temperature T_0 (Figure 3.4), it is necessary to determine the relative amount of these phases. This is done by applying *Lever Rule*. The tie line can be treated as a lever with its fulcrum at the gross composition of alloy under consideration and the total weight of each of the two phases impinging upon the lever at the respective ends where their compositions are represented. Mathematically for alloy X at temperature T_0 (Figure 3.4b), it can be said that

$$\%\alpha = \frac{X_0L_0}{\alpha_0L_0} \times 100 \tag{3.2}$$

$$\%L = \frac{\alpha_0X_0}{\alpha_0L_0} \times 100 \tag{3.3}$$

At temperature T_1, we can not apply Lever Rule, since we get 100% α solid solution.

It must be borne in mind that the diagram shown in Figure 3.4 is under equilibrium solidification of the alloys. However this condition is not realizable in practice, because it will take very very long time to make the system in equilibrium, so that the compositions of these two phases α and Liquid readjusts. Nonequilibrium solidification is characterized by an increased temperature range over which liquid and solid are present and by a composition range remaining in the solidified alloy. This is illustrated in Figure 3.5. In other words if the rate of cooling is fast, the displacement of the average solid solution compositions would be more. It is obvious from the diagram that in temperature range T_4 and T_5, the solid solution coming out shall be richer in solute i.e., B. If one looks into the microstructure, the grain of solid solutions shall be *cored*, i.e., each grain has a high melting central portion surrounded by lower melting material, such a microstructure is undesirable and must be eliminated by suitable heat treatment known as *homogenization*. The heat treatment temperature must be so selected that there is no scope of liquid phase formation.

There are certain binary isomorphus systems, where the liquidus and solidus do not vary continuously from the melting point of one component to that of the other, but, instead, pass through a minimum temperature, which lies below the melting points of both components (Figure 3.6). When this occurs, the liquidus and solidus meet tangentially. The alloys which

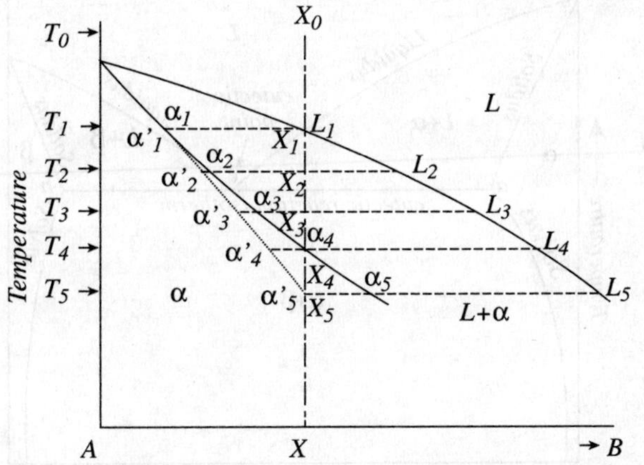

Fig. 3.5. Effect of faster cooling on the solidus curve in isomorphous solid solution system.

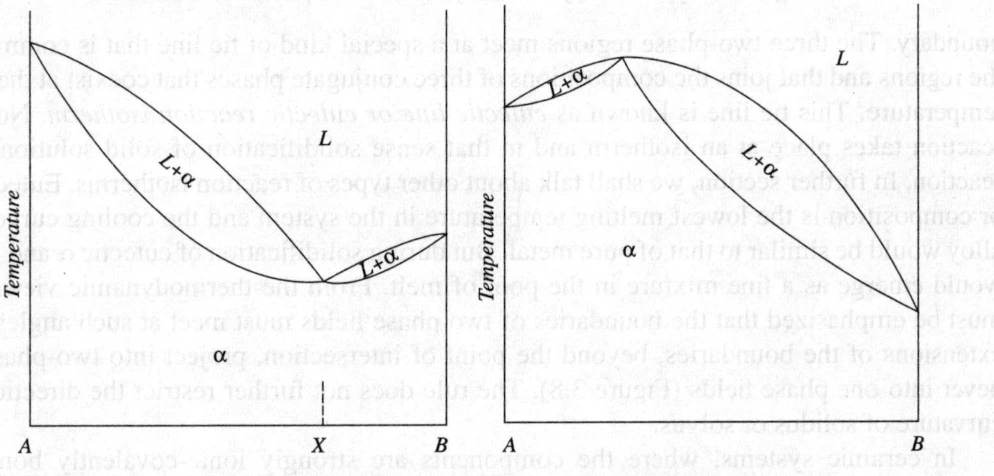

Fig. 3.6. Solid solutions showing congruent melting.

melt and solidify with the liquid and solid phases undergoing no changes in composition, are called *congruently melting alloys*. The congruently melting alloy may be treated as a component of an alloy system. Thus the system AB may be divided into two systems at composition X, one being the simple isomorphous binary system AX and the other XB.

In contrast to isomorphous solid solutions, we have terminal or primary solid solutions in which the solute concentration is limited in the solid solution (Figure 3.7). According to *Vegard's law*, the lattice parameters of primary solid solutions vary linearly with the atomic percentage of the solute element.

3.3.2 Eutectic and Eutectoid Systems

Figure 3.7 shows a schematic eutectic diagram. α and β are terminal or partial solid solutions, since they originate at a side of the diagram. It is evident that both α and β solid solutions undergo solid state transformation along solvus curve. This curve is also known as solid solution

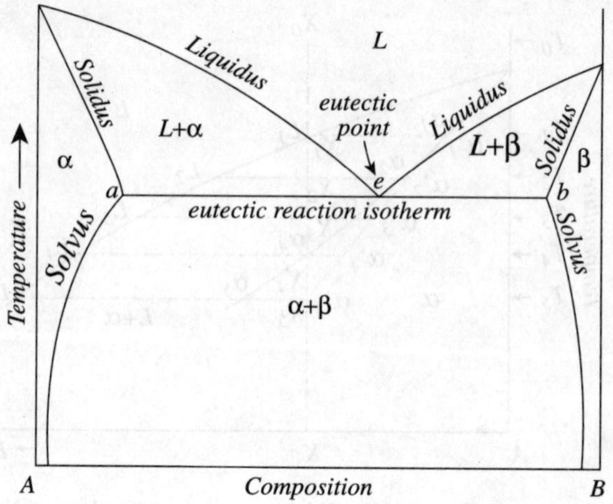

Fig. 3.7. A typical binary eutectic system showing partial solid solubility.

boundary. The three two-phase regions meet at a special kind of tie line that is common to all the regions and that joins the compositions of three conjugate phases that coexist at the eutectic temperature. This tie line is known as *eutectic line or eutectic reaction isotherm*. Note that a reaction takes place at an isotherm and in that sense solidification of solid solution is not a reaction. In further section, we shall talk about other types of reaction isotherms. Eutectic point or composition is the lowest melting temperature in the system and the cooling curve of such alloy would be similar to that of pure metal. But during solidification of eutectic α and β phases would emerge as a fine mixture in the pool of melt. From the thermodynamic view point, it must be emphasized that the boundaries of two phase fields must meet at such angles that the extensions of the boundaries, beyond the point of intersection, project into two-phase fields, never into one phase fields (Figure 3.8). The rule does not further restrict the direction of the curvature of solidus or solvus.

In ceramic systems, where the components are strongly ionic-covalently bonded, the possibility of limited solid solubility is far lower.

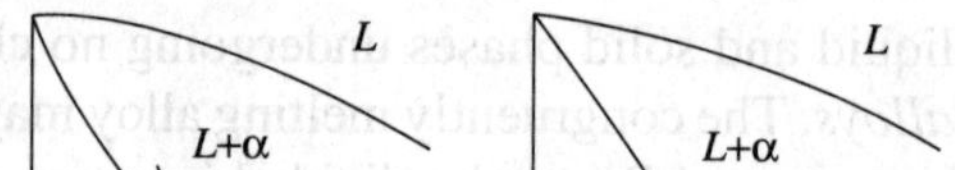

Fig. 3.8. Correct curvatures of solidus and solvus lines in the range of partial solid solutions.

As described in previous section, the lever rule is equally applied in the regions of two phases in such systems as well. The alloy on the left hand side of the eutectic composition in a binary phase diagram is known as *hypo-eutectic*, while on the right hand side is called *hypereutectic*.

As an example in the real world, we shall illustrate the binary silver-copper system, which forms partial solid solutions along with an eutectic reaction (Fig. 3.9). The melting points of Ag and Cu are 960 ° and 1083 °C respectively. Silver rich α and copper rich β partial solid solutions are designated in the figure. The maximum solid solubility of Cu in Ag is 8.8%, while for Ag in Cu is 8.0%. Eutectic composition exists at 28.5% Cu, while eutectic temperature is 779 °C. If we notice the solvus lines, it is evident that both the solubility of copper in silver (α) and the solubility of silver in copper (β) decrease with fall in temperature. Such a nature is useful in development of agehardening alloys, which shall be described in the later section. In Figure 3.9 two compositions are marked one containing 5% Cu and the other 65% Cu. In the first case as soon as solidification begins α dendrites are of composition Q. When we cool the alloy slowly, at temperature S_1 the composition of liquid shall be P_1 and that of solid solution as Q_1. With fall in temperature up to temperature R, no change in phase results. As the temperature falls slowly β phase continues to precipitate from α and both solid solutions α and β change in composition along the solvus RF and UG respectively. The changes in microstructure are depicted on the left hand side of the figure.

The alloys containing 65% Cu is a hypereutectic composition and the sequence of microstructure at various temperature is shown at the right hand side of the figure. At temperature T_1, the amount of primary β phase and liquid phase can be determined by Lever Rule. If we assume very slow cooling rate, cooling below eutectic temperature, where every thing is in solid state gives interesting observations. Since the solid solubility is varying with temperature, during cooling copper will be rejected from the α layers of the eutectic in the form of particles of the other solid solution β. These particles will join the layers of β near to those of α. Similarly silver will be precipitated from the β layers in the form of a small amount of α which will join the α - layers. The crystals of primary β will also reject some silver in the form of α, which is sometimes seen as a film surrounding the primary β grains (see the bottom most microstructure on the right hand side of Figure 3.9). When temperature is below 400 °C, there is not much change in microstructure because of the extremely low rate of diffusion in the system.

The eutectoid system is similar to eutectic with the only difference that the reaction is in entirely solid state i.e., $\gamma \underset{Heating}{\overset{Cooling}{\Leftrightarrow}} \alpha + \beta$. The significance of this diagram is in the heat treatment of steels.

3.3.3 Peritectic and Peritectoid Systems

The peritectic reaction consists upon heating in the decomposition of one solid phase into a liquid and a new solid phase:

$$L + \alpha \underset{Heating}{\overset{Cooling}{\Leftrightarrow}} \beta$$

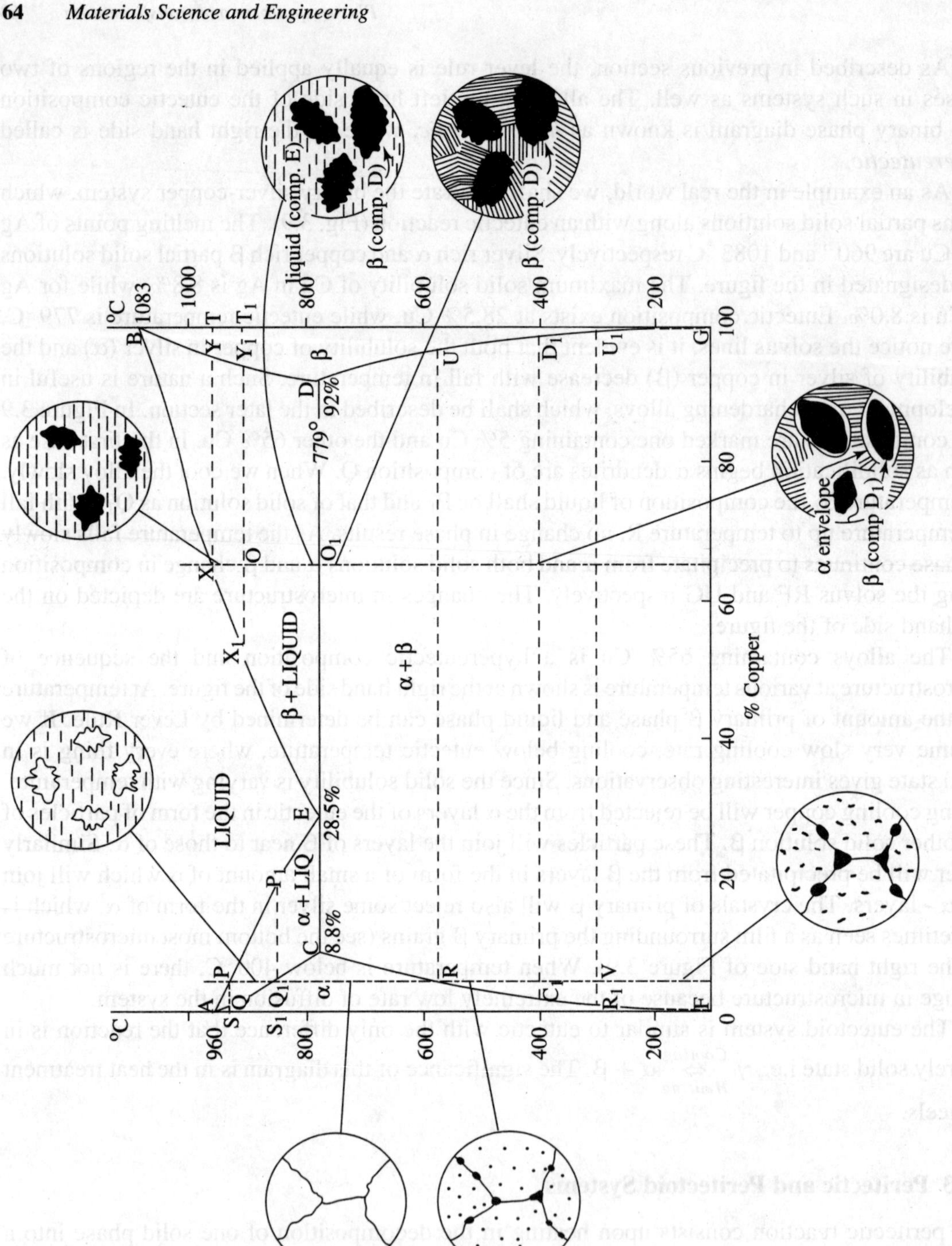

Fig.3.9. Binary phase diagram of Ag-Cu system. Also slown are the microstructural evolution in partial solid solution and hyper eutectic alloy.

Fig. 3.9.

This type of reaction is represented in Figure 3.10 where at peritectic reaction isotherm such a decomposition occurs. The β phase of peritectic composition (Point p) melts by decomposition into $L + \alpha$ of composition l and α, at the peritectic temperature. Figure 3.11 shows three alloys X, Y, Z of hypoperitectic, peritectic and hyperperitectic compositions respectively. In two phase regions, we can apply Lever Rule.

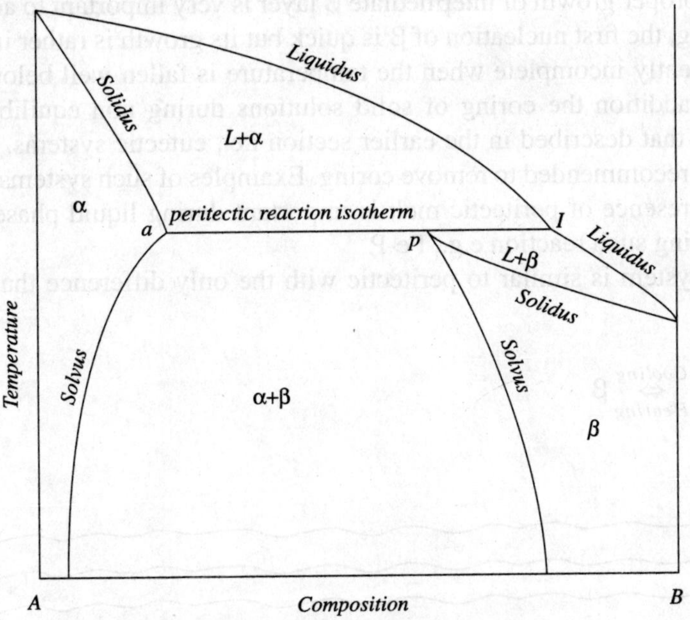

Fig. 3.10. Binary peritectic phase diagram.

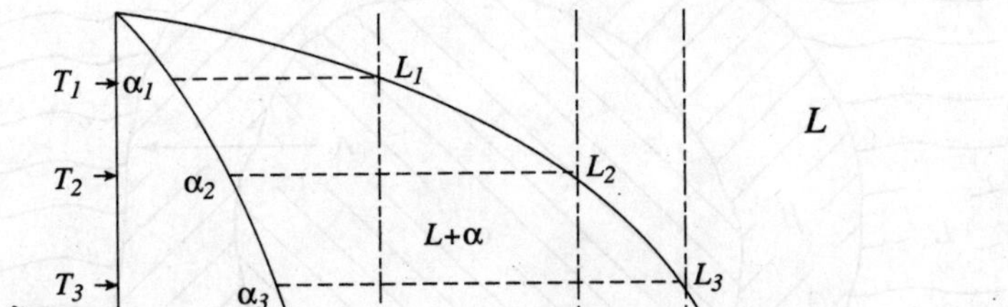

Fig. 3.11. Different alloys undergoing peritectic reaction.

In natural solidification of peritectic alloys, a much larger departure from equilibrium is felt. This is because of the fact that β phase, the reaction product, forms at the interface between solid α and liquid. This provides obstruction to diffusion, which is necessary for the completion of the reaction. Figure 3.12 shows schematically how to change α into β, since A- atoms must diffuse out of the α phase and are replaced by B-atoms, which must come from the liquid phase. In other words the proper growth of intermediate β layer is very important to achieve equilibrium. In normal cooling, the first nucleation of β is quick but its growth is rather impeded, so that the reaction is frequently incomplete when the temperature is fallen well below that of peritectic temperature. In addition the coring of solid solutions during non equilibrium cooling does occur, similar to that described in the earlier section i.e., eutectic systems. A homogenization heat treatment is recommended to remove coring. Examples of such systems are Cu-Zn, Cu-Sn, and Fe-C. The presence of peritectic melt is important during liquid phase sintering of alloy systems undergoing such reaction e.g., Fe-P.

Peritectoid system is similar to peritectic with the only difference that the reaction is in solid state i.e.,

$$\alpha + \gamma \underset{Heating}{\overset{Cooling}{\Leftrightarrow}} \beta$$

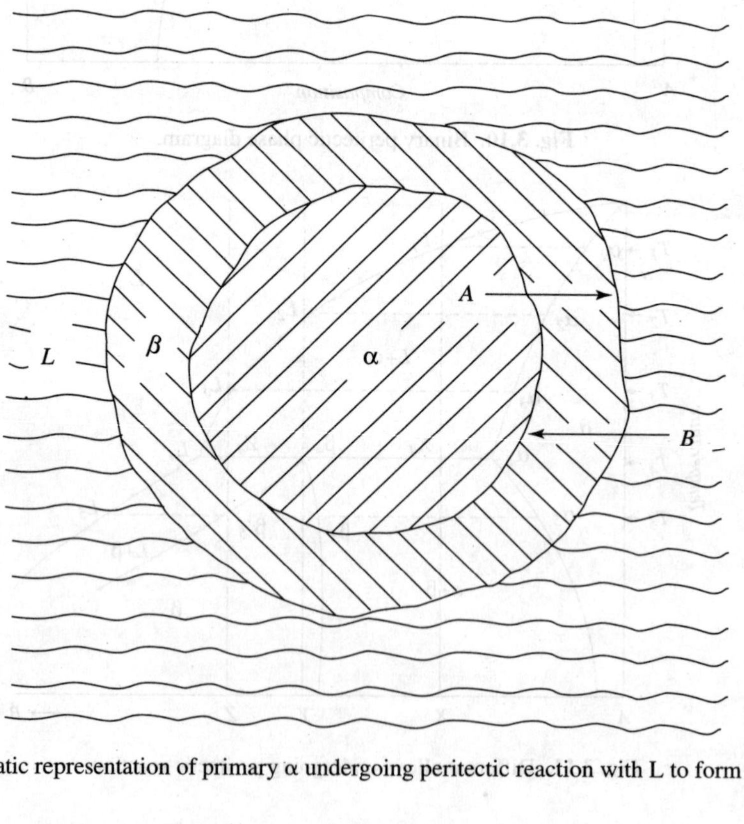

Fig. 3.12. Schematic representation of primary α undergoing peritectic reaction with L to form an envelop of β.

3.3.4 Other Binary Reactions

Apart from the four reactions described in earlier sections, there are two more reactions. They are:

$$\text{Monotectic}: \quad L_I \underset{Heating}{\overset{Cooling}{\Leftrightarrow}} \alpha + L_{II}$$

$$\text{Syntectic}: \quad L_I + L_{II} \underset{Heating}{\overset{Cooling}{\Leftrightarrow}} \beta$$

The coexistence of L_1 and L_2 signifies liquid immiscibility. A typical example is Cu-Pb system.

Figure 3.13 illustrates an imaginary binary phase diagram, where all the six types of isothermal reactions are designated.

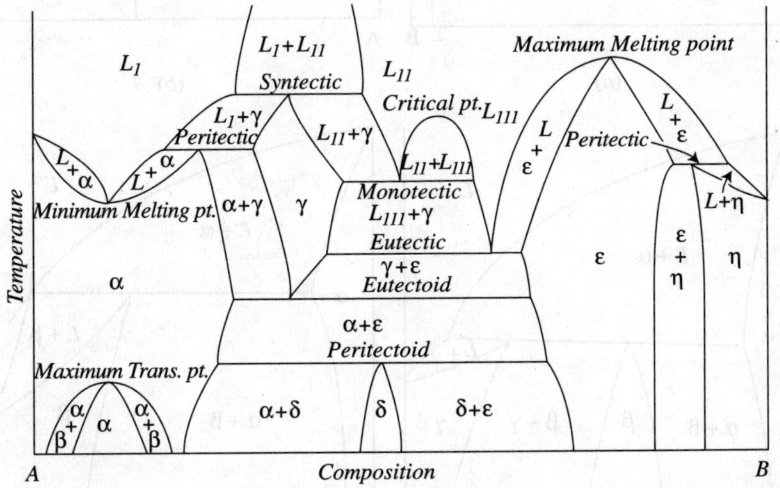

Fig. 3.13. Imaginary binary phase diagram showing various isothermal reactions.

Example 3.1. A melts at 1000 °C; B melts at 700 °C. There is a peritectic reaction: α (5%B) + L(50%B) ⇔ β (30%B) at 800 °C; a eutectic reaction: L(80%B)⇔ β (60%B)+γ (95%B) at 600 °C; and a eutectoid reaction: β (50%B) ⇔ α (2%B) +γ (97%B) at 400 °C. Complete the diagram.

Solution. See Figure Ex. 3.1.

Example 3.2. Figure Ex. 3.2 shows some incorrect phase diagrams. Indicate what are the mistakes.

Solution.

(a) System pertains to Binary eutectic. Eutectic reaction is at a fixed temperature. Hence it must be an isotherm.

(b) Peritectic alloys $L + \alpha \rightarrow \beta$ must solidify incongruently.

(c) At the isotherm, we can have one composition (peritectic), where three phase reaction takes place. $L + \alpha$ cannot transform to $\beta + \gamma$. Hence $\gamma/\beta + \gamma$ phase boundary is not

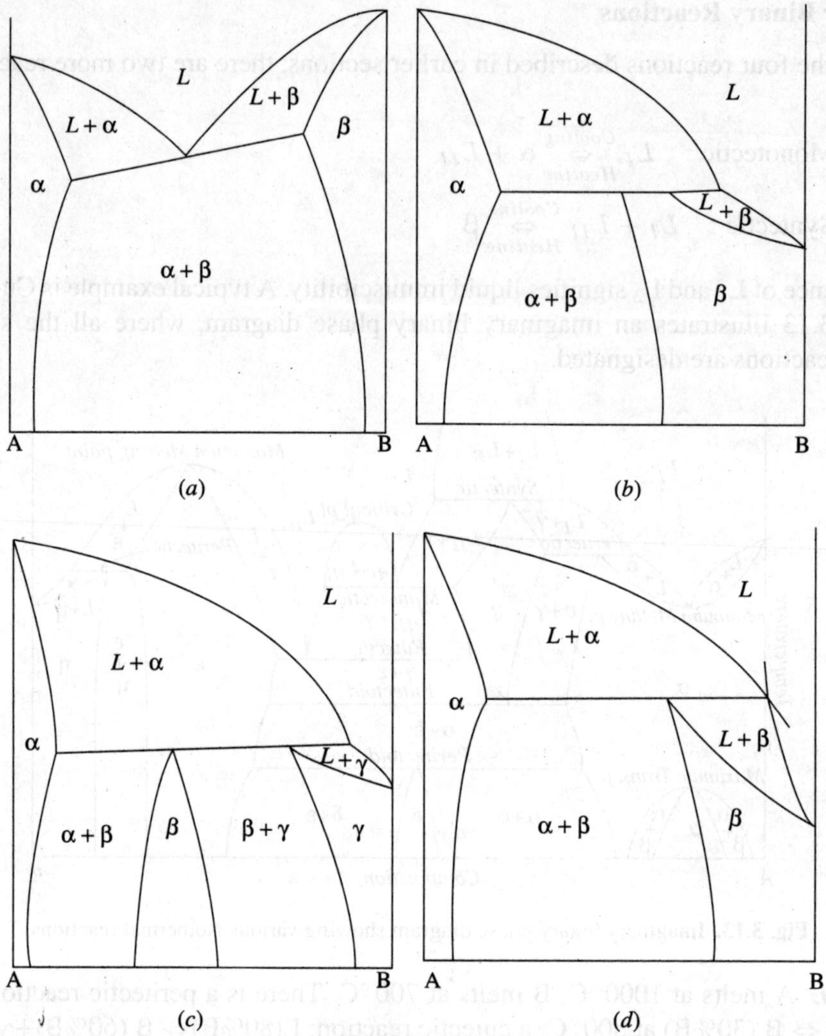

Figure Ex.3.1

tenable. So the diagram should be like in (d).

(d) The slopes of the liquidus curves must be such that after extrapolation it must extend to a two phase region. The curve showing L/L + β phase boundary needs to be modified.

Example 3.3. Two eutectic reactions in the Mg-Pb system are:

L(66.8w/o Pb) ⇔ (41.7w/o Pb) + Mg_2Pb

L(97.8w/o Pb) ⇔ Mg_2Pb + β(99.3w/o Pb)

Draw schematically Mg-Pb phase diagram. Immediately upon completion of the eutectic reaction in a 20 w/0 Mg alloy,

(i) What weight per cent of the alloy is proeutectic Mg_2Pb?

(ii) What weight per cent of the alloy is Mg_2Pb that is contained within the eutectic solid?

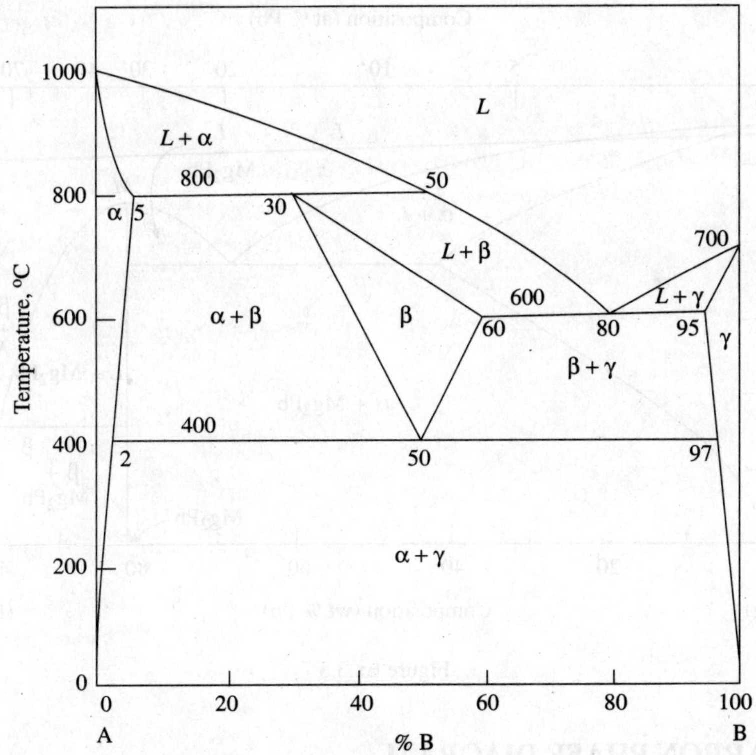

Figure Ex.3.2

(Given atomic weight of Mg = 24.31, atomic weight of Pb = 207.2, M.P. of Mg 922 K and of Pb 601 K).

Solution. The phase diagram of binary Mg-Pb is given in Figure Ex. 3.3. In the intermetallic Mg_2Pb, 2/3 atomic fraction (66.66 at.%) of Mg and 1/3 of atomic fraction (33.33 at.%) of Pb are present.

Wt.% of Pb in $Mg_2Pb = 207.2 \times 33.33 \times 100/24.31 \times 66.66 + 207.2 \times 33.32$
$= 81\%$

Therefore the first eutectic reaction can be written as:

L(66% Pb) $\Leftrightarrow$ α(41.7w/o Pb) + Mg_2Pb(81% Pb)

$$\text{(i) Wt. Proeutectic } Mg_2Pb = \frac{80 - 66.8}{81 - 66.8} \times 100 = 92.95\%$$

$$\text{(ii) Wt.\% eutectic in the given alloy (80\% Pb)} = \frac{81 - 80}{81 - 66.8} \times 100 = 7.05\%$$

$$\text{The Wt.\% } Mg_2Pb \text{ in eutectic} = \frac{66.8 - 41.7}{81 - 41.7} \times 100 \times \frac{7.05}{100} = 4.5\%$$

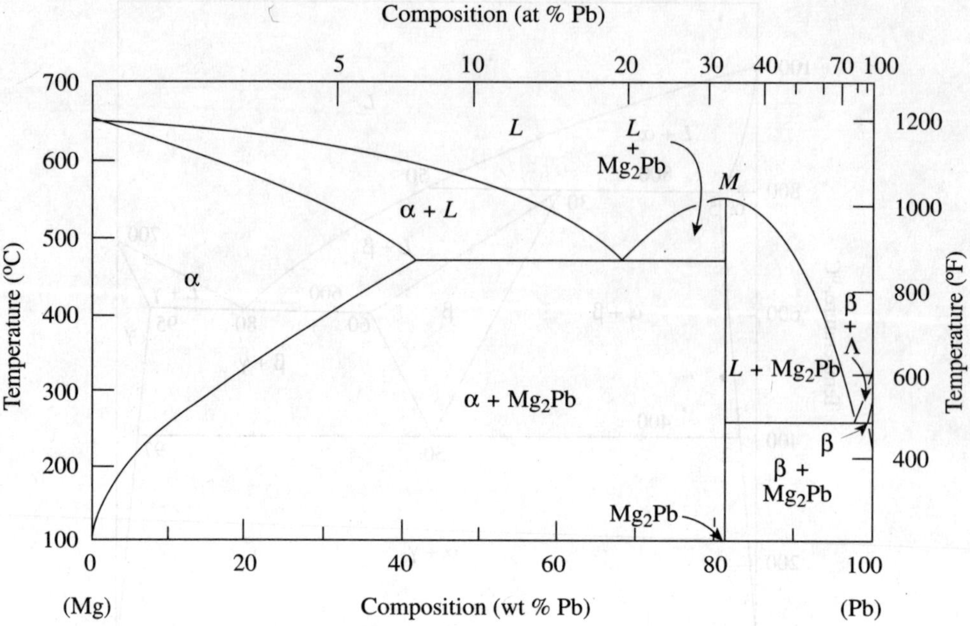

Figure Ex.3.3

3.4 IRON-CARBON PHASE DIAGRAM

Iron-carbon phase diagram is one of the most important diagram of any metallic system. As a matter of fact, it should be called iron-Fe_3C metastable diagram, as in real practice the nucleation of stable equilibrium graphite is very very difficult. Although graphite can be made to form in high carbon cast iron, it can only form in steels (< 2%C) after heating for years at temperatures above 600 °C. Cementite, Fe_3C possesses orthorhombic structure.

The Fe-C diagram (Figure 3.14) illustrates three types of reactions (1) peritectic at 1495 °C (2) eutectic at 1147 °C and (3) eutectoid at 727 °C. The maximum solubility of carbon in ferrite (α iron) is 0.022%, while in austenite (γ - iron) it is about 2%. The diagram may be divided into two portions, the left hand side pertaining to steels, while the right hand belongs to cast iron. It is obvious that melting point of cast irons is at far lower temperature than that of steels. From the diagram it is evident that near 0.8%C (exactly 0.77%C) known as eutectoid composition, the austenite transforms into an eutectoid mixture of ferrite and cementite known as *pearlite*. Figure 3.15(b) shows the mixtures of ferrite and cementite. Steels with carbon less than the eutectoid composition is known as hypoeutectoid steels and those greater than eutectoid composition (0.8% C) as hypereutectoid steels. The eutectoid temperature is also called A_1 temperature. Austenite/Austenite-Ferrite phase boundary depicts A_3 temperature, while in case of hypereutectoid steels this phase boundary designates A_{cm} temperature. It is evident that the slope of A_{cm} line is greater than A_3 line. In hypoeutectoid steels, we have primary ferrite and pearlite phases at room temperature, where as for hypereutectoid steels, they are cementite and pearlite. Cementite is noticed as a network around pearlite colonies. Since cementite is a hard

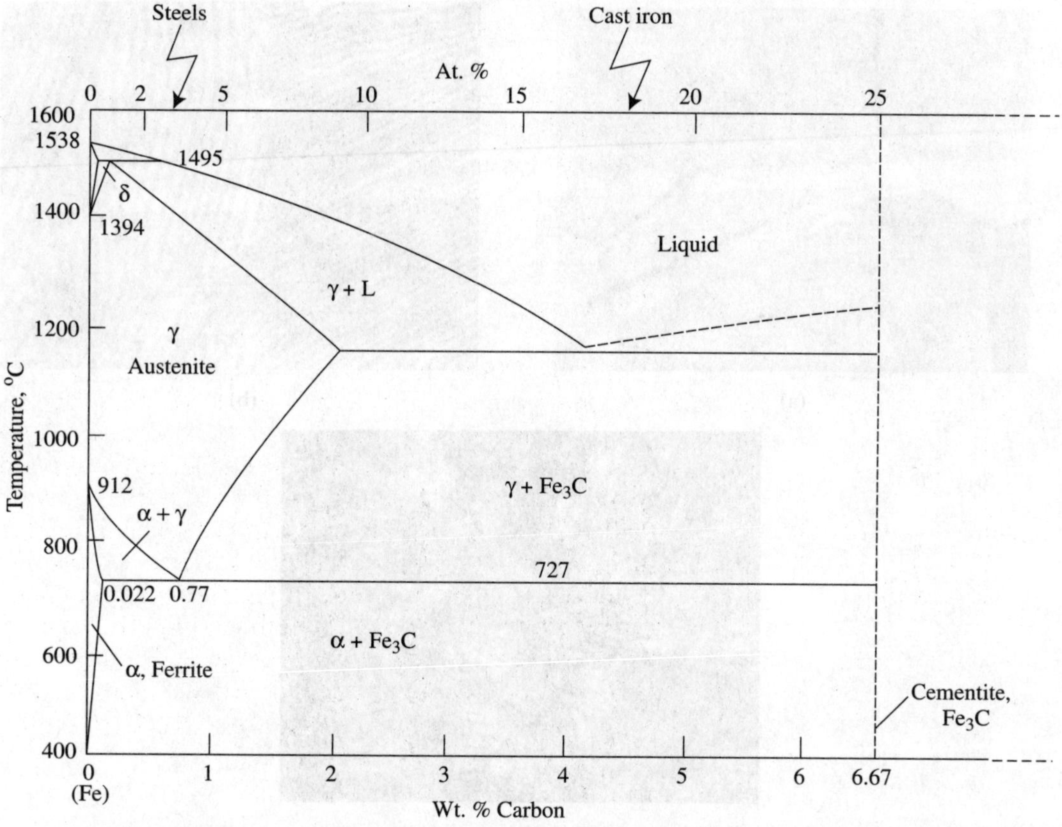

Fig. 3.14. The Fe-Fe_3C phase diagram.

compound, such steels have lower ductilities. The variation in microstructure morphology with respect to cooling rate during solid state transformation shall be described in the later section related to 'transformation'. In offeutectoid compositions during rapid cooling at temperatures below the eutectoid temperature, the percentage of proeutectoid constituent is less than the equilibrium value at the eutectoid temperature. For some of the compositions near the eutectoid composition under such cooling the proeutectoid constituent may be absent.

The microstructures of steels quite off the eutectoid composition do give rise to *Widmanstatten structure* of proeutectoid constituent i.e., ferrite or cementite. Widmanstatten plates nucleate mainly from prior austenite grain boundaries or develop from existing grain boundary allotriomorphs.

Most of the engineering steels have alloying additions and are not simple Fe-C alloys. The alloying additions do effect the polymorphic transformation temperatures A_3 ($\alpha \Leftrightarrow \gamma$ transformation) and A_4 ($\gamma \Leftrightarrow \delta$ transformation). Some elements, notably nickel, manganese, copper and cobalt, raise the A_4 temperatures and depress the A_3 as shown in Figure 3.16(i). Consequently when added to a carbon steel, these elements tend to stabilize austenite (γ) range. Since most of these elements do not form carbides, the carbon tends to remain in solid solution in austenite. Other elements e.g., chromium, molybdenum, tungsten, vanadium, aluminium, phosphorus and

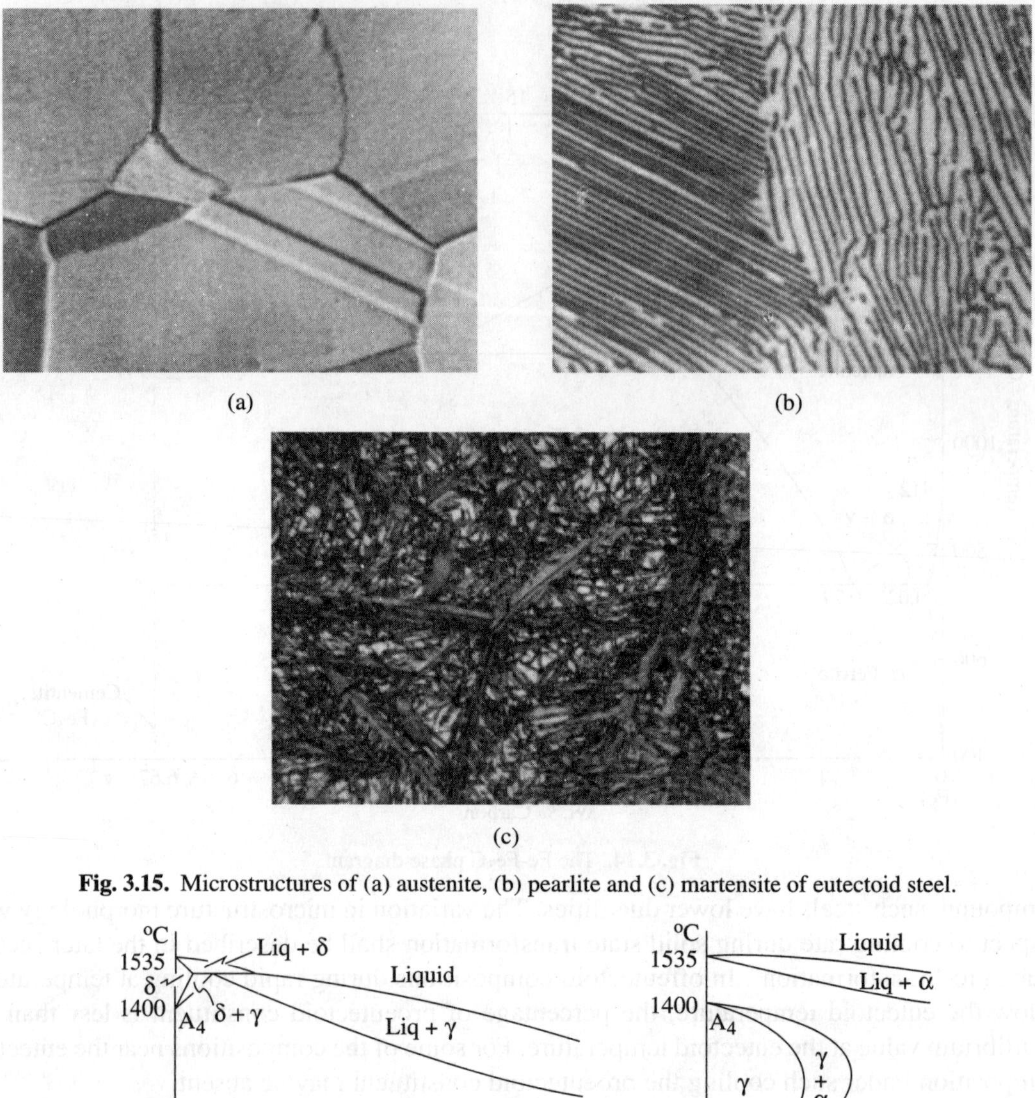

Fig. 3.15. Microstructures of (a) austenite, (b) pearlite and (c) martensite of eutectoid steel.

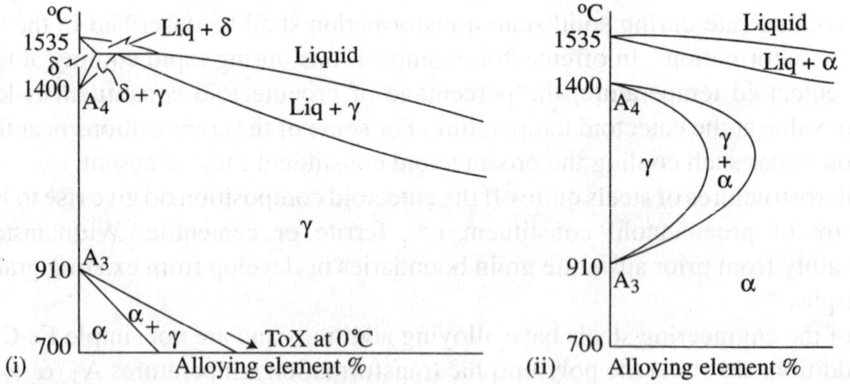

Fig. 3.16. The effects of alloy addition in Fe-C phase diagram (i) γ-stabilizer additions i.e., Ni, Mn, Cu and Co; (ii) α-stalilizer additions i.e., Cr, W, Mo, V, Al and Si.

silicon, are ferrite (α) stabilizers. By raising the A_3 temperature and depressing the A_4 these elements promote the formation of γ-loop in the phase diagram (Figure 3.16(ii)).

All alloying elements lower the eutectoid composition to values less than 0.8 w/o carbon (Figure 3.17). Manganese and nickel lower the eutectoid temperature, while chromium, tungsten, silicon, molybdenum and titanium raise the eutectoid temperature. As an example, addition of 2.5% Mn to a steel containing 0.65%C will produce a completely eutectoid i.e., pearlitic structure.

In the cast iron portion of the Fe-C phase diagram, solidification begins by liberating dendrites of austenite and that by the time the eutectic temperature (1147 °C) has been reached the microstructure will consist of austenite (γ) grains containing 2.0% C and a remaining liquid containing 4.3% C. This liquid gets solidified at 1147 °C for forming an eutectic mixture of austenite (2.0% C) and cementite (6.68% C). On further cooling both primary and eutectic austenite will change in composition by depositing particles of cementite until at 723 °C. The remaining austenite contains 0.8% C, where it transforms to pearlite. However in practice we don't get such microstructure always. The reason being that in cast irons there are other

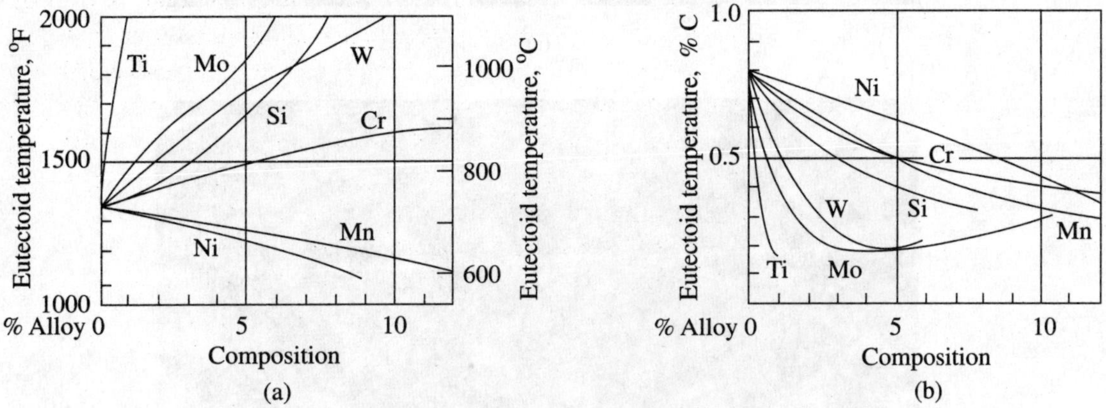

Fig. 3.17. Eutectoid shifts in steel caused by addition of alloying elements (a) shift in eutectoid temperature, (b) shift in the carbon content of the eutectoid.

constituents, which modify the microstructure. In high silicon composition (2.5%) carbide is unstable and gives rise to equilibrium graphite flakes. The presence of large amounts of graphite in the structure leads to a fractured surface having a gray appearance, hence the name *gray cast iron*. In case of low Si cast iron ($\sim$ 0.5%), cementite tends to be stable and no graphite will be formed. This type of cast iron is known as white cast iron, because of silvery white fracture surface due to cementite. Figure 3.18 shows typical microstructures pertaining to white and gray cast irons.

Example 3.4. Calculate the mass fraction of α - ferrite and cementite in pearlite.

Solution. Refer Fe-Fe_3C phase diagram (Figure 3.14).

The eutectoid reaction is

$$\gamma_{0.77C} \rightarrow \alpha_{0.022C} + Fe_3C_{6.7C}$$

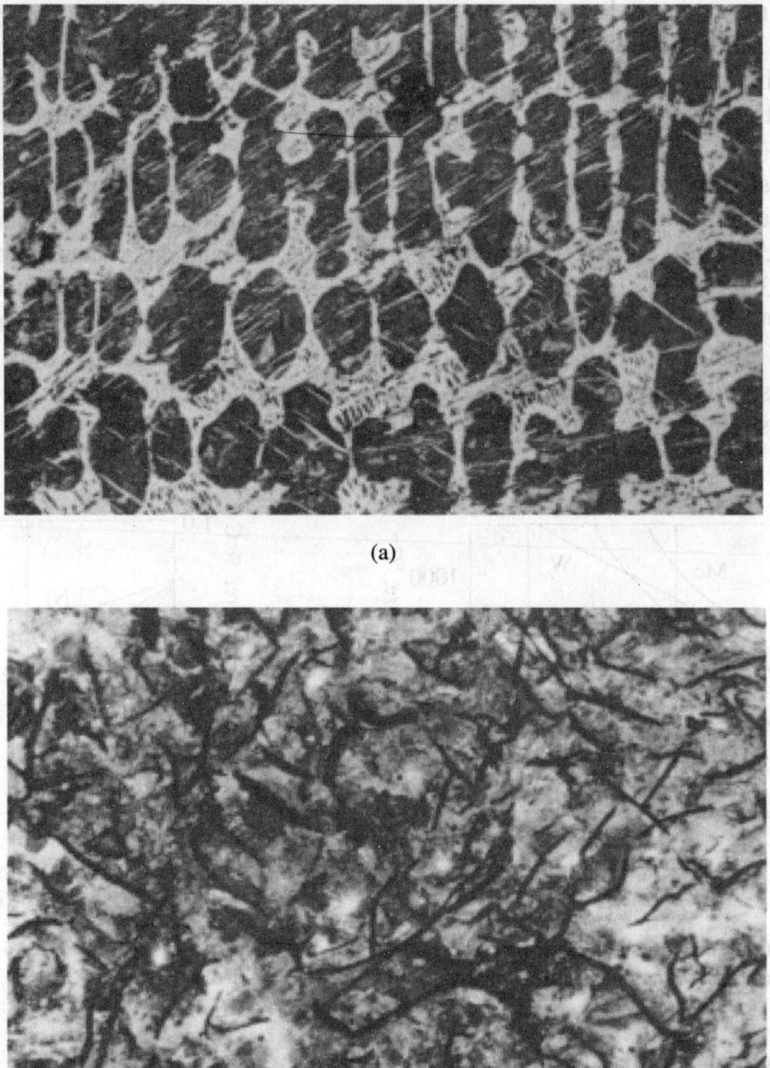

(a)

(b)

Fig. 3.18. Microstructures of (a) white cast iron and (b) grey cast iron.

Applying Lever Rule:

$$W_\alpha = \frac{C_{Fe_3C} - C_O}{C_{Fe_3C} - C_\alpha},$$

where C_O = Carbon content in pearlite

$$= \frac{6.7 - 0.77}{6.7 - 0.022} = 0.888$$

$$W_{Fe_3C} = \frac{C_O - C_\alpha}{C_{Fe_3C} - C_\alpha}$$
$$= 1 - W_\alpha = 1 - 0.888$$
$$= 0.112$$

Example 3.5. Calculate the relative thicknesses of ferrite and cementite in a completely pearlitic structure just below the eutectioid isotherm. Given the densities of ferrite and cementite as 7.87 g cm^{-3} and 7.66 g cm^{-3} respectively.

Solution. Evidently the steel pertains to eutectoid composition (0.77% C). Just below eutectoid temperature, referring the Example 3.4.

$$\text{Wt. \% of Ferrite} = 88.78\% \quad \text{(i)}$$
$$\text{Wt. \% of cementite} = 11.22\%$$

Let x = thickness of ferrite

y = thickness of cementite

Relative weight of ferrite/cementite $= 7.87x/7.66y$

We know from (i) the relative weight fraction of these phases.

Thus
$$\frac{7.87x}{7.66y} = \frac{0.8878}{0.1122}$$

or
$$\frac{x}{y} = \frac{7.70}{1}$$

Thus ferrite has 7.7 times greater thickness than the cementite.

3.5 TERNARY SYSTEMS

When the number of components is three, we have to represent the phase diagram in three dimensional model, such that each side of the equilateral trangle, represent a binary phase diagram (Figure 3.19). Unlike in binary system, any three phase equilibrium in a ternary system occurs over a temperature range. There are two kinds of 'tie elements' that appear in ternary diagram, namely, the tie-line and tie-triangle. In two dimensional plots the ternary diagrams are illustrated as isotherms as shown in Figure 3.20, pertaining to Lead-Bismuth-Antimony system.

Ternary reaction involves four-phase equilibrium. Some of the examples are:

Ternary eutectic

$$L \overset{Cooling}{\Leftrightarrow} \alpha + \beta + \gamma \quad \text{(Type I)}$$

Intermediate between eutectic and peritectic

$$L + \alpha \overset{Cooling}{\Leftrightarrow} \beta + \gamma \quad \text{(Type II)}$$

Ternary peritectic

$$L + \alpha + \beta \overset{Cooling}{\Leftrightarrow} \gamma \quad \text{(Type III)}$$

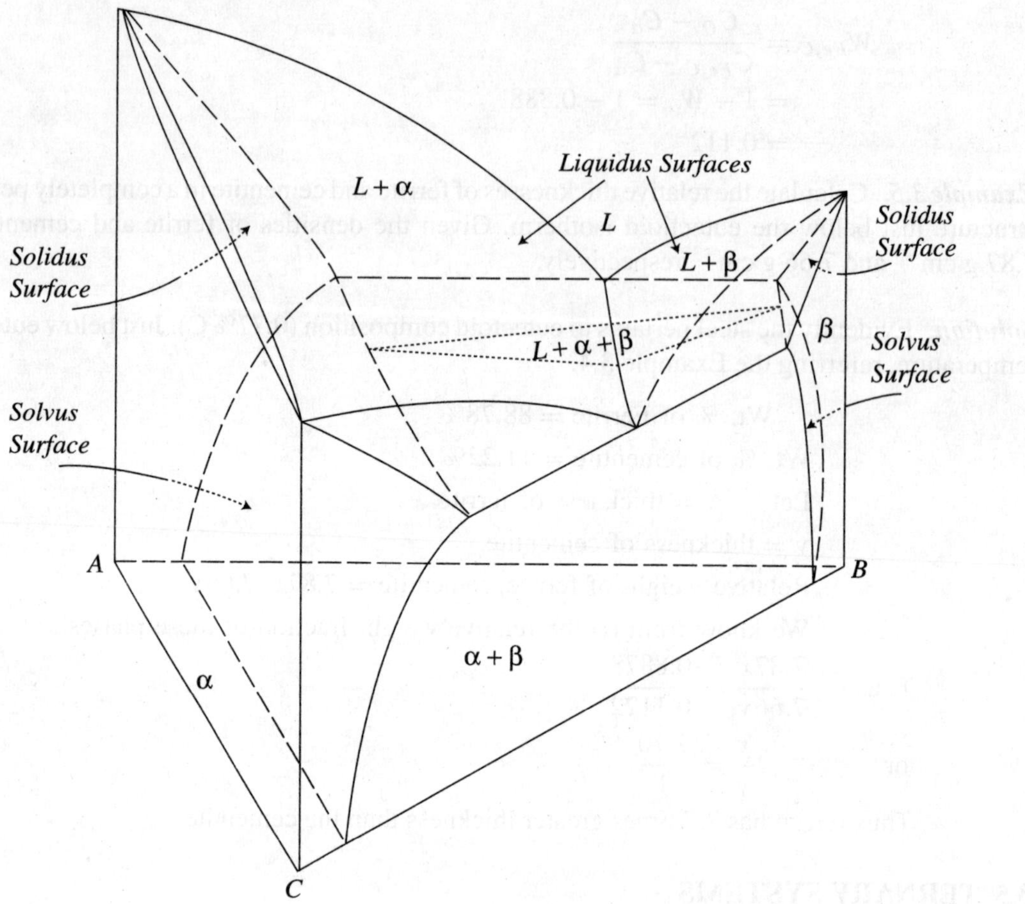

Fig. 3.19. Ternary phase diagram showing three phase equilibrium.

The detailed description of above reactions is rather beyond the scope of the present book.

For development of alloy steels (Fe-C-X) including stainless steels and other refractory ceramic systems, the study of ternary phase diagrams is a must.

3.6 CALCULATED PHASE DIAGRAMS

In case of high number of components i.e., three or more, all the possible phase diagrams have not been established experimentally, which is a very time taking operations. For this CALPHAD method (CALculation of PHAse Diagram) along with commercially available, published or self generated data bases may be used to calculate the phase diagrams, including the effects of trace additions. The software can be used to determine the stable phases present in a system as a function of composition, temperature and pressure. The stable phases are determined from the relative minimum in the Gibbs free energy term for each phase possible in the system. As the topic is beyond the scope of this book, further details are not dealt with.

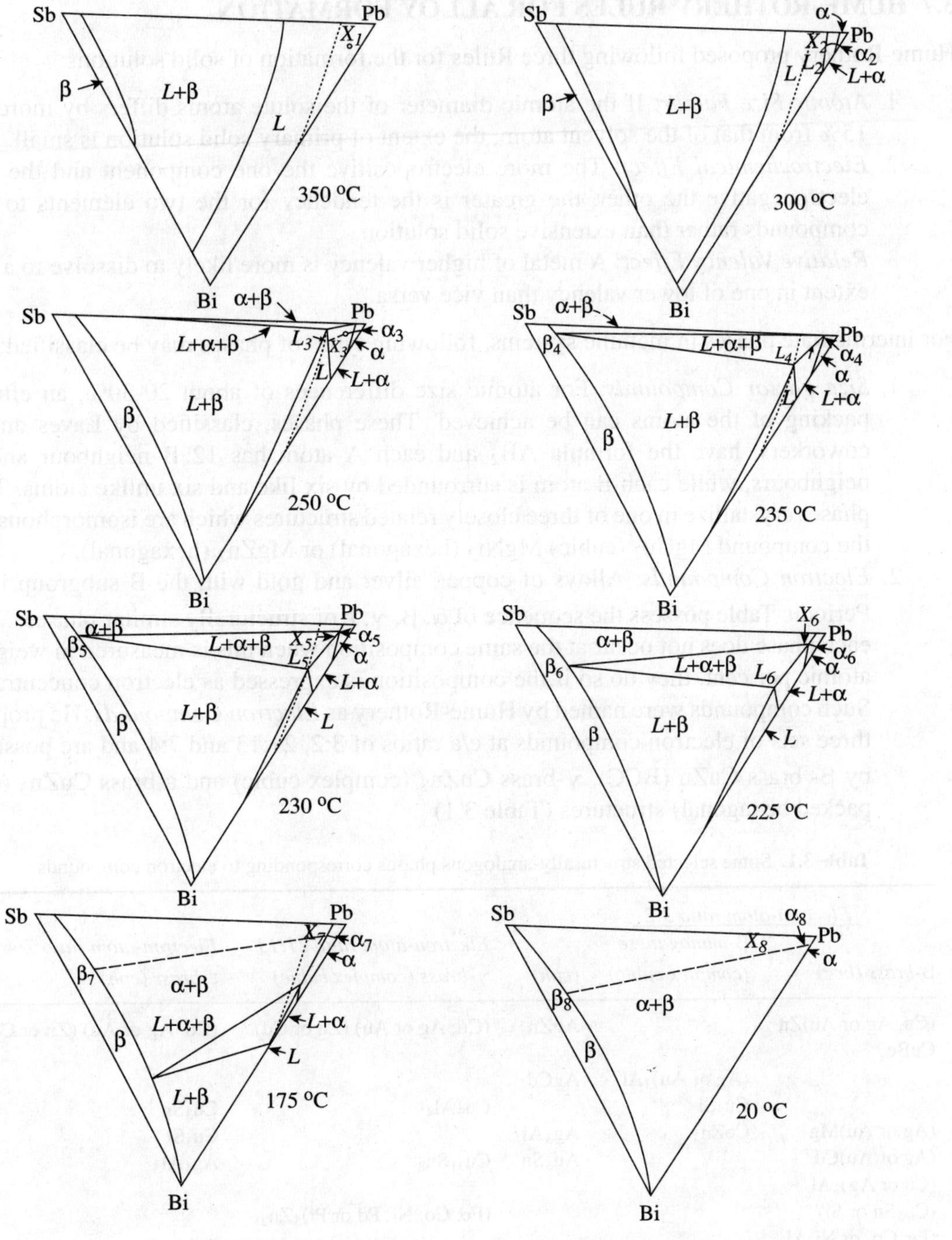

Fig. 3.20. Isotherms of the ternary system Lead-bismuth-antimony.

3.7 HUME-ROTHERY RULES FOR ALLOY FORMATION

Hume-Rothery proposed following three Rules for the formation of solid solutions:

1. *Atomic Size Factor*: If the atomic diameter of the solute atoms differs by more than 15% from that of the solvent atom, the extent of primary solid solution is small.
2. *Electrochemical Effect*: The more electropositive the one component and the more electronegative the other, the greater is the tendency for the two elements to form compounds rather than extensive solid solution.
3. *Relative Valency Effect*: A metal of higher valency is more likely to dissolve to a large extent in one of lower valency than vice versa.

For intermediate phases in metallic systems, following types of phases may be classified:

1. *Size Factor Compounds*: For atomic size differences of about 20-30%, an efficient packing of the atoms can be achieved. These phases, classified by Laves and his coworkers, have the formula AB_2 and each A atom has 12 B neighbour and 4A neighbours, while each B atom is surrounded by six like and six unlike atoms. These phases crystallize in one of three closely related structures which are isomorphous with the compound $MgCu_2$ (cubic) $MgNi_2$ (hexagonal) or $MgZn_2$ (hexagonal).
2. *Electron Compounds*: Alloys of copper, silver and gold with the B subgroup in the Periodic Table possess the sequence of α, β, γ, ε of structurally similar phases. While each phase does not occur at the same composition when this is measured in weight or atomic per cent, they do so if the composition is expressed as electron concentration. Such compounds were named by Hume-Rothery as *Electron Compounds*. He proposed three sets of electron compounds at e/a ratios of 3:2, 21:13 and 7:4 and are possessed by β- brass CuZn (BCC), γ-brass $CuZn_8$ (complex cubic) and ε-brass $CuZn_3$ (close packed hexagonal) structures (Table 3.1)

Table 3.1. Some selected structurally-analogous phases corresponding to electron compounds

Electron-atom ratio 3:2			*Electron-atom ratio 21:13*	*Electron-atom ratio 7:4*
β-brass (bcc)	*β-managanese (complex cubic)*	*(cph)*	*γ-brass (complex cubic)*	*ε-brass (cph)*
(Cu, Ag or Au)Zn		AgZn	(Cu, Ag or Au) (Zn or Cd)$_8$	(Cu, Ag or Au) (Zn or Cd)$_3$
CuBe				
	(Ag or Au)$_3$Al	AgCd		
	Cu_5Si		Cu_9Al_4	Cu_3Sn
(Ag or Au)Mg	$CoZn_3$	Ag_3Al		Cu_3Si
(Ag or Au)Cd		Au_5Sn	$Cu_{31}Sn_8$	Ag_5Al_3
(Cu or Ag)$_3$Al				
(Cu_5Sn or Si)			(Fe, Co, Ni, Pd or Pt)$_5$Zn$_{21}$	
(Fe, Co, or Ni)Al				

3. *Electrochemical Compounds*: When one element is electropositive and the other is electronegative, there is strong tendency to form such compounds. The magnesium

based compounds are probably the most common examples having the formula Mg_2 (Pb, Sn, Ge or Si). These have many features in common with salt like structure since their compositions satisfy the chemical valency laws, their range of solubility is small, and usually they have high melting point. Many of these types of compounds have crystal structures identical to chemical compounds such as NaCl and CaF_2. In this respect Mg_2X series are anti-isomorphous with the CaF_2 structure. Although these compounds follow all the chemical principles, they may often be considered as special electron compounds.

3.8 DIFFUSION IN SOLIDS

Before we go to the details of phase transformation, it is necessary to know about diffusion in solids. This is important as most of the solid state transformation occur through diffusion. Since diffusion in solids occurs by means of points defects like vacancies and interstitials, the diffusion mechanisms are classified as vacancy mechanism, interstitial mechanism, and intestitialcy mechanism (Figure 3.21). Vacancy mechanism involves the diffusing atom, which moves by jumping into a neighbouring vacant lattice site. Interstitial mechanism occurs when an atom passes from one interstitial site to one of its nearest neighbour interstitial sites without permanently displacing any of the matrix atoms. Example is carbon in α or γ iron. If the diffusing interstitial atom is almost as large as the atoms on the normal lattice site, then the lattice becomes significantly distorted and an interstitialcy mechanism of diffusion becomes operative. There are two laws for diffusion related to steady and unsteady states, known as *Fick's Laws*.

Fick's First Law - For one-dimensional steady state diffusion the simple linear law is stated as

$$J = -D\frac{\partial c}{\partial x} \tag{3.4}$$

where J, the flux or diffusion current density, is the amount of material (the net number

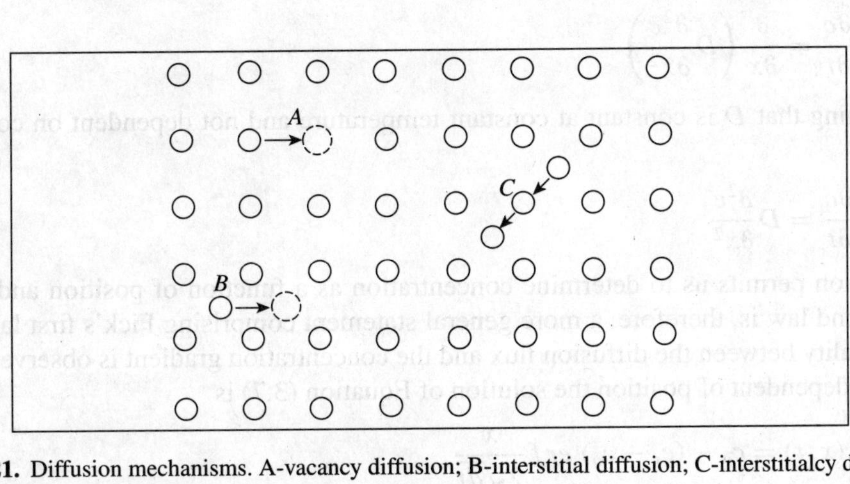

Fig. 3.21. Diffusion mechanisms. A-vacancy diffusion; B-interstitial diffusion; C-interstitialcy diffusion.

of atoms) diffusing in a unit time per unit area perpendicular to the x-axis. The proportionality constant D is the diffusion coefficient, c is the volume concentration of atoms and x is the distance along the direction in which diffusion occurs. Equation (3.4) shows that J and $\partial c/\partial x$ have opposite signs; this means that the diffusion current J is in a direction opposite to that of the concentration gradient. A concentration profile developed in the direction perpendicular to the surface as a function of the distance x is given in Figure 3.22. Equation (3.4) can be integrated to the form

$$J = -D\frac{c_2 - c_1}{\Delta x} \tag{3.5}$$

where c_1 and c_2 are concentrations at the ends of the diffusion length $\Delta x = x_2 - x_1$.

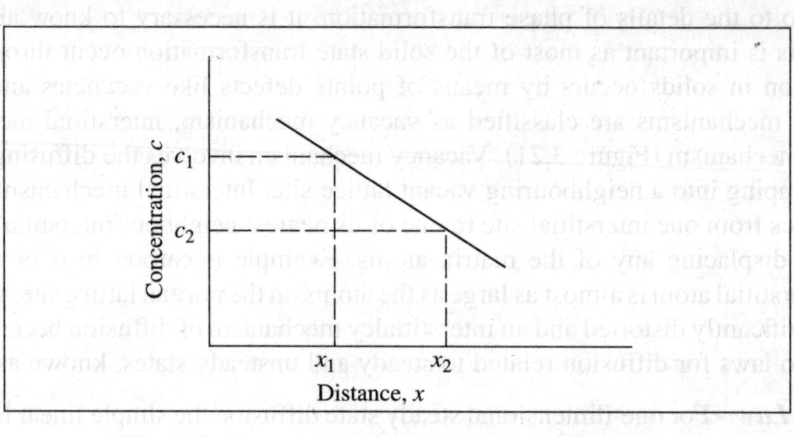

Fig. 3.22. Concentration profile in steady state diffusion.

Fick's Second Law : For unsteady state conditions there will be an accumulation of diffusible matter in a unit volume, and the concentration at any point within a solid will vary with time. Therefore $\partial c/\partial t \neq 0$. The second law of Fick's states as

$$\frac{\partial c}{\partial t} = \frac{\partial}{\partial x}\left(D\frac{\partial^2 c}{\partial x^2}\right) \tag{3.6}$$

Assuming that D is constant at constant temperature and not dependent on composition, we obtain

$$\frac{\partial c}{\partial t} = D\frac{\partial^2 c}{\partial x^2} \tag{3.7}$$

This equation permits us to determine concentration as a function of position and time. The Fick's second law is, therefore, a more general statement comprising Fick's first law, where a proportionality between the diffusion flux and the concentration gradient is observed. If D is a constant independent of position the solution of Equation (3.7) is

$$c(x, t) = c_o - (c_s - c_o)\, erf\frac{x}{2\sqrt{dt}} \tag{3.8}$$

where $c(x,t)$ is the concentration at time t and distance x from the interface, c_s is the concentration at the interface of the solid at $x = 0$, which is maintained constant, c_o is the initial concentration of solid and $\text{erf}(x/2\sqrt{Dt})$ is the probability integral or Gauss error function, which is found in any mathematical hand books. A concentration profile developed in a direction perpendicular to the surface as a function of time is shown in Figure 3.23.

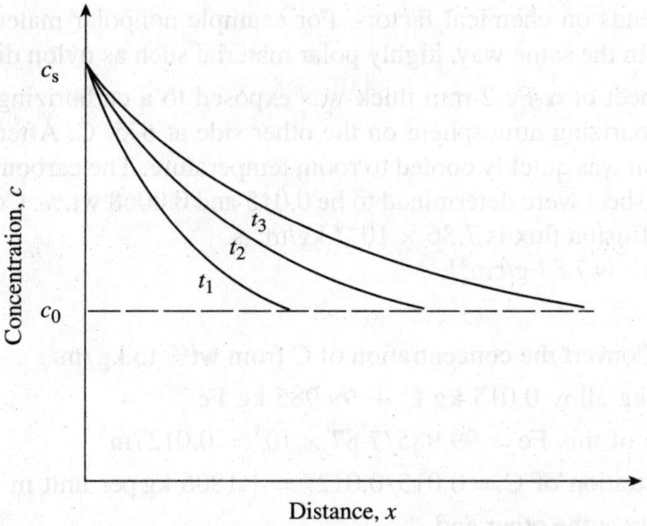

Fig. 3.23. Concentration profiles at various diffusion times $t_3 > t_2 > t_1$.

The two most important factors affecting the diffusion coefficient D are temperature and composition. Because of the activation energy term, the rate of diffusion increases with temperature according to

$$D = D_o \exp[-(E_f + E_m)/kT] \tag{3.9}$$

where D_o is a constant for a given diffusion system and E_f is the energy of formation of a vacancy and E_m is energy of migration. The sum of these two energies $Q = E_\text{f} + E_\text{m}$ is the activation for self diffusion.

Now let us consider an alloy A-B. Here the diffusion coefficients D_A and D_B are generally not equal. This inequality of diffusion was first demonstrated by *Kirkendall* using an α -brass/copper couple. He demonstrated that the lower melting point metal zinc atoms diffuse out of alloy more rapidly than copper atoms diffuse in. This effect is important in understanding the pores formation during sintering of alloys, to be described in the 'material forming' (Chapter 8).

In conclusion it can be said that the knowledge of diffusion in crystalline solids is essential in understanding elevated temperature processes like annealing, heat treatment, age hardening, sintering, surface hardening, oxidation and creep.

Now the question arises on the diffusion in polymers. The mutual diffusion of different macromolecules in the melt is an interesting example. When a layer of polymer with a low molecular weight diffuses into a layer of same polymer with high molecular weight, a feature similar to *Kirkendall's effect* is observed. The viscous bulk flow which drives this effect is equivalent to the vacancy flux in metals. Diffusion occurs where small molecules pass through the voids between the polymer molecules. The rate of diffusion will depend to a large extent on the size of the small molecules and the size and mobility of the voids. The latter depends on the state of the polymer, i.e., whether it is crystalline, glassy or rubber- like. The diffusivity will fall in the order rubber-like → glassy → crystalline. The steady state diffusion in polymers will depend on the solubility of the solute molecules in the polymer and the diffusion coefficient. The solubility depends on chemical factors. For example nonpolar material rubber dissolves nonpolar gasoline. In the same way, highly polar material such as nylon dissolves water.

Example 3.6. A sheet of α-Fe 2 mm thick was exposed to a carburizing gas atmosphere on one side and decarburizing atmosphere on the other side at 675 °C. After having reached the steady-state, the iron was quickly cooled to room temperature. The carbon concentration at the two surfaces of the sheet were determined to be 0.015 and 0.0068 wt.%. Compute the diffusion coefficient if the diffusion flux is 7.36×10^{-9} kg/m^2-s.
[Given: density of Fe is 7.87 g/cm^3].

Solution.

Steps: Convert the concentration of C from wt% to kg/m^3.

In 100 kg alloy 0.015 kg C + 99.985 kg Fe

Volume of this Fe = $99.985/7.87 \times 10^3 = 0.0127\text{m}^3$

Concentration of C = $0.015/0.0127 = 1.1806$ kg per unit m^3

Similarly at the other end

100 kg of alloy = 999.9932kg Fe + 0.0068kgC

Volume of Fe = $0.0127\, m^3$

Concentration of carbon = $0.0068/0.0127 = 0.0535$ kg

Note: volume of Fe remains unaffected by this small amount of C.

$$J = -D\frac{\Delta C}{\Delta X}$$

$$D = -J\frac{\Delta X}{\Delta C}$$

$$= 7.36 \times 10^{-9}\left[\frac{2 \times 10^{-3}}{1.1806 - 0.535}\right]$$

$$= 2.28 \times 10^{-11}\ \text{m}^2/\text{s}$$

Example 3.7. At approximately what temperature would a specimen of γ-Fe has to be carburized for 4h to produce the same diffusion result as at 1000 °C for 12h?
[Given: D_o and Q for diffusion of C in Fe 1×10^{-5} m^2/s and 32.4 kcal/mol]

Solution.

$$(Dt)_{1000\,^\circ C} = (Dt)_T$$

$$D_{1000\,^\circ C} = 1 \times 10^{-5}\ \exp\left[\frac{-32400}{1.987 \times 1273}\right] = 2.736 \times 10^{-11}\ \text{m}^2/\text{s}$$

$$(2.736 \times 10^{-11})(12 \times 3600) = D_T(4 \times 3600)$$
$$D_T = 8.207 \times 10^{-11} \ \text{m}^2/\text{s}$$
$$D_T = 1 \times 10^{-5} \exp\left[-\frac{32400}{1.987 \times T}\right] = 8.207 \times 10^{-11}$$
$$-11.71 = -32400/1.987T$$
$$T = 1392.4\,\text{K}$$
$$= 1119\,^\circ\text{C}$$

3.9 PHASE TRANSFORMATION

The phase transformation may be categorized in three types:

(i) Diffusion dependent transformation in which there is no change in the composition of the phases present. Pure metal solidification, allotropic transformation and recrystallization are examples.

(ii) This is similar to (i), but here there is some change in phase composition and often in number of phases present. The pearlite formation in eutectoid steel is a typical example.

(iii) Diffusionless transformation, where a metastable phase is produced. Martensite in steel is a typical outcome of such a transformation. Certain metals, alloys (both ferrous and nonferrous) and ceramics undergo such type of phase transformation (Table 3.2).

Table 3.2. Some metals, alloys and ceramics exhibiting martensitic transformation

Material	*Structure Change*
Co	FCC ⇔ HCP
Ti	BCC ⇔ HCP
Fe-30Ni	FCC ⇔ BCC
Au-47.5Cd	BCC ⇔ Orthorhombic
In-19Tl	FCC ⇔ FCT
Fe-C	FCC ⇔ BCT
BN	Wurtzite type ⇔ Graphite type
ZrO_2	Tetragonal ⇔ Monoclinic
SiO_2	Rhombohedral ⇔ Hexagonal

The above three types of transformations are also categorized by general name of heterogeneous transformation, which occurs by nucleation and growth. Apart from this, there are homogeneous transformations that occur simultaneously throughout the parent phase without any nucleation. Spinodal decomposition and order-disorder transformations belong to this category, which shall be described in a later section.

If a molten metal is undercooled below the equilibrium melting temperature (Figure 3.24), there is a chance that the crystal nucleated will be thermodynamically stable and will continue to grow. The change in free energy which helps the nucleus to form, can be expressed as

$$\Delta G = \frac{\Delta H}{T_m}(T_m - T) \tag{3.10}$$

where T_m = melting point

$T_m - T$ = degree of undercooling

If ΔH is the latent heat given out over unit volume of the solid formed, the total available energy is $\left(\frac{4}{3}\pi r^3\right)\frac{\Delta H}{T_m}(T_m - T)$, where r is radius of nucleus.

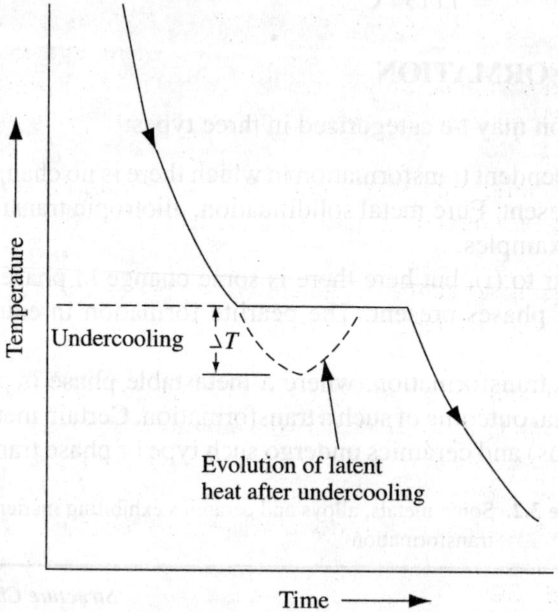

Fig. 3.24. Cooling a molten metal with an under cooling.

But this is offset by the work $4\pi r^2\gamma_{SL}$ needed to create the solid/liquid interface around the crystal. Here γ_{SL} is the solid/liquid interfacial energy. The net work needed to form the crystal is:

$$W_f = 4\pi r^2\gamma_{SL} - \frac{4}{3}r^3\Delta H\left(\frac{T_m - T}{T_m}\right) \tag{3.11}$$

If we plot these two components of work (Figure 3.25), it is obvious that the maximum value of W_f, corresponds to a critical radius r*. The critical radius r* can be obtained by differentiating equation (3.11); to give

$$\frac{dW_f}{dr} = 8\pi r\gamma_{SL} - 4\pi r^2|\Delta H|\frac{T_m - T}{T_m}$$

Since $\frac{dW_f}{dr} = 0,\ at\ r = r^*$, we obtain

$$r^* = \frac{2\gamma_{SL}T_m}{\Delta H(T_m - T)} \tag{3.12}$$

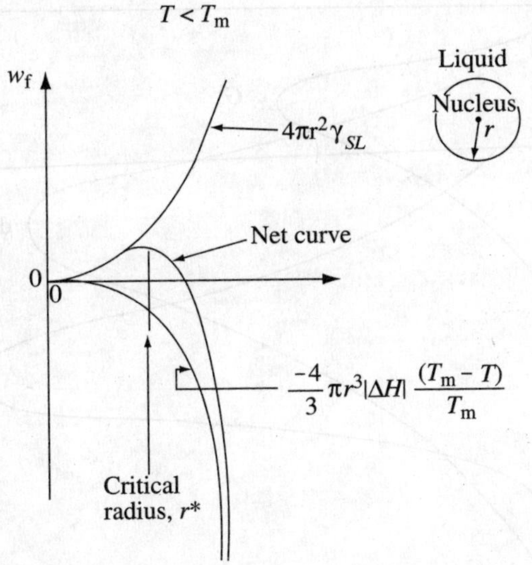

Fig. 3.25. The work needed to make a spherical nucleus.

This type of nucleation is known as *homogeneous nucleation*, i.e., atoms involved are of those of material itself. However there is another type of nucleation known as *heterogeneous nucleation*, where there is a strong tendency for the crystals to stick to the surface of the catalyst. Mould wall for ingot casting is a common example for heterogeneous nucleation. Nickel is used as a heterogeneous nucleating agent in the production of artificial diamond from graphite.

3.9.1 Transformation Kinetics

Figure 3.26 shows the temperature dependence of growth rate G along with the temperature dependence of the nucleation rate N. It is obvious that at lower temperature nucleation rate shall be greater than the growth rate. Further, the growth rate is maximum, at some intermediate degree of undercooling. The maximum in the growth rate usually occurs at a higher temperature than the maximum in the nucleation rate. To sum up, the overall transformation kinetics can be described as some function of the nucleation and growth rates:

$$\frac{dx}{dt} = f(N, G) \tag{3.13}$$

where dx/dt is the rate at which the fraction x of the new phase increases with time. Assuming that a constant rate of nucleation occurs randomly in the original phase and that the product particles grow at a constant rate as spheres, till impingment with neighbouring particles occurs, the fraction x can be derived as a function of t:

$$x = 1 - exp\left(\frac{\pi}{3}NG^3t^4\right) \tag{3.14}$$

A plot of x versus t shown in Figure 3.27 has the sigmoidal form. This is also named as S -curve. As the transformation rate is a function of both the nucleation and the growth rates,

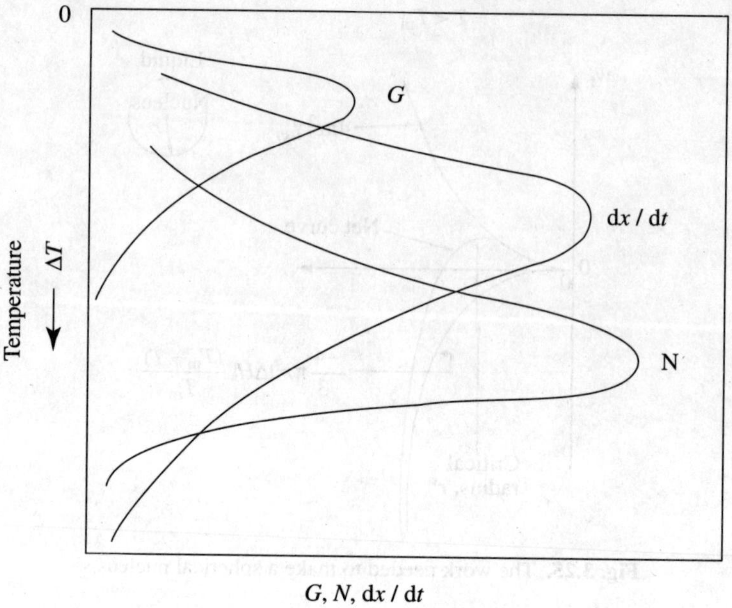

Fig. 3.26. The temperature dependence of nucleation rate N, the growth rate G and transformation rate dx/dt.

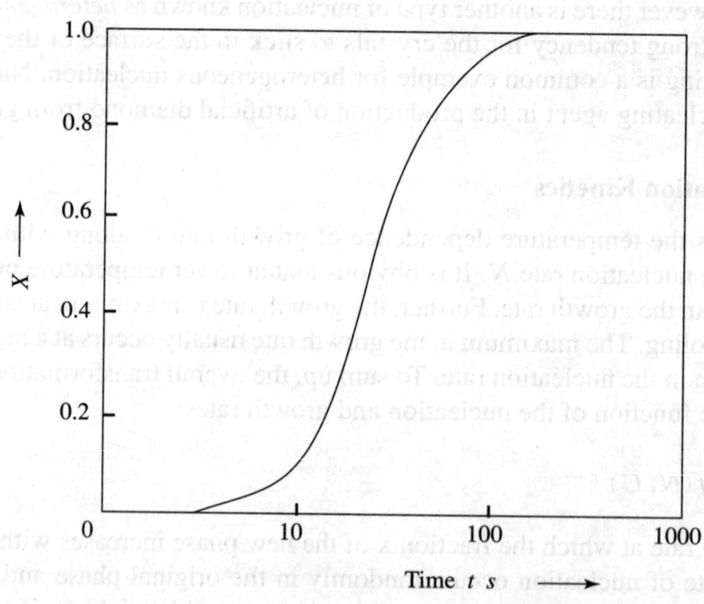

Fig. 3.27. The fraction transformed X plotted with respect to time.

the dx/dt curve has the same shape as the G and N curves. The temperature of the maximum transformation rate lies between the temperatures of maximum growth rate and maximum nucleation rate. There is another curve, known as C-curve, which is vary useful in studying the transformation in steels. This is inversely related to dx/dt curve. This is also known as T-T-T (time-temperature-transformation) curve. The nose of the C-curve corresponds to the minimum time for a specified fraction of transformation. The nose temperature is the temperature at which dx/dt is a maximum. The T-T-T diagram for a eutectoid (0.8%C) steel is shown in Figure 3.28. Here the nose temperature is 550 °C. The transformation kinetics decreases as the transformation temperature is raised above the nose, because the thermodynamic driving force for the austenite to pearlite transformation falls to zero as the temperature approaches 727 °C, i.e., A_1 temperature. The rate also decreases as the temperature is lowered below the nose of the C-curve, owing to the temperature dependence of the rate of diffusing carbon. That is at low temperatures there is enough of driving force, but here the transformation rate is controlled by the rate of carbon diffusion. Figure 3.28 is also called isothermal transformation diagram, since the transformation kinetics is studied at constant temperature, which may be varied. However care must be made to cool the steel from austenite range to the new lower temperature with

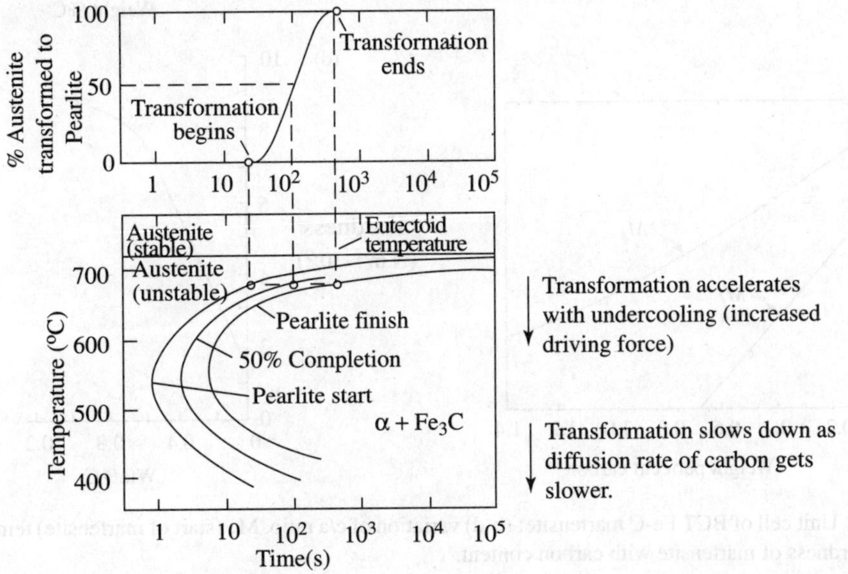

Fig. 3.28. The use of S-curve to construct a C-curve for a eutectoid steel. The specimens were austenitized and quenched to 680 °C and held for different periods.

such a rate, that metastable austenite is present prior to isothermal treatment. In case the steel is water quenched from austenite, a metastable martensite results. Such type of transformations are characterized by the fact that the reaction is (a) athermal (b) is displacive (c) is diffusionless (d) exhibits hysteresis (e) is isothermal to some degree and (f) is autocatalytic.

Martensite pertaining to Fe-C system has a body centred tetragonal structure with c/a ratio and hardness increasing with the carbon content (Figure 3.29). It is worth noting that the start and finish of martensite temperature (M_s and M_f) also drops with increase in the carbon content.

In practical industrial heat treatment, it is not possible to do the operations isothermally and the steels are cooled continuously from the stable austenite field. The cooling rates can be varied i.e., furnace cooling, air cooling, water cooling and brine solution cooling. In this order

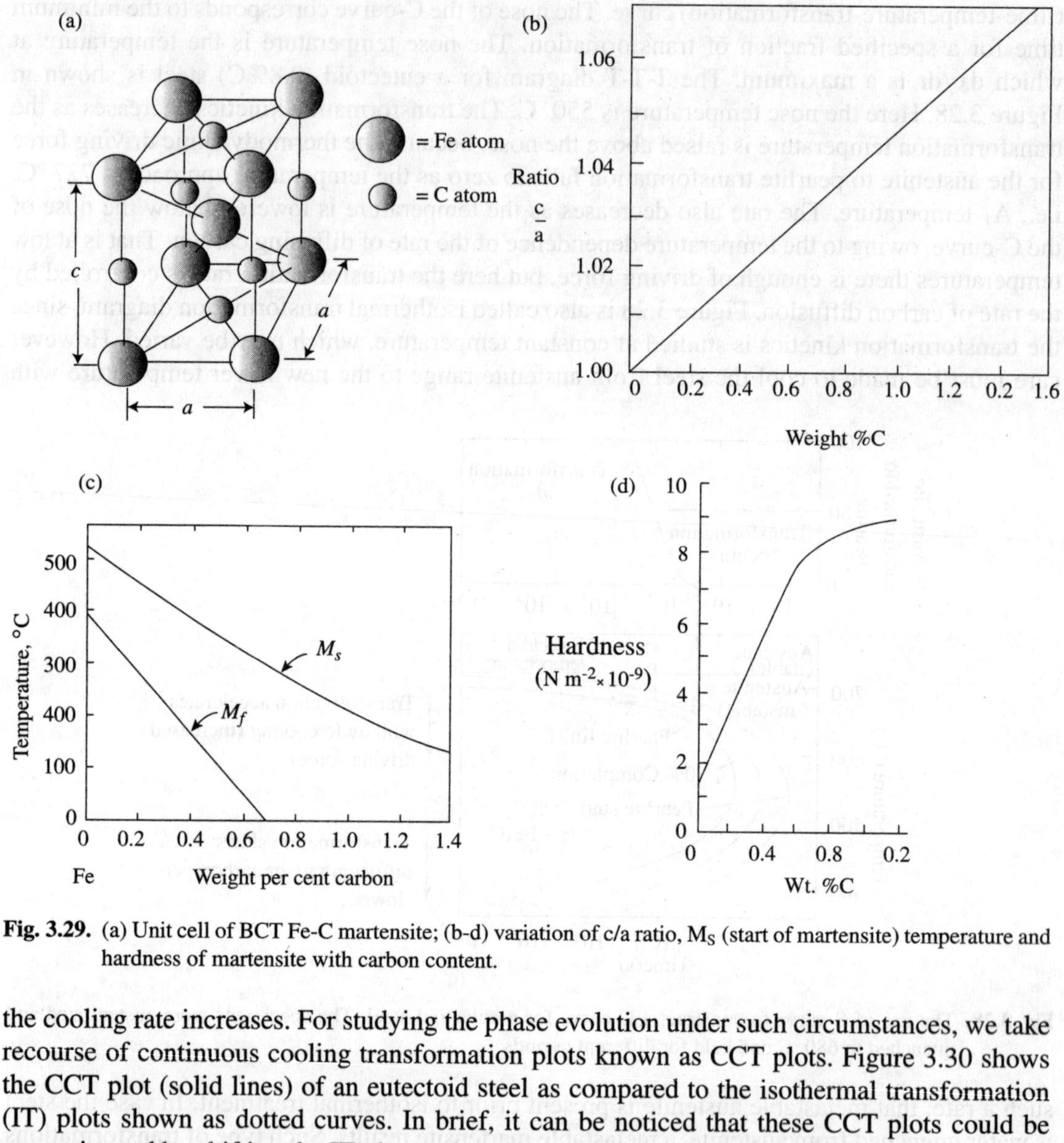

Fig. 3.29. (a) Unit cell of BCT Fe-C martensite; (b-d) variation of c/a ratio, M_S (start of martensite) temperature and hardness of martensite with carbon content.

the cooling rate increases. For studying the phase evolution under such circumstances, we take recourse of continuous cooling transformation plots known as CCT plots. Figure 3.30 shows the CCT plot (solid lines) of an eutectoid steel as compared to the isothermal transformation (IT) plots shown as dotted curves. In brief, it can be noticed that these CCT plots could be approximated by shifting the IT plots to lower temperature but larger period.

3.9.2 Homogeneous Phase Transformations

We have briefly defined such types of transformations in section 3.9. Presently we shall dwell upon them in some detail.

Spinodal Decomposition: There are a number of alloy systems, which exhibit miscibility gap in solid state. A spinodal decomposition is obtained by quenching a homogeneous solution from a temperature above such a gap. During such decompositions, regions of high and low

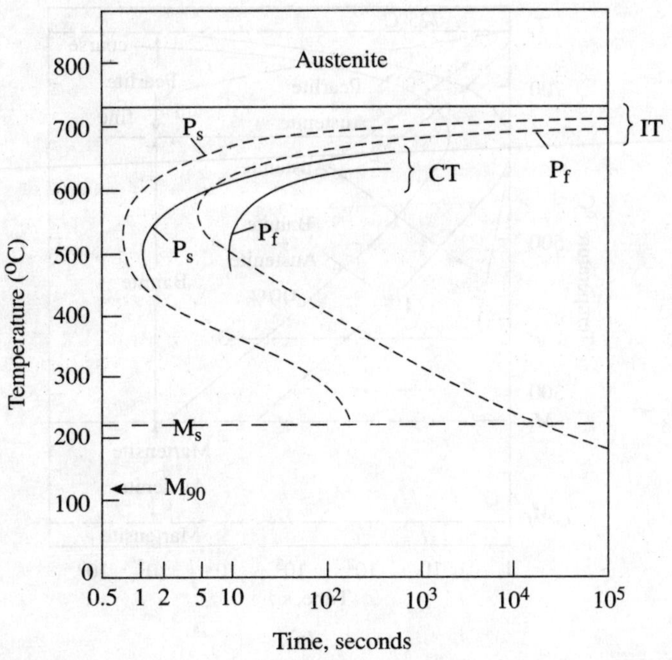

Fig. 3.30. Comparison of isothermal (IT) and continuous cooling (CT) transformation diagrams for a eutectoid steel.

concentrations are found next to each other. As a result sharp concentration builds up in the system. A typical application of such phase transformation in developing unique alloys has been illustrated in a case study in Chapter 12 (Case Study 12.28).

Order-Disorder Transformation: The ordering in a solid solution crystal may be of two types: short range order and long range order. In the long range order, as the name suggests, atoms of different components preferentially occupy lattice sites in an ordered manner over large regions. Such a structure is called super lattice structure. A completely ordered structure of this type resembles a chemical compound, in which the atomic fractions of components are in a specific ratio. Large ordered regions in long range ordered solutions may be separated by boundaries extending over one to two interatomic distance, called *anti-phase boundaries*. Anti-phase boundaries in ordered solutions are analogous to grain boundaries in polycrystalline materials. Such type of transformation is a second order transformation, since it occurs continuously and simultaneously in the whole volume. Such a transformation is of technological significance in the study of intermetallics, eg. transition metal aluminides. Some of the crystal structures of ordered intermetallics are shown in Figure 2.13 (Chapter 2).

3.10 HEAT TREATMENT OF STEELS

In earlier section we have already talked about the TTT diagrams. In this section we shall take up more applied aspects, i.e., heat treatment of steels. The full details are again recapatulated in Figure 3.31(a), which corresponds to a eutectoid composition of steel. The diagram can be

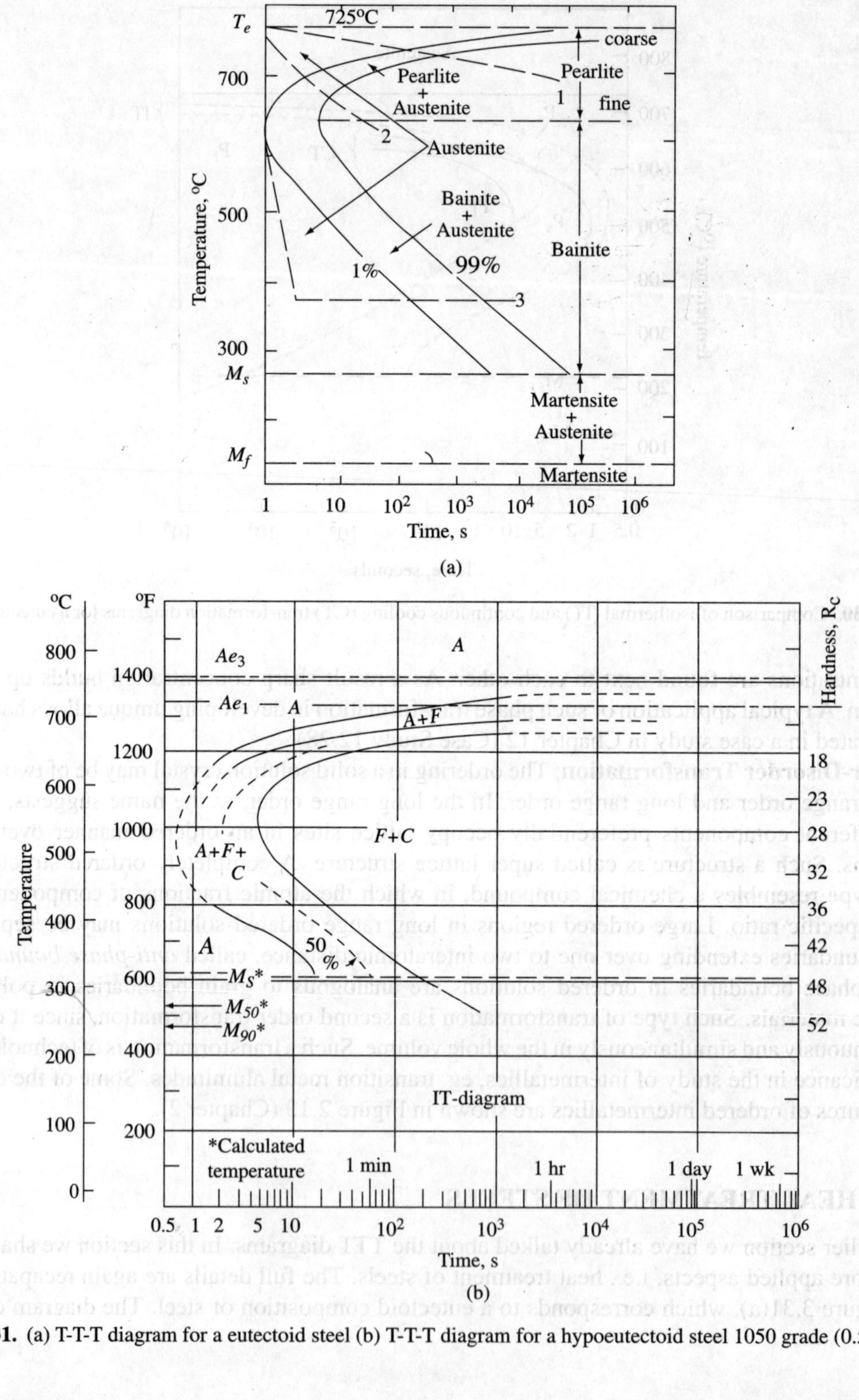

Fig. 3.31. (a) T-T-T diagram for a eutectoid steel (b) T-T-T diagram for a hypoeutectoid steel 1050 grade (0.5% C).

divided into three portions (i) pearlitic region (ii) banitic region and (iii) martensitic region. The first two pertain to diffusion transformation and the microstructures can be attained by isothermal treatment. The last category belongs to nondiffusion transformation. Pearlitic field can again be divided into coarse pearlite and fine pearlite, depending on the temperature of isothermal transformation. In Figure 3.31(a), cooling curve (i) gives rise to pearlite, cooling curve (ii) to upper bainite. But in case we wish to obtain lower bainite, we can not get it by continuous cooling, but through a quenching followed by isothermal holding. If the cooling is not interrupted, but extended up to room temperature, we will obtain martensite. Martensite formation takes place in a range of temperature from M_s (start of martrensite) to M_f (finish of martensite). In between these two temperatures, we shall have martensite and austenite, both being in metastable conditions. Bainite, like pearlite, is also a mixture of ferrite and cementite, but apart from having very fine distribution of these two phases, have a different morphology. Here the cementite particles are short needles embededed in plates of ferrite. If the metastable austenite is cooled rapidly below the bainitic range, it transforms to mertensite (α'), which possesses body centred tetragonal structure. The amount of martensite formed is a function of temperature to which the austenite is cooled and not a function of time. Martensite has plate like shape (Figure 3.15) as it occurs through an atomic level shear process. The hardness of martensite is a function of its carbon content, being maximum of R_c65 at about 0.6% C. It is an extremely hard and brittle phase and therefore has to be tempered, where the metastable phase decomposes into the more stable products of ferrite and cementite by the process of carbon diffusion. The decrease in hardness and increase in ductility can be accurately controlled, by a proper selection of tempering temperature and time. In the case of alloy steels, the carbide phase during tempering is not simple Fe_3C, but complex alloy carbides, which have even more hardness than Fe_3C both at room and elevated temperatures. This feature is important in high speed steels, where the aim is to achieve high red hardness, during severe cutting operations.

The martensite start temperature can be decreased by higher carbon or alloy content according to the equation

$$M_s(^\circ C) = 561 - 474(\%C) - 33(\%Mn) - 17(\%Cr) - 21(\%Mo) \quad (3.15)$$

If one switches to hypo- or hyper-eutectoid steels, the nature of TTT diagram is similar, except that the formation of proeutectoid ferrite or cementite must also be shown in the isothermal transformation plots. The TTT diagram for a hypoeutectoid 1050 steel (0.5%C) is shown in Figure 3.31(b). An alloy which is only partially transformed at a particular isothermal transformation temperature has a microstructure containing austenite, ferrite and cementite. If this steel in this condition is quenched to room temperature, the proeutectoid phase and the pearlite are retained but the austenite undergoes a transformation to metastable martensite.

3.10.1 Hardenability

Hardenability of a steel is a measure of the slowest rate of cooling at which the material can form martensite. The greater the hardenability of steel, the slower the rate of cooling that will result in a martensitic microstructure. In other words for a better hardenable steel, the TTT plots shift to the right hand side.

One possible test for hardenability is the method of quenched rounds. This involves quenching cylindrical samples of the steel which have various diameters. The cooling rate at the surface of the cylinders is high, while the cooling rate at any point within the cylinder is low. The cylinders are cut and hardness tests are taken along the diameter of the samples. The plot of hardness versus radius is called a U-curve. Figure 3.32 shows typical U-curves for hardenable and less hardenable steels. The microstructures of the etched surface of the rounds

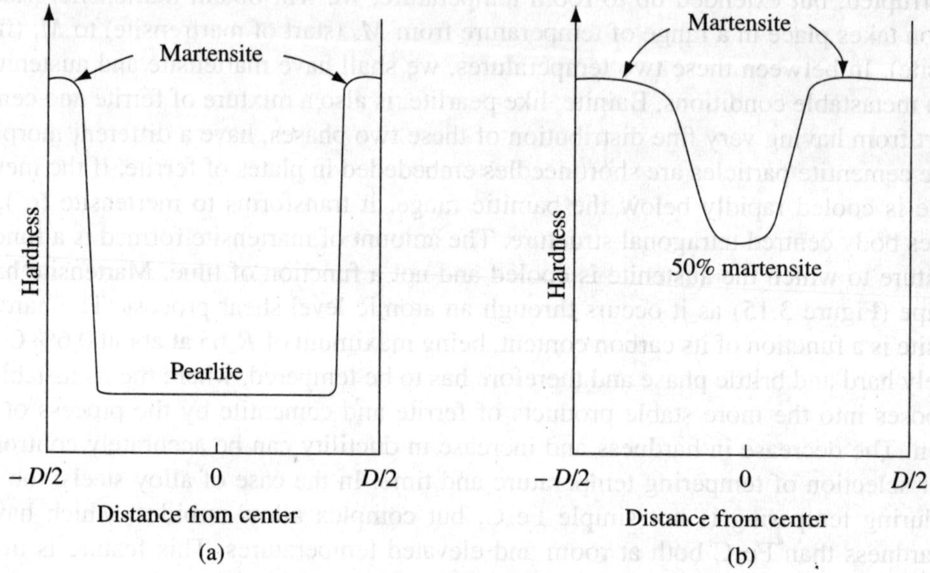

Fig. 3.32. U-curves of quenched rounds in a steel.

must be examined. Naturally martensite will form most easily near the outer edge of the cross section where the cooling rate is greatest. The slower cooling rates which predominate in the interior of the bars may be too slow to miss the nose of the CCT curve and pearlite will form. In the U-curve of the deep hardening steel (Figure 3.32b), the center section of the bar contains a mixture of 50 per cent martensite and 50% pearlite. This particular diameter of the bar is known as critical diameter. Air hardenable steel have the greatest hardenability, and water hardening steels the least. In case cracks in the steel components are formed during water quenching, this may be avoided by the use of a steel of greater hardenability. This would allow the sample to be cooled at a slower cooling rate and still insure that the central region of the part would be hardened.

Another important method of measuring the hardenability of a steel is the *Jominy end-quench* method. This technique involves removing the sample of fixed geometry from the austenitizing furnace and inserting it into an end-quench apparatus (Figure 3.33). A stream of cold water is sprayed against the bottom of the cylindrical specimen. In this way different cooling rates are obtained at different points along the length of the specimen. Since the cooling rates differ along the length of the bar, the microstructure varies from one end to the other.

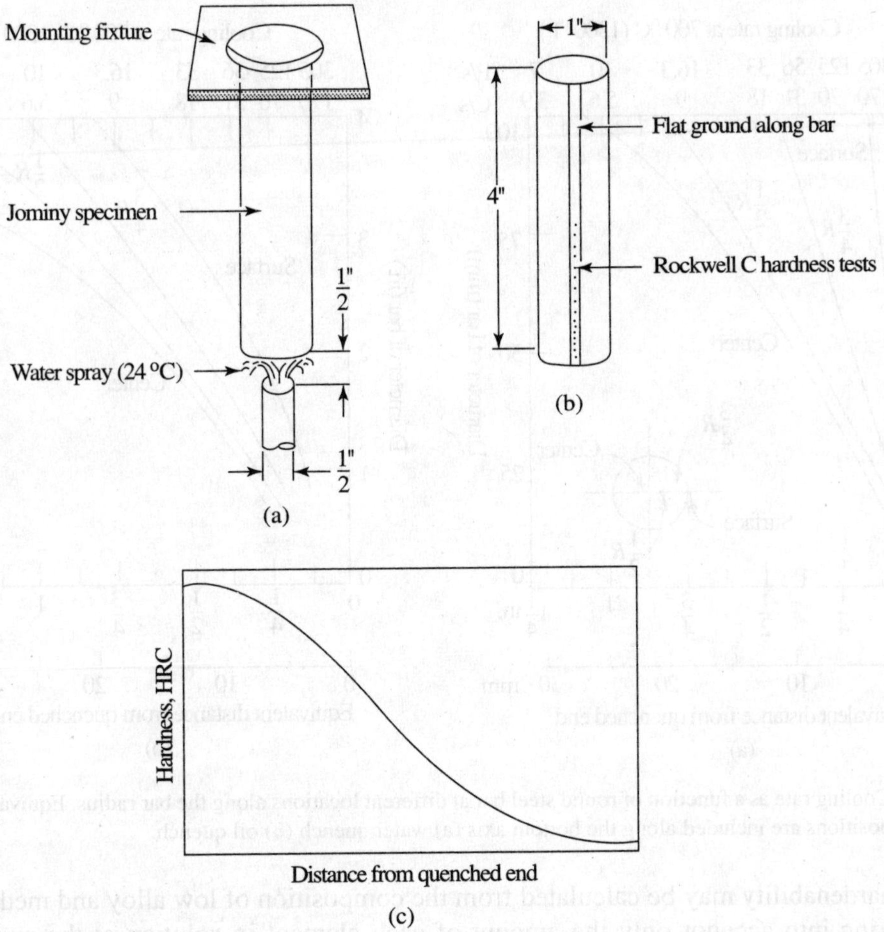

Fig. 3.33. (a) Jominy quench test fixture set up (b) marking at intervals of 1/16″ for hardness measurement (c) Hardness plot.

Subsequently the test piece is polished along the axis and hardness is measured as a function of distance from the water quenched end. Figure 3.33(c) shows a typical hardenibility plot. If the steel is very hardenable, the martensitic region extends quite far from the water-quenched end, but if the steel is only slightly hardenable, the martensitic region is very sharp and might include the face of the water quenched end. In case during continuous cooling a particular steel, bainite is also formed apart from pearlite, one must expect two plateaus in the hardenability plot, the higher hardness portion next to martensite, thus, pertains to bainite.

There are graphs available which exhibit the interconvertibility of diameter of the bar with equivalent Jominy distance. One such graph is shown in Figure 3.34. Note that any graph set is for one type of quenching medium. In case of oil quench, where the cooling rate is rather low, the curves shift to the right hand side (Figure 3.34b) as compared to those corresponding to water quenching (Figure 3.34a).

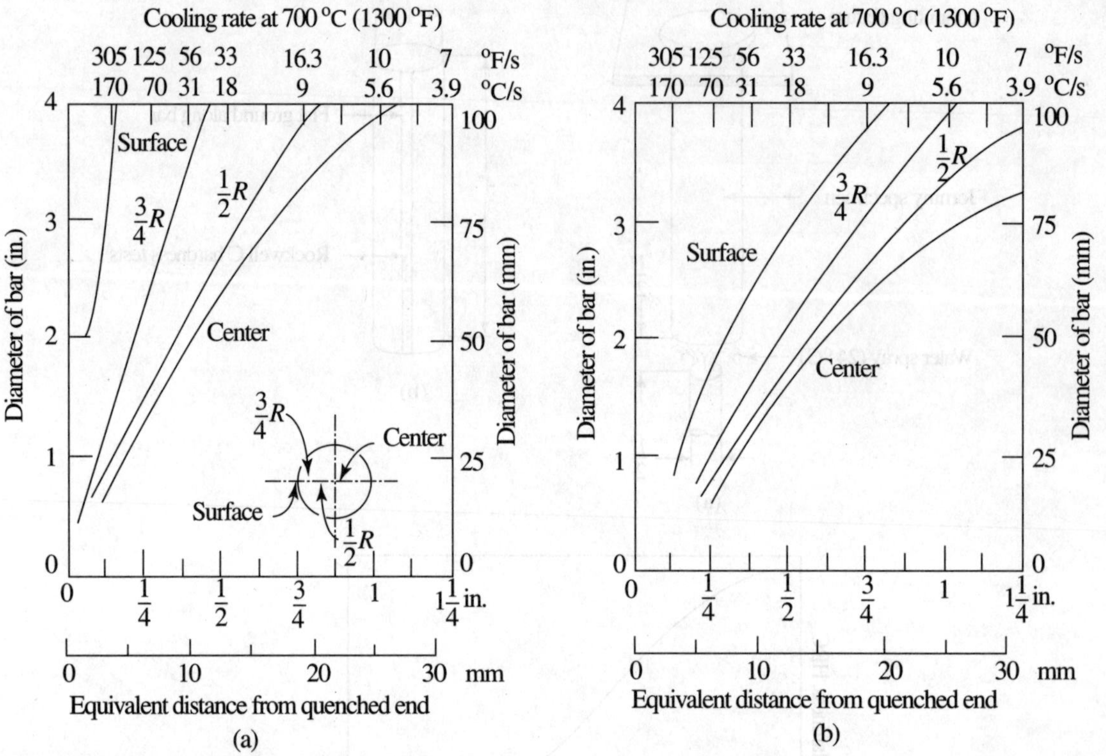

Fig. 3.34. Cooling rate as a function of round steel bar at different locations along the bar radius. Equivalent Jominy positions are included along the bottom axis (a) water quench (b) oil quench.

The hardenability may be calculated from the composition of low alloy and medium alloy steels, taking into account only the amount of each element in solution at the austenitising temperature. The austenite grain size must also be indicated. The smaller the grain size the lower is the hardenability. This is due to the fact that the total surface area of the grain boundaries increases as the grain size decreases, thereby giving rise to an increasing number of nuclei which serve as sites for pearlite formation. The question now arises how to calculate D_i-values from the chemical composition and the grain size data of the given low alloy steel. Figure 3.35 shows the 'base' hardenability characteristic for D_i. For other alloying additions, we have to read the various multiplying factors from another plot (Figure 3.36).

Example 3.8. A steel of following composition with ASTM grain size 7 is to be hardened

C	Si	Mn	Cr	Mo	
0.25	0.3	0.7	1.1	0.2	%

Calculate the D_i value of the steel.

Solution. From Figure 3.35 the D_i for 0.25%C and ASTM 7 comes out to be 0.17 in. Using Figure 3.36, we obtain following factors

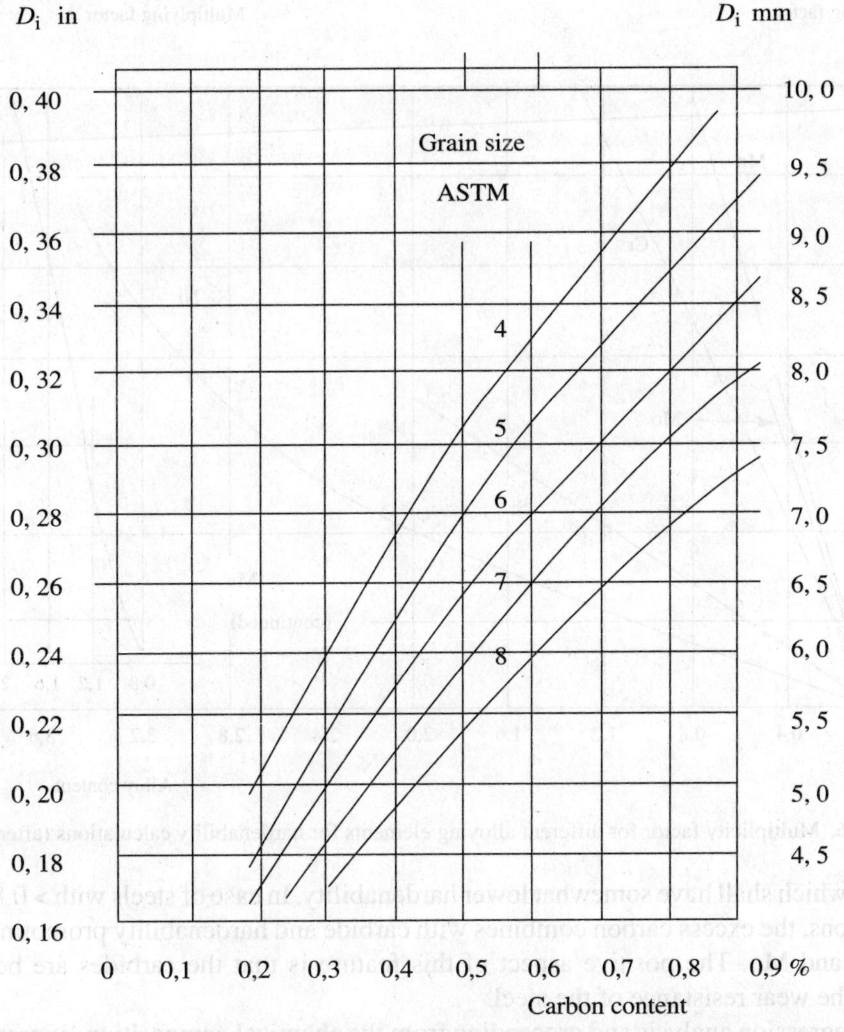

Fig. 3.35. The ideal diameter as a function of the carbon content and austenite grain size for plain carbon steels (after Grossman).

0.3Si ≈ 1.2; 0.7%Mn ≈ 3.3; 1.1%Cr ≈ 3.4; 0.2Mo ≈ 1.6;
Therefore D_i of the steel = 0.17 × 1.2 × 3.3 × 3.4 × 1.6
= 3.7 in. = 94 mm

Example 3.9. D_i values given in various plots are true assuming that that all carbides in the steel are in homogeneous solution at the austenitizing temperature. In case of steels with C> 0.8%, therefore higher austenitising temperature shall be required before quenching treatment. Is this decision correct? If not, why?

Solution. No, because higher austenitising temperature shall have two disadvantages (i) grain growth and (ii) high retained austenite. Consequently, conventional hardening temperature shall

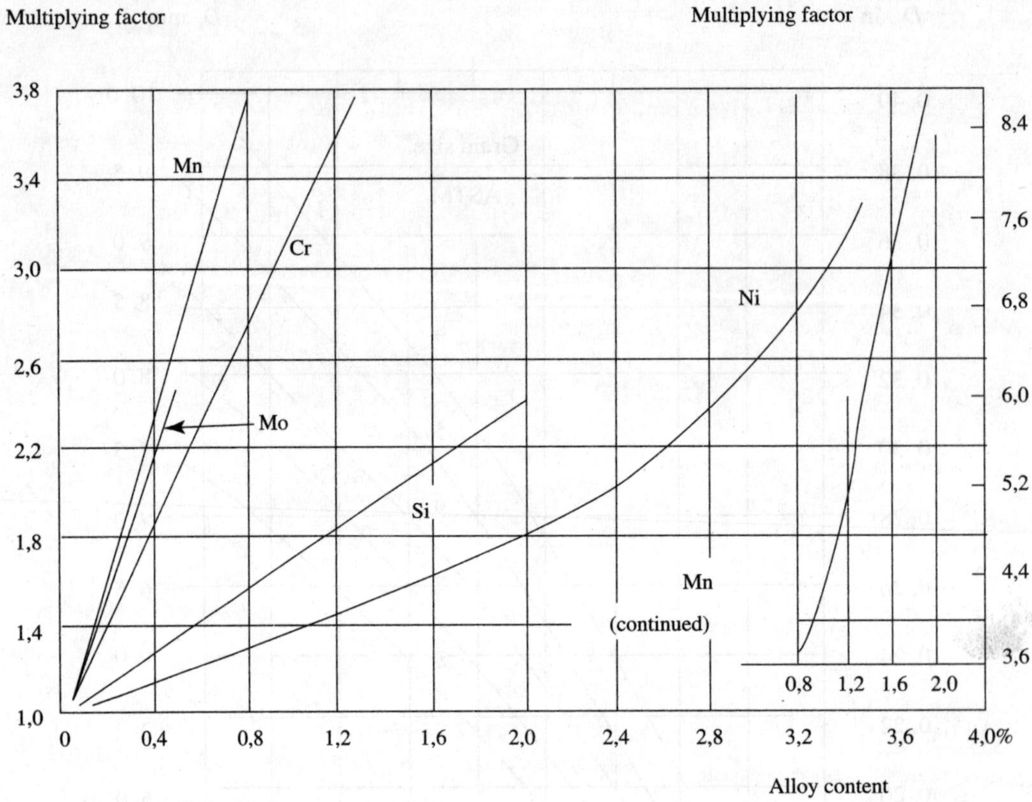

Fig. 3.36. Multiplicity factor for different alloying elements for hardenability calculations (after AISI).

be selected which shall have somewhat lower hardenability. In case of steels with > 0.8%C under such situations, the excess carbon combines with carbide and hardenability promoting elements such as Cr and Mo. The positive aspect of this feature is that the carbides are beneficial in increasing the wear resistance of the steel.

Using regression analysis and proceeding from the chemical composition, expressions have been derived that enable the hardness at different Jominy distance to be calculated direct. For a Jominy distance range of 6-80 mm, the general expression is

$$J_{6-80} = 95\sqrt{C} - 0.0028s^2\sqrt{C} + 20Cr + 38Mo + 14Mn + 6Ni + 6Si + 39V$$
$$+96P - 0.8K - 12\sqrt{s} + 0.9s - 13 \text{ HRC} \quad (3.16)$$

where J = Jominy hardness, HRC
s = Jominy distance, mm
K = ASTM grain size.

In some steels, for example a Ni-Cr-Mo low alloy steel, there may be two C-curves as shown in Figure 3.37. The advantage of such steel is that even under continuous cooling, we may get entirely bainite structure.

Factors Affecting Hardenability of Steel: The following are some of the variables in steel which affect hardenability:

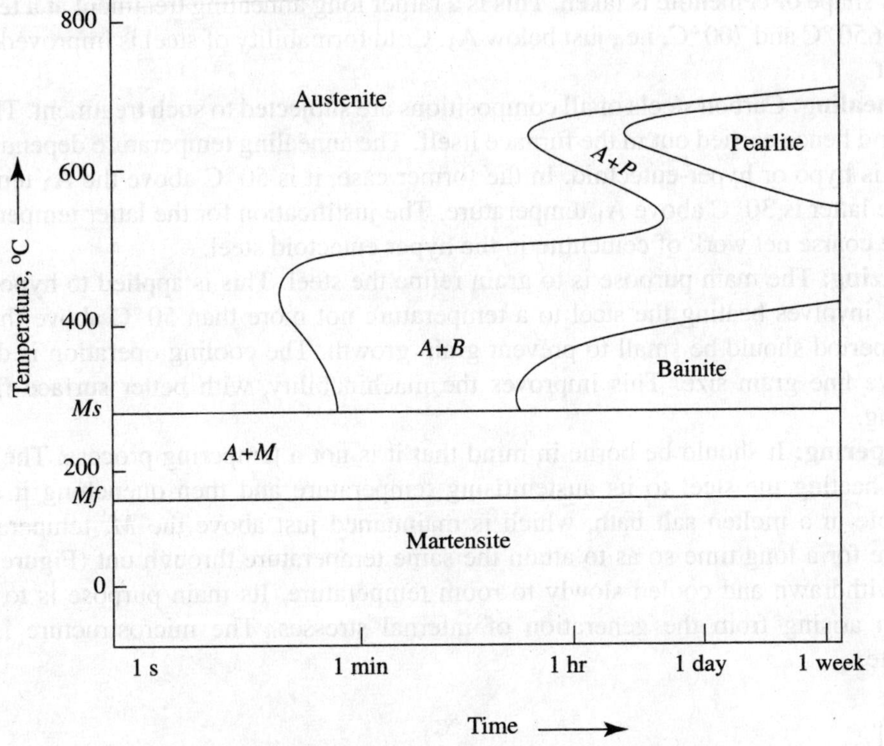

Fig. 3.37. The T-T-T diagram for Ni-Cr-Mo low alloy steel.

(a) **Carbon content:** For eutectoid composition the hardenability is maximum i.e., the incubation period in TTT diagram is the highest. At off-eutectoid composition, the incubation period is reduced because of the effects of proeutectoid constituents (ferrite or cementite) and an enhanced rate of pearlite nucleation

(b) **Alloying addition:** All the alloying elements except cobalt decrease the isothermal transformation rate. Boron retards the nucleation of proeutectoid ferrite by segregating to the grain boundaries. It, therefore, increases hardenability significantly.

(c) **Austenite grain size:** Fine austenite grains in steels provide heterogeneous nucleation sites at the grain boundaries. The reaction rate is increased and the incubation period is decreased with a decrease in the grain size.

3.10.2 Various Heat Treatment Operations

A brief description of hardening and tempering has been made in previous sections. In this section, we shall deal with following heat treatment operations.

Process Annealing: This is an inter-stage annealing operation such as might be applied to any heavily cold worked metal or alloy in order to achieve recrystallization. On industrial scale, annealing of steel at about 650 °C is generally employed, so that recrystallization will proceed more quickly and a treatment time of about one hour is sufficient.

Spheroidizing: This operation is done in order to increase machinability, where the advantage of spherical shape of cementite is taken. This is a rather long annealing treatment at a temperature between 650 °C and 700 °C, i.e., just below A_1. Cold formability of steel is improved after such treatment.

Full Annealing: Carbon steels of all compositions are subjected to such treatment. The cooling is slow and hence carried out in the furnace itself. The annealing temperature depends, whether the steel is hypo or hyper-eutectoid. In the former case, it is 50 °C above the A_3 temperatures but in the latter is 30 °C above A_1 temperature. The justification for the latter temperature is to avoid the coarse net work of cementite in the hyper eutectoid steel.

Normalizing: The main purpose is to grain refine the steel. This is applied to hypo-eutectoid steel and involves heating the steel to a temperature not more than 50 °C above the A_3. The soaking period should be small to prevent grain growth. The cooling operation is done in air to achieve fine grain size. This improves the machinability, with better surface finish after machining.

Martempering: It should be borne in mind that it is not a tempering process. The treatment involves heating the steel to its austenitising temperature and then quenching it as rapidly as possible in a molten salt bath, which is maintained just above the M_s temperature. It is held there for a long time so as to attain the same temperature through out (Figure 3.38a). It is than withdrawn and cooled slowly to room temperature. Its main purpose is to minimize distortion arising from the generation of internal stresses. The microstructure is uniform martensite.

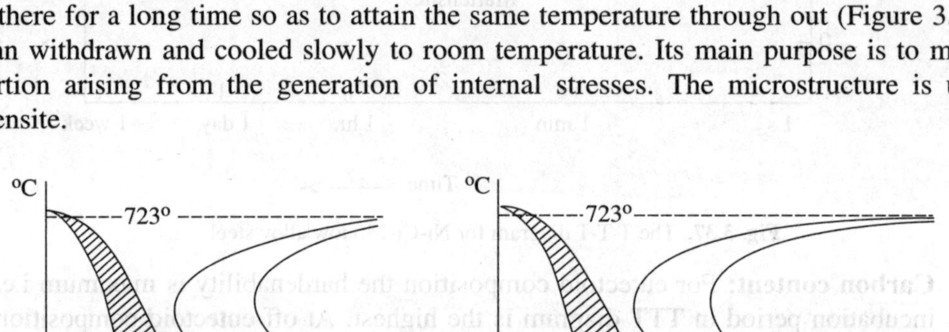

Fig. 3.38. Heat treatment cycle for (i) martempering and (ii) austempering.

Austempering: This is a true isothermal transformation to bainite. The object is to eliminate the need for a drastic water quench and the thermal shock associated with it. Figure 3.38(b) illustrates this operation. The toughness and ductility of austempered steels are often greater than those of similar steels obtained after conventional hardening and tempering to similar hardness values.

Tempering of Hardened Steel

As the hardened steels having martensitic structure are brittle, they must be given a tempering heat treatment. The tempering process involves the decomposition of martensite to the stable phases of ferrite and Fe_3C. For low carbon steels, since M_s and M_f temperatures are quite high, the martensite obtained is essentially a tempered one.

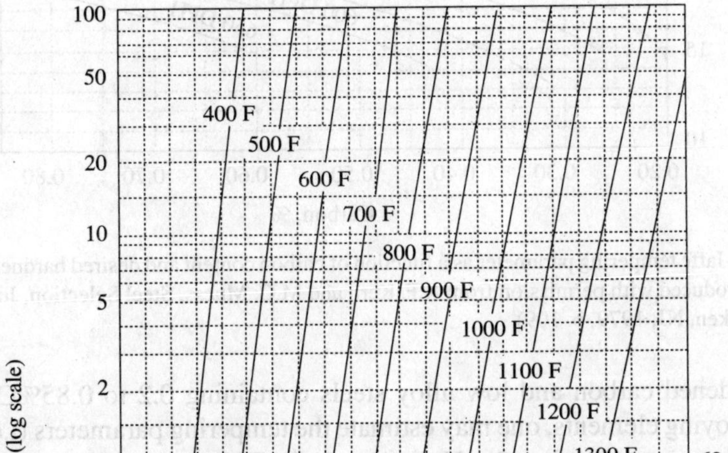

Fig. 3.39. Tempering time and temperature as a function of tempering parameter of steel (Reproduced with permission from R.F. Kern and M.C. Muess, Steel Selection, John Wiley, NY, 1979, p. 111).

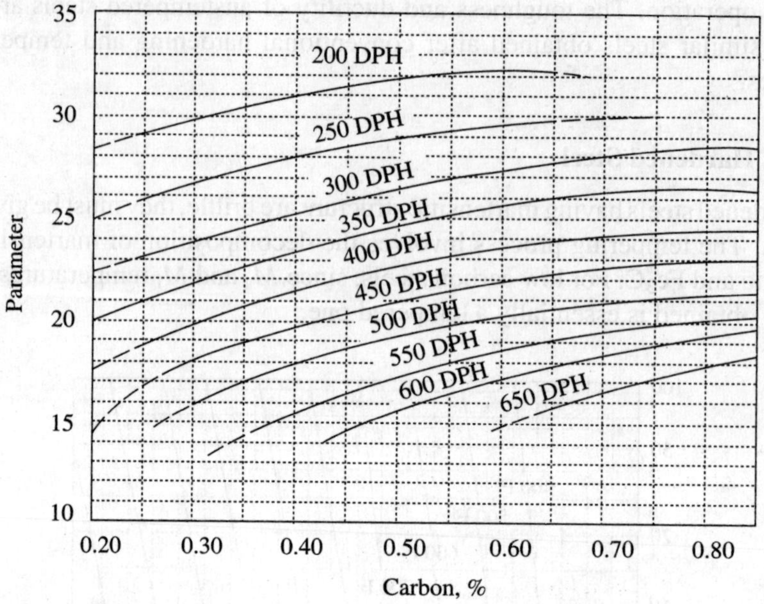

Fig. 3.40. Hollomon-Jaffe tempering parameter as a function of carbon content and desired hardness for plain carbon steel (Reproduced with permission from R.F. Kern and M.C. Muess, Steel Selection, John Wiley & Sons, Inc., Hoboken, NJ, 1979, p. 109)

For fully hardened carbon and low alloy steels containing 0.2 to 0.85%C and less than 5 per cent total alloying elements, one may estimate the tempering parameters (i.e., temperature and time at the temperature) using the *Holloman-Jaffe* (H-J) parameter, when the tempering temperature is in the range 345 °C to 650 °C. The expression for H-J parameter is

$$\text{H-J Parameter} = T(\log t + 18) \times 10^{-3} \tag{3.17}$$

where T is absolute temperature and t is time in hours. Using the data in Figures 3.39 and 3.40, we can determine the possible combination of tempering time and temperature. In low alloy steels (less than 5% alloying elements), we can still use the data related to Fe-C for tempered hardness, but with a change Δ, such that

$$\Delta_p = \sum_p wt\%\ X.F_X \tag{3.18}$$

where Δp = Change in hardness

X = The particular alloying element

F_X = Factor for the element at the particular parameter used.

The tempering hardness factors F_X for the different alloying elements with in certain composition ranges are given in Table 3.3 for varying parameters.

Example 3.10. A tempering treatment for a low alloy steel specifies 5 hrs at 500 °C. What temperature would be required to achieve the same tempering effect in 1 hour?

Solution. Mathematically, using equation (3.17)

$$T_1(\log t_1 + 18) = T_2(\log t_2 + 18)$$
$$773(\log 5 + 18) = T_2(\log 1 + 18)$$
$$\text{Therefore } T_2 = 803\ \text{K} \approx 530^\circ\ \text{C}$$

Table 3.3. Tempering hardness factors, F_X, of alloying elements added to tempered hardness of plain carbon steels (Reproduced with permission from R.F. Kern & M.C. Suess, Steel Selection, John Wiley, N.Y., 1979)

				Factor at Indicated Parameter Value			
Element	*Range*	20	22	24	26	28	30
Manganese	0.85-2.1%	36	25	30	30	30	25
Silicon	0.3-2.2%	65	60	30	30	30	30
Nickel	Up to 4%	5	3	6	8	8	6
Chromium	Up to 1.2%	50	55	55	55	55	55
Molybdenum	Up to 0.35%	40 (20)*	90 (45)*	160 (80)*	220 (110)*	240 (120)*	210 (105)*
Vanadium+	Up to 0.2%	0	30	85	150	210	150

* If 0.5−1.2% Cr is also present, use this factor.
\+ For AISI-SAE chromium-vanadium steels; may not apply when vanadium is the only carbide forming present.
Note: Boron factor is 0.

3.10.3 Case Carburizing and Carbonitriding

Low carbon steels are case hardened to provide a hard and wear resisting surface or case and also to enhance its impact strength. Because of the development of compressive stresses in the surface layers during the case hardening treatment, the fatigue strength of the steel is also increased. The details on fatigue strength are made in Chapter 5. The most common form of case hardening is through case carburizing. It is affected either in solid media (carburizing compounds), in salt baths or in gases at a temperature normally between 825 °C and 925 °C. The transport of carbon from the carburizing medium invariably takes place via a gaseous phase, usually carbon mono-oxide. These methods are (Figure 3.41):

- Direct hardening i.e., quenching straight from the carburizing medium.
- Single quenching: Heating and quenching the carburized parts after first allowing them to cool to room temperature from the carburizing treatment.
- Double quenching: This consists of a direct quench and a requench from a lower temperature. The lower temperature range 780-820 °C relates to the range of hardening for the case. This process is generally carried out to steels that have not been normalized i.e., fine grain treated. Apart from time and temperature, the depth of carburization depends on the carbon potential of the carburizing medium and on the composition of the steel. The higher the carbon potential of the medium the higher the carbon concentration at the surface of the steel, when the equilibrium has been established and the deeper the carburizing depth. According to ISO specification the depth of case hardening is defined as the distance from the surface to a plane at which the hardness is 550 HV. This standard covers depth of hardening of 0.3 mm and more.

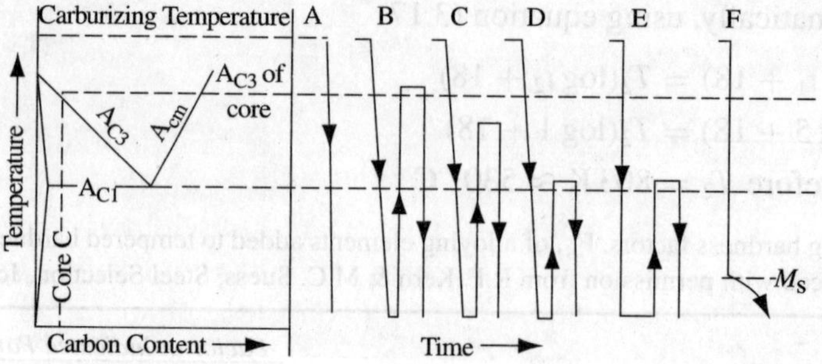

Treatment	Case	Core
A	Unrefined. Solution of excess carbide favored. Austenite retained. Distortion low to medium.	Unrefined but hardened.
B	Refined. Solution of excess carbide favored. Austenite retention promoted in highly alloyed steels.	Refined. Maximum core strength and hardness.
C	Refined. Some Solution of excess carbide.	Partially refined. Stronger and tougher than Treatment D.
D	Refined. Excess carbide not dissolved.	Unrefined. Soft and machinable.
E	Refined. Some Solution of excess carbide. Austenite retention minimized.	Low hardness. High toughness. Machinable.
F (Interrupted quench; marquenching)	Unrefined. Solution of excess carbide favored. Austenite retained. Distortion minimized.	Fully hardened.

Fig. 3.41. Various heat treatments of carburized steel

The validity of simple diffusion equation $x = k\sqrt{t}$ (x in mm and t in hours) has been established and the 'k' values are 0.34, 0.41 and 0.52 at 875, 900 and 925 °C. For example, a carburizing treatment at 900 °C for 12h would give a case depth of 1.42 mm. The full details about the carburizing atmospheres and their thermodynamics can be referred from the Metals Handbook.

From the above description, it is clear that the critical factor to determine the surface hardness of the steel is the dissolution of carbon in austenite during heat treatment. We know that if the carbon level is greater than 0.7%, the M_s temperature of the steel falls substantially, giving rise to retained austenite after quenching. In case of overcarburized steel, it is possible to control the carbon level into solution and hence the retained austenite by varying the quenching temperature. In case of other alloying additions in the steel the parameters do change. For example as the Ni content in the steel increases the value for carbon concentration corresponding to the maximum hardness is reduced. For Cr-Mn steel the maximum hardness occurs at 0.7%C and for Cr-Mo steel at 1% C.

Case carburized steels are tempered at temperatures generally around 160-220 °C.

Carbonitriding Carbonitriding is carburizing the steel in gas with simultaneous nitrogen pick up. The treatment temperature is normally 800-900 °C. Nitrogen in the form of gaseous ammonia is supplied direct into furnace or in the same feed pipe as is used by the carburizing gases. The ammonia content is usually kept between 3-8 per cent. Figure 3.42 shows a typical distribution of carbon and nitrogen on the surface layer of a carbonitrided mild steel.

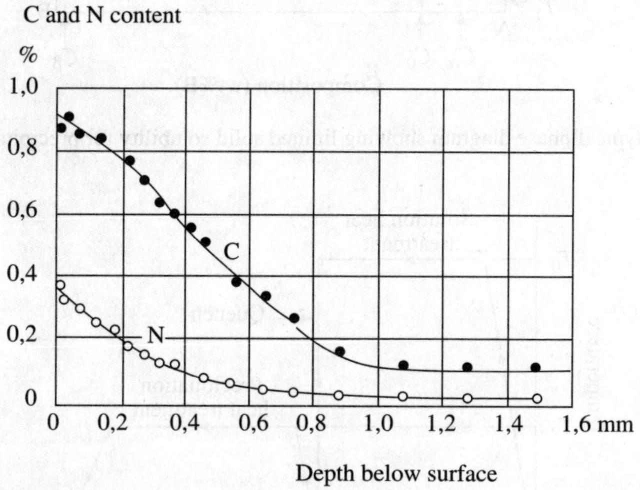

Fig. 3.42. Typical carbon and nitrogen concentration in a mild steel after carbonitriding at 860 °C.

Other types of surface hardening treatments are not described in this chapter as they are not related to phase transformation. They shall be treated separately in Chapter 11.

3.11 PRECIPITATION AND PRECIPITATION HARDENING

The essential requirement for precipitation to occur in solid solution is the decreasing solubility of a solute with decreasing temperature. This results in a supersaturated solid solution that, being unstable, tends to decompose according to the reaction.

Supersaturated α Solid Solution $\rightarrow$ Saturated α solid solution $+\beta$ precipitates

Figure 3.43 shows a phase diagram illustrating the type of alloy and conditions for precipitation hardening. In general, the second phase, β precipitate, is formed by a sequence of thermal treatments (Figure 3.44). These are solution treatment, quenching and ageing. Solution treatment is carried out at a temperature for a sufficiently long time so that complete homogenization of the solid solution takes place. During quenching treatment, supersaturated solid solution results. Here the high concentration of vacancies corresponding to the equilibrium concentration at the solution temperature is retained. These vacancies slowly precipitate or condense at, producing clusters and dislocation loops that form the sites at which precipitate particles can nucleate. Furthermore, the quenched-in vacancies enhance the

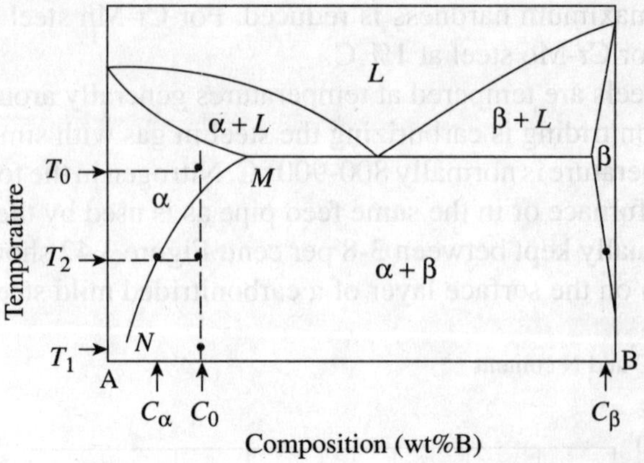

Fig. 3.43. The typical phase diagram showing limited solid solubility for precipitation hardening.

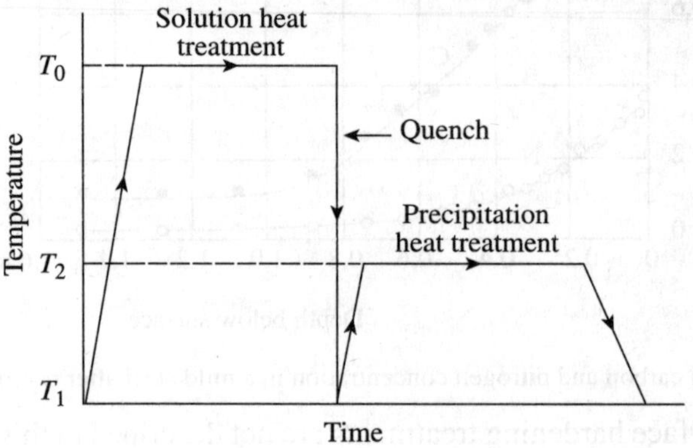

Fig. 3.44. Precipitation hardening cycle for an alloy (for temperature refer Figure 3.43)

diffusion rate of the solute, thereby promoting the formation of precipitate. The third step, called age hardening or precipitation hardening, involves reheating the alloy to an elevated temperature somewhere below the solvus line and holding at this temperature, called ageing temperature for a certain period of time to develop the necessary amount and kind of precipitate that will impart a desired hardness. One may cause the ageing process to occur at room temperature, known as natural ageing. But here to achieve a hardness peak is not possible even after a long period as the diffusion is slow at room temperature. Duralumin is a typical Al-4.5% Cu alloy, which undergoes such type of heat treatment. Here a temperature between 100 and 180 °C (Figure 3.45) is the optimum ageing temperature, where the ageing time is not too large. After reaching the peak, the hardness starts to fall. This phenomenon is called *overageing*. The fall in hardness is because of coarsening of the fine precipitate, which makes a decrease in the coherency strain. The precipitate particles, $CuAl_2$, in case of Al-Cu binary is submicroscopic and can be resolved only under electron microscope. The overaged particles could be observed under optical microscope. It is obvious that an alloy artificially aged at certain elevated temperature, would not overage during service at room temperature. However such alloys are not useful for elevated temperature service. We shall discuss this under the section 'Strengthening Mechanisms' in Chapter 5.

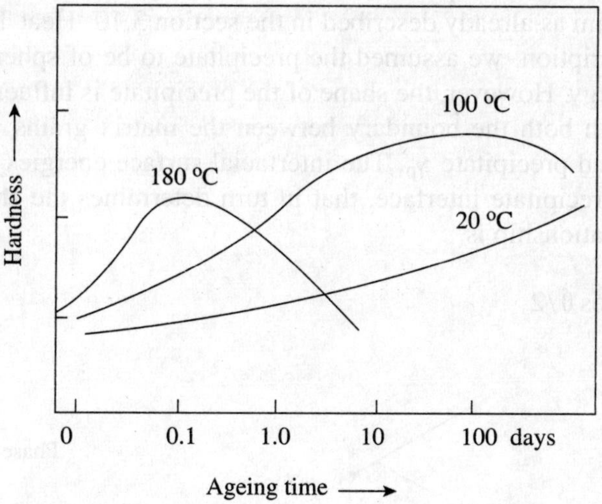

Fig. 3.45. Ageing curves for a duralumin (Al-Cu) alloy.

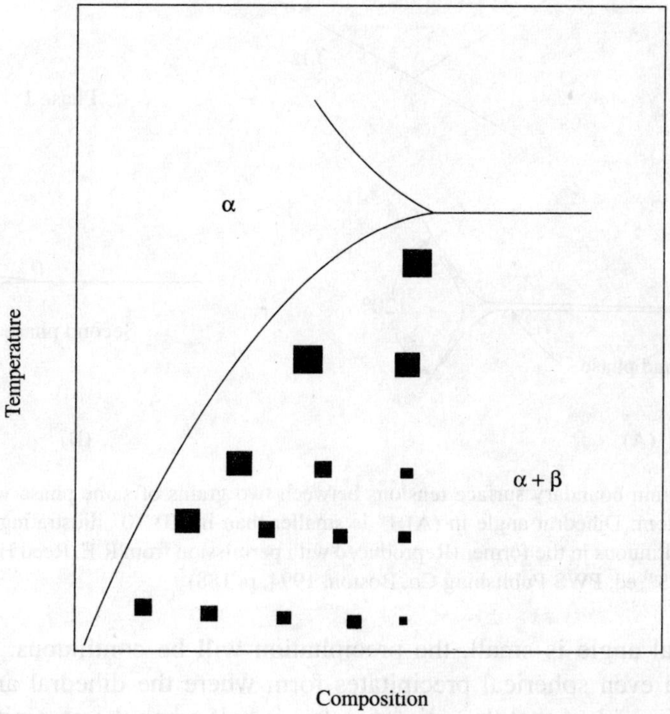

Fig. 3.46. Variation of precipitate size with ageing temperature

The ageing temperature affects the precipitate size in the manner illustrated in Figure 3.46. Low ageing temperature corresponds to high saturation and excessive nucleation of precipitates occurs, whereas at high ageing temperatures (lower supersaturation) fewer, coarser particles are formed. The kinetics of precipitation can be shown by a C-curve or Time-Temperature-Transformation diagram as already described in the section 3.10 'Heat Treatment of Steels'.

In the above description, we assumed the precipitate to be of spherical shape in order to minimize surface energy. However, the shape of the precipitate is influenced by the *interfacial energy* associated with both the boundary between the matrix grains γm and the boundary between the matrix and precipitate γ_p. The interfacial surface energies fix a *dihedral* angle θ between the matrix-precipitate interface, that in turn determines the shape of the precipitate (Figure 3.47). The relationship is

$$\gamma_m = 2\gamma_p \cos\theta/2 \tag{3.19}$$

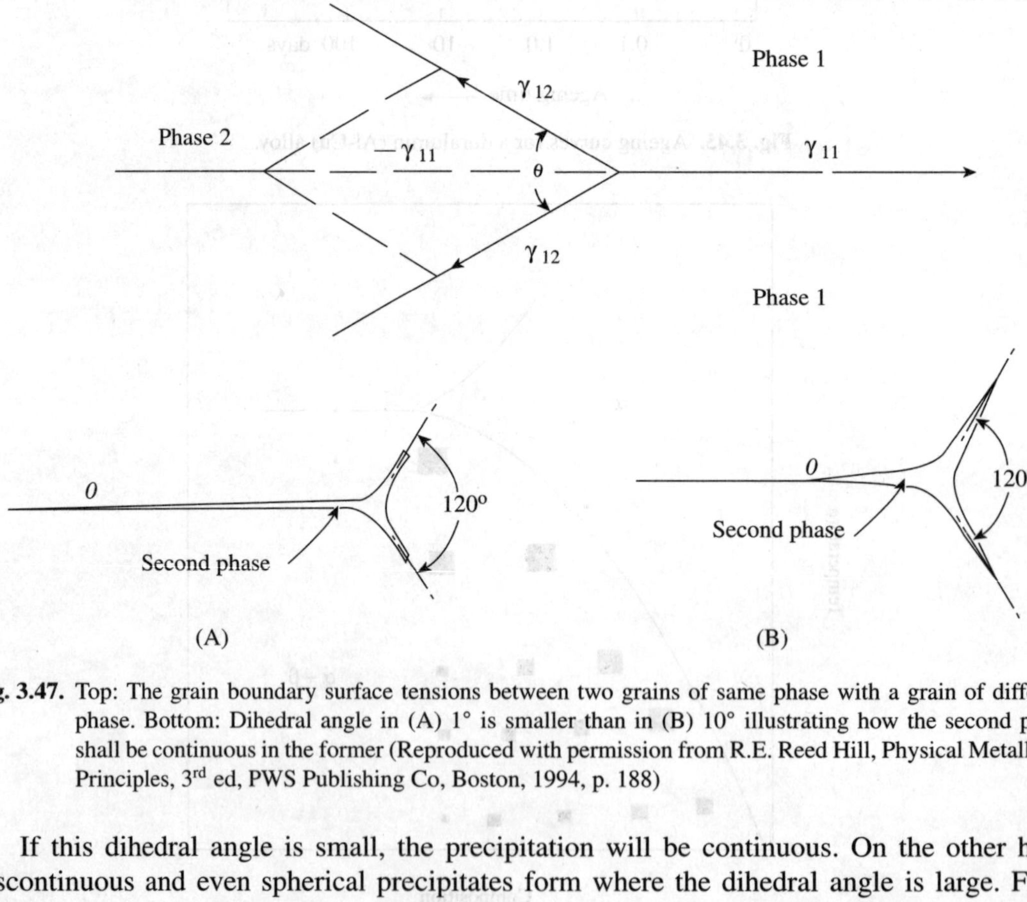

Fig. 3.47. Top: The grain boundary surface tensions between two grains of same phase with a grain of different phase. Bottom: Dihedral angle in (A) 1° is smaller than in (B) 10° illustrating how the second phase shall be continuous in the former (Reproduced with permission from R.E. Reed Hill, Physical Metallurgy Principles, 3rd ed, PWS Publishing Co, Boston, 1994, p. 188)

If this dihedral angle is small, the precipitation will be continuous. On the other hand discontinuous and even spherical precipitates form where the dihedral angle is large. From above relation, it is evident that the ratio of γ_p/γ_m is half when the precipitation is continuous i.e., θ is zero degrees.

The coherency strain described earlier introduces elastic strain energy which usually increases with square of the disregistry (misfit). The strain energy may be either shared by both phases or present primarily in one of the phases depending upon their elastic properties. Strain energy is positive and it increases the free energies of the phases concerned.

Example 3.11. Show schematically as a free energy-precipitated particle radius diagram how critical nuclei radius decreases with fall in the precipitation temperature. What basic assumption have you made in your consideration?

Solution. Assumption is that surface energy does not change with temperature. Figure Ex. 3.11 shows the schematic presentation.

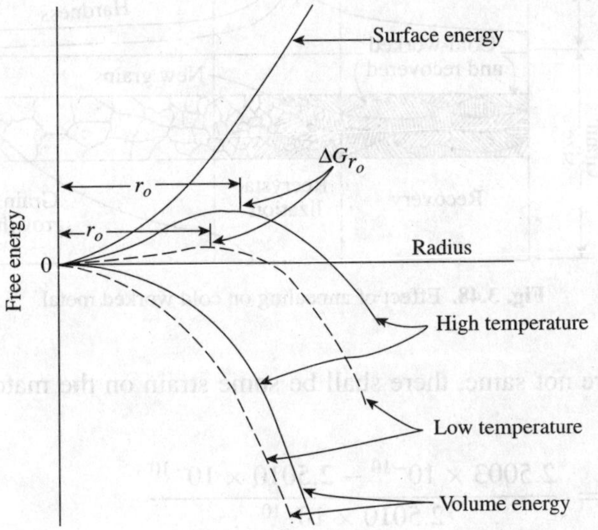

Figure Ex. 3.11

Example 3.12. During solid state transformation (001) plane of HCP cobalt is parallel to (111) plane of FCC cobalt. Both these planes are close packed planes. The HCP cobalt is likely to form coherently on FCC cobalt such that (001) HCP || {111} FCC and [100] HCP || <101> FCC. The '*a*' values of HCP and FCC cobalts are 2.5003×10^{-10} m and 3.5370×10^{-10} m. Justify whether there shall be any strain and if so how much.

Solution. It is known that the close packed planes (001) HCP and (111) of FCC have the same atomic arrangement. Therefore HCP-cobalt is likely to form coherently on FCC cobalt. The interatomic distance along the [100] direction on the (001) plane of HCP cobalt is the same as its lattice parameter '*a*' i.e., 2.5003×10^{-10} m. The interatomic distance along the <101> direction on the (111) plane of FCC cobalt is $a\sqrt{2}/2$, where '*a*' is the lattice parameter.

$$\text{Hence interatomic distance} = (3.5370 \times 10^{-2} x\sqrt{2})/2$$
$$= 2.5010 \times 10^{-10}\ \text{m}$$

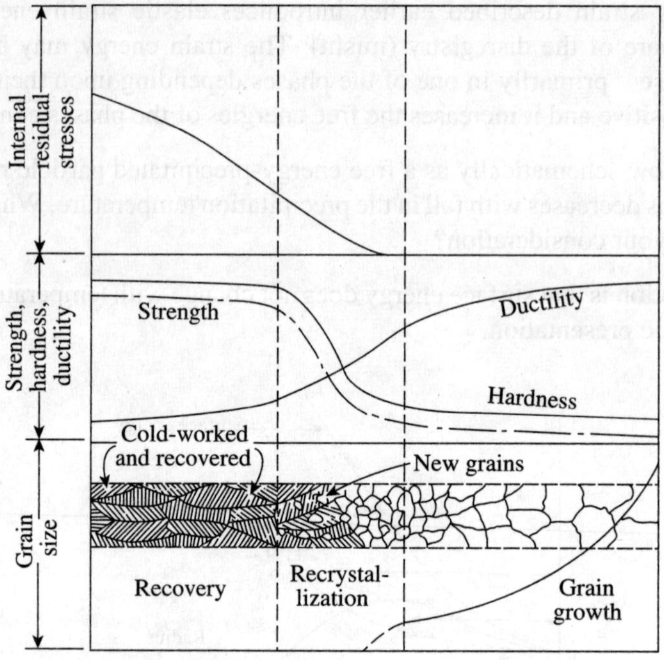

Fig. 3.48. Effect of annealing on cold worked metal.

Since these values are not same, there shall be some strain on the matching planes along the matching direction.

$$\text{Mismatch} = \frac{2.5003 \times 10^{-10} - 2.5010 \times 10^{-10}}{2.5010 \times 10^{-10}}$$

$$= 0.028\%$$

3.12 EFFECT OF ANNEALING ON COLD WORKED METALS

Heating of a cold worked metal to a high enough temperature eliminates by diffusion, most of the crystal defects and restores the properties to their original value. Such a heat treatment is referred to as annealing. During annealing, cold worked metals undergo the restoration processes of recovery and recrystallization.

Recovery is the process of releasing the internal stresses in a metal at relatively low temperature, above 0.1 T_m, where T_m is the melting point on the absolute temperature scale. During this process, relaxation of internal stresses and redistribution of more mobile dislocations occur, known as polygonization. In the previous chapter, we have discussed in 2.7.3, how low angle grain boundaries are formed from edge dislocations. Recovery although releases the internal stresses, it hardly affects the mechanical properties (Figure 3.48). This is an important method for reducing the internal stress in castings, forgings, welded joints, without lowering the strength acquired during cold working. Another name for this is *stress-relief annealing*.

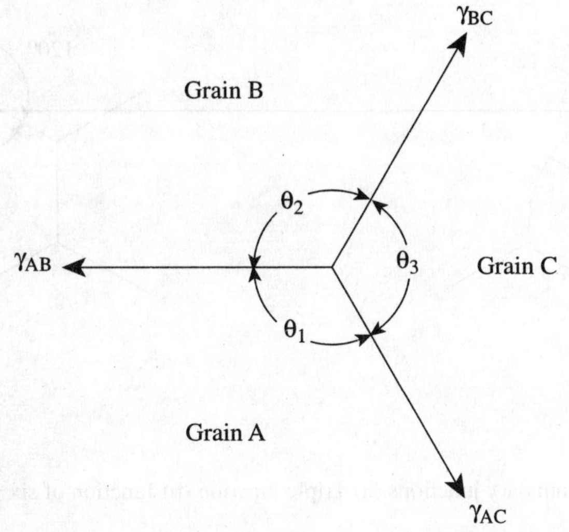

Fig. 3.49. Interfacial grain boundary energy in polycrystal.

If we heat the cold worked metals and alloys at a temperature between 0.3-0.5 T_m, recrystallization occurs, where new sets of crystals emerge. This results in considerable change in properties such as tensile strength and ductility. The cold worked microstructure i.e., elongated grains, disappears and new grains are formed. If we go on annealing further , grain growth takes place, that is, recrystallized grains grow at the expense of the smaller ones, thereby reducing the total surface energy. Figure 3.48 presents the microstructural evolution during this stage.

The shape of the grain in polycrystalline materials is based on the grain boundary energy. A polycrystalline material is never a true equilibrium structure. If all grain boundaries in a polycrystal are assumed to have the same grain boundary energy independent of boundary orientation (Figure 3.49) then at equilibrium the relation will be.

$$\frac{\gamma_{AB}}{\sin\theta_3} = \frac{\gamma_{AC}}{\sin\theta_2} = \frac{\gamma_{BC}}{\sin\theta_1} \tag{3.20}$$

If $\gamma_{AB} = \gamma_{AC} = \gamma_{BC}$ as in the case of the junction between three grain boundaries, angles θ_1, θ_2 and θ_3 are equal with a magnitude of 120°. Such generated microstructure is known as *equiaxed structure*. If the total number of boundaries around a grain is less than six, each boundary must be concave inward producing a metastable structure (Figure 3.50a). Such grain boundaries will tend to shrink and eventually disappear during grain growth stage. On the other hand larger grains having more than six boundaries will grow.

The grain growth kinetics depends on temperature and heating time. The temperature of recrystallization depends on the degree of deformation of cold worked metal. The greater the degree of work hardening is, the lower the recrystallization temperature. The addition of certain alloying element in a metal may significantly raise the recrystallization temperature. There is a critical strain value, below which there is no recrystallization. Figure 3.51 illustrates the above features graphically. At the critical strain a few recrystallized grains form and grow extremely large. This technique is sometimes used to grow single crystals. Excessive grain growth can

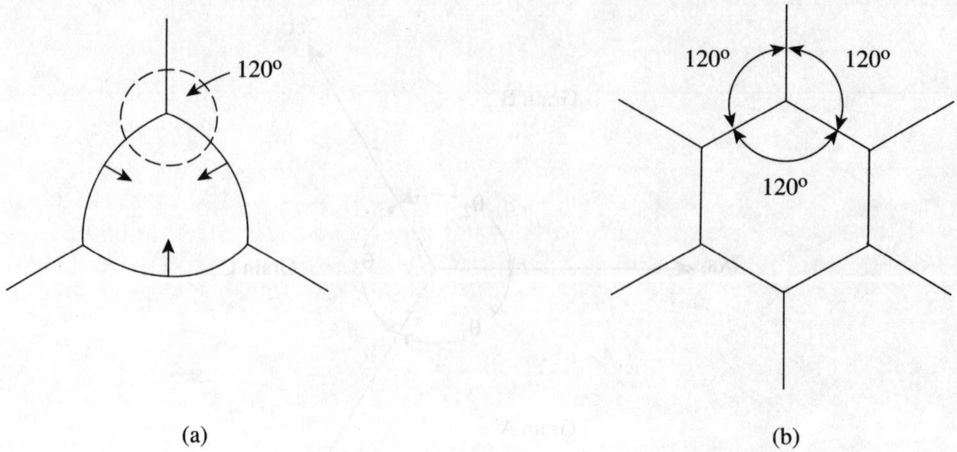

Fig. 3.50. Grain boundary junctions (a) Triple junction (b) Junction of six grain boundaries.

result in surface roughness and some times in premature failure. The study of recrystallization is important in establishing proper schedules in thermo mechanical treatment, a topic to be covered later in the Chapter 8.

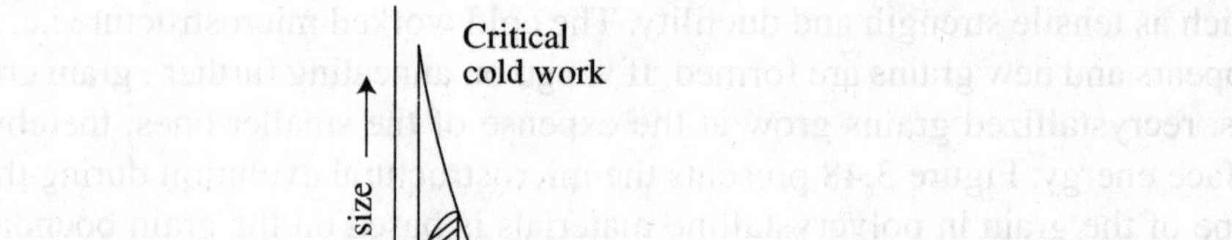

Fig. 3.51. Interrelationship between prior cold work, annealing temperature and the grain size.

The type of recrystallization that has been described so far is called *primary recrystallization.*

In some cases during the grain growth stage, a few large grains will form that grow preferentially and thereby consume the smaller recrystallized grains. This phenomenon is called *secondary recrystallization*. It is found to occur in metals that have achieved a relatively stable small grain size.

Example 3.13. Assuming that a cold worked metal recrystallizes completely in 100 hours at 260 °C, but that at 300 °C, the process is complete in only six minutes, calculate the activation energy of the recrystallization process. At what time would recrystallization be complete at 290 °C? R = 8314.3 J mol^{-1}K^{-1}.

Solution. We know

$$\text{The rate of recrystallization } \dot{R} = A \exp\left(-\frac{Q}{RT}\right)$$

$$\text{where } Q = \text{Activation energy and } A = \text{constant}$$

$$\text{Let } R_1 = \text{Recrystallization rate at temperature } T_1$$

$$R_2 = \text{Recrystallization rate at temperature } T_2$$

$$\text{Thus, } \frac{R_2}{R_1} = \exp\frac{Q}{R}\left(\frac{T_2 - T_1}{T_1 T_2}\right) \quad \text{(i)}$$

$$\text{Time } t_1 \text{ at } 260\,^\circ\text{C} = 100 \text{ hours } = 6000 \text{ minutes}$$

$$\text{Time } t_2 \text{ at } 300\,^\circ\text{C} = 6 \text{ minutes}$$

$$\dot{R}_1 \alpha \frac{1}{6000} \text{ min}^{-1}; \quad \dot{R}_2 \alpha \frac{1}{6} \text{ min}^{-1}$$

Substituting above values in equation (i), we get

$$\frac{6000}{6} = \exp\frac{Q}{8314.3}\left(\frac{603 - 533}{603 \times 533}\right)$$

$$= \exp Q\left(\frac{70}{2.672 \times 10^9}\right)$$

$$\text{Solving } Q = 2.63 \times 10^8 \text{ J k mol}^{-1}$$

$$\text{Let } \dot{R}_3 = \text{Recrystallization rate at } 290\,^\circ\text{C}$$

$$\text{and } t = \text{time required for recrystallation at } 290\,^\circ\text{C (563 K)}$$

$$\text{Again } \frac{R_3}{R_1} = \exp\frac{2637 \times 10^8}{8314.8}\left(\frac{563 - 533}{533 \times 563}\right) = \frac{6000}{t}$$

$$t = 251 \text{ minutes}$$

3.13 PHASE TRANSITION IN GLASSES

We have mentioned in the introduction of this chapter how pure metals and various polymorphs of a pure metal transforms at a constant temperature during liquid to solid or solid to solid transformation. In glasses, which are predominantly amorphous (noncrystalline) such a feature

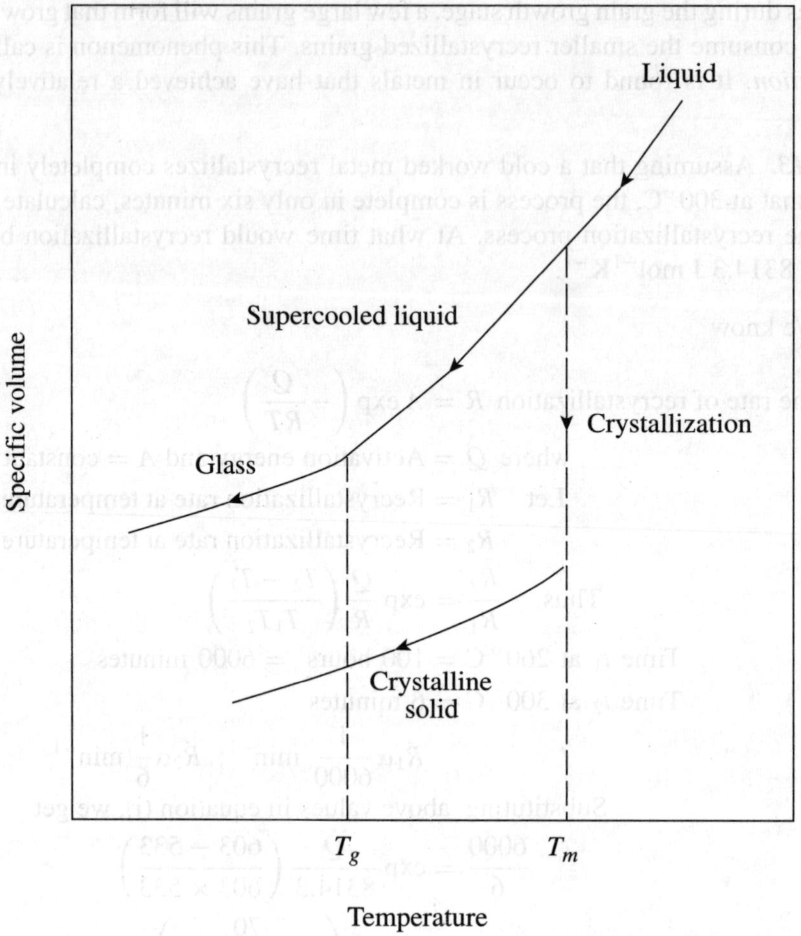

Fig. 3.52. Volume versus temperature response of crystalline and amorphous materials.

is not observed. Upon cooling, a glass becomes more and more viscous in a continuous manner with decreasing temperature. Figure 3.52 shows a plot of specific volume with temperature for both crystalline and glassy materials. Specific volume is the volume per unit weight, i.e., it is reciprocal of density. For glassy materials, volume decreases continuously with fall in temperature. A slight decrease in slope of the curve occurs at glass transition temperature Tg. Below this temperature the material is considered to be a glass. Above this critical temperature the glass is in liquid state, first a super cooled melt and then a melt.

There are various temperatures used in defining the state of the glass (Figure 3.53) These are:

Melting point: It is the temperature, where the fluidity is enough to call the phase as liquid.

Working point: It is the temperature at work the glass is easily deformed.

Softening point: It is the maximum temperature at which a glass piece may be handled without causing much dimensional changes.

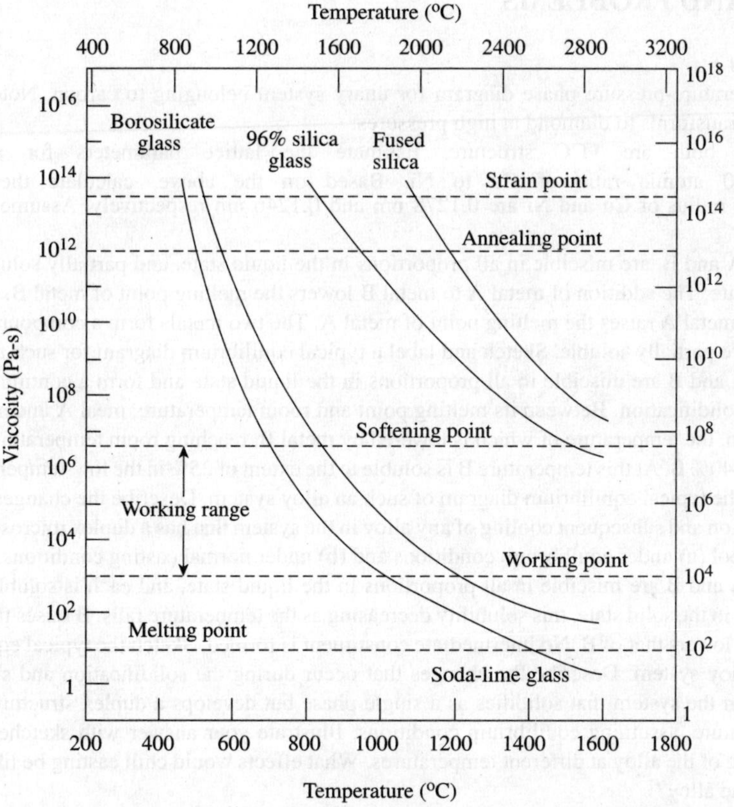

Fig. 3.53. Variation of viscosity (log scale) with temperature for different glasses (Reproduced with permission from W.D. Callister, Jr, Materials Science & Engineering 5th ed, John Wiley & Sons, Inc.,Hoboken, NJ, 2001, p. 425).

Annealing point: It is the temperature at which the diffusion is fast, so that any residual stresses may be removed within about 15 min.

Strain point: It is the temperature below which fracture of glasses occur before the onset of plastic deformation. Glass transition temperature is above the strain point.

In the above sequence of temperature, the viscosity of the glass increases.

In case of glass ceramic which is a phase mixture of amorphous and crystalline glasses, a nucleating agent e.g., titanium oxide, is added to induce crystallization or divitrification.

3.13.1 Annealing of glasses

Since glasses in general fail under thermal shock, due to residual stresses, it is often subjected to annealing, also known is thermal tempering. The temperature for this operation is above the glass transition, but below the softening point. It is then cooled to room temperature in a jet of air, or in some cases, an oil bath. The treatment offers compressive stresses on the surface, with tensile stresses on the interior. The details on the fracture in glasses shall be described in Chapter 5 'Mechanical Properties'.

QUESTIONS AND PROBLEMS

Phase Diagrams

3.1 Draw a temperature-pressure phase diagram for unary system belonging to carbon. Note carbon exists as graphite and transforms to diamond at high pressures.

3.2 Cu and Ni both are FCC structure. Estimate the lattice parameters for a solid solution with a 50/50 atomic ratio of Cu to Ni. Based on the above, calculate the density of the alloy. Atomic radius of Cu and Ni are 0.1278 nm and 0.1246 nm respectively. Assume the alloy follows Vegard's law.

3.3 Two metals, A and B, are miscible in all proportions in the liquid state, and partially soluble in one another in the solid state. The addition of metal A to metal B lowers the melting point of metal B, while the addition of metal B to metal A raises the melting point of metal A. The two metals form a compound, A_xB_y, in which both metals are partially soluble. Sketch and label a typical equilibrium diagram for such an alloy system.

3.4 Two metals A and B are miscible in all proportions in the liquid state and form a continuous series of solid solutions on solidification. Between its melting point and room temperature, meal A undergoes an allotropic transformation, the temperature of which is lowered by metal B, reaching room temperature at a composition of 60% A and 40% B. At this temperature B is soluble to the extent of 25% in the low-temperature modification of A. Sketch the typical equilibrium diagram of such an alloy system. Describe the changes that occur during the solidification and subsequent cooling of any alloy in the system that has a duplex microstructure, assuming the alloy to cool (a) under equilibrium conditions and (b) under normal casting conditions.

3.5 Two metals A and B are miscible in all proportions in the liquid state, and each is soluble in the other to a limited extent in the solid state, this solubility decreasing as the temperature falls. B raises the melting point of A, whereas A lowers that of B. No intermediate constituent is formed. Sketch the typical equilibrium diagram of such an alloy system. Describe the changes that occur during the solidification and subsequent cooling of any alloy in the system that solidifies as a single phase but develops a duplex structure while cooling to room temperature, assuming equilibrium conditions. Illustrate your answer with sketches representing the microstructure of the alloy at different temperatures. What effects would chill casting be likely to have on the structure of the alloy?

3.6 Some date on two hypothetical metals, A and B, are listed below:

	Metal A	Metal B
Melting point, °C	1, 000	700
Atomic Diameter (10^{-8}cm.)	2.9	2.6
Space Lattice Type	F.C.C.	B.C.C.
Valency	1	2

Sketch a typical equilibrium diagram which in your opinion accords with the information given above. Explain why you think the system might reasonably be assumed to have an equilibrium diagram of this type.

3.7 Some data relating to two hypothetical metals A and B are listed below: Sketch a typical equilibrium diagram

	Metal A	Metal B
Freezing point (F.P.), °C	1,000	500
Effect of one metal on F.P. of other	lowered by B	raised by A
Atomic weight	60	70
Atomic diameter (10^{-8} cm.)	2.5	2.7
Space lattice type	F.C.C.	H.C.P
Valency	1	2
Intermediate constituents	Only one based on compound AB	

which, in your opinion accords with the information given above. Explain why you think the system might reasonably be assumed to have an equilibrium diagram of this type. What type of space lattice do you think the intermediate constituent might have?

3.8 A Fe-C steel of eutectoid composition solidifies in the presence of a magnetic field. Assuming that the field has an effect on the phase equilibria, determine the number of degrees of freedom when the steel is in the austenite phase.

3.9 The mass fraction of eutectoid ferrite in an Fe-Fe_3C system is 0.71. On the basis of this information, is it possible to determine the composition of the alloy? If so, what is the composition?

3.10 For a Fe-3%C cast iron, make schematic microstructures that would be observed for conditions of vary slow cooling at the following temperatures: 1250, 1145, 700 °C. Label the phases and indicate their approximate composition.

3.11 In case of Fe-graphite system in contrast to Fe-Fe_3C system the eutectoic composition is 0.7%C and the eutectoid temperature is 738 °C. How much graphite can form in a cast iron containing 3.3% C at 740 °C?

3.12 Assume that the eutectoid composition of a 4140 (Cr-Mo) steel is reduced from 0.8%to 0.65% by the alloying additions, how much proeutectoid ferrite and how much eutectoid constituents would be expected in an austenitized and slowly cooled sample?

3.13 Describe the meaning of the term partitioning coefficient.

Diffusion

3.14 Calculate the diffusion coefficients of gold in silver at 100 °C, 500 °C and 1000 °C. The data for D_o and Q are 0.26 cm^2/s and 45500 cal/mole respectively. What you infer from these data?

3.15 The melting point of silicon and nickel are close to each other (1410 and 1453 °C respectively), but the activation energies for self diffusion are 496 kJ/mole for silicon and only 280 kJ/mole for nickel. Justify your answer.

3.16 The activation energy for diffusion in FCC nickel is only 36 kJ/mole for the interstitial solute hydrogen compared with 283 kJ/mole for nickel. Comment on the above statement.

3.17 For semiconductor development arsenic is to be diffused into silicon. Determine the time required to give a concentration of 1.5×10^{22} atoms/m^3 at a depth of 1.2 μm, if initially there was zero concentration, the surface concentration of arsenic remains constant at 5×10^{24} atoms/m^3 and the diffusion coefficient at the processing temperature is 3.0×10^{-18} m^2/s.

3.18 In Alclad, 20 mm thick duralumin sheets are covered on either side by 0.2 mm thick pure aluminium sheets. For retaining the corrosion resistance, the copper concentration at a depth of 0.1 mm from the outer surface should not exceed 0.400. How long can be material be kept at 550 °C, without damaging the corrosion resistance? Given: D_o of Cu in Al 0.25×10^{-4} m^2s^{-1}, Q of Cu in Al = 121 $kJmol^{-1}$.
The following Error Function table is also given

z	erf(z)	z	erf(z)
0.75	0.7112	0.90	0.7970
0.80	0.7421	0.95	0.8209
0.85	0.7707	1.0	0.8427

Alclad is essentially an aluminium alloy with 4 wt% Cu.

3.19 A diffusion couple, made between two bars containing pure Cu and Cu-10wt% Zn, is kept at 827 °C for 100 hours. Analysis shows a Zn concentration of 0.7 wt% at a distance of 1.0 mm from the interface. How much time will be required to obtain the same concentration (0.7 wt%) in twice the distance i.e., 2.0 mm from the interface?

3.20 What is the difference in the vacancy mechanism for diffusion in metals (Na) and ionic solids (NaCl)?

3.21 The activation energies for surface and grain boundary diffusion of thorium in tungsten are 55 per cent and 75 per cent respectively, of the activation energy for lattice diffusion of thorium in tungsten. Comment on the statement.

3.22 In a tungsten wire of 10 μm diameter, when do you envisage that the contribution of surface diffusion be greater than that from the lattice?

3.23 CaO can dissolve into ZrO_2 because Ca^{2+} and Zr^{4+} have nearly the same size. There is a 15/85 ratio of Ca^{2+}/Zr^{4+}. The charge is balanced by O^{2-} ion vacancies. What is the ratio of O^{2-} ion vacancies □ to O^{2-}?

3.24 Why is it widely believed that the allotropic transformation in titanium is a diffusionless martensitic type?

3.25 Order-disorder transitions are commonly associated with metallic systems and not with polymers. Why?

Phase Transformation in Steels

3.26 The growth rate of pearlite remains essentially constant as the growth proceeds, whereas the growth rate of preeutectoid ferrite slows down as growth proceeds. Explain.

3.27 Draw a typical cooling curve corresponding to air cooling over of TTT diagram of a hypoeutectoid steel and explain why a decrease in the slope is noticed during the transformation.

3.28 What phases are present in bainite? How does bainite differ from pearlite and martensite?

3.29 Why the pearlitic reaction in Hadfield manganese steel (1% C, 12% Mn) never goes to completion during isothermal transformation?

3.30 What is full annealing and how does it differ from process annealing?

3.31 A hypereutectoid steel containing 1 w/o C is to be given a full anneal
(a) What temperature should be used?
(b) What cooling rate should be used?
(c) Sketch the resulting microstructure after the full anneal.
(d) Calculate the weight fraction of each micro constituent present.

3.32 Describe the heat treatment of steel which would produce the following microstructure:
(a) A grain boundary network of cementite and fine pearlite.
(b) Spheroids of cementite and bainite.
(c) Spheroids of cementite in a ferritic matrix.
(d) Ferrite and martensite.
(e) Ferrite, martensite and pearlite.
Show these schedules on TTT diagram.

3.33 The strong carbide formers have low solubility in austenite. What is the implication of this on heat treatment of steels.

3.34 Why does an addition of carbon in the BCC iron lattice causes an increase in the c/a ratio of martensite lattice? How many carbon atoms are in one BCT unit cell of martensite? Give their positions.

3.35 A single crystal of martensite is more stable than the γ- iron below M_S. Why does a sample of the steel only partially transform at a given temperature between M_S and M_F to martensite?

3.36 If a 1080 steel were quenched to martensite and then tempered for a long time at say 650 °C, the microstructure would comprise spheroids of Fe_3C in a matrix of ferrite. How could this microstructure be converted to pearlite?

3.37 Distinguish between hardness and hardenability of steel. In what way do T-T-T diagrams indicate the hardenability of steel?

3.38 What simple heat treatment is widely used to improve the toughness of structural steels?

3.39 Why is not low carbon steel hardened by heat treatment?

3.40 How fast must you quench 4340 steel to prevent formation of soft spots in hardening? (Consult Metals Handbook for the TTT diagram).

3.41 The martensite transformation in steels is normally not reversible. Comment on the statement.

3.42 Why tempering invariably follows hardening operation for steel? How are tempering time and tempering temperature related to one another.

3.43 A sample of steel which has been carburized at 927 °C for 10 hrs has a carburized depth of 0.04 cm. How long must this steel be carburized at the same temperature to produce a carburized depth of 0.08 cm?

Precipitation Phenomenon

3.44 Referring Example 2.6 in Chapter 2, calculate the surface energy of copper. Given the bond energy in copper is 56.4 kJ mol^{-1}.

3.45 Alumina precipitates coherently from magnesium aluminate (MgO-Al_2O_3) such that (111) of the spinal (cubic) is parallel to the close packed based plane of alumina (Hexagonal). Show the compatibility of the arrangement of atoms on these two planes for the formation of a coherent interface.

3.46 Two solid phases α and β have tetragonal and orthorhombic structures respectively such that their lattice parameters have relationships $a_\alpha = a_\beta$ and $c_\alpha = b_\beta$. If the grain of these phases are oriented such that $(100)_\alpha \parallel (001)_\beta$ and $[010]_\alpha \parallel [100]_\beta$, justify that the interface is a coherent one with no strain whatsoever.

3.47 Draw a curve showing the relation between the two phase dihedral angle on the ratio of the two phase surface tension to the single phase surface tension.

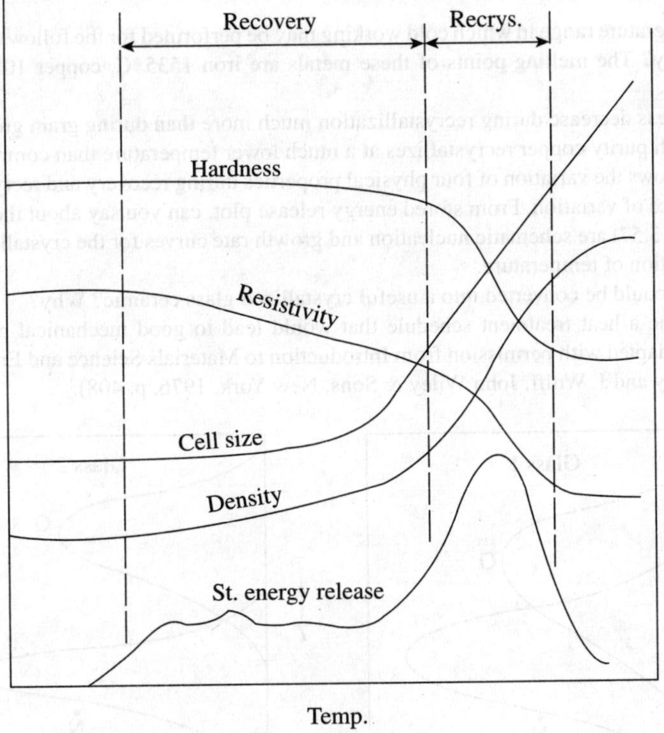

Figure P. 3.56

3.48 Because of coherency the precipitates do not coarsen rapidly. Comment on the statement.

3.49 Determine the geometry of a grain boundary triple junction in which one of the three grain boundaries has an energy of one half the average, while the other two are average.

3.50 Noticing the binary phase diagrams of the concerned systems, suggest which of the following alloys can not undergo age hardening-

(a) 95Cu-5Al (b) 95Al-5Cu (c) 95Cu-5Ni
(d) 95Ni-5Cu (e) 95Cu-5Zn (f) 95Zn-5Cu
(g) 95Mg-5Al (h) 95Al-5Mg (i) 95Cu-5Ag
(j) 95Ag-5Cu

3.51 Copper with 2% beryllium is a precipitation hardening alloy. Consult the Metals Handbook for phase diagram and describe the heat treatment schedule for obtaining maximum hardness in this alloy.

3.52 An aluminium alloy(2% Cu; 0.3% Mg) in the form of rivets is solution treated at 495 °C and used whilst it remains in this soft supersaturated solid-solution condition. It subsequently precipitation hardens at ambient temperature after the riveting operation is complete. During tests it was found that the solution-treated alloy began to harden in 4.5 hours at ambient temperature (20 °C) whilst if placed in boiling water (100 °C) the precipitation hardening process began in only 3.5 minutes. It is required to refrigerate a batch of solution-treated rivets so that the onset of precipitation hardening is delayed in order that the rivets may be stored and used over a period of four days after solution treatment. What refrigeration temperature must be used so that the rivets will not fail during the riveting process? (Adapted with permission from 'Properties of Engineering Materials', R.A. Higgins, 2nd Ed., Industrial Press Inc, N.Y., 1994, p. 266).

Recrystallization

3.53 What is the temperature range in which cold working may be performed for the following metals: iron, copper, lead and mercury? The melting points of these metals are iron 1535 °C, copper 1083 °C, lead 327 °C and mercury - 39 °C.

3.54 Why does hardness decrease during recrystallization much more than during grain growth.

3.55 Explain why high purity copper recrystallizes at a much lower temperature than commercial purity copper.

3.56 Figure P.3.56. shows the variation of four physical properties during recovery and recrystallization of a metal. Explain the nature of variation. From stored energy release plot, can you say about the purity of the metal?

3.57 Shown (Figure P.3.57) are schematic nucleation and growth rate curves for the crystallization of two different glasses as a function of temperature.

(a) Which glass could be converted into a useful crystallized glass ceramic? Why?

(b) Briefly outline a heat treatment schedule that would lead to good mechanical properties. Justify your schedule. (Adapted with permission from Introduction to Materials Science and Engineering, K.M. Ralls, T.H. Courtney and J. Wulff, John Wiley & Sons, New York, 1976, p. 408).

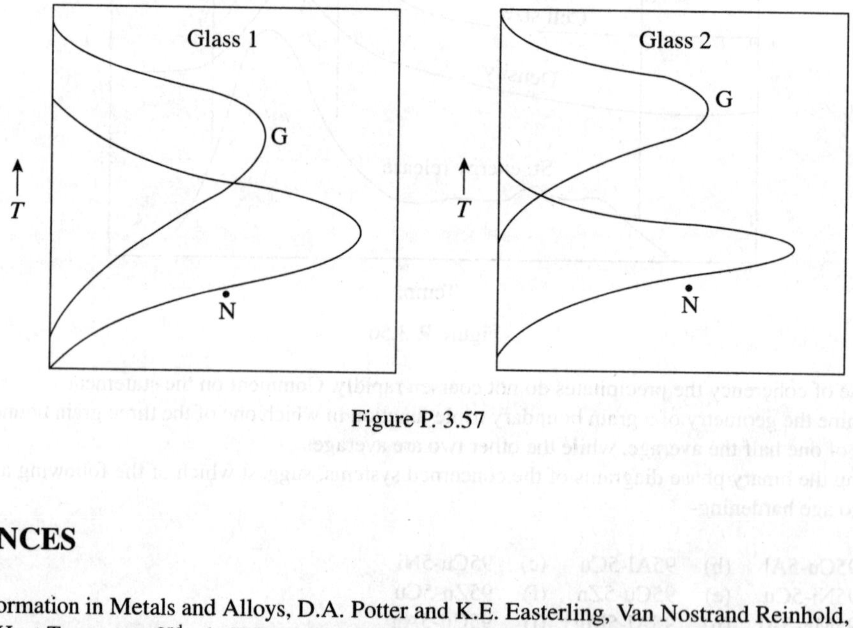

Figure P. 3.57

REFERENCES

Phase Transformation in Metals and Alloys, D.A. Potter and K.E. Easterling, Van Nostrand Reinhold, 1981.
Steel and Its Heat Treatment, 2nd ed., K.E. Thelnig, Butterworth Heinemann, Oxford, 1984.
Principles of Phase Diagrams in Materials Systems, P. Gordon, McGraw Hill, N.Y., 1968.
Steels: Heat Treatment and Processing Principles, G. Krauss, ASM International, Materials Park, Ohio, 1990.
Diffusion in Solids, 2nd ed., P.G. Shewman, TMS, Warrendale, PA, 1989.
Physical Metallurgy Handbook, A.K. Sinha, McGraw Hill, New York, 2003.
Fundamentals of Physical Metallurgy, J.D. Verhoeven, John Wiley, N.Y., 1975.
Phase Diagrams in Metallurgy, F.N. Rhines, McGraw Hill Book Co., New York, 1956.

4

PHYSICAL PROPERTIES OF MATERIALS

Practically, all properties of materials are dependent on the interatomic bonding, which on its own turn, depends on the nature of electronic structure of the material. We are aware that real crystals do contain lattice defects. All the properties of the crystalline materials can be classified into structure-insensitive and structure sensitive. Properties like elastic constants, melting point, density, specific heat, coefficient of thermal expansion are structure insensitive. On the other hand properties like electrical and magnetic ones, yield, fracture and creep strengths are structure sensitive. Table 4.1 lists materials in the descending order for different physical properties. Before going into details of various physical properties, it is worthwhile to refresh some knowledge on the electronic energy levels.

4.1 ELECTRONIC ENERGY LEVELS IN SOLID

The energy states in isolated atoms are sharply defined and occupy a certain definite energy level according to the Pauli principle. In the condensed state i.e., solid the original sharp energy levels of the individual atoms are broadened into wider regions called energy bands. The number of discrete levels in each band is same as the number of atoms. The width of these energy bands depends on the energy level to which the electrons belong and on the degree of overlapping of these energy values. The valence bands corresponding to outer shells become wider as the degree of overlap becomes greater. The empty bands, which correspond to the excited states of the outer shell electrons, are also called *conduction bands*. The forbidden band is the *energy gap* in a nonmetallic crystal is the gap between the highest valence band and the lowest conduction

Table 4.1. Physical Properties of Materials, in Descending Order

Density	*Melting point*	*Specific heat*	*Thermal conductivity*	*Thermal expansion*	*Electrical conductivity*
Platinum	Tungsten	Wood	Silver	Plastics	Silver
Gold	Tantalum	Beryllium	Copper	Lead	Copper
Tungsten	Molybdenum	Porcelain	Gold	Tin	Gold
Tantalum	Niobium	Aluminium	Aluminium	Magnesium	Aluminium
Lead	Titanium	Graphite	Magnesium	Aluminium	Magnesium
Silver	Iron	Glass	Graphite	Copper	Tungsten
Molybdenum	Beryllium	Titanium	Tungsten	Steel	Beryllium
Copper	Copper	Iron	Beryllium	Gold	Steel
Steel	Gold	Copper	Zinc	Ceramics	Tin
Titanium	Silver	Molybdenum	Steel	Glass	Graphite
Aluminium	Aluminium	Tungsten	Tantalum	Tungsten	Ceramics
Beryllium	Magnesium	Lead	Ceramics		Glass
Glass	Lead		Titanium		Plastics
Magnesium	Tin		Glass		Quartz
Plastics	Plastics		Plastics		

band that one has in mind (Figure 4.1). Only solids with partially filled energy bands are conductors. Other possibility of good conduction may be overlap with second empty valence band. In insulators, which are ionic, covalent or molecular solids, the core and valence energy bands are completely filled and do not overlap with the next empty conduction band but remain separated by a forbidden energy band, also called the energy gap. The transfer of an electron from the valence band to the next conducting band would require a high activation energy, which depends on the width of the energy gap between the two bands. This value is large for insulators at room temperature. The distinction between semiconductors and insulators is largely of a qualitative nature; the former have small energy gaps (1 or 2 eV or less), whereas the insulators have large gaps of several electron volts.

Since there are many electrons in a solid, it is not enough to know the energy spectrum for a single electron, but the distribution of electrons in the band must also be known to understand the physical properties of a solid. The energy of the highest occupied state is known as *Fermi level* and is denoted by E_F. In case of metals, this energy level is within the valence band, while in case of semiconductors this lies in the band gap region.

The behaviour of a material as a semiconductor or as an insulator also depends on the temperature. The electron concentration in the conduction band increases exponentially as the temperature is raised. For example, at low temperatures silicon makes an excellent insulator, but at very high temperature it has a conductivity approaching that of a metal (Figure 4.2). We can, therefore, say that the magnitude of the band gap and the temperature are both very important factors in determining the electrical properties of a semiconductor or an insulator.

In case of free electron, from our knowledge of basic physics we know

$$E(k) = \frac{h^2k^2}{2m} \tag{4.1}$$

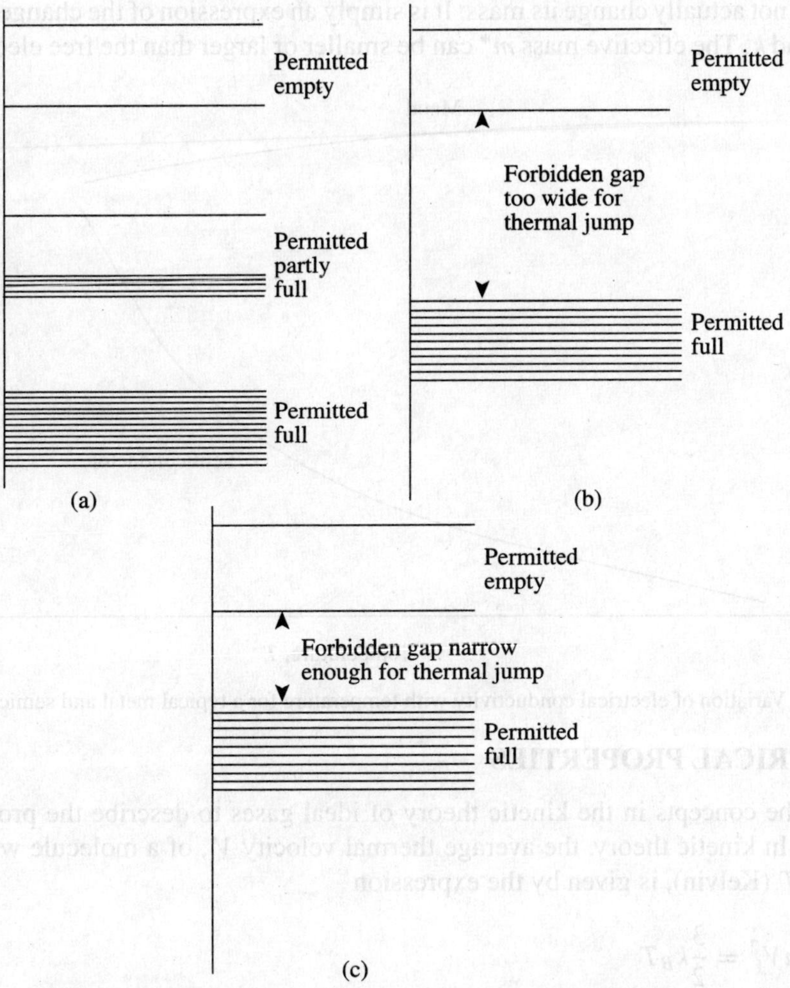

Fig. 4.1. Electron occupancy of energy bands (a) insulator (b) conductor.

where E = Energy
h = Planck's constant
k = Wave vector
m = Mass of the electrons (constant in this case)

This relationship between E and k breaks down in a crystalline solid. However, the form of the above relationship can be mentioned by using the relation

$$E(k) = \frac{h^2}{2m^*}k^2 \tag{4.2}$$

where $m^*$is an adjustable parameter, which is called effective mass of the electron. This means that any deviation from a parabolic relationship between E and k can be expressed as a change in the effective mass of the electrons at that point in k space. It must be borne in mind that the electron does not actually change its mass. It is simply an expression of the changed relationship between E and k. The effective mass m^* can be smaller or larger than the free electron mass m.

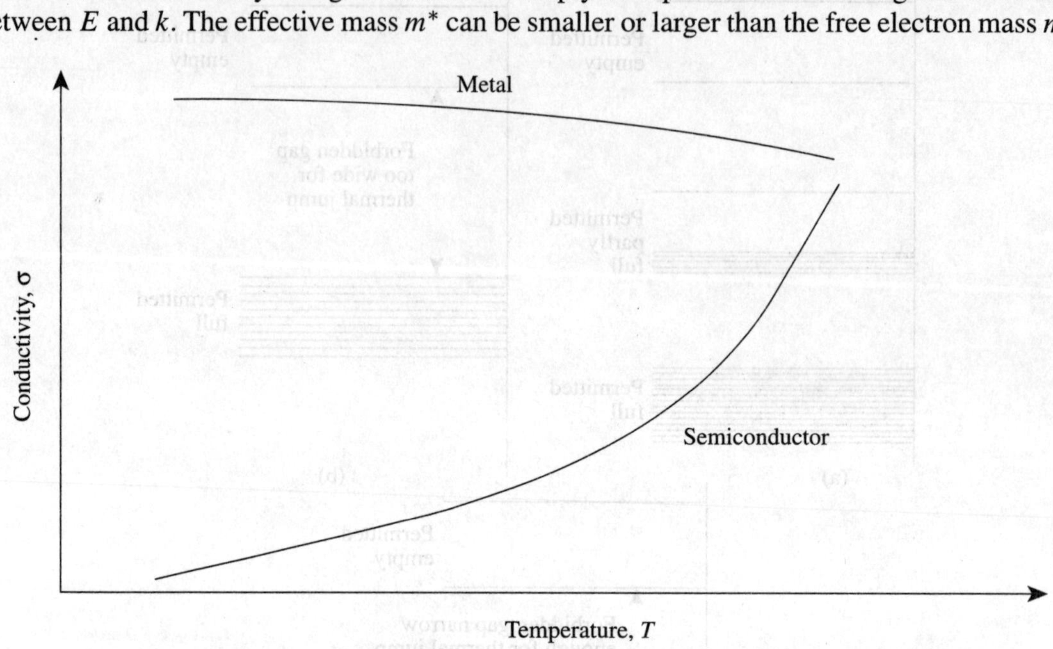

Fig. 4.2. Variation of electrical conductivity with temperature for a typical metal and semiconductor.

4.2 ELECTRICAL PROPERTIES

Drude used the concepts in the kinetic theory of ideal gases to describe the properties of the electron gas. In kinetic theory, the average thermal velocity V_t, of a molecule with mass m at temperature T (Kelvin), is given by the expression

$$\frac{1}{2}mV_1^2 = \frac{3}{2}k_BT \tag{4.3}$$

where k_B is the Boltzmann's constant. If we apply the same formula to the electrons in a metal and replace m by the electron mass m_c, then we obtain a value for the speed of the electrons at room temperature of about 10^5 ms^{-1}. Using the analogy with kinetic theory we get

$$V_t = \frac{\lambda}{\tau} \tag{4.4}$$

where λ = mean free path and τ = relaxation time i.e., the average time duration between collisions. If we assume that the mean free path is of the order of a few atomic spacing, say 1 nm, then the relaxation time is about 10^{-4}s.

The electrical conductivity of solids caused by the flow of electric charge carriers under an applied electric field is expressed by the relation

$$J = \sigma\varepsilon \tag{4.5}$$

where J = the electric current density

ε = Applied electric field

σ = Proportionality constant known as the electrical conductivity

The electric current density is also represented as

$$J = ne\bar{v}_d \tag{4.6}$$

where n = number of electrons per unit volume

$\bar{v}_d$ = The average drift velocity of the electrons attained under the electric field.

Combining equation 4.5 and 4.6, we get

$$\sigma = \frac{ne\bar{v}_d}{\varepsilon} = ne\mu \tag{4.7}$$

where μ is the mobility of an electric charge $\mu = \dfrac{\bar{v}_d}{\varepsilon}$

Electrical resistivity of a material ρ is inverse of electrical conducting σ and is expressed as:

$$\rho = \frac{RA}{l} \tag{4.8}$$

where R is the resistance of a sample of cross-sectional area A and length l. The units of resistivity are Ω m. In Figure 4.3, the resistivity of some solid materials are plotted on a logarithmic scale.

Example 4.1. The electrical resistivity of Al at room temperature is $2.62 \times 10^{-8}\Omega$. Calculate the drift velocity acquired by the free electrons in a field gradient of 100 V/m. Given lattice parameter of aluminium is 4.04 Å .

Solution. Al has a FCC structure and is trivalent.

If n = number of electrons per unit volume

$$= \frac{N}{V}$$

$$= \frac{4 \times 3}{(4.04 \times 10^{(-10)})^3} \qquad (i)$$

we know from equation 4.7

Drift velocity = electron mobility × electric field

$$= \mu\varepsilon \qquad \text{(ii)}$$

From equation 4.6, $\mu_e = \dfrac{\sigma}{n.e}$

Substituting this in equation (ii)

$$\bar{v}_d = \frac{\varepsilon\sigma}{n.e} = \frac{\varepsilon}{\rho.n.e} \quad \text{Substituting the value of n from (i)}$$

$$= \frac{100 \times (4.04 \times 10^{-10})^3}{2.62 \times 10^{-8} \times 1.6 \times 10^{-19} \times 4 \times 3}$$

$$= 0.131 \text{ ms}^{-1}$$

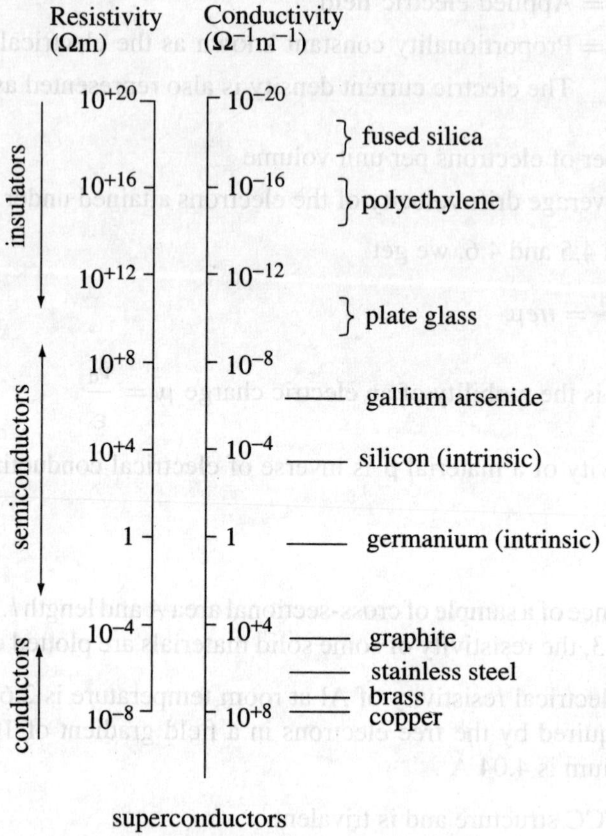

Fig. 4.3. The range of electrical resistivities and conductivities in solid materials.

4.2.1 Electrical Properties of Conductors

At absolute zero, an ideal metal should have zero electrical resistance, but since structural imperfections do occur, a residual resistivity remains. With increase in temperature, the resistivity of a metal rises.

While describing bonding in metals, it was assumed that all the valence electrons are free and form an 'electron cloud', which is uniformly distributed throughout the lattice of ion cores. For an electron to become free, it must be excited into one of the empty and available states above the Fermi energy, E_F. For metals, there are vacant energy states adjacent to the highest filled state of E_F.

With decrease in temperature, the phonon-electron collision becomes insignificant as compared to impurities, grain boundaries etc. It is, therefore, possible to increase the electrical resistance of a metal at ordinary temperatures and at the same time reduce its temperature coefficient of resistivity, by introducing many defects. An example is disordered solid solution, whose resistance may be much higher than that of either constituent. On the other hand in two phase alloys, the situation may be quite different. For example in eutectic alloys, where there are phase mixtures, the resistivity of alloy can be determined by the rule of mixture.

$$\rho = \rho_\alpha V_\alpha + \rho_\beta V_\beta \tag{4.9}$$

where V_α and V_β are volume fraction of the α and β phases.

4.2.2 Electrical Properties of Semiconductors

From Chapter 2, we know that silicon, germanium and gallium arsenide are covalent bonded with a cubic diamond structure. These are semiconductors. Here each atom has four electrons, shown as dots in Figure 4.4(a). Figure 4.4(b) shows how under the influence of enough energy, some electrons get freed from the bond. These free electrons behave like the conduction electrons in a metal. Because the number of free electrons is very small compared with the number in a metal the electrical conductivity is very much lower. The escaping electron leaves a vacant site in the structure and because the electron takes negative charge with it, the site it leaves behind has a positive charge. This is known as 'positive hole'. Other positive holes are created at other sites. When a potential difference is applied across the crystal, the negative electrons and positive holes drift independently and in opposite direction through the material. Elemental semiconductors i.e., Si, Ge are called *intrinsic semiconductor*, as all the charge carriers originate from the material itself.

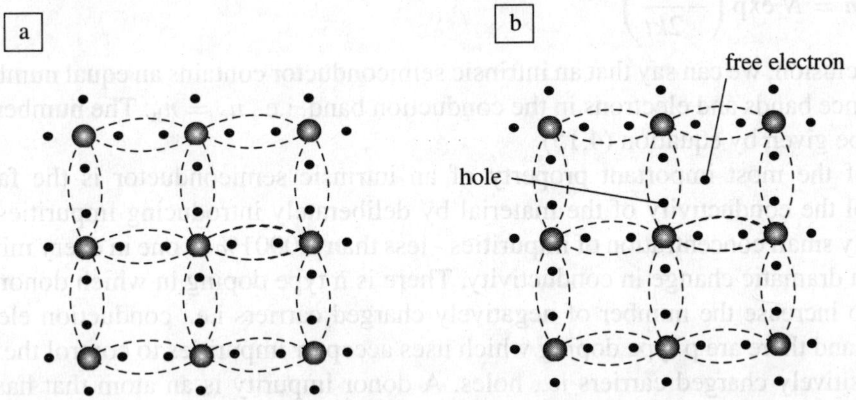

Fig. 4.4. Two dimensional representation of (a) perfect bonding in silica; (b) a valence electron escapes from a bond leaving a 'positive hole'.

Mobility of electron (μ_e) or of hole (μ_h) is expressed as

Mobility = Drift Velocity/Field Gradient

The unit of this is $m^2 v^{-1} s^{-1}$. The electrical conductivity of a semiconductor can be expressed as

$$\sigma = n_e e \mu_e + n_h e \mu_h \tag{4.10}$$

where e is the charge of electron and n_e and n_h are concentration of electron and hole per unit volume. The electrons that are excited from the top of the valence band to the bottom of the conduction band by thermal energy are responsible for conduction. The number of electrons excited across the gap can be calculated for the Fermi-Dirac probability distribution

$$P(E) = \frac{1}{1 + \exp\left[\frac{(E - E_F)}{RT}\right]} \tag{4.11}$$

The Fermi level E_F in an intrinsic semiconductor lies midway in the forbidden gap. This means that the probability of finding an electron here is 50 per cent, even though energy levels at this point are forbidden.

$$E - E_F = Eg/2 \tag{4.12}$$

where Eg is the band gap. The equation (4.11) can be simplified as

$$P(E) = \exp(-Eg/2kT) \tag{4.13}$$

we know $n = NP(E)$ (4.14)

where n = nuclear of electrons promoted across the gap

N = number of electrons available for excitations from the top of the valence band.

Substituting the value for *P(E)*, we get

$$n = N \exp\left(-\frac{Eg}{2kt}\right) \tag{4.15}$$

In conclusion, we can say that an intrinsic semiconductor contains an equal number of holes in the valence bands and electrons in the conduction band, i.e., $n_e = n_h$. The number of each of these can be given by equation (4.15).

One of the most important property of an intrinsic semiconductor is the fact that we can control the conductivity of the material by deliberately introducing impurities. In fact, a surprisingly small concentration of impurities - less than 0.0001% or one in every million atoms - leads to a dramatic change in conductivity. There is n type doping in which donor impurities are used to increase the number of negatively charged carriers i.e., conduction electrons. On the other hand there are p type doping which uses acceptor impurities to control the concentration of positively charged carriers ie., holes. A donor impurity is an atom that has one more valance electron than the atom that it replace. An example is a phophorus atom in silicon since phosphorus has five valence electrons and silicon has four. In the p-type, we have examples of aluminium as acceptor impurity in silicon, which produces an incomplete band, i.e., a hole in the valence band. Figures 4.5 and 4.6 show the schematics of n and p types of extrinsic semiconductors, along with their energy band diagram.

Much interest, of late, has been focused on gallium arsenide (GaAs) semiconductor. The main reasons for this are (a) a high electron mobility μ_e, (b) a large band gap Eg of about 1.5 eV and (c) a direct band gap. The first attribute makes the GaAs circuits faster and operating at equal or lower power than silicon circuits. The large band gap makes GaAs sensitive to wavelengths of light of 827 nm and shorter. It can therefore be used in light emitting diodes (LEDs), which emit red light. Table 4.2 compares the properties of GaAs with other elemental semiconductors i.e., Si and Ge.

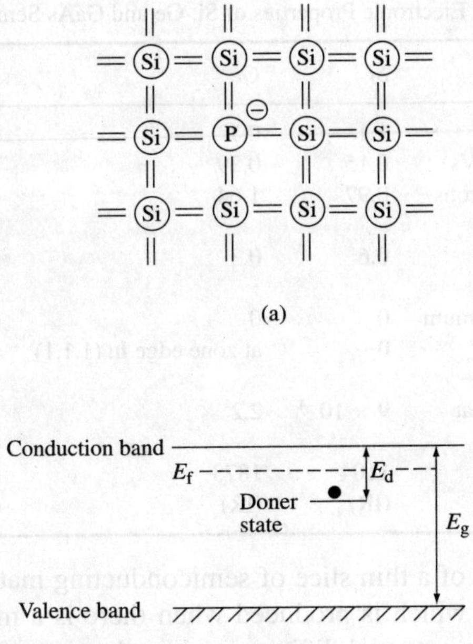

Fig. 4.5. (a) Bonds representing *n*-type extrinsic semiconductor where P is added in Si. (b) The energy band diagram for an *n*-type semiconductor. The donor state has energy below the conduction band edge. E_f represents Fermi energy mid way between E_d and conduction band edge.

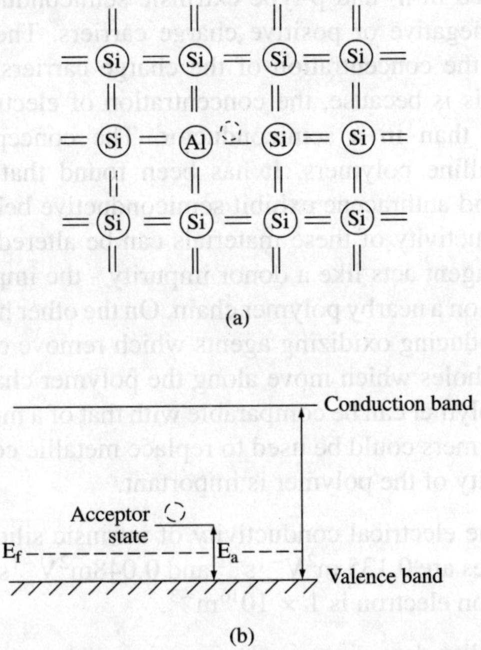

Fig. 4.6. (a) Bonds representing p-type extrinsic semiconductor, where Al is added in Si. (b) The energy band diagram for a p-type semiconductor. The acceptor state has energy above valence band edge. E_f represents Fermi energy mid way between E_a and valence band edge.

Table 4.2. Electronic Properties of Si, Ge and GaAs Semiconductors

Property	*Si*	*Ge*	*GaAs*
Band gap (eV)	1.1	0.7	1.4
Electron mobility (m^2/Vs)	0.15	0.39	0.85
Effective mass of electrons m^*/m_e	0.97	1.64	0.07
Effective mass of holes m^*/m_h	0.6	0.3	0.5
k at valence band maximum	0	0	0
k at conduction band minimum	0	at zone edge in (1,1,1)	near zone edge in (1,0,0)
Electrical conductivity at 300 K ($\Omega^{-1}m^{-1}$)	9×10^{-4}	2.2	1×10^{-6}
Absorption edge (nm)	1104 (IR)	1873 (IR)	871 (visible)

One of the applications of a thin slice of semiconducting material is for *Hall Probe*. This measures the *Hall Voltage*, which is produced when there is a magnetic field at right angles to the probe. The sign of the potential difference gives the sign of the majority charge carriers in the sample. Hall Probes are also used to measure magnetic flux density. In an intrinsic semiconductor, where the number of free electrons equals the number of positive holes, the Hall Voltage due to the negative and positive charges will more or less cancel out. It is more conveniently detected in n- and p-type extrinsic semiconductors, because they contain large excesses of either negative or positive charge carriers. The magnitude of the voltage can be used to calculate the concentration of the charge carriers. In metals, the Hall effect is very much smaller. This is because, the concentration of electrons in a metal is, at about 10^{28} m^{-3}, much higher than in a semiconductor. The concept of band theory can be equally applied to crystalline polymers. It has been found that certain organic materials, notably phthalocyanine and anthracene exhibit semiconductive behaviour. As in conventional semiconductors, the conductivity of these materials can be altered by introducing impurities. For example, a reducing agent acts like a donor impurity - the impurity becomes ionized and produces an extra electron on a nearby polymer chain. On the other hand a p-type semiconductor can be developed by introducing oxidizing agents which remove electrons from the chemical bonds and so give rise to holes which move along the polymer chains. In some instances, the conductivity of a doped polymer can be comparable with that of a metal and so another possibility is that conducting polymers could be used to replace metallic conductors in cases in which the low density or flexibility of the polymer is important.

***Example*4.2** Calculate the electrical conductivity of intrinsic silicon at 300 K given that the electron and hole mobilities are 0.135 $m^2V^{-1}s^{-1}$ and 0.048$m^2V^{-1}s^{-1}$ respectively and that the concentration of conduction electron is $1 \times 10^{16}m^{-3}$.

Solution. Using the mobility data given in the question and recognizing that for an intrinsic semiconductor, we have n=p, the conductivity due to the combined effect of the electrons and holes is given by

$$\begin{aligned}\sigma &= (ne\mu_e + p\mu_h)\\ &= ne(\mu_e + \mu_h)\\ &= (1.60 \times 10^{-19}\,\text{C})(1 \times 10^{16}\,\text{m}^{-3})[0.135\,\text{m}^2\,\text{V}^{-1}\text{s}^{-1} + 0.048\,\text{m}^2\,\text{V}^{-1}\text{s}^{-1}]\\ &= 2.93 \times 10^{-4}\Omega^{-1}\,\text{m}^{-1}\end{aligned}$$

Example 4.3. Calculate the room temperature electrical conductivity of high purity silicon doped by phosphorus to give a concentration of 10^{23} m^{-3} of charge carriers. Given electron and hole mobilities of 0.14 and 0.048 m^2 V.s.

Solution. Since P belongs to Group VA of Periodic Table, it will give rise to n-type extrinsic semiconductor. After comparing from Example 4.2, we notice that the electron concentration of intrinsic semiconductor silicon (1×10^{16} m^{-3}) is lower than the extrinsic one (10^{23} m^{-3}).

Electrical Conductivity $\sigma = ne\mu_e$

$$\begin{aligned}&= 10^{23}\,\text{m}^{-3} \times 1.6 \times 10^{-19}\,\text{C} \times 0.14\,\text{m}^2\,\text{V.s.}\\ &= 2240\Omega^{-1}\,\text{m}^{-1}\end{aligned}$$

Example 4.4. Indium phosphide (InP) has an energy gap of 1.29 eV. The electron and hole mobilities for this semiconducting compound are 0.46 and 0.15 m^{2}V^{-1}s^{-1} respectively at 300 K. Calculate the electrical conductivity? (Given $N = 5 \times 10^{25}$ m^{-3})

Solution. InP is an intrinsic semiconductor

$$n_e = n_h = n,\ T = 300\,\text{K},\ \mu_e = 0.46,\ \mu_h = 0.15,\ e = 1.6 \times 10^{19}\,\text{C}$$

We know electrical conductivity

$$\sigma = (\mu_e + \mu_h)/ne$$

$$n = N\exp(-Eg/2kT)$$

$$= 5 \times 10^{25}\exp\left[\frac{-1.29 \times 1.6 \times 10^{-10}}{2 \times 1.38 \times 10^{-23} \times 300}\right] = 7.465 \times 10^{14}\ \text{m}^{-3}$$

Therefore, $\sigma = 1.6 \times 10^{-19} \times 7.465 \times 10^{14}[0.46 + 0.15]$

$$= 7.287 \times 10^{-5}\,\Omega^{-1}\,\text{m}^{-1}$$

Semiconductor Junctions Mere individual semiconductors whether intrinsic or extrinsic are not of much use technologically. However, the combination of an n-type semiconductor with a p-type material through a rather sharply defined interface results in a device of atmost practical significance. In a real crystal, there would be a gradual change for p-type to n-type. By the term junction, we should not mean that these materials are physically joined. Such a device is generally made by growing half of the crystal with the addition of one type of dopant impurity, and completing crystal growth with the addition of the other type. The junction as boundary between the n- and p- regions cannot be distinguished optically in this type of crystal growth.

When n-type and p- type semiconductors are placed in contact, equalization of the Fermi levels occurs as electrons diffuse from the n-type material into the p-type material. As more electrons cross the junction and recombine with holes, the number of carriers in junction region is reduced. This leads to a depletion region. If we consider the diffusion of the holes, a similar result is obtained.

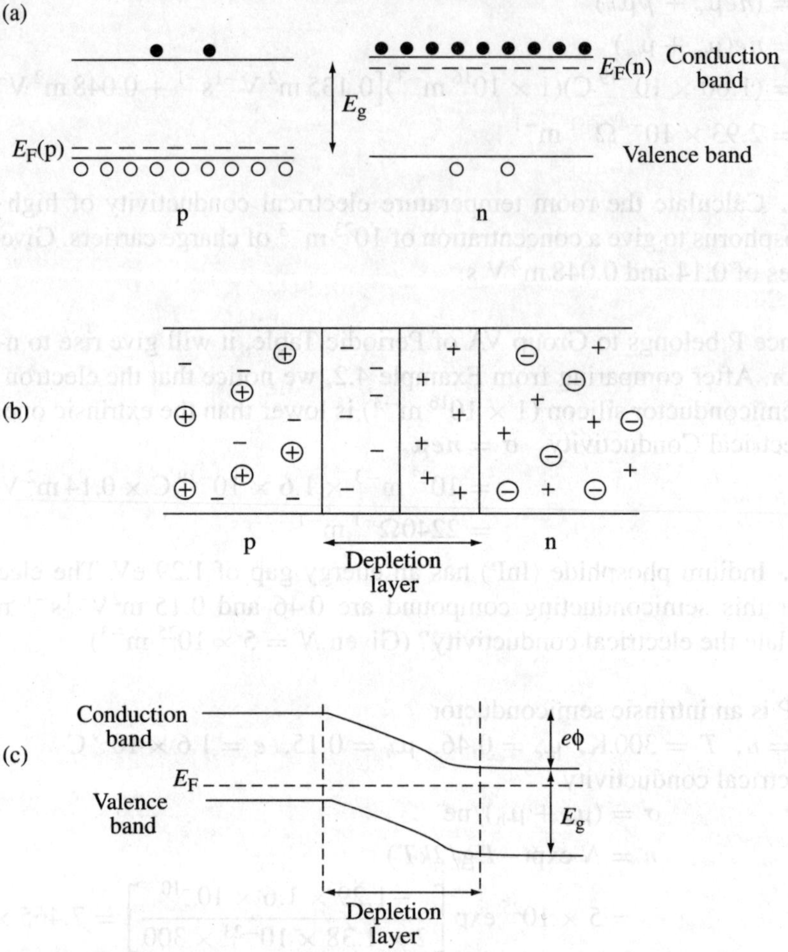

Fig. 4.7. (a) Location of Fermi energy in p-and n-type semiconductors (b) Depletion layer in a p-n junction (c) Contact potential due to charge distribution at a junction.

Figure 4.7 illustrates the above features schematically. Figure 4.7(b) shows that the process results in an accumulation of negative charge on the n-type side of the junction and positive charge on the p-type side. Since it is more difficult for the conduction electrons to move from the n-type to the p-type side, it can be shown that the energy of the conduction band edge is higher on the p-type side than on the n-type side (Figure 4.7(c)). Such a phenomenon shall give rise to contact potential ϕ. The difference in energy between the two sides of the junction is equal to $e\phi$. The quantitative analysis of the p-n junction is beyond the current discussion.

If a battery or other external source of potential is applied to the p-n junction, the type of current flow response observed depends upon the polarity of the applied field. In the forward bias the current increases exponentially with the voltage, whereas in the reverse bias, the current remains very small regardless of the magnitude of the voltage. The typical variation of current with voltage is shown in Figure 4.8. To a good approximation we can assume that a p-n junction only allows current to flow one way. So one use of p-n junction is as a rectifier, in which alternating current is changed to direct current.

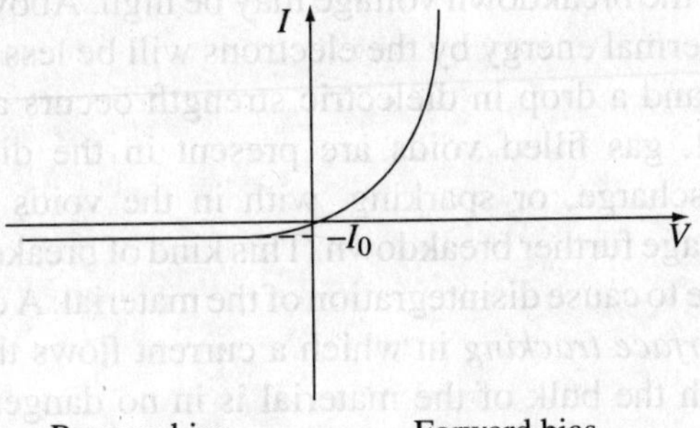

Fig. 4.8. The current-voltage variation for p-n junction showing forward and reverse bias.

Example 4.5. Sketch the energy band profile and calculate the contact potential for silicon p-n junction if the Fermi energy on the n-type side is 150 meV below the conduction band edge and the Fermi energy on the p- type side is 80 meV above the valence band edge. Band gap of silicon is 1.11 eV. (Adapted with permission from R. Turton, The Physics of Solids, Oxford University Press, Oxford, 2000, p. 180).

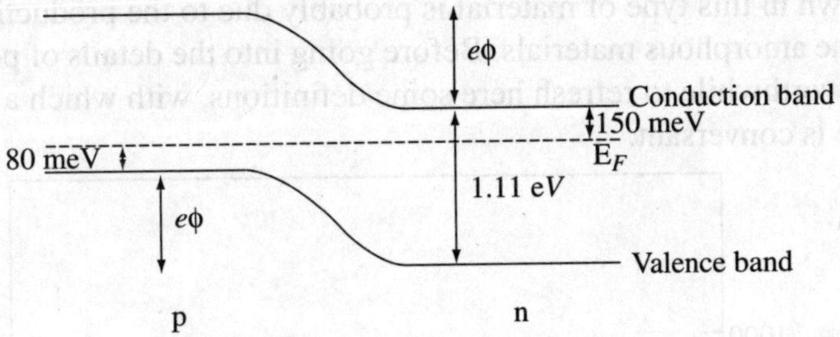

Figure Ex. 4.5

Solution. The energy band profile is as follows (Figure Ex. 4.5).

From the diagram the value for $e\phi$ is

$$e\phi = (1.11\,\text{eV}) - (0.15\,\text{eV}) - (0.08\,\text{eV})$$
$$= 0.88\,\text{eV}$$

Therefore the contact potential is 0.88 eV

4.2.3 Electrical Properties of Insulators

Electrical insulators are also called dielectric materials. These have large energy gap (5 eV or more) between the valence and conduction bands. Ordinary thermal excitation is sufficient to permit electrons to jump this gap. as they can in intrinsic semiconductors and dielectrics retain their insulating properties over a wide temperature range, provided no chemical degradation occurs when they are heated. The energy gap can be crossed, however, if a sufficiently high voltage is applied to the material. The insulator is then said to have broken down. Temperature plays an important role in dielectric breakdown. At low temperature, there will be no electrons in the conduction band and the breakdown voltage may be high. Above a particular temperature, the rate of dissipation of thermal energy by the electrons will be less than the rate at which they gain energy by collisions and a drop in dielectric strength occurs as illustrated in Figure 4.9 for polyethylene. If small, gas filled voids are present in the dielectric the material may breakdown by internal discharge, or sparking, with in the voids causing locally increased temperatures which encourage further breakdown. This kind of breakdown may be accompanied by pitting sufficiently severe to cause disintegration of the material. A common cause of dielectric breakdown is known as *surface tracking* in which a current flows through or over the surface of the material even though the bulk of the material is in no danger of breakdown. Polymers are particularly prone to this type of breakdown mechanism under conditions of high humidity. Dust attracted from the air by electrostatic charges on a dielectric can also provide paths for surface tracking. Anti-static agents can be added to these materials to reduce the tendency for this type of surface breakdown.

The avoidance of surface tracking is important because, the high temperatures generated along surface can lead to the formation of low resistance carbon soote. This will result in permanent breakdown of the insulating properties of the polymer. The breakdown feature may be applied advantageously in chalcogenide amorphous glasses, which are used as memory switches. Dielectric breakdown in this type of material is probably due to the production of a crystalline pathway through the amorphous materials. Before going into the details of polarization aspects in dielectrics, it is worthwhile to refresh here some definitions, with which a student with basic physics knowledge is conversant.

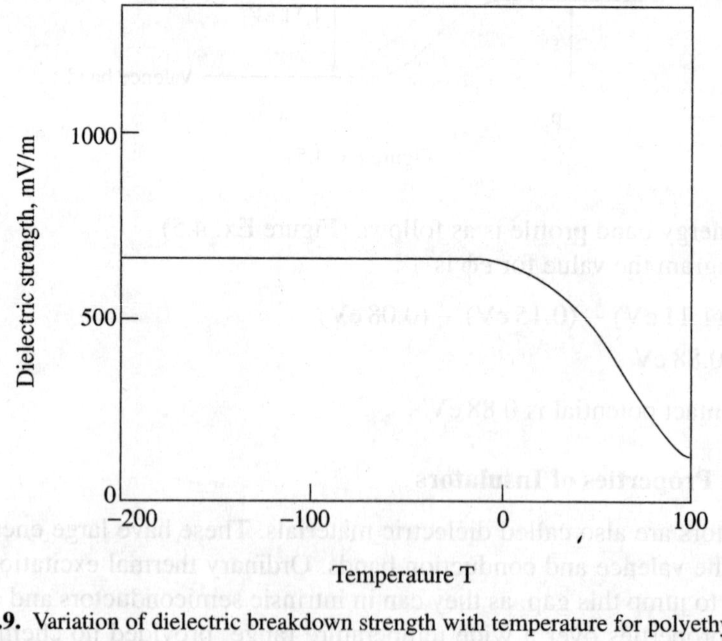

Fig. 4.9. Variation of dielectric breakdown strength with temperature for polyethylene.

When a voltage is applied across a capacitor, one plate becomes positively charged, the other negatively charged with the corresponding electric field directed from the positive to the negative. The *capacitance* C is related as:

$$C = Q/V \tag{4.16}$$

where V is the voltage applied across the capacitor. The units of capacitance are coulombs per volt or farads F.

When a parallel plate capacitor, with a plate area A and separation d, is taken into account (Figure 4.10) with vacuum (or air) between the plates, the capacitance can be written as

$$C_o = \varepsilon_o A/d \tag{4.17}$$

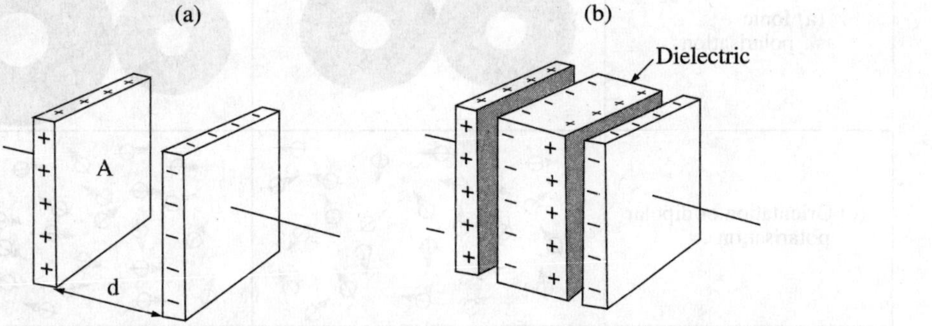

Fig. 4.10. A parellel plate capacitor in case when (a) no dielectric is present, (b) a dielectric material is placed.

The term ε_o is called *permitivity* of a vacuum, which has a value 8.85×10^{-12}F/m. However, if a dielectric material is inserted between the plates of the capacitor, we find that the electric field within the capacitor induces a surface charge on the dielectric, which increases the charge on the capacitor plates (Figure 4.10(b)). The capacitance is now given by

$$C = \frac{\sum \varepsilon_o A}{d} \tag{4.18}$$

where ε is the *dielectric constant* of the material.

$$\text{Therefore } \varepsilon = C/C_o \tag{4.19}$$

Polarization The process of dipole alignment is called *polarization*. This may be caused by either permanent or induced atomic or molecular dipole moments. There are four types of polarization (Figure 4.11):

(a) *Electronic Polarization* The electronic state of insulators is such that, all of the electrons are tied to individual atoms or groups of atoms, but they can still react to the electric field. Figure 4.11(a) shows that the electron spatial arrangement gets shifted relative to the nucleus in the opposite direction to the field. Hence the atom is said to be electronically polarized.

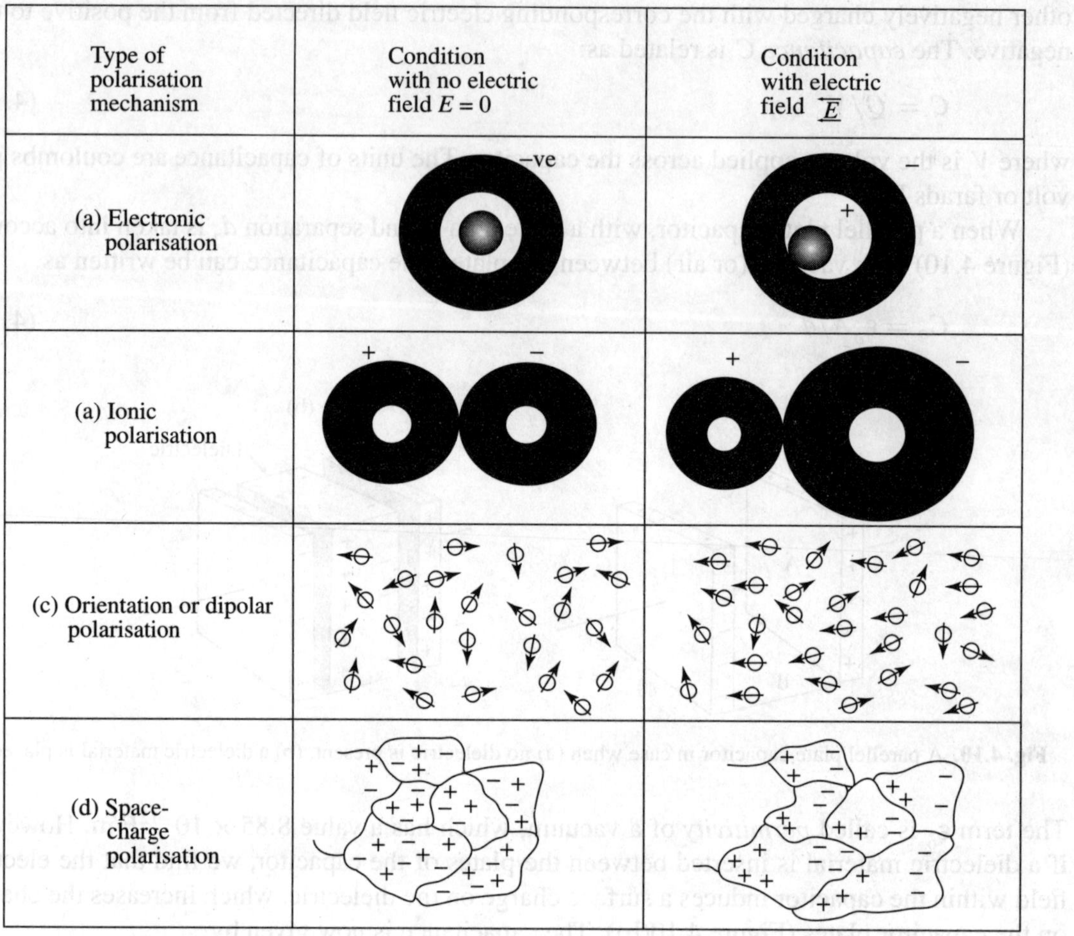

Fig. 4.11. Schematic representation of the different types of polarization mechanisms.

(b) *Ionic Polarization* This is obtained in ionic crystal, where the outer bonding electrons may be preferentially attracted towards atoms of the species in which the binding is stronger. This causes a shift of the atoms relative to one another (Figure 4.11(b)).

(c) *Orientation Polarization* In this case the dipoles are aligned in the direction of the electric field (Figure 4.11(c)). The degree of such polarization falls with increasing temperature.

(d) *Space Charge Polarization* This occurs in solid, multiphase dielectrics in which various phases or components have different dielectric constants. Any free charge carriers collect at the boundaries between the constituent materials (Figure 4.11(d)) and this results in a local distortion of the applied field.

In case of alternating current, where voltage or electric field changes direction with time, dipoles shall attempt to reorient with the field. For each polarization type, some minimum reorientation time exists. A *relaxation frequency* is, therefore, defined as the reciprocal of minimum reorientation time. When the frequency of the applied electric field exceeds its relaxation frequency, a dipole cannot keep shifting orientation direction. The dependence of dielectric constant on the field frequency is shown in Figure 4.12. It is evident that when there is a sudden drop in the dielectric constant, the polarization mechanism ceases to function.

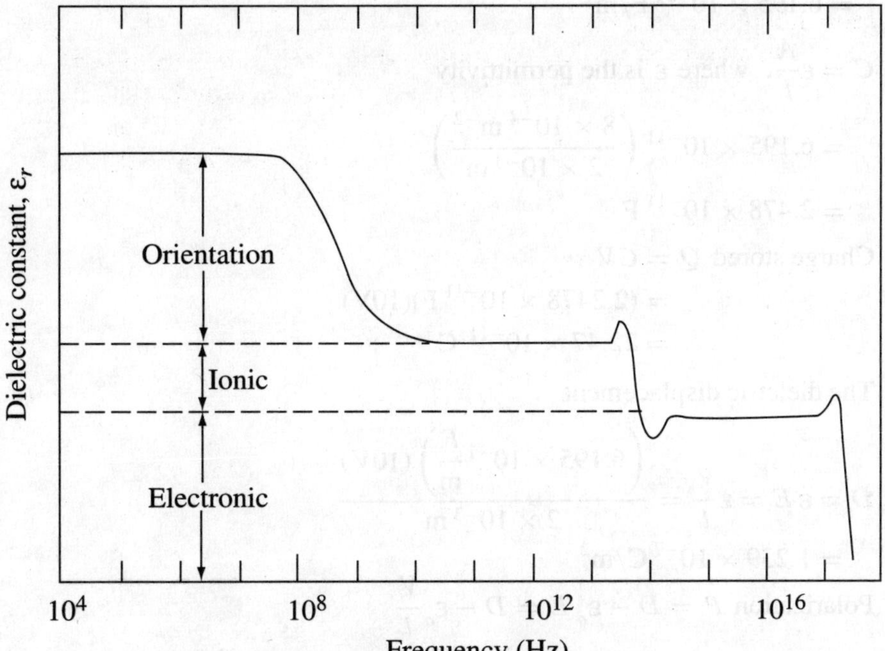

Fig. 4.12. Variation of dielectric constant with frequency of an alternating electric field.

Impurities in dielectrics can play similar role as they do in semiconductors. But, here, high doping concentrations are required, than are conventionally used in semiconductors if the device is to be operated at room temperature. One biggest advantage of using large band gap dielectric is that they are far more suitable for use in higher temperature applications. Semiconductor devices cease to operate correctly at temperature of few hundred degrees Celsius because the concentration of intrinsic carriers becomes so large that it is no longer possible to distinguish clearly between the minority and majority carriers. This is not a problem, when the band gap is much larger.

Example 4.6. A parallel-plate capacitor has an area of 8×10^{-4} m^2 and a plate separation of 2×10^{-3} m across which a potential of 10 V is applied. If a dielectric material having a constant of 7.0 is positioned within the plates, calculate

(a) The capacitance
(b) The magnitude of the charge stored on each plate
(c) The dielectric displacement D

(d) The polarization

The permittivity in vacuum $\varepsilon_o = 8.85 \times 10^{-12}$ F/m

Solution.

(a) $\varepsilon = \varepsilon_r \varepsilon_o = 7.0 \times 8.85 \times 10^{-12}$ F/m

$= 6.195 \times 10^{-11}$ F/m

$C = \varepsilon \frac{A}{l}$, where ε is the permittivity

$$= 6.195 \times 10^{-11} \left(\frac{8 \times 10^{-4}\ \text{m}^{-2}}{2 \times 10^{-3}\ \text{m}} \right)$$

$= 2.478 \times 10^{-11}$ F

(b) Charge stored $Q = CV$

$= (2.2478 \times 10^{-11}\ \text{F})(10\text{V})$

$= 22.47 \times 10^{-11}$ C

(c) The dielctric displacement

$$D = \varepsilon E = \varepsilon \frac{V}{l} = \frac{\left(6.195 \times 10^{-1} \frac{F}{\text{m}}\right)(10V)}{2 \times 10^{-3}\ \text{m}}$$

$= 1.239 \times 10^{-7}\ \text{C/m}^2$

(d) Polarization $P = D - \varepsilon_o E = D - \varepsilon_o \frac{V}{l}$

$= 1.239 \times 10^{-7}\ \text{C/m}^2 - [(8.85 \times 10^{-12})(10)]/(2 \times 10^{-3}\ \text{m})$

$= 1.129 \times 10^{-7} - 0.442 \times 10^{-7}$

$= 0.687 \times 10^{-7}\text{C/m}^2$

Application of dielectric material is too varied. The sleeving on the electrical wiring may be one of many types of dielectric material including polyethylene, polyvinyl chloride, neoprene, as well as paints and lacquers. Insulation of the winding in transformers is important. Electronic circuiting requiring the separation of conducting tracks carrying minute current and heavy duty circuit breakers for switching very large currents do require the selection of dielectric materials. Table 4.3 gives the dielectric constant values of different materials. Table 4.4 summarizes various electrical parameters in SI Units.

4.2.4 Ferroelectricity

Ferroelectrics are groups of dielectrics showing spontaneous polarization, that is polarization in the absence of an electric field. Barium titanate is the most common example of ferroelectrics. The unit cell of this crystal is already dealt in Chapter 2. A detailed investigation shows that at room temperature the oxygen ions are not quite at the centers of the cube faces, but are displaced slightly in a common direction. Since the centroids of positive and

Table 4.3. Dielectric Properties of Various Materials

Materrial	ε_r
Copper	∞
Barium titanate	7000
Distilled water	80
Inorganic glasses	6-20
Alumina	9
Polyster	4
Polystyrene	2
Air	1
Vaccum	1

Table 4.4. Primary and Derived Units for Various Electrical Parameters and Field Vectors

		SI	
Quantity	*Symbol*	*Derived*	*Primary*
Electric potential	V	volt	kg-m^2/s^2-C
Electric current	I	ampere	C/s
Electric field strength	ε	volt/meter	kg-m/s^2-C
Resistance	R	ohm	kg-m^2/s-C^2
Resistivity	ρ	ohm-meter	kg-m^3/s-C^2
Conductivity	σ	(ohm-meter)$^{-1}$	s-C^2/kg-m^3
Electric charge	Q	coulomb	C
Capacitance	C	farad	s^2-C^2/kg-m^2
Permittivity	ε	farad/meter	s^2-C^2/kg-m^3
Dielectric constant	ε_r	ratio	ratio
Dielectric displacement	D	farad-volt/m^2	C/m^2
Electric polarization	P	farad-volt/m^2	C/m^2

negative charge do not coincide, the material possesses a finite polarization even when there is no external electric field present. It is found that when stress is applied, the positive and negative ions separation changes, thus inducing piezoelectric effect. This is to be described in the next section. Thus all ferroelectric materials display the piezoelectric effect, but not vice versa.

Ferroelectric materials have been, in general, used for their piezoelectric properties. Since they have very high dielectric constant, with values of several thousand, very small size capacitors can be made using ferroelectrics. This is an important development in microelectronic circuits.

4.2.5 Piezoelectricity

In certain types of dielectric materials, the application of external stress gives rise to polarization i.e., produces electric charge on the surface of the material. This phenomenon is reversible, i.e., if the material is placed in an electric field it deforms and becomes strained. A characteristic of piezoelectric crystals is that they have a low degree of symmetry. Figure 4.13(a) illustrates this. When there is no applied stress, the ions are arranged in a pattern which has three fold symmetry. However when a stress is applied in the direction shown in Figure 4.13(b), the three fold symmetry is destroyed and the crystal acquires a net polarization.

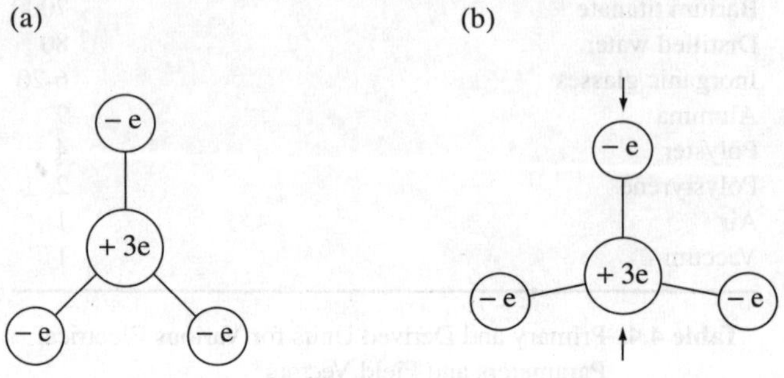

Fig. 4.13. (a) A crystal structure with three fold symmetry and zero polarization; (b) Altered symmetry, when a stress is applied. The material acquires polarization even in the absence of an electric field.

Piezoelectric materials are extensively applied in electro-mechanical transducers in single crystal form e.g., quartz. The charges which appear at the surfaces of piezoelectric materials are soon neutralized by external charges being attracted to the surface or by leakage of the charges through the associated electrons. For this reason piezoelectric transducers are not usually suitable for static measurements. Since piezoelectric constants vary with direction in an anisotropic crystal, it is important to cut the crystal in such a way so as to get maximum piezoelectric effect or to provide a transducer whose behaviour is relatively temperature independent. A recent application of piezoelectric material is in scanning-tunneling microscope.

Example 4.7. A piezoelectric material has a Young's modulus of 72×10^3 MPa. What stress is required to change its polarization from 640 C/m^2 to 645 C/m^2?

Solution. $\dfrac{\Delta\rho}{\rho} = \dfrac{645 - 640}{640} = +0.0078$

Since Q and V do not change, therefore

$$\Delta d/d = 0.0078 = \text{strain}$$

$$\text{Stress} = \text{Modulus} \times \text{strain}$$
$$= 72 \times 10^3 \times 0.0078$$
$$= 560 \text{ MPa}$$

4.3 RELATED ELECTRICAL PHENOMENA IN MATERIALS

Several electrical phenomena are associated with materials. Some of them are electron emission (thermionic, photo electric, secondary and field emission) and thermoelectricity (Seebeck, Peltier and Thomson effects). There are other phenomena such as Hall effect and superconductivity, which do take into account the role of magnetic field. These are highlighted in the sections 4.2.2 and 4.5 respectively.

4.3.1 Electron Emission

Electron can be accelerated so that they can leave the material in following ways. (a) heating (thermionic emission); (b) light (photoelectric emission); (c) electron bombardment (secondary emission) and (d) application of an intense electric field (field emission). Out of the above the photoelectric emission shall be dealt in the section 'optical properties' of this chapter in detail.

Thermionic emission When a metal that has been heated in vacuum is made electrically negative (cathode) with respect to another metal (anode), the electric current flows between the two metals. This current increases as the anode potential increases. If the anode potential becomes sufficiently high with respect to the cathode, a saturation current flows, the value of which, depends ideally only on the temperature of the cathode and the metal that is used.

If the energy input is sufficiently great, electrons can be completely removed from the materials and the voltage which is required for this to happen is called the *work function* of the material. Table 4.5 gives some examples of work function for metals. These values are represented at 0 K, when electrons completely fill the energy bands upto the Fermi level. Work function is temperature dependent. If the temperature is sufficiently high some electrons may acquire enough energy to escape from the surface of the material by *thermionic emission*.

Table 4.5. Values of the Work Function, ϕ, for Some Metals

Metal	*ϕ volts*
Calcium (Ca)	2.24
Chromium (Cr)	4.6
Cobalt (Co)	4.41
Iron (bcc) (Fe)	4.48
Nickel (Ni)	4.61
Palladium (Pd)	4.99
Platinum (Pt)	5.32
Tantalum (Ta)	4.12
Thorium (Th)	3.38
Tungsten (W)	4.54

Figure 4.14 shows the typical voltage - plate current characteristics of a diode. It is evident that at any temperature, saturation occurs (I_s). If the temperature is increased, the thermionic emission increases as shown in Figure 4.14(b). Note that the plate current, I_p, is not zero, even when the plate voltage V_p is zero. This is due to the fact that the electrons, leave the hot cathode, the more rapidly moving ones penetrating the cloud of space charge and reaching the plate even with no accelerating field. The work function is usually between 1 and 7 volts, varying with material and surface preparation. The work function ϕ represents the minimum energy that electrons have to acquire to take them from the energy level E_f (Fermi level) to outside the metal. High emission efficiency is favoured by a low work function. Most commercial electron tubes have oxide coated cathodes, because the electropositive layers decrease the potential barrier and the work function. The common commercial electropositive layer is thorium on tungsten.

Secondary Electron Emission If the surface of any solid is bombarded by electrons with velocities greater than a critical value, electrons may acquire enough energy to over come the potential barrier and escape from the solid. The secondary electrons leave the surface in the direction from which the primary arrive. The total flow of electrons in this direction consists of (a) electrons elastically reflected, (b) electrons reflected inelastically after an energy loss of

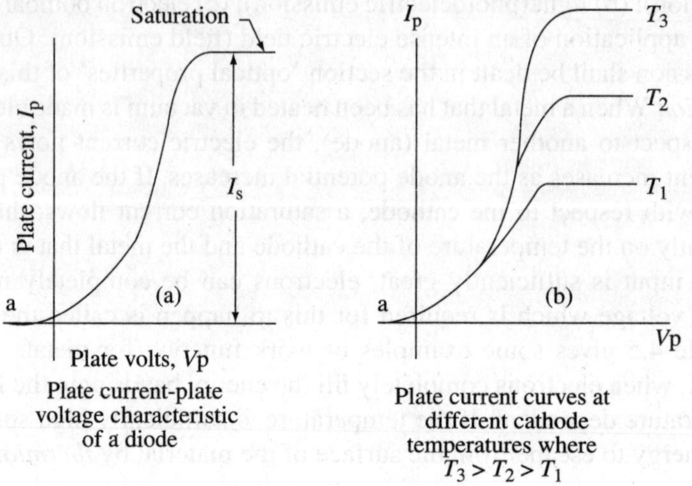

Fig. 4.14. Characteristics of a vacuum diode and the effect of temperature on thermionic emission.

some tens of electron-volts, and (c) true secondary electrons with energy independent of the primary energy. The secondary emission coefficient δ is defined as

$$\delta = \frac{a + b + c}{I}$$

where I is the primary current and a, b, c are the contributors from described above. The value, δ, reaches a maximum at about 200 to 600 V. In general, when the work function is low, the secondary emission is low. Pure metals have rather low secondary emission. Compounds, such as oxides and halides have secondary emission effect. For gas-contaminated or oxidized metals, secondary emission ranges is as high as six.

Field Emission In field emission, the application of a high positive field in the vicinity of the cathode can produce considerable emission even at room temperature. This occurs under condition of highest vacuum. Field emission explains the large emission current densities observed at the cathode spot of an arc between metallic welding electrodes.

4.3.2 Thermoelectricity

Thermoelectricity is electricity produced at a heated junction of two dissimilar conductors. The thermoelectricity covers three basic phenomena, namely Seebeck effect, the Peltier effect and the Thomson effect. *Seebeck effect* is used in thermocouples for temperature measurement above and below room temperature. The thermoelectric force is the voltage per degree of temperature difference between the junctions. A given metal is said to be thermoelectrically positive with respect to another metal if the current flows from the first metal to the second at the cold junction. The thermo emf generated by any two elements, A and B can be determined being the algebraic sum of the thermo emf of the individual elements.

Peltier effect is just the inverse of Seebeck effect and is illustrated by Figure 4.15(a). This effect is not to be confused with the Joule heating effect, which is proportional to the electrical

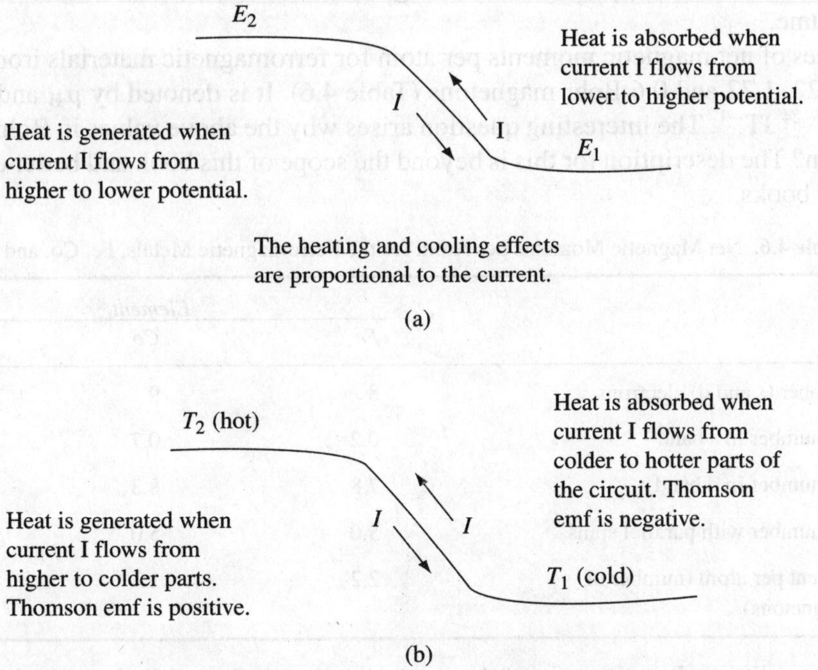

Fig. 4.15. Schematic representation of (a) Peltier effect and (b) Thomson effect.

resistivity and to the square of the currents as well as the dimensions of the conductor. The direction and magnitude of the Peltier emf depend on the metals and the temperature of the junction.

Thompson proposed that there must be a reversible absorption of heat due to the flow of current through those parts of the conductors in which a temperature gradient exists. It was shown that in some metals, heat is absorbed when current flows from colder to hotter parts of the metal and heat is generated when current flows from the hotter to colder parts (Figure 4.15(b). For other metals, the effect is reversed and in some others, the effect is negligible. The Thompson emf is generally too small and hence mostly disregarded in thermoelectric considerations.

4.4 MAGNETIC PROPERTIES

The origin of the magnetic properties in materials is related to the orbital motion of electron around nuclei and the spin of electrons. These moving electrons may be treated as tiny magnets.
Magnetic Induction B It is defined by the number of hypothetical induction lines per unit area normal to their direction. The unit is tesla (T). This is also referred as flux density.

Magnetic Field Strength H This also called as magnetic intensity or magnetizing force and like magnetic induction, are represented by lines of magnetization. It is measured in amperes per meter.

Magnetic Movement It is the product of the magnetic moment of each atom and their number in a unit volume.

The values of net magnetic moments per atom for ferromagnetic materials iron, cobalt and nickel are 2.22, 1.72 and 0.6 Bohr magnetons (Table 4.6). It is denoted by μ_B and has a value of 9.27×10^{-24} JT^{-1}. The interesting question arises why the above values in Bohr magnetons are in fraction? The description for this is beyond the scope of this book and better dealt in solid state physics books.

Table 4.6. Net Magnetic Moments per Atom for the Ferromagnetic Metals, Fe, Co, and Ni

	Element		
	Fe	*Co*	*Ni*
Total number (s and d) electrons	8	9	10
Average number in s band	0.2	0.7	0.6
Average number in d band	7.8	8.3	9.4
Average number with parallel spins	5.0	5.0	5.0
Net moment per atom (number of Bohr magnetons)	2.2	1.7	0.6

Magnetic Susceptibility It is the ratio of the magnetization to the applied magnetic field

$$\chi_m = \frac{M}{H} \tag{4.20}$$

The magnetization M, the magnetic field H and the induction B are related as

$$B = H + 4\pi M \tag{4.21}$$

$$\text{Also} \quad \mu = 1 + 4\pi\chi_m \tag{4.22}$$

By rearranging we get

$$B = H(1 + 4\pi\chi_m) \tag{4.23}$$

$$\text{Also} \quad B = \mu H \text{ or } \mu = \frac{B}{H} \tag{4.24}$$

The μ is known as *magnetic permeability* which is the measure of the ability of the material to be magnetized.

In free space $B = \mu_o H$, where μ_o is the permeability in vacuum equal to $4\pi \times 10^{-7}$ H/m.

In solid materials $\mu \neq \mu_o$and the relative permeability $\mu_r = \mu/\mu_o$.

For diamagnetic and paramagnetic material, $\mu_r = 1$.

For sake of convenience, both SI and CGS -emu units of magnetic properties are included in Table 4.7.

Table 4.7. Magnetic Units and Conversion Factors for the SI and cgs-emu systems

Quantity	*Symbol*	*SI Units*		*cgs-emu Unit*	*Conversion*
		Derived	*Primary*		
Magnetic induction (Flux density)	*B*	tesla (Wb/m^2)	kg/s-C	gauss	1 Wb/m^2 = 10^4 gauss
Magnetic field strength	*H*	amp-turn/m	C/m-s	oersted	1 amp-turn/m $= 4\pi \times 10^{-3}$ oersted
Magnetization	*M* (SI) I (cgs-emu)	amp-turn/m	C/m-s	Maxwell/cm^2	1 amp-turn/m $= 10^{-3}$ maxwell/cm^2
Permeability of a vacuum	μ_o	henry/m	kg-m/C^2	Unitless (emu)	$4\pi \times 10^{-7}$ henry/m = 1 emu
Relative permeability	μ_r(SI) μ' (cgs-emu)	Unitless	Unitless	Unitless	$\mu_r = \mu'$
Susceptibility	χ_m (SI) χ'_m (cgs-emu)	Unitless	Unitless	Unitless	$\chi_m = 4\pi\chi'_m$

4.4.1 Classification of Materials According to Magnetic Properties

All materials can be characterized as being either diamagnetic, paramagnetic, or ferromagnetic depending upon their response under the influence of magnetic field. Under major title of ferromagnetism, we have further subclasses, i.e., anitferromagnetic and ferrimagnetic. *Diamagnetic materials* do not have permanent magnetic moments, but they have only weak induced magnetic moments. Ionic solids and molecular solids are generally diamagnetic. *Paramagnetic materials* arise because the atoms or ions have an odd number of electrons that give rise to a residual permanent magnetic moment. *Ferromagnetic materials* show spontaneous magnetic moment, that exists even in the absence of an applied magnetic field. Iron, nickel, cobalt and gadolinium are examples of such materials. In such materials, interaction between adjacent atoms or ions in a crystal are strong enough to hold magnetic moments of the neighbouring atoms parallel to each other in spite of thermal agitation.

Inorganic compounds e.g., Cu_2MnAl (Heusler alloy) is a typical example of ferro-magnetic material where none of the constituent possesses such property. It was found that, the ordered form of this alloy is ferromagnetic, while the disordered form was antiferromagnetic. In the ordered form, the distance between nearest neighbour manganese atoms in the crystal structure was greater than the mean distance that was in the disordered form, and this brought about the ferromagnetism.

Antiferromagnetism arises from an antiparallel locking of spins, which results in a negative interaction between the nearest neightbours. When the magnetic moments are not equal, the antiparallel arrangement results in *ferrimagnetism* (Figure 4.16).

The inert atoms have a magnetic moment of zero because, they have all filled electron subshells. Similarly, ionic materials also have a magnetic moment of zero, because in this case, the electrons are transferred from one atom type to another so that, the resulting ions have only

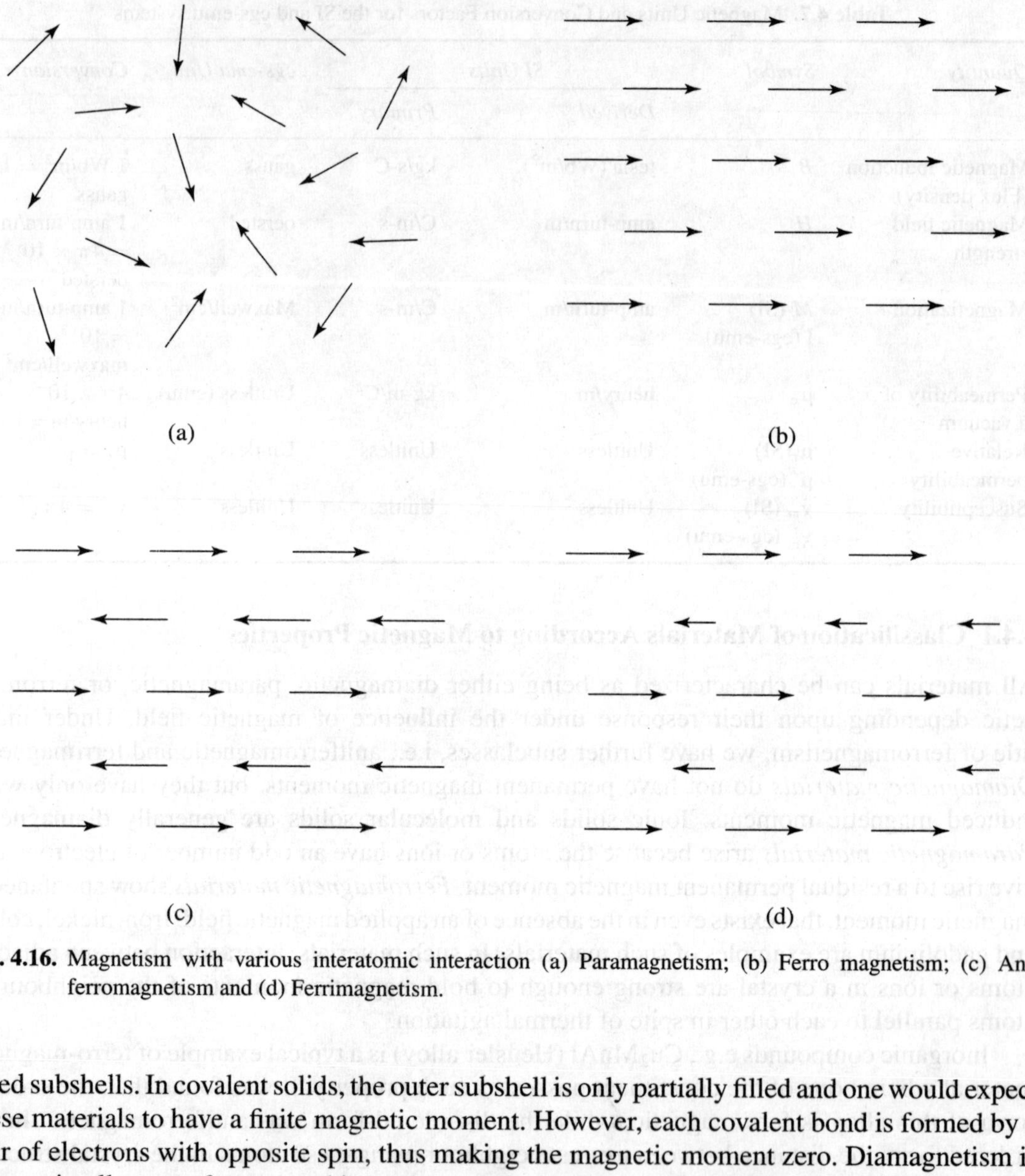

Fig. 4.16. Magnetism with various interatomic interaction (a) Paramagnetism; (b) Ferro magnetism; (c) Anti ferromagnetism and (d) Ferrimagnetism.

filled subshells. In covalent solids, the outer subshell is only partially filled and one would expect these materials to have a finite magnetic moment. However, each covalent bond is formed by a pair of electrons with opposite spin, thus making the magnetic moment zero. Diamagnetism is present in all types of atoms and it produces finite magnetic susceptibility in all solids, but the effect is very weak except in case of superconductors. Paramagnetism, therefore, is caused by the magnetic dipole moments becoming aligned with the magnetic field, whereas diamagnetism results from an induced magnetic field which opposes the applied field. Paramagnetism gives rise to positive magnetic susceptibility, while diamagnetism produces a negative susceptibility. The question now arises why transition metals and some rare earth and actinide elements show promise for ferromagnetism. The reason lies on the unfilled d- and f-electron shells in these elements. These elements give rise to which is known as *Curie paramagnetism* and under certain circumstances they can also form permanent magnets. Mathematically the magnetic susceptibility can be related with temperature as

$$\chi = \frac{C}{T} \tag{4.25}$$

where C is curie constant. This relationship is known as Curie's law. The Curie law holds true for many paramagnetic metals containing rare earth metals, but the agreement is not so good for the iron group transition metals. The reason for this is that in rare earth group elements, the incomplete 4f subshell of electrons lies inside the orbits of filled 5s and 5p subshells. On the other hand, in the iron group metals the incomplete 3d subshell of electrons is the outer most electrons in the ion. They, therefore, interact greatly with other ions,, and so the dipole moments of the individual ions are altered.

In case of ferromagnetic permanent magnets, above a temperature known as Curie temperature, the metal behaves like a paramagnet i.e., the dipoles are randomly aligned unless, they are subjected to an external magnetic field. The magnetic susceptibility of such paramagnetic material is expressed by a modified form of the Curie law, known as Curie-Weiss law

$$\chi = \frac{C}{T - \theta_c} \tag{4.26}$$

where θ_c is the Curie temperature. In iron, cobalt and nickel, the Curie temperature is well

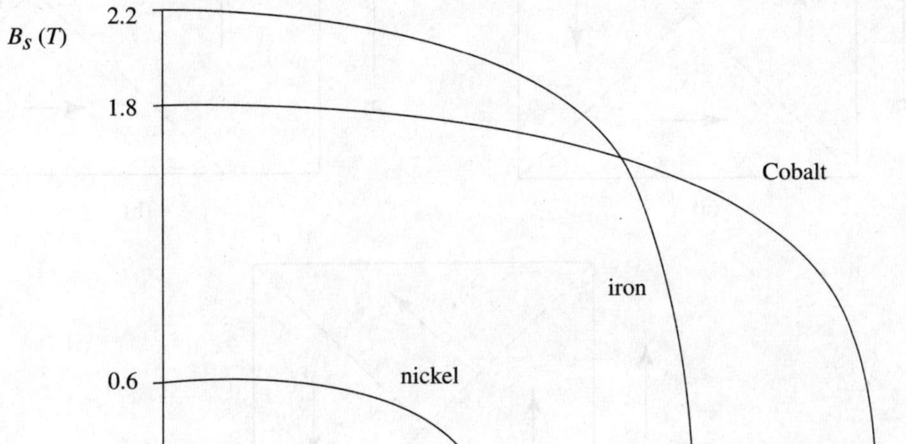

Fig. 4.17. Plot of saturation magnetization as a function of temperature for iron, cobalt and nickel. Curie temperature is marked at zero B_S.

above room temperature, but in some cases the Curie point is below room temperature. Such materials, therefore, cannot be considered useful for permanent magnets. Figure 4.17 shows the variation of saturation magnetization with temperature. Note when M_s is zero, the temperature is the Curie temperature.

Antiferromagnets also undergo a transition from an ordered phase to a paramagnetic phase at a temperature called the Neel temperature, θ_N. A large number of compounds which display Curie paramagnetism at room temperature are found to become antiferromagnetic at lower temperature.

4.4.2 Ferromagnetic Domain and Magnetization Curve

Weiss proposed that a ferromagnet is divided into a number of domains. The dipoles within each domain are mutually aligned in a common direction, but the direction of magnetization is different in neighbouring domains. In the absence of an applied magnetic field, adjacent domains may be oriented randomly in any number of several directions, called the directions of easy magnetization, which depend on the geometry of the crystal. The resultant effect of

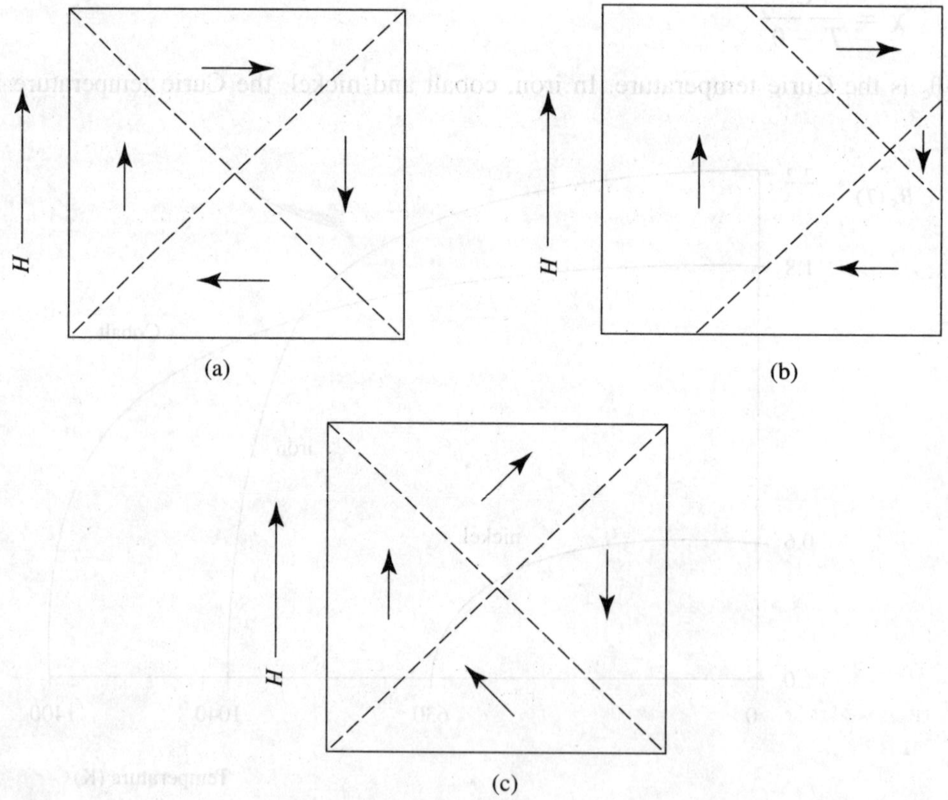

Fig. 4.18. Magnetization stages: (a) unmagnetized; (b) magnetized by domain growth; (c) magnetization by domain rotation.

all these various directions of magnetization may be zero (Figure 4.18(a)). When a magnetic field is applied, the domains most nearly parallel to the direction of the applied magnetic field grow in size at the expense of the others. This is called boundary displacement (Figure 4.18 (b). A further increase in the magnetic field causes the domains to rotate and align parallel to the applied field. The material now reaches the point of saturation magnetization and no further change will take place on increasing the magnetic field. Between the adjacent domain there is a narrow region, called a *Bloch wall* in which each dipole is slightly misaligned with the neighbouring dipoles. As a result, over a distance of typically a few hundred atomic spacings the orientation of dipoles changes from the direction of one domain to that of the adjoining domain.

The graphical presentation of changes in B with varying H is carried out by magnetization curve (Figure 4.19). Three characteristic regions can be highlighted. The initial portion has a definite slope called, the initial permeability. The middle portion is the steepest and in this

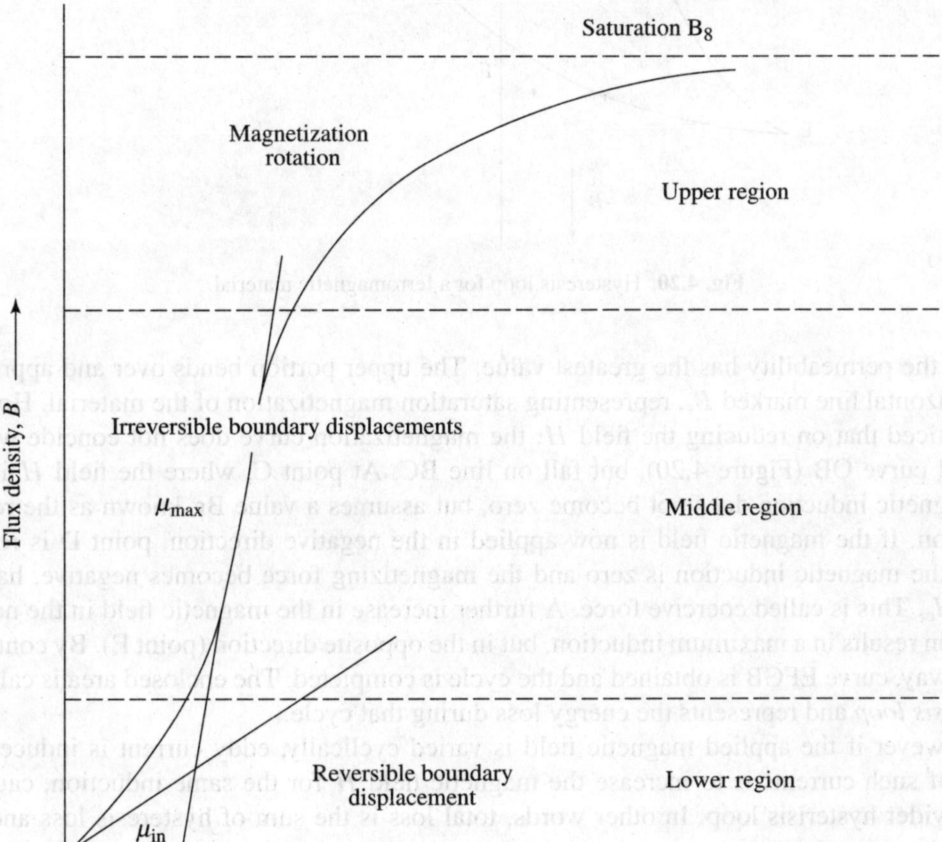

Fig. 4.19. A typical magnetization curve. μ_{in} is initial permeability; μ_{max} maximum permeability.

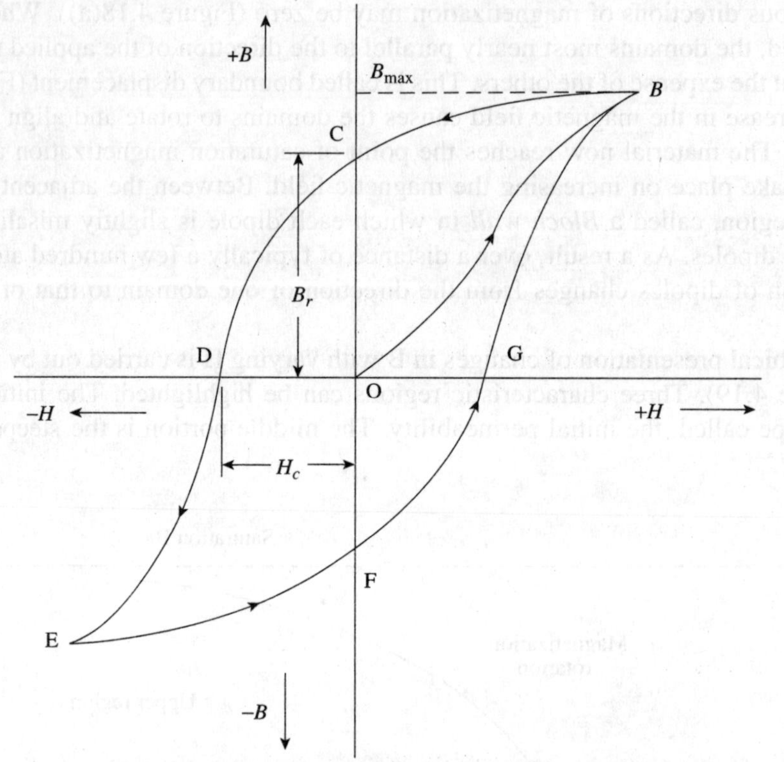

Fig. 4.20. Hysteresis loop for a ferromagnetic material.

region, the permeability has the greatest value. The upper portion bends over and approaches the horizontal line marked B_s, representing saturation magnetization of the material. However, it is noticed that on reducing the field H, the magnetization curve does not concide with the original curve OB (Figure 4.20), but fall on line BC. At point C, where the field H is zero, the magnetic induction does not become zero, but assumes a value Br, known as the residual induction. If the magnetic field is now applied in the negative direction, point D is reached, where the magnetic induction is zero and the magnetizing force becomes negative, having a value H_c. This is called coercive force. A further increase in the magnetic field in the negative direction results in a maximum induction, but in the opposite direction (point E). By continuing in this way, curve EFGB is obtained and the cycle is completed. The enclosed area is called the *hysteresis loop* and represents the energy loss during that cycle.

However if the applied magnetic field is varied cyclically, eddy current is induced. The effect of such currents is to increase the magnetic field H for the same induction, causing a much wider hysterisis loop. In other words, total loss is the sum of hysteresis loss and eddy current loss.

Almost all magnetic materials exhibit anisotropic behaviour, i.e., different magnetic strength in different crystallographic direction. Figure 4.21 illustrates this fact in case of iron nickel and cobalt single crystals.

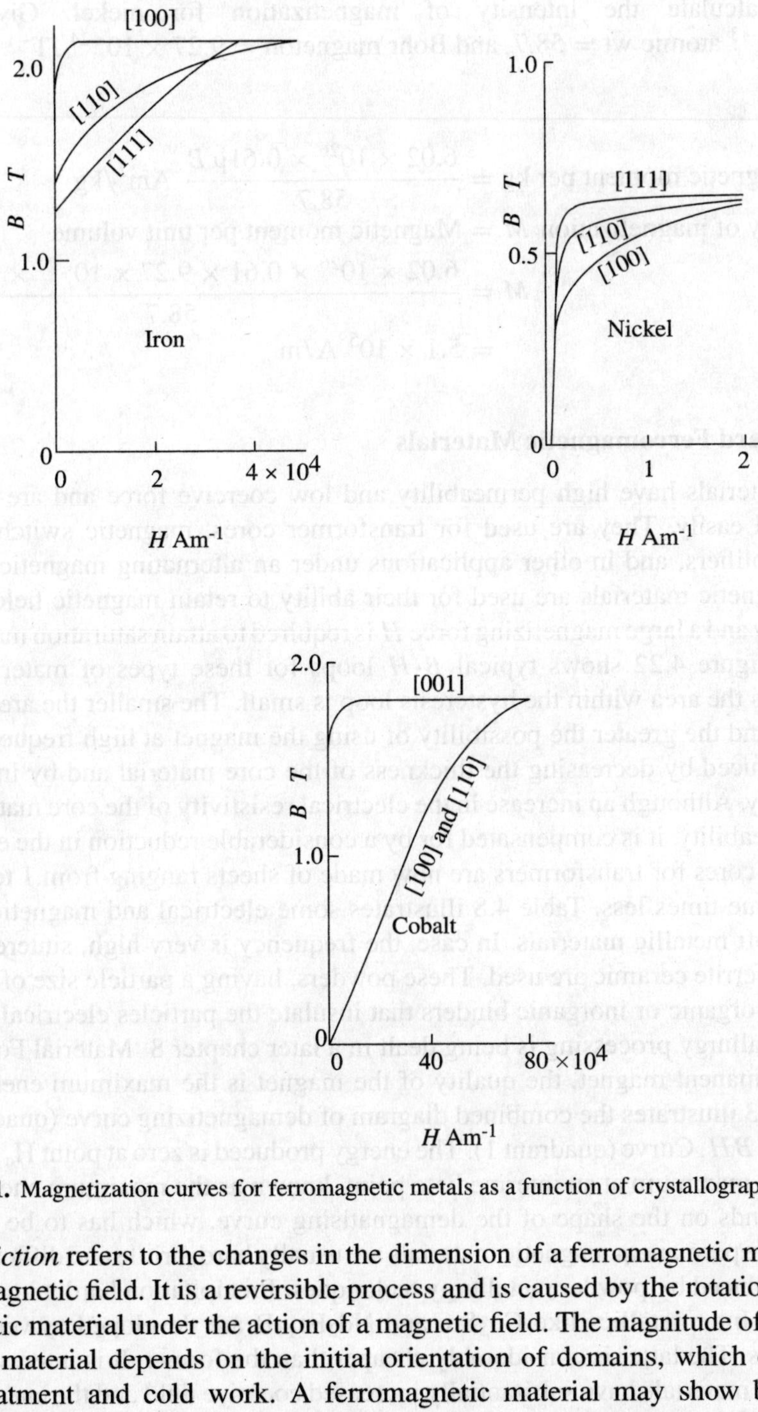

Fig. 4.21. Magnetization curves for ferromagnetic metals as a function of crystallographic orientation.

Magnetostriction refers to the changes in the dimension of a ferromagnetic material, when it is placed in magnetic field. It is a reversible process and is caused by the rotation of domains of a ferromagnetic material under the action of a magnetic field. The magnitude of magnetostriction for a given material depends on the initial orientation of domains, which can be controlled by heat treatment and cold work. A ferromagnetic material may show both positive and negative magnetostriction. By suitable alloying one may modify the degree of magnetostriction. Magnetostrictive transducers are typical devices using this type of phenomenon.

Example 4.8. Calculate the intensity of magnetization for nickel. Given density $= 8.8 \times 10^3$ kg m^{-3} atomic wt = 58.7, and Bohr magneton $= 9.27 \times 10^{-24}$ JT^{-1}

Solution.

$$\text{Magnetic moment per kg} = \frac{6.02 \times 10^{26} \times 0.61\mu B}{58.7} \text{ Am}^2/\text{kg}$$

$$\text{Intensity of magnetization } M = \text{Magnetic moment per unit volume}$$

$$M = \frac{6.02 \times 10^{26} \times 0.61 \times 9.27 \times 10^{-24} \times 8.8 \times 10^3}{58.7}$$

$$= 5.1 \times 10^5 \text{ A/m}$$

4.4.3 Soft and Hard Ferromagnetic Materials

Soft magnetic materials have high permeability and low coercive force and are magnetized and demagnetized easily. They are used for transformer cores, magnetic switching circuits and magnetic amplifiers, and in other applications under an alternating magnetic field. Hard or permanent magnetic materials are used for their ability to retain magnetic fields. Here the permeability is low and a large magnetizing force H is required to attain saturation magnetization of the material. Figure 4.22 shows typical B-H loops for these types of materials. In soft magnetic materials the area within the hysteresis loop is small. The smaller the area, the lower the power losses and the greater the possibility of using the magnet at high frequencies. Eddy current can be reduced by decreasing the thickness of the core material and by increasing its electrical resistivity. Although an increase in the electrical resistivity of the core material lowers its magnetic permeability, it is compensated for by a considerable reduction in the eddy current losses. Laminated cores for transformers are now made of sheets ranging from 1 to 0.025 mm in thickness or some times less. Table 4.8 illustrates some electrical and magnetic properties of magnetically soft metallic materials. In case, the frequency is very high, sintered materials either of metal or ferrite ceramic are used. These powders, having a particle size of 2 to 10 μm are mixed with an organic or inorganic binders that insulate the particles electrically from one other. Powder metallurgy processing is being dealt in a later chapter 8 'Material Forming'.

In case of permanent magnet, the quality of the magnet is the maximum energy product BH_{max}. Figure 4.23 illustrates the combined diagram of demagnetizing curve (quadrant 2) and the energy product BH. Curve (quadrant 1). The energy produced is zero at point H_c and at point Br and it reaches a maximum at an intermediate point, known as the maximum energy product BH_{max}. This depends on the shape of the demagnatising curve, which has to be determined experimentally. The permanent magnetic materials are usually hard, brittle and difficult to shape. These are also produced by powder metallurgy techniques. Precipitation hardening type alloys are also available. Intermetallics like Co_5Sm and $Nd_2Fe_{14}B$ are developed as very powerful permanent magnets. The latter is considerably cheaper than the former. It is interesting to note that, these two new materials have substantially increased coercive field and the energy product, but the remanant magnetization has changed very little and generally comparable with that of pure iron. Table 4.9 summarizes the properties of some major hard magnetic materials.

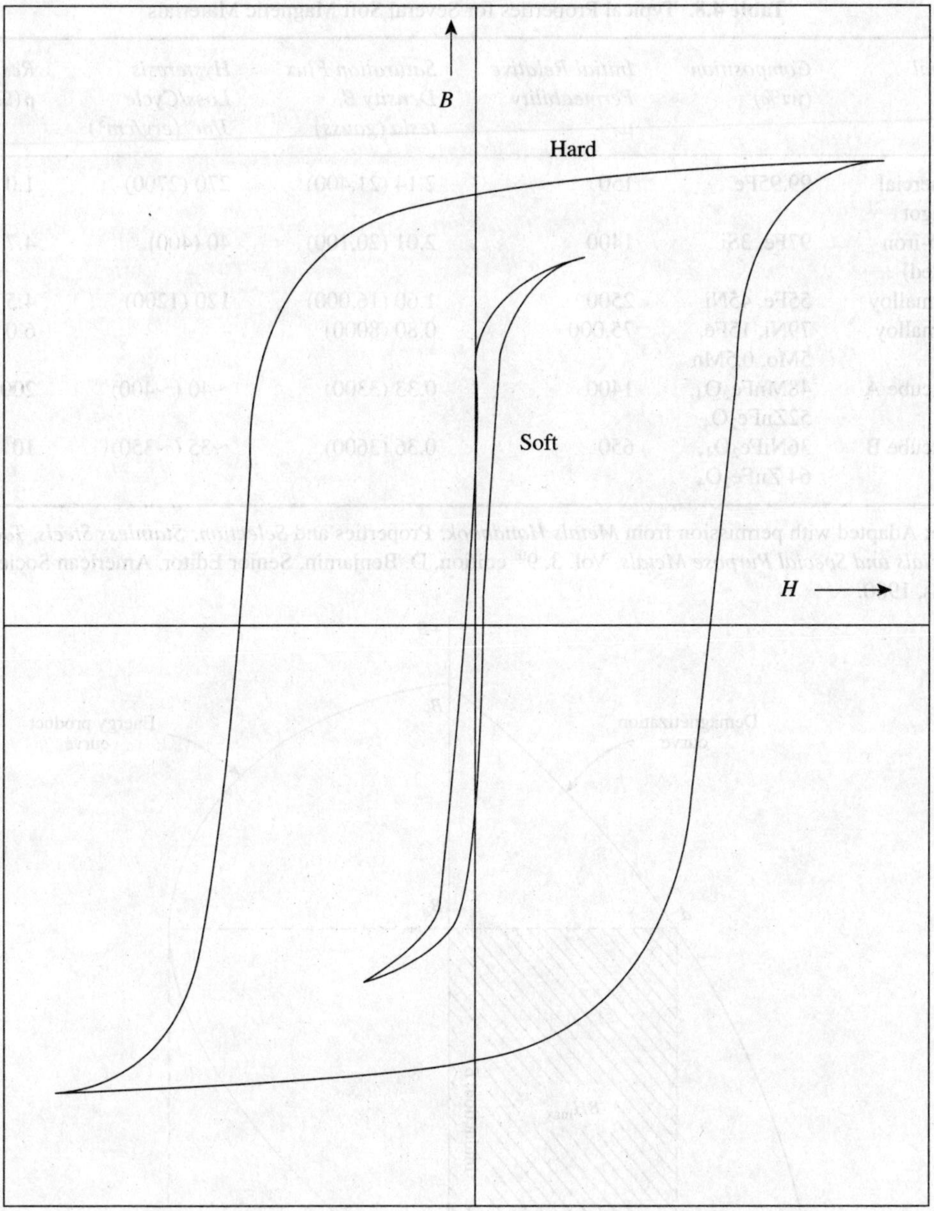

Fig. 4.22. Schematic magnetization curves for soft and hard magnetic materials.

4.4.4 Ferrimagnetic Materials

It has been mentioned earlier that ferrites are ceramics. It should not be confused with the BCC ferrite phase of pure iron. Like metallic magnets, these ceramic magnets are also divided into soft and hard ferrites.

Table 4.8. Typical Properties for Several Soft Magnetic Materials

Material	*Composition (wt%)*	*Initial Relative Permeability* μ_i	*Saturation Flux Density B, tesla (gauss)*	*Hysteresis Loss/Cycle J/m³ (erg/cm³)*	*Resistivity* $\rho(\Omega\text{-}m)$
Commercial iron ingot	99.95Fe	150	2.14 (21,400)	270 (2700)	1.0×10^{-7}
Silicon-iron (oriented)	97Fe, 3Si	1400	2.01 (20,100)	40 (400)	4.7×10^{-7}
45 Permalloy	55Fe, 45Ni	2500	1.60 (16,000)	120 (1200)	4.5×10^{-7}
Supermalloy	79Ni, 15Fe, 5Mo, 0.5Mn	75,000	0.80 (8000)	-	6.0×10^{-7}
Ferroxcube A	48$MnFe_2O_4$, 52$ZnFe_2O_4$	1400	0.33 (3300)	~40 (~400)	2000
Ferroxcube B	36$NiFe_2O_4$, 64 $ZnFe_2O_4$	650	0.36 (3600)	~35 (~350)	10^7

Source: Adapted with permission from *Metals Handbook:* Properties and *Selection: Stainless Steels, Tool Materials and Special Purpose Metals,* Vol. 3, 9th edition, D. Benjamin, Senior Editor, American Society for Metals, 1980.

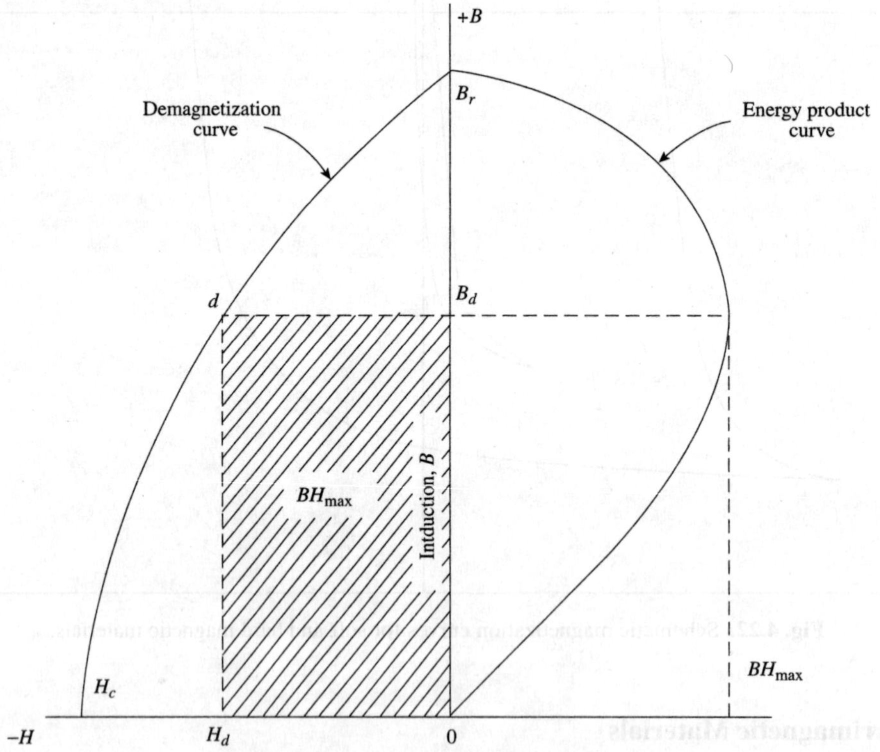

Fig. 4.23. Representation of maximum energy product BH_{max} in a permanent magnet.

Table 4.9. Typical Properties for Several Hard Magnetic Materials

Material	*Composition (wt%)*	*Remanence* B_r *tesla (gauss)*	*Coercivity* H_c *amp-turn/m (Oe)*	$(BH)_{max}$ kJ/m^3 *(MgOe)*	*Curie Temperature* T_c °C	*Resistivity* ρ (Ω-m)
Tungsten steel	92.8 Fe, 6W, 0.5Cr, 0.7 C	0.95 (9500)	5900 (74)	2.6 (0.33)	760	3.0×10^{-7}
Cunife	20 Fe, 20 Ni, 60 Cu	0.54	44,000 (550)	12 (1.5)	410	1.8×10^{-7}
Sintered alnico 8	34 Fe, 7Al, 15 Ni, 35 Co, 4 Cu, 5 Ti	0.76 (7600)	125,000 (1550)	36 (4.5)	860	-
Sintered ferrite 3	$BaO\text{-}6Fe_2O_3$	0.32 (3200)	240,000 (3000)	20 (2.5)	450	$\sim 10^{-4}$
Cobalt rare earth 1	$SmCo_5$	0.92 (9200)	720,000 (9000)	170 (21)	725	5.0×10^{-7}
Sintered neodymium-iron-boron	$Nd_2Fe_{14}B$	1.16 (11,600)	848,000 (10,600)	255 (32)	310	1.6×10^{-8}

Source: Adapted from *ASM Handbook: Properties and Selection: Nonferrous Alloys and Special-Purpose Materials*. Copyright ©1990 by ASM International, permission of ASM International Materials Park, OH.

Let us now examine the atomic arrangement in Fe_3O_4, a typical ferrite. In this structure iron cations occupy two types of positions. For one, the coordination number is 4 (tetrahedral coordinations) and for other, the coordination number is 6 (octahedral coordinations). With this inverse spinel structure, half the trivalent (Fe^{3+}) ions are situated in octahedral positions, the other half in tetrahedral positions. The divalent Fe^{2+} ions are all located in octahedral positions. Figure 4.24 shows how the spin moments of all the Fe^{3+} ions in the octahedral positions are aligned parallel to one another. However, these are directed opposite to the Fe^{3+} ions placed in the tetrahedral positions, which are also aligned. The total effective magnetic moment is offered by the Fe^{2+} ions. The saturation magnetization of such ferromagnetic material is therefore, the product of the net spin magnetic moment for each Fe^{2+} ion and the number of Fe^{2+} ions.

Cubic ferrites with different compositions can be made by adding metallic ions that substitute some iron in the crystal structure. The general formula is $M^{2+}O^{2-}(Fe^{3+})_2(O^{2-})_3$. M^{2+} represent divalent ions like Fe^{2+}, Ni^{2+}, Mn^{2+}, Co^{2+} and Cu^{2+}. Each of these ions offer different net spin magnetic moment. Instead of one metal, even two metals may be mixed e.g., $(Mn, Mg)Fe_2O_4$ in which the ratio of these metal atoms may be varied.

Other examples of ferrites are hexagonal ferrites. The chemical formula may be represented by $AB_{12}O_{19}$, in which A is a divalent metal such as barium, lead or strontium and B is a trivalent metal such as aluminium, gallium, chromium or iron.

The garnets have a general chemical formula $M_3Fe_5O_{12}$, where M represents rare earth ion such as samarium, europium, gadolinium or yttrium. The crystal structures are rather complex.

Ferrites with square magnetic loops are used as memory or logic devices in computers, as switching devices and in information storage.

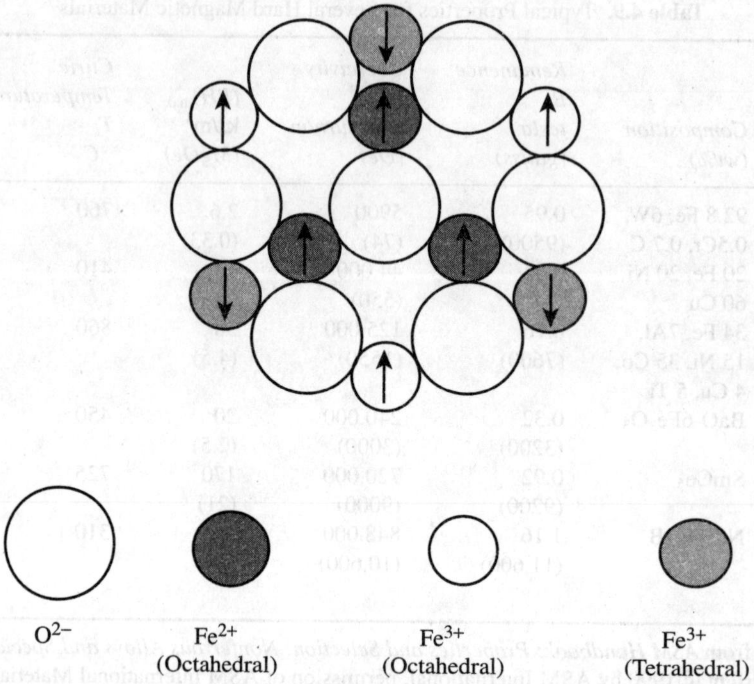

Fig. 4.24. Schematic diagram showing the spin magnetic moment configuration for Fe^{2+}, Fe^{3+} ions in Fe_3O_4.

Example 4.9. Calculate the saturation magnetization of magnetite Fe_3O_4, if it has a unit cell of lattice parameter 0.839 nm, each unit cell containing eight molecules of Fe_3O_4. Given Bohr magneton = 9.27×10^{-24} Am^2.

Solution. In Fe_3O_4, when the iron atom is ionized to give the Fe^{2+} ion there will be 4 unpaired electrons left after the loss of the two 4s electrons. When the iron atom is ionized to give the Fe^{3+} ion there will be 5 unpaired electrons left after the loss of two 4s and one 3d electrons. In the magnetite unit cell, eight Fe^{2+} ions occupy eight octahedral sites and sixteen Fe^{3+} are divided so that eight occupy octahedral sites and eight tetrahedral sites (refer section 2.4.3 for more details).

The magnetic dipoles from the electrons in the octahedral sites are in the opposite direction to that given by the electrons in the tetrahedral sites. Thus the magnetism is entirely due to the eight Fe^{2+} ions since the Fe^{3+} ions cancel each other out.

$$\text{Magnetic dipole moment for each } Fe^{2+} \text{ ion} = 4\mu_B$$

$$\text{Magnetic dipole moment for a unit cell} = 32\,\mu_B$$

$$= 32 \times 9.27 \times 10^{-24}$$

$$= 2.97 \times 10^{-22} \text{ Am}^2$$

$$\text{The magnetism per cubic meter} = \frac{2.97 \times 10^{-22}}{(0.839 \times 10^{-9})^3} \text{ TA/m}$$

$$= \frac{4\pi \times 10^{-7} \times 2.97 \times 10^{-22}}{(0.839 \times 10^{-9})^3} \text{ T}$$
$$= 0.63 \text{ T}$$

4.4.5 Magnetic Storage

Magnetic storage, such as that found in a computer hard drive, relies on the magnetic properties created by an electron's spin. However, if an electron's spin can be harnessed within a semiconductor, the potential exists to create entirely new ways of computing and signal processing that will greatly increase speed and data storage densities. The exploitation of an electron's spin to carry information, rather than its charge, often is referred to as *spintronics*. Spintronics would provide the basic properties required for advanced technologies, such as on-chip integration of magnetic storage and electronic processing functions and quantum computing, which depends on coherent spin states to transmit and store information. A material is permanently magnetic if the majority of its electrons spin in the same direction. In order to be practical, spintronics will need to use semiconductors that maintain their magnetic properties at room temperature. This is a challenge because most magnetic semiconductors lose their magnetic properties at temperatures well below room temperature, and would require expensive and impractical refrigeration in order to work in an actual computer. These properties were achieved in a crystalline oxide film for example anatase titanium dioxide that is infused with a small amount of cobalt, a magnetic impurity. This magnetic semiconductor material was created using a synthesis method called molecular beam epitaxy.

4.5 SUPERCONDUCTIVITY

We have described in section 4.2.1, how with decrease in temperature, electrical resistivity of a metal falls down. However, there are limited number of metals and compounds, which loose their electrical resistivity before absolute zero is reached and become *superconductors*. The cause for such phenomena is related to interaction between electrons and lattice vibrations. The transition temperature T_c, at which the phenomenon occurs varies. It is 0.4 K for titanium, 1.17 K for aluminium, 18.2 K for Nb_3Sn, while 92 K for $YBa_2Cu_7O_7$. At present, we can get superconductors operating at liquid nitrogen temperature.

The superconductivity must not be taken as an extension of normal electrical conductivity. The materials which become superconductors are not necessarily good conductors, when they are normal metals. As a matter of fact, the best electrical conductors copper and silver, do not superconduct even at temperatures of 0.0001 K. Bardeen, Cooper and Schrieffer formulated their theory known as BSC theory for super conductivity, which confirmed that electrons in a superconductor do not behave in the same way as those in normal metal.

The superconducting state can be destroyed by the application of an external magnetic field. In the absence of any external magnetic field, an increase in the current passing through the conductor can itself create a sufficient change in magnetic field to restore the resistivity of the conductor.

When a yttrium barium copper oxide ($YBa_2Cu_3O_7$) cooled to liquid nitrogen temperature is placed in a magnetic field, the sample start floating. This occurs because the magnetic field causes current to flow in the superconductor. As there is no resistance, the charges are in a state of perpetual motion. These currents in the superconductor set up a magnetic field which exactly opposes the applied field. The result is a strong repulsive force, thus levitating the sample. This effect is known as *Meissner effect*. This principle has been used by Japanese to introduce high speed trains (552 km/hour), running without much friction.

Another application of superconductors is in SQUID (Superconducting quantum interference device), which can be used as an extremely sensitive device for measuring magnetic fields. This has great application in biomedical engineering. Superconductors may also be used in cellular telephone systems as filters.

Let us now consider the effects of magnetic field. The maximum magnetic field that a material can tolerate and still remain superconducting is zero at T_c, but increases to a maximum value B_c as the temperature is reduced to absolute zero. The critical magnetic field, B_T, at temperature T is generally found to obey the relationship

$$B_T = B_c\left[1 - \left(\frac{T}{T_c}\right)^2\right] \qquad (4.27)$$

The above equation is represented by the curve shown in Figure 4.25 as the base curve in the three dimensional presentation.

Since there are three independent parameters which affect the superconductivity, it is worthwhile to represent them by a three dimensional graph (Figure 4.25). For all combinations of temperature, magnetic field and current density corresponding to points below the critical surface the material behaves as a superconductor, and for all other combinations the material is a normal one.

BSC theory predicts that at absolute zero the band gap in a supercondctor is given by

$$E_g = 3.5k_B T_c \qquad (4.28)$$

where k_B is Boltzmann's constant, and T_c the critical temperature. For a typical superconductor the energy gap is a few meV, which is typically a thousand times smaller than the band gap in a semiconductor.

From Figure 4.25 it is evident that as either the temperature approaches T_c or the applied magnetic field approaches B_c, the magnitude of the critical current density approaches zero. J_c is not strictly an intrinsic property of the superconductor, but depends on the crystalline state of the material i.e., crystallite size, defect density etc.

Example 4.10. Can the Meissner effect be explained from a classical view point if we treat the superconductor as a material which has zero resistivity?

Solution. No, it cannot be explained using classical electromagnetism because there is no change in the external magnetic field, and therefore no induced current.

Example 4.11. Lead has $T_c = 7.19$ K and B_c (O) $= 0.0803$T. Calculate the critical magnetic field at 6.0K.

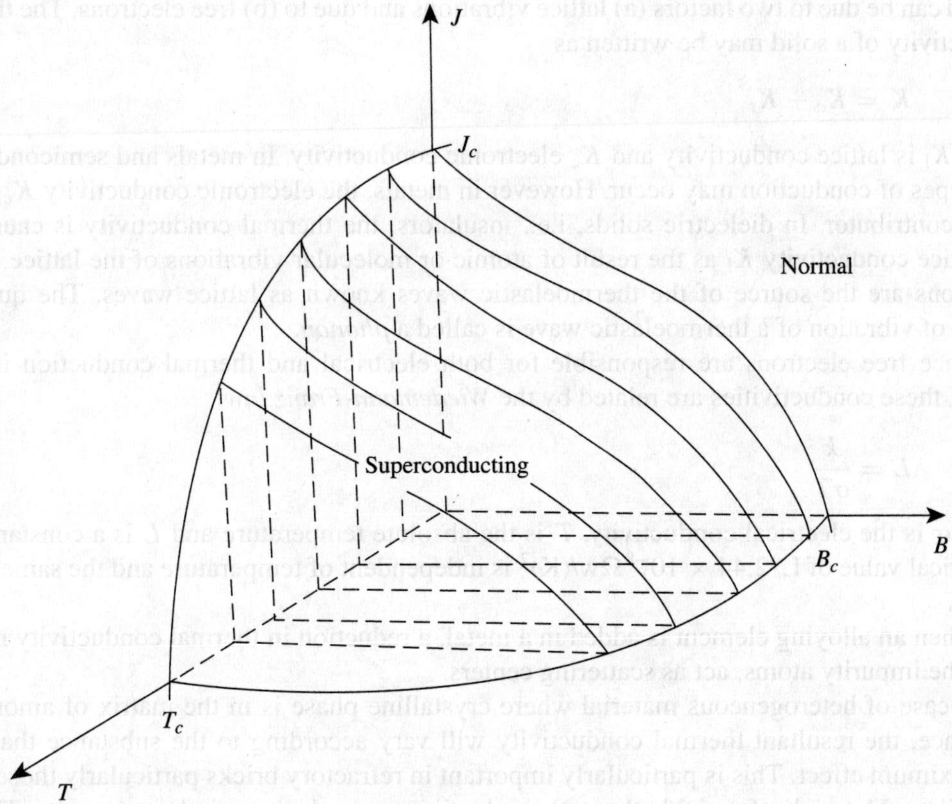

Fig. 4.25. Superconducting and normal states, showing the effect of temperature, magnetic field B and current density J.

Solution.

$$\begin{aligned}
\text{We know } H_C(O) &= B_C(\text{O})/\mu_o \\
&= 0.0803\text{T}/1.257 \times 10^{-6}\ \text{A/m} \\
&= 6.39 \times 10^4\ \text{A/m}
\end{aligned}$$

$$\begin{aligned}
H_C(T) &= H_C(O)\left[1 - \frac{T^2}{T_C^2}\right] \\
&= (6.39 \times 10^4\ \text{A/m})\left[1 - \frac{(6\ \text{K})^2}{(7.19\ \text{K})^2}\right] \\
&= 1.98 \times 10^4\ \text{A/m}
\end{aligned}$$

4.6 THERMAL CONDUCTIVITY

A temperature gradient in a material gives rise to heat flux transport. The mathematical treatment can be taken similar to Fick's laws of diffusion, described in section 3.8. Thermal conduction in solid can be due to two factors (a) lattice vibrations and due to (b) free electrons. The thermal conductivity of a solid may be written as

$$K = K_c + K_t \tag{4.29}$$

where K_l is lattice conductivity and K_c electronic conductivity. In metals and semiconductors both types of conduction may occur. However in metals, the electronic conductivity K_c, is the major contributor. In dielectric solids, i.e., insulators, the thermal conductivity is caused by the lattice conductivity K_l as the result of atomic or molecular vibrations of the lattice. These vibrations are the source of the thermoelastic waves known as lattice waves. The quantum energy of vibration of a thermoelastic wave is called a *phonon*.

Since free electrons are responsible for both electrical and thermal conduction in pure metals, these conductivities are related by the *Wiedemann-Franz law*.

$$L = \frac{k}{\sigma T} \tag{4.30}$$

where σ is the electrical conductivity, T is the absolute temperature and L is a constant. The theoretical value of L, $2.44 \times 10^{-8} \Omega w/(K)^2$ is independent of temperature and the same for all metals.

When an alloying element is added in a metal, a reduction in thermal conductivity results, since the impurity atoms, act as scattering centers.

In case of heterogeneous material where crystalline phase is in the matrix of amorphous substance, the resultant thermal conductivity will vary according to the substance that exert the maximum effect. This is particularly important in refractory bricks particularly those made from silica. Materials of variable thermal conductivity may also vary in heat capacity. The rate at which the temperature rises in them is important in applications of thermal insulation to buildings and plants. A good insulator does not necessarily remain cold but if its heat capacity is low, the material is soon saturated and does not go on absorbing heat from the source. The parameter used to describe this is the *thermal diffusivity* α, the ratio of thermal conductivity to heat capacity

$$\alpha = \frac{k}{\rho C_p} \tag{4.31}$$

ρ is the density, C_p heat capacity per mole. Metals have a high value of 'α' because their thermal conductivity is high. Insulating materials are frequently porous. This gives a low value for both k and ρ C_p, although 'α' may not be any smaller than that for denser material. Ceramic products are generally porous, as they are fabricated through powder processing. Heat transfer across pores is ordinarily slow and inefficient. Internal pores normally contain still air, which has an extremely low thermal conductivity.

Next question arises as to what happens to heat abstraction when a cold and a hot surface are placed in contact. The *contact coefficient* 'b' is defined by the square root of the product of conductivity and heat capacity.

$$b = \sqrt{k\rho C_p} \tag{4.32}$$

Good insulating materials, therefore, have low values of '*b*', since both *k* and ρ C_p are low. Such materials are applied to floor covering, upholstery and handles. The value of '*b*' for metals is remarkably independent of temperature, although *k* and ρ C_p are both temperature dependent for metals.

In case of polymers, both vibration and rotation of the chain molecules contribute in thermal conductivity. It is also a function of the crystallinity in the polymer. A polymer with a highly crystalline and ordered structure will have a greater thermal conductivity than the equivalent amorphous material. Like ceramics, the insulative property of a polymer may be further increased by introducing small pores, which are ordinarily introduced by foaming during polymerization.

Polycrystalline metals have much lower thermal conducting than single metal crystals because of the scattering effect of electrons along grain boundaries. Hence the thermal conductivity of commercial metals depends on the grain size.

4.7 THERMAL EXPANSION

Practically all solids expand when heated and contract when cooled. Thermal expansion of solid, like heat capacity and other thermal properties, has its origin in the lattice vibration, the intensity of which increases as the temperature rises. The thermal coefficient of expansion is related to the specific heat and to the melting point of the solids. Figure 4.26 shows the relationship between atomic number and coefficient of linear thermal expansion. Here the case

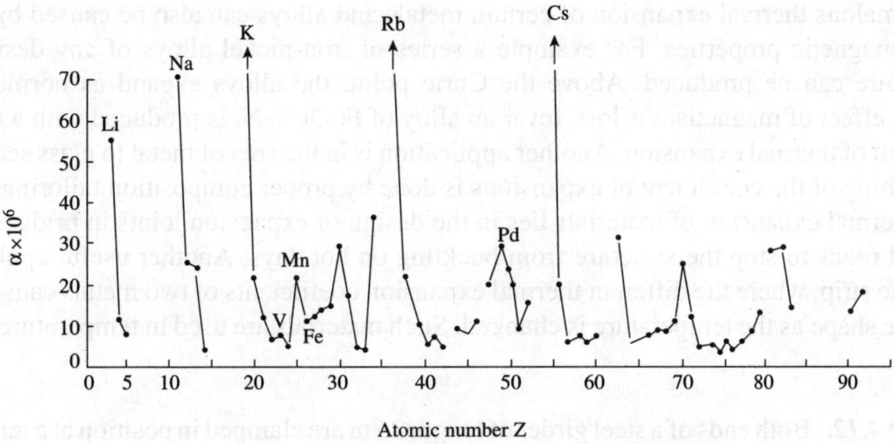

Fig. 4.26. Variation of linear coefficient of expansion with atomic number.

of transition metals is interesting where there are hybrid states between 4s and 3d. The third long period is even more complex by the rare earths and the introduction of 4f electrons.

Thermal expansion response is isotropic in those crystal structures, which are isotropic. In case of HCP metals i.e., Zn, along the basal plane, the coefficient of thermal expansion is minimum along the close packed direction, and actually becomes negative near the absolute zero.

We know that linear and volumetric expansion are given as

$$\frac{\Delta l}{l_o} = \alpha_l \Delta T \tag{4.33}$$

and

$$\frac{\Delta V}{V_0} = \alpha_v \Delta T \tag{4.34}$$

and Δl and ΔV_v are change in length and volume and l_o and V_o are initial length and volume of the material.

If we go through the energy vs. interatomic spacing plot pertaining to a crystal (*see* section 2.2), it is obvious that the thermal expansion is due to the asymmetric curvature of the potential energy well, rather than the increased atomic vibration with rising temperature. In strongly bonded solids like covalent solids, the potential energy well plot is deeper and more narrow. This gives rise to a smaller value of the coefficient of thermal expansion. With regard to temperature dependence, the magnitude of the coefficient of thermal expansion increases with increase in temperature.

Phase transformation in metals or ceramics will give rise to a change in the coefficient of expansions, thus introducing volumetric change. The sudden change in volume during transformation results in poor thermal shock resistance. A good example is poor shock resistance of silica brick in a temperature range room temperature to 600 °C. On the other hand amorphous silica has a very low and uniform coefficient of thermal expansion. It is possible, by making changes in compositions of more complex ceramics and glasses, to produce materials with coefficients of thermal expansion varying over a wide range, from positive to negative values. The anomalous thermal expansion of certain metals and alloys can also be caused by changes in their magnetic properties. For example a series of iron-nickel alloys of any desired curie temperature can be produced. Above the Curie point, the alloys expand as normal metals, since the effect of magnetism is lost. Invar an alloy of Fe-36% Ni is produced with a minimum coefficient of thermal expansion. Another application is in the area of metal to glass seals, where the matching of the coefficient of expansions is done by proper composition tailoring. Another use of thermal expansion of materials lies in the design of expansion joints in bridges, railway lines and roads to stop the structure from buckling on hot days. Another useful application is bimetallic strip, where the different thermal expansion coefficients of two metals cause the strip to change shape as the temperature is changed. Such materials are used in temperature sensitive switch.

Example 4.12. Both ends of a steel girder of length 25 m are clamped in position at a temperature of 5 °C. Calculate the change in length that occurs when the temperature is raised to 30 °C and determine the stress that is produced in the girder. If the yield stress of the steel is 5.0×10^7 Nm^{-2}, is the thermal stress sufficient to cause a permanent deformation of the girder?

Given the linear coefficient of expansion of steel is 1.1×10^{-5} K^{-1} and Young's modulus is 2.1×10^{11} Nm^{-2}.

Solution.

$$\Delta L = L_o \alpha \Delta T$$

$$= (1.1 \times 10^{-5}\,K^{-1})(25\,m)(25\,K)$$
$$= 6.875 \times 10^{-3}\,m$$

We know that

$$\text{Strain } \varepsilon = \frac{\Delta L}{L_o} = \frac{6.875 \times 10^{-3}\,m}{25}$$
$$= 2.75 \times 10^{-4}$$
$$\text{Also stress} = \text{Young's modulus} \times \text{strain}$$
$$= (2.1 \times 10^{11}\,Nm^{-2})(2.75 \times 10^{-4})$$
$$= 57.75 \times 10^{6}\,Nm^{-2}$$

Since the value exceeds the yield stress, it will produce a permanent deformation.

4.7.1 Thermal Shock Resistance

A sudden change in temperature of a material may give rise to volume change, which may cause cracking. Thermal shock is a combination of (a) expansion or contraction constraints (b) temperature gradient due to thermal conductivity, and (c) phase transformations. The result of such thermal shock is the generation of residual tensile stresses over the surface. A ductile material generally withstands severe thermal shock much better than brittle materials of comparable strength and thermal properties. This is the reason, why ceramics are least prone to thermal shock resistance. Thermal fatigue may also cause cracking, where repeated thermal cycling is involved. The resistance of thermal shock is measured by determining the maximum temperature difference that can be tolerated during a quench without causing any cracking. Some of the values for ceramics are 950 °C for Sialon, 500 °C for partially stabilized zirconia, 350 °C for SiC and 200 °C for alumina and ordinary glass.

The reason for stabilizing zirconia by CaO or MgO is to avoid polymorphism, thereby getting a stable cubic phase at all temperatures. This offers an enhancement in thermal shock resistance of zirconia. Cutting tools are also subjected to thermal shock depending on the severity of cutting. This property becomes significant in such materials. Low elastic modulus and high brittle strength favourably affect the thermal shock resistance of the material. The ratio of brittle strength to modulus ($\sigma b/E$) should be as high as possible. Mathematically thermal shock resistance (TSR) can be expressed as

$$TSR = \frac{\sigma_b K}{E\alpha} \tag{4.35}$$

where σ_b = The brittle strength of the material

α = The thermal coefficient of expansion

E = The Young's modulus

K = The thermal conductivity

From the above expression, it is obvious that a change in α values is a good controller for developing thermal shock resistance materials. For example, soda-lime glass is prone for thermal shock. Reducing the CaO and Na_2O contents, while at the same time adding B_2O_3 to develop borosilicate (Pyrex) glasses is a good solution to increase thermal shock resistance. This glass is ideal for a kitchen oven heating and cooling cycles.

4.8 OPTICAL PROPERTIES

Optical properties like thermal properties are related to the interaction of a material with radiation in the form of wave or particles of energy. The spectrum of the electromagnetic radiation covers a vary broad range of frequency from very high energy gamma radiation to long radio waves, as shown in Figure 4.27. The electromagnetic radiation propagates through space with the speed c equal to the product of their frequency ν and λ :

$$c = \lambda\nu \tag{4.36}$$

The degree of interaction of the electromagnetic waves with matter depends on wavelength and frequency and on the characteristic properties of materials for example, dielectric constant. When light strikes a material, it can undergo one of the three transitions. For a given wavelength λ, an incident radiation may be reflected, absorbed or transmitted, depending on the characteristic properties of the medium. The total incident radiation is, therefore, the sum of these effects

$$A_\lambda + T_\lambda + R_\lambda = 1 \tag{4.37}$$

where A_λ is the fraction of the light absorbed, T_λ is the fraction transmitted and $R\lambda$ is the amount reflected. No solid reflects or transmits all of the incident radiation, but a certain amount of it is also absorbed. If the radiation of intensity I_o is incident on a solid of thickness x, intensity of the transmitted portion of light I is given by

$$I = I_o e^{-\beta x} \tag{4.38}$$

where β is the absorption coefficient, characteristic of a particular solid or medium.

Most dielectrics that are electric insulators are transparent to visible radiation, since their energy band is usually greater than the energy necessary to cause band-to-band transition, resulting in the formation of electron-hole pairs. Thus, for transparency, a dielectric should have an energy gap beyond the visible spectral range, which is from about 1.7 to 3.5 eV. For example, diamond is completely transparent, even for ultraviolet radiation, because its energy gap is 5.6 eV. A transparent material becomes translucent and then opaque if light is scattered at internal imperfections such as grain boundaries and dispersed crystallites in the amorphous matrix. The presence of impurities may lower the energy gap of the crystal imparting a colour to the transparent crystal, for example Cr^{3+} in Al_2O_3. The colour arises because of partial absorption of white light and selective transmission of the other portion of the spectrum. Silicon and germanium with their low energy gaps (1.11 and 0.67 eV respectively) are opaque upto 1.1 and 1.8μ m radiation, but beyond this range, they become transparent. Metals are opaque to all electromagnetic radiation from very low frequency to the middle region of ultraviolet frequency, beyond which they become transparent. This is due to the fact that in metals electrons at the Fermi level can absorb energy, readily passing to higher energy states in the band.

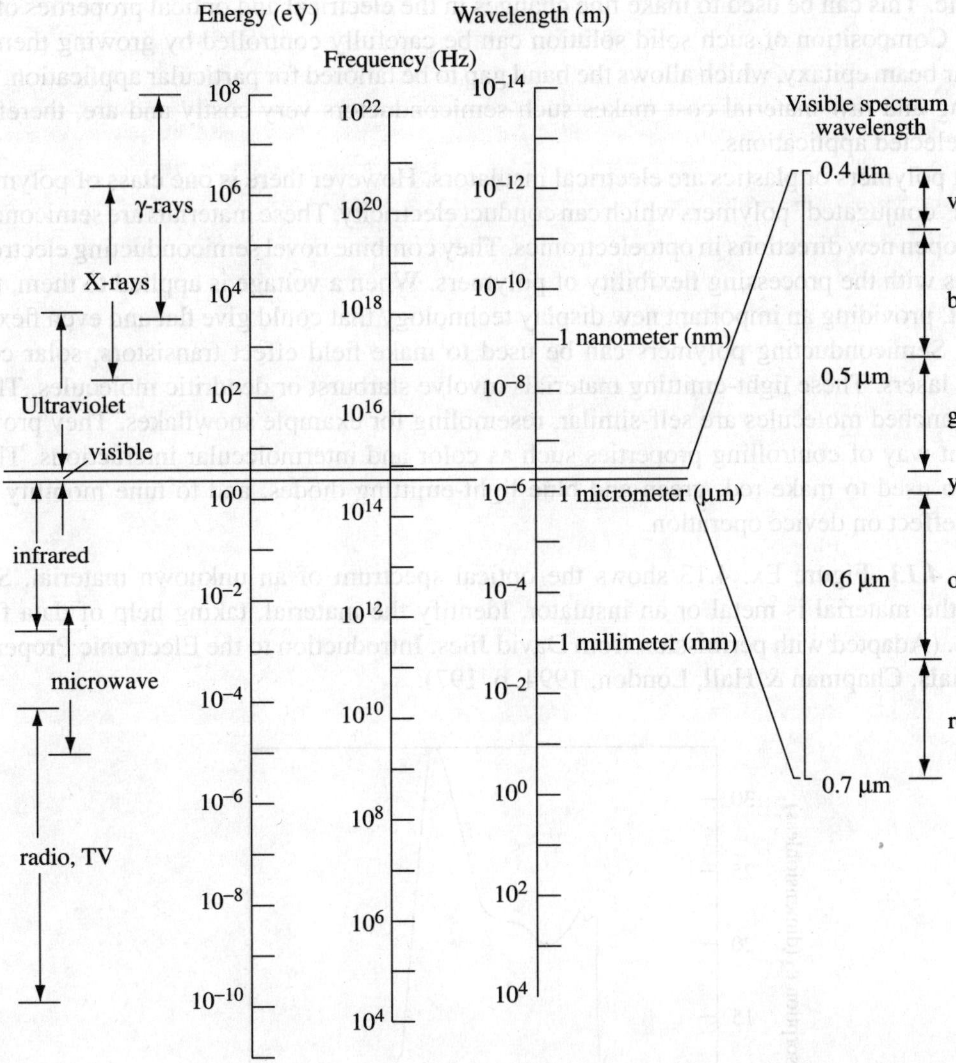

Fig. 4.27. The spectrum of electromagnetic radiation.

In earlier section, we have mentioned about the technological significance of GaAs, as it is highly efficient in detecting and generating light. This is derived from its direct band gap. This means that it can be used in fabricating lasers and in optical communications devices for computers. In this application, information is transmitted by photons instead of by electrons. These optical computers are much faster than conventional computers. Table 4.2 illustrates the comparative data for silicon, germanium and GaAs.

Another interesting property of GaAs is that it can be mixed with other components to form solid solutions. For example in GaAs aluminium or indium can be substituted for some or all of the gallium. In the same way, phosphorus or antimony can be substituted for some or all of the arsenic. This can be used to make fine changes in the electrical and optical properties of the material. Composition of such solid solution can be carefully controlled by growing them by molecular beam epitaxy, which allows the band gap to be tailored for particular application. The processing and raw material cost makes such semiconductors very costly and are, therefore, used in selected applications.

Most polymers or plastics are electrical insulators. However there is one class of polymers, known as "conjugated" polymers which can conduct electricity. These materials are semiconductors, and open new directions in optoelectronics. They combine novel semiconducting electronic properties with the processing flexibility of polymers. When a voltage is applied to them, they emit light, providing an important new display technology that could give flat and even flexible displays. Semiconducting polymers can be used to make field effect transistors, solar cells, and even lasers. These light-emitting materials involve starburst or dendritic molecules. These highly branched molecules are self-similar, resembling for example snowflakes. They provide an elegant way of controlling properties such as color and intermolecular interactions. These have been used to make red, green and blue light-emitting diodes, and to tune mobility and study its effect on device operation.

Example 4.13. Figure Ex. 4.13 shows the optical spectrum of an unknown material. State whether the material is metal or an insulator. Identify the material, taking help of data from Table 4.2. (Adapted with permission from David Jiles, Introduction to the Electronic Properties of Materials, Chapman & Hall, London, 1994, p. 197).

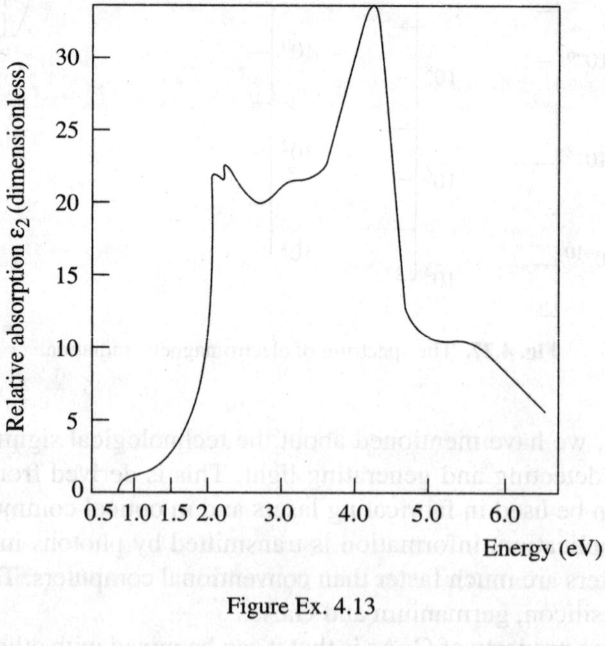

Figure Ex. 4.13

Solution. The optical spectrum is that of an insulator or semiconductor because of the low absorption at long wave lengths (low energies) and higher absorption at short wavelengths (high energies). The absorption edge, which corresponds to the energy at which electrons can just begin to cross the band gap occurs in this material at about 0.75 eV. From Table 4.2 it is obvious that the material is Germanium.

4.8.1 Refractive Index

When light is transmitted into the transparent solid, there is a decrease in its velocity, which results in its bending at the interface. This phenomenon is called *refraction*. The *refractive index* (n) of a material is defined as the ratio of the velocity in a vacuum c to the velocity in the medium ν

$$\text{or} \quad n = c/\nu \tag{4.39}$$

The phenomenon of refraction is related to electronic polarization at the relatively high frequencies for visible light. There is a relation between the refractive index and the dielectric constant ε as

$$n \approx \sqrt{\varepsilon} \tag{4.40}$$

The size of the atoms or ions has influence as the electronic polarization. Larger the atom or ion, the greater is the electronic polarization, the slower the velocity and the greater the refractive index. The refractive index of a typical soda-lime glass is approximately 1.5. Addition of large barium or lead ions (as BaO and PbO) to a glass, therefore, increases n value. For example, highly leaded glasses containing 90% PbO have a refractive index of about 2.1. In case the crystal is not isotropic, the index depends on the crystallographic direction.

The change in refractive index from medium to medium has been successfully exploited in the development of fiber optic glasses. Figure 4.28 shows the cross-section of a typical fiber glass. The outer layer is a protective coating. The core and the surrounding cladding are made of high purity silica glass modified by the controlled addition of vary small amount of other substances. Pure silica has a refractive index of 1.45 at 1 μm. Very small quantities of boron oxide (B_2O_3) can be added to lower the index, while the addition of GeO_2 causes it to increase. If we see the refractive index profile (Figure 4.28(b)), it is obvious that the change is sharp at the interface between the core and the cladding. Such fibers are known as a *step - index fiber*. However, there is a big disadvantage in such fiber, as it becomes difficult to resolve the separate pulses when they reach the receiving end of the fiber. This effect is called multipath dispersion. There are two ways to avoid this problem. Firstly we can make the diameters of the fiber very small, so that only one beam, with a path down the center of the core, is propagated. This type of fiber is termed as *monomode*. The other way is to tailor the refractive index of the fiber so that the light travels faster in the outer layers of the core than it does inside. It has been established that the percentage transmission is maximum at about 1.55 μm wavelength, making this the preferred wavelength for a signal. This wavelength corresponds to infrared region. This is the reason, why InGaAsP laser emitters are more suited than GaAlAs laser which has a wavelength range 0.82 μm to 0.9 μm for transmission.

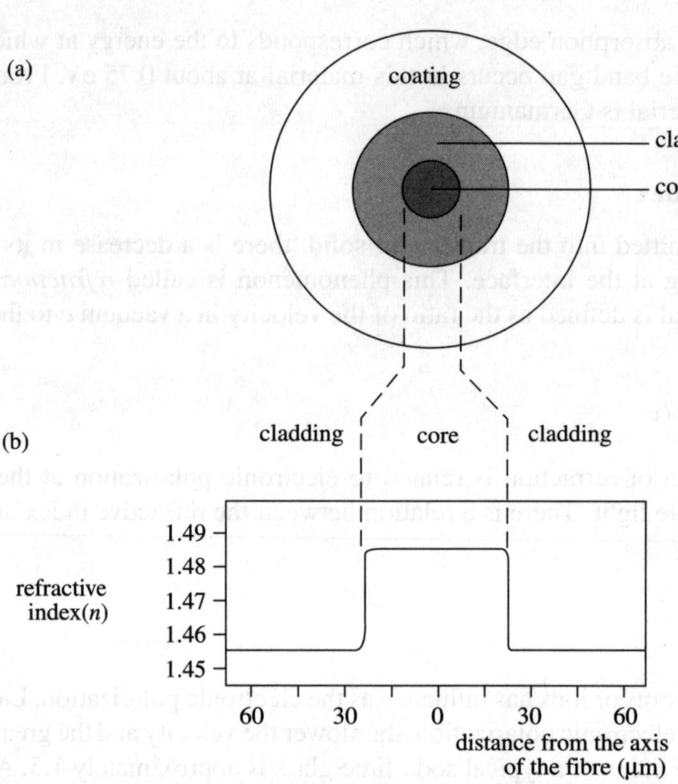

Fig. 4.28. (a) An optical fiber in cross-section showing the core, cladding and coating. (b) Refractive index profile in a step index fiber.

QUESTIONS AND PROBLEMS

4.1 Why the density of ceramics are generally lower than that metals?

4.2 Certain physical properties of Carborundum (SiC), naphthalene ($C_{10}H_8$) and Cryolite (Na_3AlF_6) can be predicted on the basis of the atomic structure of the constituent elements and the types of interatomic bonding suggested by the formulae. Enumerate these properties and explain the reason of the predictions.

4.3 Arrange the following inorganic compounds in the probable order of increasing melting temperatures, giving your reasons: KCl, CaO, B_2O_3.

Electrical Properties of Conductors

4.4 What is the mobility of electrons in copper with an electrical conductivity of 5.5×10^7 S/m if it has a current density of 2.0×10^6 A/m? Copper has 8.0×10^{28} electrons per cubic meter, each with a charge of 1.6×10^{-9}C.

4.5 Aluminium has a resistivity of $2.5 \times 10^{-8}\,\Omega$m. What will be the resistance of an aluminium wire with a length of 1 m and a cross sectional area of 2 mm^2?

4.6 Which of the following solids might be expected at 300 K to be good conductors, semiconductors or insulators? Aluminium energy gap 0 eV, aluminium nitride, energy gap 6.3 eV, gallium antimonide 1.95 eV.

4.7 Why does a 50/50 Cu-Ni alloy have an electrical conductivity that is much smaller than that of either pure copper or pure nickel?

4.8 Electrical resistivity of a pure copper sample and that of a Cu-2%Ni alloy at room temperature are 2×10^{-8} ohm m and 4×10^{-8} ohm m respectively. Calculate the residual resistivity for each per cent of Ni addition in pure copper.

4.9 A two phase eutectic alloy contain 10% of the α-phase and 90% the β-phase by volume. If the electrical resistivities are 1.7×10^{-8} ohm m and 7.0×10^{8} ohm m respectively, calculate the resistivity of the alloy.

4.10 Show that the electrical conductivity of a continuous composite along the direction parallel to axis takes the form

$$\sigma_c = V_1\sigma_1 + V_2\sigma_2$$

and along the transverse direction

$$\frac{1}{\sigma_c} = \frac{V_1}{\sigma_1} + \frac{V_2}{\sigma_2}$$

Will the thermal conductivity of the composite take a similar type of equation? Comment.

Electrical Properties of Semiconductor

4.11 The electrical conductivity of a material is given by the equation $\sigma = n\mu e$, where n is the electron concentration and μ is its mobility. Discuss the temperature dependence of conductivity of a semiconductor with that of a metal.

4.12 Compare the known electronic properties i.e., band gap, electron mobility, conductivity, optical properties etc. of Silicon, Germanium and Gallium Arsenide. Suggest engineering applications where GaAs has a distinct advantage over the others. (Refer Table 4.2).

4.13 Determine the concentration of conduction electrons in a sample of silicon if one in every million silicon atoms is replaced by a phosphorus atom. Molar mass of silicon 0.028 $kgmol^{-1}$ and density of silicon 2300 kgm^{-3}.

4.14 Explain how germanium, which is a semiconductor under normal condition, becomes a metal under a pressure of 120 k bar (12GPa).

4.15 What element(s) could be added to InSb to make an n-type semiconductor?

4.16 When phosphorus is added as a dopant to pure silicon, do the following quantities increase, decrease or remain the same? Explain your answer
(a) The hole concentration
(b) The mobility of electron
(c) The conductivity
(d) The temperature coefficient of conductivity

4.17 Extrinsic germanium used for transistor has the resistivity of 0.017 ohm m and an electron hole concentration from impurities of 1.6×10^{21} holes/m^3. (a) What is the mobility of the holes in the germanium? (b) What impurity atoms could be added to the germanium to create holes?

4.18 Explain why the Hall Voltage tends to be much bigger in semiconductor than in metals.

Electrical Properties of Insulators

4.19 When do ceramics make better electrical insulators than plastics?

4.20 If the type of plastic used for insulating electric wires has a breakdown field of 5×10^{7} Vm^{-1}, determine the minimum thickness required to insulate a domestic household supply (assuming a voltage of 240V). Compare the figure with the actual thickness of insulation in practice, which is typically about 0.5 mm.

4.21 The capacitance of a parallel plate capacitor increases from 0.085 μF to 0.195 μF when a sheet of polythene is placed between the plates. What is the dielectric constant of this polymer?

4.22 A charge of 2×10^{-10}C is to be stored on each plate of a parallel plate capacitor having an area of 800 mm^2 and plate separation of 4 mm.
(a) What voltage is required of a material having a dielectric constant of 3.5 when positioned within the plates?
(b) What voltage would be required if a vacuum is used?
(c) What are the capacitance for parts (a) and (b)?
(d) Compare the dielectric displacement for part (a)
(e) Calculate the polarization for part (a).

4.23 When a single crystal of $BaTiO_3$ is polarized by application of an electric field along various crystallographic directions, are there easy and hard directions as with ferromagnetic materials? Explain.

Related Electrical Phenomena

4.24 Explain why anodes in electronic tubes are often covered with a black deposit of sputtered metal or deposited carbon?

4.25 Explain how the thermo emf of a thermocouple shall be influenced by (a) elastic strain, (b) cold work, (c) chemical composition, (d) structural change and (e) magnetization.

Magnetic Properties

4.26 Explain the differences in relative permeability between diamagnetic, paramagnetics and ferromagnetic materials.

4.27 On the same set of B-H coordinates, draw schematically a hysteresis loop for iron at room temperature, 1020 K and 1060 K. (The Curie temperature for iron is 1043 K).

4.28 Given a piece of iron, describe possible treatments that would make it a soft or hard magnetic material.

4.29 Explain why wide domain walls are favored by exchange interactions.

4.30 Iron saturates at a flux density of 2.18 T. What is the average number of electrons per atom in the solid state that appear to contribute to this magnetization, if the iron has a BCC structure with a lattice parameter of 0.287 nm? Given Bohr magneton= 9.27×10^{-24} Am^2.

4.31 Assume that nickel consists entirely of Ni^{2+} ions, each have eight 3d electrons. Determine the magnetic moment per nickel ion for a sample of nickel given that the magnetization of nickel is 4.85×10^5 JT^{-1}m^{-3}, the density of nickel is 8.9 kg m^{-3} and the molar mass of nickel is 0.058 kg mol^{-1}. Given the spin magnetic moment of a single electron is 9.27×10^{-24} JT^{-1}.

4.32 Compare the known saturation magnetization of iron, cobalt and nickel in bulk with the known magnetic moment of the free atoms or ions and comment on the result.

4.33 Iron alloyed with 30-35% Co can be made into a very fine powder with particle size approximately 10 nm. If the powder particles are coated with a thin layer of antimony and then pressed to high density in a magnetic field, the resulting metal can be magnetized to produce a permanent magnet of very high coercivity. Explain the reaosons.

4.34 Determine the saturation magnetic flux density for the ferrite $NiFe_2O_4$ if the resultant moment is due to eight Ni^{2+} ions per unit cell. Lattice parameter of ferrite is 0.83 nm.

4.35 Magnetic glasses have been developed for use in transformer cores. How does their microstrucure make the core more energy efficient?

4.36 Does a tape recorder head made of ferrite need to be laminated? Explain.

Superconductivity

4.37 A strong magnet may be made by winding a coil of superconducting wire.

(a) Why must the coil be cooled for use as a magnet?

(b) What advantages does such a magnet have over other strong electromagnets?

(c) Suggest two different uses for such a super conducting magnet.

4.38 To what temperature must mercury be cooled in magnetic field of 14000 A/m for it to be superconductive. Given $B_c(O) = 0.0411$ Tesla, T_c=4.15K.

Thermal Expansion

4.39 Explain thermal expansion using a typical potential energy curve.

4.40 For small thermal strain, show that the coefficient of volume expansion is three times the coefficient of linear expansion.

4.41 If a rod of aluminium 40 cm long is heated from 15 to 85 °C, while its ends are maintained rigid, determine the type and magnitude of stress that develops. Assume that at 15 °C, the rod is stress free. What about the stress magnitude if a rod of 1 m long is used? If the original rod is cooled from 15 to −15 °C, what type and magnitude of stress will result? Modulus of elasticity for Al = 6.9×10^4 MPa. Coefficient of linear expansion for Al = 23.6×10^{-6}°C^{-1}.

Optical Properties

4.42 Explain in terms of band theory, why insulators can be transparent to visible light.

4.43 Silicon is somewhat transparent in the infrared. In what application is it desirable to have a p-n junction near to the surface?

4.44 Given that the shortest wavelength of light that is visible to the eye is about 380 nm, determine the value of the largest band gap which is able to absorb visible light. Given $h = 6.63 \times 10^{-34}$ Js and $c = 3.00 \times 10^8$ ms^{-1}.

4.45 A sample of Ge exhibits intrinsic conductivity at 300 K. If the absorption edge is at 1771 nm, estimate the temperature rise that will result in a 30 per cent increase in electrical conductivity.

4.46 Explain why silicon is opaque, whereas diamond is transparent.

4.47 Gallium arsenide has a band gap of 1.43 eV. If all of the gallium atoms are replaced with indium, the resulting crystal of indium arsenide has a band gap of 0.36 eV. Assuming that the band gap varies linearly between these two extremes, calculate the percentage of indium that must be used in alloy of indium gallium arsenide in order to emit light with a wavelength of 1 μm.

4.48 The GaAlAs semiconductor laser was designed for optic fiber applications. It emits in the range 0.82 μm to 0.9 μm. Explain why for optical communication purposes, this is not ideal.

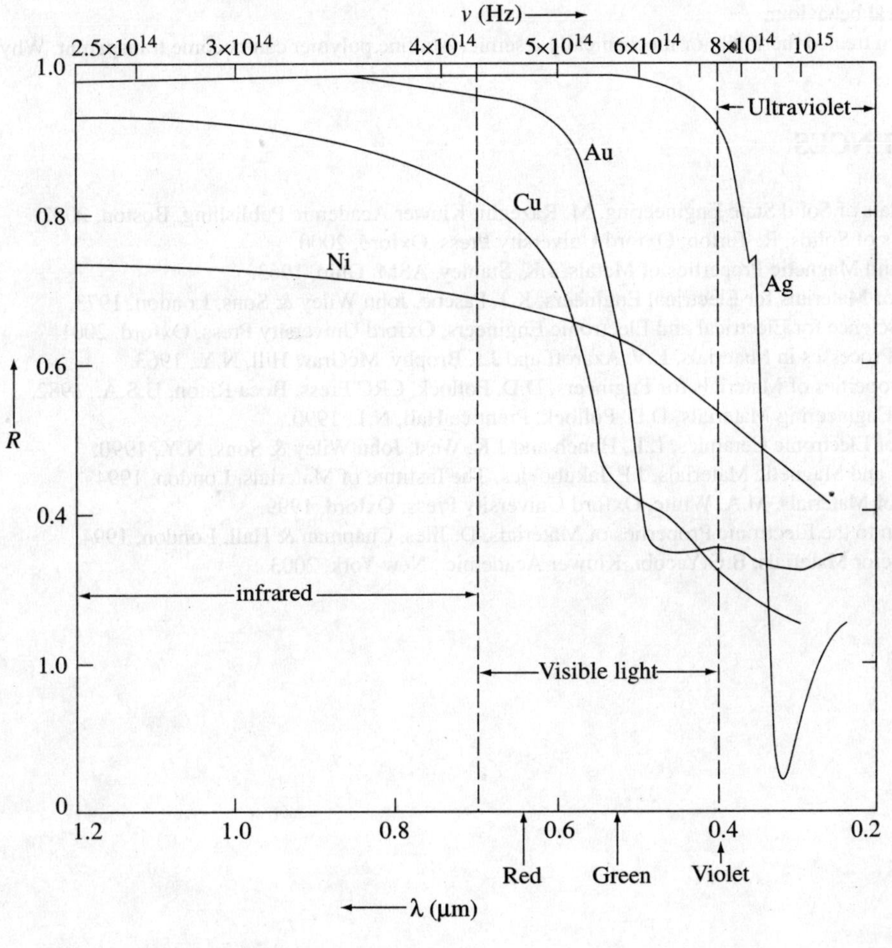

Figure P. 4.51

4.49 High capacity telephone links operate at wavelengths in the range 1.3 μm to 1.55 μm. Suggest reasons for this choice.

4.50 A semiconductor can be illuminated with electromagnetic radiation. Can superconductor be also affected likewise?

4.51 The figure P 4.51 shows the wavelength and frequency dependence of reflectivity for copper, silver, gold and nickel. On this basis discuss rationale of colours of these metals under different conditions. Why the behaviour of nickel is so different than other three metals? (Adapted with permission from K.M. Ralls, T.H. Courtney and J. Wulff, Introduction to Materials Science and Engineering, John Wiley & Sons, N.Y., 1976, p. 624).

4.52 Pure Al_2O_3, which is colourless, forms a complete series of solid solutions with Cr_2O_3, which is green. For solid solutions containing less than about 8 mole % Cr_2O_3, a red color is observed (this composition range includes ruby), and the molar volume remains close to the value for Al_2O_3 (25.58 cm^3/mol). At higher Cr_2O_3 contents, the solid solution becomes progressively greener, and the molar volume increases toward the value for Cr_2O_3 (29.09 cm^3/mol). In case of ruby, it has been observed experimentally that, heating causes it to become green in color, and subsequent cooling restores the red color. Suggest some qualitative reasons for the behaviour cited above.

4.53 Semicrystalline long-chain polymers usually are white and exhibit translucency if not too thick, whereas totally noncrystalline long-chain polymers are transparent if pure. Discuss the reasons for this difference in optical behaviour.

4.54 When treated thermally or mechanically a semicrystalline polymer can become transparent. Why?

REFERENCES

Fundamentals of Solid State Engineering, M. Razeghi, Kluwer Academic Publishing, Boston, 2002.
The Physics of Solids, R. Turton, Oxford University Press, Oxford, 2000.
Electrical and Magnetic Properties of Metals, J.K. Stanley, ASM, Ohio, 1963.
Properties of Materials for Electrical Engineers, K.J. Pascoe, John Wiley & Sons, London, 1973.
Materials Science for Electrical and Electronic Engineers, Oxford University Press, Oxford, 2001.
Electronic Processes in Materials, L.V. Azaroff and J.J. Brophy, McGraw Hill, N.Y., 1963.
Physical Properties of Materials for Engineers, D.D. Pollock, CRC Press, Boca Raton, U.S.A., 1982.
Physics of Engineering Materials, D.D. Pollock, Prentice Hall, N.J., 1990.
Principles of Electronic Ceramics, L.L. Hench and J.K. West, John Wiley & Sons, N.Y., 1990.
Magnetism and Magnetic Materials, J.P. Jakubovics, The Institute of Materials, London, 1994.
Properties of Materials, M.A. White, Oxford University Press, Oxford, 1999.
Introduction to the Electronic Properties of Materials, D. Jiles, Chapman & Hall, London, 1994.
Semiconductor Materials, B.G.Yacobi, Kluwer Academic , New York, 2003

5

MECHANICAL PROPERTIES OF MATERIALS

The mechanical behaviour of solids shows the relationship between its response to an applied load or force. The important mechanical properties are strength, hardness, ductility, stiffness and fracture toughness. Elastic constants are structure–insensitive, while other mechanical properties such as yield, fracture and creep strengths are structure–sensitive.

Table 5.1 ranks various materials with descending order for different mechanical properties.

Table 5.1. Relative Mechanical Properties of Various Materials at Room Temperature, in Decreasing Order. Metals are in their Alloy Form

Strength	*Hardness*	*Toughness*	*Stiffness*	*Strength/Density*
Glass fibers	Diamond	Ductile metals	Diamond	Reinforced plastics
Graphite fibers	Cubic boron nitride	Reinforced plastics	Carbides	Titanium
Kevlar fibers	Carbides	Thermoplastics	Tungsten	Steel
Carbides	Hardened steels	Wood	Steel	Aluminum
Molybdenum	Titanium	Thermosets	Copper	Magnesium
Steels	Cast irons	Ceramics	Titanium	Beryllium
Tantalum	Copper	Glass	Aluminum	Copper
Titanium	Thermosets		Ceramics	Tantalum
Copper	Magnesium		Reinforced plastics	
Reinforced thermosets	Thermoplastics		Wood	
Reinforced Thermoplastics	Tin		Thermosets	
Thermoplastic	Lead		Thermoplastics	
Lead			Rubber	

5.1 THEORETICAL STRENGTH OF CRYSTALLINE MATERIALS

The theoretical stress for slip in a crystalline material may be calculated by assuming a periodic variation of the shear stress opposing movement during the slip process (Figure 5.1).

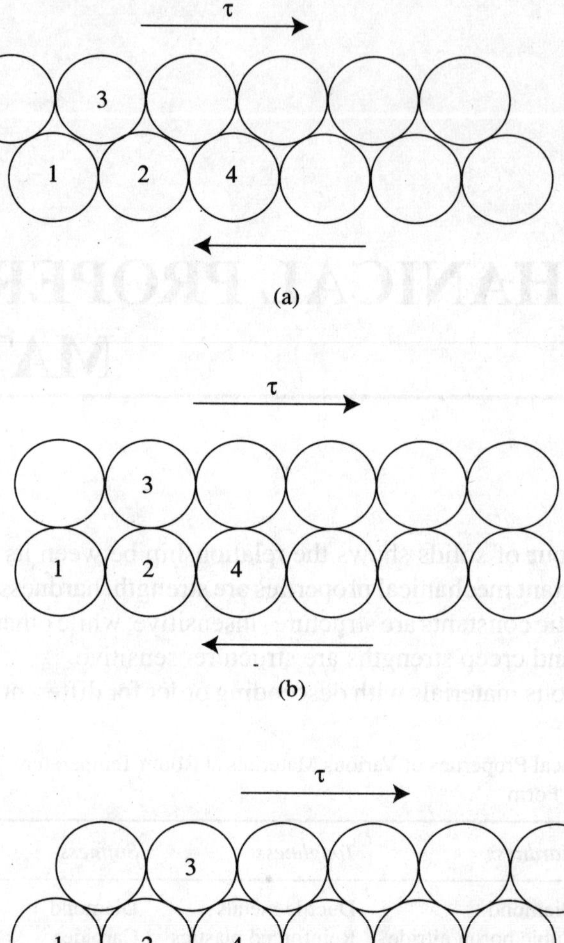

Fig. 5.1. Slip at the atomic level.

The shear stress in all the cases (a), (b), and (c) is zero because the atoms are in a set of normal sites. The variation of shear stress between two normal sites may be expressed as

$$\tau = \tau_o \sin\left(\frac{2\pi x}{d}\right) \tag{5.1}$$

where τ_o = maximum value of stress encountered
x = displacement
d = interatomic distance

Figure 5.2 shows a schematic of the equilibrium positions of two rows of close-packed atoms. For values of x which are small compared to d, that is $x/a << 1$, the shear strain γ is given by

$$\gamma = x/d \tag{5.2}$$

and the Hooke's law may be rewritten as

$$\tau = Gx/d \tag{5.3}$$

where G is the shear modulus.
Equating (5.1) and (5.3), we get

$$\tau_o \sin\left(\frac{2\pi x}{d}\right) = \frac{Gx}{d} \tag{5.4}$$

For values of x which are small compared to d, the approximation may be made such that

$$\sin\left(\frac{2\pi x}{d}\right) = \frac{2\pi x}{d} \tag{5.5}$$

Substituting equation (5.5) into equation (5.4) and rearranging, we obtain

$$\tau_o = \frac{G}{2\pi} \approx \frac{G}{6} \tag{5.6}$$

Based on the values of shear modulus of different crystalline materials, the theoretical shear stress for plastic deformation would be of the order of millions of MPa. This has not been achieved in real crystals except the whiskers. These are grown and have diameters of few micrometers. The whisker crystals also deforms elastically to a much greater extent than ordinary single crystals. The reason for this anomalous behaviour is that there are only few dislocations in them. Usually the only defect present in a whisker is a single screw dislocation parallel to the axis of the whisker.

The theoretical estimation of yield shear stress is $G/30$. Unlike plastic material, in brittle material, the fracture occurs at the elastic limit. In general, the fracture stress of a brittle material is about 0.1Y, where Y is the Young's modulus. The fracture of brittle materials at such a low stress levels, cannot be accounted by the presence of dislocation, which shall be elaborated in later sections related to plastic deformation. The detailed description of fracture and fracture mechanics shall be made in section 5.13.

5.2 PLASTICITY AND ROLE OF DISLOCATION

We have described in Chapter 2 line defects i.e., dislocations. For example edge dislocations (extra half plane) movement in a crystal would give rise to the slip to occur. This would be accomplished by breaking the bond to the right of the half plane (Figure 5.3(a)) and attaching the bottom of it to the half plane so that the atomic picture is as shown in Figure 5.3(b). Since the

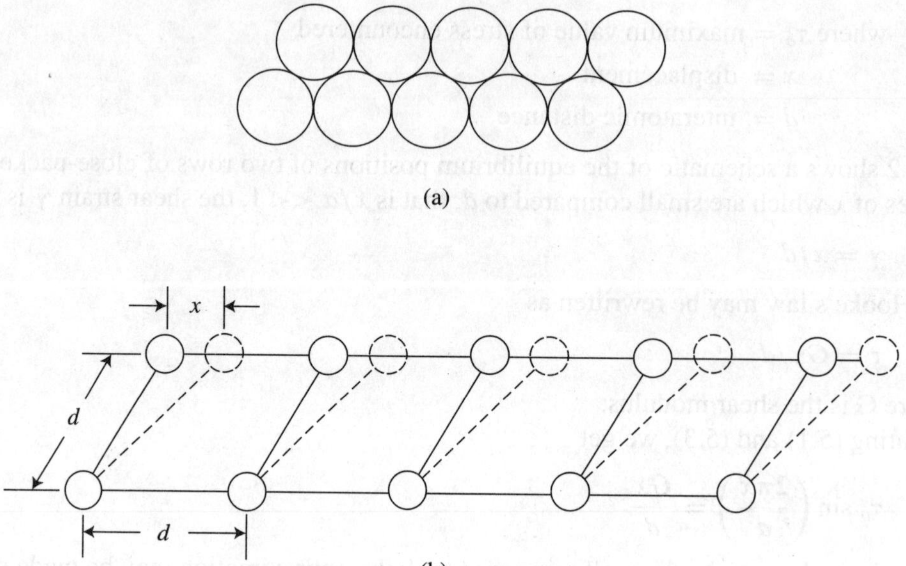

(a)

(b)

Fig. 5.2. Simultaneous displacement of atomic planes.

crystal is only as strong as its weakest link, the crystal will flow plastically as soon as the stress level reaches a value sufficient to cause the dislocations to move. With more and more plastic deformation the density of dislocations in the crystal increases. The simplest type of source for dislocations is known as *Frank-Read Source*.To operate this source the force applied must be sufficient to overcome the restoring force on the dislocation line due to its line tension. Such source continues to generate dislocations until there are so many dislocations present that they interfere with one another. For each dislocation generated, a higher value of stress is needed to produce the next loop because of their mutual interference and thus work hardening or strain hardening occurs.

The shear stress needed to activate a Frank-Read source is given as

$$\tau = \frac{Gb}{d} \tag{5.7}$$

where b = Burgers vector; G = Shear modulus; d = Distance between pinning points

Edge dislocations of opposite sign on the slip plane attract each other and subsequently annihilate each other. On the other hand, dislocations of like sign on the same slip plane will repel each other. However, when unlike dislocations are on closely spaced neighbouring slip planes, complete annihilation cannot occur. In this situation they combine to form a row of point defects.

The force between two parallel screw dislocations is simple to evaluate, since the stress field is radially symmetrical. The force between two screw dislocations is a radial force which depends only on the distance of separation r. The force F is given by the expression

$$F = \text{stress} \times b$$

$$= Gb^2/2\pi r \tag{5.8}$$

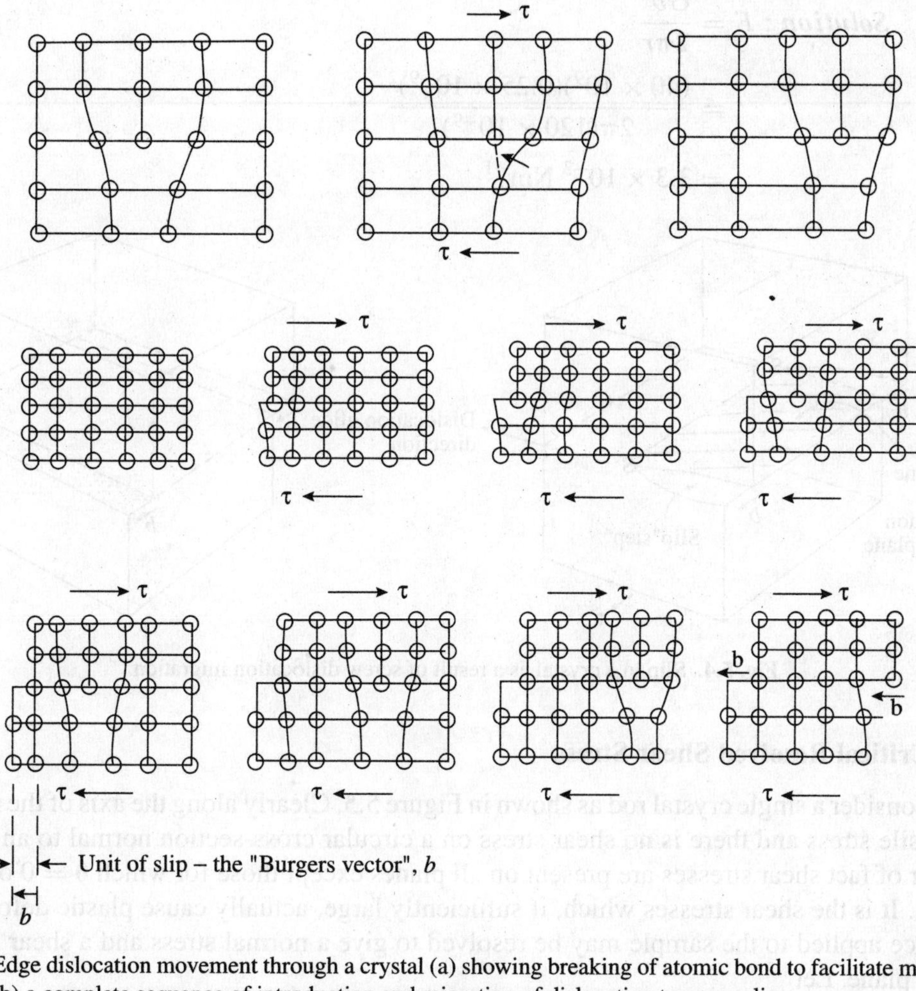

Fig. 5.3. Edge dislocation movement through a crystal (a) showing breaking of atomic bond to facilitate movement; (b) a complete sequence of introduction and migration of dislocation to cause slip.

The force is attractive for dislocations of opposite sign (antiparallel screw) and repulsive for dislocations of the same sign (parallel screws).

Figures 5.3 and 5.4 illustrate the sequence of migration of edge and screw dislocations in a crystal to cause slip to occur. In HCP structures, the slip is more easily maintained in one plane. However in BCC and FCC structures, other slip systems are easily activated. The rotation and direction of the slip plane will easily put other systems in a favourable position. In HCP crystals with $c/a > 1.633$, slip occurs mainly in the basal plane, while the pyramidal and prism planes are preferred in crystals with $c/a < 1.633$.

***Example* 5.1**: Two parallel and straight screw dislocations in copper single crystal are separated by a distance of 120 nm. Calculate the total force on each dislocation. Given Shear modulus = 40 GPa and Burgers Vector is 0.25 nm.

$$\textbf{Solution}: F = \frac{Gb^2}{2\pi r}$$

$$= \frac{(40 \times 10^9)(0.25 \times 10^{-9})^2}{2\pi(120 \times 10^{-9})}$$

$$= 3.3 \times 10^{-3}\ \text{Nm}^{-1}$$

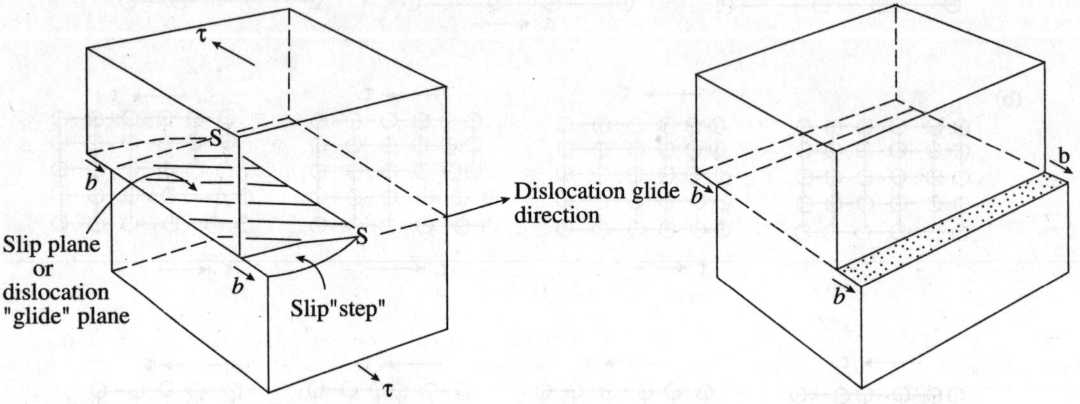

Fig. 5.4. Slip in a crystal as a result of screw dislocation migration.

5.2.1 Critical Resolved Shear Stress

Let us consider a single crystal rod as shown in Figure 5.5. Clearly along the axis of the rod there is a tensile stress and there is no shear stress on a circular cross-section normal to an axis. As a matter of fact shear stresses are present on all planes except those for which $\theta = 0$ or $\theta = 90$ degrees. It is the shear stresses which, if sufficiently large, actually cause plastic deformation. The force applied to the sample may be resolved to give a normal stress and a shear stress on the slip plane. Let

$A_o =$ Area of specimen cross section

$P =$ Applied tensile load

If the normal to the slip plane makes an angle θ with the axis of the specimen, and the slip direction makes an angle ϕ with the specimen axis, then the area of the slip plane

$$A = \frac{A_o}{\cos\theta} \tag{5.9}$$

The stress on this plane is

$$S = P/A \tag{5.10}$$

and the shear stress resolved in the slip direction is

$$\tau = \frac{P}{A}\cos\phi = \frac{P}{A_o}\cos\theta\cos\phi \tag{5.11}$$

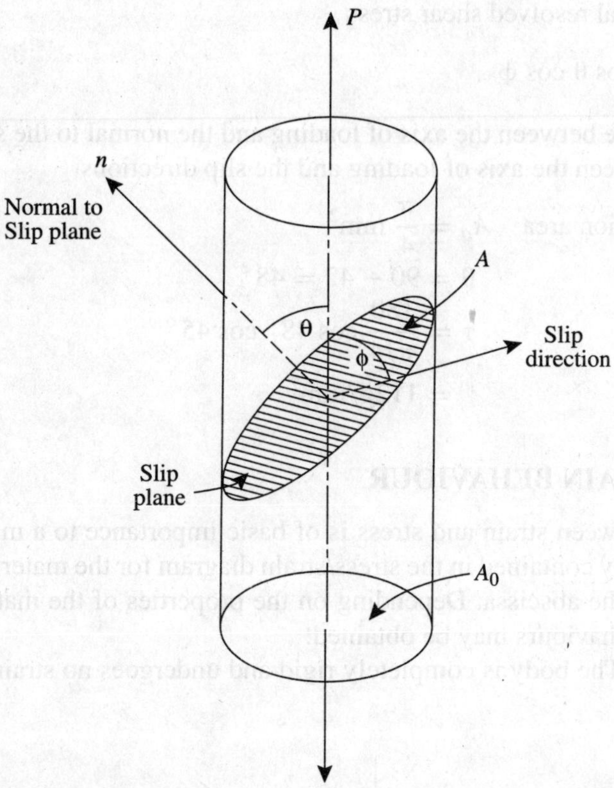

Fig. 5.5. Schematics of critical resolved shear stress in a single crystal specimen.

From equation (5.11), the maximum value of τ occurs when both θ and φ are 45 ° and is equal to

$$\tau_{max} = \frac{P}{2A_o} \tag{5.12}$$

Since the normal stress on the cross section is

$$\sigma = \frac{P}{A_o} \tag{5.13}$$

then

$$\tau_{max} = \frac{\sigma}{2} \tag{5.14}$$

The critical resolved shear stress of a single crystal of a given material is, therefore fixed. At this stress the plastic flow would begin, i.e., the yield stress of the material. The actual shape of the stress-strain curves for single crystal depends on the number of slip systems in the crystal.

***Example* 5.2**: A single crystal of zinc is 1 mm in diameter. The angle between the slip plane and the wire axis is 42 ° and the angle between the operative slip direction and the wire axis is 45 °. Yielding begins at a tensile load of 186 g. Calculate the critical resolved shear stress.

Solution: The critical resolved shear stress

$$\tau = \frac{P}{A_o} \cos\theta \cos\phi$$

where θ = Angle between the axis of loading and the normal to the slip plane
ϕ = Angle between the axis of loading and the slip directions

$$\text{Cross section area} \quad A_o = \frac{\pi}{4}\ \text{mm}^2$$

$$\theta = 90 - 42 = 48^\circ$$

$$\tau = \frac{180}{\pi/4} \cos 48^\circ \cos 45^\circ$$

$$= 112\text{g/mm}^2$$

5.3 STRESS-STRAIN BEHAVIOUR

The relationship between strain and stress is of basic importance to a materials engineer. The information is usually contained in the stress/strain diagram for the material, where stress is the ordinate and strain the abscissa. Depending on the properties of the material being tested the following type of behaviours may be obtained:

Rigid behaviour: The body is completely rigid and undergoes no strain under applied stress (Figure 5.6(a)).

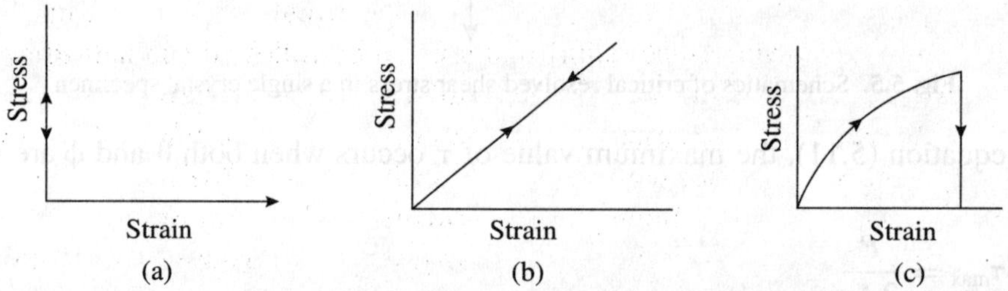

Fig. 5.6. Stress-strain diagrams for pure (a) rigid, (b) elastic (c) plastic materials.

Elastic behaviour: Here the body undergoes deformation under the action of the applied stress, but once the stress is removed the deformation disappears and the body reverts to its original shape and dimension (Figure 5.6(b)).

Plastic behaviour: In this type of behaviour, the body undergoes deformation or strain under the action of an applied stress and the strain is permanent and persists after the removal of the applied stress (Figure 5.6(c)).

In all cases, failure will occur if the applied stress is progressively increased. The behaviour of actual material consists of a combination of these individual modes of deformation (Figure 5.7).

Figure 5.8 illustrates a typical stress-strain plot for a ductile metal under tension. The sample is a single crystal and the illustrations show what is happening along the curve. We shall now highlight various properties obtained from the tensile test.

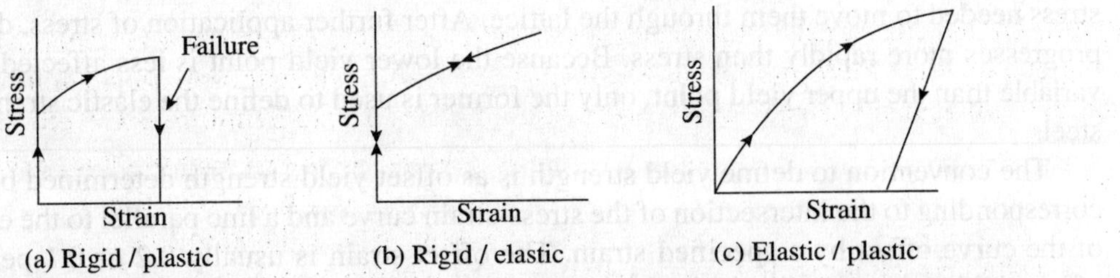

Fig. 5.7. Stress-strain diagrams for (a) rigid/plastic, (b) rigid-elastic and (c) elastic-plastic materials.

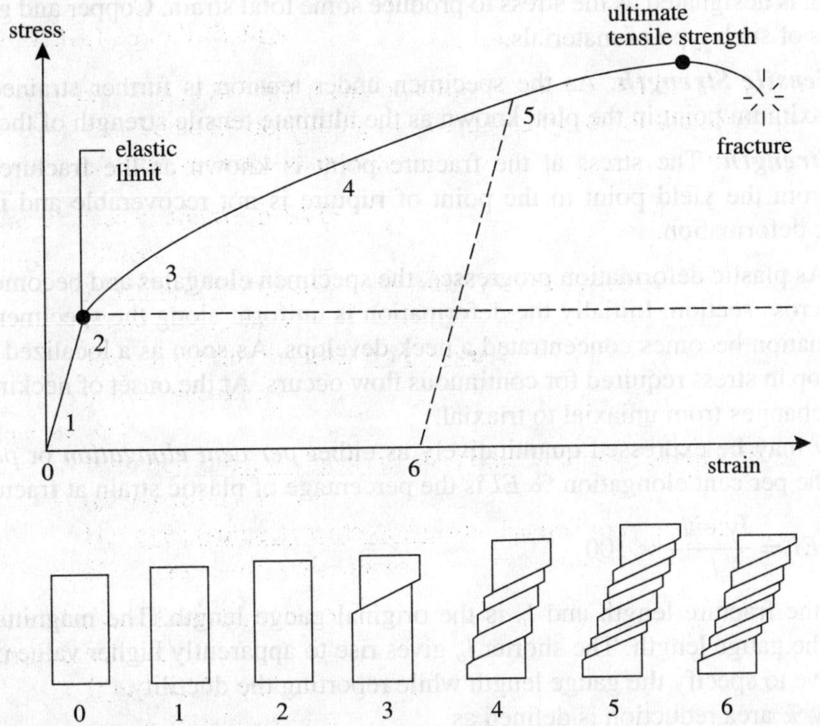

Fig. 5.8. A typical stress-strain curve for a metal. The lower part of the Figure shows slip producing plastic deformation at various points along the curve.

Elastic limit: Up to point 2 in the figure there is straight line relationship where the Hooke's law applies. The slope of this initial straight line portion the curve is a measure of the stiffness of the material. The ratio of stress to strain is known as the modulus of elasticity in tension or *Young's modulus*. Elastic limit is the greatest stress a material is capable of withstanding without producing a permanent set of deformation upon release of that stress. For ductile materials the elastic limit is close to the proportional limit and may be considered as having the same value.
Yield Strength: At point 3, the yield point or yield strength, the onset of slip i.e., plastic deformation occurs. In mild steel there are two yield strengths, the upper yield strength corresponds to the stress necessary to free dislocation and the lower yield strength relates to the

stress needed to move them through the lattice. After further application of stress, deformation progresses more rapidly than stress. Because the lower yield point is less affected by the test variable than the upper yield point, only the former is used to define the elastic strength in mild steel.

The convention to define yield strength is as offset yield strength determined by the stress corresponding to the intersection of the stress-strain curve and a line parallel to the elastic point of the curve offset by a specified strain. The offset strain is usually 0.2 or 0.1 per cent. The offset yield strength is also termed as *proof stress*. This property obtained by an offset method is very useful to designers. In cases, where there is no linear portion in the stress-strain plot, yield strength is designated as the stress to produce some total strain. Copper and gray cast iron are examples of such type of materials.

Ultimate Tensile Strength: As the specimen under tension is further strained, the stress reaches a maximum point in the plot, known as the ultimate tensile strength of the material.

Fracture Strength: The stress at the fracture point is known as the fracture stress. The elongation from the yield point to the point of rupture is not recoverable and is a measure of the plastic deformation.

Ductility: As plastic deformation progresses, the specimen elongates and becomes smaller in diameter or cross section. Initially the deformation is uniform along the specimen length, but when deformation becomes concentrated a neck develops. As soon as a localized neck begins to form, a drop in stress required for continuous flow occurs. At the onset of necking, the stress distribution changes from uniaxial to triaxial.

Ductility may be expressed quantitatively as either *per cent elongation* or *per cent area reduction*. The per cent elongation $\% El$ is the percentage of plastic strain at fracture or

$$\% El = \frac{l_f - l_o}{l_o} \times 100 \tag{5.15}$$

where l_f is the fracture length and l_o is the original gauge length. The magnitude of $\% El$ depends on the gauge length. The shorter l_o gives rise to apparently higher value of $\% El$. So, it is imperative to specify the gauge length while reporting the ductility.

Similarly % area reduction is defined as

$$\% AR = \frac{A_o - A_f}{A_o} \times 100 \tag{5.16}$$

where A_o is the original cross sectional area and A_f the cross sectional area at fracture point. Both the values i.e., $\% El$ and $\% AR$ in a material are not the same. Brittle materials are generally classified, when the fracture strain is less than 5%.

Deformation in a polycrystalline material is more complex because such material cannot behave like a single crystal. Slip does occur but in ways that allow the grains to remain in contact. Here a number of families of slip plane may be operative. For polycrystalline materials one has to prepare a standard test piece from a round bar or a flat. Figure 5.9 shows the standard size test pieces, while Table 5.2 shows the standard dimensions that can be used. Test pieces

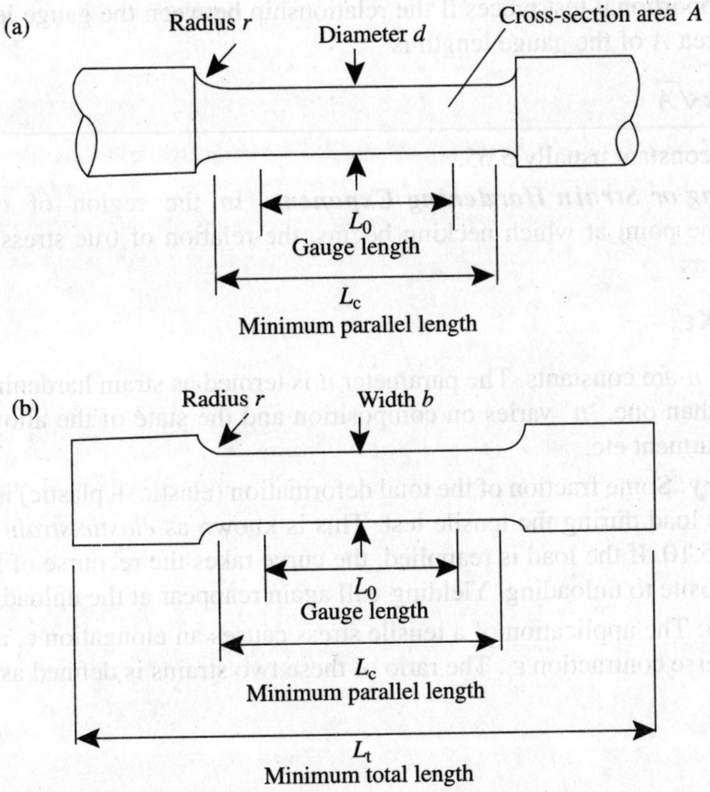

Fig. 5.9. Standard test pieces for tensile testing (a) round (b) flat.

Table 5.2. Dimensions of Standard Test Pieces

Flat Test Pieces

b mm	$L_o$mm	$L_c$mm	$L_f$mm
20	80	120	140
12.5	50	75	87.5
Round Test Pieces (Proportional)			
d mm	A mm^2	$L_o$mm	$L_c$mm
20	314.2	100	110
10	78.5	50	55
5	19.6	25	28

Note: K= 5.85

Reproduced with permission from W. Bolton, Engineering Materials Technology, 3rd Ed., Butterworth-Heinemann, Oxford, 1998, p. 46.

are said to be proportional test pieces if the relationship between the gauge length L_o and the cross sectional area A of the gauge length is

$$L_o = k\sqrt{A} \tag{5.17}$$

where k is a constant usually 5.65.

Work Hardening or Strain Hardening Exponent: In the region of onset of plastic deformation to the point at which necking begins, the relation of true stress-true strain may be approximated as

$$\sigma_T = K\varepsilon_T^n \tag{5.18}$$

where K and n are constants. The parameter n is termed as strain hardening exponent and has a value less than one. 'n' varies on composition and the state of the alloy i.e., prior cold working, heat treatment etc.

Elastic Recovery: Some fraction of the total deformation (elastic + plastic) is recovered after the release of the load during the tensile test. This is known as *elastic strain recovery* and is shown in Figure 5.10. If the load is reapplied, the curve takes the recourse of linear portion in the direction opposite to unloading. Yielding will again reappear at the unloading stress level.

Poisson's Ratio: The application of a tensile stress causes an elongation ε_l along the tensile axis and a transverse contraction ε_t. The ratio of these two strains is defined as Poisson's ratio

$$v = \frac{\varepsilon_l}{\varepsilon_t} \tag{5.19}$$

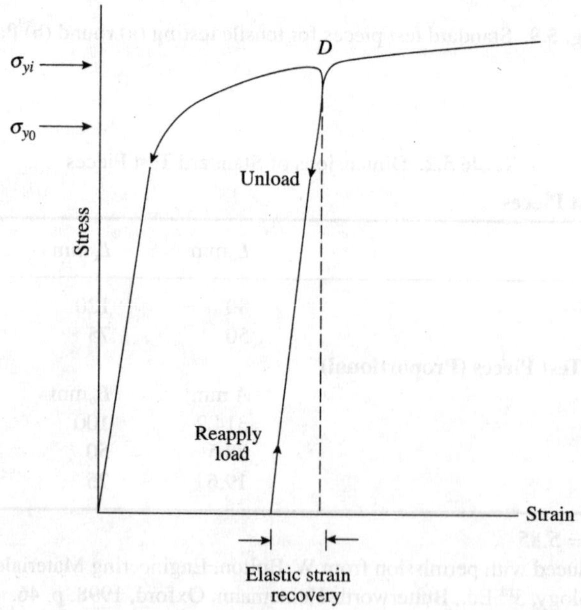

Fig. 5.10. Schematic stress-strain curve plot showing elastic strain recovery and strain hardening (σ_{yo} = initial yield strength, σyi = yield strength after releasing the load at point D and then reloading).

The Poisson's ratio for metals is around 0.3, for polymers and rubber it is between 0.4 and 0.5, and for ionic solids it around 0.2.

Resilience: It is the capacity of the material to absorb energy, when it is deformed elastically. ***Modulus of resilience*** **(U_r)** is the strain energy per unit volume required to stress a material from an unloaded state upto the point of yielding. It is given by the expression after calculating the area in the stress-strain plot as

$$U_r = \frac{\sigma_y^2}{2E} \tag{5.20}$$

where σ_y is the yield stress and E is the Young's Modulus. Its unit is joules per cubic meter.

5.3.1 True Stress/True Strain

The plot of stress-strain shown in Figure 5.8 is also known as engineering stress-strain curve, since the stress calculated is based on the original cross sectional area of the specimen before any deformation and, therefore, does not take into account the decrease in the area at the neck. It is more scientific to represent a true stress-true strain relationship. *True stress* σ_T is defined as the load per unit instantaneous cross sectional area A_i over which deformation takes place, i.e.,

$$\sigma_T = \frac{F}{A_i} \tag{5.21}$$

True strain ε_T is defined by

$$\varepsilon_T = \ln\frac{l_i}{l_o} \tag{5.22}$$

Keeping in view the volume constancy factor during deformation, i.e.,

$$A_i l_i = A_o l_o \tag{5.23}$$

So the relation reduces to

$$\sigma_T = \sigma(l + \varepsilon) \tag{5.24}$$

$$\varepsilon_T = ln(l + \varepsilon) \tag{5.25}$$

Equation (5.21) and (5.22) are valid only to the onset of necking. Beyond this point true stress and strain is computed from actual load, cross sectional area and gauge length measurement. As the actual stress components are complex during necking, it is noticed that a corrected stress may be used, which is lower than the true stress computed. Figure 5.11 shows a typical corrected True Stress/True strain plot.

Temperature and strain rates have opposite effects on stress-strain plots. Increasing the strain rate (or decreasing the temperature) leads to higher stress levels, but lower values of strain. Figure 5.12 illustrates the above features.

Example 5.3: A structural steel specimen of diameter 1.25 cm and gauge length of 5 cm was used in a standard tension test. The load at the lower yield point was found to be 3700 kg. The maximum and fracture loads were 6450 and 4500 kg, respectively and the gauge length at fracture was 6.5 cm. The diameter of the fracture cross-section was found to be 1 cm. Total

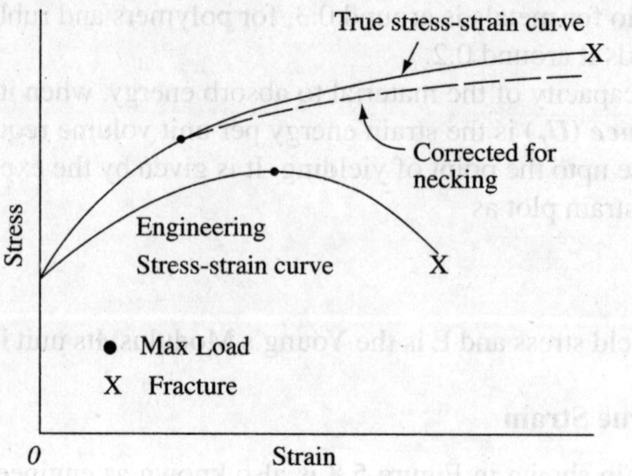

Fig. 5.11. Comparison of engineering and true stress-strain curves. Also included is the corrected true stress-true strain curve.

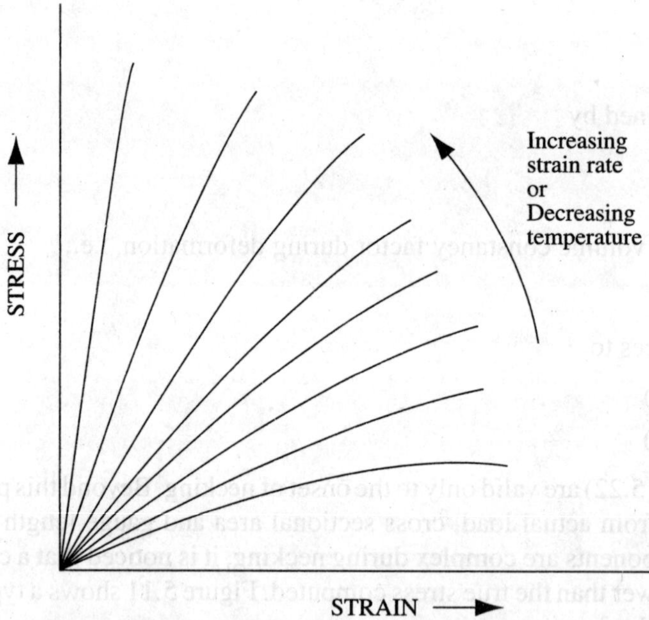

Fig. 5.12. Effect of strain rate and temperature on stress-strain curves.

elongation at a load of 1800 kg was 0.034 mm. A plot of the stress-strain diagram showed that the initial straight-line part of the diagram passed through the point of zero stress and strain. Determine: (a) the lower yield point, (b) modulus of elasticity, (c) the modulus of elastic resilience, (d) ultimate stress, (e) fracture stress, (f) the percentage elongation, (g) the percentage reduction in area and (h) the static toughness value.

Solution: The original cross section area of specimen

$$A_o = \frac{\pi(1.25)^2}{4} = 1.23\,\text{cm}^2 = 123\text{ mm}^2$$

(a) Then, the lower yield stress becomes

$$S_{yp} = \frac{P_y}{A_o} = \frac{3700 \times 9.81\,\text{N}}{123\,\text{mm}^2} = 295.1\,\text{MPa} \quad (1\text{N/mm}^2 = 1\text{MPa})$$

(b) The modulus of elasticity O is determined as

$$E = \frac{S}{\varepsilon} = \frac{PL_o}{A_o\, e} = \frac{1800 \times 9.81 \times 50}{123 \times 0.034} \frac{\text{N}}{\text{mm}^2} \frac{\text{mm}}{\text{mm}}$$

$$= 211.1\text{ GPa}$$

(c) From Eq. (5.20) the modulus of resilience is:

$$U_y = \frac{S_y^2}{2E} = \frac{(295.1 \times 10^6)^2 (\text{N/m}^2)^2}{2 \times 211.1 \times 10^3 \times 10^6 (\text{N/m}^2)}$$

$$= \frac{87084 \times 10^{12}}{2 \times 211.1 \times 10^9}$$

$$= 206.3 \times 10^3\text{ N/m}^2$$

(d) The ultimate stress is

$$S_u = P_u/A_o = (6450 \times 9.81)\,\text{N}/123\text{ mm}^2 = 514.4\text{ MPa}$$

(e) The fracture stress becomes

$$S_f = P_f/A_o = 4500 \times 9.81\text{ N}/123\text{ mm}^2 = 358.8\text{ MPa}$$

(f) From Eq. (5.15) the percentage elongation (or fracture strain) is

$$\%El = [(L_f - L_o)/L_o] \times 100 = [(6.5 - 5.0) \times 5.0] \times 100 = 30\%$$

(g) The area at fracture is

$$A_f = \frac{\pi \times 10^2}{4} = 78.5\text{ mm}^2$$

Then from Eq. (5.16) the percentage reduction in area becomes

$$\%AR = \frac{A_o - A_f}{A_o} \times 100 = \frac{123 - 78.5}{123} \times 100 = 36.2\%$$

(h) Static toughness (T_o) can be given by the area of the stress/strain curve upto the fracture point.

$$T_o = S_u \varepsilon_f = 514.4 \times 0.3\text{ MPa}$$

$$= 154 \times 10^6\text{ m.N/m}^3$$

$$= 154\text{ MPa}$$

5.3.2 Strain Ageing

This type of response is related to the yield-point phenomenon, in which the strength of the material is increased and the ductility is decreased by heating at a relatively low temperature after cold working. Figure 5.13 illustrates how after ageing a cold worked low carbon steel at 130 °C, the enhanced yield point is observed. The reappearence of the yield point is due to diffusion of interstitial atoms like carbon and nitrogen during the ageing period to form an atmosphere of such atoms around the dislocations. Technologically it is necessary to remove strain ageing in deep drawing steels, because the reappearance of yield point gives rise to surface markings or 'stretcher strains' on the steel. The conventional method to remove such a defect is to deform the steel to a point by roller levelling or skin pass rolling and draw it immediately before it can age.

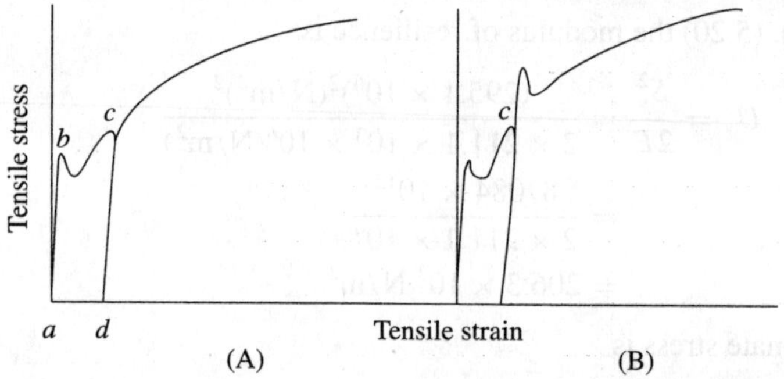

Fig. 5.13. Strain ageing in two cases (A) load remained at point C and the specimen reloaded within a short period (hours); (B) load removed at point C and specimen reloaded after a long period of time (months).

Strain ageing is also associated with the occurrence of the serrations in the stress-strain curve, known as the *Portevin - Le Chatelier effect*. Blue brittleness observed in plain carbon steel in the temperature interval of 500-650 K is due to accelerated strain-ageing.

5.3.3 Effect of Strain Rate

Strain rate increases flow stress. Also, the strain rate dependence of strength increases with increasing temperature. A general relationship between flow stress and strain rates at constant strain and temperature is expressed as

$$\sigma = \left| C(\dot{\varepsilon})^m \right|_{\varepsilon . T} \tag{5.26}$$

The exponent m is defined as *strain – rate sensitivity*. It can be evaluated from the slope of plot of log σ with log $\dot{\varepsilon}$. A more accurate method is to measure the change in flow stress brought about by a change in strain rate $\dot{\varepsilon}$ at constant strain e and temperature T (Figure 5.14). Mathematically

$$m = \left(\frac{\partial \ln \sigma}{\partial \ln \dot{\varepsilon}}\right)_{\varepsilon . T} = \frac{\log (\sigma_2/\sigma_1)}{\log\left(\frac{\dot{\varepsilon}_2}{\dot{\varepsilon}_2}\right)} \tag{5.27}$$

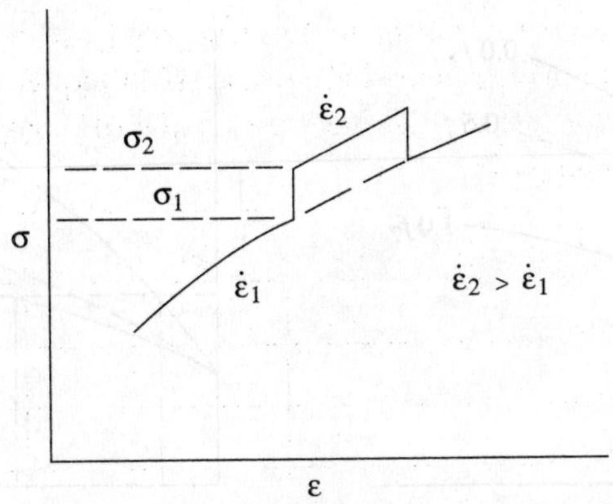

Fig. 5.14. Strain rate change test to determine strain rate sensitivity.

Strain rate sensitivity of metals is low (< 0.1) at room temperature, but it increases with temperature, particularly at temperatures greater than the half of the absolute melting point ($> 0.5T_m$). The significance of strain-rate sensitivity is obvious in case of super plastic materials. Superplasticity is associated with extensive elongation between 100 and 1000 per cent, without any necking. The maximum value of superplasticity, i.e., $m = 1$, is noticed in hot glass, which can be drawn from the melt into long glass fibers. The material is said to behave as Newtonic viscous fluid. Superplastic metals and alloys have a very small grain size or interphase spacing of the order of 1 μm. Two phase structures like eutectic/eutectoid are advantageous in maintaining a fine grain size during testing. It is established that a fine equiaxed grain structure and grain boundary sliding are common prerequisites for superplastic deformation. In polycrystals, triple junction points obstruct the sliding process and give rise to a low m value. In nutshell it can be concluded that superplasticity is controlled by grain boundary diffusion. Superplastic alloys such as Zn-Al eutectoid, Pb-Sn eutectic, Al-Cu eutectic etc have 'm' values approaching unity at elevated temperature. Testing at high temperature and low strain rates promotes superplastic behaviour. The industrial application of superplasticity is in super plastic forming of complex shaped automobile and aerospace structures.

5.3.4 Tensile Properties of Two Phase Alloys

Two phase alloys may have aggregated type of microstructure in which grain size of both the phases are similar or dispersed type of microstructure, in which the other phase is present as particles. The latter type is of greater interest as the strength of the alloy shall be dictated by the size, shape, number and distribution of the second phase and the interfacial energy and interfacial bonding between the phases. In case the contributions of each phase are independent, the strength of the alloy will be the weighted average of the strengths of the individual phases. Estimation of flow stress of two phase alloy may be done on the basis of equal strain or equal stress basis. The average stress and strain values of the alloy can be given as

$$\sigma_{avg} = f_1\sigma_1 + f_2\sigma_2 \tag{5.28}$$

and $$\varepsilon_{avg} = f_1\varepsilon_1 + f_2\varepsilon_2 \tag{5.29}$$

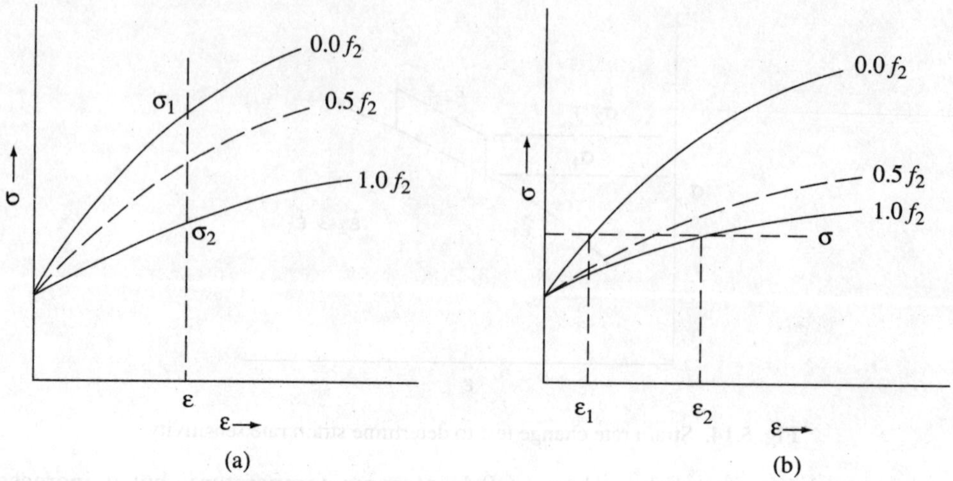

Fig. 5.15. Estimate of flow stress of two phase alloys (a) equal strain; (b) equal stress.

where f_1 and f_2 are volume fractions of phase 1 and phase 2 respectively. Figure 5.15 shows the flow curve for a 0.5 volume fraction alloy on the basis of equal stress, equal strain hypothesis. It must be borne in mind that these hypotheses are mere approximate ones. In the aggregate structure, the second phase block slips so that plastic deformation is not uniform in the matrix. Strain in the matrix is localized and is increased over the average strain of the specimen. The details of the particulate strengthening shall be described in section 5.14.

5.3.5 Tensile Behaviour of Ceramics and Polymers

Elastic moduli (E) of ceramics are generally greater than those of metals. This is related to higher stiffness of ionic/covalent bonds in ceramics. Since ceramics have rather low density (ρ) values, the E/ρ ratio i.e., specific moduli are high. This is the main reason why ceramic or glass fibers are used in composites. Because of the utmost brittleness, it is generally difficult to do tensile test on ceramics. For this bending stress or compressive stress is preferred. For plastically deformable solids (metals/thermoplastics) the tensile and compressive strengths have same values. However for ceramics this is not so. The general relation is

$$\sigma_c = 15\sigma_T \tag{5.30}$$

where σ_C and σ_T are compressive and tensile strengths respectively. The elastic moduli of ceramics are strongly dependent on porosity. Ceramics are porous due to their fabrication and one should be aware of the effect of porosity. This is also true in case of powder metallurgically processed metals and alloys. For relatively low porosity the elastic modulus relation is

$$E = E_o(1 - 1.9p) \tag{5.31}$$

where E_o is the modulus of fully dense metals and p is the volume fraction porosity. In general it can be said that the porosity has an effect on elastic modulus equal to approximately double the volume of pores.

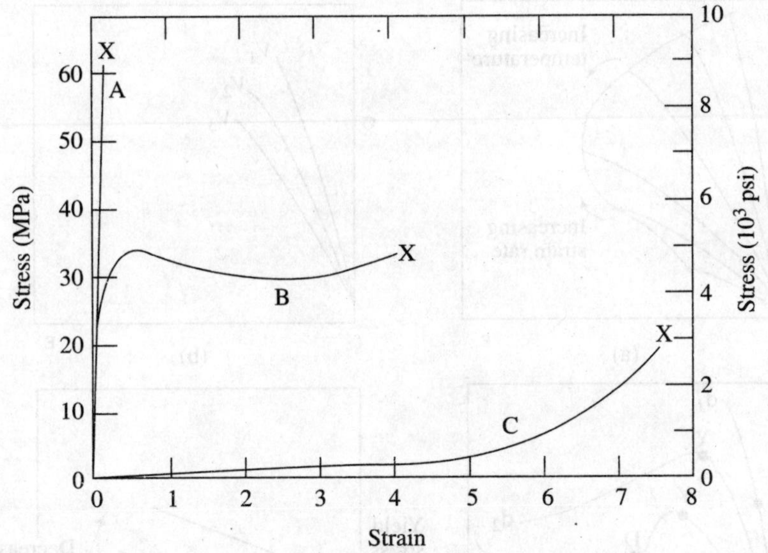

Fig. 5.16. The stress-strain curves for (A) brittle, (B) plastic and (c) highly elastic polymers (Reproduced with permission from W.D. Callister, Jr, Materials Science & Engineering 3rd ed, John,Wiley & Sons, Inc.,Hoboken, NJ, 1994, p. 474).

It has been discussed in Chapter 2, how polymers have varied structures. The wide spectrum of structure affects the mechanical properties. Figure 5.16 shows three types of plots of polymers under tensile loading. Curve A corresponds to a brittle or glassy polymer, while curve C corresponds to totally elastic materials like rubber. Curve B shows an intermediate behaviour of a plastic material similar to that of metal. However this should not be generalized, as the mechanical properties are highly sensitive to the rate of deformation, the temperature and the environment. Semicrystalline polymer shall response according to curve B in the Figure 5.16.

In a semicrystalline polymer, the crystals are embedded in a matrix of amorphous polymer whose properties depend on the ambient temperature relative to its glass transition temperature. Thus, the overall elastic properties of the semicrystalline polymer can be predicted by treating the polymer as a composite material with stiff crystals embedded in a more compliant amorphous matrix. The yield stress of semicrystalline polymers is linearly related to the fraction of crystallinity, and it increases sharply as the thickness of the crystalline region increases. The molecular weight does not seem to have any systematic effect. This shows that only the crystalline regions deform irreversibly.

Various variables in tensile mechanical properties of polymers are highlighted in Figure 5.17. Figure 5.17c has three plots designated as d_1, d_2 and d_3. Curve d_1 shows yielding at point Y, but not as pronounced as in case of metals. In d_2 an unstable neck forms at point Y, followed by inhomogeneous yield response. In d_3, a neck forms at point Y followed by some strain hardening, which stabilizes the neck, giving rise to a drawing out of the necked region.

In case of amorphous polymers the modulus of elasticity shall depend on the temperature. In case the temperature is below glass-transition temperature the polymers shall be stiff but brittle. On the other hand at a temperature above glass-transition temperature, the modulus

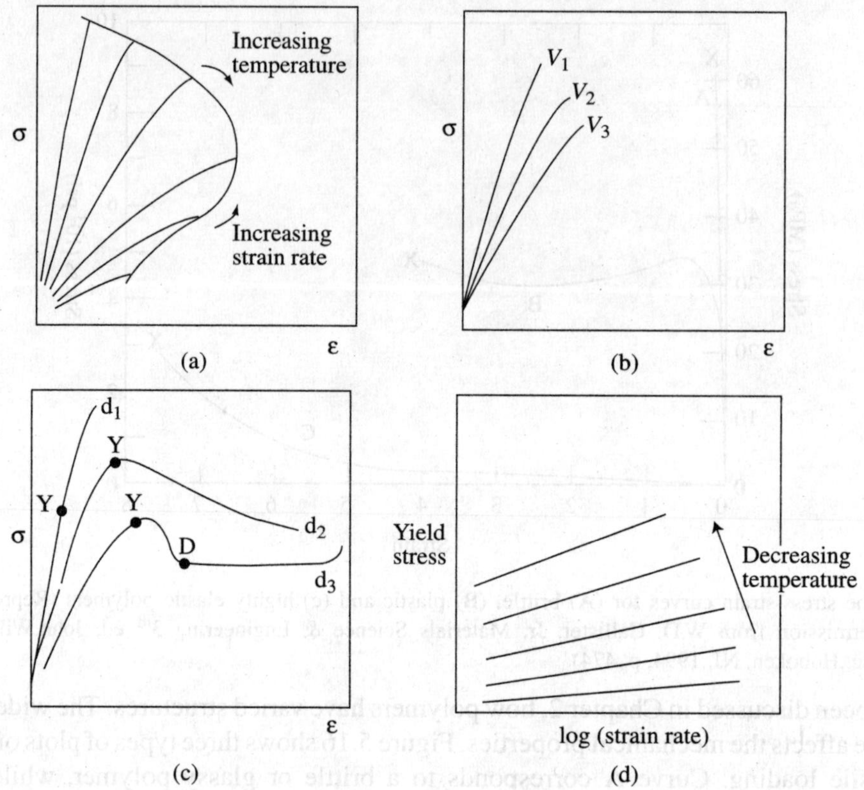

Fig. 5.17. Stress-strain plots of polymers showing
(a) effect of temperature and strain rate
(b) effect of strain rate in brittle polymer ($V_1 > V_2 > V_3$)
(c) different types of ductile behaviour
(d) effect of strain rate as tensile yield stress of an amorphous polymer at different temperatures. (Reproduced with permission from C. Hall, Polymer Materials, An Introduction for Technologists and Scientists, 2nd ed, Wiley -VCH,Weinheim, 1989, p. 79).

shall be low and the rubbery region is observed. However it may be remembered that even at room temperature testing, both types of behaviour are noticed depending on the composition and hence molecular structure of the polymer. Rubber is elastic at room temperature, but if cooled to liquid nitrogen temperature is quite brittle. Most polymers become rubbery at some temperature, the exception being heavily cross-linked thermosets, which decompose before they reach their glass transition temperature.

Many linear polymers draw at room temperature. The draw-strengthening is widely used to produce high strength fibers and films. An example is nylon made by melt spinning and drawing (draw ratio 4). The drawn fiber has a modulus and strength some eight times greater than that of the bulk, unoriented, polymer. Another example is polymer of PS type, but it draws well at higher temperatures. At room temperature during tensile loading it cracks. The cracks are easily visible as white streaks or as general whitening, when cheap injection moulded articles are bent.

The cracks are precursor to fracture. However such polymer could be appreciably strained under compression. The phenomenon is known as shear banding. Within each band a finite shear has taken place. As the number of bands increases, the total over all strain accumulates.

Polymers because of their molecular structure often exhibit an elastic and viscoelastic behaviour. The time dependent elastic behaviour is called *anelasticity*. For metals and ceramics this component of elasticity is rather negligible, but in some polymeric materials this effect may be significant. In amorphous polymers the behaviour is glassy at low temperature, a rubbery at intermediate temperature and a pure viscous at a still higher temperature. The behaviour at intermediate temperature is known as *viscoelasticity*. Figure 5.18c shows a typical strain-time relation for a viscoelastic polymer. This means that viscoelastic behaviour of polymer is dependent on both time and temperature.

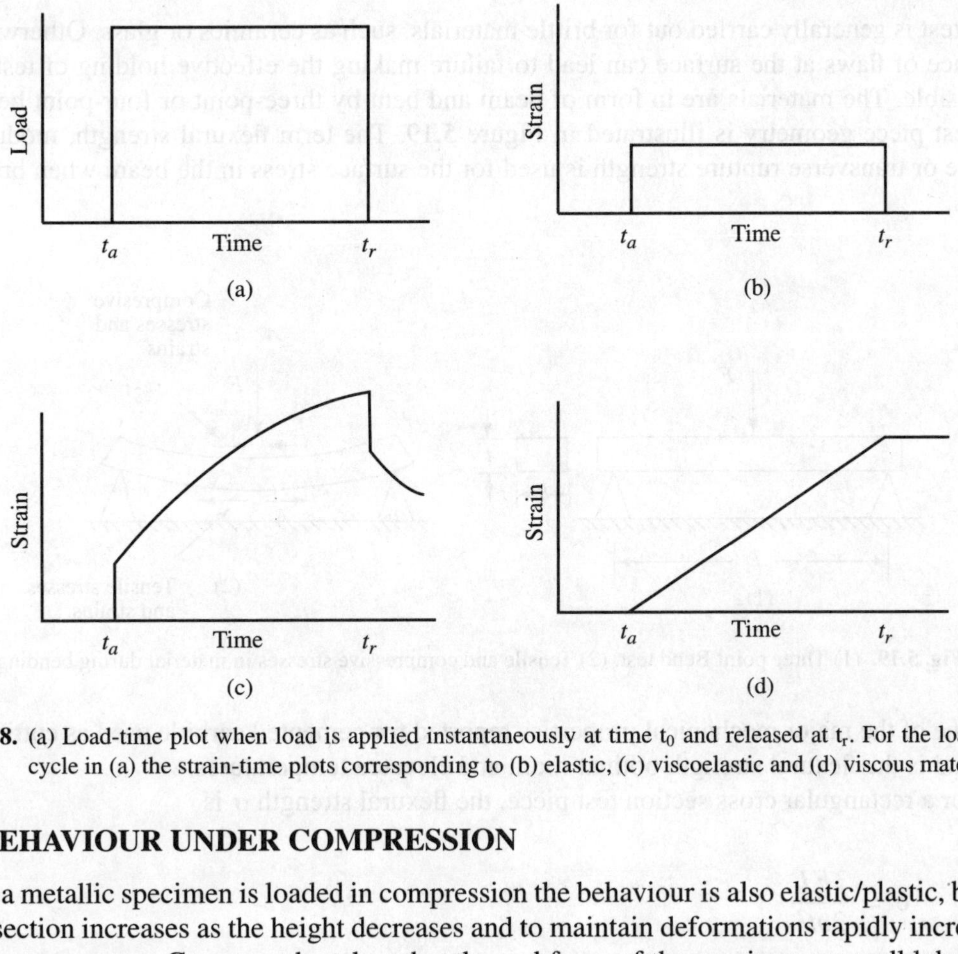

Fig. 5.18. (a) Load-time plot when load is applied instantaneously at time t_0 and released at t_r. For the load-time cycle in (a) the strain-time plots corresponding to (b) elastic, (c) viscoelastic and (d) viscous materials.

5.4 BEHAVIOUR UNDER COMPRESSION

When a metallic specimen is loaded in compression the behaviour is also elastic/plastic, but the cross section increases as the height decreases and to maintain deformations rapidly increasing loads are necessary. Care must be taken that the end faces of the specimen are well lubricated with the help of solid lubricant. This is in order to avoid frictional energy loss, otherwise the tensile and compression test plots of true stress/true strain shall not be identical. The yield point in a metal is usually determined by means of the tensile test and not compression test.

When the results are considered relative to industrial deformation processes (Chapter 8), only the plastic deformation portion of the curve is of interest, giving flow stress/deformation rather than stress-strain diagrams.

The engineering stress-engineering strain curve under compressive loading is concave, whereas under tensile loading it is convex.

Because cracks and flaws tend to close up in compression, brittle materials tend to be much stronger in compression than in tension. For example tensile strength of alumina is 210 MPa with a compressive strength of 2625 MPa.

5.5 BEHAVIOUR UNDER BENDING

Bend test is generally carried out for brittle materials, such as ceramics or glass. Otherwise the presence of flaws at the surface can lead to failure making the effective holding of test piece impossible. The materials are in form of beam and bent by three-point or four-point bending. The test piece geometry is illustrated in Figure 5.19. The term flexural strength, modulus of rupture or transverse rupture strength is used for the surface stress in the beam when breaking occurs.

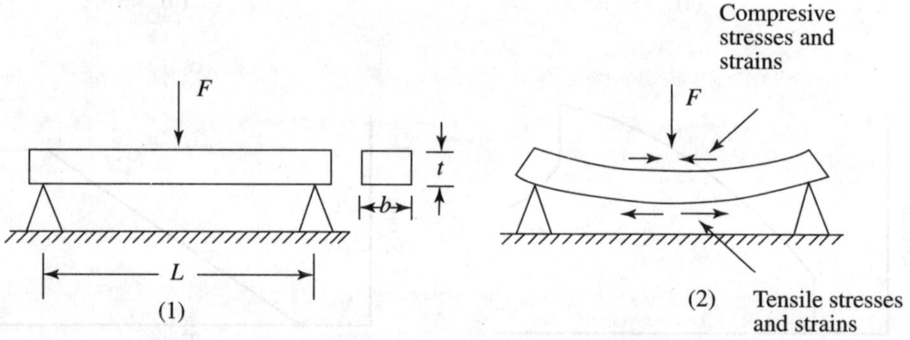

Fig. 5.19. (1) Three point Bend test, (2) Tensile and compressive stresses in material during bending.

One of the major mechanical properties reported for cemented carbide used as cutting tool material is the flexural strength both at room and elevated temperatures.

For a rectangular cross section test piece, the flexural strength σ is

$$\sigma = \frac{3FL}{2bt^2} \tag{5.32}$$

where b = breadth of the section; t = thickness of the specimen, F = load, L = distance between supports to the test piece.

Generally, the flexural strength values are about twice their tensile strength.

5.6 BEHAVIOUR UNDER SHEAR AND TORSION

Figure 5.20 shows the schematic deformation by shear and torsion. The latter is a variation of pure shear. The shear stress τ is expressed as

$$\tau = F/A_o \quad (5.33)$$

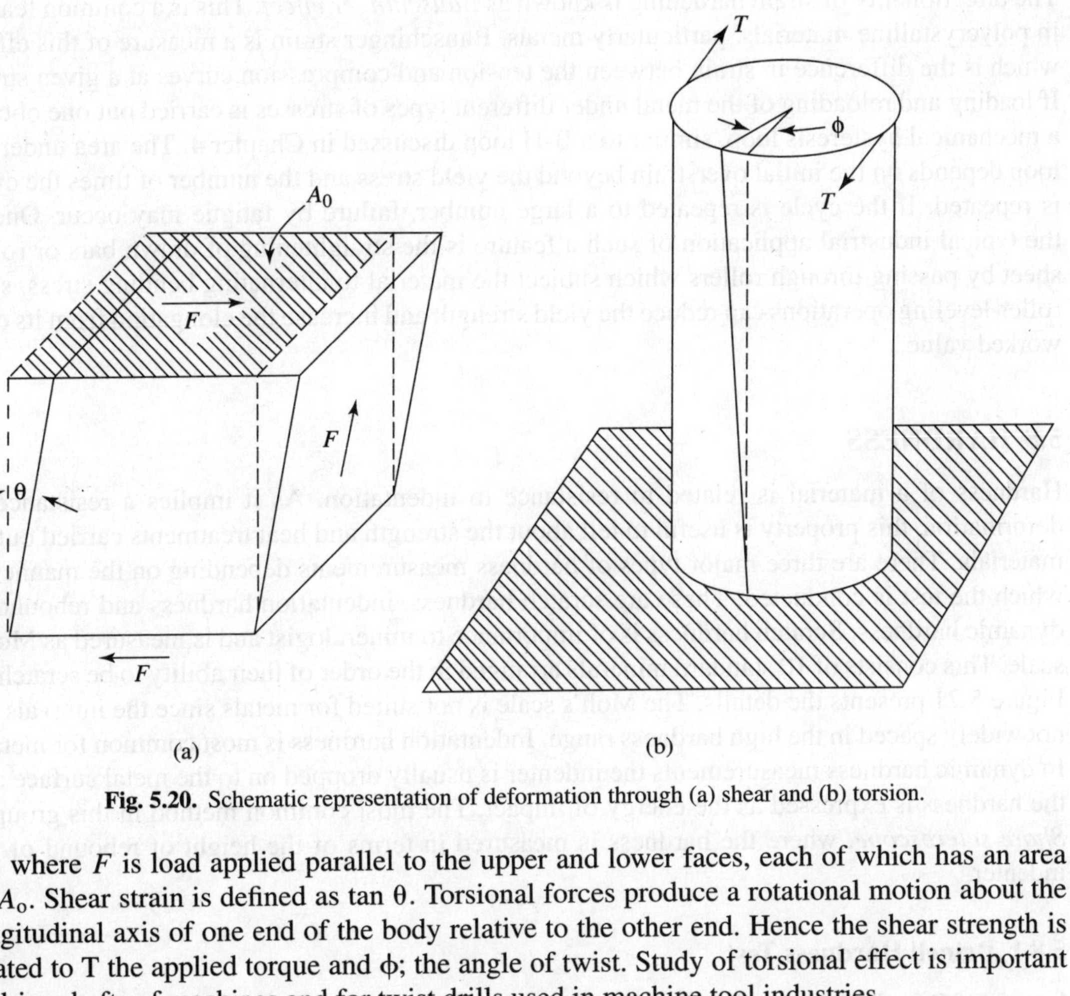

Fig. 5.20. Schematic representation of deformation through (a) shear and (b) torsion.

where F is load applied parallel to the upper and lower faces, each of which has an area of A_o. Shear strain is defined as tan θ. Torsional forces produce a rotational motion about the longitudinal axis of one end of the body relative to the other end. Hence the shear strength is related to T the applied torque and ϕ; the angle of twist. Study of torsional effect is important in drive shafts of machines and for twist drills used in machine tool industries.

The shear modulus of a material G, which is defined as the shear stress divided by the shear strain, is related to the Young's modulus Y, though the Poisson's ratio (ν) as

$$G = Y/2(1 + \nu) \quad (5.34)$$

In engineering applications bolts, rivets and drive keys are loaded under shear in such a manner as to cleve the material in half. A shear test can be performed in a tensile machine with special grips replacing the tensile specimen. Double shear is a standard test for metals. It is not common to use polymer or ceramics as shear loaded members in machines, so their shear properties are seldom reported in hand books. Roughly speaking the shear strength of a metal is about 0.4 times the tensile strength.

5.7 BAUSCHINGER EFFECT

The directionality of strain hardening is known as *Bauschinger effect*. This is a common feature in polycrystalline materials, particularly metals. Bauschinger strain is a measure of this effect, which is the difference in strain between the tension and compression curves at a given stress. If loading and reloading of the metal under different types of stresses is carried out one obtains a mechanical hysteresis loop, similar to a B-H loop discussed in Chapter 4. The area under the loop depends on the initial overstrain beyond the yield stress and the number of times the cycle is repeated. If the cycle is repeated to a large number, failure by fatigue may occur. One of the typical industrial application of such a feature is the straightening of drawn bars or rolled sheet by passing through rollers which subject the material to alternating bending stress. such roller-leveling operations can reduce the yield strength and increase the elongation from its cold worked value.

5.8 HARDNESS

Hardness of a material is related to resistance to indentation. As it implies a resistance to deformation, this property is useful to tell about the strength and heat treatments carried out on materials. There are three major types of hardness measurements depending on the manner in which the test is conducted. These are scratch hardness, indentation hardness and rebound or dynamic hardness. Scratch hardness is of importance to mineralogist and is measured as Moh's scale. This consists of 10 standard minerals arranged in the order of their ability to be scratched. Figure 5.21 presents the details. The Moh's scale is not suited for metals since the intervals are not widely spaced in the high hardness range. Indentation hardness is most common for metals. In dynamic hardness measurements the indenter is usually dropped on to the metal surface and the hardness is expressed as the energy of impact. The most common method in this group is *Shore scleroscope*, where the hardness is measured in terms of the height of rebound of the indenter.

5.8.1 Brinell Hardness Test

In case of Brinell test, a hardened steel ball is pressed for 10 to 15 seconds on to the surface of the material by standard load. After the load and ball have been removed the diameter of the indentation is measured.

Brinell hardness number (BHN) is given as

$$\text{BHN} = \text{Applied Load/Surface area of indentation}$$

$$= \frac{P}{(\pi D/2)\left(D - \sqrt{D^2 - d^2}\right)} \tag{5.35}$$

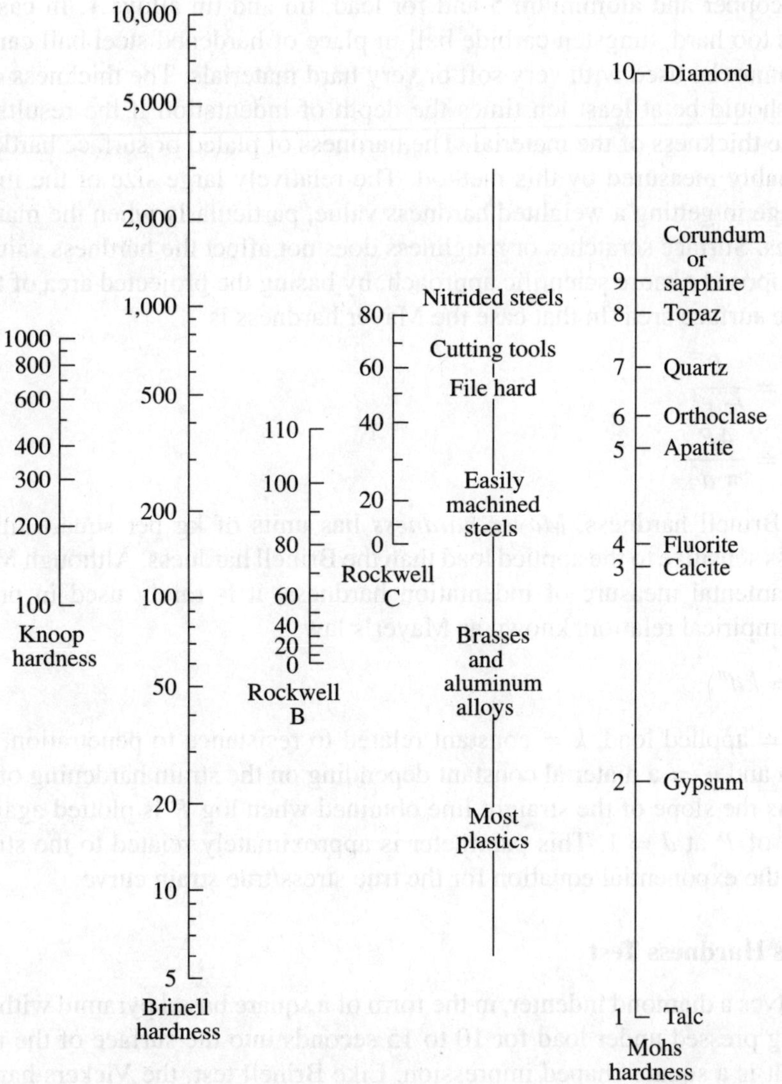

Fig. 5.21. Comparison of several hardness scales (Reproduced with permission from W.D. Callister, Jr, Materials Science & Engineering 6th ed, John Wiley & Sons, Inc., Hoboken, NJ, 2003 p. 139).

where P is the applied load in kg, D = diameter of ball in mm, d = diameter of indentation in mm. The diameter D of the ball used and the magnitude of the applied load (P) are chosen for the British Standard, to give P/D^2 values of 1, 5,10 or 30 with the diameters of the balls being 1, 2, 5 or 10 mm. In principle, the same value of P/D^2 should give the same hardness value, regardless of the diameter of the ball used. It is necessary for the impression to have a diameter of between $0.25D$ and $0.50D$ if accurate values of the hardness are to be obtained. The P/D^2 value is chosen to fit the materials concerned, the harder the material the higher the value used. For steels and cast irons the value used for P/D^2 is 30, for copper and aluminium alloy 10, for copper and aluminium 5 and for lead, tin and tin alloys 1. In case the material to be tested is too hard, tungsten carbide ball in place of hardened steel ball can be used. The Brinell test cannot be used with very soft or very hard materials. The thickness of the material being tested should be at least ten times the depth of indentation if the results are not to be affected by the thickness of the material. The hardness of plated or surface hardened materials cannot be reliably measured by this method. The relatively large size of the impression may be an advantage in getting a weighted hardness value, particularly when the material has local heterogeneities. Surface scratches or roughness does not affect the hardness values.

Mayer proposed a more scientific approach, by basing the projected area of the impression rather than the surface area. In that case the Mayer hardness is

$$P_m = \frac{P}{\pi r^2} = \frac{4P}{\pi d^2} \tag{5.36}$$

Like the Brinell hardness, *Mayer hardness* has units of kg per square millimeter. This hardness is less sensitive to the applied load than the Brinell hardness. Although Mayer hardness is more fundamental measure of indentation hardness, it is rarely used in practice. Mayer proposed an empirical relation, known as Mayer's law

$$P = kd^{n'} \tag{5.37}$$

where P = applied load, k = constant related to resistance to penetration, d = diameter of indentation and n' = a material constant depending on the strain hardening of material. The parameter n' is the slope of the straight line obtained when log P is plotted against log d, and k is the value of P at $d = 1$. This parameter is approximately related to the strain hardening coefficient in the exponential equation for the true stress/true strain curve.

5.8.2 Vickers Hardness Test

This test involves a diamond indenter, in the form of a square based pyramid with an apex angle of 136°, being pressed under load for 10 to 15 seconds into the surface of the material under test. The result is a square shaped impression. Like Brinell test, the Vickers hardness number (HV) is obtained by dividing the applied load by the surface area (mm^2) of the indentation.

$$\text{Area of the indentation} = \frac{d^2}{2\sin\theta/2} = \frac{d^2}{1.854}$$

where d is the mean diagonal length and the apex angle 2 θ is equal to 136°.

$$\text{Therefore } HV = \frac{1.854P}{d^2} \tag{5.38}$$

The biggest advantage in Vickers test is that the hardness value is independent of the magnitude of the load used. Typically a load of 30 kg is used for steels and cast irons, 10 kg for copper alloy, 5 kg for pure copper and aluminium alloys, 2.5 kg for pure aluminium and 1 kg for lead, tin and tin alloys. Vickers hardness testing is very common in research, but in industry Rockwell type tests are generally more preferred, as the latter is the quick one.

5.8.3 Rockwell Hardness Test

This test utilizes the depth of indentation, under constant load, as a measure of hardness. This depth is directly indicated by a pointer on a calibrated scale. The test uses either a diamond cone or a hardened steel ball as the indenter. There are a number of Rockwell scales (Table 5.3),

Table 5.3. Rockwell Hardness Scales

Scale	*Indenter*	*Load kg*	*Typical Applications*
A	Diamond	60	Extremely hard materials, e.g., tool steels
B	Ball 1.588 mm dia.	100	Softer materials, e.g., Cu alloys, Al alloys, mild steel
C	Diamond	150	Hard materials, e.g., steels, hard cast irons, alloy steels
D	Diamond	100	Medium case hardened materials
E	Ball 3.175 mm dia.	100	Soft materials, e.g., Al alloys, Mg alloys, bearing metals
F	Ball 1.588 mm dia.	60	As E, the smaller ball being more appropriate where inhomogeneities exist
G	Ball 1.588 mm dia.	150	Malleable irons, gun metals, bronzes
H	Ball 3.175 mm dia.	60	Soft aluminium, lead, zinc, thermoplastics
K	Ball 3.175 mm dia.	150	Aluminium and magnesium alloys
L	Ball 6.350 mm dia.	60	Soft thermoplastics
M	Ball 6.350 mm dia.	100	Thermoplastics
R	Ball 12.70 mm dia.	60	Very soft thermoplastics

Note: The diameter of the balls arise from standard sizes in inches, 1.588 mm being 1/16 in, 3.175 mm being 1/8 in, 6.350 mm being 1/4 in, and 12.70 mm being 1/2 in.

Reproduced with permission from W. Bolton, Engineering Materials Technology, 3rd Ed., Butterworth – Heinemann, Oxford, 1998, p. 59.

the scale being determined by the indenter and the additional force used. In any reference to the results of a Rockwell test the scale letter must be quoted. For metals the B and C scales are the most common. In case of hard materials like cemented carbides A scale is used. For polymers, Rockwell R scale is generally selected. A variation of the Rockwell test is used for thin sheet, this test being referred to as the Rockwell superficial hardness test. Smaller loads are used and the depth of indentation which is correspondingly smaller is measured with a more sensitive device.

Many times for engineering materials conversion of hardness becomes necessary. Various Handbooks give the conversion data for steels and the ultimate tensile strength equivalents. However, these conversions must be taken as a rough guide, as different hardness scales are meaningful only through experimental verification.

There is an approximate relationship between hardness values and tensile strengths.

$$TS = k \times \text{Hardness} \tag{5.39}$$

where k is a constant for a particular material. Thus for annealed steels the tensile strength in MPa is about 3.54 times the Brinell hardness value and for quenched and tempered steel, this value is 3.24. For copper alloys this factor is 5.6, while for aluminium alloys it is 4.2.

5.8.4 Microhardness Test

Microhardness tests are carried out to measure the hardness of individual constituents of a microstructure. Another use is the measurement of hardness gradient of a very thin carburized steel surface, where Vickers microhardness test is common. The Knoop indenter is a modification over the Vickers, as it has a diamond shaped indentation with the long and short diagonals in the approximate ratio of 7:1 resulting in a state of plane strain in the deformed region. The Knoop hardness number is expressed as

$$\text{KHN} = \frac{P}{A_p} = \frac{P}{L^2C} \tag{5.40}$$

where P is applied load in kg, A_p is unrecovered projected area of indentation, mm^2, L = length of long diagonal, mm and C = constant for each indenter supplied by each manufacturer. The biggest advantages for such a test method are efficient testing of a thin layer and testing of brittle solids, since here the tendency for fracture is proportional to the volume of the stressed material. The surface of the specimen in case of microhardness testing must be carefully prepared similar to metallographic polishing. Work hardening of the surface during polishing can influence the results. Hot microhardness tests are very convenient method for studying creep deformation.

5.9 HOT TORSION TEST

This test is extremely useful for studying hot working conditions, which shall be dealt in detail in Chapter 8. Because the torsion test piece is not subjected to necking, as in tension, or barreling, as in compression, it is possible to carry out the test to a large plastic strains. While necking is not possible in torsional deformation, some torsional plastic instability in the form of a band of intense localized flow can occur. Figure 5.22 shows a typical torque-twist curves for a nickel based super alloy Inconel 600.

5.10 IMPACT TOUGHNESS

It is a measure of the ability of a material to absorb energy upto fracture. Specimen geometry and the type of load application are important in its determination. For dynamic loading (Izod or Charpy test), and when a notch of specified geometry is present, the measurement obtained is known as *notch toughness*. In case of static test, the toughness may be evaluated from the results of a tensile stress-strain curve upto the point of fracture. The unit of toughness is energy per unit volume of material. In qualitative sense, toughness is a combination of both strength and ductility.

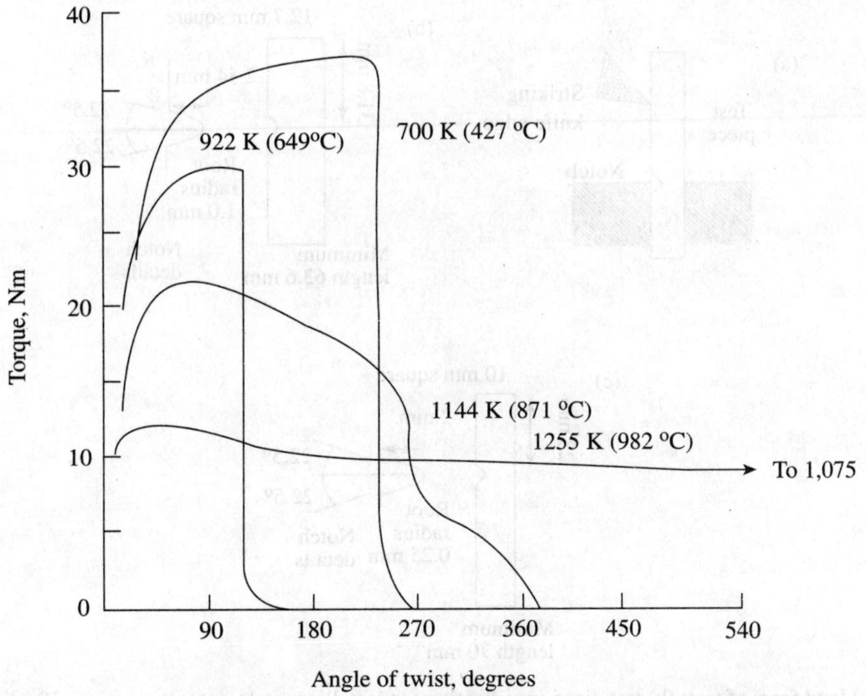

Fig. 5.22. Torque-twist curves at different temperatures for Inconel 600 nickel based superalloy (shear strain rate 2.5 s^{-1}) (After E. Shapiro and G.E.Dieter, Metall. Trans., Vol.1, 1970, p.1712).

Both Izod and Charpy tests involve the same type of measurement but differ in the form of the test pieces. Both involve a pendulum swinging down for a specified height to hit the test piece and fracture it. Figures 5.23 and 5.24 show test pieces for Izod and Charpy testing. It must be noted that there is no direct relationship between the impact energies given for breaking test pieces of different sizes or different notches with the same test method. The impact energy value for a material is affected by factors like temperature, the speed of impact, anisotropy in the microstructure. For plastics, we generally use Izod tests. Machines of varying capacities are available with different hammer sizes. The test is very important to study ductile-brittle transitions in materials at different temperatures.

5.10.1 Ductile-Brittle Transition

With respect to temperature, metals and alloys, particularly having BCC structures, behave differently. At elevated temperature, the material is tough showing ductile fracture, while at low temperature it is generally brittle. With brittle fracture there is little plastic deformation prior to fracture and so little energy is required to break the test piece. There is a sharp or, some time, diffused range of temperature when ductile to brittle transition takes place (Figure 5.25). This temperature is known as *ductile-brittle transition temperature*. From Figure 5.25 it is noticed that FCC metals, particularly with medium or low strength, have high notch toughness values.

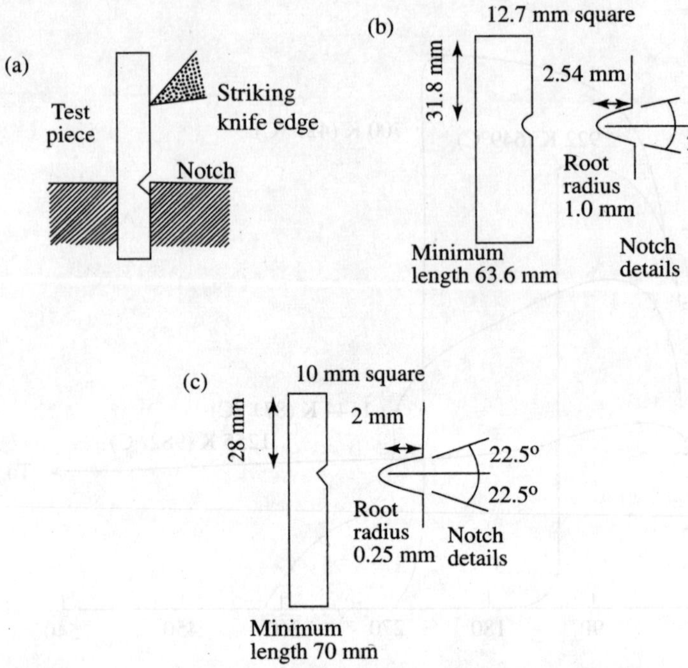

Fig. 5.23. (a) Izod form of test; (b) test piece specification for metallic metals; (c) test piece specification for plastic (Reproduced with permission from W. Bolton, Engineering Materials Technology, 3rd ed, Elsevier ,Oxford, 1998, p. 50).

Low strength BCC materials exhibit ductile-to-brittle transition. The transition is very sharp in very low carbon steels, say 0.01C. Hexagonal structure metals like Be and Zn and some ceramics also show a ductile-to-brittle nature. At low temperature the fracture occurs by cleavage, while at high temperature the fracture is a ductile one. The lower curve in Figure 5.25 corresponds to high strength FCC materials. The low value of notch toughness is evidently interlinked with low ductility due to the high strength of such materials.

The aim of designer is to select such compositions, where this transition temperature is on lower and lower side, so that the structure may perform tough duties. With a brittle metal, the fractured surfaces are crystalline in appearance, while in case of ductile materials, the surfaces are rough and fibrous in appearance. With ductile failure there is a significant reduction in the cross sectional area of the test piece, while for brittle fracture the reverse is true. The attention towards investigation started during World War II, when a number of ships and tankers met catastrophic failure both in rough sea or when anchored at dock. Most of the failure occurred during the winter months. Engineers doubted that the failures might have been because of faulty welding, but this feature was found to be absent.

In metals, the ductile-brittle transition occurs at 0.1 to 0.2 of the absolute melting point T_m, while in ceramics, the transition occurs at about 0.5 to 0.7 T_m. There is further subclassification of this critical temperatures into fracture-appearance transition temperature, ductility transition temperature and nil-ductility temperature.

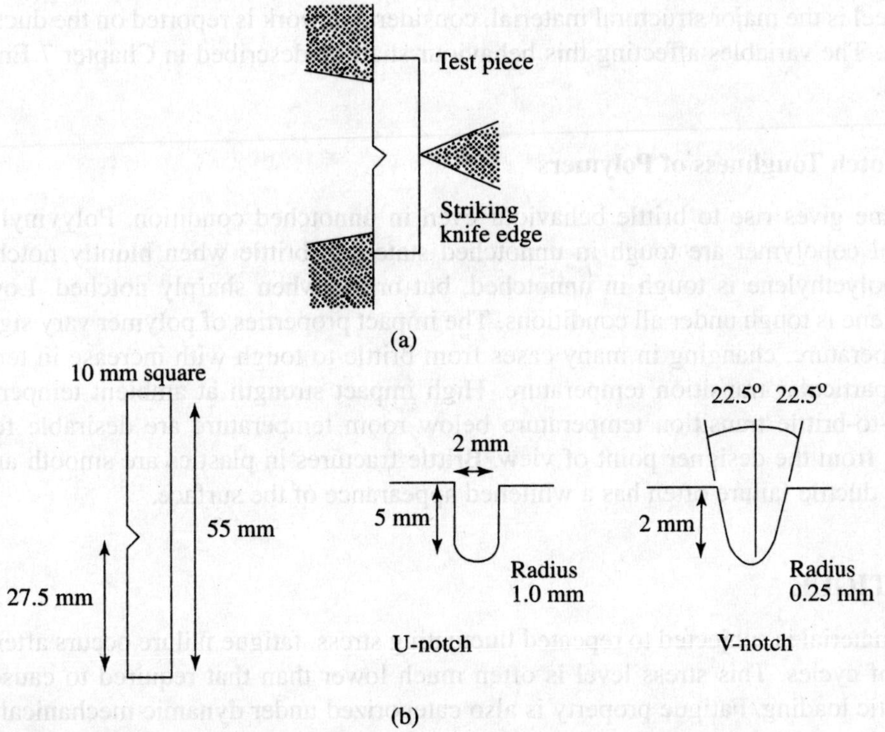

Fig. 5.24. (a) Charpy impact test (b) Geometry for charpy test pieces (Reproduced with permission from W. Bolton, Engineering Materials Technology, 3rd ed, Elsevier, Oxford, 1998, p. 50).

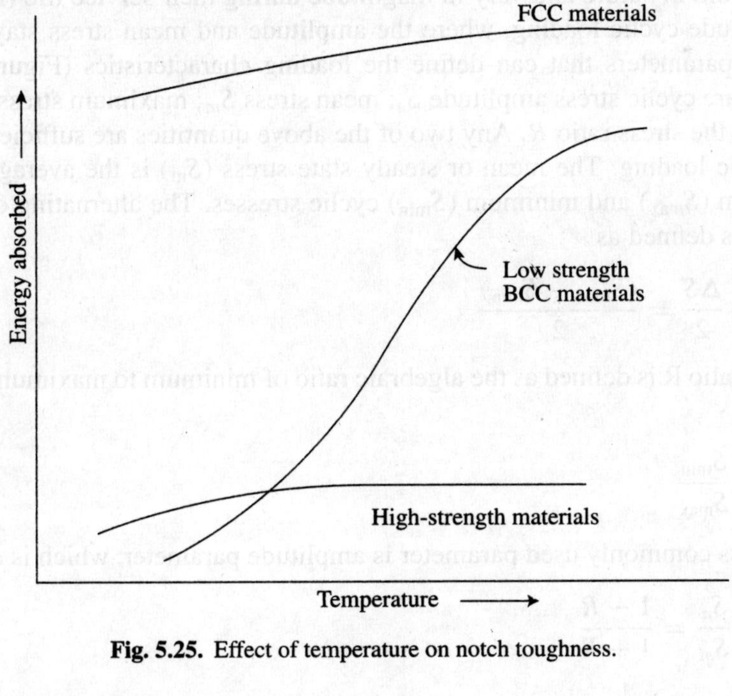

Fig. 5.25. Effect of temperature on notch toughness.

As steel is the major structural material, considerable work is reported on the ductile-brittle transition. The variables affecting this behaviour shall be described in Chapter 7 Engineering Materials.

5.10.2 Notch Toughness of Polymers

Polystyrene gives rise to brittle behaviour even in unnotched condition. Polyvinyl Chloride and acetal copolymer are tough in unnotched state and brittle when bluntly notched. High density polyethylene is tough in unnotched, but brittle when sharply notched. Low density polyethylene is tough under all conditions. The impact properties of polymer vary significantly with temperature, changing in many cases from brittle to tough with increase in temperature at some particular transition temperature. High impact strength at ambient temperature and a ductile-to-brittle transition temperature below room temperature are desirable features in polymers from the designer point of view. Brittle fractures in plastics are smooth and glassy, while the ductile failure often has a whitened appearance of the surface.

5.11 FATIGUE

When a material is subjected to repeated fluctuating stress, fatigue failure occurs after a certain number of cycles. This stress level is often much lower than that required to cause fracture under static loading. Fatigue property is also categorized under dynamic mechanical property. The treatment of fatigue can be considered into engineering factors, metallurgical factors and fine structural factors. Typical forms of cyclic loading that occur in real structural materials are almost random in nature and vary in magnitude during their service life (Figure 5.26a). In constant amplitude cyclic loading, where the amplitude and mean stress stay constant, there are five stress parameters that can define the loading characteristics (Figure 5.26b). These characteristics are cyclic stress amplitude S_a; mean stress S_m; maximum stress S_{max}; minimum stress S_{min}; and the stress ratio R. Any two of the above quantities are sufficient to completely define the cyclic loading. The mean or steady state stress (S_m) is the average algebraic sum of the maximum (S_{max}) and minimum (S_{min}) cyclic stresses. The alternating or variable stress amplitude, S_a, is defined as

$$S_a = \frac{\Delta S}{2} = \frac{(S_{max} - S_{min})}{2} \tag{5.41}$$

The stress ratio R is defined as the algebraic ratio of minimum to maximum cyclic stresses, i.e.,

$$R = \frac{S_{min}}{S_{max}} \tag{5.42}$$

Another less commonly used parameter is amplitude parameter, which is expressed as

$$A = \frac{S_a}{S_m} = \frac{1 - R}{1 + R} \tag{5.43}$$

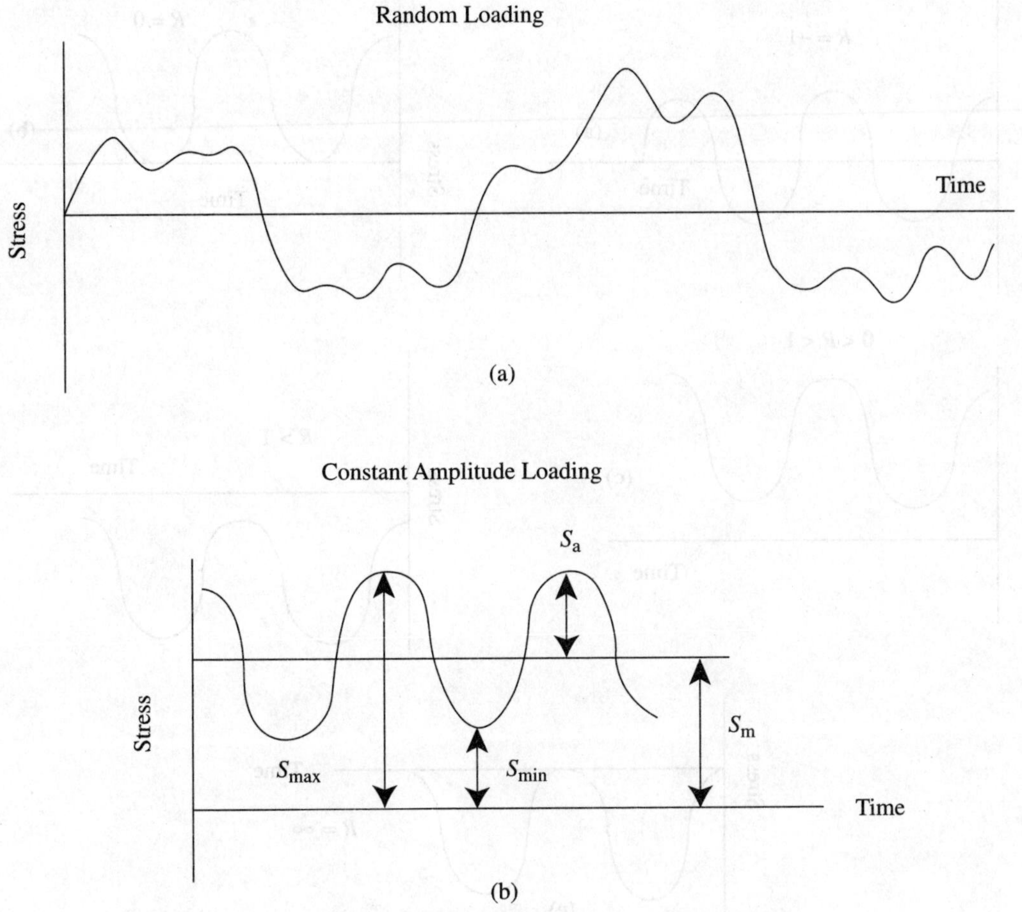

Fig. 5.26. Illustration of (a) random and (b) constant amplitude loading in fatigue.

R is widely used to distinguish different constant amplitude cyclic loading conditions in fatigue analysis. Figure 5.27 shows typical constant amplitude loading cycles. The simplest case of $R = -1$, is a rotating bending test where a constant moment to a rotating cylindrical specimen is provided. Here the maximum and minimum stresses are equal but of opposite sign. In real engineering, the problem may be complex. For example, *R* may be zero in case of the pressurization and depressurization cycle of a pressurized tank. In other situation, R may be less than 1, example being a preloaded bolt subjected to cyclic tensile stresses. Another situation may arise where *R* is greater than 1, example being a plate with a hole that has undergone a sleeve cold expansion or mandrelizing process and subjected to fluctuating load. Another example is a satellite orbiting around the earth, at a certain position in its orbit not exposed to the sun, the induced thermal stress is compressive. The induced compressive stress can be thought of as a compressive sustained load. The mechanical cycling together with the sustaining compressive load will create a load cycling.

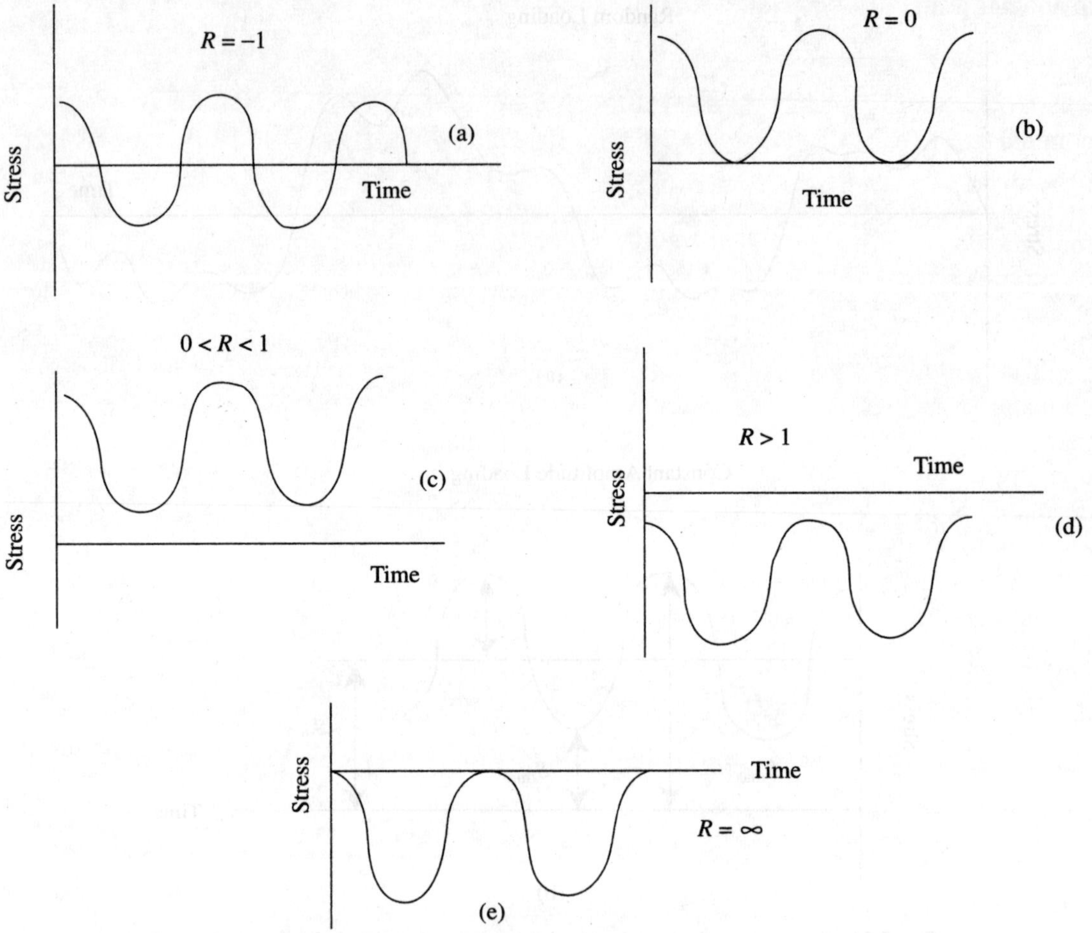

Fig. 5.27. Typical constant amplitude loading cycles.

The most convenient way to express fatigue response of a material is through the S-N curve, where S is stress (better term is stress amplitude S_a) and N is the number of cycles to failure. The stress values are usually nominal stresses. Most determination of the fatigue properties of materials are made in completly reversed bending where the mean stress S_m is zero. Nominal or average stress S is calculated from load or moment or their combination as a matter of convenience and is only equal to σ (stress at a point). For simple axial loading, the stress σ is same everywhere and so is equal to the nominal stress $S(S = P/A$, where A is the cross sectional area). Similarly for bending and notched specimens, nominal stress S must be calculated.

When there is notch in the test piece, the relationship has to be modified and we meet across a term *elastic stress concentration factor* k_t, such that $k_t = \sigma/S$. This means that it is the ratio of the local notch (point) stress σ to the nominal (average) stress S. In other words on S-N plot, the effect of notch is to reduce the stress amplitude corresponding to any given life by the factor k_t. It is observed that the actual test data lie above this estimate. This anomaly gives rise to another term k_f (*fatigue notch factor*) such that

$$k_f = \sigma_{ar} / S_{ar} \tag{5.44}$$

where σ_{ar} is the completely reversed stress for smooth test piece, and S_{ar} completely reversed nominal stress for notched test piece. In case the notch has a large radius at its tip, the k_f is equal to k_t.

The extent to which the stress concentration effect under fatigue conditions approaches that for static conditions is given by the *notch sensitivity factor q*; such that

$$q = \frac{k_f - 1}{k_t - 1} \tag{5.45}$$

Thus q is always less than 1. Notch sensitivity is a very complex factor depending not only upon the material but also upon the grain size, a finer grain size resulting in a higher value of q than a coarse grain size. It also increases with section size and tensile strength. It also depends upon the sensitivity of notch and type of loading. Under some circumstances it is possible to decrease the fatigue life by increasing the tensile strength.

Fatigue behaviour is also classified into two categories (i) high cycle fatigue and (ii) low cycle fatigue. In the former case the stress is elastic, but the material deforms plastically in a highly localized way. At higher stresses, it is very difficult to interpret only in terms of stress. For the low cycle fatigue ($N < 10^4$ or 10^5 cycles) tests are conducted with controlled cycles of elastic plus plastic strain in stead of controlled load or stress cycles.

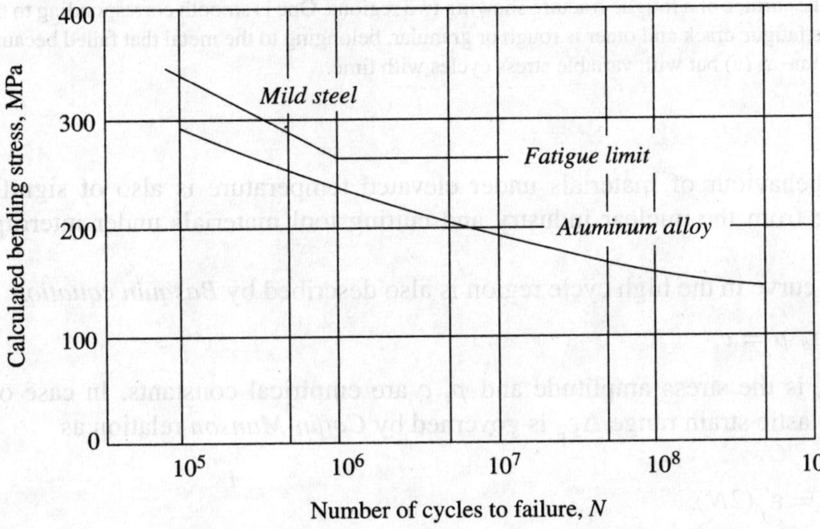

Fig. 5.28. Typical fatigue curves for ferrous and nonferrous materials.

Figure 5.28 shows two types of fatigue curves. In one case i.e., mild steel the curve takes a horizontal position at certain limiting stress. The limiting stress is called *fatigue limit* or *endurance limit* of the material. Nonferrous metals like aluminium, magnesium and copper alloys do not exhibit any true fatigue limit. In such cases, an arbitrary number of cycles, for example 10^8 cycles, is taken to report the *fatigue strength* value. It must be borne in mind that the fatigue has a statistical nature and therefore, tests must be carried out on a large number of test pieces say 10 or more.

There are two distinct regions in the general appearance of a typical fatigue fracture (Figure 5.29). The first is relatively smooth area, though which the fatigue crack has spread slowly. This area usually has concentric marks about the points of origin of the crack which correspond to the positions at which the crack was stationary for some time. The second portion of the fractured surface is typically rough transcrystalline fracture. This portion is responsible for the catastrophic failure. It is observed that the lower the ductility of the material, the less well defined are the striations.

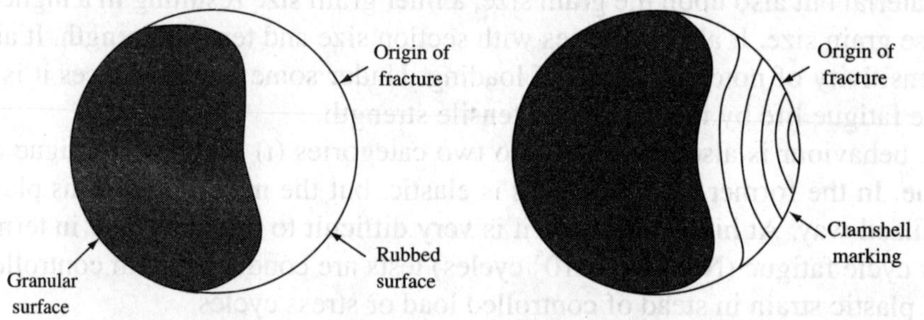

Fig. 5.29. (a) The surface of a fatigue fracture showing two regions. One is smooth corresponding to the slow spread of the fatigue crack and other is rough or granular, belonging to the metal that failed because of overload. (b) same as (a) but with variable stress cycles with time.

Fatigue behaviour of materials under elevated temperature is also of significance. The examples are from the nuclear industry and cutting tool materials under interrupted cutting conditions.

The S-N curve in the high cycle region is also described by *Basquin equation*

$$N\sigma_a\, p = c \tag{5.46}$$

where σ_a is the stress amplitude and p, c are empirical constants. In case of low-cycle fatigue, the plastic strain range $\Delta\varepsilon_p$ is governed by *Coffin-Manson* relation as

$$\frac{\Delta\varepsilon_p}{2} = \varepsilon'_f (2N)^c \tag{5.47}$$

where $\Delta\varepsilon_p/2 =$ plastic strain amplitude, ε'_f is the fatigue ductility coefficient defined by the strain intercept at $2N = 1$. It is equal to true fracture strain for many metals.

$2N =$ Number of strain reversals to failure

$c =$ Fatigue ductility exponent, which varies between –0.5 and –0.7 for many metals. The smaller value of c results in large, values of fatigue life.

Various empirical relations have been proposed that take into account the effect of mean stress on fatigue life. Some of these are:

Goodman's relationship which assumes a linear effect of mean stress between $\sigma_m = 0$ and σ_{UTS}

$$\sigma_a = \sigma_o \left[1 - \frac{\sigma_m}{\sigma_{UTS}}\right] \tag{5.48}$$

Gerber's relationship which assumes a parabolic effect of mean stress between $\sigma_m = 0$ and σ_{UTS}

$$\sigma_a = \sigma_o \left[1 - \left(\frac{\sigma_m}{\sigma_{UTS}}\right)^2\right] \tag{5.49}$$

Soderberg's relationship which assumes a linear effect of mean stress $\sigma_m = 0$ and σ_y

$$\sigma_a = \sigma_o \left[1 - \frac{\sigma_m}{\sigma_y}\right] \tag{5.50}$$

In all these relationship, σ_m is mean stress, σ_a is the fatigue strength in terms of stress amplitude when $\sigma_m \neq 0$, σ_o is the fatigue strength in terms of stress amplitude when $\sigma_m = 0$, σ_{UTS} is the monotonic ultimate tensile strength and σ_y is the monotonic yield stress. Figure 5.30 shows the three relationships schematically. Experimental data by majority fall between Gerber and Goodman lines. Goodman diagram can be considered as a conservative estimate of the mean stress effect. It is worth noting that all the above relationships involve uniaxial stresses. In real life, the situation may not be as simple.

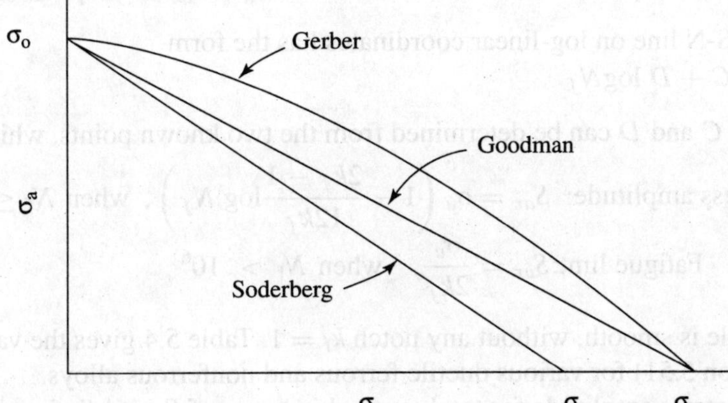

Fig. 5.30. Gerber, Goodman and Soderberg diagrams showing mean stress effect on fatigue life.

5.11.1 Estimation of S-N Curves

Estimates of fatigue limits can be used as part of a procedure for estimating entire S-N curves. The most common method is one described by Collins. Estimated S-N curves are useful in preliminary calculations and gives rough estimation of fatigue resistance. Collins estimated the completely reversed S-N curve for ductile steels and irons with σ_u below about 1200 MPa. Figure 5.31 illustrates this method. Draw a straight line on log-linear coordinates between the ultimate tensile strength (σ_u) at one cycle and half of this value ($\sigma_u/2$) at one million cycles. If a notch is present, the same procedure is used except the stress value at $N_f = 10^6$ cycles is divided by k_f.

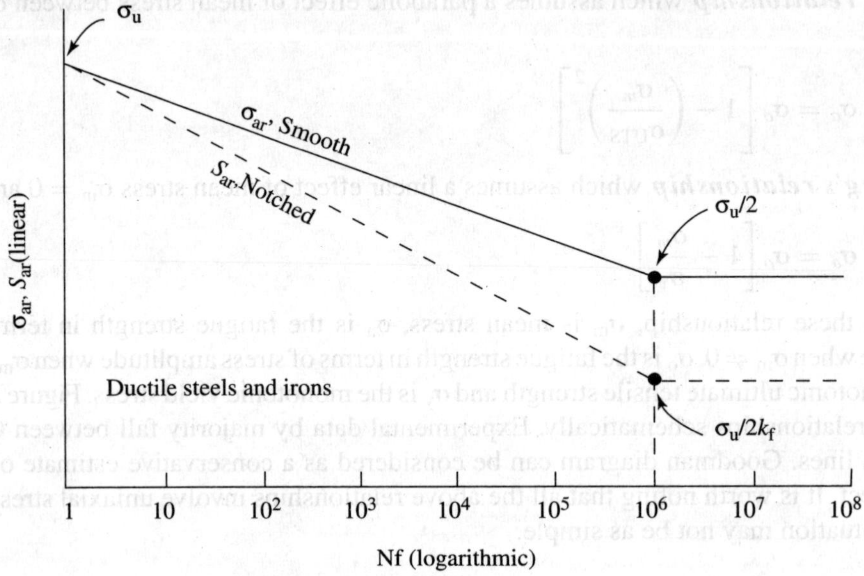

Fig. 5.31. Estimation of completely reversed S-N curves for smooth and notched test pieces according to Collins.

The straight S-N line on log-linear coordinates has the form

$$S_{ar} = C + D \log N_f \tag{5.51}$$

The constant C and D can be determined from the two known points, which leads to

$$\text{Nominal stress amplitude: } S_{ar} = \sigma_u\left(1 - \frac{2k_f - 1}{12k_f}\log N_f\right), \quad \text{when } N_f \leq 10^6 \tag{5.52}$$

$$\text{Fatigue limt:} S_{ar} = \frac{\sigma_u}{2k_f}, \quad \text{when } N_f > 10^6 \tag{5.53}$$

In case the sample is smooth, without any notch $k_f = 1$. Table 5.4 gives the values of constant C and D (Equation 5.51) for various ductile ferrous and nonferrous alloys.

The nominal stress amplitude for total reversal in terms of S_a and S_m is expressed as

$$S_{ar} = \frac{S_a}{1 - \dfrac{k_{fm} S_m}{\sigma_u}} \tag{5.54}$$

where k_{fm} is the ***fatigue notch factor for the mean stress***. For brittle material $k_{fm} = k_f$; while for ductile material $k_{fm} = 1$.

Table 5.4. Constants for Stress-Life Curves for Various Ductile Engineering Metals, from Tests at Zero Mean Stress on Unnotched Axial Specimens

Material	*Yield*	*Ultimate*	$\sigma = C + D \log N_f$	
	σ_o, *MPa*	σ_u, *MPa*	C	D
Steels				
AISI 1015 (normalized)	227	415	545	−69.6
AISI 4142 (Q & T, 450 HB)	1584	1757	1529	−148
AISI 4340 (aircraft quality)	1103	1172	1247	−137
Nonferrous alloys				
2024-T4 Al	303	476	624	−69.9
Ti-6Al-4V (solution treated and aged)	1185	1233	1393	−157

***Example* 5.4**: A ground surface round bar of AISI 4340 steel (ultimate tensile strength $\sigma_u = 1172$ MPa) has a circumferential groove. Estimate the completely reversed S-N curve and also determine the life for cyclic loading under bending at a nominal stress amplitude $S_a = 250$ MPa with mean of $S_m = 300$ MPa. Assume $k_f = 2.29$.

Solution: Recalling equations (5.52) and (5.53) we have

$$S_{ar} = \sigma_u \left(1 - \frac{2k_f - 1}{12k_f} \log N_f\right), \quad N_f \leq 10^6$$

$$S_{ar} = \frac{\sigma_u}{2k_f}, \quad (N_f > 10^6)$$

Substituting the values of σ_u and k_f, we get

$$S_{ar} = (11732\,\text{MPa})\left[1 - \frac{2(2.29) - 1}{12(2.29)} \log N_f\right]$$
$$= 1172(1 - 0.130 \log N_f)\,\text{MPa} \qquad (N_f \leq 10^6)$$

Also fatigue limit $S_{ar} = 1172/2 \times 2.29 = 256\ \text{MPa}\,(N_f > 10^6)$

We can now plot the S-N curve, similar to Figure 5.31.

Since this steel is considered as ductile, k_{fm} in equation (5.54) is unity. Therefore

$$S_{ar} = \frac{S_a}{1 - \dfrac{S_m}{\sigma_u}} = \frac{250}{0.74}\ \text{MPa}$$
$$= 336\ \text{MPa}$$

This value is above the estimated fatigue limit $S_{ar} = 256$ MPa and so it corresponds to a finite life. This value, therefore, can be used with the S-N equation to determine N_f.

$$\log N_f = \frac{1}{0.130}\left(1 - \frac{S_{ar}}{1172}\right)$$
$$= \frac{1}{0.130}\left(1 - \frac{336}{1172}\right)$$

Therefore $N_f = 3.5 \times 10^5$ cycles.

5.11.2 Cumulative Fatigue Damage

The effects of fatigue are cumulative and it is difficult to predict the fatigue life of components which work under varying conditions of stress. The assessment of such damage becomes particularly important when one deals with materials in which the fatigue properties are measured by a finite endurance limit rather than by a fatigue limit. Let a certain stress amplitude S_{a1} is applied for a number of cycles N_1, where the number of cycles to failure from the S-N curve for σ_{a1} is N_{f1} (Figure 5.32). The fraction of the life used is then N_1/N_{f1}. Let another stress amplitude σ_{a2} corresponding to N_{f2} on the S-N curve be applied for N_2 cycles. An additional fraction of the life N_2/N_{f2} is then used. The *Palmgren-Miner rule* states that fatigue failure is expected when such life fractions sum to unity. In other words

$$\frac{N_1}{N_{f1}} + \frac{N_2}{N_{f2}} + \frac{N_3}{N_{f3}} + \ldots = \Sigma \frac{N_i}{N_{fi}} = 1 \tag{5.55}$$

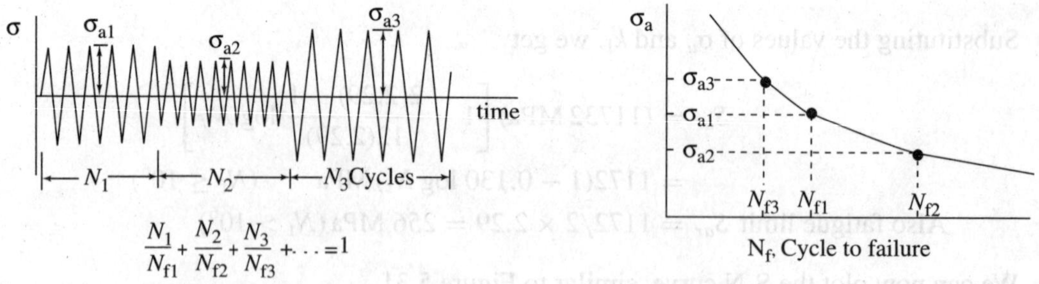

Fig. 5.32. Use of the Palmgren-Miner rule for life prediction for variable amplitude loading which is completely reversed.

***Example* 5.5**: The values of the endurance limits at various stress amplitude levels for low-alloy constructional steel fatigue specimens are given below:

σ_a (MN/m^2)	N_f(cycles)
550	1500
510	10050
480	20800
450	50500
410	12500
380	275000

A similar specimen is subjected to the following programme of cycles at the stress amplitudes stated; 3000 at 510 MN/m^2, 12000 at 450 MN/m^2 and 80000 at 380 MN/m^2, after which the sample remained unbroken. How many additional cycles would the specimen withstand at 480 MN/m^2 prior to failure? Assume zero mean stress conditions. (Adapted from E.J. Hearn, Mechanics of Materials, 2nd Ed., Pergamon Press, Oxford, 1985, p.890).

Solution: From Palmgren-Miner's Rule, equation (5.55), with X the required number of cycles:

$$\frac{n_1}{N_1}+\frac{n_2}{N_2}+\frac{n_3}{N_3}+...etc=1$$

Therefore,

$$\frac{3000}{10050}+\frac{12000}{50500}+\frac{80000}{275000}+\frac{X}{20800}=1$$

$$0.2985+0.2376+0.2909+\frac{X}{20800}=1$$

$$\frac{X}{20800}=1\quad 0.8270$$

Therefore, X = 3598 cycles.

5.11.3 Factors Affecting Fatigue Life

Stress concentration: Fatigue strength is highly reduced by the introduction of stress raiser such as a notch, or hole, fillets, key ways, screw threads, and press fits. Such stress concentration can be controlled by proper design, machining and fabrication. Other sources for stress raisers are surface roughness, porosity, inclusions, local overheating during grinding and decarburization. To study such aspect notch fatigue test is carried out.

Size effect: In laboratory testing, the test pieces are small in size. It is difficult to translate the data on large machine part. It is generally noticed that the fatigue strength of large members is lower than that of small specimens. This is known as size effect.

Surface effects: We have talked earlier about the role of surface roughness. Significant improvements in fatigue properties can result from the formation of harder and stronger surfaces on steel parts by case hardening i.e., carburizing or nitriding. This causes favourable compressive residual stresses on the surface. The residual stresses are measured by x-ray methods. A vary convenient method to impart residual compressive stress on the surface is by surface rolling with contoured rollers and by shot peening. The former is done on fillets of crankshafts and the bearing surface of railroad axles, while the latter is common for small size components. Another way to introduce residual compressive stress is by quenching of steel from the tempering temperature. The shot peened surfaces are generally polished to further improve fatigue property.

Metallurgical Factors: The grain size has great effect on fatigue life in the low stress, high cycle condition. Quenched and tempered microstructures of steels offer good fatigue properties. However austempered structure at the same hardness results in better fatigue life than a quenched and tempered structure. In general there appears to be a relationship between fatigue and tensile properties. At the same tensile strength level, the spheroidal carbide microstructure is preferred than the lamellar cementite in pearlite. With enhanced strain aging, introduced by the presence of interstitial atoms; the fatigue limit of the steel is raised. Inclusions in forging grade steels, which are subjected to mechanical workings exhibit anisotropy in fatigue behaviour. Transverse fatigue properties are one of the most structure-sensitive mechanical properties.

Environmentalfactors: Chemical environment may accelerate the initiation and growth of fatigue cracks. Corrosion pits act as stress raisers. Material dissolves more near the crack tip in a hostile environment. Moisture and gases in air, specially at high temperatures are also contributory factors.

Temperature: In general fatigue resistance of metals below room temperature is greater than above the room temperature. With increase in temperature, creep-fatigue interaction sets in. In general, the higher the creep strength of a metal, the higher its high temperature fatigue strength.

Fatigue failure can also be predicted by fluctuating thermal stresses under conditions where no stresses are produced by mechanical causes. If failure occurs after repeated applications of thermal stress of a lower magnitude, it is called *thermal fatigue*. Austenitic stainless steel is sensitive to this phenomenon, because of its low thermal conductivity and high thermal expansion. Thermal fatigue parameter is defined as $\sigma_f K / E\alpha$, where σ_f is the fatigue strength at the mean temperature, K is the thermal conductivity, E is elastic modulus and α is the linear thermal coefficient of expansion. Higher is this parameter, better the thermal fatigue resistance.

Table 5.5 summarizes the static and fatigue strength of some important engineering materials.

5.12 CREEP

Deformation that occurs with time is called creep strain. Creep is important in engineering applications involving high temperature, such as steam turbine plants, jet and rocket engines and nuclear reactors. Sagging of tungsten filament in incandescent bulbs is another common example of creep. Crystalline materials i.e., metals and ceramics, creep in the temperature range of 0.3 – 0.6 T_m. Large creep strains can occur in polymers; and glasses above the particular glass transition temperature do creep. Often, in engineering components, multiaxial stresses may occur. As the creep data and constants are based on uniaxial tests, complicated methodology has to be applied. It is beyond the present discussion to include this topic.

Let us now examine a typical creep curve, which is a plot between strain and time under constant load (Figure 5.33). There is an initial instantaneous occurrence of elastic and also plastic strain followed by the gradual increase of creep strain. The strain rate $\dot{\varepsilon} = d\varepsilon/dt$, i.e., the slope of strain and time plot is relatively high. However $\dot{\varepsilon}$ decreases and often becomes approximately constant, at which point the *primary or transient creep* stops and the *secondary*

Table 5.5. Comparison of Static and Fatigue Strengths of Some Engineering Materials

Material	*Tensile Strength (MPa)*	*Endurance Limit (MPa)*	*Endurance Ratio*
Ferrous alloys			
AISI 1010, normalized	364	186	0.46
1025, normalized	441	182	0.41
1035, normalized	539	238	0.44
1045, normalized	630	273	0.43
1060, normalized	735	315	0.43
1060, oil Q, tempered	1295	574	0.44
3325, oil Q, tempered	854	469	0.55
4340, oil Q, tempered	952	532	0.56
8640, oil Q, tempered	875	476	0.54
9314, oil Q, tempered	812	476	0.59
302, annealed	560	238	0.43
316, annealed	560	245	0.44
431, quenched, tempered	798	336	0.42
ASTM 20 gray cast iron	140	70	0.50
30 gray cast iron	210	102	0.49
60 gray cast iron	420	168	0.40
Nonferrous alloys			
AA 2011-T8	413	245	0.59
2024, Annealed	189	91	0.48
6061-T6	315	98	0.31
6063-T6	245	70	0.29
7075-T6	581	161	0.28
214 As cast	175	49	0.28
380 Die-cast	336	140	0.42
Phosphor bronze, annealed	315	189	0.60
hard drawn	602	217	0.36
Aluminum bronze, quarter hard	581	206	0.35
Incoloy 901, at 650 °C	980	364	0.37
Udimet 700, at 800 °C	910	343	0.38
Reinforced plastics			
Polyester – 30% glass	123	84	0.68
Nylon 66 – 40% glass	200	62.7	0.31
Polycarbonate – 20% glass	107	34.5	0.32
40% glass	131	41.4	0.32

or steady stage sets in. After end of the secondary stage, ε increases in an unstable manner as rupture failure approaches. This portion of plot is called *tertiary stage*. In this final stage, the deformation becomes localized, by the formation of a neck as in a tension test, or voids may form inside the material or both may occur. Figure 5.34 shows the effect of applied stress on the creep curve at constant temperature. It is apparent that a creep curve with three clear stages will be found for only certain combinations of stress and temperature. A similar family of curves

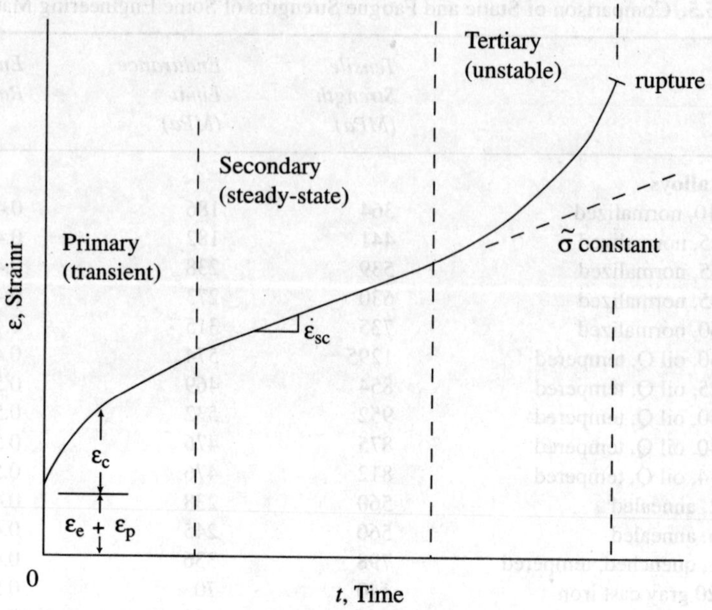

Fig. 5.33. Schematics of strain-time behaviour during creep under constant engineering stress.

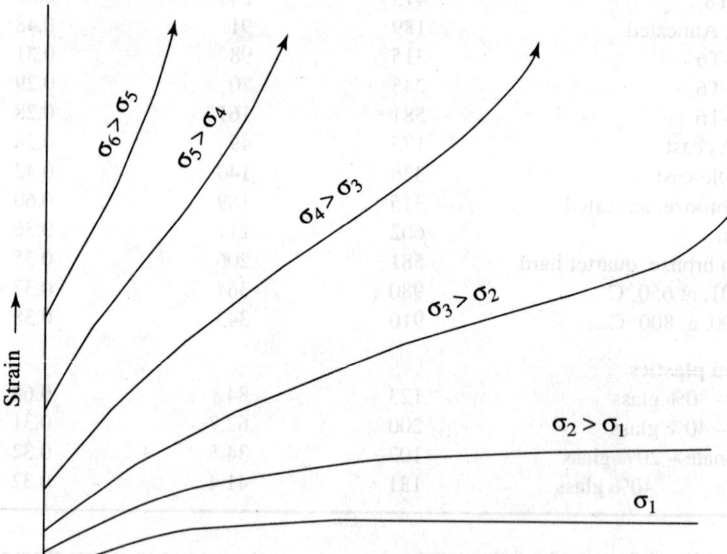

Fig. 5.34. Schematic representation of the effect of stress on creep curves at constant temperature.

is obtained for creep at constant stress for different temperatures. The higher the temperature, the greater the creep rate. The minimum creep rate is an important design parameter derived from the creep curve. Two standards are common, first the stress to produce a creep rate of 1 per cent/10,000 h and the second stress for creep rate of 1 per cent/100,000 h. The first standard is a major requirement for jet engine alloys, while the latter are for steam turbine.

The minimum creep rate is highly temperature sensitive and can be related to the temperature by the same type of exponential equation as diffusion

$$\frac{\Delta \varepsilon}{\Delta t} = A e^{Q_c/RT} \tag{5.56}$$

where Q_c is activation energy in creep, A is a constant. Similarly the relation between the minimum creep rate and stress at constant temperature can be given by

$$\frac{\Delta \varepsilon}{\Delta t} = B \sigma^{n'} \tag{5.57}$$

where B and n' are constants characteristic for each material

Another important test is the stress-rupture test, which is basically similar to the creep test except that the test is invariably carried out to the failure of the material. Higher loads are used with the stress-rupture test than in creep test, and hence the creep rates are higher. In the creep test the total strain is often less than 0.5 per cent, while in the stress rupture test the total strain may be around 50 per cent. Figure 5.35 shows a typical plot of stress-rupture data. A straight line is usually obtained for each test temperature. Changes, if any, in the slope of the stress-rupture plot is due to structural changes occurring in the material.

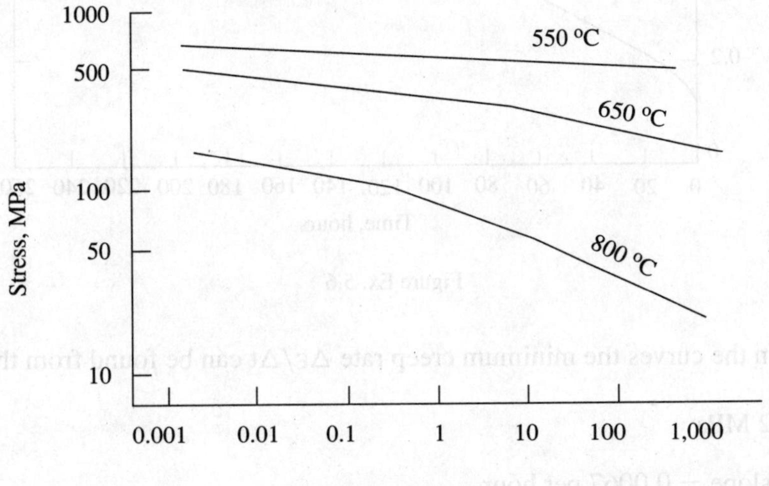

Fig. 5.35. A typical plot of strain-rupture time data.

***Example* 5.6**: Creep curves of an aluminum alloy is given in Figure Ex.5.6 at a constant temperature of 180 °C and for various stresses. Calculate the minimum creep rate for the stress 60 MPa.

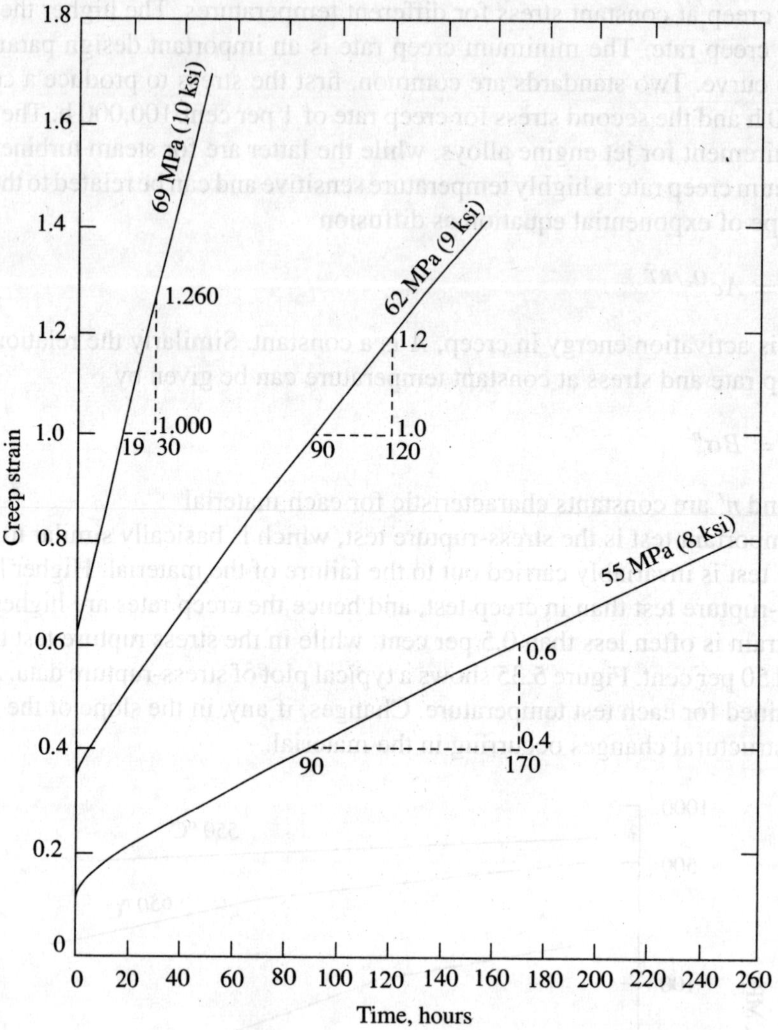

Figure Ex. 5.6

Solution: From the curves the minimum creep rate $\Delta\varepsilon/\Delta t$ can be found from the slope of the curve.

At stress 62 MPa

The slope = 0.0067 per hour

At stress 55 MPa

The slope = 0.0025 per hour

$$\text{We know } \frac{\Delta\varepsilon}{\Delta t} = B \times \sigma^{n'}$$

Substituting above numerical values of minimum creep rate

$$0.0067 = B \times 62^{n'} \quad \text{(for stress of 62 MPa)}$$
$$0.0025 = B \times 55^{n'} \quad \text{(for stress of 55 MPa)}$$

Solving above two equations

$$n' = 8.1 \text{ and } B = 2 \times 10^{-17}$$

Therefore minimum creep rate at 60 MPa

$$\text{i.e.,} \quad \frac{\Delta\varepsilon}{\Delta t} = 2\times 10^{-17} \times 60^{8.1}$$
$$= 5.1\times 10^{-3} \text{ per hour}$$

5.12.1 Mechanism of Creep Deformation and Rupture

There is a lot of similarity in diffusion and creep mechanism. Atoms, vacancies, dislocations and molecules within a solid material move in a time-dependent manner during creep. The phenomenon gets enhanced at higher temperatures. The mechanism can be classified broadly in two categories, which are diffusion flow and dislocation creep. Diffusional flow can occur at low stress but requires relatively high temperature. This mechanism involves the movement of vacancies in the crystal lattice. If the vacancies move through the crystal lattice, the behaviour is called *Nabarro-Herring Creep*. The strain rate is approximately proportional to the stress and inversely proportional to the square of the average grain size.

$$\varepsilon \propto \frac{\sigma}{d^2} \tag{5.58}$$

On the other hand, if the vacancies instead move along the grain boundary, the behaviour is termed *Coble creep*. Here the dependence on stress is

$$\varepsilon \propto \frac{\sigma}{d^3} \tag{5.59}$$

Both these types of diffusional flow are essentially viscous processes. In addition, as any diffusional process is thermally activated, the process may be expressed through Arhenius equation

$$\varepsilon = Ae^{-Q/RT} \tag{5.60}$$

where Q is activation energy. This is a measure of the energy barrier that must be overcome for atomic/molecular motion to occur. A combined equation can be written to show the effects of both stress and temperature as

$$\varepsilon = A_1\sigma e^{-Q/RT} \tag{5.61}$$

where A_1 is a new coefficient, which depends on the material. The activation energy Q like in equation (5.56), may change if the physical mechanism is altered due to an enough shift in temperature or stress.

Another mechanism for creep is dislocation creep, which is operative at high stress. In this process dislocation climb process is involved. The creep deformation in this case is time dependent, because the climb process is time dependent. As grain boundaries are not a major factor, the strain rate is not strongly affected by grain size, but the resistance to climb process is such that there is a strong stress dependence.

$$\varepsilon \propto \sigma^m \qquad (5.62)$$

where m varies with material and test conditions but typically of the order of 5.

Another term for such type of creep is 'power-law' creep. The power-law creep can be demonstrated through two consequences. Firstly, at constant stress, strain accumulates with time. Ice in the Antarctic ice cap and in glaciers spreads and flows by power-law creep. Secondly, at constant displacement, stress relaxes with time. As an example bolts in turbine casing etc. must be regularly tightened. It must be emphasized that the over tightening of bolt does not do any good since the time interval between retightening falls rapidly as stress increases.

Ashby displayed conveniently various creep operating mechanisms on a deformation-mechanism map in which the appropriate stress or equivalent stress compensated by modulus on a log scale is plotted against homologous temperature (T/T_m) for a material with a fixed grain size. Effect of alloying addition can be also gauged by the comparing such maps.

Earlier it has been mentioned that voids along grain boundaries during creep occur due to creep cavitations. The enlarging and joining of grain boundary or other voids then cause cracks and can progress to fracture. In case the temperature is sufficiently high in a ductile and relatively pure metal, a process known as dynamic recrystallization can occur. Large deformation are then possible, and failure occurs due to necking

5.12.2 Creep in Polymers

At temperature below the glass transition temperature T_g of a given polymer, creep effects are relatively small. Above T_g the effect is significant. For a primarily crystalline thermoplastic such as polyethylene, viscous flow occurs at temperature substantially above T_g because of easier sliding of molecular chains. The process becomes easier in linear polymers if the molecular chains are shorter, and also by the absence of obstacles to sliding, such as bulky side groups on the molecules or cross-linking between the chain like molecules. The creep response is more complicated at intermediate temperatures, only slightly above T_g, where the behaviour is leathery or rubbery. Here, sliding of the molecular chains is more difficult and they are more easily entangled with one another. Such entanglement offers more creep resistance. Such obstacles to sliding of molecules also offer a memory effect.

In section 5.3.5 we have introduced the concept of viscoelasticity in polymers. A fixed but long-term load on viscoelastic polymer can lead to creep. Conversely, a fixed but long –term strain can lead to *stress relaxation*. It is, therefore, the decay of stress by creep, but with the

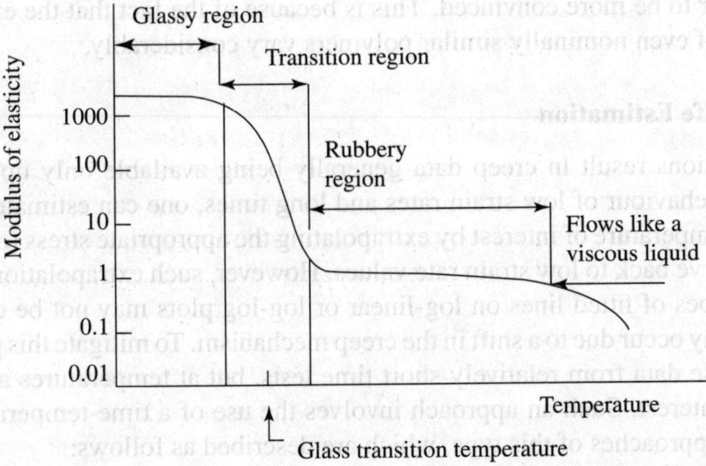

Fig. 5.36. Effect of temperature on the modulus of elasticity of an amorphous polymer.

displacement remaining constant. The total shear displacement r can be written as

$$r = r_e + r_v = \frac{\tau}{G} + \left(\frac{\tau}{\eta}\right)^t \tag{5.63}$$

where r_e = elastic shear strain, r_v = viscous flow, τ = shear stress, G = shear modulus, η = viscosity and t = time. The mechanisms of stress relaxation and creep are very much related, except that as the value of r_v increases, r_e decreases to maintain the constant total displacement. If the strain does not change with time, i.e., if $r/t = 0$, the above equation can be written as

$$\frac{\tau}{t} = -\frac{G\tau}{\eta} \tag{5.64}$$

Differentiating the above equation

$$\frac{d\tau}{dt} = -\frac{\tau}{\left(\frac{\eta}{G}\right)} \tag{5.65}$$

$$\frac{d\tau}{\tau} = -\frac{dt}{\left(\frac{\eta}{G}\right)}$$

or $$\ln \tau - \ln \tau_o = -t/(\eta/G) \tag{5.66}$$

where τ_0 = initial stress and τ = stress at time t. The ratio of viscosity to shear modulus (η/G) is called the, *relaxation time* λ.

Stress-strain relationships suggested by simple linear rheological models, which include time dependency are very useful for polymers. But more complex behaviours might also occur. For design applications with polymers, generally it is assumed that the behaviour follows linear viscoelasticity. Many times the users have to generate laboratory data for the specified lot of the material, in order to be more convinced. This is because of the fact that the exact composition and processing of even nominally similar polymers vary considerably.

5.12.3 Creep Life Estimation

Test time limitations result in creep data generally being available only upto 1000 hrs. For estimating, the behaviour of low strain rates and long times, one can estimate the creep stress for the service temperature of interest by extrapolating the appropriate stress (σ) with respect to strain rate (ε) curve back to low strain rate values. However, such extrapolations are not strictly valid, as the shapes of fitted lines on log-linear or log-log plots may not be constant. Sudden slope changes may occur due to a shift in the creep mechanism. To mitigate this problem, another approach is to use data from relatively short time tests, but at temperatures above the service temperature of interest. Such an approach involves the use of a time-temperatures parameter. There are three approaches of this type, which are described as follows:

Sherby – Dorn (S – D) Parameter: Sherby-Dorn parameter is expressed as

$$P_{SD} = \log t_r - 0.217Q(1/T) \quad (5.67)$$

where t_r is the rupture time, Q is the activation energy for creep, T temperature in absolute K.

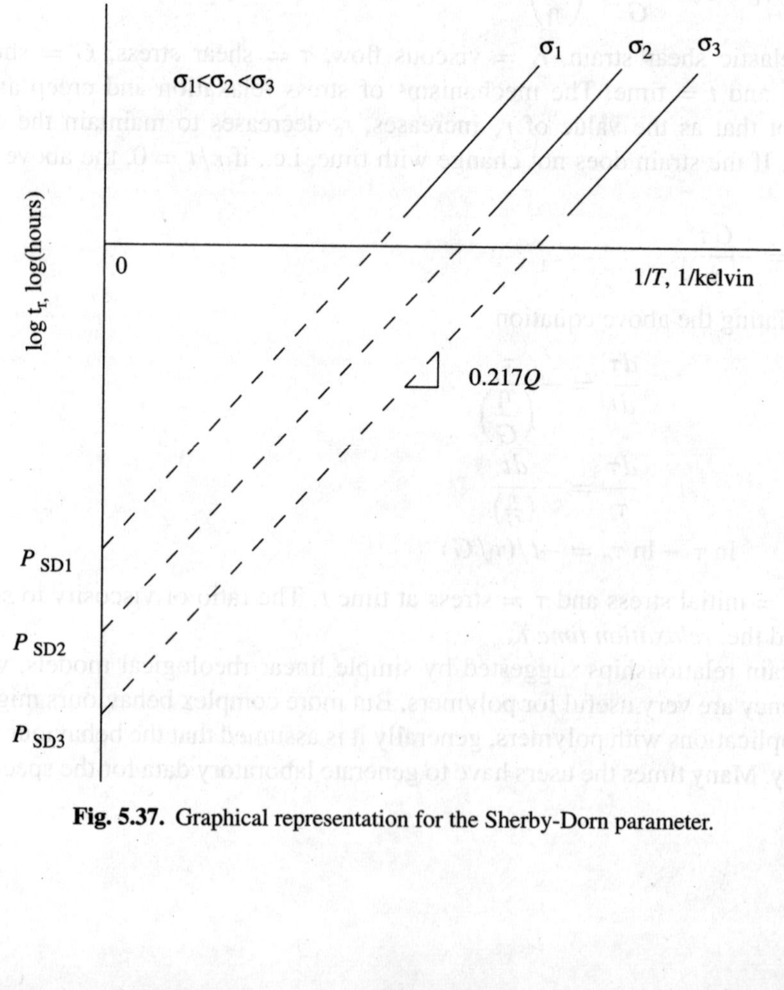

Fig. 5.37. Graphical representation for the Sherby-Dorn parameter.

The values of activation energy Q for use with the S-D parameter can be obtained by plotting creep-rupture data or coordinate axes of log t_r versus $1/T$ (Figure 5.37). A family of parallel straight lines is obtained, one for each value of stress. The lines here is a common slope $0.217Q$. Each intercept as $1/T = 0$ gives the value of P_{SD} for that stress. Table 5.6 gives some values of constants for time-temperature parameters for different metallic alloys. It is useful to identify a particular value of creep strain, such as 1 per cent or 2 per cent that is considered to represent failure. S-D parameter can be used for this situation too, the only difference being that t_r is replaced by t_f, the time to reach the strain of interest.

Table 5.6. Some Values of Constants for Time-Temperature Parameters[1].

Material	*Sherby-Dorn*	*Larson-Miller*	*Manson-Haferd*	
	Q, cal/mole	C	T_a, K	*log* t_a
Various steels and stainless steels	≈9000	≈20	-	-
Pure aluminum and dilute alloys	≈36,000	-	-	-
S-590 alloy (Fe base)	85,00	17	172	20
A-286 stainless steel	91,000	20	367	16
Nimonic 80A (Ni base)	91,000	18	311	16
1Cr-1Mo-0.25V steel	110,000	22	311	18

Note: [1]Values given are for temperatures T in kelvin (K) and rupture time t_r in hours.

Larson – Miller (L – M) Parameter: L-M parameter is expressed by the equation

$$P_{LM} = 0.217Q = T(\log T_r + C) \tag{5.68}$$

where C is a constant.

The value of C is obtained by an extrapolated intercept on a plot of log t_r versus $1/T$ (Figure 5.38). A family of straight lines is obtained for various stress values, with all of these lines having a common intercept at $1/T = 0$ of log $t_r = -C$. The slopes are the values of P_{LM} corresponding to each stress. The whole treatment is similar to S-D parameter, except that the activation energy is assumed not to be constant but to vary with stress. A plot of stress versus P_{LM} of a particular alloy is plotted, in order to estimate rupture times. The L-M parameter may also be used to estimate time t_f, corresponding to a limiting creep strain prior to rupture. If a set of stress/P_{LM} plots of various alloys are plotted, for a fixed L-M parameter, higher the value of stress infers a better creep resistant material.

Manson – Haferd (M – H) Parameter: In this case log t_r vs. T is plotted instead of $1/T$ as done in previous two parameters i.e., S-D and L-M parameters. A family of straight lines for various stresses is assumed to converge at point (T_a log T_a) as shown in Figure 5.39. The values of both T_a and log t_a are considered to be material constants. For any given stress, the M-H parameter is given by the inverse of the slope of the line.

$$P_{M-H} = \frac{T - T_a}{\log t_r - \log t_a} \tag{5.69}$$

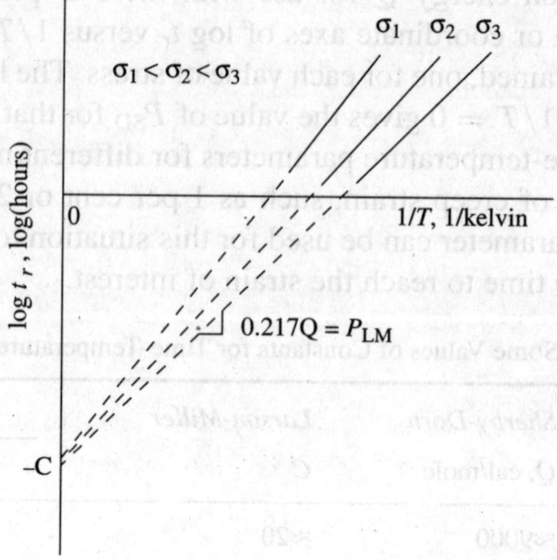

Fig. 5.38. Graphical representation for the Larson-Miller parameter.

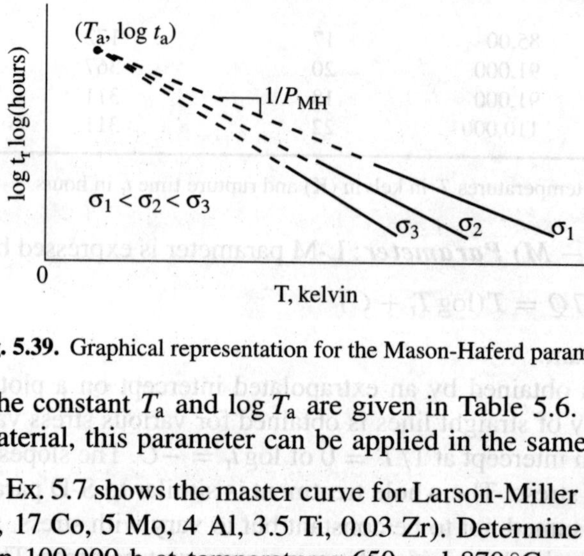

Fig. 5.39. Graphical representation for the Mason-Haferd parameter.

Some values of the constant T_a and $\log T_a$ are given in Table 5.6. Once T_a and $\log t_a$ are known for a given material, this parameter can be applied in the same manner as the others.

***Example* 5.7**: Figure Ex. 5.7 shows the master curve for Larson-Miller Parameter for Astroloy (0.02C, 0.02B, 15 Cr, 17 Co, 5 Mo, 4 Al, 3.5 Ti, 0.03 Zr). Determine the stress required for failure of this alloy in 100,000 h at temperatures 650 and 870 °C respectively. Assume the Larson-Miller constant $C = 46$.

Solution:

$$\text{At } 650\,^\circ\text{C (923 K)}, P = 923\,(\ln 10^5 + 46) = 53.1 \times 10^3$$

$$\text{At } 870\,^\circ\text{C (1143 K)}, P = 1143\,(\ln 10^5 + 46) = 65.7 \times 10^3$$

From the master plot of this material given in Figure Ex. 5.7, we find the stresses corresponding to the values of the parameters. The values are

$$\text{At } 650\,^\circ\text{C}, \ \sigma = 500 \text{ MPa}$$

$$\text{At } 870\,^\circ\text{C}, \ \sigma = 85 \text{ MPa}$$

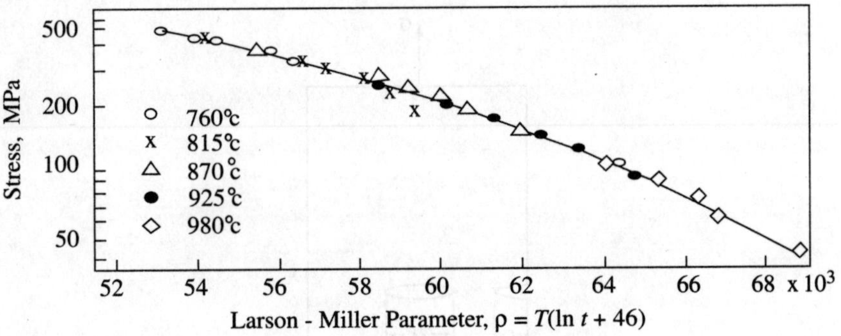

Figure Ex. 5.7

5.13 FRACTURE AND FRACTUE TOUGHNESS

In earlier sections we have mentioned about the fracture of metals during tensile loading and of a brittle solid during impact loading. Fracture takes place in two stages i.e., crack initiation and its propagation. Fracture can be classified into ductile and brittle fractures. The former is characterized by appreciable plastic deformation prior to and during the propagation of the crack. On the other hand, the brittle fracture is characterized by a rapid rate of crack propagation with no gross deformation. Much study on fracture has been carried out on the brittle materials. In the present section, we shall deal with fracture caused by uniaxial tensile stress. However in practice the situation may be much more complex, like fatigue, creep etc. These are dealt in their respective sections.

Griffith proposed that a brittle material contains a number of fine cracks which produce a stress concentration so that the theoretical strength is reached in localized regions at a nominal stress, much below the theoretical strength. Figure 5.40 shows the schematic of the crack model. The effect of both types of cracks i.e., in interior (length $2a$) and at surface (length a), on the fracture behaviour is the same.

Let us assume that the crack is of elliptical shape. The total elastic strain energy that has been released per unit volume by the crack is equal to two times the energy for either upper or lower crack surface. Therefore

$$\text{Total elastic strain energy} = 2 \times \frac{1}{2}\sigma \times \frac{\sigma}{E} \times \pi a^2 t \tag{5.70}$$

where σ = applied stress; E = Modulus of elasticity, a = half the crack length and t = thickness of the material. On the other hand the energy needed to crack the surfaces of the crack is the product of total surface area of crack and the surface energy i.e.,

$$\text{Total Energy needed} = 2 \times 2a \times t \times \gamma = 4at\gamma$$

$$\text{Net change in energy } U = \frac{\pi\sigma^2 a^2 t}{E} - 4at\gamma$$

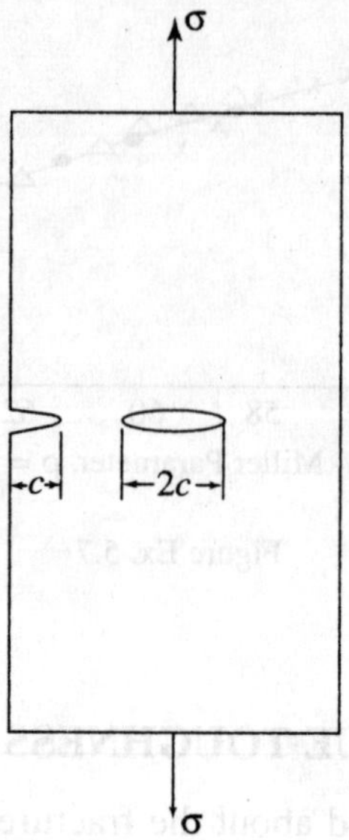

Fig. 5.40. Griffth model.

The energy change rate with respect to a

$$\frac{dU}{da} = \frac{2\pi\sigma^2 at}{E} - 4t\gamma$$

$$\text{At equilibrium} \frac{2\pi\sigma^2 at}{E} = 4t\gamma$$

$$\text{i.e.,} \qquad \sigma = \left(\frac{2\gamma E}{\pi a}\right)^{1/2} \tag{5.71}$$

$$\approx \left(\frac{\gamma E}{2a}\right)^{1/2} \tag{5.72}$$

The above relation is true for pure brittle materials like glass, ceramics and brittle polymer. However in many cases some plastic deformation can be associated. In that situation equation (5.71) is modified as

$$\sigma = \left[\frac{2E(\gamma + \gamma_p)}{\pi a}\right]^{1/2} \tag{5.73}$$

In most metals plastic deformation energy γ_p is considerably greater than surface energy γ. Generally the term $(\gamma + \gamma_p)$ is replaced by G_c, a term representing the energy needed for fracture per unit area of crack extension, so that the equation (5.73) becomes the Orowan-Irwin relationship.

$$\sigma = \sqrt{\frac{EG_c}{\pi a}} \tag{5.74}$$

The term G_c can be presented in a number of ways. These are (a) work of fracture, (b) critical strain energy release rate or (c) toughness which is a measure of the ability of a material to absorb energy. Its dimension is Jm^{-2}.

Figure 5.41 shows various modes of fracture. When the mode of deformation is tension, it is known as *opening mode*. It is also called as fundamental mode of fracture or simply mode-I. Shear mode is classified as in-plane shear mode or mode-II. This is also called as *sliding mode*. Mode-III is *out-of-plane* shear mode. It is also called *tearing mode*. By K_{IC} one means the parameter in mode–I.

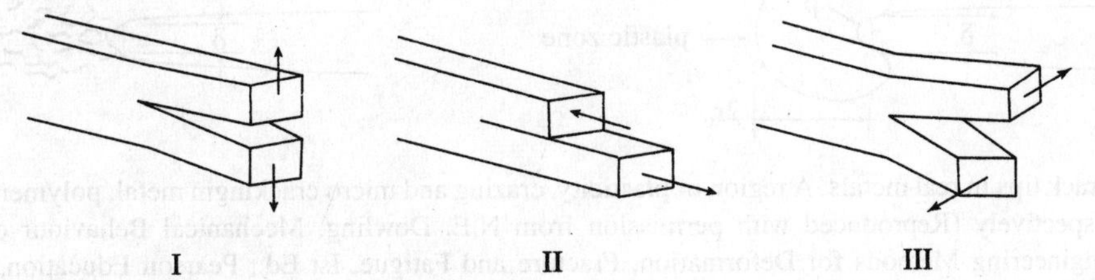

Fig. 5.41. The three opening modes associated with crack growth under I: tensile mode; II: in plane shear mode and III: anti plane shear mode.

***Example* 5.8**: A refractory component made out of MgO must not fail when a tensile stress of 13.5 MPa is applied. Determine the maximum allowable surface crack length if the surface energy and elastic modulus of MgO are 1 J/m^2 and 210 GPa respectively.

Solution:

$E_{MgO} = 210$ GPa,

Let a = length of surface crack

$$a = \frac{2E\gamma}{\pi\sigma_c^2} = \frac{2 \times 21 \times 10^{10} \times 1}{\pi(13.5 \times 10^6)^2}$$

$= 0.73$ mm

5.13.1 Crack Tips in Real Materials

From the description in the earlier section, it is evident that the crack morphology is dependent on the nature of real material. An infinite stress cannot exist in a real material (Figure 5.42). In ductile materials, such as many metals, large plastic deformation occurs near the crack tip. Here some blunting of crack occurs, because of large plastic deformation. The stress at the tip is no longer infinite and the crack near the tip gets opened by a finite amount δ. The δ is known as *crack tip opening displacement* (CTOD). In case of some polymers, elongated voids having a fibrous structure get developed. This is known a *craze zone*. In case of ceramics (brittle materials), numerous tiny cracks known as microcracks near the crack tip may develop.

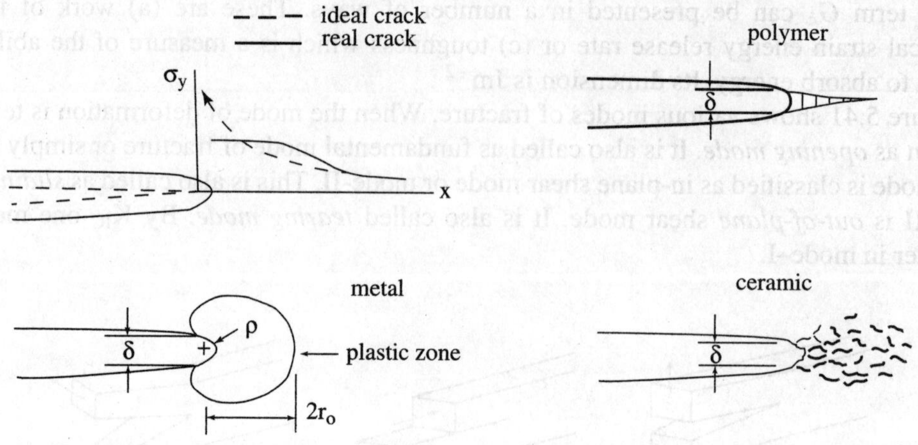

Fig. 5.42. Crack tips in real metals. A region of plasticity, crazing and micro crackingin metal, polymer and ceramic respectively (Reproduced with permission from N.E. Dowling, Mechanical Behaviour of Materials: Engineering Methods for Deformation, Fracture and Fatigue, Ist Ed., Pearson Education, Inc., Upper Saddle River, NJ, 1993, p. 280).

5.13.2 Fracture Toughness

The significant question arises on the effect of cracks on strength. Here we encounter a term known as *stress intensity factor K*. This characterizes the severity of the crack and depends on crack size, stress and geometry. So far, the material is assumed to behave in a linear – elastic manner. *Fracture toughness* of a material is a mechanical property, which signifies the resistance of brittle fracture, as long as the K is below a critical value K_C.

The values of K_C for a material depends on temperature, loading rate and thickness of the material under test. Figure 5.43 shows the relationship of K_C with the thickness of the test piece of the concerned material. High values of K_C occur with thin sheets and decreases as the sheet thickness is increased to become almost constant at large thickness. The lower limiting value of the critical stress intensity factor is called the *plane strain fracture toughness* and is denoted by K_{IC}. This value is used in design for all but the very thin sheets. The lower the value of K_{IC}, the less tough the material is. The SI unit of K_{IC} is MPa $m^{1/2}$. Fracture toughness can be as low as 0.5–1 MPa $\sqrt{m}$ for glass and as high as 150 MPa $\sqrt{m}$ for alloy steel.

In the earlier section, from equation (5.74) we know that $\sigma\sqrt{\pi a} = \sqrt{EG}$, which indicates that fast fracture will occur when a material is stressed to a value σ and the crack reaches some critical size 'a'. Alternately when a material containing a crack is subjected to some critical stress and crack length for fast fracture, $\sqrt{EG}$ for the material is constant. Here E is Young's modulus and G is the strain energy release rate. The term $\sigma\sqrt{\pi a}$ given a symbol K, is known as stress intensity factor. Fast fracture will then occur when $K = K_C = \sqrt{EG_c}$. K_C is known as *critical stress intensity factor* or more commonly the *fracture toughness parameter*.

Normal manufactured components are of finite size and generally cracks grow from a free surface. The stress intensity factor is modified to

$$K = \sigma Y \sqrt{a} \tag{5.75}$$

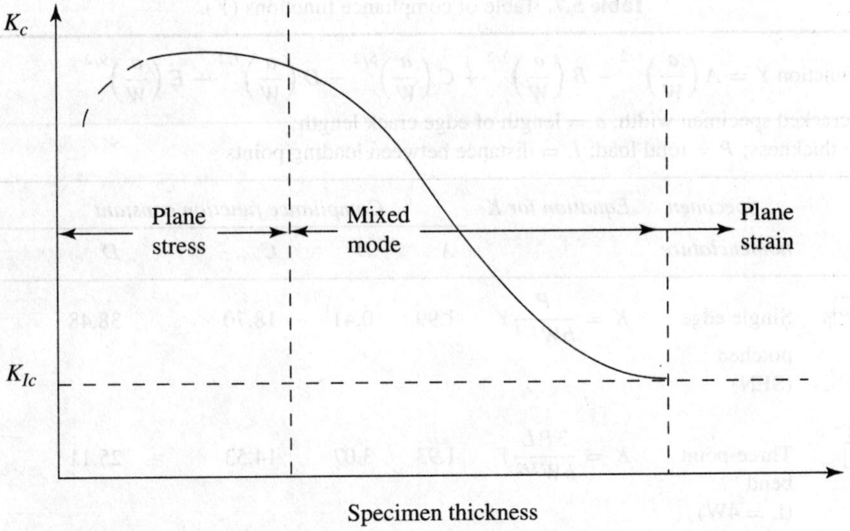

Fig. 5.43. Effect of specimen thickness on the critical stress intensity factor.

where a = length of the edge crack, and Y is a dimensionless correction factor often termed a '*compliance factor*'. Y is a polynomial of the ratio a/W, where W is the uncracked width in the crack plane.

In terms of specimen geometry for tensile loading

$$K = \frac{P}{bW^{1/2}}Y \tag{5.76}$$

where P = load, b = specimen thickness, W = width of the specimen. Y for different common laboratory specimen geometries are given in Table 5.7.

In laboratory, fracture toughness testing involves test pieces with sharp notches being strained until the crack propagates and the test piece fails. The problem in obtaining test pieces is producing the sharp notches. This is done by taking a test piece with a machined notch and then using a standardized precracking procedure involving loading with an alternating stress i.e., fatigue loading. Figure 5.44 shows the dimensions of typical test pieces for three point bending and tensile loading, according to British Standards BS7448. For the test a strictly increasing load is applied to a test piece, which causes the notch to open. The maximum value of load (P) is recorded before breaking occurs as a result of the crack propagation.

Ashby proposed the property chart of fracture toughness versus strength, which can be referred in any Materials Handbook. It must be borne in mind that while designing a structure, care should be taken that it yields before it breaks. For a minimum detectable crack size of 2a this condition is given by $(K_{IC}/\sigma_y) \geq \sqrt{\pi a}$, where σ_y is the yield stress. The safest material is the one with the greatest value of K_{IC}/σ_y.

***Example* 5.9** An aircraft component is fabricated from an aluminium alloy that has a plane strain fracture toughness of 40 MPa$\sqrt{m}$. It has found that fracture results at a stress of 300 MPa, when the maximum (or critical) internal crack length is 4 mm. For the same component, will

Table 5.7. Table of compliance functions (Y).

Compliance function $Y = A\left(\frac{a}{W}\right)^{1/2} - B\left(\frac{a}{W}\right)^{3/2} + C\left(\frac{a}{W}\right)^{5/2} - D\left(\frac{a}{W}\right)^{7/2} + E\left(\frac{a}{W}\right)^{9/2}$

with W = uncracked specimen width; a = length of edge crack length;
b = specimen thickness; P = total load; L = distance between loading points

Specimen geometry	*Specimen nomenclature*	*Equation for K*	*Compliance function constant*				
			A	B	C	D	E
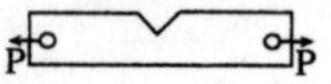	Single edge notched (SEN)	$K = \frac{P}{bW^{1/2}}Y$	1.99	0.41	18.70	38.48	53.85
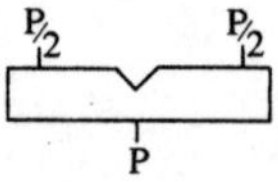	Three-point bend (L = 4W)	$K = \frac{3PL}{bW^{3/2}}Y$	1.93	3.07	14.53	25.11	25.80
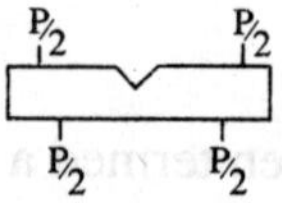	Four-point bend	$K = \frac{3PL}{bW^{3/2}}Y$	1.99	2.47	12.97	23.17	24.80
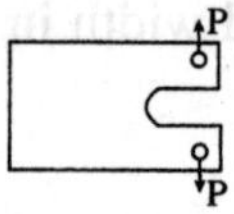	Compact tension (C.T.S)	$K = \frac{P}{bW^{1/2}}Y$	29.60	185.50	655.70	1017.0	638.90

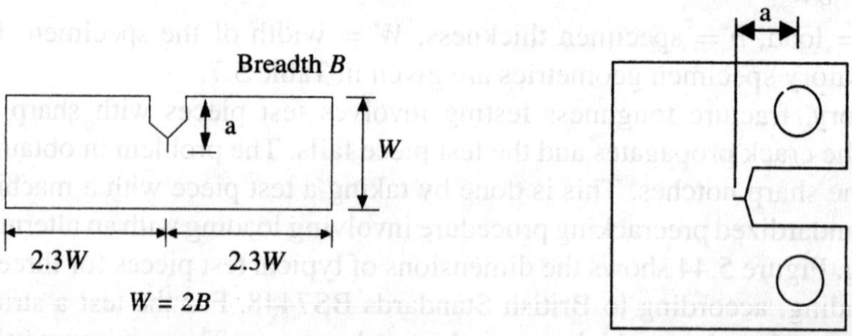

Fig. 5.44. Fracture toughness test pieces for three point bending (left) and for tensile loading (right) (Reproduced with permission from W. Bolton, Engineering Materials Technology, 3rd ed, Elsevier, Oxford, 1998, p. 54).

fracture occur at a stress level of 260 MPa, when the maximum internal crack length is 6 mm. Justify your answer.

Solution.

Given $K_{IC} = 40\ \text{MPa}\sqrt{\text{m}}$

(i) $\sigma = 300$ MPa $a_c = 4/2 = 2$ mm

(ii) $\sigma = 260$ MPa a $= 6/2 = 3$ mm

$$Y = \frac{K_{IC}}{\sigma\sqrt{\pi a}} = \frac{40}{300\sqrt{\pi \times 2 \times 10^{-3}}}$$
$$= 1.68$$

For (ii)

$$Y\sigma\sqrt{\pi a} = K_c = 1.68 \times 260\sqrt{\pi \times 3 \times 10^{-3}}$$
$$= 42.4 \text{ MPa } \sqrt{m}$$
$$\text{Here} K_C > K_{IC}$$

i.e., Fracture shall take place.

***Example* 5.10**: A 9 mm thick tension panel 10 cm wide containing an edge crack of 1 mm yielded at a load of 150 kN. However at a load of 120 kN, another panel of the same material cracked into two pieces when the crack was 5 mm long. Calculate the yield stress and fracture toughness of the material. Given compliance factor = 1.12.

Solution:

$$\text{Yield stress} = (150 \times 10^3)/300 = 500 \text{ MPa}$$

At fracture the nominal stress is

$$\sigma = (120 \times 10^3)/300 = 400 \text{ MPa}$$

$$\text{Now } K_{IC} = 1.12\sqrt{\pi a}$$
$$= 1.12 \times 400 \times \sqrt{\left(\frac{\pi a}{1000}\right)} = 56.16 \text{ MPa}\sqrt{m}$$

5.13.3 Fracture Toughness and Fatigue Life

We have earlier mentioned that fatigue failure is identified by three steps (1) crack initiation, (2) crack propagation and (3) final failure. The fatigue life N_f, can be taken as

$$N_f = N_i + N_p \tag{5.77}$$

where N_i = number of cycles for cracks initiation and N_p = number of cycles for crack propagation.

The contribution by the third step is insignificant, since it occurs very rapidly. The question, now, arises as to the possibility of applying fracture mechanics concept to assess the maximum crack length that may be tolerated without causing fatigue failure. The rate of crack growth, da/dN can be related to stress intensity factor 'K' in the form

$$\frac{da}{dN} = A(\Delta K)^m \tag{5.78}$$

where a = crack length, A and m are constants for a material. These may also depend on environment, frequency and the stress ratio R. The value of m generally ranges from 1 to 6. The stress intensity factor ΔK is

$$\Delta K = K_{max} - K_{min} \tag{5.79}$$

From equation (5.75), we get

$$\Delta K = Y\Delta\sigma\sqrt{\pi a} = \gamma(\sigma_{max} - \sigma_{min})\sqrt{\pi a} \tag{5.80}$$

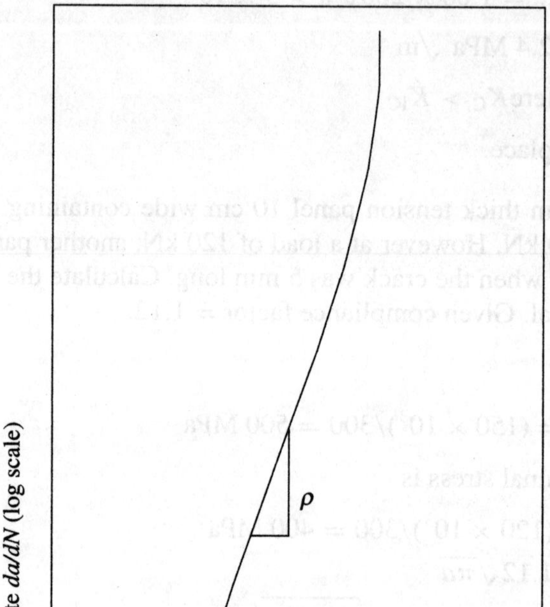

Fig. 5.45. Schematic representation of fracture crack growth behaviour.

Figure 5.45 illustrates schematically the fatigue crack growth rate response of material, and is divided into three regions. In region II, the curve is practically linear. The slope and intercept of the straight line region correspond to the values of m and log A respectively (equation 5.78).

The equation (5.78) can be written as

$$dN = \frac{da}{A(\Delta K)^m}$$

After integrating we get

$$N_f = \int_{a_o}^{a_c} dN = \int_{a_o}^{a_c} \frac{da}{A(\Delta K)^m} \tag{5.81}$$

where N_f = fatigue life, a = initial crack length, a_c = critical crack length, determined from fracture toughness tests. Substituting the value of ΔK from equation (5.75), we get

$$N_f = \int_{a_o}^{a_c} \frac{da}{A(Y\Delta\sigma\sqrt{\pi a})^m} \tag{5.82}$$

Therefore, it must be borne in mind that it ignores the time taken to initiate the crack and also for final failure. The N_f thus obtained be treated as a mere estimated one and not the exact ones.

***Example* 5.11**: A cylindrical steel pressure vessel with yield strength of 360 MPa is subjected to cyclic loading with a maximum internal pressure of 1.5 MPa. The critical stress intensity factor of the steel is, $K_{IC} = 20$ MPa m$^{1/2}$. If the radius of the vessel is 1 m, estimate the minimum thickness of the vessel wall to avoid fast fracture. Assume, $\alpha = 0.72$ for semi-elliptical surface cracks parallel to cylinder axis. Suppose a residual tensile stress of 80 MPa is developed in the steel walls during the fabrication of the pressure vessel. What would be its effect? Assume constant $A = 2.1 \times 10^{-7}$ and $m = 2.25$.

Solution: Hoop stress or Tangential stress (σ)

$$\sigma = \frac{Pr}{t} \qquad \text{(assuming : r >> t)}$$

$$\sigma = \frac{1.5 \times 1}{t} = \frac{1.5}{t}\ \text{MPa}$$

The hoop stress, σ is normal to crack surface

$$K = \alpha\sigma\sqrt{\pi a} = 0.72 \times \frac{1.5}{t}\sqrt{\pi a}$$

Critical crack length a_c when $K = K_{IC}$, is given by the expression

$$20 = 0.72 \times \frac{1.5}{t}\sqrt{\pi a_c}$$

$$a_c = 109\ t^2$$

To avoid fast fracture (i.e., leak occurs rather than fast fracture),

$$a_c > t, t > \frac{1}{109}, \text{ or } t > 0.0092 \text{ m i.e., } t > 9.2 \text{ mm}$$

$$\text{Thus, for } t = 9.2 \text{ mm}, \sigma = \frac{1.5}{0.0092} = 163 \text{ MPa} < \text{Yield stress of 360 MPa}$$

In the presence of residual tensile stress of 80 MPa

$$20 = 0.72 \times \left(\frac{1.5}{t} + 80\right) \times \sqrt{\pi a_c}$$

Putting $a_c = t$ and solving for t

$$\frac{1}{\pi}\left(\frac{20}{0.72}\right)^2 = \left(\frac{1.5}{t} + 80\right)^2 \times t; \text{ or } 6400\, t^2 - 5.6t + 2.25 = 0$$

$$\text{Thus } t = \frac{5.6 \pm \sqrt{(5.6)^2 - 4 \times 6400 \times 2.25}}{2 \times 6400}$$

Only imaginary solutions of t exist.

Thus in the presence of residual stress of 80 MPa, it is not possible to design against fast fracture.

$\frac{da}{dN} = A(\Delta K)^m$, Given $A = 2.1 \times 10^{-7}$ and $m = 2.5$.Take 'ΔK' in MPa $m^{1/2}$ and 'a' in m

Integrating above

$$\int_o^{N_f} dN = \int_{a_i}^{a_f} \frac{da}{A(Y\Delta\sigma\sqrt{\pi a})^m}$$

$$\text{or} \quad N_f = \frac{2}{A(Y\Delta\sigma\sqrt{\pi})^m(m-2)}\left[\frac{1}{a_i^{m-2/2}} - \frac{1}{a_f^{m-2/2}}\right]$$

$$\Delta K = Y(\sigma_{max} - \sigma_{min})\sqrt{\pi a} = Y\Delta\sigma\sqrt{\pi a}$$

where

a_i = initial crack length

a_f = final crack length at failure

In the present case, $a_f = t = 0.0092$ m

(Other cases: $a_f = a_c$)

N_f = Number of cycles to failure

$\sigma_{max} = 1.5/0.0092 = 163$ MPa; $\sigma_{min} = 0$; therefore $\Delta\sigma = 163$ MPa

$$N_f = \frac{2}{2.1 \times 10^{-7}(0.72 \times 163 \times \sqrt{\pi})^{2.25} \times 0.25}\left[\frac{1}{(0.1 \times 10^{-3})^{0.125}} - \frac{1}{(0.0092)^{0.125}}\right]$$

$$N_f = 231.82 \times (3.162 - 1.797) \approx 316 \text{ cycles}$$

The pressure vessel is expected to fail in 316 cycles if a minimum value of $t = 0.0092$ m is used.

5.13.4 Statistics of Brittle Fracture

In case of brittle solids there is a statistical variation of strength. The distribution of crack lengths in such solids would have effects. A large sample will fail at a lower stress than a small one, because it is more likely that it will contain one of the larger defects. There is, therefore, a volume dependence of the strength. Weibull defined the survival probability $P_s(V_o)$ as the fraction of identical samples, each of volume V_o, which survives to a tensile stress σ . According to him

$$P_s(V_o) = \exp\left\{-\left(\frac{\sigma}{\sigma_o}\right)^m\right\} \tag{5.83}$$

where σ_o and m are constants. Figure 5.46 shows the Weibull distribution function. When $\sigma = 0$ all the samples survive and $P_s(V_o) = 1$. As σ increases more and more samples fail and $P_s(V_o)$ decreases. Large stresses cause virtually all the samples to break, so $P_s(V_o) \to 0$ as $\sigma \to \infty$. The constant 'm' is the indicator showing how rapidly the strength falls as one reaches σ_o. This is called the *Weibull modulus*. The lower m, the greater the variability of strength.

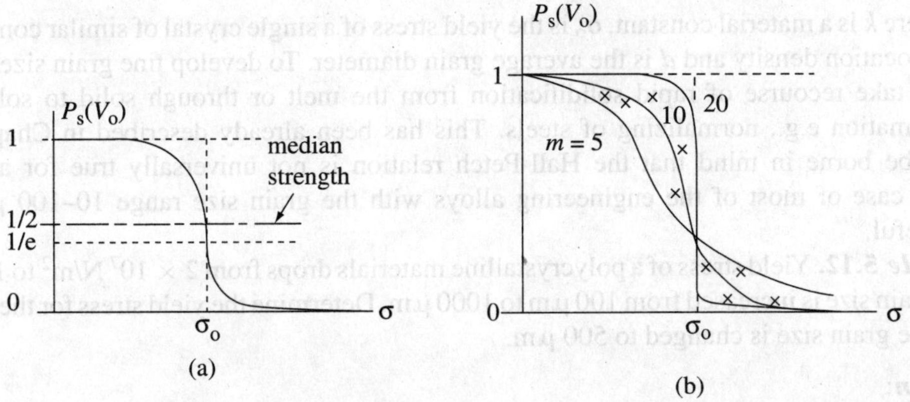

Fig. 5.46. (a) Weibull distribution function; (b) change in survival probability with change in Weibull modulus 'm'.

Many times the environment e.g., moisture does have effect on crack propagation in ceramics making it a time dependent phenomenon. If the standard test which was used to measure tensile strength σ_{TS} takes a time t (test), then the stress which the sample will support safely for a time t is

$$\left(\frac{\sigma}{\sigma_{TS}}\right)^n = \frac{t(\text{test})}{t} \tag{5.84}$$

where n is the slow crack growth exponent.

5.14 STRENGTHENING MECHANISMS

There are various mechanisms for strengthening a metal or alloy. Many times the mechanisms are cumulative. In this section we shall highlight some of the strengthening mechanisms.

Point Defect Hardening: When there is excess concentration of either vacancies or interstitials, a significant change in strength is observed. A large increase in the yield stress may occur just after quenching. Similarly, irradiation by high energy particles (e.g., neutron) may produce irradiation hardening.

Work Hardening: With increase in plastic deformation, the dislocation density increases, causing more and more work hardening. It must be borne in mind that this contribution is lost, when the material is heated. The shear yield strength (τ) of a crystal containing a network of dislocation of density ρ is given by:

$$\tau = \alpha Gb(\rho^{1/2}) \tag{5.85}$$

where G is the shear modulus, b Burgers vector of dislocation and α is a constant, about 0.2.

Grain Size: In polycrystalline materials the role of grain size is significant. The tensile yield strength (σ_y) of polycrystalline material is higher the smaller the grain size. These parameters are related to the *Hall-Petch* equation.

$$\sigma_y = \sigma_o + k_y d^{-1/2} \tag{5.86}$$

where k is a material constant, σ_o is the yield stress of a single crystal of similar composition and dislocation density and d is the average grain diameter. To develop fine grain size material we can take recourse of rapid solidification from the melt or through solid to solid phase transformation e.g., normalizing of steels. This has been already described in Chapter 3. It should be borne in mind that the Hall-Petch relation is not universally true for any grain size. In case of most of the engineering alloys with the grain size range 10–100 μm, it is very useful.

***Example* 5.12.** Yield stress of a polycrystalline materials drops from 2×10^7 N/m^2 to 10^7 N/m^2 when grain size is increased from 100 μm to 1000 μm. Determine the yield stress for the material when the grain size is changed to 500 μm.

Solution:

Hall-Petch relation is expressed as

$$\sigma_y = \sigma_I + kd^{-0.5}$$

So,

$$2\times 10^7 = \sigma_I + k(100 \times 10^{-6})^{-0.5}$$

$$1 \times 10^7 = \sigma_I + k(1000 \times 10^{-6})^{-0.5}$$

Solving we get

$$k = 146.2 \times 10^3$$

$$\sigma_I = 0.538 \times 10^7 \text{ Nm}^{-2}$$

Therefore, substituting above values

$$\sigma_{y500} = 0.538 \times 10^7 + 146.2 \times 10^3(500 \times 10^{-6})^{-0.5}$$
$$= 1.192 \times 10^7\ \text{Nm}^{-2}$$

Solute Hardening: We have earlier described two types of solid solutions i.e., interstitial and substitutional. Both contribute to rise in local stresses within the lattice, which raise the yield stress of the alloy. The degree of strengthening depends on the concentration of the solute atoms in the alloy and also upon the magnitude of the local misfit strains associated with the individual solute atoms.

The increase in shear yield stress $\Delta\tau_y$ due to solute strengthening is expressed as

$$\Delta\tau_y = G\varepsilon^{3/2}bC^{1/2} \tag{5.87}$$

where G is the shear modulus, b Burgers vector of dislocation, C the concentration of solute atoms and ε is the solute atom mismatch parameter.

***Example* 5.13**: Figure Ex. 5.13 shows the effect of dilute additions of silicon and manganese in strengthening the ferrite phase in a low carbon steel. Calculate the effect of 2% Si and 1% Mn on the increase in the yield stress of the above spring steel.

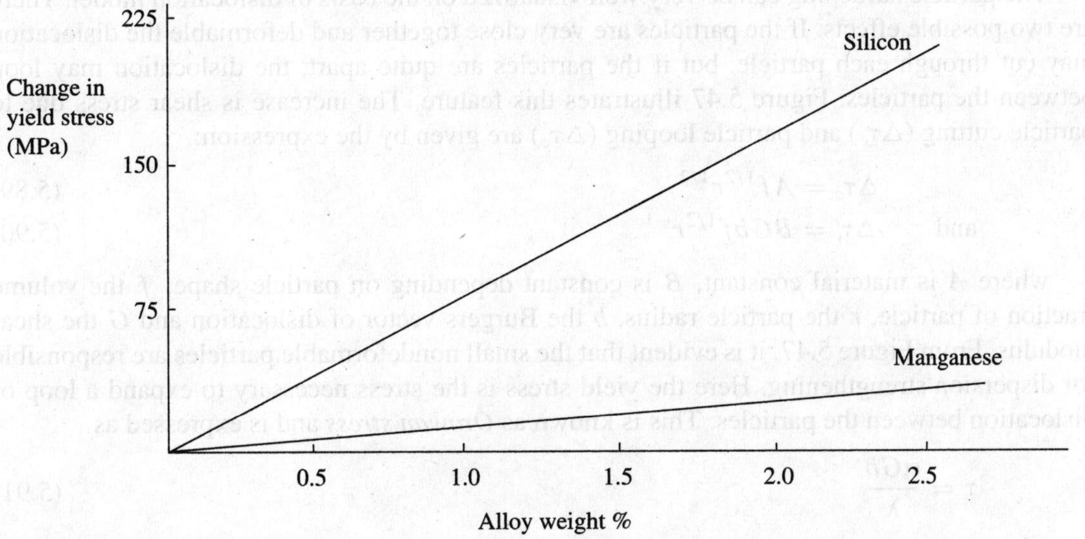

Figure Ex. 5.13

Solution: From the Figure Ex. 5.13 it is apparent that 2 wt% Si and 1 wt% Mn addition in steel enhances the yield stress by 169 and 15 MPa respectively.

The cumulative effect of these solutes can be expressed as:

$$\Delta\tau^2 = \Delta\tau_{Si}^2 + \Delta\tau_{Mn}^2$$
$$= 28561 + 225$$

$$= 28786$$

$$\text{Therefore } \Delta\tau = 170 \text{ MPa}$$

Particle hardening: In this broad category there are two mechanisms i.e., precipitation hardening and dispersion hardening. In many systems phase transformation in the solid state occurs, where the solid solubility decreases with fall in temperature. The details of precipitaion hardening has been given in Chapter 3. During *precipitation* or *age hardening* coherent precipitates impart strengthening. On the other hand, by producing a fine stable particles (10–100 nm) dispersion in an alloy, the yield stress of the material is enhanced. This is known as *dispersion hardening*. The degree of strengthening depends on the distribution of particles in the matrix, volume fraction, average particle diameter and mean interparticle spacing. These factors are all interrelated. The particles act in two ways to retard the dislocation motion. The particles either may be cut by the dislocations or the dislocations are forced to bypass them. A critical parameter of the dispersion of particles is the interparticle spacing λ. This is expressed as

$$\lambda \frac{4(1-f)r}{3f} \tag{5.88}$$

where f is the volume fraction of spherical particles (assumed) of radius r.

The particle hardening can be very well visualized on the basis of dislocation model. There are two possible effects. If the particles are very close together and deformable the dislocation may cut through each particle, but if the particles are quite apart, the dislocation may loop between the particles. Figure 5.47 illustrates this feature. The increase is shear stress due to particle cutting ($\Delta\tau_c$) and particle looping ($\Delta\tau_e$) are given by the expression:

$$\Delta\tau_c = Af^{1/2}r^{1/2} \tag{5.89}$$

and

$$\Delta\tau_l = BGbf^{1/2}r^{-1} \tag{5.90}$$

where A is material constant, B is constant depending on particle shape, f the volume fraction of particle, r the particle radius, b the Burgers vector of dislocation and G the shear modulus. From Figure 5.47, it is evident that the small nondeformable particles are responsible for dispersion strengthening. Here the yield stress is the stress necessary to expand a loop of dislocation between the particles. This is known as *Orowan stress* and is expressed as

$$\tau = \frac{\alpha G b}{\lambda} \tag{5.91}$$

where α is a constant, G is the shear modulus, λ the interparticle separation and b the Burgers vector of dislocation. Figure 5.48 illustrates the Orowan mechanism of dispersion hardening.

Example 5.14: An oxide dispersion strengthened aluminium alloy has unshearable dispersoids of average diameter 200 nm and a shear yield strength of 30 MPa. What would be the shear yield strength if the microstructure consisted of dispersion of the same volume fraction of 50 nm diameter oxide particle?

Solution: Since the dispersoids are responsible of dispersion strengthening, here Orowan mechanism is operative. We know that for nonshearable particles $\tau \propto r^{-1}$, (refer to Equation

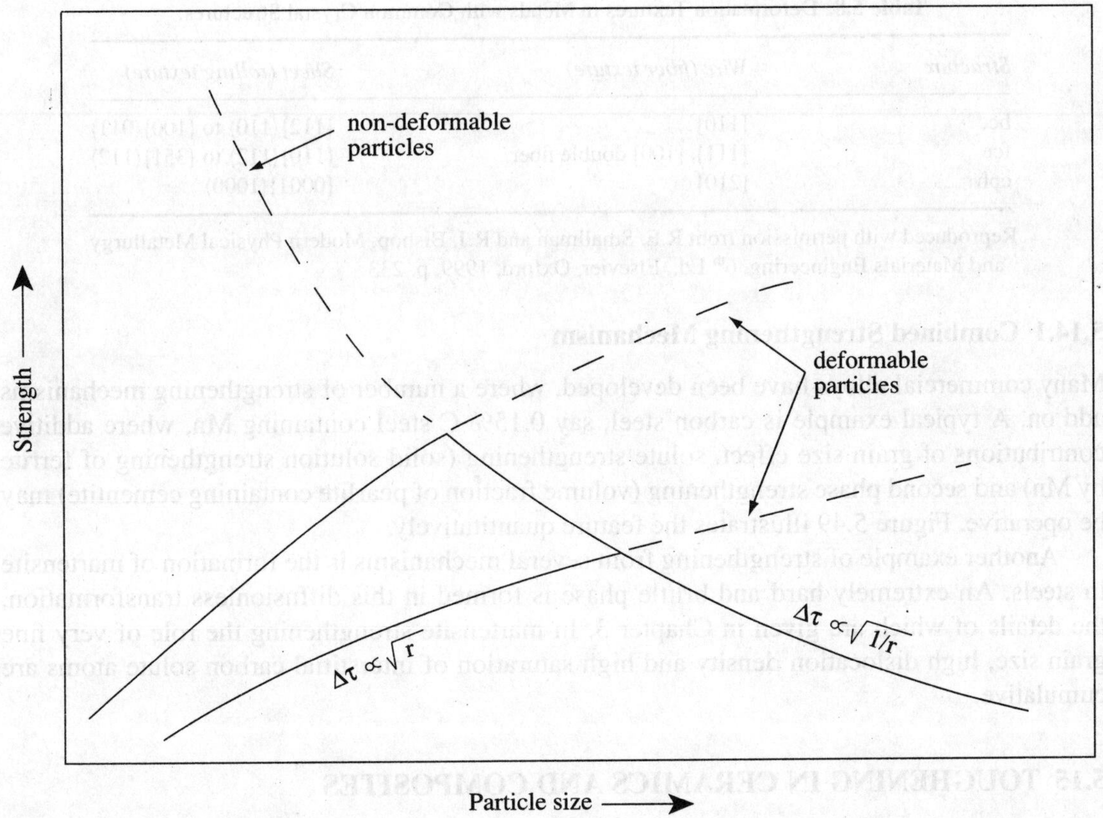

Fig. 5.47. Variation of strength with particle size, defining the deformable and nondeformable particle regimes.

5.90) where r = radius of the dispersoid. In other words

$$\frac{\tau}{30} = \frac{100}{25}$$

$$\text{Therefore, } \tau = 30 \times 4$$

$$= 120 \text{ MPa}$$

Texture hardening: When a polycrystalline metal is plastically deformed extensively (~ 90% reduction in area), the individual grains tend to rotate into a common orientation. Thus preferred orientation is developed. This study is carried out by X-ray analysis, the description of which is beyond the scope of current discussion. Table 5.8 give the information for deformation textures in metals having different crystal structures. The deformation texture cannot, in general, be eliminated by an annealing operation even when such a treatment causes recrystallization. Instead, the formation of a new annealing texture results, which is related to the deformation texture by standard lattice rotations. Application of texture hardening lies in the use of suitably textured sheet for high biaxial strength e.g., pressure vessels etc. Chapter 8 related to deformation forming shall describe this effect of texture in much more details.

Table 5.8. Deformation Textures in Metals with Common Crystal Structures.

Structure	*Wire (fiber texture)*	*Sheet (rolling texture)*
bcc	[110]	$\{112\}\langle 1\bar{1}0\rangle$ to $\{100\}\langle 011\rangle$
fcc	[111], [100] double fiber	$\{110\}\langle 112\rangle$ to $\{3\bar{5}1\}\langle 112\rangle$
cph	[210]	$\{0001\}\langle 1000\rangle$

Reproduced with permission from R.E. Smallman and R.J. Bishop, Modern Physical Metallurgy and Materials Engineering, 6th Ed., Elsevier, Oxford, 1999, p. 233.

5.14.1 Combined Strengthening Mechanism

Many commercial alloys have been developed, where a number of strengthening mechanisms add on. A typical example is carbon steel, say 0.15% C steel containing Mn, where additive contributions of grain size effect, solute strengthening (solid solution strengthening of ferrite by Mn) and second phase strengthening (volume fraction of pearlite containing cementite) may be operative. Figure 5.49 illustrates the feature quantitatively.

Another example of strengthening from several mechanisms is the formation of martensite in steels. An extremely hard and brittle phase is formed in this diffusionless transformation, the details of which are given in Chapter 3. In martensite strengthening the role of very fine grain size, high dislocation density and high saturation of interstitial carbon solute atoms are cumulative.

5.15 TOUGHENING IN CERAMICS AND COMPOSITES

Internal flaws are major sources for weakening of glass and ceramics. We have described earlier how once the fracture toughness is known, the fracture stress or the stress at which the crack will propagate can be obtained. The stress at failure may be less than the yield stress. Flaws in almost all silicates and oxide based ceramics and glasses invariably grow at stress levels lower than the calculated stress that produces a catastrophic fracture. Such a crack growth is often environmentally assisted and is produced by a reaction between the shared bonds at the top of the crack with water or other molecules.

A simplest method to strengthen a ceramic is through careful processing so as to eliminate as much as possible, flaws in the material. Application of simple equation for fracture mechanics $K_{IC} = \sigma\sqrt{\pi a}$, would suggest that, for a common ceramic having a fracture toughness of 4 MPa m$^{1/2}$, a reduction in flaw size from 1 to 0.1 mm has the effect of increasing the maximum tensile stress that the ceramic will withstand from 16 to 56 MPa.

A unique type of toughening, often witnessed in ceramic systems is *transformation toughening*. A typical example is zirconia (ZrO_2) which possesses monoclinic, tetragonal and cubic polymorphs with increasing temperature. When zirconia is cooled through 1000 °C the accompanying volume change causes ceramic shattering. However, if a small amount of Yttria, CaO or MgO is added to the ZrO_2, the tetragonal phase can be stabilized down to room temperature. The particle size of stabilizer is critical (0.05 to 1 μm size). The ability of tetragonal precipitates to get transformed to monoclinic zirconia on application of a stress or passage of a

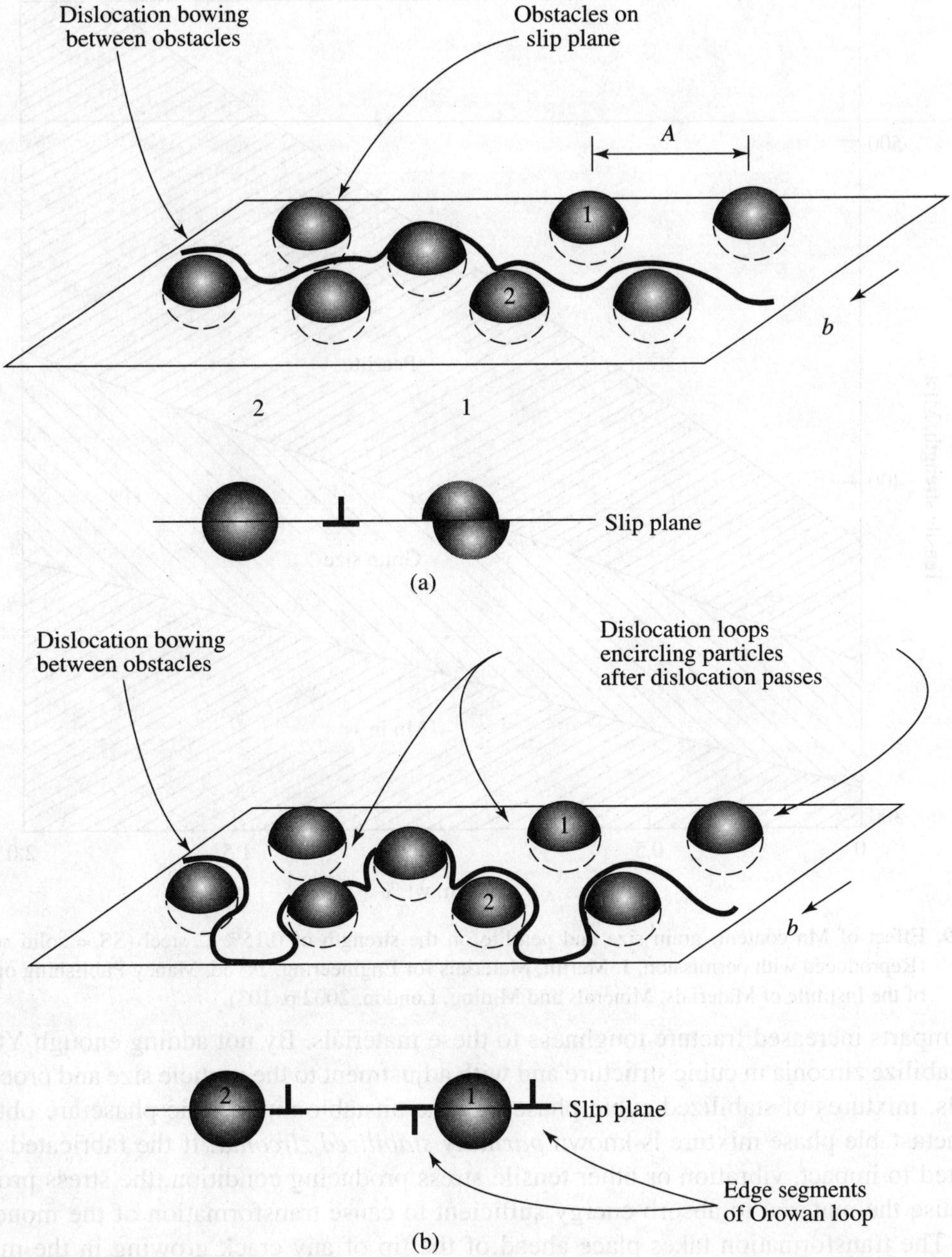

Fig. 5.48. Dislocation-particle interaction in two phase alloys (a) bowing and cutting through precipitates, (b) dislocations by passing particles behind after leaving behind loops.

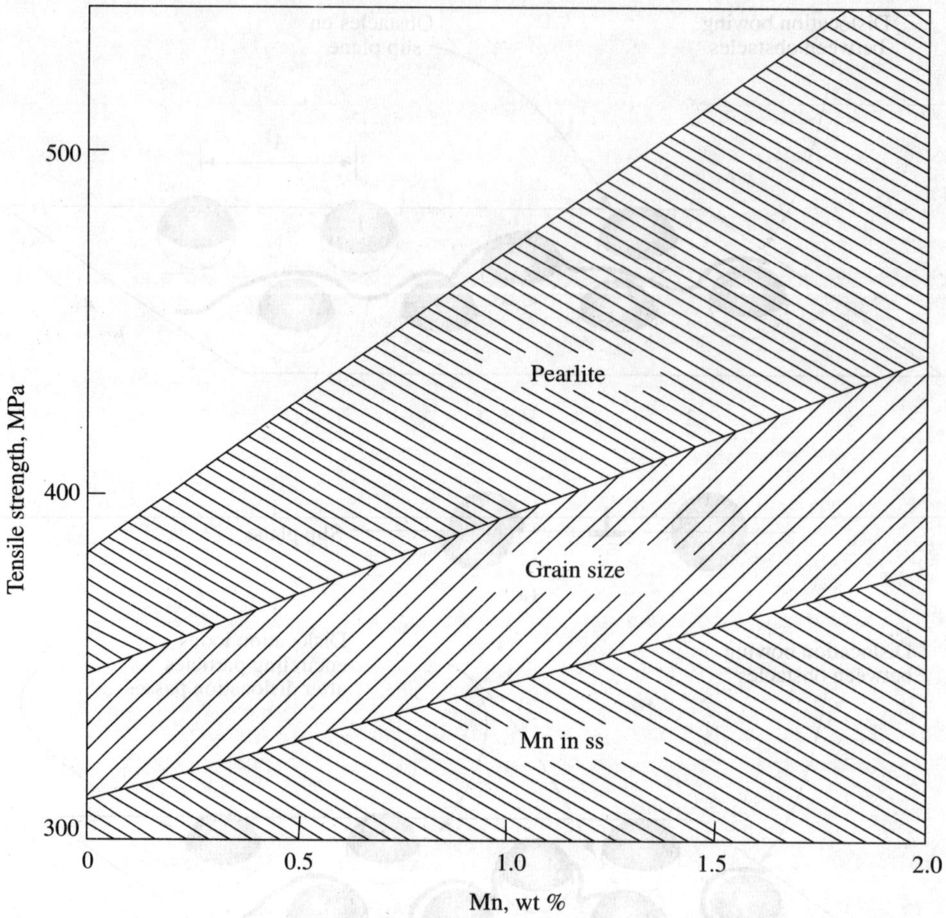

Fig. 5.49. Effect of Mn content, grain size and pearlite on the strength of 0.15% C steel (SS = solid solution) (Reproduced with permission, J. Martin, Materials for Engineering, 2nd ed, Maney Publishing on behalf of the Institute of Materials, Minerals and Mining, London, 2002 p. 103).

crack imparts increased fracture toughness to these materials. By not adding enough Yttria to fully stabilize zirconia in cubic structure and with adjustment to the particle size and processing controls, mixtures of stabilized cubic phase and the unstable monoclinic phase are obtained. This metastable phase mixture is known *partially stabilized zirconia.* If the fabricated part is subjected to impact, vibration or other tensile stress producing condition, the stress produced will cause the material to absorb energy sufficient to cause transformation of the monoclinic phase. The transformation takes place ahead of the tip of any crack growing in the material (Figure 5.50) resulting in a decrease in the stress level and in turn causing the crack to propagate and hence an increase in the toughness of the ceramic material. As with other materials, such ceramics can be heat treated in a variety of ways to modify their microstructures for particular applications.

Another method of toughening glass and ceramics is by subjecting the surface under compression. The role of compressive stress in enhancing the fatigue life has been described

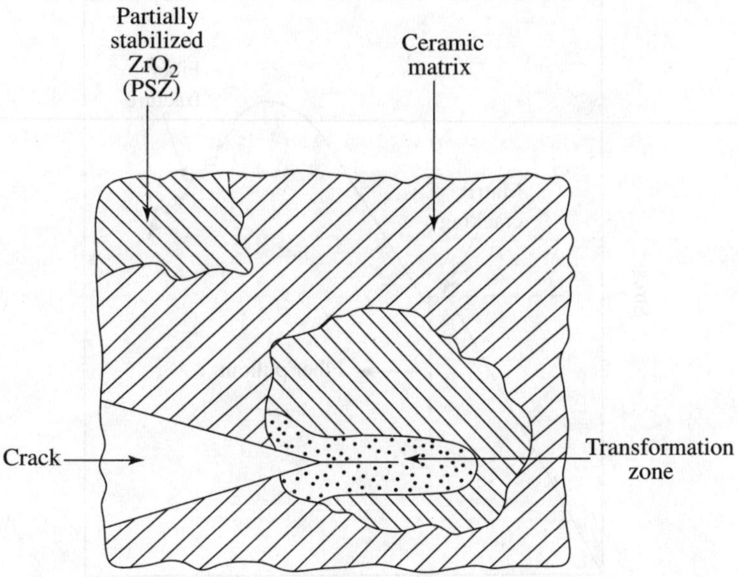

Fig. 5.50. Formation of a transformation zone surrounding the tip of a crack due to polymorphic transformation of ZrO_2 (Reproduced with permission, J.A. Jacobs and T.F. Kilduff, Engineering Materials Technology: Structures, Processing, Properties and Selection, 5th ed, Pearson Education , Inc., Upper saddle River, NJ, 2005).

earlier. The surface compression can be achieved by ion exchange or thermal treatment. In the former the larger ions are substituted for the smaller ions, which after cooling introduces compressive stress over the surface. In case of thermal treatment, a glassy material is cooled rapidly after heating, so as to develop solid hard case around the still somewhat viscous interior. This process imparts a compressive prestress. This method is commonly adopted in the toughening of tempered glass and composite material like prestressed concrete. In case of fiber reinforced composites, a multistep process is used to achieve this prestressing. First the fiber are tensioned mechanically, second, they are encapsulated by the matrix material, which is allowed to solidify. Third the tensioning force is released, causing the fibers to contract towards their original length, thus placing the matrix under a compressive stress. During application when the external tensile stress is applied at the composite, the compressive stress in the matrix must be exceeded before the matrix can actually feel the presence of tensile stress.

In fiber reinforced ceramic matrix composites, the toughness is modified by varying the interfacial bond between fibers and matrix materials. In such composites the bond between the fibers and matrix must be weakened to allow the fiber to absorb more of the energy that is initiating or propagating a crack. This is opposite to what we do in bonding with a polymeric matrix, which calls for an increase in the interfacial bonding. The main purpose why the silicon fibers are coated with carbon before they are used in a matrix of silicon carbide, is to reduce the interfacial bonding between the fiber and the matrix material. By doing this the fibers are debonded from the matrix material and pull out before breaking. Figure 5.51 shows a typical stress-strain plot of such a composite, where fiber pull out and matrix cracking is indicated.

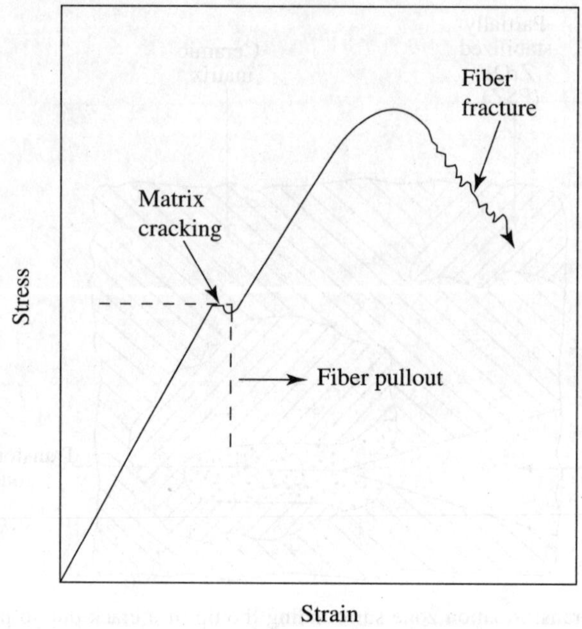

Fig. 5.51. Stress-strain diagram of continuous fiber reinforced ceramic matrix composite exhibiting increased toughness.

Each of these events takes energy away from that required to propagate a crack, thus leading to increased toughness. Another contribution to toughening may be the blockage of crack path by the fibers, thus diverting the crack to run along the fiber-matrix interface.

If we compare the stress-strain plot for brittle fibers and ductile matrix (Figure 5.52), it is evident that the modulus of elasticity of fiber is very high as compared to that of the matrix. The stress-strain plot of the composite in the longitudinal direction is also shown in the figure. It is clear that the onset of composite failure begins as the fibers start fracturing at a strain level of E_f. It is worth noticing that the composite failure is not catastrophic.

In fiber reinforced composites, attention must be paid on the fiber length. For strengthening to occur there is a critical length l_c, which is expressed as

$$l_c = \frac{\sigma_f d}{\tau_c} \tag{5.92}$$

where σ_f = ultimate tensile strength of the fiber, d = diameter of the fiber and τ_c is the fiber-matrix bond strength or shear yield strength of the matrix. In case the fiber length is far far greater than critical length (generally $l > 15\ l_c$), the fibers are termed as continuous. In case the length is less than critical, the fiber is known as *short fire* or *discontinuous fiber*. In this case the load transfer is not through the fiber. Under such circumstances, the composite is essentially a particulate composite.

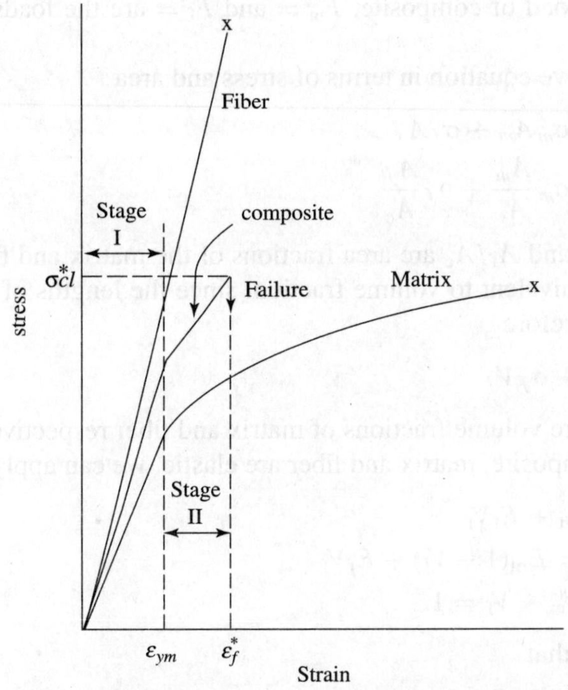

Fig. 5.52. Schematic stress-strain curve for an aligned fiber reinforced composite subjected to longitudinal tensile loading (Reproduced with permission from W.D. Callister, Jr, Materials Science & Engineering 5th ed John Wiley & Sons, Inc., NJ, 2001,p. 531).

The rule of mixture equations predict that the modulus of elasticity of the particulate (E_c) composite must fall between an upper bound given by and a lower bound by

$$E_c = E_m V_m + E_p V_p \tag{5.93}$$

$$E_c = \frac{E_m E_p}{V_m E_p + V_p E_m} \tag{5.94}$$

where E and V are elastic modulus and volume fraction respectively. m and p designate matrix and particle respectively. Such composites are also known as large-particle composites in contrast to dispersion strengthened systems described in section 5.14. The typical examples of the former are filler filled polymers and concretes.

Continuous fibers in composites are invariably aligned, while the discontinuous fibers may be aligned or randomly oriented. The mechanical properties of composites containing aligned fibers are highly anisotropic. Let us assume that a load is applied on a fiber composite along the longitudinal direction in order to achieve deformation. In case the matrix-fiber interfacial bonding is good, the degree of deformation in both matrix as well as fiber shall be the same. The condition is known as isostrain condition. Here

$$F_c = F_m + F_f \tag{5.95}$$

where F_c = total load of composite, F_m = and F_f = are the loads shared by matrix and fiber respectively.

Converting the above equation in terms of stress and area

$$\sigma_c A_c = \sigma_m A_m + \sigma_f A_f \tag{5.96}$$

or $$\sigma_c = \sigma_m \frac{A_m}{A_c} + \sigma_f \frac{A_f}{A_c} \tag{5.97}$$

The terms A_m/A_c and A_f/A_c are area fractions of the matrix and fiber phase respectfully. These fractions are equivalent to volume fraction, since the lengths of composite, matrix and fiber are all equal. Therefore

$$\sigma_c = \sigma_m V_m + \sigma_f V_f \tag{5.98}$$

where V_m and V_f are volume fractions of matrix and fiber respectively.

Assuming that composite, matrix and fiber are elastic, we can apply mixture rule as

$$E_c = E_m V_m + E_f V_f \tag{5.99}$$

or $$E_c = E_m(1 - V_f) + E_f V_f \tag{5.100}$$

since $V_m + V_f = 1$.

It can be also deduced that

$$\frac{F_f}{F_m} = \frac{E_f V_f}{E_m V_m} \tag{5.101}$$

where F denotes load in longitudinal direction.

In case the fibers are discontinuous, the expression for the rule of mixtures for the elastic modulus takes a modified form as

$$E_c = K E_f V_f + E_m V_m \tag{5.102}$$

where K is known as *fiber efficiency parameter*. This depends on the V_f and the E_f/E_m ratio. Its magnitude is in the range of 0.1 to 0.6.

***Example* 5.15**: A continuous fiber reinforced epoxy based composite is to be developed containing 30 volume fraction of fibers. The minimum longitudinal modulus of elasticity and tensile strength of the composites must be 5.5×10^4 and 1000 MPa respectively. Among glass, carbon and aramid fibers, which are possible candidates and why? Given modulus of elasticity of epoxy = 3.1×10^3 MPa and its tensile strength is equal to 69 MPa. For glass fiber $E_f = 72 \times 10^3$ MPa and $(TS)_f = 3.5 \times 10^3$ MPa. For carbon $E_f = 325 \times 10^3$ MPa, and $(TS)_f = 4 \times 10^3$ MPa. For aramid $E_f = 124 \times 10^3$ MPa and $(TS)_f = 3.5 \times 10^3$ MPa.

Solution:

Here $V_m = 0.7$, $V_f = 0.3$, $E_m = 3.1 \times 10^3$ MPa and $(TS)_m = 69$ MPa

Therefore for **glass fiber reinforced composite**

$$E_c = E_m V_m + E_f V_f$$

$$= (3.10 \times 10^3 \text{MPa}) \times (0.7) + (72 \times 10^3 \text{ MPa})(0.3)$$
$$\approx 2.4 \times 10^4 \text{ MPa}$$

Not suited, since the value is less than the specified minimum.

For **carbon fiber reinforced composite**

$$E_f = 325 \times 10^3 \text{ MPa and } (\text{TS})_f = 3.5 \times 10^3 \text{ MPa}$$

Therefore

$$E_c = (3.1 \times 10^3 \text{ MPa}) \times (0.7) + (325 \times 10^3 \text{ MPa})(0.3)$$
$$= 9.9 \times 10^4 \text{ MPa}$$

This is suited since the value is greater than the specified minimum.

For **aramid fiber reinforced composite**

$$E_f = 124 \times 10^3 \text{ MPa and } (\text{TS})_f = 3.5 \times 10^3 \text{ MPa}$$

$$\text{Thus } E_c = (3.1 \times 10^3 \text{ MPa}) \times (0.7) + (124 \times 10^3 \text{ MPa})(0.3)$$
$$\approx 3.9 \times 10^4 \text{ MPa}$$

Not suited, since the value is lower than the specified minimum.

Further, for **carbon fiber reinforced composite**

$$(\text{TS})_c = (\text{TS})_m V_m + (\text{TS})_f V_f$$
$$= (69 \text{ MPa})(0.7) + (4 \times 10^3 \text{ MPa})(0.3)$$
$$\approx 1.24 \times 10^3 \text{ MPa}$$

The value is greater than the specified minimum of 1000 MPa.

QUESTIONS AND PROBLEMS

5.1 Many physical and mechanical properties of materials are simple limiting values beyond which a material undergoes drastic changes in the behaviour. List some of them.

5.2 A view parallel to the layers of (1) talc, (2) muscovite mica, and (3) kaolinite can be represented schematically.
(a) Justify why these minerals cleave as they do.
(b) Explain why talc is much softer than mica.

Strength of Solid

5.3 If the critical resolved shear stress for yielding in copper single crystal is 0.4 MPa, calculate the tensile stress required to cause the yielding when the tensile stress axis is [001].

5.4 The axis of a cylindrical single crystal of copper lies in the (111) plane. Show that, when a tensile stress σ is applied along the axis of the crystal, the resolved shear stress τ in any direction on this plane is zero.

5.5 Nickel with a FCC lattice has a lattice parameter of 3.6Å. A pure edge dislocation with Burgers vector $b = (a/2)[101]$ lies on the $(11\bar{1})$ plane of its single crystal. Is the dislocation sessile or glissile? Why? What is Burgers vector for this dislocation?

5.6 Calculate the energy stored in J/m^3 in a specimen of copper, that was cold worked to a dislocation density of 10^{16}m/m^3. Assume that dislocations here 50% edge and 50% screw in character. Given $G = 42$ GPa, $b = 2.5$Å, ν=1/2.

5.7 Although dislocation can exist in polymeric crystals, they do not play major role in deformation of the polymers. Justify the above statement.

Stress-Strain Behaviour

5.8 Describe the difference among metals, polymers, ceramics and silica glasses in relation to their tensile strength, ductility and stiffness.

5.9 The relatively large elastic deformation encountered in case of rubbery or elastomeric materials is due to stretching of atomic bonds. Is this statement correct? Discus your answer.

5.10 Show schematically how the modulus of elasticity of an amorphous polymer depends on the glass transition temperature. Assume that the molecular weight was kept constant.

5.11 Explain the difference between engineering stress and true stress and the difference between engineering strain and true strain.

5.12 The offset yield stress and the proportional limit stress are both used to characterize the beginning of nonlinear behaviour in a tensile test. Why is the offset method generally preferable?

5.13 Define 'resilience' and 'toughness' of a solid. Draw the engineering stress-strain curves for two materials one having a higher toughness and lower resilience in comparison to the other.

5.14 Shown in Figure P.5.14 a stress strain diagram for a polycrystalline metal.

(a) Suppose the dislocation density of the unstrained metal was 10^7 dislocations per cm^2. Make a qualitative estimate of dislocation density at the Hooke's law stress, *A*, at the yield stress, *B*, and at two flow stresses, *C* and *D*.

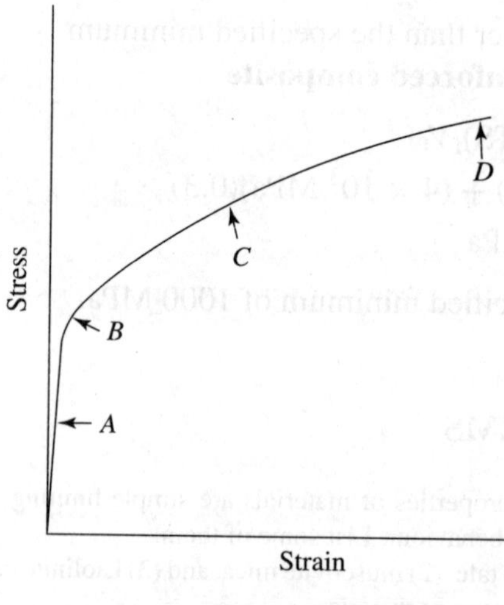

Figure P. 5.14

(b) According to dislocation theory what happens in the metal as the stress goes past the yield point *B*.

(c) As the stress increases from point *B* to *C* to *D* the metal gets work hardened. How does dislocation theory account for this increased strength?

(d) Could you determine the critical resolved shear stress from the above stress-strain diagram for a polycrystalline metal? If so, explain how; if not, explain why not. (Adapted with permission from J.D. Verhoeven, Fundamentals of Physical Metallurgy, John Wiley & Sons New York, 1975, p. 125.)

5.15 Consider two I-beams of the same size, loaded in the center by a concentrated load and supported at either end. One I-beam is made of aluminium and supports a load which produces on elastic deflection of 2 mm. If the second I-beam is made of magnesium, how much will it deflect if the same load is applied and produces only elastic action?

5.16 At extremely low temperature, how the strength and ductility of copper would behave?

5.17 What is the effect on properties of notching tensile specimens of brittle and ductile metals? What is the effect of notch on stress distribution?

5.18 In compression testing unlike tensile testing one does not meet across necking, but barelling. Explain the reasons. What is the effect of barelling on the nature of true stress-true strain curves under compressive loading?

5.19 Both glass and superplastic crystalline alloys can be pulled in tension to extremely high strain. Is the deformation mechanism identical? Justify your answer.

5.20 Both ends of a steel girder of length 20 m are clamped in position at a temperature of 10 °C. Calculate the change in length that occurs when the temperature is raised to 40 °C and determine the stress that is produced in the girder as a result. If the yield stress of the girder is 5.0×10^7 Nm^{-2}, is the thermal stress sufficient to cause a permanent deformation of the girder? Given linear coefficient of expansion for steel 1.1×10^{-5} K^{-1}, Young's Modulus for steel = 2.1×10^{11} Nm^{-2}.

5.21 Why the agreement between the calculated and experimental values of thermal shock resistance is much better when the term tensile strength is replaced by the modulus of rupture?

5.22 The K, n and m values for a stainless steel sheet are 1140 MPa, 0.35 and 0.01 respectively. A test piece has initial width, thickness and gauge length of 12.5, 0.45 and 50 mm respectively. Determine the increase in load when the extension is 10% and the extension rate of the gauge length is increased from 0.5 to 50 mm/minute. (Reproduced with permission from Z. Marciniak et al., Mechanics of Sheet Metal Forming, 2nd ed., Elsevier/Butterworth-Heinemann, Oxford, 2002, p. 13)

5.23 Explain why inclusion shape control increases the transverse ductility and toughness but has little effect on the longitudinal properties.

5.24 How much torque is needed to produce a shear stress of 80 MPa on the surface of a 40 mm diameter shaft with a polar moment of inertia equal to 25.2×10^{-8} m^4?

5.25 A steel bar of length 1 m and diameter 2 cm projects horizontally from a building. Given that the elastic limit is reached when a mass of 1000 kg is suspended from the free end of the bar, determine the yield shear stress based on a theoretical approach and explain why the values differ. Calculate the angle of shear corresponding to the elastic limit. Modulus for rigidity for steel is 84×10^9 Nm^{-2}. (Adopted with permission from Richard Turton, The Physics of Solids, Oxford University Press, Oxford, 2000, p. 86).

5.26 Prove mathematically that the maximum shear stress in a tensile specimen is half the applied tensile stress.

5.27 Why ceramics are usually much stronger in compression, followed by bending and then by tension?

5.28 Quasicrystals are long range ordered but nonperiodic. Describe the mechanical behaviour of such materials from the view point of (a) high isotropy of isocahedral symmetry, (b) glide vs. climb mechanism for dislocations, (c) twinning, (d) yield point occurrence and (e) effect of temperature e.g., the material may compete with superplastic.

Hardness

5.29 With the Vickers hardness test a 30 kg load gave for a sample of steel an indentation with diagonals having mean length of 0.53 mm. What is the hardness?

5.30 With the Brinell hardness test a sample of cold worked copper with a 1 mm diameter ball and 20 kg load give an indentation of diameter 0.63 mm. What is the hardness?

5.31 Compare the relative advantages and disadvantages of the Rockwell and Brinell hardness tests.

5.32 Give reasons why a quasicrystalline material is generally harder than a crystalline material of the same composition.

5.33 Show schematically how you would expect the hardness of the alloys to be related to their chemical composition of alloy system in *Question 3.7*.

Fatigue

5.34 Which of the following are true for a completely reversed cycling? Which ones are true for zero to tension cycling?

(a) $R = 0$

(b) $R = -1$

(c) $\sigma_a = \sigma_{max}$
(d) $\Delta\sigma = \sigma_{max}$
(e) $\sigma_m = \sigma_a$
(f) $\sigma_m = 0$

5.35 **The Figure P.5.35 shows S-N plot for unnotched specimens of aluminium alloy. What is the limitation of this type of plot and how would you avoid it?**

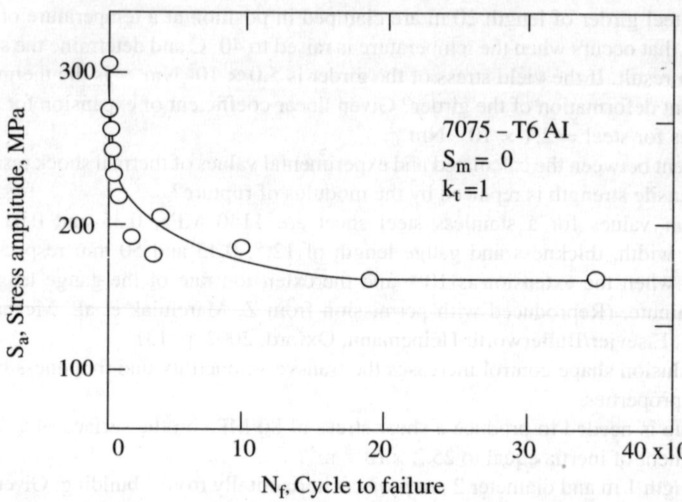

Figure P. 5.35

5.36 **The fatigue behaviour of a specimen under alternating stress conditions with zero mean stress is give by the expression:**

$$\sigma_r^a . N_f = K$$

where σ_r is the range of cyclic stress, N_f is the number of cycles to failure and K and a are material constants. It is known that $N_f = 10^6$ when $\sigma_r = 300$ MN/m^2 and $N_f = 10^8$ when $\sigma_r = 200$ MN/m^2. Calculate the constants K and a and hence the life of the specimen when subjected to a stress range of 100 MN/m^2. (Reproduced with permission from E.J. Hearn, Mechanics of Materials, Vol. 2, 2nd ed., Elsevier, Oxford, 1985, p. 887).

5.37 **Using the data for 2024 T4 Al alloy in Table 5.6, plot the S-N curve on log-linear scale covering N_f from 1 to 10^6.**

5.38 **Predict using the Goodman diagram for a steel having a fatigue limit of 320 MPa and a tensile strength of 740 MPa. Whether the falling stress amplitudes are likely to lead to fatigue failure (a) 100 MPa at a mean stress of 200 MPa (b) 150 MPa at a mean stress of 400 MPa?**

Creep

5.39 **The relaxation time for creep in a thread of a plastic product is 80 hours. How much time is required for the thread to lose half of its stress at a fixed strain?**

5.40 **Discuss the role of following on creep resistance (a) alloy composition; (b) grain size; (c) prior cold work.**

5.41 **Nimonic 80 A has $C = 20$ and a Larson-Miller parameter of 22800 for a stress of 300 MPa. What will be the time to rupture at 800 °C?**

5.42 **A rubber band, made without cross linking and used for tight fastening, has been left for a year or more. Do you expect the rubber band to return to the original dimensions after unfastening? If not, why?**

5.43 **Why does high temperature creep have the same activation energy as self diffusion?**

Fracture and Fracture Toughness

5.44 Draw stress-strain curves on the same scale for a metal exhibiting (a) brittle fracture (b) brittle fracture with limited propagation; (c) ductile fracture.

5.45 Make a list of the different types of fracture that occur in single crystals. Indicate how grain boundaries add to the fracture modes, when the material is polycrystalline.

5.46 Describe the development of a cup-and-cone fracture in a tensile specimen.

5.47 Sketch schematically the stress pattern at the surface and center of a through hardening steel bar of 100 mm diameter, when quenched in water after austenitizing. If we increase the diameter of the bar, is the risk of cracking more? If so, why?

5.48 Because alloy steel has better hardenability than plain carbon steel, under hardening treatment the former gets through hardened, while the latter is shallow hardened. Describe which steel will fail first during bending and why? Show by schematic sketches the stress distribution in both types of steels during bending.

5.49 Calculate the stress required to break a bar of steel by brittle fracture assuming that the most serious surface cracks are of length 1 μm and have a tip radius of curvature of 1.0 nm. Young's Modulus of steel $= 210 \times 10^9$ Nm^{-2}, and yield stress $= 200 \times 10^6$ Nm^{-2}. Explain why steel undergoes plastic deformation at room temperature.

5.50 In case of diamond, which plane would cleave more easily, {100} or {111}? Explain.

5.51 At room temperature Steel *A* has a higher Charpy value as compared to another steel *B*. At –20 °C the relation is just the reverse. Are you sure steel *A* is suited better than steel *B* in engineering applications? Is the above information sufficient to a design engineer?

5.52 Can the glass transition temperature of a long-chain polymer be considered to be the ductile-brittle transition temperature? Explain.

5.53 Why does gray cast iron fracture normal to the specimen axis in tension and at about 45 ° to the specimen axis in compression? Compare this with the fractures in brittle polymer in tension and compression.

5.54 Determine the size of the internal flaws believed to be in a brittle glass if it fractures under a tensile stress of 50 MPa. Given elastic modulus $E = 70$ GPa and surface energy $\gamma = 0.3$ Jm^{-2}.

5.55 Why the thickness of hard coatings over a substrate are always kept to a lower value in order to minimize failure?

5.56 Explain how the notch impact test of a metal can be applied in real engineering situation.

5.57 An aluminium alloy structural component having a crack has a yield strength of 480 MPa and a plane strain fracture toughness $K_{IC} = 25$ MPa m$^{1/2}$. Using a safety factor of 2, estimate the critical crack half length.

5.58 Explain what is the limitation of linear elastic fracture mechanics. What refinement would you suggest to treat the material where the plastic zone at the crack tip is present?

Strengthening Mechanisms

5.59 Is there any advantage of strengthening by precipitation over that by dispersion of strong inert particles?

5.60 T D Nickeel (Thoria dispersed nickel) contains 2 vol % ThO_2 particles dispersed in nickel. What effect would there be for splitting the particles in half to double the number of particles, while maintaining the same volume fraction?

5.61 Why grain size strengthening contribution remains essentially constant during tempering of a steel?

5.62 Can all strengthening mechanisms be explained in terms of dislocation? That is, are there plastic flow mechanisms that do not involve dislocations?

5.63 N.J. Grant and coworkers found that reductions in the boron constent of stainless steel Met Glass foils promotes improved ductility upon crystallization by reducing the boride phase separation volume fraction. Thermomechanical processing of Fe-25Cr-20Ni – 10 at % B gave following mechanical properties (Ultimate tensile strength): 0.2 μm grain size. 1450 MPa; 0.5 mm grain size 1240 MPa, 2 μm grain size, 900 MPa. Does this fit a Hall-Petch type relation? What would be the ultimate tensile strength of a 0.01 μm grain size material obtained after processing? (Adapted with permission from L.E. Murr, Industrial Materials Science and Engineering, Marcel Dekker, New York, 1984, p. 571).

5.64 Give three mechanisms of strengthening of polymeric materials.

5.65 An inorganic glass can be strengthened by the following treatment. After the glass is heated above the softening point, its surface is cooled rapidly by blasts of air or an oil bath, while the interior remains at a higher temperature. When the temperature becomes uniform throughout, the resulting tempered glass exhibits a greater strength than would be realized with more uniform, slower cooling. Why?

Mechanical Properties of Composites

5.66 You want to develop continuous fiber reinforced metal matrix composites based on austenitic stainless steel and zinc respectively. In which system relatively larger critical volume fraction of fiber shall be required for composite strengthening? Give reasons.

5.67 What is the ratio of the longitudinal modulus of elasticity to the transverse modulus for a composite with continuous aligned fibers constituting 50% of the volume, if the tensile modulus of the fibers in 50 times that of the matrix?

5.68 A continuous fiber reinforced composite is tested under transverse loading. Derive an expression for the elastic modulus of composite, if the elastic moduli and volume fractions of matrix and fibers are known.

5.69 What is the difference in the strength of the continuous fiber reinforced composites of the following types (a) glass fiber-polyester resin composites and (b) glass fiber – epoxy resin composites. Does the density of the components have effect on composite development. Given $E_{\text{glass}} = 76$ GPa, $E_{\text{Polyester}} = 4.41$ GPa and $E_{\text{Epoxy}} = 2.41$ GPa.

5.70 Glass fibers provide longitudinal reinforcement for a nylon. The fiber diameter is 20 μm and the volume fraction is 0.45. (a) what fraction of the load is carried by the glass (b) what is the stress in glass when the average stress in the composite is 14 MPa?

5.71 A 1060 steel wire (1 mm in diameter) is coated with copper (combined diameter = 2 mm). The two materials have elastic moduli 205×10^3 MPa and 110×10^3 MPa respectively and yield strengths of 280 MPa and 140 MPa respectively. (a) If this composite is loaded in tension, which metal will yield first? (b) How much load can the composite carry in tension without plastic deformation? (c) What is the modulus of elasticity of the composite material? (Adapted with permission from L.H. Van Vlack, Materials for Engineering, Addison-Wesley, Reading, USA, 1981, p. 391).

5.72 Discuss why the failure under longitudinal loading in a continuous fiber reinforced composite is not catastrophic.

5.73 Why in a composite strengthened by discontinuous parallel fibers, should the fiber length be made as long as possible?

5.74 Why unlike elastic moduli, it is difficult to predict the strength of a composite by a simple rule of mixture relationship?

5.75 Sketch stress-strain plots for concrete in tension and compression. Is there any difference and if so why?

5.76 Why in particulate reinforced composites the Young's modulus is much less than predicted by the rule of mixture? Why rule of mixture is valid for continuous fiber reinforced composites?

5.77 One difficulty with using porcelain coatings on steel is the stress buildup in the porcelain as a result of the difference in thermal expansion. Compute the compressive stress in the porcelain when it is cooled to 25 °C from 425 °C. Assume that it is stress free at 425 °C and that both components behave elastically. Porcelain has $E = 750$ kbars and $\nu = 0.25$. $\alpha_{\text{Steel}} = 11 \times 10^{-6}\ \text{K}^{-1}$, $\alpha_{\text{Porcelain}} = 4 \times 10^{-6}\ \text{K}^{-1}$ [1 bar = 0.1 MPa]. (Adapted with permission from A.L. Ruoff, An Introduction to Materials Science, Prentice-Hall, N.J., 1972, p. 502).

5.78 Al-4% Cu alloy is available to precipitation hardening type heat treatment. Using phase diagram determine the volume fraction of precipitate phase ($CuAl_2$). Also, estimate the size of the precipitate particles (assuming spherical in shape) and the interparticle spacing. The yield strength of the alloy was 600 MPa. Given, shear modulus G=27.6 GPa, Burgers Vector **b** = 0.25 nm, density of α - phase = 2700 kg m^{-3} and the density of precipitate phase = 4430 kg m^{-3}. (Adapted with permission from G.E. Dieter, Mechanical Metallurgy, McGraHill Book Co., New York, 1988, p. 219)

REFERENCES

Mechanical Behaviour of Materials, N.E. Dowling, Prentice Hall, N.J., 1993.

Deformation and Fracture of Solids, R.M. Caddell, Prentice Hall, N.J., 1980.

Mechanics of Materials, Vol. 2, 2nd ed., E.J. Hearn, Pergamon Press, Oxford, 1985.

Deformation and Fracture Mechanics of Engineering Materials, 3rd ed., R.W. Hertzberg, John Wiley & Sons, N.Y., 1989.

Brittle Fracture in Steel Structures, ed. G.M. Boyd, Butterworths, London, 1970.

Mechanical Metallurgy, G.E. Dieter, McGraw Hill Book Co., N.Y., 1988.

Mechanical Behaviour of Materials, M.A. Meyers and K.K. Chawla, Prentice Hall Inc., N.J., 1999.

Principles of Mechanical Metallurgy, I. Le May, Arnold, London, 1981.

Fatigue and Fracture Mechanics of High Risk Parts, B. Farahmand, Chapman and Hall, N.Y., 1997.

6

ELECTROCHEMICAL PROPERTIES, CORROSION AND DEGRADATION

The study of electrochemical properties of materials is essential in the following areas:

- Direct chemical generation of electricity by batteries and fuel cells
- Electroplating (which includes electrorefining) and electrolysis
- Corrosion

To understand these phenomena we must understand the electrolytic cell. When electrodes are of different materials say copper and zinc, we call this a *dissimilar electrode cell*. An electrochemical cell is always comprised of four parts: two electrodes, an external electrical conductor and an electrolyte. Some cells also contain a semi permeable membrane which allows one or more ions to diffuse through, but does not permit the transfer of the solvent or a gas. The electrolyte may be liquid or a solid. Let us consider the electrolysis of NaCl in these two conditions.

Electrolysis of Molten Salt: Let us consider that the two electrodes are inert. If the electrodes are connected externally to a direct current power supply that is capable of impressing a suitable voltage, the sodium cations will migrate towards the negative electrode (cathode), where they can combine with electrons supplied by the external circuit to form neutral sodium atoms. Since the cell temperature is above the melting point of sodium, the product of this reaction is liquid sodium. Simultaneously, chloride ions migrate to the positive electrode (anode) where each gives up one electron and is converted into a neutral chlorine atom. The chemical reaction taking place in the cell may be written as two separate reaction (Half-Cell reaction).

$$Na^+ + e^- \rightarrow Na\ (l) \qquad \text{Cathode reaction} \qquad (6.1)$$

$$Cl^- \rightarrow 1/2\ Cl_2\ (g) + e^- \qquad \text{Anode reaction} \qquad (6.2)$$

In chemical sense, the reduction reaction occurs at cathode, while oxidation reaction at the anode. By oxidation one should not conclude the combination of the species with oxygen, but the loss of one or more electrons amounts as if the reaction species is oxidized.

The above two half cell reactions (6.1 and 6.2) may be combined to give the overall cell reaction for electrolysis of molten NaCl:

$$Na^+ + Cl^- \rightarrow Na\ (l) + 1/2\ Cl_2\ (g) \qquad (6.3)$$

It is worth noting that electrons participate in half cell reactions, but not in the overall cell reaction.

Electrolysis of Aqueous Salt Solution: Before going into details of an aqueous salt solution, let us consider a neutral water molecule, which undergoes limited dissociation into hydrogen ion and a hydroxyl ion. The equilibrium reaction is:

$$H_2O \Leftrightarrow H^+ + OH^- \qquad (6.4)$$

At any given temperature and pressure, the degree to which this process occurs is constant. By applying the Law of Mass Action for dilute solution of a salt we can say that

$$[H^+].[OH^-] \rightarrow \text{Constant} \qquad (6.5)$$

The constant measured experimentally for above equation is 10^{-4} under standard conditions, that is 25 °C and 1 atmosphere. Water is considered neutral since the concentration of hydrogen and hydroxyl ions are equal. In case the concentration of hydrogen ions exceeds the concentration of hydroxyl ions, the solution is termed *acidic*. On the other hand, if the concentration of hydroxyl ions exceeds the concentration of hydrogen ions, the solution is termed *alkaline*. The accepted method of defining acidity is by means of a term called pH. This is measured on a scale from 0 to 14 and is defined as:

$$pH \rightarrow -\log\ [H^+] \qquad (6.6)$$

The term in square bracket is the concentration of hydrogen ions expressed as a molarity. Thus a concentration of 10^{-8} M when substituted in equation (6.6) gives a pH of 8. The pH of pure water is thus 7 under standard condition. If temperature goes up (>25 °C), the pH of water falls.

Electrolysis of aqueous solution of NaCl is more complex than that for molten NaCl. This is because of the presence of water, which offers possible alternative oxidation and reduction reaction such as:

$$2H_2O + 2e^- \rightarrow H_2(g) + 2OH^- \qquad \text{Reduction reaction}$$

$$H_2O \rightarrow 1/2\ O_2\ (g) + 2e^{-1} \qquad \text{Oxidation reaction} \qquad (6.7)$$

In relatively concentrated NaCl aqueous solution, Cl^- is oxidized more easily than H_2O and H_2O is reduced more easily than Na^+. The overall cell reaction can therefore be given by

$$2Cl^- + 2H_2O \rightarrow Cl_2(g) + H_2(g) + 2OH^- \quad (6.8)$$

Example 6.1. In case of electroplating copper over an electrode of pure copper (other electrode being of impure copper), the electrolyte being aqueous solution of copper sulphate with some sulphuric acid, write down the cell reaction. How would you calculate the deposition rate?

Solution.

$$2Cu^{2+} = 4e^- \rightarrow 2Cu\,(s) \qquad \text{Cathode reaction}$$

$$2H_2O \rightarrow 4H^+ + (O_2)(g) + 4e^- \qquad \text{Anode reaction}$$

The overall cell reaction is

$$2Cu^{2+} + 2H_2O \rightarrow 2Cu\,(s) + 4H^+ + O_2(g)$$

The rate of copper deposition equals to the rate of overall cell reaction, which can be calculated in terms of the cell current:

Deposition rate (g/s) = I(A) × atomic weight (g/mol)/ nF

where n = Number of Faradays to deposit one mole

F = Faraday constant = 96490 C/mol

Faraday is equal to the magnitude of charge on one mole (i.e., 6.022×10^{23}) of electrons. In the case of copper, $n = 2$, and atomic weight = 63.54 g/mol; therefore

Copper deposition rate (g/s) ≈ $(3.3 \times 10^{-4}) \times I$(A)

In case we wish to determine the deposition rate per unit cathode area, I in the above equation is to be substituted by current density J = I/Area of cathode surface.

Example 6.2:. What is the pH value of a solution with molarity of 2×10^{-4}?

Solution:. From equation (6.6), we know that

$$\begin{aligned} pH &= -\log[H^+] \\ &= -\log(2 \times 10^{-4}) \\ &= -[\log 2 + \log(10^{-4})] \\ &= -(0.3–4) \\ &= 3.7 \end{aligned}$$

i.e., the solution is acidic in nature

6.1 ELECTROCHEMICAL AND GALVANIC CELLS

An electrochemical cell is a combination of type

Electronic conduction (metal)	Ionic conduction (electrolyte)	Electronic conduction (metal)

in which electrochemical processes may occur with the passage of electric current. If, the electrochemical cell consumes current from an external current source, and, in effect, stores chemical energy, it is named as *electrolytic cell.* On the other hand if the electrochemical cell generates electric energy, under the consumption of chemical energy, it is called a *galvanic cell.* In the next section, we shall describe this in a bit more detail.

A galvanic cell is formed when two dissimilar metals are connected electrically while both are immersed in an electrolytic solution. The study of such a cell is important to understand galvanic corrosion, where one of the electrode corrodes preferentially. Another cell, which works on a similar principle is *concentration cell,* where differing solution concentrations are present on the same metal surface. A typical example of galvanic corrosion is when steel is coupled to stainless steel. Another name for such a corrosion is *contact corrosion,* where the preferentially corroded alloy is the anode and the unattacked alloy is the cathode.

The question now arises which metal corrodes preferentially (i.e., which is anode), for which the galvanic series chart for a particular environment is required. Table 6.1 shows a typical series

Table 6.1. A Portion of the Galvanic Series in Sea Water

Corroded End (Anodic, or Least Noble)
Magnesium
Magnesium alloys
Zinc
Galvanized steel
Aluminum and its alloys
Mild steel
Cast iron
50% Pb–50% Sn alloy
Nonpassivated 304 (18% Cr–8% Ni) stainless steel
Nonpassivated 316 (918% Cr–8% Ni–3% Mo) stainless steel
Lead
Tin
Muntz metal
Naval brass
Yellow brass
Admirality brass
Aluminum bronze
Red brass
Copper
Silicon bronze
70-30 Copper-nickel
Passivated nickel
Passivated inconel
Monel
Passivated 304 (18% Cr – 8% Ni) stainless steel
Passivated 316 (18% Cr–8% Ni–3% Mo) stainless steel
Protected End (cathodic, or Most Noble)

in case of sea water. The metal higher on the list will be the anode. For environment other than sea water, the galvanic series may differ considerably.

A galvanic cell possesses a limited capacity for supplying electrical energy because of the finite amount of reactants. Batteries are, therefore, often rated in terms of ampere-hours. Other energy producing cells e.g., fuel cells do not have limitations of batteries. They shall be described later.

It must be borne in mind that the Galvanic Series should not be confused with the emf series described in Section 6.1. The emf series is a list of half cell potentials proportional to the free energy changes of the corresponding reversible half cell reactions for standard state conditions. The Galvanic Series gives only tendencies for galvanic corrosion and it does not give any information about the rate of attack i.e., kinetics. It, thus, gives qualitative indications of the likelihood of galvanic corrosion but quantitative predictions are rather impossible.

6.2 STANDARD ELECTRODE POTENTIALS

The transfer of current from an electrode to an electrolytic (ionic) conductor is possible only through an electrode process, i.e., a chemical reaction, in which free electrons take part.

When the voltage in a cell is measured under standard conditions (i.e., temperature = 25 °C, pressure = 1 atm and electrolyte concentration = 1 molar) and vanishingly small current, it is called the *standard state potential* (ΔV°). Since an overall cell reaction can be written as the sum of two half-cell reactions, each half-cell can be considered to have a half-cell potential such that the difference between the cathode and anode potentials equals the cell voltage:

$$V_c^\circ - V_a^\circ = \Delta V^\circ \tag{6.9}$$

The cathode potential is algebraically larger than the anode potential.

Individual electrode potentials are measured relative to a *standard hydrogen electrode (SHE)*. The hydrogen electrode is arbitrarily assigned a half-cell potential of zero when the inert platinum electrode is immersed in an acid solution of unit hydrogen ion concentration (i.e., 1 molar H^+) at 25 ° C and hydrogen gas at 1 atmosphere is bubbled over its surface. In other words, a half-cell potential of zero is taken for the reaction $2H^+ + 2e^- \Longleftrightarrow H_2$ under standard conditions and this serves as a reference for the determination of other half-cell or electrode potentials. Figure 6.1 illustrates a typical copper-hydrogen galvanic cell. Table 6.2 gives the *standard electrochemical potential* values of a number of half-cell reactions.

Electrochemical potentials may be reported directly in terms of the secondary reference or a value may be calculated in terms of the SHE. Table 6.2 summarizes the potential values for common reference electrodes. Conversion of potential from one reference electrode scale to another can be done. For example an electrode which measures – 0.341 V vs. SCE (Saturated Calomel Electrode) would measure only – 0.1 V vs. SHE. The saturated calomel electrode is the most commonly used secondary electrodes for laboratory purpose. It consists of mercurus chloride Hg_2Cl_2, mixed in a mercury paste on a pool of liquid mercury in contact with a saturated KCl solution.

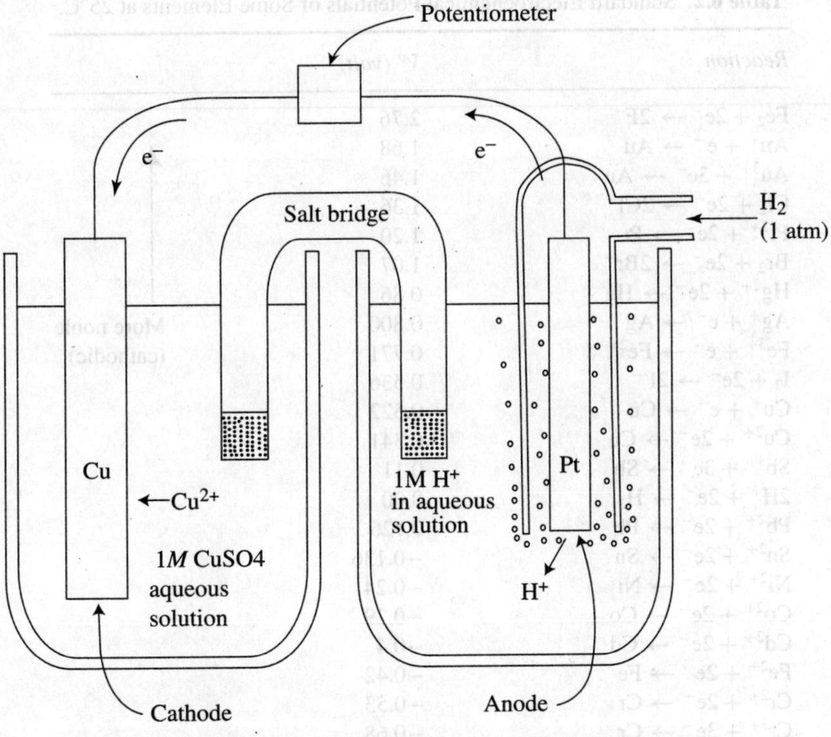

Fig. 6.1. A copper-hydrogen galvanic cell.

6.3 POLARIZATION

All electrochemical reactions proceed only at a finite rates. In case, excess electrons with their negative charges accumulate at the metal/solution interfaces due to the reason that, the reaction is not fast enough to accommodate all the available electrons, a negative potential change (*cathodic polarization*) results. Similarly, a deficiency of electrons in the metal liberated at the interface produces a positive potential change called *anodic polarization*. Anodic polarization, thus, represents a driving force for corrosion by the anodic reaction.

Elecrochemical polarization is of three types:

- Concentration polarization
- Resistance polarization
- Activation polarization

Concentration polarization is caused by a deviation of the concentration on the electrode surface from that of the bulk solution. As an example, when copper is deposited on the cathode during copper refining, a lower copper ion concentration at the cathode results than in the bulk solution. On the other hand, at the anode, copper is dissolved, resulting in a higher concentration than in the bulk.

Table 6.2. Standard Electrochemical Potentials of Some Elements at 25°C.

Reaction	*V° (volt)*	
$Fe_2 + 2e^- \rightarrow 2F^-$	2.76	
$Au^+ + e^- \rightarrow Au$	1.68	↑
$Au^{3+} + 3e^- \rightarrow Au$	1.46	
$Cl_2 + 2e^- \rightarrow 2Cl^-$	1.36	
$Pt^{2+} + 2e^- \rightarrow Pt$	1.20	
$Br_2 + 2e^- \rightarrow 2Br^-$	1.07	
$Hg^{2+} + 2e^- \rightarrow Hg$	0.86	
$Ag^+ + e^- \rightarrow Ag$	0.800	More noble
$Fe^{3+} + e^- \rightarrow Fe^{2+}$	0.771	(cathodic)
$I_2 + 2e^- \rightarrow 2I^-$	0.536	
$Cu^+ + e^- \rightarrow Cu$	0.522	
$Cu^{2+} + 2e^- \rightarrow Cu$	0.341	
$Sb^{3+} + 3e^- \rightarrow Sb$	0.11	
$2H^+ + 2e^- \rightarrow H_2$	0.00	
$Pb^{2+} + 2e^- \rightarrow Pb$	0.126	
$Sn^{2+} + 2e^- \rightarrow Sn$	−0.136	
$Ni^{2+} + 2e^- \rightarrow Ni$	−0.24	
$Co^{2+} + 2e^- \rightarrow Co$	−0.29	
$Cd^{2+} + 2e^- \rightarrow Cd$	−0.4	
$Fe^{2+} + 2e^- \rightarrow Fe$	−0.42	
$Cr^{2+} + 2e^- \rightarrow Cr$	−0.53	
$Cr^{3+} + 3e^- \rightarrow Cr$	−0.68	
$Zn^{2+} + 2e^- \rightarrow Zn$	−0.762	
$Ti^{2+} + 2e^- \rightarrow Ti$	−1.63	
$Al^{3+} + 3e^- \rightarrow Al$	−1.67	More active
$Mg^{2+} + 2e^- \rightarrow Mg$	−2.36	(anodic)
$Ca^{2+} + 2e^- \rightarrow Ca$	−2.71	
$K^+ + e^- \rightarrow K$	−2.92	
$Rb^+ + e^- \rightarrow Rb$	−2.96	
$Cs^+ + e^- \rightarrow Cs$	−2.97	
$Li^+ + e^- \rightarrow Li$	−3.04	
$Na^+ + e^- \rightarrow Na$	−2.71	↓

Table 6.3. Potential Values for Common Secondary Reference Electrodes. Standard Hydrogen Electrode included for reference.

Name	*Half-Cell Reaction*	*Potential V vs. SHE*
Mercury-Mercurous Sulfate	$HgSO_4 + 2e^- = Hg + SO_4^{2-}$	+0.615
Copper-Copper Sulfate	$CuSO_4 + 2e = Cu + SO_4^{2-}$	+0.318
Saturated Calomel	$Hg_2Cl_2 + 2e^- = 2Hg + 2Cl^-$	+0.241
Silver-Silver Chloride	$AgCl + e^- = Ag + Cl^-$	+0.222
Standard Hydrogen	$2H^+ + 2e^- = H_2$	+0.000

Resistance (or Ohmic) polarization is due to an ohmic resistance in a film, e.g., an oxide film on the electrode surface, causing an ohmic potential drop. The passivity of metal surface to be described in the next section may be considered as a special case of an anodic ohmic polarization. The ohmic polarization disappears, when the current is broken.

Activation polarization is caused by a certain slow step in the electrode process requiring an activation energy for over coming the reaction hindrance. The activation polarization or *overpotential* is strongly dependent on the composition of the solution, particularly its content of anions and inhibitors. This type of polarization is independent of stirring of the solution, but decreases rapidly with rise in temperature.

When a metal e.g., Zinc is corroding in an acidic solution, both anodic and cathodic half-cell reactions occur simultaneously on the surface, as follows:

$Zn \rightarrow Zn^{2+} + 2e^-$ Anodic reaction (Oxidation) (6.10)

$2H^+ + 2e^- \rightarrow H_2$ Cathodic reaction (Reduction) (6.11)

Figure 6.2 shows the plot between potential and current density for each of the half-cell reaction. The two half-cell electrode potentials eZn/Zn^2 and eH^+/H_2 cannot coexist separately on an electrically conductive surface. In other words, each must polarize or change potential to a common intermediate value E_{corr}, known as *corrosion potential*. This is also known as *mixed potential* since, it is a mixture of the half-cell electrode potentials for two reactions i.e., anodic and cathodic. At E_{corr} the rates of the anodic and cathodic reactions are equal. Here the rate of anodic dissolution is identical to the corrosion rate i_{corr}.

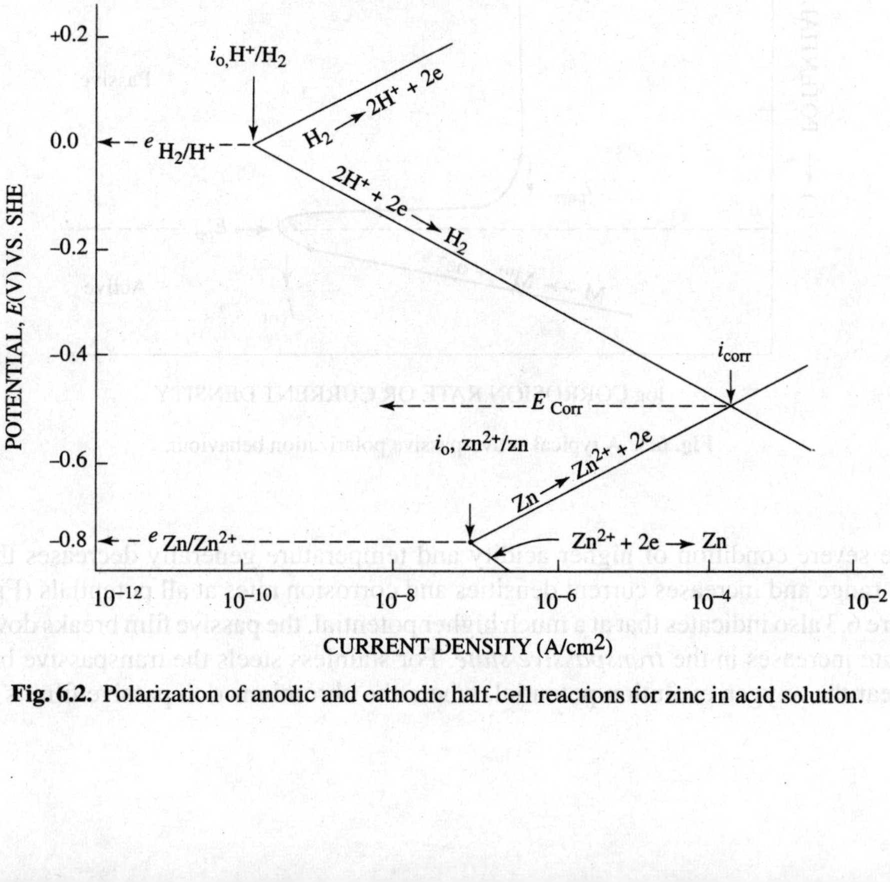

Fig. 6.2. Polarization of anodic and cathodic half-cell reactions for zinc in acid solution.

6.4 PASSIVITY

For most active metals the rate of corrosion increases as the concentration of a dissolved oxidizing agent increases. However, contrary to this, certain active metals such as chromium, iron, nickel, titanium and many of their alloys become passive or more noble in behaviour when exposed to suitable oxidizing agents at high enough concentrations. The active-passive transition exhibited by these metals is caused by the formation of a surface film of hydrated oxide that interferes with the anode reactions and drastically lowers the corrosion rate. Figure 6.3 illustrates the existence of some critical potential E_{pp}, such that below E_{pp} the metal corrodes at a relatively high rate. Depending on the oxidizing power of the solution, an alloy may exist in the *passive state* above E_{pp} or in the *active state* below it. As an illustration, type 304 stainless steel is passive in aerated but active in deaerated salt water. Passivity must not be considered as the ultimate solution for corrosion resistance. In case the passive film is brittle, its failure may result in localized form of corrosion, including pitting, crevice corrosion or stress corrosion cracking. The details of these shall be highlighted in the later section.

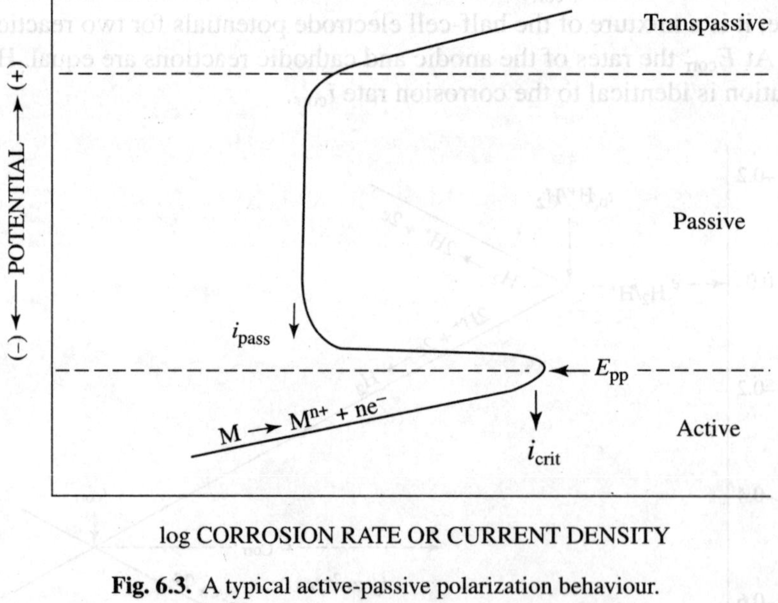

Fig. 6.3. A typical active-passive polarization behaviour.

More severe condition of higher acidity and temperature generally decreases the passive potential range and increases current densities and corrosion rates at all potentials (Figure 6.4).

Figure 6.3 also indicates that at a much higher potential, the passive film breaks down and the anodic rate increases in the *transpassive state*. For stainless steels the transpassive breakdown occurs near the oxygen evolution potential, where the chromium-rich passive film is unstable.

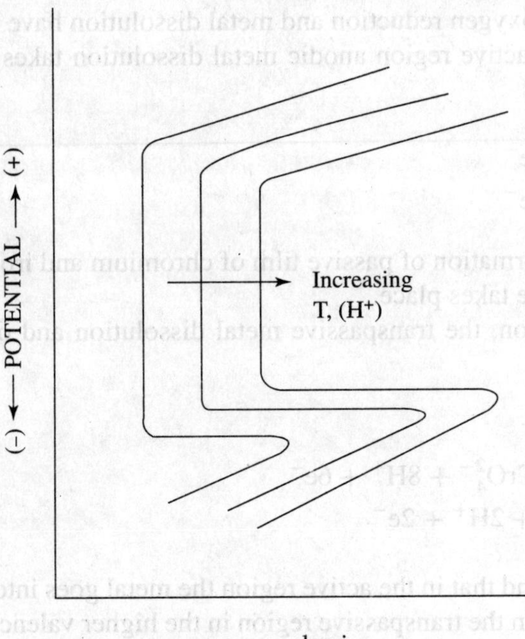

Fig. 6.4. Effect of increasing acid concentration and temperature on passivity.

Example 6.3. Figure Ex. 6.3 shows the quasi stationary polarization curves of 13% Cr steel in following media (a) 0.2 M H_2SO_4 and (b) 0.2 M Na_2SO_4. Discuss the special features.

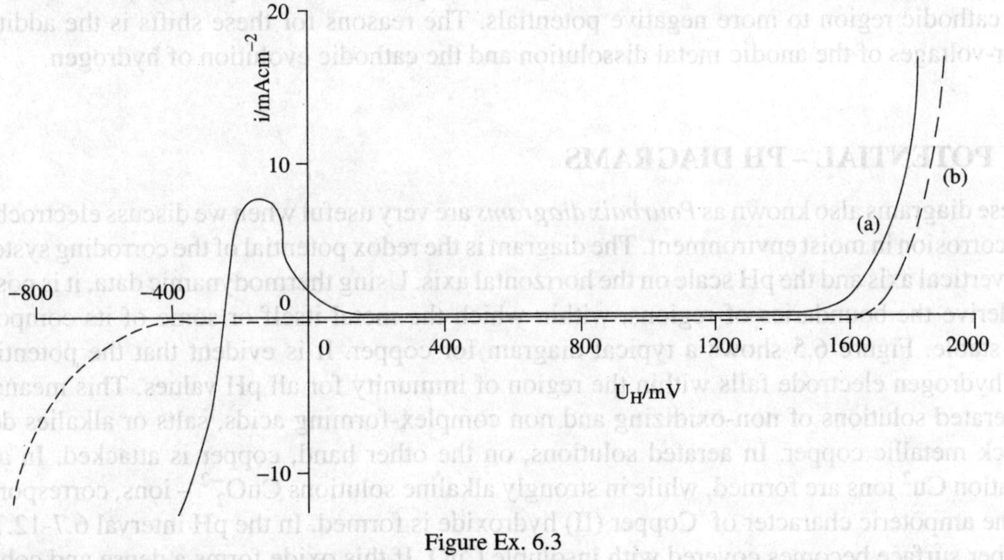

Figure Ex. 6.3

Solution. The curve (a) can be described in two parts.

Cathodic current: In this region hydrogen evolution takes place predominantly

$(H^+ + e^- \rightarrow 1/2H_2)$. The oxygen reduction and metal dissolution have been neglected.

Anodic current: In the active region anodic metal dissolution takes place predominantly, according to

$$Fe \rightarrow Fe^{2+} + 2e^-$$
$$Cr \rightarrow Cr^{3+} + 3e^-$$

In the passive region, formation of passive film of chromium and iron oxides and the slow corrosion in the passive state takes place.

In the transpassive region, the transpassive metal dissolution and oxygen evolution take place according to

$$Fe \rightarrow Fe^{3+} + e^-$$
$$Cr + 4H_2O \rightarrow CrO_4^{2-} + 8H^+ + 6e^-$$
$$H_2O \rightarrow 1/2O_2 + 2H^+ + 2e^-$$

It should be borne in mind that in the active region the metal goes into solution in the lower valence oxidation state and in the transpassive region in the higher valency oxidation state. The beginning of the transpassive region is termed the breakdown potential.

In electrolytes which cause pitting, the breakdown potential is displaced in a negative dimension and is termed pitting potential.

The curve (b) pertains to neutral solution. As compared to curve (a), in this case no active region is noticed. This is due to the fact that the passive region extends to strongly negative potentials. In addition, the transpassive region is displaced to more positive potentials and in the cathodic region to more negative potentials. The reasons for these shifts is the additional over-voltages of the anodic metal dissolution and the cathodic evolution of hydrogen.

6.5 POTENTIAL – PH DIAGRAMS

These diagrams also known as *Pourbaix diagrams* are very useful when we discuss electrochemical corrosion in moist environment. The diagram is the redox potential of the corroding system as the vertical axis and the pH scale on the horizontal axis. Using thermodynamic data, it is possible to derive the boundaries of regions, within which the metal itself or some of its compounds are stable. Figure 6.5 shows a typical diagram for copper. It is evident that the potential of the hydrogen electrode falls within the region of immunity for all pH values. This means that deaerated solutions of non-oxidizing and non complex-forming acids, salts or alkalies do not attack metallic copper. In aerated solutions, on the other hand, copper is attacked. In acidic solution Cu^2 ions are formed, while in strongly alkaline solutions CuO_2^{-2} – ions, corresponding to the ampoteric character of Copper (II) hydroxide is formed. In the pH interval 6.7-12.7, the copper surface becomes covered with insoluble CuO. If this oxide forms a dense and coherent film, the attack stops rapidly. At pH values close to the borderline between the areas of corrosion and passivity, incomplete coverage and local attacks are likely to occur.

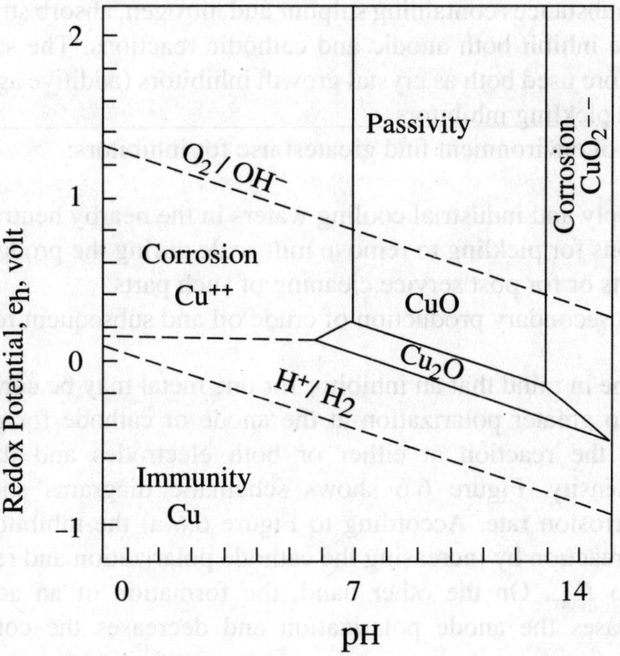

Fig. 6.5. Potential-pH diagram for copper in an aqueous solution containing 10^{-6} gram-atoms copper per litre. Temperature 25 °C.

Example 6.4:. Write down some typical reactions applicable in the following conditions:

(a) Independent of either pH or potential
(b) Dependent on pH
(c) Dependent on potential
(d) Dependent on both pH and potential

Solution.

(a) $CO_2(\text{aqueous}) + H_2O \rightarrow H_2CO_3(\text{aqueous})$
(b) $Fe^{2+} + H_2O \rightarrow Fe(OH)_2 + 2H^+$
(c) $Fe^{2+} \rightarrow Fe^{3+} + e$
(d) $Fe^{2+} + 3H_2O \rightarrow Fe(OH)_2 + 3H^+ + e$

6.6 INHIBITORS

We have mentioned about this under activation polarization. Inhibitors in general are negative catalysts. These are substances which increase overvoltage. Some anions, such as phosphate, chromate and nitrate, by forming insoluble salts or oxides with the metal exert a specific influence on anodic polarization. Thus they act as powerful anodic corrosion inhibitors. Others, particularly organic substances containing sulphur and nitrogen, absorb strongly on all the metal surface and therefore inhibit both anodic and cathodic reactions. The same types of organic substances are therefore used both as crystal growth inhibitors (additive agent) in electroplating and as corrosion and pickling inhibitors.

Following types of environment find greatest use for inhibitors:

1. Natural, supply and industrial cooling waters in the nearby neutral (pH 5-9) range.
2. Acid solutions for pickling to remove mill scale during the production and fabrication of metal parts or for post service cleaning of such parts.
3. Primary and secondary production of crude oil and subsequent refining processes.

It should be borne in mind that an inhibitor for one metal may be corrosive to others.

Inhibitors lead to greater polarization at the anode or cathode for given current density by interfering with the reaction at either or both electrodes and thus by reducing the corrosion current density. Figure 6.6 shows schematic diagrams showing the effect of inhibitors on the corrosion rate. According to Figure 6.6(a) the inhibitor interferes with the hydrogen reduction reaction by increasing the cathode polarization and reducing the corrosion current from I_{corr} to I'_{corr}. On the other hand, the formation of an adsorved layer on the anode surface increases the anode polarization and decreases the corrosion current from I_{corr} to I'corr.

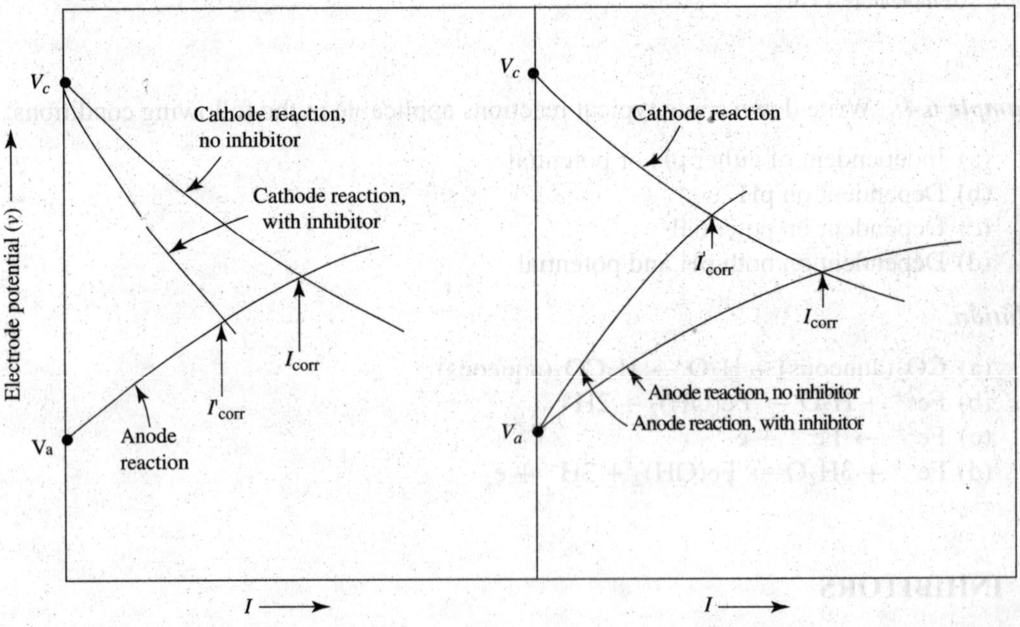

Fig. 6.6. Schematic diagram showing the effect of inhibitors on the corrosion rates: (a) An inhibitor that interferes with hydrogen reduction; (b) An inhibitor that forms adsorbed layer on the anode surface, thus increasing the anode polarization.

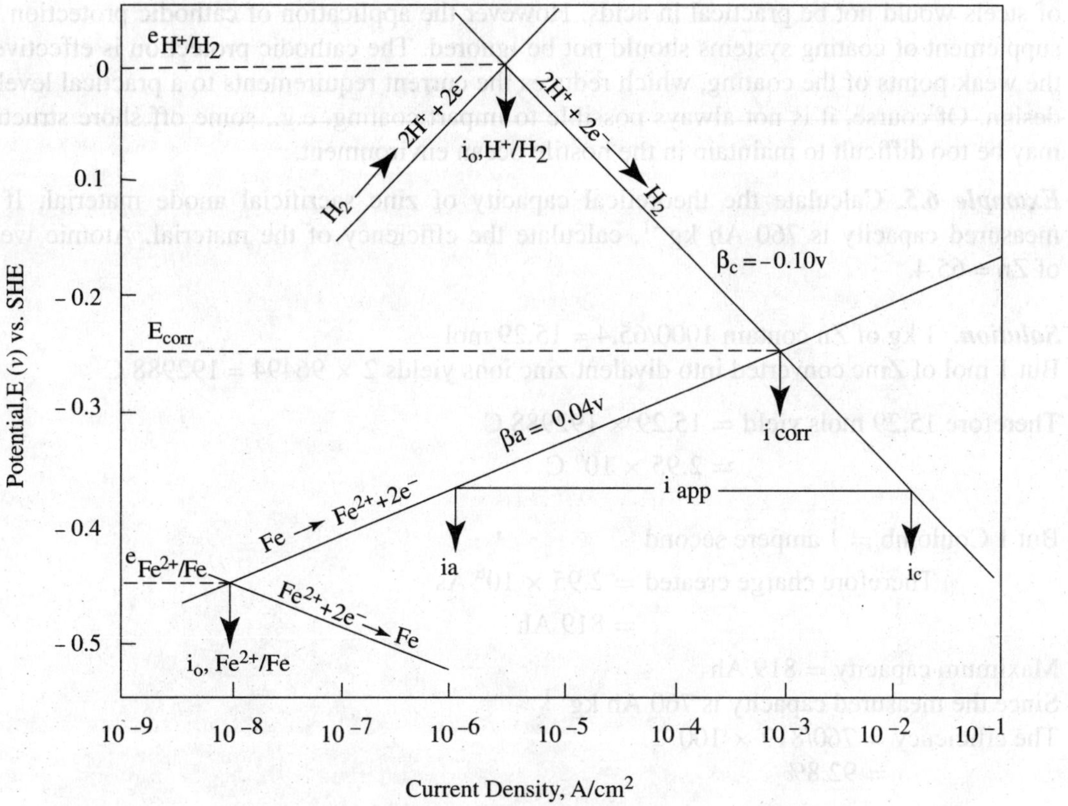

Fig. 6.7. Cathodic protection with an impressed current density for steel in an acidic solution.

6.7 CATHODIC PROTECTION

This is the second way of limiting the rate of corrosion by making a metal's potential cathodic relative to the corrosion potential. This can be achieved chemically through the appropriate galvanic coupling or electrically through an applied current. For example, a steel pipe buried in the ground can be protected by magnesium or zinc anodes that are electrically connected to the pipe. Such active anodes corrode preferentially to the steel and, therefore, are called *sacrificial anodes*. Galvanised steel i.e., coated with zinc is protected similarly. Even when the anodic zinc coating is punctured, galvanized steel corrodes only slowly in normal environments because the corrosion current is spread over a large anode area. The combination of an inert electrode and an external power supply that impresses a current produces the same effect as galvanic coupling to a sacrificial anode. The applied current $I_{app} = I_{cathodic} - I_{anodic}$. Figure 6.7 shows a polarization diagram for steel in acid solution. This is similar to that for zinc given in Figure 6.2. The cathodic polarization curve gives an idea of the impressed current density, I_{app} for cathodic protection to any required level of cathodic polarization. In case the current requirement is high, coatings are usually applied to decrease the current requirements for cathodic protections. The coatings generally is not effective for aggressive acid solutions. Thus, cathodic protection of steels would not be practical in acids. However the application of cathodic protection as a supplement of coating systems should not be ignored. The cathodic protection is effective on the weak points of the coating, which reduces the current requirements to a practical level for design. Of course, it is not always possible to impart coating, e.g., some off shore structures may be too difficult to maintain in the hostile ocean environment.

Example 6.5. Calculate the theoretical capacity of zinc sacrificial anode material. If the measured capacity is 760 Ah kg^{-1}, calculate the efficiency of the material. Atomic weight of Zn = 65.4.

Solution. 1 kg of Zn contain 1000/65.4 = 15.29 mol
But 1 mol of Zinc converted into divalent zinc ions yields 2 × 96494 = 192988 C

$$\text{Therefore } 15.29 \text{ mols yield} = 15.29 \times 192988 \text{ C}$$
$$= 2.95 \times 10^6 \text{ C}$$

But 1 Coulomb = 1 ampere second

$$\text{Therefore charge created} = 2.95 \times 10^6 \text{ As}$$
$$= 819 \text{ Ah}$$

Maximum capacity = 819 Ah
Since the measured capacity is 760 Ah kg^{-1}
The efficiency = 760/819 × 100
= 92.8%

6.8 TYPES OF CORROSION AND THEIR PREVENTION

After the description of the electrochemical fundamentals, we can now appreciate the details of various forms of corrosion and their possible prevention. It must, however, be borne in mind that not all forms of corrosions are due to electrochemical factors. Figure 6.8 shows the schematics of various forms of corrosions.

6.8.1 Uniform Corrosion (Oxidation)

Such type of corrosion is common when the corrosive environment is equally accessible to all parts of the metal surface and the metal itself is metallurgically and compositionally uniform. The latter demands are too idealistic to be met in practice. Atmospheric corrosion is the most common example in this category of corrosion.

A combination of high humidity, high average temperature, and the presence of industrial pollutants or air-entrained sea salt increases atmospheric corrosion rates. The maximum atmospheric corrosion, therefore, is evident in marine tropical or semitropical environments. The rusting of iron or steel is electrochemical in nature, such that iron anodically dissolves. In cathodic reaction, the ferric iron in FeOOH (rust) is reduced to ferrous iron in Fe_3O_4 (Magnetite) at the interface between the two. In case the anodic reaction becomes localized, the type

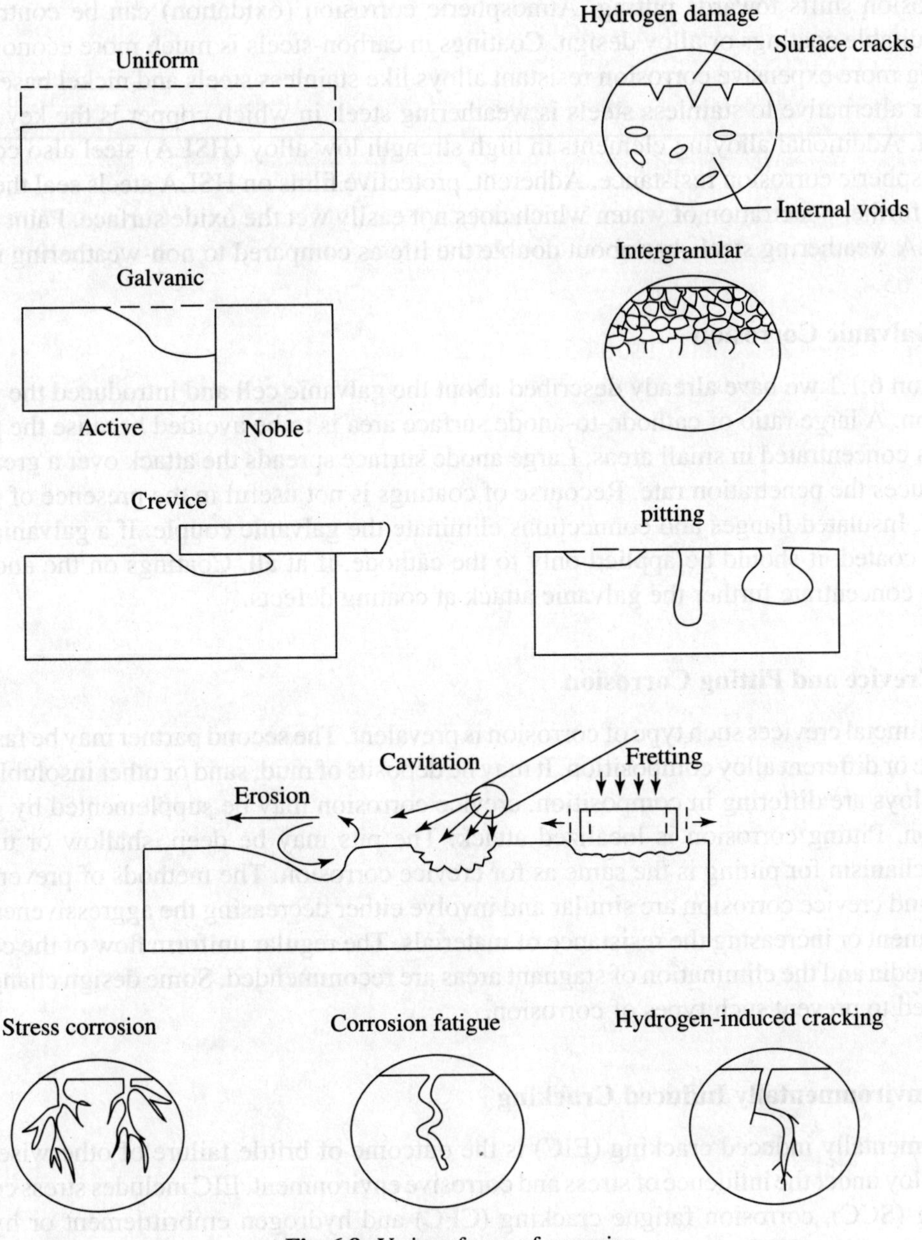

Fig. 6.8. Various forms of corrosion.

of corrosion shifts towards pitting. Atmospheric corrosion (oxidation) can be controlled by either suitable coatings or alloy design. Coatings in carbon-steels is much more economic than selecting more expensive corrosion resistant alloys like stainless steels and nickel based alloys. Cheaper alternative to stainless steels is weathering steel, in which copper is the key alloying element. Additional alloying elements in high strength low alloy (HSLA) steel also contribute in atmospheric corrosion resistance. Adherent, protective films on HSLA steels seal the surface against further penetration of water, which does not easily wet the oxide surface. Paint coatings on HSLA weathering steels has about double the life as compared to non-weathering steels.

6.8.2 Galvanic Corrosion

In Section 6.1.1 we have already described about the galvanic cell and introduced the galvanic corrosion. A large ratio of cathode-to-anode surface area is to be avoided because the galvanic attack is concentrated in small areas. Large anode surface spreads the attack over a greater area and reduces the penetration rate. Recourse of coatings is not useful in the presence of galvanic couples. Insulated flanges and connections eliminate the galvanic couple. If a galvanic couple is to be coated, it should be applied only to the cathode, if at all. Coatings on the anode only serve to concentrate further the galvanic attack at coating defects.

6.8.3 Crevice and Pitting Corrosion

In metal-metal crevices such type of corrosion is prevalent. The second partner may be fastener of the same or different alloy composition. It may be deposits of mud, sand or other insoluble solids. If the alloys are differing in composition, crevice corrosion may be supplemented by galvanic corrosion. Pitting corrosion is localized attack. The pits may be deep, shallow or undercut. The mechanism for pitting is the same as for crevice corrosion. The methods of prevention for pitting and crevice corrosion are similar and involve either decreasing the aggressiveness of the environment or increasing the resistance of materials. The regular uniform flow of the corrosive liquid media and the elimination of stagnant areas are recommended. Some design changes may be needed to prevent such types of corrosion.

6.8.4 Environmentally Induced Cracking

Environmentally induced cracking (EIC) is the outcome of brittle failure of otherwise ductile metal/alloy under the influence of stress and corrosive environment. EIC includes stress corrosive cracking (SCC), corrosion fatigue cracking (CFC) and hydrogen embrittlement or hydrogen induced cracking (HIC). In real situations more than one of the above may contribute in metal failure. Table 6.4 summarizes the comparison of the above mentioned crackings. This should be treated as a general guidelines.

Stress Corrosion Cracking is related to a relatively low constant tensile stress of an alloy exposed to a corrosive environment. One of a typical classical example of this type of cracking is the season cracking of brass in ammonia bearing environment. Caustic cracking associated with even explosion of carbon steel steam engine boilers is another example. Another example

Table 6.4. Characteristics of Environmentally Induced Cracking[a].

Characteristic	*Stress Corrosion Cracking*	*Corrosion Fatigue Cracking*	*Hydrogen Induced Cracking*
Stress	Static tensile	Cyclic with some tensile	Static tensile
Aqueous corrosive environment	Specific to the alloy	Any	Any
Temperature increase	Accelerates	Accelerates	Increases to RT, then decreases
Pure metal	More resistant	Susceptible	Susceptible
Crack morphology	TG or IG branched sharp tip	TG unbranched blunt tip	TG or IG unbranched sharp tip
Corrosion products in the crack	Absent (usually)	Present	Absent (Usually)
Crack surface appearance	Cleavage like	Beach marks and/or striations	Cleavage like
Cathodic polarization	Suppresses (usually)	Suppresses	Accelerates
Near maximum strength level	Susceptible, but HIC often predominates	Accelerates	Accelerates

[a]RT: room temperature, TG: transgranular, IG: intergranular
HIC: hydrogen induced cracks

is stainless steels under hot aqueous chloride solutions. Prevention of SCC may be done by following possible ways:

- Redesigning, so as to eliminate tensile stress in critical parts.
- Stress relief annealing or shot preening
- Lowering of the oxidizing agents in the environment
- Changing chemistry of the alloy system
- Cathodic protection may stop SCC, but hastens the hydrogen induced cracking

CorrosionFatigueCracking is caused by fluctuating stress in a corrosive environment. The failure of carbon-steel boiler tube is a typical example. Another example is the cracking of buried carbon steel pipeline carrying natural gas due to the alternating stress caused by vibrations near compressor stations. The factors mentioned above for SCC failure prevention are also valid in this case, with an additional factor of reduction or removal of cyclic stresses.

HydrogenInducedCracking (HIC) is caused by hydrogen diffusing into the lattice of the alloy. This may occur during corrosion, electroplating, cleaning and pickling or cathodic protection of the alloy. HIC effects are often reversible. HIC is common in iron alloys (BCC structure). However, FCC metals like copper, aluminium and nickel and their alloys are more resistant to HIC. Many times cold working of the alloy accelerates HIC. Some of the method for preventing HIC are

- Removing the hydrogen source
- Inert coating
- Use of inhibitors
- Stopping or reducing cathodic protection

- Annealing
- Redesigning of the equipment
- Replacing BCC ferritic steels to FCC alloys. However, this is not necessarily a full-proof solution.

6.8.5 Hydrogen Damage

Unlike HIC, here the hydrogen level is high and the damage is irreversible. One of the common example is reaction of hydrogen with the carbide phase in the steel resulting in decarburization, voids and blisters. In case of reactive metals e.g., titanium, zirconium, uranium and thorium, the hydrogen damage is through the formation of brittle hydrides.

6.8.6 Intergranular Corrosion

One of the typical examples of this type of corrosion is in case of austenitic stainless steel, which after heat treatment depletes the grain boundaries of chromium. In the temperature range of 425-815 °C, chromium carbides precipitate at grain boundaries, thus depleting the grain boundary and nearby area of chromium. In case the chromium content in the steel is below 10%, this effect is predominant. Although the chromium carbide is resistant, the adjacent chromium depleted area gets considerably corroded. Propensity of precipitation is also known as *sensitization*. Figure 6.9 shows the sensitization diagram for Type 304 stainless steel for various carbon compositions.

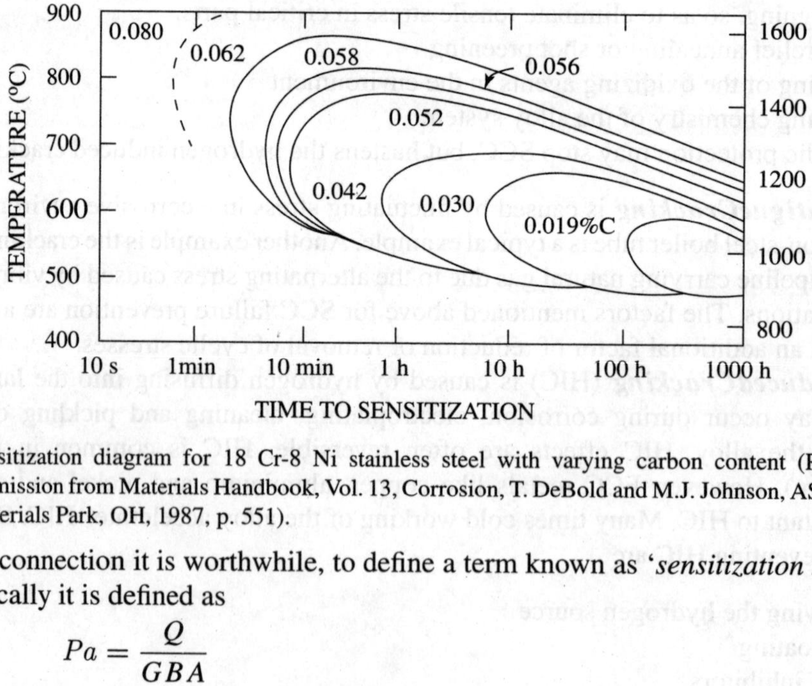

Fig. 6.9. Sensitization diagram for 18 Cr-8 Ni stainless steel with varying carbon content (Reproduced with permission from Materials Handbook, Vol. 13, Corrosion, T. DeBold and M.J. Johnson, ASM International, Materials Park, OH, 1987, p. 551).

In this connection it is worthwhile, to define a term known as '*sensitization number*' 'Pa'. Mathematically it is defined as

$$Pa = \frac{Q}{GBA} \tag{6.12}$$

where Q = charge in Coulombs, C

GBA = Grain boundary area, cm^2

GBA is calculated from the expression

$$GBA = A_s[5.095 \times 10^{-3} exp(0.34^7 x)] \quad (6.13)$$

where A_s = sample area in cm^2

x = ASTM grain size at 100*X* magnification

The most common methods to prevent intergranular corrosion are

- Solution annealing the steel above 815 °C, followed by rapid cooling (water quenching)
- Low carbon steel modifications
- Stabilized stainless steels containing niobium and titanium. These stabilizers react with carbon above 815 °C to form the respective carbides, thus preventing sensitization.

6.8.7 Dealloying

One of the typical examples in this category of corrosion is dezincification of brass (Cu-Zn alloy). Zinc is strongly active and easily leaches out of brass. Another example of dealloying is graphitic corrosion of gray cast iron. This occurs in buried cast iron pipes after a lapse of many years. Low zinc containing red brass (<15% Zn) is generally immune to dezincification, Tin addition in $\alpha - \beta$ brass castings prevents the attack in the alpha phase and produces alloy fit far marine environments. Dezincification of single phase α brass (Cu-28Zn-1Sn) is inhibited by small addition of P, As, Cr and Sb.

6.8.8 Erosion Corrosion and Fretting

This type of corrosion results due to the combination of corrosive fluid and its high velocity. In case the velocity is high, erosion of the metallic surface takes places. The fresh surface after erosion meets an accelerated corrosive attack. In case the fluid contains hard particles e.g., sand as in case of slurries, the effect of erosion-corrosion is still more significant. Erosion-corrosion is a common problem in steel pipe with flowing steam carrying condensate water droplets. *Cavitation* is a special case of erosion-corrosion when the velocity is very high. Examples are turbine blades, pump impellers, ship propellers and in tube and pipes where large pressure changes take place. Erosion-corrosion can be prevented by either design change or proper material selection. To prevent cavitation, careful design is required to minimize pressure drops across surfaces subject to high velocity liquid flow. Proper operation of pumps and other equipments is required.

Fretting is a modified version of erosion-corrosion, but in the vapour phase. The erosion is caused by repeated small movement, usually from vibration, between the corroding metal and another contacting solid under load. The erosive action abrades surface oxide film, eventually leading the surface for further oxidation. The oxide debris between the mating parts further enhances the failure. To prevent fretting one must be careful in interrupting the required surface contact and motion. It is more controlled by the abrasiveness of the wear debris and wear resistance of the materials, rather than the selection of more corrosion-resistant materials.

6.8.9 Biocorrosion

The involvement of microorganisms in the deterioration and destruction of materials can be characterized by three different terms, namely, biofouling, biodeterioration and biocorrosion or microbiologically induced corrosion. Biofouling generally refers to adherence of micro- and macro-organisms onto material surfaces in marine and fresh water systems leading to formation of fouled layers. Deterioration of nonmetallic materials such as glass, concrete, cement, rubber, wood, plastics and some organic and inorganic fluids in the presence of microorganisms is termed *biodeterioration*. Corrosion of metals and alloys induced by the activities of microorganisms is defined as *bio-or microbial corrosion*.

In microbial corrosion processes, aggressive environments are generated by micro-organisms as well as several interaction factors participating in the mechanisms. The micro-organisms often make a contribution without being solely responsible for the failure of materials.

6.9 CORROSION RATE UNITS

The corrosion rate in USA is expressed more commonly as *mils* (1 ml = 0.001 in) *penetration per year* (MPY). This is calculated as

$$\text{MPY} = 534W/DAT \tag{6.14}$$

where W = weight loss in milligrams, D = density in g cm^{-3}, A = Area in square inches and T = time in hours. Other units for W, D, A and T change the value of the constant 534. In research, many times the data are expressed as weight loss per unit area per unit time, i.e., mdd (milligrams per square decimeter per day). For conversion 1 MPY = 1.44 (mdd)/specific gravity.

Another unit of expressing corrosion is mm/year. In this case, the constant multiplying factor is 87.6 instead of 534 and the units for other parameters are the same as for MPY, except that the area A is given in square centimeter.

Another way of expressing corrosion rate (CR) is in terms of mol/m^2.s and is given by the expression

$$\text{CR} = i/n.F \tag{6.15}$$

where i = current density i.e., current per unit surface area

n = number of electrons associated with the ionization of each metal atom

F = 96500 C/mol

Table 6.5 gives the corrosion rate comparison in various units.

Example 6.6. A plain carbon steel test specimen of dimension 3 in × 4 in × 0.125 in with a 0.25 in hole for suspending in solution is exposed for 120 hours in an acid solution and loses 250 milligrams. Calculate the corrosion rate in mpy and mdd respectively. Assume the density of steel = 7.87 g mm^{-3}.

Table 6.5. Comparison of MPY with Equivalent Metric-Rate Expressions.

Relative Corrosion Resistance[a]	*mpy*	*mm/yr*	*μm/yr*	*nm/h*	*pm/s*
Outstanding	<1	<0.02	<25	<2	<1
Excellent	1-5	0.02-01	25-100	2-10	1-5
Good	5-20	0.1-0.5	100-500	10-50	20-50
Fair	20-50	0.5-1	500-1000	50-150	20-50
Poor	50-200	1-5	1000-5000	150-500	50-200
Unacceptable	200+	5+	5000+	500+	200+

[a]Based on typical ferrous and nickel alloys. Rates greater than 5 to 20 mpy are usually excessive for more expensive alloys, while rates above 200 mpy are sometimes acceptable for cheaper materials (e.g., cast iron) of thicker cross section.

Solution:. We know from equation (6.14),

Corrosion rate = 534 W/DAT mpy, where T = 120 hours and W = 250 milligram

$$\text{Area A} = 2(3\times 4) + [2(3+4)]0.125 + \pi(0.125)(0.25) - 2\left(\frac{\pi}{4}\times 0.25^2\right)$$
$$= 25.75\text{in}^2$$

Therefore Corrosion Rate $= 534 \times 250/(7.87 \times 25.75 \times 120)$
$= 5.49$ mpy

Since 1 mpy = 1.44 mdd/sp. gravity

The corrosion rate = 1 mdd

6.10 HIGH TEMPERATURE OXIDATION

6.10.1 Thermodynamics of Oxidation

Most metals readily oxidize because the change in free energy ΔG is negative for oxide formation. The variation of the standard free energy change with absolute temperature for a number of metal oxides is shown in Figure 6.10. The noble metals which are easily reduced occur at the top of the diagram and the more reactive metals at the bottom. The numerical value of ΔG° for oxidation reactions decreases with increase in temperature i.e., the oxides become less stable. Most of the ΔG versus T lines therefore slope upwards. Exception to the positive slope of the ΔG versus T line occurs for carbon oxidation to CO or CO_2. In both cases the oxide product is gaseous and thus also has a high free energy. The carbon monooxide reaction $2C + O_2 \rightarrow 2CO$ is favoured at high temperatures and, therefore, carbon is a very effective reducing agent, having a greater affinity for oxygen than most oxides. For many oxides ΔG° tends to zero at some elevated temperature. This is known as the standard dissociation temperature when the oxide is in equilibrium with the pure element and oxygen at 1 atm pressure (p) according to expression

$$\Delta G = \Delta G^\circ + RT \ln P \tag{6.16}$$

At equilibrium $\Delta G = 0$, the relationship becomes

$$P_\circ = \exp \frac{\Delta G^\circ}{RT} \tag{6.17}$$

If the pressure is lowered below this value the oxide will dissociate. On the other hand if the pressure is raised above this equilibrium value, the oxide is stable. The diagram shown in Figure 6.10 is also known as *Ellingham diagram* and shows the relative thermodynamic stability of the indicated oxides. For more details of thermodynamics readers may consult the book by D.R. Gaskell (Introduction to Metallurgical Thermodynamics, McGraw Hill, New York, 1983).

Similar prediction can be made for the atmospheres resulting from mixtures of water vapour and hydrogen and carbon monooxide and carbon dioxide (Figure 6.10).

6.10.2 Kinetics of Oxidation

During oxidation the oxygen molecules are first adsorbed on the metal surface and dissociates into atomic state, which eventually chemically combine with surface atoms of the metal. This is known as *chemisorption*. After the build up of a few adsorbed layers, the oxide is nucleated epitaxially on the grains of metal at favourable sites, such as dislocations and impurity atoms. The nucleated oxide grains and the whole metal surface is oxidized. The oxide layer can be considered as a polycrystalline aggregate. In case the oxide layer is porous the reaction would continue. As often the film is not porous, the oxidation process involves diffusion through the oxide layer. In the case of oxygen-oxide surface the contribution is by both electrons and metal ions. On the other hand, at metal-oxide interface, oxygen ions have to diffuse through the oxide and electrons migrate in the opposite direction to complete the reaction.

Figure 6.11 shows the different oxidation behaviour in metals. These are:

Logarithmic growth: This is during the initial stage of growth at low temperatures.

Parabolic growth: This occurs at intermediate temperatures e.g., 250-1000°C in iron, according to parabolic law $x^2 \propto t$. The oxidation here is a thermally activated process and the ions thermally migrate through the oxide film. The kinetics would depend on the nature of the defect structure of the oxide lattice.

Parabolic break away growth: As the oxide scale grows according to parabolic law, the interface stress level increases and eventually the oxide layer can break away. The break away parabolic law is also called para linear and is common in oxidation of titanium when the oxide reaches a critical thickness.

Linear growth: This is common in metals like U, W and Ce, where the linear rate process is associated with an interface reaction converting a thin protective oxide layer to a non-protective porous oxide.

It has been noticed that, if the oxide film grows at the oxide/metal interface, it remains in the compressive state and most probably protective. Pilling and Bedworth proposed a ratio defined

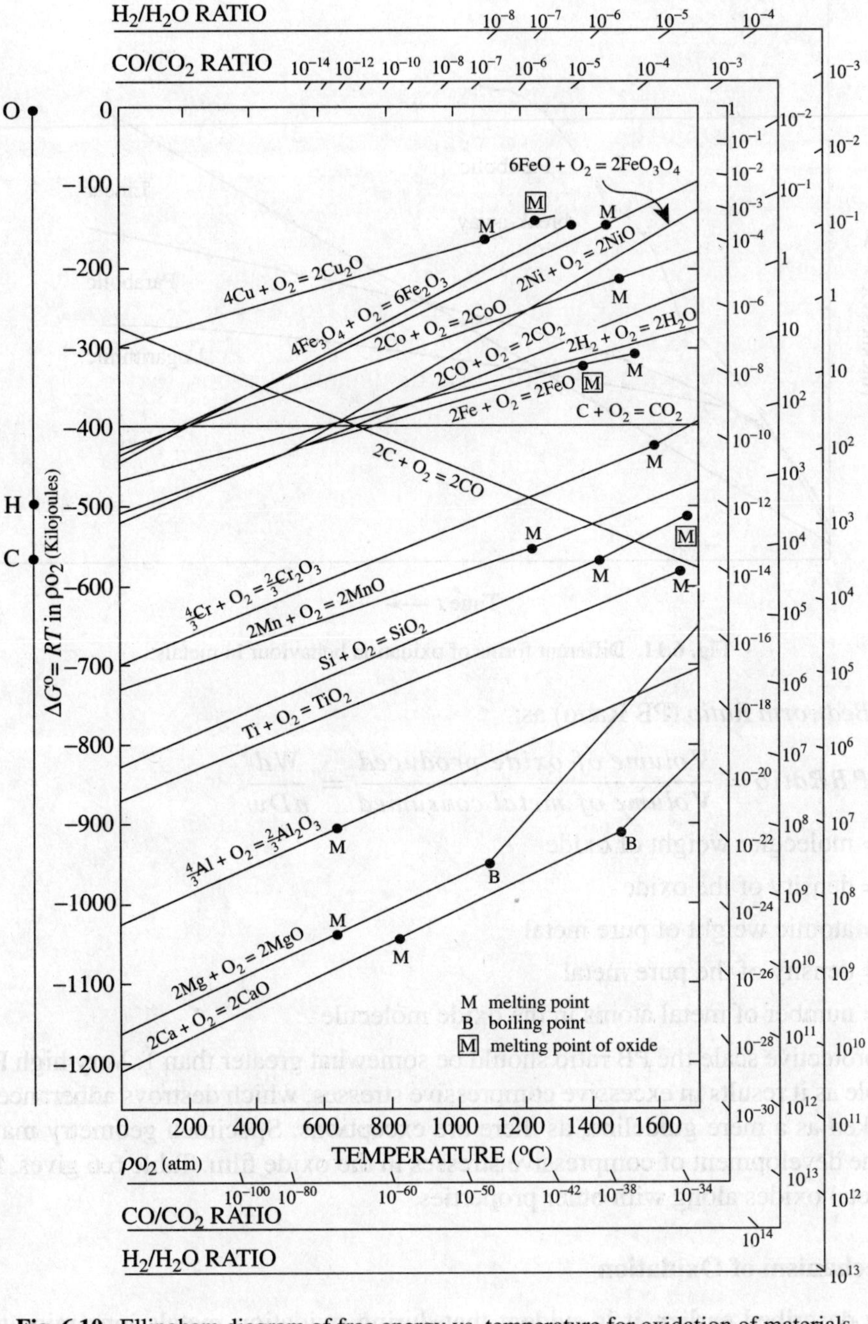

Fig. 6.10. Ellingham diagram of free energy vs. temperature for oxidation of materials.

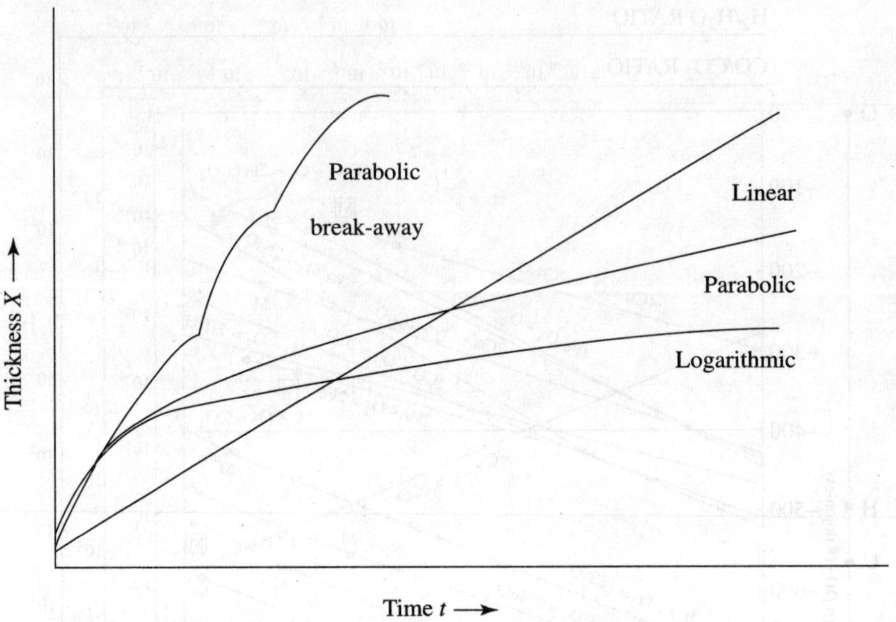

Fig. 6.11. Different forms of oxidation behaviour in metals.

as *Pilling-Bedworth Ratio* (PB Ratio) as:

$$PB Ratio = \frac{Volume\ of\ oxide\ produced}{Volume\ of\ metal\ consumed} = \frac{Wd}{nDw} \tag{6.18}$$

where W = molecular weight of oxide

D = density of the oxide

w = atomic weight of pure metal

d = density of the pure metal

n = number of metal atoms in the oxide molecule

For a protective scale the PB ratio should be somewhat greater than 1. Very high PB ratio is not desirable as it results in excessive compressive stresses, which destroys adherance. PB ratio must be taken as a mere guideline, as there are exceptions. Specimen geometry may have an effect on the development of compressive stresses in the oxide film. Table 6.6 gives, PB values of some metal oxides along with other properties.

6.10.3 Mechanism of Oxidation

As already described earlier, it is evident that during oxidation metals ions migrate out and oxygen ions inwards through the ion lattice. Electrons must migrate out wards through the oxide film in order to maintain electro-neutrality. In some cases for example iron, during high temperature oxidation two or more oxide films form (e.g., FeO, Fe_3O_4 and Fe_2O_3). Figure 6.12 illustrates this feature. Most metal oxides are nonstoichiometric, such that they either show

Table 6.6. Properties of Metal Oxides.

Metal	*Oxide*	*PB Ratio*	*Protectiveness*[a]	*Oxide Type*[b]
Aluminum	Al_2O_3	1.28	P	*n*
Calcium	CaO	0.64	NP	*n*
Cadmium	CdO	1.42	NP	*n*
Cobalt	Co_2O_3	2.40	P	*p*
Copper	Cu_2O	1.67	P	*p*
Chromium	Cr_2O_3	2.02	P	*p*
Iron	FeO	1.78	P	*p*
Magnesium	MgO	0.81	P	*n/p*
Manganese	MnO_2	2.37	P?	*n*
Molybdenum	MoO_3	3.27	NP	*n*
Nickel	NiO	1.70	P	*p*
Lead	PbO	1.28	NP	*p*
Silicon	SiO_2	2.15	P	*n*
Tantalum	Ta_2O_5	2.47	NP	*n*
Titanium	Ti_2O_3	1.76	NP	*n/p*
Uranium	UO_2	1.97	NP	*p*
Tungsten	WO_3	1.87?	NP	*n*
Zinc	ZnO	1.58	NP	*n*
Zirconium	ZrO_2	1.57	P	*n*

[a]P: Protective; NP: Not Protective
[b]n: n-type semiconductor, metal excess oxide
p: p-type semiconductor, metal deficit oxide

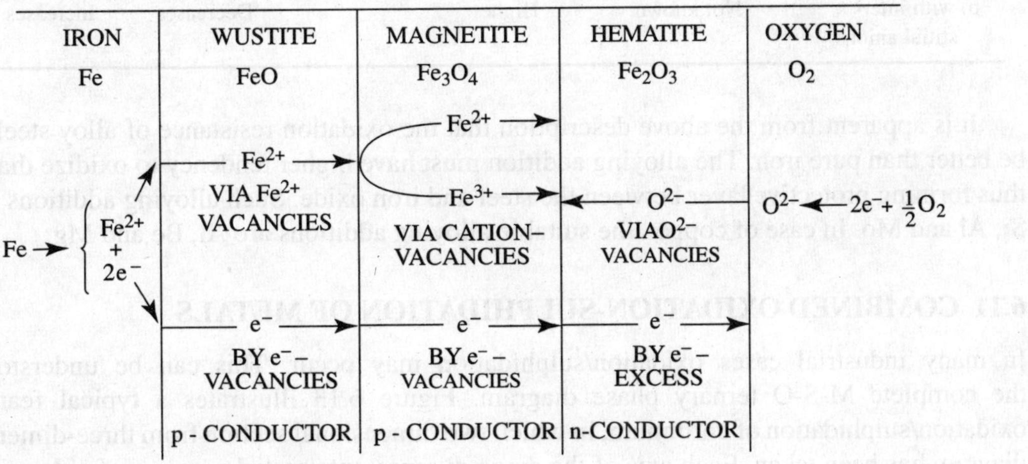

Fig. 6.12. Schematic representation of phases and conditions of diffusion in the oxidation of iron above 570 °C (Adapted with permission from An Introduction to Corrosion and Protection of Metals, G.A. Wranglen, 1972).

excess or deficit of metal. In other words, they have semiconduction behaviour. This has been highlighted in section 4.2.2. n-type semiconductor is present due to intersitial cation or anion vacancies. On the other hand p-type semiconduction takes place due to cation vacancies. Out of electron conduction and ion conduction, the rate controlling step in oxidation is due to ionic conduction. The introduction of foreign metal does influence the oxidation kinetics. For example the electron conductivity of NiO is increased by an addition of Li_2O. This feature is utilized in the positive electrode plates of alkaline Ni-Fe storage batteries. Experimentally, it is found that the oxidation of nickel is slower in the presence of Li_2O. The role of valency of the alloying metal in relation to that of the base metal is governed by the *Hauff's Valency or Wagner-Hauffe's rules*, which is demonstrated in Table 6.7.

Table 6.7. Influence of the Valency of an Alloying Metal on the Oxidation Rate of the Basic Metal. Hauffe's Valency Rules.

Type of semi-conductor oxide	*Typical oxides*	*The valency of the alloying metal in relation to that of the basic metal*	*Change of electron conductivity*	*Change of ionic conductivity and oxidation rate*
1. n-conductors (metal excess)				
a) with interstitial cations	ZnO, CdO	Lower	Decreases	Increases
b) with anion vacancies	Al_2O_3, TiO_2, Fe_2O_3, ZrO_2	Higher	Increases	Decreases
2. p-conductors (metal deficit)				
a) with cation vacancies)	NiO, FeO, Cu_2O, Cr_2O_3, Fe_3O_4	Lower	Increases	Decreases
b) with interstitial anions	Not known	Higher	Decreases	Increases

It is apparent from the above description that the oxidation resistance of alloy steels shall be better than pure iron. The alloying addition must have higher tendency to oxidize than iron, thus forming protective layer between the steel and iron oxide. Such alloying additions are Cr, Si, Al and Mo. In case of copper, the suitable alloying additions are Al, Be and Mg.

6.11 COMBINED OXIDATION-SULPHIDATION OF METALS

In many industrial cases oxidation/sulphidation may occur. This can be understood by the complete M-S-O ternary phase diagram. Figure 6.13 illustrates a typical feature of oxidation/sulphidation of chromium, in which two dimensional section from three-dimensional diagram has been taken. Each axis of the diagram represents partial pressure of either gaseous phases. In Figure 6.13(b) the reaction path way shows more oxidation, while in Figure 6.13(c) the pathway shows the existence of Cr_2O_3 and CrS depending on the values of the partial pressures. It is worth mentioning that the diffusion of S_2 in chromium is much faster than O_2. Figure 6.13(d) shows a typical case in which there is oxide stability at low P_{S_2}, but the pathway

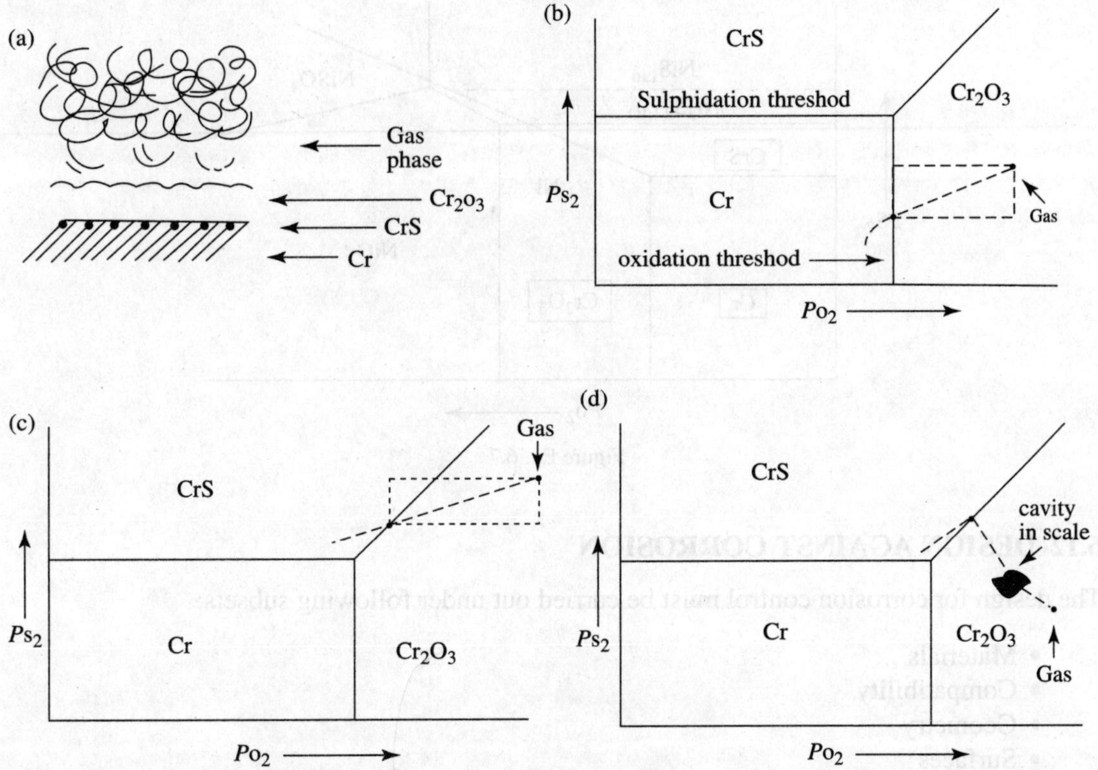

Fig. 6.13. Reaction paths for oxidation and sulphidation of chromium (Reproduced with permission from Modern Physical Metallurgy & Materials Engineering, R.E. Smallman and R.J. Bishop, 6th ed.,Elsevier, Oxford, 1999, p. 381).

direction changes with change in the partial pressures. Some times the oxide scale may crack, thus forming voids. Such condition enhances activity of S_2 and thus sulphidation.

Example 6.7:. A 80 Ni-20Cr alloy (nichrome) is subjected to the combined oxidation – sulphidation effect. Describe the oxidation/sulphidation features. Note Ni-NiS forms an eutectic reaction at a relatively low temperature of 645 °C. (Reproduced with permission from R.E. Smallman and R.J. Bishop, Modern Physical Metallurgy and Materials Engineering, 6th ed., Elsevier, Oxford, 1999, p. 382).

Solution:. Figure Ex. 6.7 is the super imposition of isothermal sections from Cr-S-O and Ni-S-O systems. From this it is evident that Cr_2O_3 is much stabler than NiO. CrS scale is formed preferentially over NiS. Although NiS is formed much later than CrS; the formation of the Ni-NiS eutectic phase plays an havoc, since the diffusivity in the melt phase is much faster. This infers that the corrosive effect in such a combined atmospheric would be severe.

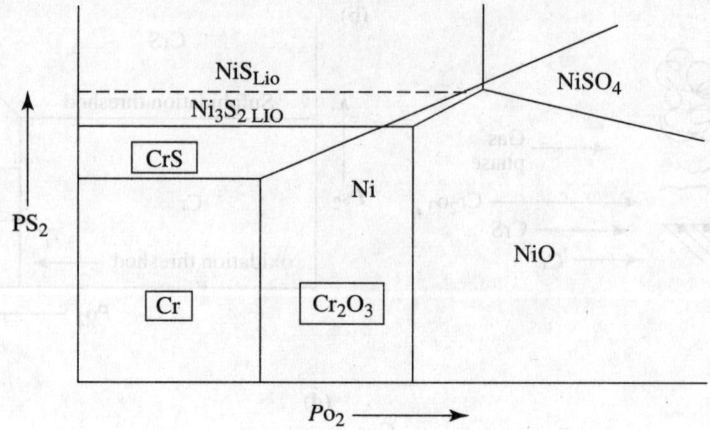

Figure Ex. 6.7

6.12 DESIGN AGAINST CORROSION

The design for corrosion control must be carried out under following subsets:

- Materials
- Compatibility
- Geometry
- Surfaces
- Protection
- Maintainability

It is not, however, essential that all points must be considered in each design. Only those points on parts appertaining to the particular design under consideration may be called upon. The objective is to enhance the ultimate practical corrosion-control reliability of the product as much as it is possible to achieve by selection, reconciliation, elimination and optimum compromise of all known corrosion-control factors.

Some of the methods for the prevention of corrosion have been indicated in section 6.8. In this section, we are more particular about the change in the design of construction to minimize/prevent corrosion. Specifics of the corrosion of the welded joints are to be described in the chapter 10 related to 'Materials Joining'. Some of the general design rules to minimize corrosion are:

1. ***Simplify the form***: The greater the angles, edges, corners and internal surfaces in a construction, more difficult becomes surface treatment. In addition, a complicated construction has a large surface, subjected to the corrosive medium. Tubular sections are, therefore, preferred against L, T and U sections.
2. ***Avoid Residual Moisture***: The sections in a construction should be so arranged that the moisture is not retained in them. Storage containers should be so designed that they can be completely emptied and cleaned. Supporting steel poles should be

placed on concrete foundations. Steel roof must have sufficient slope to allow a rapid drainage of rain water. Condensed moisture formed at the contact of a hot gas and a cold metal should be avoided. The most preferred alternative is to properly select the working temperature, so that condensation does not occur. Other alternative is to have double walled gas ducts as chimneys. Many times, the insulating materials may be having dissolvable salts e.g., chlorides and sulphates, which promote rapid corrosion. Provisions should be made for draining off condensed moisture. However, care must be taken to see that in addition to drainage hole, the construction must be properly ventilated.

3. ***Location of Plant***: Plant location must be made in such a manner that the flue dust is not falling towards the plant because of the prevailing wind.
4. ***Plant Design***: Sometimes the various sections of the plant must be considered from the design viewpoint. For example, steam heating may be introduced for vapour lines to prevent or delay condensation. In chemical plants, proper valving must be done to prevent from the corrosive to the non-corrosive areas.

6.13 BATTERIES

In earlier sections related to corrosion, it is highlighted how the electric current generated by corrosion cells was wasted. In this section, it shall be shown how the current can be harnessed in batteries. In batteries at the anode (the negative terminal) electrons are produced and at the cathode (the positive terminal) electrons are absorbed. The liquid electrolyte used historically is generally replaced by a paste or a solid electrolyte. Table 6.8 lists some of the common types of batteries. In principle, all primary batteries can be recharged but due to various reasons, this does not usually work. Two common systems where the battery is reversible are lead/sulphuric acid and nickel/cadmium batteries. The former is accumulator battery used in cars.

There is another type of cell known as *'Fuel Cells'*. In this, the chemical reaction providing the power does not involve either of the electrodes. These cells are still very expensive.

Two common fuel cell reactions are

$$2H_2 + O_2 \rightarrow 2H_2O \tag{6.19}$$

and

$$C + O_2 \rightarrow CO_2 \tag{6.20}$$

A fuel cell device comprises an anode electrode (exposed to fuel),electrolyte,and a cathode electrode (exposed to oxidant).The electrolyte separates anode and cathode electrodes and facilitates the ionic transport required for the oxidation of fuel.Electron flow in external circuit produces usable electrical power.Several individual cells are connected in series to form a 'stack' to obtain usable cell voltage and power. Table 6.9 gives informations on various types of fuel cells and their applications. Among all the fuel cell technologies ,solid oxide fuel cell technology offers the advantage of all solid-state construction,multifuel operational capability, high electrical conversion efficiency ,and a simpler balance of plant. Case Study 12.43 in Chapter 12 gives a detailed desciption of a solid oxide fuel cell and materials involved therein.

Table 6.8. Batteries in common use (adapted from Tables 1-2, 6-2 and others of Handbook of Batteries by D.Linden, by kind permmission of McGraw-Hill).

Battery	*Anode*	*Cathode*	*Electrolyte*	*Reaction*	*Voltage*	*Energy density ($Wh\ kg^{-1}$)*	*Typical applications*
Primary							
Dry Leclanche	Zn	MnO_2	NH_4Cl+ $ZnCl_2$	$Zn + 2MnO_2 \rightarrow ZnO + Mn_2O_3$	1.5	85	Torch; radio; tape recorder; CD player; calculator
Alkaline	Zn	MnO_2	KOH	$Zn + 2MnO_2 \rightarrow ZnO + Mn_2O_3$	1.5	125	Ditto (better, more expensive)
Mercury	Zn	HgO	KOH	$Zn + HgO \rightarrow ZnO + Hg$	1.3	470	Hearing aid; camera
Silver	Zn	Ag_2O	KOH	$Zn + Ag_2O + H_2O \rightarrow Zn(OH)_2 + 2Ag$	1.6	500	Watch; camera
Lithium	Li	Several e.g., I_2	e.g., solid LiI	e.g., $2Li + I_2 \rightarrow 2LiI$	2.8	800	Heart pacemaker
Fuel cell							
H_2/O_2	H_2	O_2 (or air)	e.g., KOH (low temp. version)	$2H_2 + O_2 \rightarrow 2H_2O$	1.2	3570	Spaceship
Secondary							
Lead-acid	Pb	PbO_2	H_2SO_4	$Pb + PbO_2 + 2H_2SO_4 \rightarrow 2PbSO_4 + 2H_2O$	2.0	70	Automobile
Nickel-cadmium	Cd	NiOOH	KOH	$Cd + 2NiOOH + 2H_2O \rightarrow 2Ni(OH)_2 + 2H_2O + Cd(OH)_2$	1.2	80	Lap-top computer; electric shaver

Table 6.9. Charge carriers and fuel cell operation temperature.

Fuel cell type	*Electrolyte*	*Operating Temperature*	*Charge Carrie r*	*Application* & fuel Comments
AFC	Potassium Hydroxide	~ 80°C	OH^-	Space Pure H_2, CO, CO_2 intolerant
MCFC	Molten Carbonates	~ 650°C	$CO_3^=$	Stationary Power Fuel flexibility
PAFC	Phosphoric Acid	~ 200°C	H^+	Stationary Power, Transportation Relatively pure H_2
PEMFC	Ion Exchange Membrane	~ 50°C	H^+	Transportation Pure H_2, CO intolerant
SOFC	Solid metal Oxide	600-1000°C	$O^=$	Stationary Power, APU Fuel flexibility

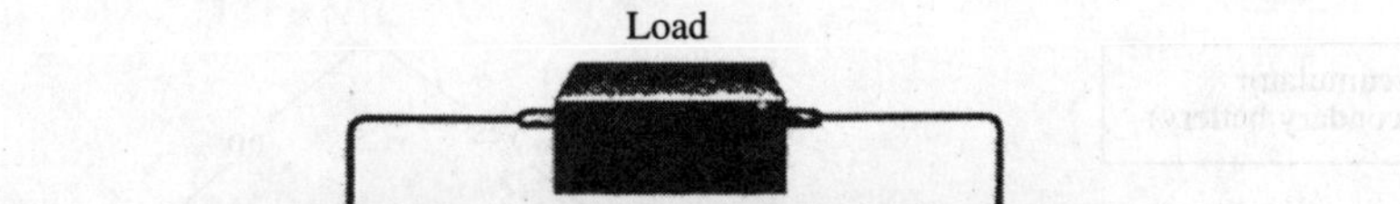

Fig. 6.14. Schematic of a hydrogen-oxygen fuel cell.

Figure 6.14 shows a schematic view of Hydrogen-Oxygen fuel cell. The reaction involved in the latter reaction is similar to a coal or oil fired power station producing heat which drives a turbine to produce electricity. The fuel cells are very efficient and convenient. The only limiting factor is that the cost of the resulting electricity is still very high. A lot of material development research is underway for the fuel cell components.

The main electrochemical aspects covered in the present chapter are conveniently summarized in Figure 6.15.

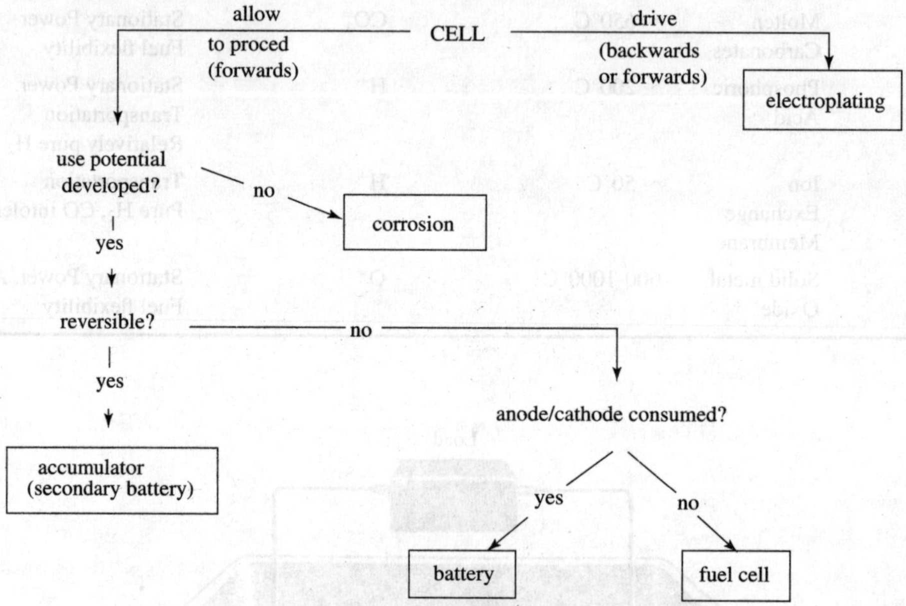

Fig. 6.15. A classification of the main processes and devices in electrochemistry (Reproduced with permission from Materials Science for Electrical and Electronic Engineers, I.P. Jones, Oxford University Press, Oxford, 2001, p. 162).

6.14 CORROSION OF CERAMICS AND GLASSES

Ceramics are generally corrosion resistant in almost all environments, specially at room temperature. Corrosion of ceramic materials generally involves chemical dissolution in contrast to electrochemical processes described in connection with metallic corrosion. However, ceramics must resist attacks by molten metals, salts, slags and glasses. In high tech energy efficient technologies, the applications of ceramics are very common and in this connection hot corrosion of ceramics is to be considered seriously.

Some of the attacks on ceramics can be classified as

- Attack by molten glasses
- Attack by molten salts
- Galvanic corrosion of contact between refractory and glass
- Attack by molten metals, example $4Al_2O_3 + 3Mg \rightarrow 3MgAl_2O_4 + 2Al$
- Attack by aqueous media, e.g., weathering of silicates

- Corrosion by gas, e.g., oxidation of non-oxide ceramics i.e., silicon carbide and nitride. Another example is reduction of silica by hydrogen at elevated temperature to the mono-oxide, which is highly volatile above 300 °C.
- Attack due to interdiffusion between two dissimilar solids in contact. When an interface reaction layer is formed, additional reaction can take place only by the diffusion of chemical species through this layer.
- Porosity induced attack. As ceramics are not fully dense, porosity plays a role such that more porous the material the more corrosion is resulted.

Both time and temperature affect the corrosion phenomenon in ceramics. In case of former, mechanism may change due to crystallization of amorphous layers, alteration of crystalline phases as diffusion progresses, cracking due to crystallization and spalling. In the latter case i.e., temperature, the mechanism of corrosion may change with increase in temperature due to crystallization of amorphous reaction layers, polymorphic transitions, melting of crystalline layers, vaporization of various species in the layer, cracking, etc.

Like molten metal, molten glass also attacks certain ceramics, example being attack of Al_2O_3 refractory by molten silica in glass melting furnace, giving rise of mullite $3Al_2O_3.2SiO_2$ formation. The corrosion takes place due to the diffusion of Na^+ ions from the molten silicate glass into the glassy phase of the refractory. To mitigate this corrosion problem alumina-zirconia-silica refractories are used in soda-lime glass melting furnaces.

Any glassy phase, if present, say in silicon nitride ceramics gets attacked by an acidic environment.

As far as glasses are concerned, their corrosion resistance is a function of structure, which is determined by the composition. Glasses can be soluble under a wide range of pH values from acids to bases, including water. Water-sodium solution silicates form the basis of the soluble silicate industry that supplies products for the manufacture of cements, adhesives, cleansers and flocculants. At $pH<5$, ion exchange is the predominant mechanism and at $pH>9$, matrix dissolution is predominant. Between pH 5 and 9, the corrosion is minimum. The attack of water vapour or glass is known as *weathering*. In electronic industries, sealing glasses and glass envelops have been developed that resist attack by alkali or mercury vapours.

6.15 DEGRADATION OF POLYMERS

Degradation normally refers to swelling, dissolving or breaking of internal polymer chains. Many plastics emit poisonous gases when burned. Some times it is desirable for polymers to deteriorate or be biodegradable so that the natural environment can break down the material, thus reducing solid waste. When polymers are exposed to sunlight, heat and other conditions, they do get oxidized. During oxidation the number of cross links in polymer is increased, thereby reducing the ability to strech. They thus become hard and crystalline. Polyethylene when exposed to high temperature in oxygen become brittle. Other example is polyvinyl chloride, which becomes coloured when exposed to high temperatures, although this does not affect its mechanical properties.

During swelling of polymer due to exposure to certain liquids, the solute diffuses and absorbed within the polymer. The increase in chain separation results in a reduction of the secondary intermolecular bonding forces, thus making the polymer softer and more ductile. Some times the liquid solute lowers the glass transition temperature, causing the polymer to become rubbery and weak. Polytetrafluoroethylene is resistant to both acidic and alkaline attack. Some synthetic polymers swell through water absorption, which causes deterioration and provides a good environment for damaging microorganisms.

QUESTIONS AND PROBLEMS

Electro-chemical Basics

6.1 What is the cell potential if the electrodes are a standard Ag electrode and a standard Zn electrode?

6.2 Under standard conditions, what is the concentration of hydrogen and hydroxyl ions in a solution of pH 8.4?

6.3 For iron corroding in acidified aerated seawater, write the probable anodic and cathodic reactions.

6.4 A single piece of metal placed in an electrolyte can act as its own anode, cathode and electrical connection. Comment on the statement.

6.5 Based on the following data sketch the polarization plots for alloys A and B

	E_{corr} V	I_{corr}, A	E_{pp}, V	I_{pass}, A	F_{Tr}, V
AlloyA	−0.4	1×10^{-6}	0	1×10^{-5}	+0.7
AlloyB	−0.2	7×10^{-7}	+0.3	1×10^{-6}	+1.2

Also comment on the following:
(a) Which will be more corrosion resistant in reducing conditions?
(b) Which will be more easily penetrated by dissolved oxidizers?
(c) Which is more corrosion resistant in strongly oxidizing solutions?
(d) Which will be more corrosion resistant in the passive state.?
(e) Which would be more easily protected by anodic protection?

6.6 Why does a crack tip act as the anode?

6.7 Discuss how the cathodic charging of steel introduces hydrogen and causes crack nucleation.

6.8 What limits the use of the emf series in predicting actual corrosion behaviour?

6.9 In a copper electro-purification cell, 1 Coulomb of electricity is passed. If the copper cathode has an area of 10 mm^2, what thickness of copper is deposited? [The electronic charge $e = 1.602 \times 10^{-19}$, the lattice parameter of copper = 0.3616 nm].

6.10 Write down the most likely cathodic reaction in case of copper corroding in an aerated hydrochloric acid solution.

6.11 The corrosion potential of carbon steel is measured as –0.540 V vs. SCE. What shall be the corrosion potential measured vs. SHE?

6.12 Explain why galvanostatic anodic polarization is unsuitable to active active-passive alloys.

6.13 In fuel, cell reducing and oxidizing species are used. Determine the number of electrons in the following overall reactions:
(a) $H_2 \rightarrow H_2O$
(b) $CO \rightarrow CO_2$
(c) $CH_4 \rightarrow CO_2 + H_2O$
(d) $CH_4 \rightarrow CO + H_2O$

Corrosion of Metals

6.14 Why can tin be used in contact with final products whereas lead cannot be used?

6.15 What precautions should be observed regarding the use of inhibitors for corrosion prevention?

6.16 Explain why water is selected to cool automobile engines. What is the effect of using more than the recommended amount of antifreeze in the automobiles cooling system in cold countries?

6.17 What determines whether a riveted lap joint corrodes at the extremities of the joint or deep within the joint?

6.18 Would aluminium be a suitable metal for (a) a roof on a house from the atmospheric corrosion view part, (b) a tank to contain strong detergent solution? Explain your answer.

6.19 Galvanised iron will not rust if it is scratched. Would Zinc coated titanium corrode if it is scratched?

6.20 What may be the reason for enhanced corrosion by differential aeration?

6.21 Suggest methods for preventing corrosion of an offshore oil platform. The structure is made of steel and there are three critical areas with corrosion rates in mm per year:
(a) The immersed zone, 0.56 mm/year
(b) The splash zone, 1.23 mm/year
(c) The atmospheric zone, 0.11-0.22 mm/year

6.22 How does the de-aeration of water in a boiler reduce corrosion?

6.23 A ship was undergoing weld repairs drawing current from DC source from a shore based generator. After some time it was noticed that the hull got thinned down. Give reason for such a failure.

6.24 Rank the following materials from the corrosion view point in sea water, 10% H_2SO_4, 8% alkaline solutions respectively: Low carbon steel, galvanized steel, Gray cast iron, 4-6% Cr steel, 18-8 stainless steel, monel, nickel, copper, red brass (85 Cu-15 Zn), Aluminum bronze, nickel-silver (65 Cu, 18 Ni, 17 Zn), aluminium and Duralumin. (Rank 1: Rapid attack, 5: unlimited service).

6.25 Mention salient features in selecting a stainless steel from the view point of (a) service environment and (b) design of the equipment.

Oxidation

6.26 Using Ellingham diagram determine the free energy for oxidation of chromium at 300°C in an atmosphere with oxygen partial pressure of 10^{-4} atms.

6.27 When iron rusts it forms a chemical compound with oxygen. Iron has two valences of 2 and 3 respectively, while oxygen has a valency of 2. Give two possible ratios of iron/oxygen atoms in rust. Also suggest two possible formulae for rust.

6.28 Weight gain data for pure nickel at 900°C is as follows. Determine whether the data best follow parabolic or cubic oxidation kinetics.

Exposure time, min.	400	520	640	750	830	1000	1100	1200
Weight gain, mg/cm^2	1.511	1.73	1.87	2.09	2.21	2.37	2.51	2.79

6.29 F.W. Young in 1956 while working at Oak Ridge National Laboratory in USA machined a sphere from a copper single crystal (cylindrical in shape), followed by mechanical and anodic electrolytic polishing. He then studied oxidation in air involving optical interference patterns. Explain why he chose a sphere for his study? [Hint: Oxidation kinetics on different crystal faces was the advantage.]

6.30 High temperature applications of intermetallics in air would require oxidation resistance. Comment on some possible series of such oxidation resistance in intermetallics.

Design Against Corrosion

6.31 In section 6.12, while describing design against corrosion, the significance of proper design of construction was emphasized. The following configurations of the constructions (Figure P. 6.31) could be modified for better corrosion resistance. Sketch the preferred configuration giving justification. (Adapted with permission from Design and Corrosion Control, V.R. Pludek, The Macmillan Press Ltd., London, 1977).

6.32 In 1985 the suspended ceiling of a swimming pool in Uster, Switzerland made out of stainless steel crashed down into the pool and killed 12 bathers. It was found that type 304 stainless steel was used as fasteners. Comment on the solution.

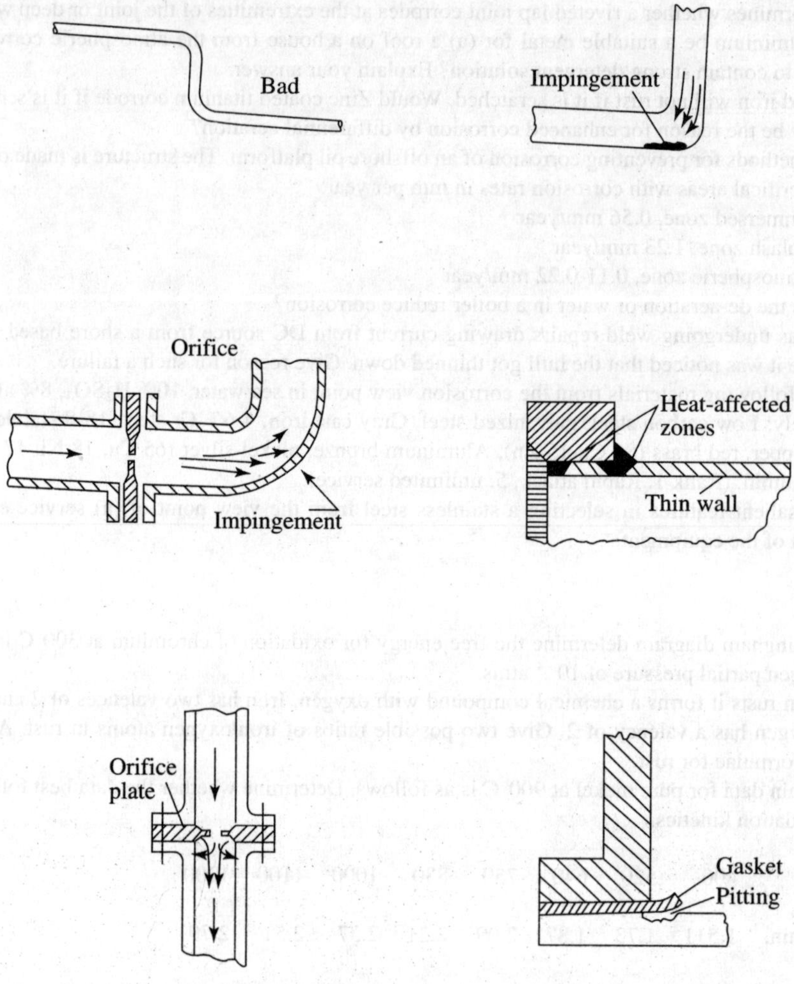

Figure P. 6.31 continued

Corrosion of Ceramics and Glasses

6.33 Discuss critically the corrosion of 60% MgO magnesite-chrome refractory in contact with molten high iron oxide containing slag in an electric furnace steel making.

6.34 Si_3N_4 is often sintered using sintering aids. Such a ceramic was subjected to corrosive effect of HCl aqueous solutions at 70°C. Will the corrosion mechanism be the same for <1M and >5M HCl solutions? If the amorphous phase at grain boundary is negligible, what shall be its effect on dissolution?

6.35 During manufacturing of float glasses, the glass is floated on to a bed of molten tin in a chamber containing a reducing gas (N_2-10% H_2). All commercial flat glasses contain some iron present over the surface in the reduced ferrous state. However the presence of metallic nickel in undesirable as it causes rejection. Assuming a maximum temperature not greater than 1100°C, what partial pressure of oxygen should be maintained. [Refer Ellingham Diagram].

6.36 Carbon fibers when admixed with carbonate in reinforced steel structures show better mechanical properties, but a fall in corrosion resistance in either $Ca(OH)_2$ or NaCl environment. What is the reason for this fall?

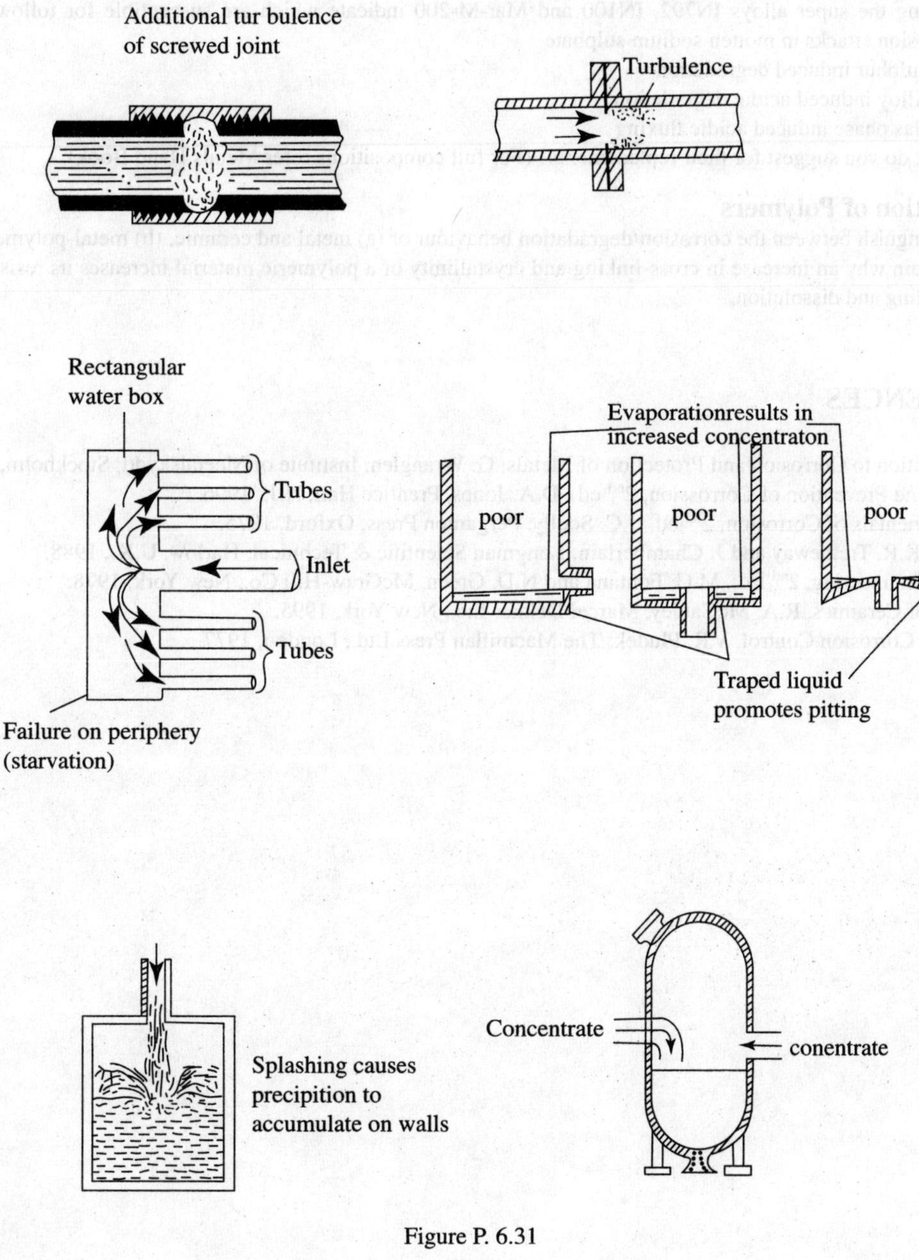
Additional tur bulence
of screwed joint
Turbulence
Rectangular
water box
Tubes
Inlet
Tubes
Failure on periphery
(starvation)
Evaporationresults in
increased concentraton
poor
poor
poor
Traped liquid
promotes pitting
Splashing causes
precipition to
accumulate on walls
Concentrate
conentrate

Figure P. 6.31

6.37 Among the super alloys IN792, IN100 and Mar-M-200 indicate which are susceptible for following hot corrosion attacks in molten sodium sulphate
(a) Sulphur induced degradation
(b) Alloy induced acidic degradation
(c) Gas phase induced acidic fluxing
What do you suggest for their replacements? (For full compositions refer Metals Hand Book).

Degradation of Polymers

6.38 Distinguish between the corrosion/degradation behaviour of (a) metal and ceramic, (b) metal-polymer.

6.39 Explain why an increase in cross-linking and crystallinity of a polymeric material increases its resistance to swelling and dissolution.

REFERENCES

An Introduction to Corrosion and Protection of Metals, G. Wranglen, Institute of Meetalskydd, Stockholm, 1972.
Principles and Prevention of Corrossion, 2nd ed., D.A. Jones, Prentice Hall, N.J., 1996.
The Fundamentals of Corrosion, 2nd ed., J.C. Scully, Pergamon Press, Oxford, 1975.
Corrosion, K.R. Tretheway and J. Chamberlain, Longman Scientific & Technical, Harlow, U.K., 1988.
Corrosion Engineering, 2nd Ed., M.G. Fontana and N.D. Green, McGraw-Hill Co., New York, 1978.
Corrosion of Ceramics, R.A. McCauley, Marcel Dekker Inc., New York, 1995.
Design and Corrosion Control, V.R. Pludek, The Macmillan Press Ltd., London, 1977.

7

ENGINEERING MATERIALS

In Chapter 1 (Introduction) classification of engineering materials into systems like metals, ceramics, polymers and composites has been highlighted. In the current chapter the same is described in detailed subsets. Various alloys important from the application viewpoints shall be indicated. Wherever possible the type of phase equilibrium and its effect on the resultant properties shall be suggested. The idea is that students appreciate composition-properties-application relationship. The processing aspects of engineering materials shall be elaborated in the future chapters. The present chapter is not aimed in discussing the philosophy and practice of materials selection. This aspect i.e., materials design, requires a detailed discussion in its own right. This has been carried out in Chapter 12.

METALLIC AND INTERMETALLIC MATERIALS

7.1 FERROUS ALLOYS

Among ferrous alloys, steels are the most common. Steels are variously categorized as carbon, low alloy steels containing alloy additions less than 5%, while high alloy steels have greater quantities. Carbon steels are generally classified as low, medium and high. Low carbon steel, also called mild steel, has less than 0.3% carbon. It is generally used for common industrial products, such as bolts, nuts, sheet plates, tubes and machine components that do not require high strength. Medium carbon steel has 0.3% to 0.6% carbon. It is generally used in applications requiring higher strengths than those using low carbon steels, such as machinery, automotive and agricultural equipment parts (e.g., gears, axles, connecting rods, crank shafts), rail road equipment and parts for metal working machinery. High carbon steel has more than 0.6% carbon. It is generally used for parts requiring strength, hardness and wear resistance such as cutting tools, cable, music wire, springs, cutlery etc.

For alloy steels, several elements are added to the plain carbon steels to impart various properties such as hardenability, strength, hardness, toughness, wear resistance, workability and machinability. Generally, the higher the percentage of these elements in steels, the higher the particular properties that they impart. The categories of alloying additions stabilizing the ferrite and austenite phases has already been described in Chapter 3.

There are a number of specifications for steels, but the most widely used are those of the American Iron and Steel Institute (AISI) and the Society of Automotive Engineers (SAE). Table 7.1 gives the summary of both AISI and SAE designations of carbon and alloy steels. The differences in listings occur as a result of differences in determining eligibility for listing. AISI uses production tonnage as the basis for including a steel. SAE includes a steel if it is used in significant quantity by two users as if it has unique engineering characteristics.

Table 7.1. SAE-AISI System for Carbon and Low-alloy Steels.

*Numerals and digits**	*Type of steel and nominal alloy content.*
Carbon steels	
10XX	Plain carbon (Mn 1.00 max)
11XX	Resulfurized
12XX	Resulfurized and rephosphorized
15XX	Plain carbon (max Mn range: 1.00-1.65)
Manganese steels	
13XX	Mn 1.75
Nickel steels	
23XX	Ni 3.50
25XX	Ni 5.00
Nickel-chromium steels	
31XX	Ni 1.25; Cr 0.65 and 0.80
32XX	Ni 1.75; Cr 1.07
33XX	Ni 3.50; Cr 1.50 and 1.57
34XX	Ni 3.00; Cr 0.77
Molybdenum steels	
40XX	Mo 0.20 and 0.25
44XX	Mo 0.40 and 0.52
Chromium-molybdenum steels	
41XX	Cr 0.50, 0.80 and 0.95; Mo 0.12, 0.20, 0.25 and 0.30
Nickel-chromium-molybdenum steels	
43XX	Ni 1.82; Cr 0.50 and 0.80; Mo 0.25
43BVXX	Ni 1.82; Cr 0.50; Mo 0.12 and 0.25; V 0.03 min
47XX	Ni 1.05; Cr 0.45; Mo 0.20 and 0.35
81XX	Ni 0.30; Cr 0.40; Mo 0.12
86XX	Ni 0.55; Cr 0.50; Mo 0.20
87XX	Ni 0.55; Cr 0.50; Mo 0.25
88XX	Ni 0.55; Cr 0.50; Mo 0.35
93XX	Ni 3.25; Cr 1.20; Mo 0.12
94XX	Ni 0.45; Cr 0.40; Mo 0.12
97XX	Ni 0.55; Cr 0.20; Mo 0.20
98XX	Ni 1.00; Cr 0.80; Mo 0.25

Table 7.1. *continues.*

*Numerals and digits**	*Type of steel and nominal alloy content.*
Nickel-molybdenum steels	
46XX	Ni 0.85 and 1.82; Mo 0.20 and 0.25
48XX	Ni 3.50; Mo 0.25
Chromium steels	
50XX	Cr 0.27, 0.40, 0.50 and 0.65
51XX	Cr 0.80, 0.87, 0.92, 0.95, 1.00 and 1.05
Chromium (bearing) steels	
50XX	Cr 0.50
51XX	Cr 1.02 C 1.00 min
52XX	Cr 1.45
Chromium-vanadium steels	
61XX	Cr 0.60, 0.80 and 0.95; V 0.10 and 0.15 min
Tungsten-chromium steel	
72XX	W 1.75; Cr 0.75
Silicon-manganese steels	
92XX	Si 1.40 and 2.00; Mn 0.65, 0.82 and 0.85; Cr 0 and 0.65
High-strength low-alloy steels	
9XX	Various SAE grades
Boron steels	
XXBXX	B denotes boron steel
Leaded steels	
XXLXX	L denotes leaded steel
Vanadium steels	
XXVXX	V denotes vanadium steel

*The XX in the last two digits of these designations represents the insertion of carbon content (in hundredths of a per cent).

A mention of terminology for steel needs to be made. The term 'grade' denotes chemical composition. The 'type' indicates the deoxidation practice, for example, killed, semikilled, or rimmed. The term 'class' indicates other attributes such as strength or surface finish. In ASTM specification, however, the terms are used interchangeably.

With few exceptions, the SAE-AISI system uses a four-digit number to designate carbon and alloy steels, specified to chemical composition ranges. There are certain types of alloy steels which are designated by five digits (numerals). Table 7.1 shows the abbreviated listing of four-digit designations of the SAE-AISI carbon and alloy steels. The first digit, 1, of this designation indicates a carbon steel; i.e., carbon steels comprise 1XXX groups in the SAE-AISI system and are subdivided into four different series due to the variance in certain fundamental properties among them.

The major alloying elements in an alloy steel are indicated by the first two digits of the designation (Table 7.1). Thus, the first digit 2 denotes a nickel steel; 3 a nickel-chromium steel; 4 a molybdenum, chromium-molybdenum, nickel-molybdenum, or nickel-chromium-molybdenum steel; 5 a chromium steel; 6 a chromium-vanadium steel; 7 a tungsten-chromium steel; 8 a nickel-chromium-molybdenum steel; and 9 a silicon-manganese steel or a nickel-chromium-molybdenum steel. In the case of a simple alloy steel, the second digit represents the approximate percentage of the predominant alloying element. For example, 2520 grade indicates a nickel steel of approximately 5% Ni (and 0.2% carbon).

The last two digits of four-numeral designations and the last three digits of five-numeral designations indicate the approximate carbon content of the allowable carbon range in hundredths of a per cent. For example, 1020 steel indicates a plain-carbon steel with an approximate mean 0.20% carbon. Similarly, 4340 steels are Ni-Cr-Mo steels and contain an approximate mean of 0.40% carbon varying within an allowable carbon range and 51100 steel is a chromium steel with an approximate mean of 1.00% carbon.

Carbon and alloy steel designations showing the letter B inserted between the second and third digits indicate that the steel has 0.0005 to 0.003% B. Likewise, the letter L inserted between the second and third digits indicates that the steel has 0.1 to 0.35% Pb for enhanced machinability. Sometimes, the prefix M is used for merchant- quality steels, and the suffix H is used to comply with standard hardenability requirements. In alloy steels, the prefix letter E is used to designate steels which are produced by the basic electric furnace process with special practice.

In another system United Numbering System (UNS) number, which is a designation of chemical composition and not a specification, a number is assigned to each chemical composition of the standard carbon and alloy steel grades for which controlling limits have been established by the SAE-AISI. The UNS designation consists of a single-letter prefix followed by five numerals (digits). The letters denote the broad class of alloys; the numerals define specific alloys within that class. The prefix letter G signifies standard grades of carbon and alloy steels; the prefix letter H indicates standard grades that meet certain hardenability requirements; the prefix T includes tool steels, wrought and cast; the prefix letter S relates to heat- and corrosion-resistant steels (including stainless steel), valve steels, and iron-base superalloys; the prefix letter J is used for cast steels (except tool steels); the prefix letter K identifies miscellaneous steels and ferrous alloys; and the prefix W denotes welding filler metals. The first four digits of the UNS number usually correspond to the standard SAE-AISI designations, while the last digit (except zero) of the five- numeral series denotes some additional composition requirements, such as boron, lead, or non-standard chemical ranges.

Ductility and Brittleness in Structural Steels: In Chapter 5 (Mechanical Properties), we have described the ductile-brittle behaviour of steels at a sharply defined temperature or in a range of temperature. It is worthwhile to dwell upon the various factors affecting the ductile-brittle transition temperature in steels. The major factors are:

Composition: In steels, significant changes in transition temperature results from the variation in the manganese and carbon content. The transition temperature is increased about 14 °C for each increase of 0.1 per cent carbon. This transition temperature is lowered about 5 °C for each increase of 0.1 per cent manganese. The Mn : C ratio should be at least 3 : 1 for satisfactory notch toughness. O, P, Si, Mo in steel are known to increase the transition temperature, while nickel lowers the temperature. The very notorious effect of oxygen in steel was a challenge for steel makers to produce good quality killed steels.

Grain Size: A lowering in grain size results in a decrease in the transition temperature. A normalized steel microstructure is advantageous. Use of lowest possible finishing temperature for hot rolling of steel plates and spray cooling lower the transition temperature. Microalloyed steels with niobium and vanadium additions are preferred, as they prevent grain growth.

The transition temperature is generally high in thick hot rolled steel plates as compared to the thin plates. This is related to the fine uniform pearlite in the latter.

Orientation of Woked Products: At high energy levels, a distinctive effect of the orientation of notch has been noticed.

Strain and Quench ageing: Strain ageing has been found to contribute in raising the transition temperature, more than the mere cold working. Quench ageing of steels results in a relatively lower impact toughness loss as compared to strain ageing.

Microstructure: Tempered martensite in a plain carbon steel is an ideal microstructure. However in case of alloy steels this cannot be generalized, as in the tempering temperature range of 200-300 °C, the precipitation of cementite platelet causes embrittlement. Suitable alloying by silicon mitigates this problem. Another variable like tempering period could be judiciously used for obtaining desirable microstructure.

7.1.1 High Strength Low Alloy (HSLA) Steels

These are primarily structural steels used in large buildings, bridges, ships, oil and gas pipelines, automobiles and pressure vessels. They are sometimes called microalloyed steels, since they contain minute quantities of niobium, titanium, vanadium, and/or molybdenum. They contain less than 0.2% carbon, such a low carbon content being necessary for welding operations. These steels have been developed as a compromise between the convenient fabrication characteristics and low cost of plain carbon steels, and the high strength of heat treated alloy steels. The steels have excellent strength (ultimate tensile strength 415-700 MPa, 15-25% Elongation). Their production can entail

- Alloying with small amount of strong carbide and nitrides forming elements.
- Controlled rolling
- Controlled cooling
- Inclusion shape control.

Niobium microalloying is much effective as the solubility of Nb(C,N) in austenite is less than the solubility of V (C, N). In most cases, all Nb, C and N are in solution at the start of the hot rolling of austenite, but precipitation occurs during the rolling as the temperature of the steel drops. The precipitate particles hinder the growth of austenite grains, and at still lower temperatures, the particles inhibit recrystallization of the deformed austenite grains. For the same microalloy content niobium is most effective in refining the ferrite grain size followed by titanium and vanadium. A typical composition has maximum of 0.21 C, 1.35 Mn, 0.04 P, 0.05 S, 0.9 Si and niobium in the range of 0.005 to 0.1 per cent. Figure 7.1 compares the conventional and advanced thermomechanical properties of low alloy steels. In case of microalloyed steels the process involves either accelerated cooling or direct quenching immediately after the steels pass the last hot rolls to produce bainite and/or martensitic structure and then tempering.

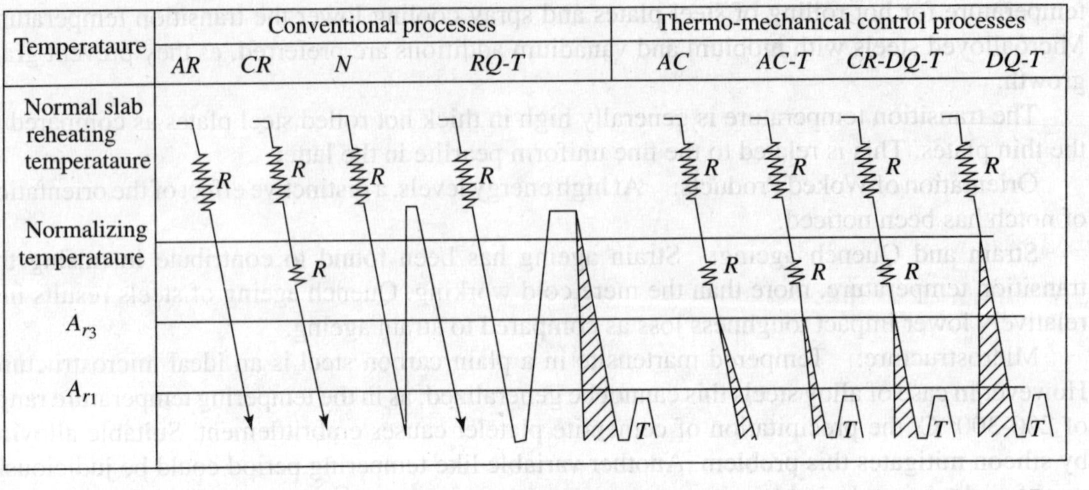

Fig. 7.1. Schematic of conventional and thermo-mechanical controlled rolling processes of steel plates: AR - as rolled; CR - control-rolled; N - rolled and normalized; RQ - T rolled quenched and tempered; AC control - rolled accelerated cooling; AC - T as AC followed by tempering; CR - DQ - T CR plus direct quenched and tempered; DQ - T rolled, direct quenched and tempered.

Another category of HSLA steel is dual-phase steels, which are heat treated to form a mixed microstructure of ferrite and martensite. This has a low yield strength but a high work hardening rate leading to excellent formability. A typical dual phase steel could be an HSLA steeel containing about 1.5% Mn and 0.5% Si strengthened by an addition of about 0.05% V. The steel is hot rolled, cold rolled, then continuously annealed at a temperature in the $\alpha + \gamma$ region (~760 °C), then cooled so that the islands of austenite transform to martensite or lower bainite. The transformation produces a dispersion of hard particles in a matrix of ferrite and the resulting tensile strength follows the rule of mixture.

7.1.2 Medium and High Alloy Steels

Tool and die steels are alloy steels (Table 7.2) designed for high strength, impact toughness and wear resistance at room and elevated temperatures. They are commonly used in forming and machining of metals. One of a typical example of high alloy steel is high speed steel, which contains major alloying elements such as Cr, Mo, W and V. When these steels are quenched and tempered, alloy carbides form in the high temperature range giving rise to secondary hardening. A particular name of 'High Speed Steel' has been coined as the steel maintains its hardness during high speed machining operations, upto 550 °C, which represents the maximum operating temperature for these steels.

There are two basic types of high-speed steels; the molybdenum type (M series) and the tungsten type (T series). The M series steels contain upto about 10% Mo, with Cr, V, W and Co as other alloying elements. The T series steels contain 12% to 18% tungsten, with Cr, V and Co as other alloying elements. As compared with the T-series steels, the M-series steels generally have higher abrasion resistance, have less distortion in heat treatment and are less expensive.

Table 7.2. Compositions and Applications of Various Types of Tool steels.

Type	*Typical compositions*	*Applications*
Cold worked tool steels		
W: Water hardening	0.6-1.4% C	Chisels, drills, files, lathe cutting tools
O: Oil hardening	0.8-1.5% C 0.5-1.0% Cr, W, Mn	Shear blades, wide variety of room-temperature cutting tools
A: Air hardening	1-2% C, Cr, Mn, Mo, V, and Ni to approx. 3% total	Punches, forming dies, blanking and trimming dies
S: Shock resistant	0.5% C, 0.5-1.8% Cr, 0.1-1.5% Mo	Repetitive-wear applications, rivet sets, shear blades, punches
Hot-worked tool steels		
H Type		
H110-H19	0.3-0.5% C, 3-5% Cr	Shearing knives
H20-H39	0.2-0.5% C, 8-19% W	Extrusion and forcing dies for brass
H40-H59	0.5-0.7% C, 4-6% Mo	Hot-worked steel tools
High-speed T	0.65-1.6% C 0.9-5% V 12-19% W	High speed tool steels , both T and M types, used for heavy-duty work calling for high hot hardness and abrasion resistance; hot forging and extrusion dies and mandrels, cutting tools
High-speed M	0.7-1.5% C 1-6% W 4-11% Mo	

Another example of high alloy steels is stainless steel. It requires a minimum of 10.5% Cr for maintaining corrosion resistance. The stainless property is achieved through the formation of a thin adherant passive chromium oxide surface film. Stainless steels can be divided into five categories; the austenitic, which accounts for well over half of the use of all stainless steels; martensitic; ferritic, precipitation hardening; and duplex stainless steels.

Ferritic Stainless Steels: These steels can be even treated as of binary Fe-Cr alloys containing between 12 and 30% Cr. The upper limit is dictated by the σ phase, which is very brittle and must be avoided. Ferritic steels on the high end of the chromium content scale have good corrosion resistance but relatively low strength compared to other stainless steels. Since they do not contain nickel, they are less expensive than the austentic.

Martensitic Stainless Steels: We have already learnt about the martensite formation in plain C-steel after quenching from the austenite region. From the Fe-Cr phase diagram one can notice that the maximum content of chromium in the austenite range at elevated temperature is 12.7%. But as carbon is austenite stabilizer, with its presence a larger chromium contents are possible. Such steels have high strength and lack ductility. They are used for cutlery, surgical instruments, springs and applications that involve high strength and corrosion resistance.

Austenitic Stainless Steels: These steels are a good compromise between the ferritic and martensitic stainless steels and are extensively used. The stability of austenite phase in stainless steels can be understood from the *Schaeffler diagram* (Figure 7.2). The equations used to calculate the chromium and nickel equivalents are shown in the axes of the diagram. The ferrite stabilizing elements similar to chromium are molybdenum, silicon, and niobium, while the

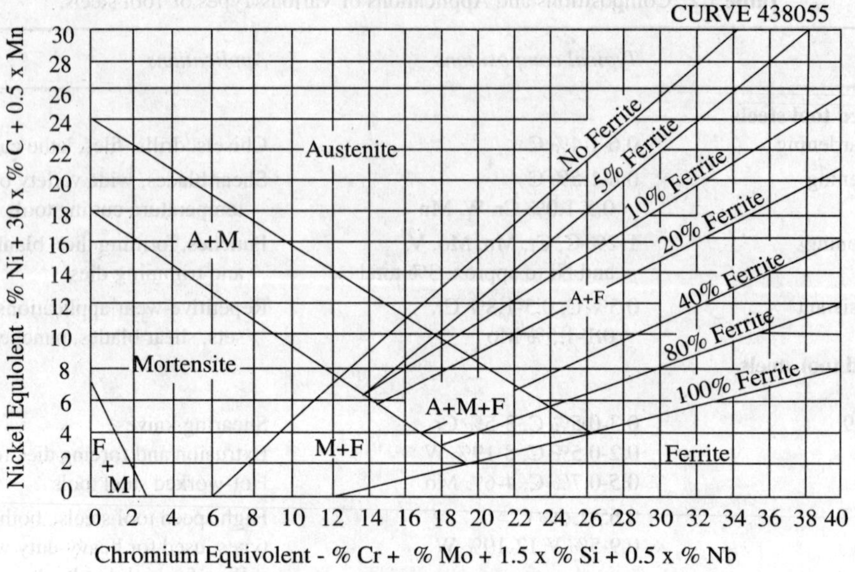

Fig. 7.2. Schaffler constitution diagram for stainless steels.

austenite stabilizing elements similar to nickel are manganese, carbon and nitrogen. By the addition of nickel in quantities on the order of 8%, the γ transformation temperature, which in pure iron is 910 °C, can be retained at room temperature. Most alloys of austenitic stainless steels contain 16-25% Cr, 7-20% Ni and the balance iron. Their strengths are in between those of the ferritic and martensitic steels, but their cold working strengths of around 760 MPa with ductilities of 50% RA are achievable.

Most austenitic stainless steels are susceptible to *sensitizations* when heated in the range of about 425-870 °C or slowly cooled through this temperature range. The chromium being strong carbide former combines with carbon in the steel and hence is not available to form the passive Cr_2O_3 surface film. In order to avoid this effect, special addition like titanium is made, which is still a stronger carbide former than chromium. This remedy is known as *stabilization*. Another solution is to have as little amount of carbon in the steel, which is expressed as the letter 'L' after the grade of the steel.

Precipitation-Hardening (PH) Stainless Steels: These steels attain very high strength levels, in excess of 1380 MPa with good ductilities of 50-70% RA. As the name suggests, the strengthening is because of precipitation of intermetallics like Ni_3Al, Ni_3Ti or Ni_3Mo. These steels are susceptible to hydrogen enbrittlement, but this can be overcome to some extent by overageing the strength level of say 1170 MPa.

Duplex Stainless Steels: From the Schaffler diagram, one can optimize the composition of ferritic-austenitic duplex stainless steels. The main advantage of these steels is their good corrosion resistance in salt water, making them suitable for marine applications. Chromium contents are about 25%. They have relatively high solubility for molybdenum and a 2% Mo content results in good corrosion resistance to chloride environments.

Tables 7.3 – 7.5 give the specification of various grades of stainless steels.

Table 7.3. Compositions of Selected Austenitic, Martensitic and Ferritic Stainless steels.

AISI type no.	*C*	*Mn*	*Cr*	*Ni*	*Others*
			Nominal composition, %		
Austenitic					
301	0.15 max	2.0	16-18	6.0-8.0	
302	0.15 max	2.0	17-19	8.0-10	
304	0.08 max	2.0	18-20	8.0-12	
304L	0.03 max	2.0	18-20	8.0-12	
309	0.20 max	2.0	22-24	12-15	
310	0.25 max	2.0	24-26	19-22	
316	0.08 max	2.0	16-18	10-14	2-3 Mo
316L	0.03 max	2.0	16-18	10-14	2-3 Mo
321	0.08 max	2.0	17-19	9-12	(5 X% C) Ti min
347	0.08 max	2.0	17-19	9-13	(10 X% C) Nb-Ta min
Marensitic					
403	0.15 max	1.0	11.5-13		
410	0.15 max	1.0	11.5-13		
416	0.15 max	1.2	12-14		0.15S min
420	0.15 min	1.0	12-14		
431	0.20 max	1.0	15-17	1.2-2.5	
440A	0.60-0.75	1.0	16-18		0.75 Mo max
440B	0.75-0.95	1.0	16-18		0.75 Mo max
440C	0.95-1.20	1.0	16-18		0.75 Mo max
Ferritic					
405	0.08	1.00	11.50/14.50	-	Al 0.10/0.30
409	0.08	1.00	10.50/11.75	-	Ti 6 X C min/0.75 max
429	0.12	1.00	14.00/16.00		
430	0.12	1.00	16.00/18.00	-	
434	0.12	1.00	16.00/18.00	-	0.75/1.25 Mo
436	0.12	1.00	16.00/18.00	-	0.75/1.25 Mo + Ta 5 X C
442	0.20	1.00	18.00/23.00	-	
446	0.20	1.50	23.00/27.00	-	N 0.25

Si 1% in each grade of ferritic stainless steel.

Table 7.4. Compositions of Selected Precipitation Hardening Stainless steels.

AISI Type no.	*Trade name producer*	*C*	*Cr*	*Ni*	*Mo*	*Cu*	*Al*	*Ti*	*Others*
Martensitic grades									
635	Stainless W (U.S. Steel)	0.06	16.75	6.25			0.2	0.8	
630	17.4 PH (Armco Steel)	0.04	16.0	4.2		3.4			0.25 Nb
. . .	PH 13-8 Mo (Armco Steel)	0.04	12.7	8.2	2.2		1.1		
Semiaustenitic grades									
631	17.7 PH (Armco Steel)	0.07	17.0	7.1			1.2		
633	AM 350 (Allegheny Ludlum)	0.05	16.5	4.25	2.75				0.10 N
Austenitic grades									
600	A-286 (Allegheny Ludlum)	0.05	14.75	25.25	1.30		0.15	2.15	0.30 V, 0.005 B

Table 7.5. Compositions of Selected Duplex Stainless steels.

Trade name/producer	*Nominal composition, %* *C*	*Cr*	*Ni*	*Mo*	*Others*
Al 2205 (Allegheny Ludlum)	0.02	22.0	5.5	3.0	0.15 N
7-Mo PLUS (Carpenter Technology Corp.)	0.02	26.0	4.7	1.4	0.20 N
Ferralium 255 (Cabot Corp.)	<0.08	25.5	5.5	3.0	>0.1 N, 1.3-1.4 Cu
DPI (Sumitomo)	<0.03	18.5	4.75	2.75	
U45 (Creusot-Loire steels)	<0.03	22.0	5.75	3.0	<0.2 N

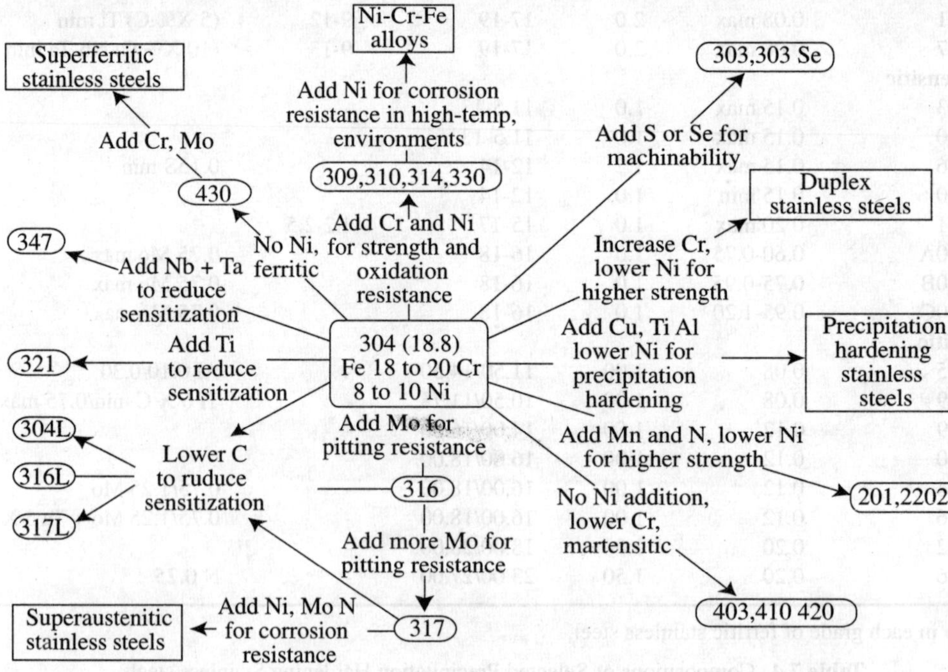

Fig. 7.3. Compositional and principle linkages in the stainless steel family.

Figure 7.3 provides a useful summary of some of the composition and property linkages in the stainless steel family.

Table 7.6 summarizes the minimum tensile properties of different grades of stainless steels and their possible applications.

Maraging Steels: To call these series of alloys as steels is a misnomer, since carbon in such steels is undesirable. Major alloying element in such alloy is nickel (~18%) and molybdenum (5%). The As quenched Fe-Ni martensite is of lath type containing a high density of dislocations. Its formation is to provide a very large number of nucleation sites for the precipitation of intermetallic compounds Ni_3Mo (orthorhombic) and FeTi (tetragonal). Unlike Fe-C martensite, Fe-Ni martensite can be obtained both athermally and isothermally. The martensite before

Table 7.6. Room-Temperaturre Properties and Applications of Stainless steels.

Type	*Condition*	*Minimum *Y.S. MPa U.T.S. MPa*	*R.A. %*	*Applications*
Martensitic 410	Quenched and tempered	800 1000	60	High-strength and elevated-temperature applications; not suitable for cryogenic uses
Martensitic 431	Quenched and tempered	750 1000	65	Similar to 410; both have acceptable resistance to SCC at hardness below R_c 22
Ferritic 430	Annealed bar	205 450	45	Popular general purpose where welding is not required; low toughness after welding
Ferritic 409	Annealed plate	205 380	-	General-purpose construction steels for automotive exhaust systems and light applications
Ferritic 405	Annealed bar	205 415	45	Toughness satisfactory after welding
Austenitic 304	Annealed bar	205 515	50	Good formability; use 304L (low carbon) for welding structures; general-purpose corrosion resistant
	Cold-worked Bar 1/4 hard	515 860	-	High-strength, corrosion-resistant structural parts; railway passenger cars, truck trailers, missiles, cryogenic
Austenitic 316	Annealed bar	205 515	5	Mo added to increase pitting, general corrosion resistance to chloride environments
Austenitic 347	Annealed bar	205 515	5	Stabilized with Nb and Ta to reduce sensitization in range 480 °C to 813 °C
PH 17-4 martensitic	Quenched and aged, 1025 °C	1000 1070	45	Weldable, strong, corrosion resistance, varied applications for small forged precision cast parts; aerospace, hydrogen embrittles
PH 13-8 Mo martensitic	Quenched and aged, 1025 °C	1210 1275	45	Highest-strength stainless steels, aerospace, fasteners, cooler sections of aircraft engines, susceptible to hydrogen embrittlement
Duplex ferralium 225	Annealed	550 760	65	Good corrosion resistance to chlorides, marine applications

* Y.S. – yield strength, U.T.S. – ultimate tensile strength.

ageing is comparatively soft about 30 R_c, so the steel can be formed before ageing. The basic composition of this alloy is essentially 18% Ni-8% Co-5% Mo. The yield strength is varied between 1375 and 2410 MPa by raising the titanium content from 0.2% to 1.4%. To ensure high toughness, the content of the residual elements P, S, C and N must be kept as low as possible, which requires vacuum melting. The processing involves ingot heating to about 1250 °C with hot rolling being finished at about 920 °C. The alloys are subsequently austenitized at 850 °C, cooled to room temperature, then aged 3 h at 480 °C. The strengthening effect of cobalt arises through short range ordering. Nickel is rejected from the ordered Fe-Co regions, and this effects the size of the Ni_3Mo particles by changing the local solubility of molybdenum. The regions depleted in nickel serve as nucleation sites for precipitation of FeTi. Overageing occurs by recovery of the martensite dislocation structure and formation of austenite. The fall in toughness is encountered by precipitation or segregation at prior austenite grain boundaries of Ti(C,N) or Ti_2S. To avoid such embrittlement the alloy must be solution treated at above 980 °C, followed by rapid cooling. Table 7.7 gives four standard compositions of maraging steels.

TRIP (Transformation-Induced Plasticity) Steels: Such steels are strengthened by strain induced transformation of austenite. Such transformed martensite by locally increasing the rate of work hardening, inhibits necking; thereby increasing the uniform elongation. TRIP steels use this phenomenon, but in austenite that has been strengthened by thermo-mechanical processing (similar to ausforming) or by thermal cycling to produce a high density of dislocation in the austenite. Table 7.7 includes the composition of such steels. However these steels suffer from relatively high cost, difficulty of larger plastic deformation at low temperature and problems with joining. Because of these limitations such steels are for special applications as flat rolled products or wire.

Table 7.7. Typical Compositions of Maraging and TRIP steels.

Steel	*C*	*Mn*	*Si*	*Cr*	*Ni*	*Mo*	*Others*
Maraging steels							
200	0.03 max	0.10 max	0.10 max		18.0	3.25	8.5 Co, 0.20 Ti, 0.10 Al
250	0.03 max	0.10 max	0.10 max		18.0	4.90	8.0 Co, 0.40 Ti, 0.10 Al
300	0.03 max	0.10 max	0.10 max		18.0	4.90	9.0 Co, 0.65 Ti, 0.10 Al
350	0.01 max	0.10 max	0.10 max		17.5	3.75	12.5 Co, 1.80 Ti, 0.15 Al
TRIP steel							
	0.25	2.5	2.0	9.0	9.0	5.0	
	0.25	1.7	2.0	10.0	8.8	5.5	
	2.0	2.0	2.0	13.5	8.8	3.0	

7.1.3 Cast Irons

In Section 3.4, a mention has been made of grey and white cast irons. These are alloys of iron that usually contain between 2.5 and 4% C and from 1 to 3% Si, which promote the formation of graphite in flaky form. Cast irons are very fluid, when molten and have good casting properties, but the castings usually have relatively low impact resistance and ductility, which limit their use. The carbon equivalent value (CE) is an index which combines the effect of Si and P upon the eutectic of Fe-C and is expressed as

$$CE = \text{Total}\%C + \frac{(\%Si + \%P)}{3} \tag{7.1}$$

This determines how close a given composition of iron is to the theoretical iron-carbon eutectic (4.3% C), and therefore, how much free graphite is likely to be present for a given cooling rate. Figure 7.4 shows the relationship between CE value, ultimate tensile and section size (i.e., cooling rate) and the resultant microstructure. It is seen that cast iron is formed at low CE value and rapid cooling, while grey cast iron as CE value and the section size increase.

As ductility of white or grey cast iron is low, proper heat treatment is desirable for achieving malleable cast iron or nodular cast iron. In the former case, when the cast iron is held at

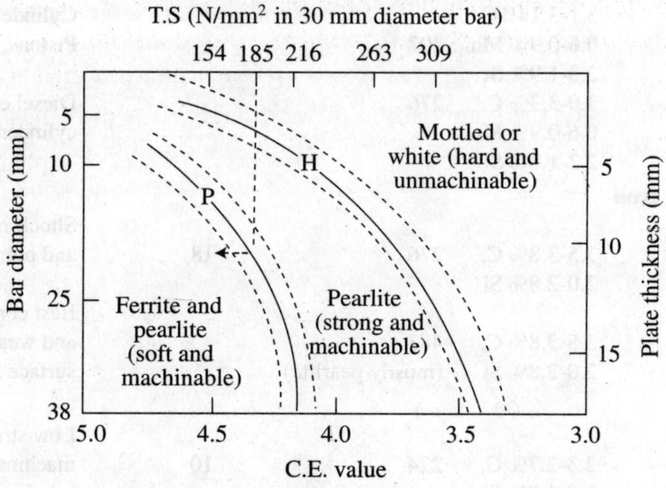

Fig. 7.4. Various stable phases in cast irons of different carbon equivalent values and section thicknesses (Reproduced with permission from Materials for Engineering, J.W. Martin, 2nd ed., Maney Publishing, on behalf of Institute of Materials, Minerals and Mining, London, 2002, p. 111).

950-1000 °C, the cementite decomposes into iron and carbon. This carbon in graphite form is known as 'temper carbon'. The graphite is in the aggregate form and not in flake form and hence the mechanical properties are much improved. In case of nodular or spheroidal graphite (SG) cast iron, graphite spheroids in the cast state is obtained by the addition of cerium and/or magnesium to the cast iron melt. SG cast iron can be made with a pearlite matrix or ferrite with appropriate heat treatment. Such cast irons have good ductility (Table 7.8).

Applications of ferritic grade malleable cast iron include automotive power trains, and hand tools such as pipe wrenches. Applications for the stronger and harder pearlitic malleable cast iron include parts that require high surface hardness (upto HRC 60 or HB 163-209), such as bearing surfaces on automobiles, trucks and heavy machinery. Applications for nodular iron include crankshafts, rocker arms and pistons in internal combustion engines; cast gears, pumps and ship propellers and caustic handling equipment. It is more expensive than grey iron in terms of weight, but its specific strength makes ductile iron economical. However, it lacks the damping ability and thermal conductivity of grey cast iron.

Like alloy steels, cast irons can also be alloyed to increase the strength, hardness, hardenability or corrosion resistance. The common alloying elements are chromium, molybdenum or nickel. Table 7.9 lists approximate ranges of alloy content for various types of alloy cast irons. One of the common applications of alloy cast iron is for brake drums and discs for automobiles.

Table 7.8. Types, Properties and Applications of Cast Irons.

Type	*Composition (wt. %)*	*Yield Strength (MPa)*	*Elongation %*	*Application*
Gray Cast Iron				
UNS No. F10006	3.1-3.14%C, 0.6-0.9% Mn, 2.3-1.9% Si	207	-	Cylinder blocks, Pistons, brake drums, clutches
UNS No. F10008	3.0-3.3% C, 0.6-0.9% Mn, 2.2-1.8% Si	276	-	Diesel engine castings, liners, cylinders, pistons, campshafts
Ductile Nodular Cast Iron				
UNS No. F32800(ferrite matrix)	3.5-3.8% C, 2.0-2.8% Si	276	18	Shock-resistant parts, valves and pump bodies
UNS No. F34800	3.5-3.8% C, 2.0-2.8% Si	483 (mostly pearlite)	3	Best combination of strength and wear resistance; can be surface hardened, crank
Malleable Cast Iron				
UNS No. F22200 ASTM A338	2.3-2.7% C, 1.0-1.7% Si	224	10	Low stress parts required, good machinability, steering-gear housings, brackets, carriers
ASTM A 602 M5503	2.2-2.9%C, 0.9-1.9% Si	379 (tempered martensite)	3	For machinability and response to induction hardening
White Cast Iron				
Class I, D	2.5-3.6% C, 1.3% Mn, 1.0-2.2% Si, 5-7% Ni, 7-11% Cr	Hardness 500-600 Brinell, tempered martensite (white)	-	For abrasion resistance

Source: Metals Handbook, vol. 1, ASM International, 10th ed., 1990.

7.2 COPPER ALLOYS

Pure copper is used for its high conductivity, corrosion resistance, and attractive appearance in electrical engineering, plumbing, architecture, and for domestic gas and water piping. Phosphorus deoxidized copper is used for ease of welding. Arsenical copper (~0.4 per cent) has higher strength at ~200-300 °C for chemical engineering applications. Small additions of Ag, Cd, and Cr allow the copper to retain its good conductivity but give it a higher softening temperature.

In beryllium copper (2% Be) precipitation gives higher strength but poorer conductivity. This is very good for springs, diaphragms, flexible bellows, and for non-sparking hand tools, e.g., for mines and explosives factories.

In Section 3.7 while describing Hume-Rothery rules for alloy formation, we have talked about electron compounds. Copper forms 3/2 electron compounds (β phase) with Zn, Al and Sn. Figure 7.5 shows the three binary diagrams i.e., Cu-Zn, Cu-Al and Cu-Sn. It is clear that in

Table 7.9. Ranges of alloy content for various types of alloy cast irons composition, wt%(a).

Description	*TC (b)*	*Mn*	*P*	*S*	*Si*	*Ni*	*Cr*	*Mo*	*Cu*
Abrasion-Resistant White Irons									
Low-C white iron(d)	2.2 to 2.8	0.2 to 0.6	0.15	0.15	1.0 to 1.6	1.5	1.0	0.5	(e)
High-C, low-Si white iron	2.8 to 3.6	0.3 to 2.0	0.3	0.15	0.3 to 1.0	2.5	3.0	1.0	(e)
Malleable white iron	2.2 to 2.5	0.3 to 0.5	0.15	0.15	1.0 to 1.6	...	...	...	...
Martensitic nickel chromium iron	2.5 to 3.7	1.3	0.3	0.15	0.8	2.7 to 5.0	1.1 to 4.0	1.0	...
Martensitic nickel high chromium iron	2.5 to 3.6	1.3	0.1	0.15	1.0 to 2.2	5 to 7	7 to 11	1.0	...
Martensitic chromium molybdenum iron	2.0 to 3.6	0.5 to 1.5	0.1	0.06	1.0	1.5	11 to 23	0.5 to 3.5	1.2
High chromium iron	2.3 to 3.6	0.5 to 1.5	0.1	0.06	1.0	1.5	23 to 28	1.5	1.2
Corrosion Resistance Iron									
High Silicon iron(f)	0.4 to 1.1	1.5	0.15	0.15	14 to 17	...	5.0	1.0	0.5
High Chromium iron	1.2 to 4.0	0.3 to 1.5	0.15	0.15	0.5 to 3.0	5.0	12 to 35	4.0	3.0
Nickel chromium gray iron(g)	3.0	0.5 to 1.5	0.08	0.12	1.0 to 2.8	13.5 to 36	1.5 to 6.0	1.0	7.0
Nickel chromium ductile iron(h)	3.0	0.7 to 4.5	0.08	0.12	1.0 to 3.0	18 to 36	1.0 to 5.5	1.0	....
Heat-Resistance Gray Iron									
Medium Silicon iron(j)	1.6 to 2.5	0.4 to 0.8	0.3	0.1	4.0 to 7.0	....	...	...	...
High chromium iron	1.8 to 3	0.3 to 1.5	0.15	0.15	0.5 to 2.5	5.0	15to 35	...	...
Nickel chromium iron(g)	1.8 to 3	0.4 to 1.5	0.15	0.15	1.0 to 2.75	13.5 to 36	1.8 to 6.0	1.0	7.0
Nickel Chromium silicon iron(k)	1.8 to 2.6	0.4 to 1.0	0.10	0.10	5.0 to 6.0	13 to 43	1.8 to 5.5	1.0	10.0
High aluminium iron	1.3 to 2.0	0.4 to 1.0	0.15	0.15	1.3 to 6.0	...	20 to 25 Al	...	...
Heat Resistance Ductile Iron									
Medium silicon ductile iron	2.8 to 3.8	0.2 to 0.6	0.08	0.12	2.5 to 6.0	1.5	...	...	...
Nickel Chromium ductile iron(h)	3.0	0.7 to 2.4	0.08	0.12	1.75 to 5.5	18 to 36	1.75 to 3.5	1.0	...

(a) Where a single value is given rather than a range, that value is a maximum limit. (b) Total carbon. (c) CP, coarse pearlite; M, martesite; A, austenite; F, ferrite. (d) May be produced from a malleable-iron base composition. (e) Cu may replace all or part of the Ni. (f) such as Duriron, Durichlor 51, Superchlor. (g) Such as Ni-resist austenitic iron (ASTM A436). (h) Such as Ni-Resist austenitic ductile iron(ASTM A439). (j) Such as Silal. (k) Such as Nicrosilal

Reproduced with permission from Metals Handboook, 9th ed, vol. 1, ASM International, 1978, p. 76.

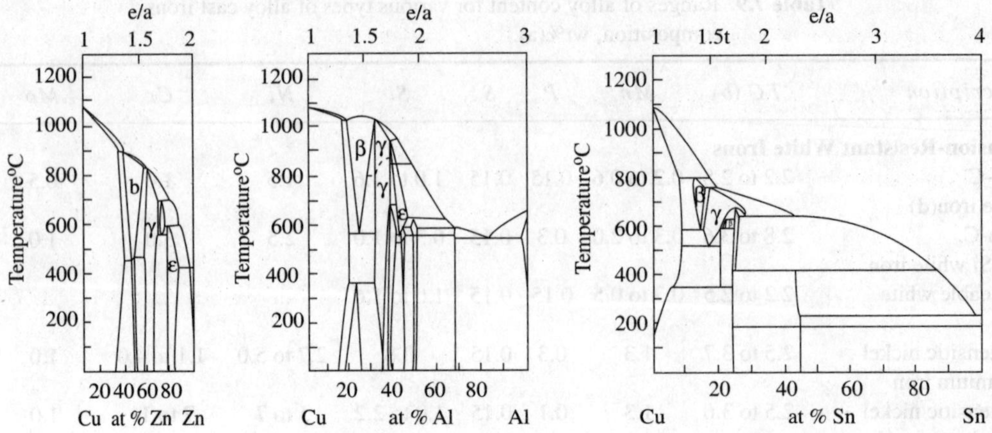

Fig. 7.5. Cu-Zn, Cu-Al and Cu-Sn phase diagrams. The three diagrams have been drawn with different length of the X-axis in order to have the same scale for the e/a values.

terms of atomic composition, smaller quantities of Al and Sn are sufficient to produce the same structure given by a certain quantity of Zn. Presence of second intermediate phase in copper substantially affects the properties. A brief account of Cu-Zn, Cu-Al and Cu-Sn alloys are given below.

7.2.1 Brasses

Zinc is added to copper to raise the strength. α-brasses are tough and cold-workable (upto 30% Zn), β-brasses are strong and brittle, but are very ductile when hot (above 30% Zn). Cu/2% Zn is *cap copper* for explosive-shell priming caps. It is soft and ductile. Cu/5-15% Zn alloys are *gilding metals*, used in sheet, strip and wire form for jewellery and decorative purposes, for architectural uses. Cu/30% Zn is known as 70/30 or cartridge brass: for shells, lamp-bulb sockets, cigarette lighters, radiators, head-light reflectors, etc. This alloy has high cold-formability. The single phase α-bronze exhibits the unique mechanical properties such as strength and ductility increase at the same time, as the zinc content is increased (Figure 7.6). 70/30 + 2% Sn is Admiralty brass which has high corrosion resistance and is used for condenser tubes in marine and other boilers. Aluminium brass is 76 Cu/22 Zn/2 Al. This has a self-healing film of Al_2O_3 which makes the alloy excellent for abrasive corrosion conditions. Muntz metal or 60/40 brass is used for sheathing, condenser end plates, forgings, castings, etc. High-tensile brasses, (α + β), are usually 60/40 + Al, Fe, Mn, Ni, Sn. These have good strength, toughness and corrosion resistance, and are used for propellers, rudders, autoclaves, non-sparking tools. They are resistant to cavitation erosion. Other β brasses with about 50% Zn are used only for brazing (MP ≈ 870 C).

7.2.2 Bronzes

Cu/9 per cent Al bronze has a tenacious oxide film and is corrosion resistant at high temperatures. The alloy has moderate strength and is used in jewellery and heat exchangers. Like steel this system also undergoes eutectoid transformation.

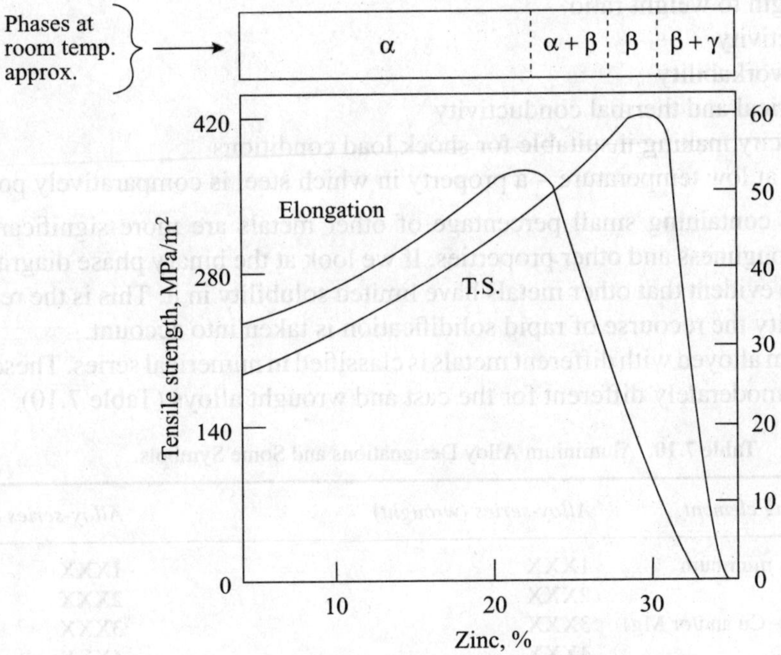

Fig. 7.6. Mechanical properties of α-brasses.

α-bronze: Cu – 8% Sn bronze is cold-workable to spring temper. 8% Sn + 0.5% P is *phosphor bronze* which is used for wire products and springs. Cu/20% Sn is bell metal, a hard and brittle alloy which is sonorous. Cu/30% Sn is *speculum metal*, a lustrous white, hard, brittle alloy polishable to high finishes for mirrors, e.g., for telescopes.

Porous and impregnated bronzes are made by powder metallurgy. They are impregnated with graphite, oils, or polymers to give self-lubricated bearings. Or they are used for filters or deicers for leading edges of aircraft wings.

88 Cu/10 Sn/2 Zn are *gunmetals*. These are casting bronzes with high strength and fluidity, which are used for valve bodies, ornamental work, slip rings.

7.2.3 Cupronickels

Copper and nickel forms continuous solid solution. The whole range of Cu/Ni alloys are hot and cold ductile, strong alloys, but perhaps the best known is 80 Cu/20 Ni. With a little iron added, they have good corrosion resistance in sea water. They are used for condenser tubes and coinage. Cu/30% Ni/0.1 Fe, Mn is *Monel*, which is one of the best of all materials for steam valves and for corrosion resistance in chemical engineering plant.

7.3 ALUMINIUM ALLOYS

Aluminium has many attractive properties. Some of them are:

- Lightness
- Resistance to corrosion

- High strength to weight ratio
- High reflectivity
- Excellent workability
- High electrical and thermal conductivity
- High elasticity making it suitable for shock load conditions
- Toughness at low temperature – a property in which steel is comparatively poorer.

Aluminium alloys containing small percentage of other metals are more significant as they provide strength, toughness and other properties. If we look at the binary phase diagrams based on aluminium, it is evident that other metals have limited solubility in it. This is the reason why to enhance solubility the recourse of rapid solidification is taken into account.

Pure aluminium alloyed with different metals is classified in numerical series. These numeric classifications are moderately different for the cast and wrought alloys (Table 7.10).

Table 7.10. Aluminium Alloy Designations and Some Symbols.

Principal alloying element	*Alloy-series (wrought)*	*Alloy-series (cast)*
Aluminium 99% minimum	1XXX	1XXX
Copper	2XXX	2XXX
Manganese (Si + Cu and/or Mg)	3XXX	3XXX
Silicon	4XXX	4XXX
Magnesium	5XXX	5XXX
Magnesium and Silicon	6XXX	-
Zinc	7XXX	7XXX
Tin	-	8XXX
Other elements	8XXX	9XXX
Unused series	9XXX	6XXX
Symbol	**Meaning**	
O	Annealed	
F	As-fabricated	
H	Strain-hardened	
W	Unstable followed by solution heat treatment	
T	Heat-treated	

The 1XXX series, which indicates a commercially pure aluminium, has a minimum of 99.00 per cent aluminium. The last two digits of a number in this series indicate the minimum percentage of aluminium in hundredths. The second digit from the left indicates the special control of the given number of impurities present. Integers 1 through 9 indicate the number of impurities that are controlled.

The 2XXX to 8XXX series indicate the alloying elements in aluminium as shown in Table 7.10. The first digit to the left indicates the alloy type, and the second digit the modifications. The original alloy is designated when the second digit from the left is zero. As indicated, if the second digit is 1 through 9, it designates the number of impurities that are controlled, which are registered with the Aluminium Association. The last two digits serve no purpose other than possibly to identify the different alloy in the group as registered. The 9XXX series is used while new alloys are being developed. Once the new alloy loses its experimental status, it loses its 9XXX designation. While it is being commercially evaluated, one of the above four-digit designations is assigned and prefixed with an X. The prefix is dropped when the new material joins other standard aluminium alloys.

Aluminium alloys fall into two general categories: those that are *heat-treatable* and those that are *non-heat-treatable*. In general, the 2000, 4000, 6000 and 7000 series may be strengthened by heat treatment. The 3000 and 5000 series cannot. These series depend upon the alloying of manganese and magnesium for their strength. They may be further strengthened by strain-hardening procedures or by cold-working.

A system of letters and numbers to indicate *temper* has also been approved by the American Standards Association. In this system the basic temper has been designated with a letter, whereas the modification of this temper is indicated with a number. The list in Table 7.10 of symbols express the following.

- EC, *electrical conductor*. Indicates aluminium used in the electrical field. It contains a minimum of 99.45 per cent aluminium, with traces of boron and copper, and a very close control of impurities.
- O, *annealed, recrystallized*. Indicates the softest temper.
- F, *as-fabricated*. Indicates the as-cast condition with no guarantee that the properties of the material will be other than those that result from the normal fabricating processes. In casting, it refers to the as-cast condition. In wrought materials, there is no intent to guarantee the properties of the material.
- H, *strain-hardened*. Refers to the aluminium products that do not have their mechanical properties affected by heat treatment alone. The mechanical properties may be changed by cold-working and subsequent heat treatment to produce partial softening. This H is always followed by two or more digits. The first digit designates the heat-treating process, and the second, or subsequent, digits, the degree of cold-working.
- H1, *strain-hardened material*. A *second-digit* range of 1 through 8 indicates the degree of hardness in *eights*. The digit 1 indicates 'dead' softness. The digit 8 indicates full hardness following a full anneal. Sometimes the digit 9 is used to indicate"extrahard" temper. Thus a half-hard material would have the number –H14 (4/8 = 1/2); a quarter-hard material would have the number –H12 (2/8 = $\frac{1}{4}$); etc.
- H2, *strain-hardened and partially annealed*. Indicates a method for achieving desirable physical and mechanical properties by first work-hardening the material and then reducing the hardness by partial annealing. The H indicates "strain hardening", and the 2 indicates partial annealing. A second numeral indicates the degree of hardness in the end product. Thus the 6 in –H26 indicates 6/8 or 3/4-hardened end product.
- H3, *strain-hardened and stabilized*. Strain-hardened alloys containing magnesium, when heat-treated at low temperatures, will achieve structural stability. Once stabilized, these materials will have better ductility than before they were processed. The 3 in –H34 indicates this type of processing. The 4 indicates a 4/8, or half-hardened end product.

- W, conditions following solution heat treating indicates the condition of alloy subject to natural age hardening after a fixed period of time. Since the age-hardening process is continuous, the time period following solution heat-treating is specified. Thus 6061-W-8 hours has specific properties as a result of solution heat-treating and precipitation hardening after 8 hours.
- T, *thermally treated products*. Indicates a stable tempered state with or without cold-working. Age-hardening may take place after the indicated operations. The digit that follows the T temper indicates a specific operation or operations. Any variation of properties from that indicated by the first digit requires a second digit that is assigned and registered with the Aluminium Association.
- T1, *hot-worked, cooled, and naturally aged to a stable condition*. Applied to those products that have been cast or hot extruded and whose strength properties are increased by room temperature ageing.
- T2, *annealed (cast products only)*. The purpose here is to increase ductility and volumetric stability.
- T3, *solution heat-treated and cold-worked*. Cold-working, when applied to wrought aluminium, is for the purpose of increasing strength after solution heat treatment. The process is allowed to proceed by age-hardening.
- T4, *solution heat treatment and natural ageing to a stable condition*. The same as T3, except that no cold-working is required.
- T5, *artificially aged only*. Applies when there is rapid cooling from elevated temperature but no solution heat treatment has taken place.
- T6, *solution heat treatment and then artificial ageing*. Applies when there is no cold-working after solution heat-treating.
- T7, *solution heat treatment and stabilizing*. Applies to materials where the temperature and time required for maximum hardness are exceeded and where there is control of grain growth and residual stress.
- T8, *solution heat treatment, cold-working, and artificial ageing*. Applies when intermediate cold-working is needed to improve the strength.
- T9, *solution heat treatment, artificial ageing, and cold-working*. Cold-working after solution heat-treating and artificial ageing to improve strength.
- T10, *partial solution heat treatment, artificially aged, and cold-worked*. Used when the cold-working improves strength.

Like wrought aluminium and its alloys, the number designation for aluminium castings and ingots, is also based on four digits, 1XXX. For aluminium castings and ingots the 1 indicates commercially pure aluminium, as seen in Table 7.10. The second two digits indicate the percentage of 1 per cent beyond 99.00 per cent pure aluminium. Castings should be indicated by the number 0 after the decimal point, and ingots by the number 1 after the decimal point. A letter X used as a *prefix* for experimental alloys and other letters would be used arbitrarily, assigned, and registered.

For alloys (2XXX, 3XXX, etc.) the first digit designates the major alloy; the next two digits are assigned when the material is registered. The 0 after the decimal point indicates a casting; a 1 indicates an ingot with the same composition as had been assigned to the casting (with some specified exceptions), and a 2 indicates a variation from the 0 category.

1XXX series alloys possess excellent electrical and thermal conductivity and resistance to atmospheric corrosion. This series is non-heat-treatable and the hardness is achieved by cold-working or alternately by temper annealing. 2XXX series respond to heat treatment. It possesses mechanical properties equal to or even better than those of mild steel. Alloy 2024 of this series is the most common aircraft alloy. Biggest limitation is poor corrosion resistance. This series of alloys, therefore, are used with surface cladding of high purity aluminium or 6XXX series alloys. Welding properties are rated poor. 3003 (Mn 1.00-1.5%) is the most common grade. It has good corrosion resistance, response to welding, and cold workability, the major applications are domestic appliances like tubes, washing machine, cooking range etc. 3004 alloy contains magnesium in addition to manganese. 3XXX series of alloys are non-heat treatable.

4XXX series aluminium alloys are non-heat treatable. In 5XXX series Al-alloys, magnesium, promotes strengthening, even better than manganese. This series has excellent cryogenic properties in addition to good response to welding. With the increase of the magnesium content, the formability drops down. The alloy has good resistance to saline water and marine atmosphere. This series of alloys finds wide application in buildings, ship making, chemical applications like desalination of sea water, canning industry, transport section and general engineering applications.

6XXX series alloys are used in solutionised condition followed by natural or artificial ageing. The series is normally used for extrusion. In mechanical properties the alloys of 6XXX series are inferior only to those of 2XXX and 7XXX series, but they have good corrosion resistance. This series of alloy is most versatile and find wide applications in making panels, coaches, pipes, structural frames, boat masts, sport gears, and in transport industries.

7XXX series of aluminium alloys are heat treatable and possess highest mechanical properties in the family of Al-alloys. Modification by smaller addition of copper and chromium improves the mechanical properties. The main alloy 7075 grade is used in aerospace, artillery and weapons. It is also used for making sport goods, nuts, bolts etc.

Alloying lithium to aluminium decreases density, while increasing modulus, due to the formation of the δ' phase Al_3Li. Although Al-Li alloys exhibit high modulus, they have a lower fracture toughness and a higher level of anisotropy than alloys currently being used. Al-Li (1.9-2.6), Cu (2.4-3.0), Mg (0.25%) composition is specified as 2090 grade alloy. Rapid solidification processing combined with thermomechanical operations offer further enhanced properties. Potential airframe application include: fuselage skins, floor beams, seat frames and seat tracks.

Table 7.11 summerizes the mechanical properties of various grades of aluminium alloys, pertaining to cast and wrought categories in different conditions. Table 7.12 summerizes the corrosion behaviour of some major class of wrought aluminium alloys.

Table 7.11. Typical mechanical properties of some cast and wrought aluminum alloys.

Alloy	*Condition*	*Density (Mg m-3)*	*Young's Modulus(GPa)*	*Proof stress (MPa)*	*UTS(MPa)*	*Elongation (% in 50 mm)*
Cast						
Al-11.5Si (LM6)	Sand cast	2.65	71	65	170	8
Al-5Mg-0.5Mn (LM5)	Sand cast	2.65	71	100	160	6
Al-6Si-4Cu-0.2 Mg-1Zn(LM21)	Sand cast	~2.7	71	130	180	1
Wrought						
Al-4.5Cu1.5Mg-0.6 Mn (2024)	Peak aged	2.77	73	395	475	10
Al-1.25Mn (3103)	Annealed	2.74	69	65	110	40
	Work hardened			185	200	7
Al-2Mg-0.3 Mn (5251)	Annealed	2.69	70	60	180	20
	Work hardened			215	270	4
Al-0.5Mg-0.5Si (6063)	Peak aged	2.70	71	210	245	20
Al-5.6Zn-2.5Mg-1.6 Cu-0.25Cr (7075)	Peak aged	2.80	72	500	570	11
Al-2.5Li-1.3 Cu-0.95Mg-0.1Zr (8090)	Peak aged	2.55	77	436	503	5

Reproduced with Permission from J.W. Martin, Materials for Engineering, 2nd ed., Maney, Leeds, 2002, p. 347, 352.

Table 7.12. Corrosion Behaviour of Wrought Aluminium Alloys.

Alloy Class	*Typical Temper**	*Alloying Elements*	*Corrosion Resistance*** *General*	*Pitting*	*Exfoliation*	*SCC*
1XXX	All	Natural impurities	E	E	E	I
2XXX	T3,T4,	Cu	F	P	P	VS
	T8		F	P	F	R
3XXX	All	Mn, Mn + Mg	E	E	E	I
4XXX	All	Si	F	G	G	G
5XXX	Most	Mn, G, Cr	E	G	G	I-R
6XXX	All	Mg, Si	E	G	E	I
7XXX	T6	Zn, Mg, Mn, Cu	F	F	F-P	S-VS

* T3, T4, T6 - age hardened; T8 - overaged;
**E - excellent; G - good; F - fair; P - poor; I - immune; R - resistant; S -susceptible; VS - very susceptible.

7.4 MAGNESIUM ALLOYS

Magnesium having a hexagonal structure, possesses limited ductility. It is, therefore, more difficult to deform it to shape, requiring higher temperatures to enhance its deformability. Because of this, magnesium alloys are used predominantly as castings. Magnesium has a higher strength/density ratio than Al, but is not as responsive to strengthening as Al alloys which are usually stronger on a strength for strength basis. Mg alloys have poor wear resistance and low fatigue resistance. Unlike Al alloys, however, corrosion resistance is poor because there is no passive layer. Magnesium is strengthened with aluminium to give precipitation hardening alloys and with zinc to give solution-hardening alloys. Magnesium alloys are best used for high strength/density uses. They have good casting properties and machinability. They are used for aircraft engine parts, automotive housings, portable machinery and in Magnox fuel cans.

The ASTM has standardized the designation of the alloys of magnesium, using a four-part system. The designation as follows:

1. The first part of the number consists of two code letters, which indicate the two main alloys. The first letter indicates the alloy with the higher percentage; the second the alloy with the next higher percentage. The letter code is as follows:

A–Aluminium	L–Beryllium
B–Bismuth	M–Manganese
C–Copper	N–Nickel
D–Cadmium	P–Lead
E–Rare earth	Q–Silver
F–Iron	S–Silicon
H–Thorium	T–Tin
K–Zirconium	Z – Zinc

2. The second part consists of two numbers corresponding to the rounded-of percentages to the nearest whole number of the two principal alloys. If the last digit of the percentage is a 5, then the nearest even whole number is used.
3. The third part consists of a letter indicating how many alloys have been developed that have the same percentages of the first two alloys. Thus an A would mean that the percentages developed under step 2 are the first alloy of magnesium which has the indicated percentages. All letters of the alphabet are used except I and O.
4. The fourth part consists of a letter and a number that indicates the temper heat treatment. This system is the same as that used for aluminium, i.e.,
   O - Annealed
   F - As-fabricated
   H - Strain-hardened
   W - Unstable followed by solution heat treatment
   T - Heat-treated.

7.5 NICKEL ALLOYS

Nickel is one of the most important metals, 65 per cent of Ni production is used for alloying in steels. But Ni has good corrosion resistance and 15 per cent is used in Ni-rich alloys for this purpose, e.g., Monel. It has high electrical resistance, Ni-chrome (80/20 Ni/Cr) is much used for heaters. Magnetic properties are very good, e.g., Permalloy. (Ni + Fe, Cr, Mo) has high permeability and low saturation. Of particular interest is the alloy *Invar* (Fe/35 Ni) which has almost zero thermal expansion. It is used for metrology standards, pendulums, balance wheels, packing sleeves to minimize thermal expansion stresses, glass/metal seals (in lamps and valves, etc.). Also much used is the alloy *Elinvar* (Fe/35 Ni/12 Cr) which is a constant modulus alloy used for hair springs, springs in balances and weighing machines diaphragms and Bowden tubes.

Gas turbines almost all use Ni-based superalloys for blades, vanes and discs at the hot end. Examples are the *Nimonics* (Ni/Cr/Al/Ti/Fe) and *Inconels* (Ni/Cr/Fe), both with small amounts of C, Co, Mo, W. These are all basically Ni/Cr solid-solution alloys, precipitation-hardened with $Ni_3(Al,Ti)$ plus various carbides. They are used in cast or wrought form, but great care is required to produce good quality castings. They are usually vacuum melted and cast. They are usable at temperatures upto 0.75 of the melting point. They have high alloy phase stability, which means no transformations occur during service, but they must be given oxidation resistant coatings. Table 7.13 gives the composition of some major nickel based commercial super alloys.

TD Nickel is a nickel with a fine dispersion of ThO_2 particles in a heavily-wrought structure. The alloy retains its strength to much higher temperatures than the superalloys, but unfortunately has much lower strength at lower temperatures.

Table 7.13. Nominal wt,% compositions of nickel base superalloys.

	C	*B*	*Cr*	*Co*	*Mo*	*W*	*Al*	*Ti*	*Zr*	*Hf*	*V*	*Nb*
Udimt 700	0.06	0.025	15	17	5	-	4	3.5	-	-	-	-
Low C Astroloy	0.02	0.02	15	17	5	-	4	3.5	0.03	-	-	-
Rene 95	0.07	0.01	13	8	3.5	3.5	3.5	2.5	0.05	-	-	3.5
AF2-1DA	0.04	0.015	12	10	2.8	6.5	4.6	2.8	0.10	-		-
AF-115	0.05	0.02	10.7	15	2.8	5.9	3.8	3.9	0.05	0.8	-	1.7
MAR-M 247	0.07	0.015	8.3	10	0.7	10	5.5	1	0.06	1.5	-	-
IN-100	0.07	0.02	12.4	18.5	3.2	-	5	4.4	0.06	-	0.8	-
Merl 76	0.02	0.02	12.4	18.5	3.2	-	5	4.3	0.06	0.8	-	1.7

7.6 TITANIUM ALLOYS

Titanium exists in two crystalline states: the low temperature α phase, which has a HCP structure and c/a ratio of 1.58 stable below 883 °C and the high temperature βphase, which is stable above this temperature and has BCC structure. The phase composition and the transformation temperature can be modified by addition of alloying elements. Titanium metal under rapid cooling undergoes martensitic transformation.

Table 7.14. Classification of Titanium Alloys.

Alpha	Alpha-Beta
Ti-0.2 Pd	Ti-8 Mn
Ti-5 Al-2.5 Sn	Ti-6 Al-4 V
Ti-8 Al-1 Mo-1 V	Ti-7 Al-4 Mo
Ti-6 Al-2 Nb-1 Ta-1 Mo	Ti-3 Al-2.5 V
Ti-6 Al-2 Sn-4 Zr-2 Mo-0.08 Si	Ti-6 Al-6 V-2 Sn
Ti-5 Al-6 Sn-2 Zr-0.8 Mo-0.25 Si	Ti-4.5 Al-5 Mo-1.5 Cr
	Ti-6 Al-2 Sn-4 Zr-6 Mo
	Ti-5 Al-2 Sn-2 Zr-4 Cr-4 Mo
	Ti-6 Al-2 Sn-2 Zr-2 Cr-2 Mo-0.25 Si
Beta	
Ti-10 V-2 Fe-3 Al(Ti-10-2-3)	
Ti-13 V-11 Cr-3 Al(Ti-13-11-3)	
Ti-15 V-3 Cr-3 Al-3 Sn(Ti-15-333)	
Ti-3 Al-8 V-5 Cr-4 Mo-4 Zr(Beta-C)	

Titanium is the heaviest of 'light' metals but it has the highest melting point. It is rarely used pure, because of its poor creep resistance, low strength and the oxidation problem.

A range of titanium alloys has been developed to extend and improve the properties of the pure metal. Table 7.14 gives the major classifications of titanium alloys, while Figure 7.7 shows the typical binary phase diagrams of the different alloy classifications. Beta (β-alloys) or near-beta alloys are highly heat treatable. They are capable of high strengths, fair creep resistance, and good formability. Alpha-beta ($\alpha + \beta$) titanium alloys are typically heat treatable and some are weldable. Their strength levels are medium to high and their formability is high. The creep strength, however, is not as high as in most alpha alloys. α-alloys are generally weldable and not heat treatable. They have medium strength, good notch toughness and good creep resistance at elevated temperatures.

Metastable beta is a non-equilibrium phase which can be developed in $\alpha + \beta$ alloys that contain sufficient amounts of beta-stabilizing elements. In these alloys the M_s temperature is suppressed to below room temperature and the beta phase is retained at room temperature on rapid cooling or quenching from the $\alpha + \beta$ phase field. The metastable conditions of the beta phase make it possible to age the alloy to high strength levels. Like aluminium alloys, titanium base alloys are also amenable to rapid solidification processing in order to provide unique properties.

Ti-6 Al-4 V alloy ($\alpha - \beta$ alloy) is extensively used and is the work horse of titanium alloys like mild steel among steels. The existance of two phases α and β is due to the fact that the alloy contains α-stabilizing element Al and β-stabilizer vanadium. In case of binary Ti-6 Al alloy, the $\beta(\alpha + \beta)$ and $(\alpha + \beta)/\alpha$ transformation take place at $\sim 1010\,°C$ and $970\,°C$respectively. The introduction of V at constant Al concentration, although it has a comparatively small influence on the position of the $\beta/(\alpha + \beta)$ transus, produces a rapid decrease in the $(\alpha + \beta)/\alpha$ transus. These effects are shown in Figure 7.8, which plots transformation temperature versus vanadium concentration for two fixed levels of aluminium. One of the major advantage of this system

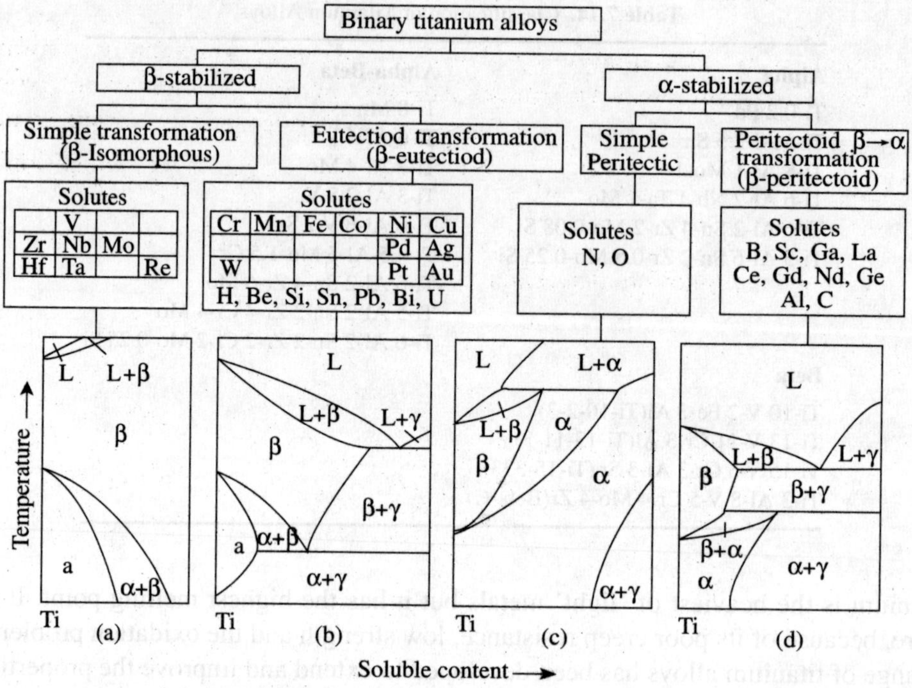

Fig. 7.7. Classification of binary titanium alloys: alloying elements and effects on phase diagram.

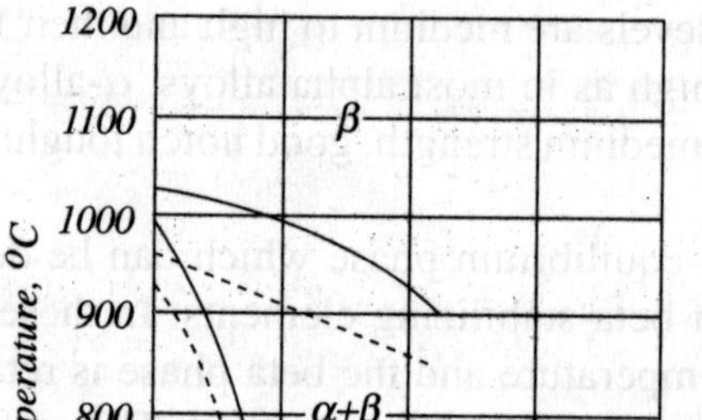

Fig. 7.8. Vertical sections of the Ti-Al-V versus temperature : equilibrium phase at 4 and 7 wt % Al (Reproduced with permission from the Physical Metallurgy of Titanium Alloys, E.W. Collings, ASM International, Ohio, 1984, p. 62).

is that a wide range of processing techniques are applied so as to produce a wide range of microstructures and hence the resultant mechanical properties.

Ti/8 Al/1 Mo/1 V (α-alloy) is used for supersonic transport (Mach 2.7 – Mach 3.0), and is applied for compressor blading for jet engines. Cost is high but life may be long. Titanium alloys have excellent corrosion resistance in aqueous environments, again because of the passivation layer. They are useful for chemical engineering plant, and are a substitute for stainless steel where weight-saving is important.

7.7 PRECIOUS METALS

Precious metals may be classified into four groups: (1) silver and its alloys, (2) gold and its alloys, (3) platinum alloys and the platinum group, (4) palladium and its alloys. In the pure state, precious metals are soft, have good malleability, are corrosion resistant, and have good electrical and thermal conductivity. The melting temperature is low enough so that alloys of silver may be used as solders at 580 °C, and high enough so that when alloyed with osmium it melts at temperatures greater than 2990 °C.

Silver. Silver may be 99.99 per cent pure or commercially pure at 99.9 per cent. Eutectic silver (72 per cent silver, 28 per cent copper) has a low melting point (772 °C) and may be used as a solder. Sometimes nickel, cadmium, palladium, zinc, tin, or phosphorus is added to achieve special purposes. Because of its resistivity to oxidation, it is suitable for electrical connections and contacts. When plated, the base metal underneath may oxidize and cause peeling, especially when used at high temperatures. It has electrical and thermal conductivity almost the same as those of copper. It is not suitable for use as electrical contacts when the voltage is below 20 V, or in low-level audio circuits because of the noise generation.

Silver-clad copper, iron, nickel, or brass is used in the electrical field as conductors. Glass, mica and ceramics may be coated with silver and used for electrical conducting parts. In some instances silver coats are used as a conducting base for electroplating. Electroplating of silver on a nickel silver base is used for tableware.

The very good physical properties and the slight change in dimension that takes place when silver-tin-mercury solidifies and cools make it useful for dental work.

Sterling silver has high reflectivity, making it suitable for jewelry and cutlery. It is also used in photography because of the photosensitivity of silver crystals. Because of its uniform and high reflectivity, it is also used to coat glass in the manufacture of mirrors. It also finds many uses in the chemical field.

Gold. In its commercially pure state, gold is used for dental and decorative work. It is used in fuses to protect electric furnaces, as a film on glass to filter light, for decorative purposes on glass, and as a high-melting solder. In the electrical field, because of its resistance to corrosion, it is used to line wave guides and as a coating on grid wires, where low-noise contacts are required and where low and stable resistances are required. Gold-silver-copper alloys are used for jewelry, decorative, and dental purposes. The balancing, or addition, of elements changes the colour of the alloy.

The combination of 70 per cent gold with 30 per cent platinum produces a high-melting-point solder at 1446 °C. Gold-palladium-iron alloys have high resistivity and are used as potentiometer wires. The combination of 49.5 per cent gold, 40.5 per cent palladium, and 10 per cent iron yields the highest resistivity.

Platinum. Pure platinum is used in resistance thermometers and thermocouples. It is alloyed with other elements to develop properties such as high corrosion resistance and hardness so that it may be used in precision electrical contacts, brushes, and potentiometers. It is also used in the pure, alloyed or clad state for jewelry and decorative purposes. It is an outstanding catalyst and therefore finds wide use in the chemical industry.

When alloyed with palladium, it is used for jewelry and electrical contacts. When alloyed with rhodium, it is used in high-temperature applications where oxidation is liable to take place. Platinum-rhodium thermocouples are used to measure and control temperatures upto 1928 °C.

Other elements used as alloys with platinum are iridium, ruthenium, nickel, tungsten, and cobalt. Combined with iridium, platinum is used for high-temperature applications in the chemical and electrical fields. It is also used for jewelry. The platinum-nickel alloy develops good high-temperature strength. Oxide-coated wires of platinum-5 per cent nickel find their greatest use in electron tubes. Where oxidation is undesirable at high temperature, this alloy should not be used. When alloyed with tungsten, it is used in potentiometers where wear resistance and low noise are required, such as in power tube grids and spark plug contacts. It is an excellent alloy for making strain gauges because of its good high-temperature physical properties. This alloy should not be used where oxidation can take place.

Palladium. This metal has about the same physical property characteristics as platinum. It is lighter and less dense than platinum. Because it remains relatively free from tarnish, it is used in electrical contacts for audio transmission where freedom from noise is required. It may be hot- or cold-worked or fabricated into very thin leaf. When alloyed with silver, it is used for electrical contacts. When alloyed with copper, it is used primarily for electrical contacts that are to operate in circuits in the milliampere range, and for slip rings where a hard material is used for brushes.

Palladium-silver-copper and the palladium-silver-gold alloys are used for electrical contacts that are subject to wear. Both are subject to age-hardening and are used in dental work. Platinum-silver-gold alloys are used for high-strength brazing. The latter alloy may also be clad to other metals. It is easily worked and is very corrosion resistant. When alloyed with ruthenium, it is used in jewelry and for electrical contacts.

Iridium. This element has a very high specific gravity of 22.5. Commercially pure iridium is used in extrusion dies in the processing of high-temperature-melting glass. It is a highly corrosion resistant material and possesses very high temperature tensile strength.

Osmium. This element has the second highest specific gravity of 22.57. It is very hard metal and has high wear and corrosion resistance. Because of this high wear resistance, it is used in phonograph needles, pens, electrical contacts, and bearings in instruments. It is very hard (600 BHN) and has the highest melting point (2712 °C) of all the precious metals.

Rhodium. Since this metal maintains its high uniform reflectivity, it is used for mirrors. It is also used on jewelry articles because of its non-tarnish characteristic. It is sometimes applied to electrical contact surfaces.

7.8 REFRACTORY METALS AND THEIR BASED ALLOYS

Refractory metals such as W, Mo, Cr, Ta, Nb, V may be used for prosthetics (e.g., Ta, Cr); vacuum or inert gas furnace parts (W, Mo); liquid metal (Na,K) containment in fast breeder reactors(Nb, V); capacitors (Ta); chemical engineering plant (Cr, Ta); superconductors (Nb alloys). Other uses are air-breathing engines, aerospace-(aerodydnamic heating resistance in non-oxidising atmospheres); containment of flames (rocket-nozzle liners); and in fine wire and fiber form for composite materials requirements.

Refractory metals, particularly tungsten and molybdenum are processed through powder metallurgy route, since their melting points are very high. The details of such processing are given in Chapter 8. Table 7.15 gives the mechanical properties data of tungsten and molybdenum in different conditions.

Table 7.15. Mechanical Properties of Tungsten and Molybdenum.

	State	*Tungsten*	*Molybdenum*
Hardness, HV	Sintered	200-250	150-160
	Hammered	350-500	200-240
	Sheet, rolled	450-500	250-350
	Sheet, annealed (recrystallized)	260-380	250-300
Tensile strength, MPa	Sintered	100-150	
	Hammered	400-1500	500-1000
	Wire, drawn	1800-4100	1000-2500
	Wire, annealed	900-1100	800-1100
Elongation, %	Wire, drawn	1-4	2-5
	Wire, annealed	-	10-20
Elastic modulus, GPa	At 20 °C	390	355
Shear modulus, GPa	At 20 °C	150	134

TZM is a molybdenum based alloy with about 0.5% Ti and 0.08% Zr. The carbon content in the alloy does not exceed 0.01 – 0.2%. It offers significant advantages over molybdenum metal in respect to strength and resistance to recrystallization. The interesting properties like high thermal conductivity, high stiffness, low thermal expansion, high hot hardness and excellent resistance to many molten metal, has led to its applications in die-casting and cores, hot working dies, mandrels, heating element and boats, apart from the wide range of applications in aerospace industry.

One of the major application of tungsten is in heavy alloy penetrators used as armour piercing cores. The alloy is essentially a metal-metal composite processed through powder metallurgy route. A typical composition is 90 W-7 Ni-3 Fe, which has a hardness of 41 R_c , Young's modulus 310 GPa tensile strength 1200 MPa and % elongation 7%. It is possible to machine heavy alloy with a great degree of accuracy. The key factor in penetrator performance is not whether the penetrator will yield and deform, but the manner in which the deformation and inversion process occurs. Impact-induced stresses set up shock waves, the reflection of which within the penetrator can cause fracture. Attempts have been made to develop a 70 vol. %

tungsten-30 vol. % uranium (depleted) penetrator material by infiltrating uranium into a porous tungsten perform.

Major application of another refractory metal, niobium, is for superconducting magnets. One of the typical alloy for this purpose is Nb-47 wt. % Ti, which is commercially available ($J_c = 3 \times 10^9$ Amperes/meter^2 at 5 Tesla and at 4.2 K). Fabrication of such superconductor is in the form of multifilaments. This shall be described in Chapter 8. Another alloy is Nb_3Sn compound superconductor (A15 type structure). Grain boundaries are responsible to a great extent for the high value of critical current J_c in this material. Grain size can be made as small as 30 nm by suitable high treatment. Such materials are finding applications in generation of high magnetic fields, in loss-less electrical power transmission and storage, in magnetic sensors, in electromagnetic radiation detectors as well as in high speed digital signal and data processing.

7.9 INTERMETALLICS

Of late, aluminide intermetallics (Ni_3Al, NiAl, Fe_3Al, TiAl, Ti_3Al) and silicide intermetallic ($MoSi_2$) have drawn much attention for high temperature structural applications. Similarly, rare-earth intermetallics (RCo_5, R_2Co_{17}, $R_2Fe_{14}B$) have been developed for high strength permanent magnets.

Some properties of aluminide intermetallics are summarized in Table 7.16. They possess much lower densities as compared to the existing alloys. They contain sufficient amount of aluminium to form thin film of alumina (Al_2O_3) in oxidizing environment which are compact and protective. The ordered structures exhibit attractive elevated temperature properties. However, those aluminides suffered from their brittleness at room temperature and high brittle-to-ductile transition temperatures. In aerospace industry, still the real impact of intermetallic has not been felt. However, a number of other general applications have been successfully found out. Products in cast shape are more in common than the wrought ones, although powder metallurgy processing has brought a new change.

Table 7.16. Important Aluminides and Their Properties.

Alloy	*Crystal Structure*	*M.P. (°C)*	*Density (g/cm^3)*	*Y.S. (MPa)*	*Room Temperature Ductility*	*Creep Limiting Temp. (°C)*
Ni_3Al	$L1_2$	1,390	7.50	250 - 500.	2 - 50$^+$	760
NiAl	B_2	1,640	5.86	250 - 475	2	~1200$^\#$
Fe_3Al	DO_3	1,540	6.72	385 - 392	2 - 12$^\$$	~700
FeAl	B_2	1,250	5.56	360 - 380	2 - 17$^\$$	~827
TiAl	Al_0	1,460	3.91	400 - 650	1 - 4	1,000
Ti_3Al	DO_{19}	1,600	4.20	700 - 900	2 - 10	760

* 0.2 wt. % B

$^+$ 0.1 wt. % B

$^\#$ Alloyed with Nb

$^\$$ Depends on atmosphere.

Table 7.17. Physical and mechanical properties of Ni_3Al-based cast alloy IC-221M.

Property	*Temperature, °C*							
	Room	200	400	600	800	900	1000	1100
Density (g/cm^3)	7.86	-	-	-	-	-	-	-
Hardness (R_c)	30	-	-	-	-	-	-	-
Microhardness (DPH)	260	270	280	290	280	230	120	-
Modulus (GPa)	200	190	174	160	148	139	126	114
Mean Coeff. of thermal expansion (10^{-6}/°C)	12.77	13.08	13.72	14.33	15.17	15.78	16.57	-
Thermal Conductivity (w/m K)	11.9	13.9	16.7	20.3	25.2	27.5	30.2	-
0.2% tensile yield strength (MPa)	555	570	590	610	680	600	400	200
Ultimate tensile strength (MPa)	770	800	850	850	820	675	500	200
Total tensile elongation (%)	14	14	17	18	5	5	7	10
10^2 h Rupture strength (MPa)	-	-	-	-	252	124	55	28
10^3 h Rupture strength (MPa)	-	-	-	-	172	83	36	18
10^4 h Rupture strength (MPa)	-	-	-	-	124	55	24	11
Charpy impact hardness (J)	40	40	40	35	15	10	-	-
Fatigue strength 10^6 cycle life (MPa)	-	-	-	630	-	-	-	-
Fatigue strength 10^7 cycle life (MPa)	-	-	-	550	-	-	-	-

Source: J.H. Westbrook and R.L. Fleischer, Intermetallic Compounds, vol. 3, John Wiley & Sons, Chichester, 2002, p. 515.

The physical and mechanical properties of a typical Ni_3Al based cast alloy IC-221 M are given in Table 7.17. This is cast intermetallic-based alloy that is approved in ASTM specifications (Designation A 1002-99). This alloy possesses following attractive properties.

- Excellent resistance to oxidation at temperature ≥900 °C,
- Excellent room temperature ductility for nickel-rich compositions,
- Good high temperature strength upto temperatures approaching 1100 °C,
- Resistance to carburization,
- An increase in yield strength with temperature that peaks at about 850 °C,
- Possibility of creating non-conductive and wear resistant surfaces by preoxidation at 1100 °C in air that results in the Al_2O_3 formation.

Table 7.18 gives a summary of various general applications of Ni_3Al.

Another important intermetallic is $MoSi_2$, which possesses a high congruent melting point of 2030 °C and excellent high-temperature oxidation resistance. Although brittle at low temperatures, it has a brittle-ductile transition of approximately 1000 °C, $MoSi_2$ is thermodynamically stable with potential ceramic reinforcement such as SiC, Si_3N_4, ZrO_2, Al_2O_3, TiB_2 and TiC, and may also be alloyed with other high melting point silicides such as WSi_2. Due to its low electrical resistivity (15 μΩ cm at room temperature) it can be electrodischarge machined, which is a significant benefit for the low-cost fabrication of components. Since the thermal expansion coefficient of $MoSi_2$ is closer to metals, it can be easily joined with metals.

$MoSi_2$ is the most stable in the temperature range 1000-1600 °C, when a glassy SiO_2 surface layer is formed during oxidation. At low temperatures (600-800 °C) $MoSi_2$ exhibits a low

Table 7.18. Current and potential applications of Ni_3Al-based alloys.

Component	*Application/Objectives*	*Replacing*
Thin-wall tubes and statically cast return bends	Radiant burner tubes in heat treating furnaces • To increase life via higher creep ,oxidation and carburizing resistance	Cast stainless steel such as HU, HT, HK etc.
Thick-wall tubes and pipe and statically cast trunions	Furnace rolls in austenitizing and hydrogen annealing furnaces • To increase life through elimination of blisters , non-sticking of iron oxide to the roll and higher creep and high-temperature wear resistance	Cast stainless steels such as HU and HP modified
Statically cast trays and fixtures	Trays and fixtures for holding components during carburization and annealing processes • To increase life from higher creep strength and better resistance to carburization and oxidation	Cast stainless steel such as HU, HT, HK etc
Statically cast die blocks	Dies for the hot-forging process • To increase life due to its peak in yield strength at 850 °C and good oxidation resistance	H-13 and other die steels
Statically cast tubes hangers	Hangers to support tubes operating at high temperatures in chemical plants • To increase life due to high creep strength and oxidation resistance	SUPERTHERMS™ and other high-temperature alloys
Statically cast components	Variety of applications for heat-treating furnaces • To increase life by higher creep strength, oxidation and carburization resistance	Alloy steel and cast stainless steels
Preoxidized sheet	Electronic circuit board applications • The electrically insulating surface with a metallic core will absorb more heat from the circuit boards than monolithic Al_2O_3 boards. This will extend the electronic boards capability for higher power capacity	Monolithic Al_2O_3 boards
Cast-shaped rods	Industrial furnace heating elements • Increase the life time of heating elements by preventing their oxidation and sagging	Fe-Cr-Al alloy heating elements
Ni_3Al powder	Binder for tungsten and chromium carbide as tool and die materials • Improves hardness and wear resistance and aqueous corrosion resistance in certain acid solutions	Expensive cobalt that is currently used as the binder material
Cast hot-pressing dies	Dies for hot pressing of permanent magnet materials • Improved die material because of excellent chemical compatibility and peak yield strength at the hot pressing temperature of ~850 °	IN-718 dies

Source: J.H. Westbrook and P.L. Fleischer, Intermetallic Compounds, vol. 3, John Wiley & Sons, Chichester, 2002, p. 512.

oxidation resistance. The so called '$MoSi_2$-pest' reaction destroys the material within several hours. However, this type of destructive oxidation occurs only in samples which have not been oxidized at higher temperatures. $MoSi_2$ shows good resistance to attack by metal melts, the only exception being aluminium, which attacks it to form $MoAl_2$.

Among rare-earth intermetallics RCo_5 is the oldest one, which is widely applied as permanent magnet. Sm-Co alloy (<81 at %) solidifies as the primary phase $SmCo_5$; while at higher cobalt contents the compound forms by a peritectic reaction between the primary phase Sm_2O_{17} and the melt. The formation of Sm_2O_{17} in $SmCo_5$ should be carefully avoided, as the former greatly reduces the coercive force. Samarium can be partly replaced by Pr, La, Ce and mischmetal.

The most important and recent development in the area of rare-earth intermetallic permanent magnets has been of $Nd_2Fe_{14}B$. The bonded rare-earth cobalt magnets have been almost entirely replaced by bonded Nd-Fe-B magnets. The schematic phase diagram of a Nd-Fe-B ternary is shown in Figure 7.9. The composition of the alloy is slightly hypostoichiometric of the compound $Nd_2Fe_{14}B$ in order to compensate for oxidation during processing. The Nd-rich grain boundary phase is an eutectic one. Excess oxidation of the alloy composition during processing results in a shift in the effective composition to the Fe-rich side of the phase diagram. The sintering temperature is 1100 °C for 20-90 min and the density of the finished magnets is 7.4 g cm^{-3}. They are heat treated for 1-3 h at temperatures between 600 °C and 900 °C in order to maximize the coercivity. Table 4.9 (Chapter 4) includes the magnetic and electrical properties of both of the above categories of rare-earth intermetallics.

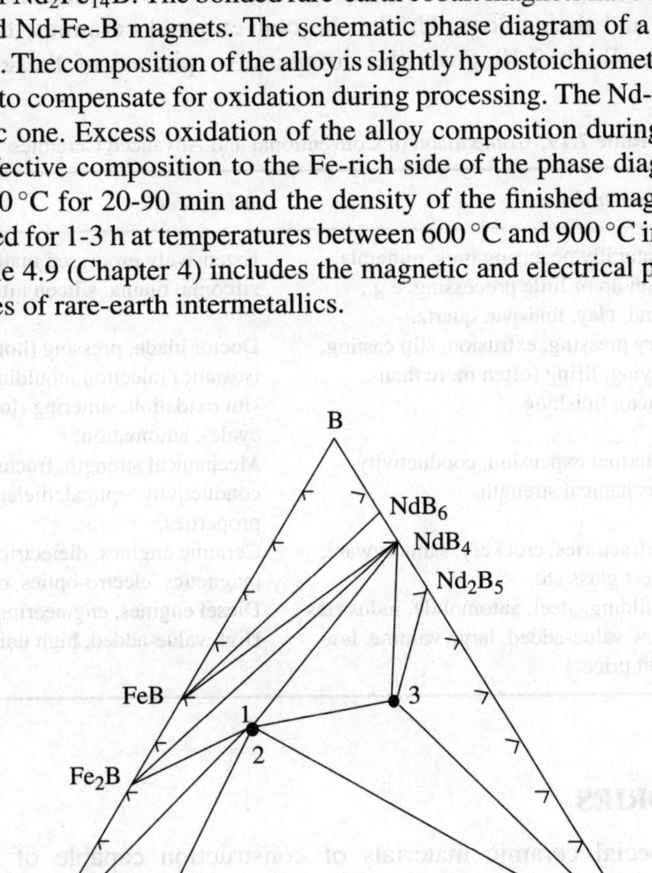

Fig. 7.9. Nd-Fe-B ternary phase diagram.

CERAMIC MATERIALS

Ceramics are inorganic, non-metallic materials that are processed or used at high temperatures. They include a broad range of silicates, metallic oxides and combination of silicates and metal oxides. Crystal structures of ceramics have already been dealt in Chapter 2. Ceramics can be grouped into three broad divisions – clay products, refractories and glasses – according to their common characteristic features. Closely related to ceramics in chemical composition are inorganic cements that are used as binding materials to produce concrete, mortars and similar products.

Another classification of ceramics is as conventional and advanced ceramics. In the latter, the input of materials science is much larger. The former category includes refractories, sanitary ware, sheet glass, building materials, crockery etc. Advanced ceramics include pure oxide and nonoxide ceramics and their alloys. Unlike the conventional ceramics, they are extensively processed materials. Table 7.19 gives the comparative picture of these two categories of ceramics.

Table 7.19. Comparison of Conventional and Advanced Ceramics.

Characteristics	*Conventional*	*Advanced*
Raw material	Naturally occurring ores, minerals with no or little processing, e.g., sand, clay, feldspar, quartz.	Extensively processed materials, e.g., alumina, zirconia, titania, silicon nitride, sialon.
Processing	Dry pressing, extrusion, slip casting, drying, firing (often more than once), finishing.	Doctor blade, pressing (hot, iso-static and hot , isostatic) injection moulding, C.V.D., controlled in situ oxidation, sintering (low temperature, fast cycle), automation.
Properties	Thermal expansion, conductivity, mechanical strength.	Mechanical strength, fracture toughness, electrical conductivity, optical, dielectric and magnetic properties.
Application	Refractories, crockery, sanitaryware, sheet glass etc.	Ceramic engines, dielectrics and piezoelectrics, magnetics, electro-optics, opto-electronics.
Markets	Building, steel, automobile, industries.	Diesel engines, engineering industry, electronics.
Value	Low value-added, large volume, low unit price.	High value-added, high unit costs.

7.10 REFRACTORIES

Refractories are special ceramic materials of construction capable of withstanding high temperature in various industrial processes and operations. The main bulk of the commercial refractories comprises complex solid binders consisting of high melting oxides or a combination of oxides of elements such as silicon, aluminium, magnesium, calcium and zirconium, with small amounts of other elements present as impurities. One of the most widely used refractories is based on alumina-silica compositions; varying from nearly pure silica, through a wide range of alumina-silicates to nearly pure alumina.

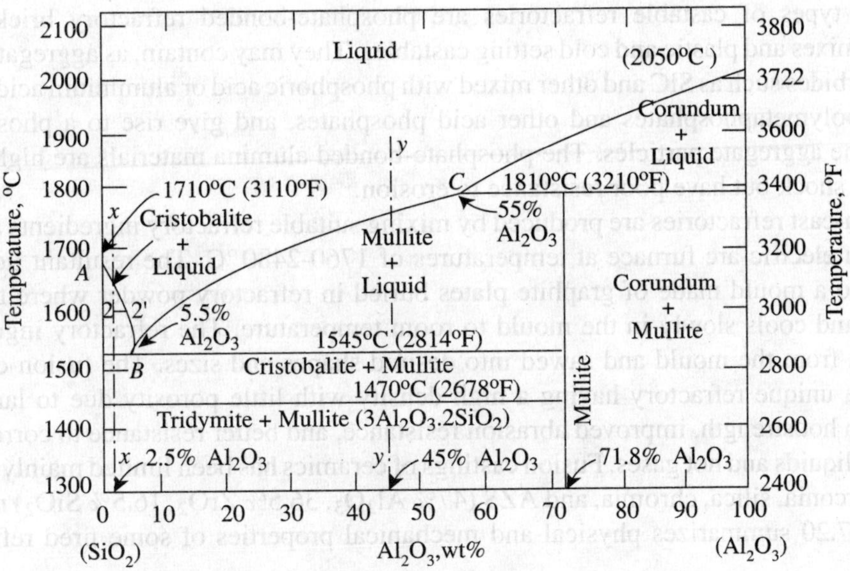

Fig. 7.10. Phase diagram of the system Al_2O_3-SiO_2.

Figure 7.10 shows a typical binary phase diagram of silica-alumina system, in which a eutectic is formed. Most commercial alumina-silica refractories always contain a certain amount of impurities such as basic oxides of iron, calcium, magnesium and smaller amounts of alkaline metal oxides. These impurities greatly affect equilibrium relations between the two major components, silica and alumina. The fluxing action of the basic oxides follows the order $MgO < CaO < FeO < N_2O < K_2O$, the latter two being much more effective than the former three. The compositions from 20% to 40% alumina are known as fire clay refractories, which are classified as super duty, high duty, medium-duty and low-duty bricks, depending on their alumina content and degree of firing. Mullite refractories are prepared by fusing alumina and silica materials in any required proportion or by calcining sillimanite, a naturally occurring mineral of the composition $Al_2O_3.SiO_2$ (63% Al_2O_3 by weight). In the range of composition of 63.5% to 71.8% alumina, corundum appears due to the dissociation of mullite at 1810 °C. Compositions above 71.8% alumina yield a solid phase consisting of mullite and corundum only. With the increasing amount of alumina the refractory consists mainly of crystalline corundum bonded with a glassy matrix formed from the molten impurities.

There is another grade of refractories known as castable and fused refractories. These are made by mixing refractory aggregate of suitable grading, such as alumina-silicates or high alumina, with a refractory high alumina cement and water to desired consistency. The mix is then either cast, rammed, gunned or sprayed into shape and permitted to set until it becomes hydraulically bonded. On subsequent heating to high temperature this cementitious bond is dehydrated and replaced by a refractory bond that is developed between the matrix and aggregate particles. Castable refractories have been used in applications where abrasion resistance at elevated temperatures is required and as a protective barrier against corrosive attack by hot gases and liquids that are highly detrimental to other structural materials.

Other types of castable refractories are phosphate-bonded refractory bricks, mortars, ramming mixes and plastic and cold setting castables. They may contain, as aggregate refractory oxides, carbides such as SiC and other mixed with phosphoric acid or aluminium acid phosphate, or alkali polymetaphosphates and other acid phosphates, and give rise to a phosphate bond between the aggregate particles. The phosphate-bonded alumina materials are highly resistant to thermal shock but have poor resistance to erosion.

Fusion-cast refractories are produced by mixing suitable refractory ingredients and melting them in an electric-arc furnace at temperatures of 1760-2480 °C. The resultant liquid is then poured into a mould made of graphite plates buried in refractory powder where the material solidifies and cools slowly in the mould to room temperature. The refractory ingots are then withdrawn from the mould and sawed into desired shapes and sizes. The fusion-cast process produces a unique refractory having a high density with little porosity due to large isolated voids, high hot strength, improved abrasion resistance, and better resistance to corrosive attack by molten liquids and hot gases. Fusion castings of ceramics has been limited mainly to alumina, mullite, zirconia, silica, chromia, and AZS (47% Al_2O_3, 36.5% ZrO_2, 16.5% SiO_2) refractories.

Table 7.20 summarizes physical and mechanical properties of some fired refractories in brick form.

Table 7.20. Some physical and mechanical properties of fired refractory bricks.

	Magnesia ($95\%MgO$)	*Chrome* ($30\%Cr_2O_3$)	90% *Alumina*	70% *Alumina*	*Zircon*
Bulk density, g/cm^3	2.8-2.95	3.06-3.14	2.9-2.96	2.53-2.6	3.6-3.72
Porosity, %	15-19	16-20	14-18	17.5-21.5	19-23
Cold Crushing strength, MPa(ksi)	48-70 (7-10)	35-55 (5-8)	62-95 (9-14)	27-48 (4-7)	48-76 (7-11)
Modulus of rupture, MPa(ksi)	17-24 (2.5-3.5)	14-21 (2-3)	17-21 (2.5-3)	7.6-11 (1.1-1.6)	15-23 (2.2-3.3)
Reheat test, % permanent linear change after heating to:					
1600 °C (2910 °F)	...	...	...	+3.5-6	...
1650 °C (3000 °F)	...	...	...	...	0.0
1725 °C (3135 °F)	−0.2 to 1.0	...	+0.1-1.0	...	...
Load test at 170 kPa (25 ksi);					
Withstands load to temperature, °C (°F)	1620 (2950)	1400 (2550)	1760 (3200)	1450 (2640)	1600 (2910)

7.11 PURE OXIDE CERAMICS

Alumina is the most widely used ceramic and it possesses low thermal conductivity, a comparatively high thermal expansion coefficient, high hardness and chemical stability. It can exist in different forms, some amorphous (hydrous and anhydrous) and some as crystalline oxides or hydrates. A third group comprises beta variations with small amounts of alkali or alkaline earth ions lodged into the structure.

Zirconia is another most important oxide ceramic. There are three polymorphs of zirconia. Monoclinic zirconia is stable upto 1150 °C, when it transforms to tetragonal. At temperatures above 2300 °C the cubic form exists. Tetragonal-monoclinic phase change is of technological importance, since it is accompanied by a volume expansion of about 4%. This transformation is of the martensitic diffusionless type and if this is allowed to develop in an uncontrolled manner then spontaneous failure of zirconia ceramic can occur. Control of particle size, composition and distribution of the zirconia allows the properties to be engineered. In Chapter 5, we have described how the mechanical properties of zirconia containing ceramics can be improved because of the impeding of the crack growth. Figure 7.11 illustrates how the amount of ZrO_2 particles in alumina and its size control the flexural strength and fracture toughness values.

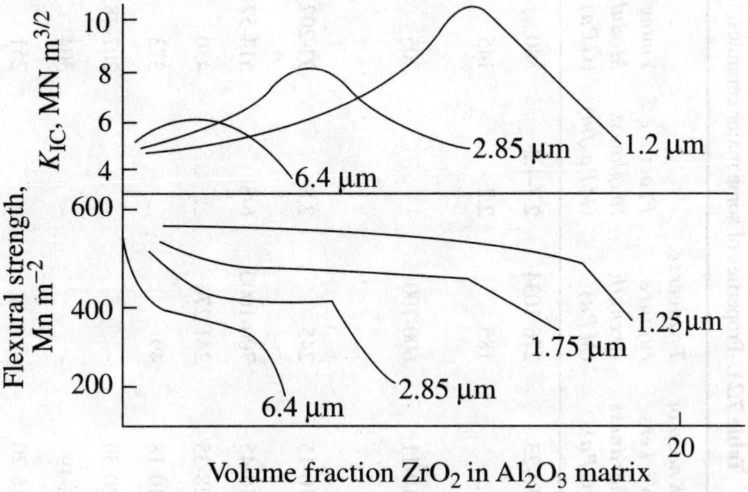

Fig. 7.11. Effect of particle size and volume fraction of ZrO_2 on toughness and strength of alumina (Reproduced with permission from Sintered Metallic and Ceramic Materials, G.S. Upadhyaya, John Wiley & Sons Ltd, Bognor Regis, West Sussex, U.K., 2000, p. 504).

A relatively recent development is partially stabilized zirconia (PSZ), which has high strength and toughness and performs more reliably than zirconia. It is obtained by doping the zirconia with oxides of calcium, yttrium or magnesium. Typical applications include dies for hot extrusion of metals, zirconia beads for grinding and dispersion media, for aerospace coatings etc. The coefficient of thermal expansion of PSZ is only about 20% lower than that of cast iron, and its thermal conductivity is about one third that of the ceramics. Consequently it is very suitable for heat engine components, such as cylinder liners and valve bushings. New development to further improve upon the properties of PSZ include transformation toughened zirconia, which has higher toughness than that of PSZ, because of dispersed tough phases in the ceramic matrix. Table 7.21 includes properties of some oxide ceramics.

Table 7.21. Properties of some major ceramics.

Material	*Crystal Structure*	*Theoretical Density* (g/cm^3)	*Knoop or Vickers hardness* (GPa)	*Tranverse rupture strength* (MPa)	*Fracture toughness* $(MPa\sqrt{m})$	*Young's Modulus* (GPa)	*Poisson's ratio*	*Thermal expansion* $(10-6/K)$	*Thermal conductivity* $W/m.K$
Al_2O_3	Hexagonal	3.97	18-23	276-1034	2.7-4.2	380	0.26	7.2-8.6	27.2
Mullite	Orthorhombic	2.8	...	185	2.2	145	0.25	5.7	5.2
Partially stabilized ZrO_2	Cubic monoclinic, teragonal	5.7-5.75	10-11	600-700	-	205	0.23	8.9-10.6	1.8-2.2
Fully stabilized	Cubic	5.56-6.1	10-15	245	2.8	97-207	0.23-0.32	13.5	1.7
TiB_2	Hexagonal	4.5-4.54	15-45	700-1000	6-8	514-574	0.09-0.13	8.1	65-120
TiC	Cubic	4.92	28-35	241-276	...	430	0.19	7.4-8.6	33
Cr_3C_2	Orthorhombic	6.7	10-18	49	...	373	...	9.8	19
SiC	α, Hexagonal	3.21	20-30	-	-	207-483	0.19	4.3-5.6	63-155
Si_3N_4	α, Hexagonal	3.18	8-19	-	-	304	0.24	3.0	9-30
TiN	Cubic	5.43-5.44	16-20	-	-	251	-	8.0	24

7.11.1 Magnetic Oxides

Magnetic ceramics are complex oxides and can be divided into three classes: spinel or cubic ferrite, garnets or rare earth ferrites and magneto plum-bites or hexagonal ferrites (Table 7.22). Table 7.23 gives electrical and magnetic properties of some spinel ferrites. In the last category, substitutional solid solution of various divalent cations for Ba^{2+} can occur in the lattice. A variation of Ba^{2+} content is also possible in the BaO.6 Fe_2O_3 lattice before the equimole BaO : Fe_2O_3 phase is formed. These features offer the compositional flexibility (Figure 7.12) to tailor the magnetic properties of barium ferrite by substituting a cation, such as Co^{2+} with magnetic moments, for nonmagnetic Ba^{2+} ions.

Table 7.22. Classes of Magnetic Ceramics.

Structure	*Composition*	*Applications*
Spinel (cubic ferrites)	1 MeO : 1 Fe_2O_3 (where MeO = transition metal oxide, e.g., Ni, Co, Mn, Zn)	Soft magnets
Garnet (rare earth ferrites)	3 Me_2O_3 : 5 Fe_2O_3 (where Me_2O_3 = rare earth metal oxide, e.g., Y_2O_3, Gd_2O_3)	Microwave devices
Magnetoplumbite (hexagonal ferrites)	1 MeO : 6 Fe_2O_3 (where MeO = divalent metal oxide from group IIA; e.g., BaO, CaO, SrO)	Hard magnets

Table 7.23. Properties at 20 °C of Some Spinel Ferrites.

	Mn − Zn	*Mn − Zn*	*Mn − Zn*	*Ni − Zn*	*Ni*
Composition (vol. %)					
MnO	27	34	30		
NiO				32	50
ZnO	20	14	15	18	
Fe_2O_3	53	52	55	50	50
Initial Permeability (μ_i)	2000	1500	650	110	20
Coercivity H_c (Oe)	0.2	0.25	0.7	4	14
Curie temperature	160	220	210	380	440
DC electrical resistivity	20	20	150	10^6	10^6
Applications	Low-power wide band transformers	High-power transformers	Medium-frequency inductors	High-frequency inductors	Misc.

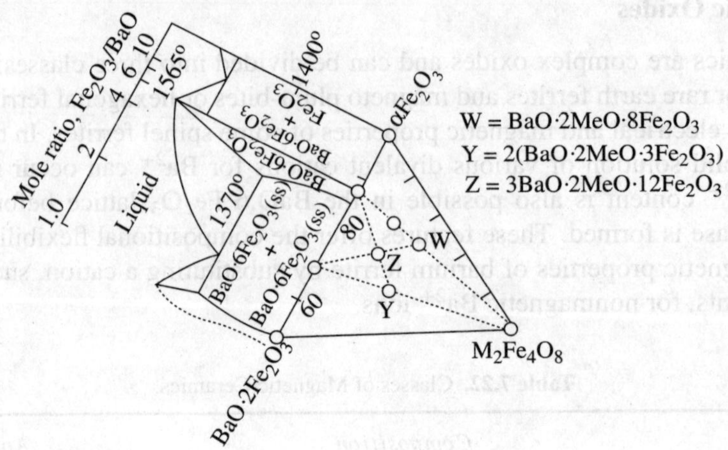

Fig. 7.12. Phase relations between $BaO.Fe_2O_3$, $\alpha - Fe_2O_3$ and hexagonal ferrites.

7.11.2 Ferroelectric Oxides

The PZT ceramics are oxide alloys of lead, zirconium and titanium. This is the most commonly used ferroelectric. By varying the composition and microstructure of PZT, it is possible to control their properties. Figure 7.13 shows the equilibrium phase diagram for the system. Much technical attention is paid to the ceramic compositions near the morphotropic boundary, separating the tetragonal and rhombohedral behaviours. PZT as such are p-type extrinsic semi-conductors. The most important additive to PZT to control this conductivity is lanthanum oxide, La_2O_3. This additive oxide brings extra oxygen in the lattice, thus giving more free electrons which converts Pb^{4+} ions to Pb^{2+}. In this manner holes are filled and p-type conductivity is lowered. In case one intends to promote n-type conduction, the additive is VO_3, which is substituted for TiO_2 or ZrO_2 rather than for PbO. In addition, certain additives (e.g., Co^{3+}) pin down the ferroelectric domain walls, which reduce the ageing rate of poled ceramics and increase their coercive fields. Poling signifies paraelectric to ferroelectric transition. Other additives like Ba, Sr, La affect in lowering the Curie temperature.

Lanthanum doped lead zirconate-titanate (PLZT) ceramics ($Pb_{1-1.5y}La_yTi_{1-z}Zr_zO_3$) have good electro-optical characteristics, which results in their use in light modulation or image storage. Its effect on PZT are as follows:

- It fills oxygen vacancies in PZT, making domain walls mobile so that the polarization of poled ceramics can quickly stabilize.
- It provides spare electrons to fill holes and suppresses p-type conductivity.
- It reduces Curie temperature to an extent that can enable paraelectric effects at room temperature.

La contents <5% are usually used for piezoelectric and >6% for electro-optic devices.

7.11.3 Oxide Superconductors

Another application of oxide ceramic is as high T_c super-conductors. $YBa_2Cu_3O_{7-y}$(123) compound is a typical example. Other superconducting oxides are based on bismuth/thallium

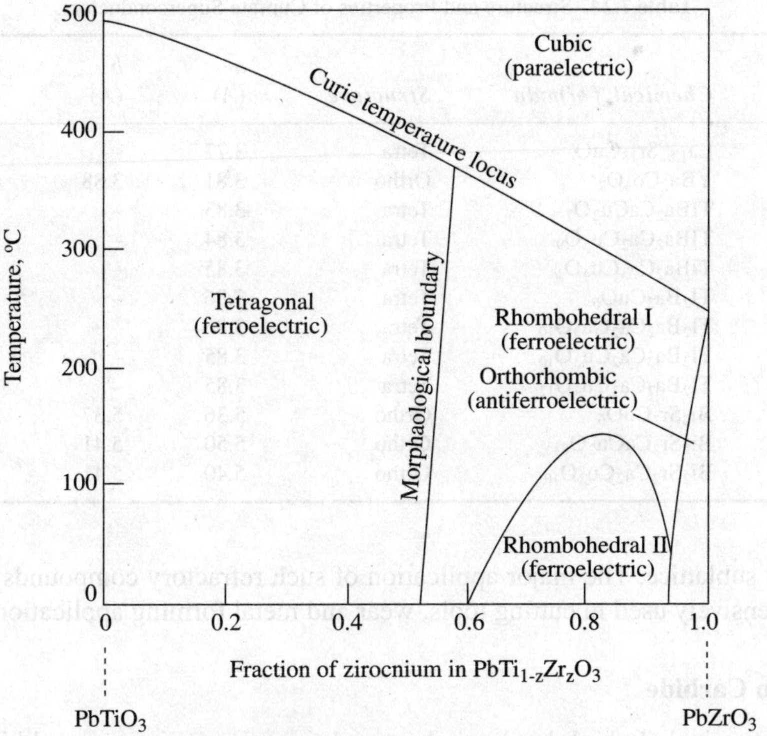

Fig. 7.13. Equilibrium phase diagram for PZT.

and copper oxides. Unlike conventional oxide ceramics, high T_c oxide superconductors are extremely sensitive to their conditions of processing and dopants. Oxygen stoichiometry and associated defects play a key role in influencing the temperature of such materials as well as the rates at which they approach equilibrium during processing. Table 7.24 gives a brief summary of the crystal structures and properties of such high T_c superconductors.

One of the limitations of using high T_c superconductors is what is known as *flux creep*. High T_c materials show thousand of time worse current instability than low T_c materials, due to the instability of the vortex lattice in these materials on the introduction of magnetic field, which results in flux creep. This prevents the use of these materials at sufficiently high temperature and magnetic fields.

7.12 NON-OXIDE CERAMICS

7.12.1 Refractory Compounds

Carbides, nitrides and borides of early transition metals are commonly called refractory compounds. These are well known for their unique properties such as high melting point, hardness and other interesting physical and mechanical properties (Table 7.21). Such non-oxide ceramics mostly have a metallic lustre and electrical and thermal conductivities of the same order as metals. Refractory carbides and nitrides have homogeneity ranges as a result of vacancies in

Table 7.24. Structure and Properties of Cuprate Superconductors.

Simplified notations	*Chemical formula*	*Structure*	*a* (*A*)	*b* (*A*)	*c* (*A*)	T_c (*K*)
214	$La_{1.82}Sr_{18}CuO_4$	Tetra	3.77	- -	13.25	38
123	$YBa_2Co_3O_7$	Ortho	3.81	3.88	11.63	90
TI-1212	$TIBa_2CaCu_2O_7$	Tetra	3.83	–	12.68	80
TI-1223	$TIBa_2Ca_2Cu_3O_9$	Tetra	3.84	–	15.88	105
TI-1234	$TIBa_2Ca_3Cu_4O_{11}$	Tetra	3.85	–	19.10	120
TI-2201	$TI_2Ba_2CuO_6$	Tetra	3.86	–	23.24	15
TI-2212	$TI_2Ba_2Ca_2Cu_3O_{10}$	Tetra	3.86	–	29.39	108
TI-2223	$TI_2Ba_2Ca_2Cu_3O_{10}$	Tetra	3.85	–	35.60	125
TI-2234	$TI_2Ba_2Ca_3Cu_4O_{12}$	Tetra	3.85	–	43.00	105
BI-2201	$Bi_2Sr_2CuO_6$	Ortho	5.36	5.37	24.62	6
BI-2212	$Bi_2Sr_2CaCu_2O_8$	Ortho	5.50	5.41	30.78	81
BI-2223	$Bi_2Sr_2Ca_2Cu_3O_{10}$	Ortho	5.40	5.41	37.18	110

the non-metal sublattice. The major application of such refractory compounds are in cermets, which are extensively used in cutting tools, wear and metal forming applications.

7.12.2 Silicon Carbide

Its characteristics include high hardness, heat and corrosion resistance and high-temperature strength. These properties make it promising as the main material for structural components such as high-temperature furnace components (heating elements, core tubes, refractory bricks, etc.). These applications, however, do not necessarily require high-density sintering bodies. The development of SiC components as structural materials with high density was mainly begun with the objective of application to gas turbine engines. Application of SiC as structural materials can be broadly divided into two categories, (1) abrasion- and corrosion-resistant components and (2) heat-resistant components.

Table 7.21 includes the physical and mechanical properties of silicon carbide (hexagonal).

Liquid phase sintering has been extensively applied in order to obtain dense SiC ceramics. It is observed that a 3 : 5 ratio of Y_2O_3 and Al_2O_3 of additives imparts optimum densification in the β-SiC powder. The advantage of using a liquid phase is that the silicon carbide, which is not easily sintered could be sintered to high densities (>95% theoretical) at temperatures well below 2000 °C.

7.12.3 Silicon Nitride

In both its structural modifications, α and β, silicon nitride (Si_3N_4) has a unique combination of properties: it is strong, hard, wear-resistant, stable to temperatures above 1800 °C, oxidation-resistant and due to its low coefficient of thermal expansion, it has excellent resistance to thermal shock. It is also less than half as dense as steel. The greatest impetus to development of silicon nitride for engineering applications was its selection by the US Department of Defence in 1971 as the material for the ceramic gas turbine, specified to run at 1370 °C. Not only would this be more efficient than the nickel-based superalloy engine running at about 1050 °C, it could use poorer fuels and would cause less environmental pollution. The major problem in the widespread use of silicon nitride is that the desirable properties of silicon nitride are achieved only in fully dense material and in this form silicon nitride is difficult to fabricate into precisely shaped components.

In reaction bonded silicon nitride, the silicon powder mass is shaped, partially nitrided and then machined to product before fully nitriding. The material is porous and has a relatively low strength but finds some applications in metal handling, for example, as thermocouple sheaths. Post-sintered reaction bonding closes the pores and raises the strength to a level approaching that of hot-pressed silicon nitride. Table 7.21 includes the physical and mechanical temperatures of silicon nitride (α-hexagonal).

Other methods of consolidation for silicon nitride are gas pressure sintering, hot pressing, sintering followed by hot isostatic pressing and hot isostatic pressing after encapsulation of powder. Silicon nitride ceramics are heterogeneous, multicomponent materials characterized by the inherent properties of the crystalline modifications α and β of Si_3N_4 and the significant influence of the densification additives. With a view to ability of the α and β modification to form solid solutions α-Si_3N_4 (αss) and β-Si_3N_4 (βss) solid solutions can be distinguished. Each group contains engineered materials with interesting properties for special applications. Phase relations and microstructures determine the properties decisively. Composition of the phases, the distribution of the grains, their aspect ratio and the grain boundary phase are pronounced microstructural features. The formation of the microstructure strongly depends on the one hand on the quality of the Si_3N_4 starting powders, which closely is related to the chemistry of the production process, and on the other on the liquid phase sintering as the most important step in the densification route.

7.12.4 Boron Carbide

Boron carbide, a covalent ceramic, has a density of 2.52 g cm^{-3} and a melting point of 2450 °C. It does not decompose upto its melting point and vaporizes by the preferential loss of gaseous boron. Boron carbide is characterized by a relatively wide gap in its forbidden band, a low thermal conductivity and a high thermoelectric power. This is useful for nuclear applications due to its high neutron absorption cross-section, as it provides a high concentration of boron atoms in a strong and refractory form and is relatively easy to shape. After diamond and cubic boron nitride, boron carbide is the hardest industrial material known. Another application is in high-temperature (upto 2727 °C) thermocouples in combination with graphite. It is also used in lightweight body armour.

7.12.5 Boron Nitride

There is a marked similarity between BN and carbon, as they are isoelectronic. Boron nitride is known to exhibit three polymorphic modifications; hexagonal graphite type (h-BN), cubic zinc blende type (c-BN) and wurtzite type (w-BN). Boron nitride is used extensively as a solid lubricant, as a chemically resistant container, and as a dielectric in electronic applications.

Mechanical and chemical stability and the high electrical resistivity of hexagonal boron nitride at a high temperature (2500 °C) makes it suited as a candidate for use as microwave windows in re-entry vehicle communication systems. However, this form of BN sublimes at 2500 °C. Other applications are: masks for X-ray lithography, crucibles for aluminium evaporation and for molecular beam epitaxy and powder and bonded abrasives and cutting metals.

7.12.6 Aluminium Nitride

Aluminium nitride is a highly stable covalent ceramic with the unusual combination of high thermal conductivity comparable to that of metals and high electric insulation comparable to the best insulators. In addition, it has excellent high-temperature strength, oxidation and thermal shock resistance and resistance against attack by liquid metals.

The strength of AlN at room temperature is greater than that of reaction-bonded silicon nitrides, but decreases when the temperature increases and reaches a stable value between 800 °C and 1400 °C. At 1400 °C the strength is of the same order as that of other nitrides. The highest strength value (328 MN m^{-2}) can be obtained with composition of 75 AlN-25 Y_2O_3. AlN replaces BeO advantageously for heat-sink applications in electronic devices, since the latter is toxic. AlN is known to offer translucency for light in the visible and the infrared spectrum at wavelengths between 0.5 μm and 3 μm.

7.13 GLASSES

In Chapter 2, we have already described the structure of glasses, while Chapter 5 included the toughening mechanism of glasses. In the current section, a descriptive picture of various types of glasses shall be given. The major constituents of glasses are glass former oxides (B_2O_3, SiO_2, GeO_2, Al_2O_3, V_2O_5, P_2O_5, Sb_2O_5, ZrO_2). They can produce glassy structure on their own. There are other additives too. These are designated as either intermediate oxides (TiO_2, ZnO, PbO, Al_2O_3, CdO, BeO, ZrO_2, ThO_2) or modifier oxides (PbO, MgO, PbO_2, ZnO, BaO, CaO, Na_2O, Li_2O, K_2O, CdO, SnO_2, Ge_2O_5, La_2O_3). A modifier is an oxide that is incapable of forming a glass on its own, but is frequently added in large quantities to the glass formers to alter or modify the structures and properties of the glass. An intermediate oxide is incapable of forming a glass by itself, but is capable of substituting for a glass former in the glassy amorphous network. In other words, its behaviour is intermediate between those of a glass former and a modifier. Table 7.25 gives the composition of commercial types of glasses. Some physical and mechanical properties are summarized in Table 7.26.

As an example of smart glass, the development of *photochromic glass* is worth mentioning. This glass was discovered at Corning Glass by Stookey. The concept was based on the light triggered chemistry of photography. By dispersing silver and copper halides into the glass formula, such glasses were developed. As light energy hits the glass, an electron from the copper ion is released and picked up by the silver ions, producing neutral silver atoms that block light just as they do in photographic exposures. Once the light is removed the electrons are liberated from the silver ions and return to the copper ion, which does not produce light-blocking atoms.

Table 7.25. Commercial Glass Types.

Type	Major Components, % SiO_2	Al_2O_3	CaO	Na_2O	B_2O_3	MgO	PbO	*Other*	*Comments*
Fused silica	99								Very low thermal expansion, very high viscosity
96% silica(vycor)	96				4				Very low thermal expansion, very high viscosity
Pyrex type	81	2		4	12				Low thermal expansion, low ion exchange
Containers	74	1	5	15		4			Easy workability, High durability
Plate	73	1	13	13					High durability
Window	72	1	10	14		2			High durability
Lamp bulbs	74	1	5	16		4			Easy workability
Lamp stems	55	1		12			32		High resistivity
Fiber (E-glass)	54	14	16		10	4			Low alkali
Thermometer	73	6		10	10				Dimensional stability
Lead glass tableware	67			6			17	K_2O 10	High index refraction
Optical flint	50			1			19	BaO 13 K_2O 8 ZnO 8	Specific index and dispersion values
Optical crown	70			8			10	BaO 2 K_2O 8	Specific index and dispersion values

Table 7.26. Some Properties of Glasses.

Property	*Units*	*Silica glass*	*Soda lime glass*	*Boro- silicate glass*	*Glass ceramic*
Modulus of rupture	MPa	70	50	55	>110
Compressive strength	MPa	1700	1000	1200	~1200
Young's modulus	GPa	70	74	65	92
Thermal expansion coefficient	10^{-6} K^{-1}	0.62	7.8	3.2	1.0
Thermal conductivity	$W\ m^{-1}\ K^{-1}$	1.8	1.8	1.5	2.4
Toughness(G_c)	$J\ m^{-2}$	~1	~1	~1	

Reproduced with permission from Materials for Engineering, J.W. Martin, 2nd ed., Maney Publishing on behalf of the Institute of Materials, Minerals and Mining, London, 2002.

Glass-ceramics are important group of materials which contain ceramics (crystalline) and glasses (amorphous). The ceramic phase is obtained after a controlled nucleation and growth process from a glass matrix containing nucleating agents. Glass-ceramics are by definition at least 50 per cent crystalline (ceramics) by volume and are generally more than 90 per cent crystalline. The nucleating agents can be dispersed phases of a metal, titanate, zirconate or fluorides, which are structurally incompatible with the matrix and highly unstable as glasses.

Depending on the composition and heat treatment, grain sizes can range from <0.1 μm to >10 μm. Unlike ceramics that are formed via sintering, glass ceramics are fully densified, free of porosity and their uniform microstructure ensures that their physical properties are reproducible. They are, therefore, stronger than their parent glasses and can be designed to meet a wide range of thermal expansion requirement with excellent thermal shock resistance, dielectric properties and chemical durability. Table 7.27 gives the composition, crystalline phases and some properties of Pyroceram 9606.

Table 7.27. Composition, Properties and Application of Cordierite-based Pyroceram Glass-ceramic.

Composition, wt %	
SiO_2	56.1
Al_2O_3	19.8
MgO	14.7
CaO	0.1
TiO_2	8.9
Sb_2O_3	0.3
Fe_2O_3	0.1
Properties:	
Crystalline phases	Cordierite Cristobalite Rutile Mg-dititanate
Coefficient of thermal expansion (10-700 °C) $K^{-1} \times 10^{-6}$	4.5
Modulus of rupture, MPa	250
Fracture toughness, MPa $\sqrt{m}$	2.2
Young's modulus, GPa	120
Thermal conductivity, W/m . K	38
Hardness, HK_{100}	700
Dielectric constant at 8.6 GHz	5.5
Loss tangent at 8.6 GHz	0.0003
Softening temperature, °C	>1300

Corning Glass Works specifications.

POLYMERIC MATERIALS

A brief introduction of polymers has been done in Chapter 1 (Introduction). Figure 7.14 shows the various classification of polymeric materials, of which plastics are most common. Plastics are again divided into two groups – thermoplastic and thermosetting. Each one of these again subdivided into commodity and engineering plastics. Commodity plastics generally have lower cost, and they are the most used plastics. Engineering plastics include all others. Tensile mechanical properties of various types of polymers are given in Figure 7.15.

Some of the general features and applications of these plastics arranged in alphabetical order are described below.

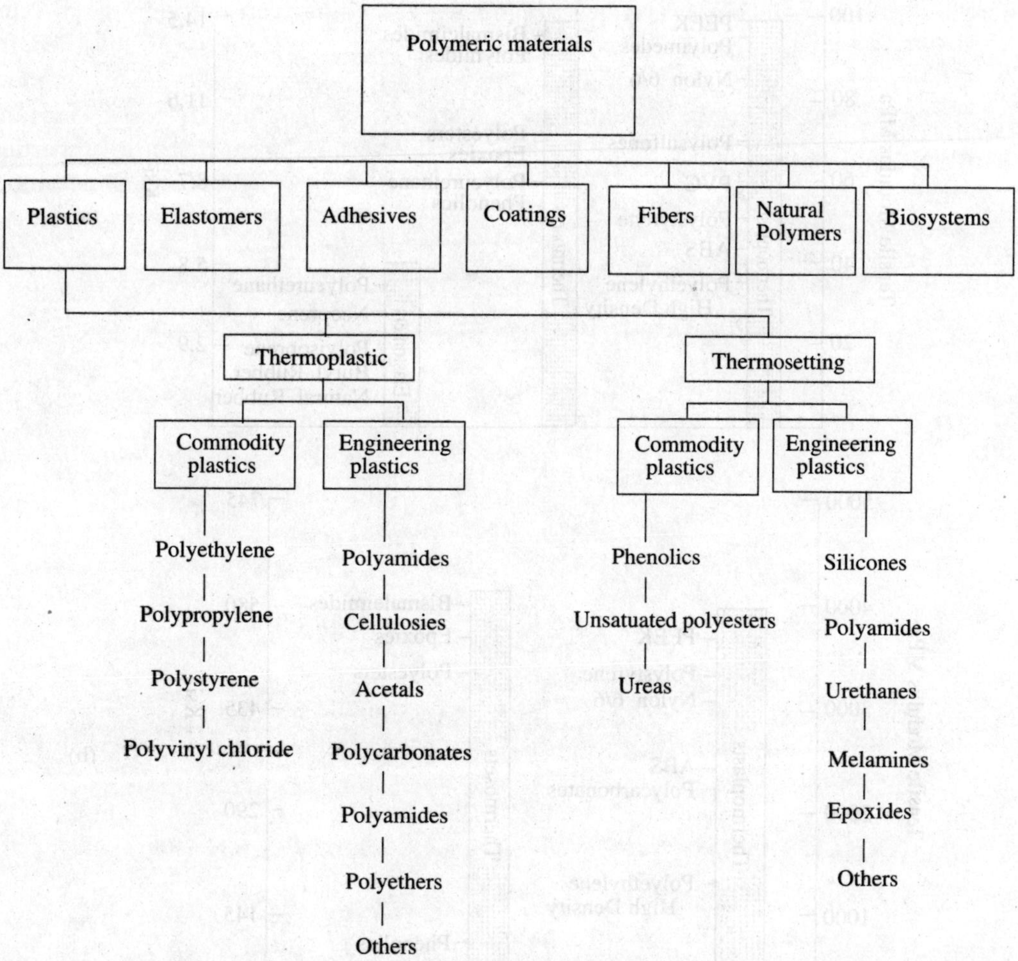

Fig. 7.14. Polymeric materials and some of the important thermoplastic and thermo-setting plastic families.

7.14 THERMOPLASTIC POLYMERS

7.14.1 ABS (Acrylonitrile/Butadiene/Styrene Copolymer) SAN (Styrene/Acrylonitrile Copolymer)

Butadiene and acrylonitrile improve the poor impact properties of polystyrene. Varying the proportions of the three monomers gives a range of impact and strength properties as needed for specific purposes. They have good strength and resistance to chemical attack in water, air, salt, alkalies, alcohols and hydrocarbons (except ketones). They are usable between −40 °C and +95 °C.

ABS is used for radio and TV cabinets, camera casings, fans, automobile facias and exterior trim. SAN is a transparent plastic and very good for lighting fitments and diffusers, TV implosion shields.

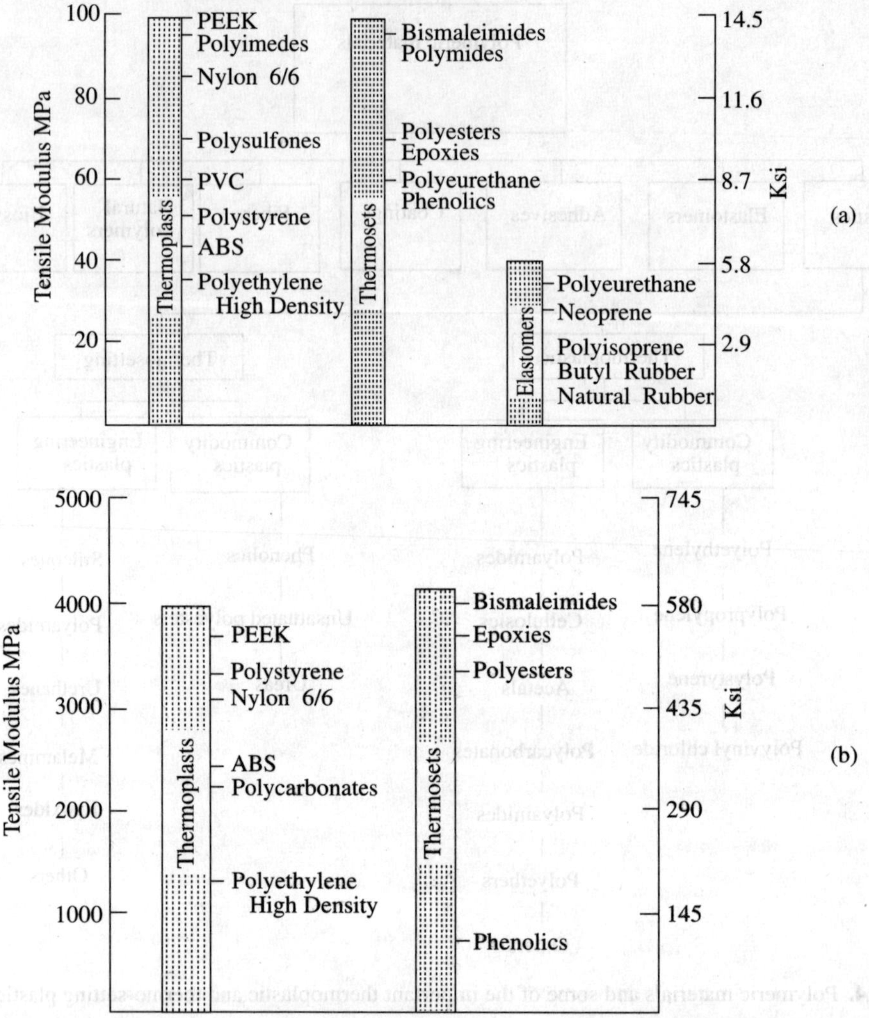

Fig. 7.15. Room temperature tensile strength (a) and tensile modulus (b) of various types of polymers.

7.14.2 Acetals

These are polymers of formaldehyde. They are tough and rigid and they can stand in for metals in many low temperature applications (i.e., Zn, Al, Cu, even steel). The main characteristics are:

- Properties are retained over wide temperature range (−40 to +120 °C)
- Strong, rigid: good impact strength and fatigue endurance
- High dimensional stability and low moisture absorption
- Low coefficient of friction (~ 0.2) and good abrasion resistance
- Resistant to most organic solvents (attacked by strong acids and alkalies)
- Can be fabricated by injection moulding and extrusion.

Applications are: car door handles, bearings, bushings, instrument housings, domestic taps, gears, valves, agitators for washing machines, moulded casings for TV sets, electric drill casings, fridge mouldings, typewriter parts, insulators, push buttons, switch parts, zippers, door knobs, cutlery handles, camera casings, pens, light diffusers, pipes, aerosol bottles, shoe heels, toys.

7.14.3 Acrylics

These are thermoplastic polymers of methylmethacrylate, e.g., PMMA (Perspex or Plexiglass) and are supplied as small beads for moulding, or as sheet, rods, tubes, etc. (extruded) and as cast sheet. The major characteristics are:

- Transparency good enough for optical use (92 per cent transmission)
- Good weathering properties
- Good resistance to weak acids, alkalies, etc. but not concentrated acids and alkalies
- Maximum service temperature is 80 °C
- Good mechanical properties (about 70 MN m^{-2})
- Easily coloured
- Surface hardness is close to that of aluminium – i.e., reasonable wear resistance
- Not easily ignited.

Applications are *Q* light fittings, windscreens for cars and aeroplanes, wall lights, goggles, control panels, machine safety guards, oxygen tents, hypodermics, decorative uses of all types, bathroom fittings, baths, sinks, cutlery handles, wine glasses, telephones, radio and TV receivers, drawing instruments, electrical covers, mirrors, prisms and lenses, particularly non-shattering lenses, e.g., for contact lenses.

7.14.4 Fluorocarbons

Thermoplastics such as polytetrafluorethylene (PTFE) comes under this category. The main characteristics are:

- Chemically inert
- Unique electrical properties (nontracking)
- Coefficient of friction ~0.001(lowest known value)
- Very low water absorption.

Applications are in chemical, food and pharmaceutical industries for non-contaminating containers, pipes, gaskets, valve components, diaphragms, release agent in baking equipment, bearings, sleeves, insulators, etc.

7.14.5 Polyamides

These are the thermoplastics of the Nylon family. The main features are:

- Tough, high strength, resistant to abrasion, low creep and good impact resistance
- Self-lubricating
- Deterioration of properties in sunlight

- High H_2O absorption
- Excellent chemical resistance
- Adequate electrical characteristics
- Crystalline MP = 264 °C – therefore can be sterilized in boiling H_2O or steam
- Low density

Some major applications are: Injection-moulded components, self-lubricated cams, gears, bearings, bushes, fasteners, curtain runners and fittings, cables and wire ropes, bottles, tubes, film for packaging.

7.14.6 Polycarbonates

These are the thermoplastics based on bisphenol-A (e.g., Lexan). The major characteristics are:

- Tough, high impact properties
- Good dimension stability
- Low H_2O absorption and expansion coefficient
- Heat distortion temperature about 275 °C
- Good electrical properties
- Good weathering resistance
- Rigid and transparent

Applications are insulated drill casings, shaver casings, glazing and electrical fittings, lighting fittings, covers, fan grills, baby bottles, combs, water meter housings, camera parts, centrifuge tubes.

Polypropylene and polyethylene: These are tonnage, cheap thermoplastics used for domestic and industrial containers, electrical components, valves, pumps, toys, shoe heels, sanitary ware, bottles, films for packaging, pipes for corrosive fluids.

Polystyrene: Similar in most respects to acrylics. But also used in foam form for heat insulation and buoyancy.

PVC (Polyvinyl chloride): Largest tonnage plastic used today. Major applications are for raincoats, boots, shoes, toys, bottles, gloves, electrical covers, gramophone records, cable sheathing, flexible pipes, containers, steel coating for chemical resistance.

7.15 THERMOSETTING POLYMERS

7.15.1 Aminos

These are thermosets, such as melamine-formaldehyde or urea-formaldehyde. The major characteristics are:

- Hard, rigid, good abrasion resistance
- Cheap
- High resistance to deformation
- Very high electrical resistance
- Good low temperature properties resistant to embrittlement

- Service upto 100 °C
- Self-extinguishing if ignited.

However they exhibit pronounced shrinkage in moulding, which is an important design problem.

The major applications are: electrical fittings, toilet seats, decorative trays and lampshades, saucepan handles, ash trays, knobs, moulded table-ware (Melamine).

7.15.2 Epoxides

These are condensation products of epichlor-hydrin + complex substances. Catalytically-cured, sometimes without heat. Usually used with additives, e.g., powder fillers or glass fiber. The major characteristics are:

- Excellent adhesion (e.g., for glues)
- Good resistance to chemicals
- Low shrinkage in curing
- Outstanding electrical properties – high resistance, low power factor and dielectric loss
- Good impact properties when properly cured.

The major applications are: electrical and electronic encapsulation where pressure cannot be applied, bushings, laminates for casings in electrical equipment, cable jointing, surface coatings for domestic and industrial use (e.g., lining of chemical tanks), flexible coatings for wire, etc., boats, jigs and tools in metal forming, moulds, filament winding, for ablative shields in rockets, tanks, pressure vessels, etc., flooring and road surfacing, wall cladding.

7.15.3 Phenolics

These are the oldest family of thermosetting, artificial plastics (i.e., Bakelites). Some of the characteristics are:

- Very cheap – plentiful raw materials
- Strong, with good impact properties
- Very little creep, rigid at ordinary temperature ($E \sim 7\ GN\ m^{-2}$)
- Very good electrical insulation and heat insulation
- Good chemical resistance
- Easy to mould, even in complicated shapes.

The common applications are: electrical insulation, water pump and valve components, heat insulation (e.g., electric iron handles, cooker knobs, etc.). Mostly used with fillers and pigments, e.g., cloth and paper or mica for good insulation in electrical work.

7.16 NATURAL POLYMERS

Natural rubber (Polyisoprene): Elastic modulus is controllable by varying the number of cross-links introduced. Highly incompressible, even under high loads, which makes it ideal for compression springs, and it can sustain large elastic strains without failure.

Rubbers rapidly stiffen below −20 °C and become brittle: it is most important to add antioxidants because of serious ozone attack leading to brittleness. This is aggravated by exposure to ultra-violet or electron radiation. Rubbers absorb large quantities of liquids (e.g., oils, water) leading to swelling.
Summary: Table 7.28 summarizes the roles of additives and modifiers commonly used in polymeric materials.

All the polymer applications described so far are based on mechanical and thermal properties of polymers. Of late, new applications of polymers have emerged in electronic devices. Some transducers are based on piezoelectric behaviour of polymers e.g., polyvinylidene fluoride (PVF_2). In such materials there is a built-in polarization due to charge separations. Any stress alters this polarization resulting in an electrical signal. Since polymer can be fabricated in large areas, these sheets can be very effective as burglar or fire alarms. Polymer spheres (~ 10 μm diameter) form the black or coloured toner particles in xerography to help transfer images from an original to a copy. Some progress has been made in polymer display devices.

COMPOSITE MATERIALS

In Chapter 1 (Section 1.14) a brief introduction of composite materials has already been done. Table 7.29 summarise the listing of various matrix and reinforcement materials and the resultant modification of the overall properties. As shown in the table the reinforcement is either in particle form or fiber form. In fiber form glass/ceramic and polymers are used. Fine wires of metallic materials are also used. For example steel wires are used as a radial steel reinforcement in automobile tyres and in wire-wound high pressure hoses. Table 7.30 gives density and mechanical properties data for some major glass, ceramic and polymer (Kevlar) fibers.

7.17 METAL MATRIX COMPOSITES

In this, category are systems like metal-metal, metal-ceramic and metal-polymer. A typical example of metal-metal composite is that of tungsten based heavy alloys, already described earlier in this chapter under Section 7.8. Another application is composites like Cu-W, Ag-W, W-Cu etc. which are extensively used as electric contact materials.

In the category of metal-ceramic composites, examples are refractory compound based composites containing metal/alloys as binders. Cemented carbides, WC-Co system is a typical example, which is processed through powder metallurgy route. From the pseudobinary phase diagram of refractory monocarbides (TiC, ZrC, HfC, VC, NbC, TaC, WC) with Fe, Co, Ni, it is evident that they invariably undergo eutectic reactions, which can be advantageously used for the liquid phase sintering. Carbon content in cemented carbides is very critical as deviation from the stoichiometric carbon content causes the precipitation of a third phase, either the 'η' phase (M_6C) or free graphite, and both deteriorate mechanical properties (Figure 7.16)

The ISO classification divides all grades of cemented carbides into three colour-coded groups: straight tungsten carbide grades (letter K, red), highly alloyed carbide grades (letter

Table 7.28. Additives and Modifiers for Polymeric Materials to control Mechanical, Surface and Chemical Properties.

Additive type	*Purpose*	*Examples*
Mechanical property modifiers		
Impact modifiers	Improves fracture toughness and impact strength	Polybutadiene rubber, EVA copolymers, SIS block copolymers
Plasticizers	Improves flexibility and toughness, reduces modulus	Phthalates, adipates, trimelliates, ricinolates, acetoxy stearates, sebecates
Nucleating agents	Control the rate of crystallization	Carbon fiber, glass fiber, fabrics
Reinforcements	Increase strength	(cotton, canvas), mica
Fillers	Controls shrinkage or reduces resin cost	Wood flour, titanium dioxide, mica, alumina, calcium carbonate, polymer or glass spheres, silicas, titanates, ground nut shells or rice husks, clays, talc
Surface property modifiers		
Internal lubricants	Improves tribological properties, prevents polymer from adhering to processing equipment	Silicon, waxes, stearates, fatty acid amides, fluoro-dispersions, glycerides, petrolatum
External lubricants	Improves tribological properties	PTFE, silicone, molybdenum disulfide, graphite, waxes, stearates
Slip and antiblocking agents	Reduces the tendency of films and sheets from sticking together	Precipitated silica, aluminium silicates, fatty acids, stearates
Antistatic agents	Reduces static charge buildup	Pyrogenic alumina, metal powders, polyether block amides, carbon, alkali salts
Wetting agents	Stabilizes dispersions of fillers	
Antifogging agents	Disperses moisture droplets on films	Fatty acid esters of glycerol, sorbitan, fatty alcohols
Chemical property modifiers		
Antioxidants/heat stabilizers	Prevents thermal or oxidative degradation: chain scission, cross linking, crazing, gelation deploymerization	Hindered phenols, mercaptobenzoididi-zoles, organophosphites, organotins
UV stabilizers	Mitigates ultraviolet light degradation	Hindered amines, hydroxylphenyl benzotriazoles, benzophenones
Flame retardants, synergists	Suppresses flammability	Antimony oxide, borates, reactive bromates, intumescent phosphates
Smoke suppressants	Suppresses smoke	Aluminum trihydrate, magnesium hydroxide, zinc oxide
Antistats	Reduces static charge buildup	Pyrogenic alumina, metal powders, polyether block amides, carbon, alkali salts
Biocides, fungicides, preservatives	Reduces microbial activity, mildew, and fungus formation	

P, blue) used mainly for machining steel, and moderately alloyed grades of carbide (letter M, yellow), which are multipurpose and may be used on steels, complex high-temperature alloys, cast iron, etc. Each grade within a grouping is given a number between 01 and 50, representing a range from maximum hardness to maximum toughness. Typical applications are given for

Table 7.29. Different Matrices, Reinforcements and Properties of Composite Materials (Adapted from the Principles of Material Selection for Engineering Design, P.L. Margonon, Prentice-Hall, USA, 1990, p. 702).

Matrix material	*Reinforcement material*	*Properties modified*
Metal	Metal, ceramic, carbon, glass	Elevated temperature strength Electrical resistance Thermal stability
Ceramic	Metallic and ceramic particles and fibers	Elevated temperature strength Chemical resistance Thermal resistance
Glass	Ceramic fibers and particles	Mechanical strength Temperature resistance Chemical resistance Thermal stability
Polymers Thermosets, Thermoplastics	Carbon, glass, organic fibers, glass beads, flakes, ceramic particles, metal wires	Mechanical strength Elevated temperature strength Chemical resistance Antistatic Electrical resistance EMF shielding Flexibility Wear resistance Energy absorption Thermal stability

Table 7.30. Mechanical Properties of Some Reinforcing Fibers.

Material	*Density* ρ ($kgm^{-3} \times 10^3$)	*Young's Modulus E (GPa)*	*Tensile Strength* σ (*GPa*)	*Fiber Radius r* (μ*m*)
E-glass fibers	2.56	76	1.4 - 2.5	10
Carbon fibers (high modulus)	1.75	390	2.2	8.0
Carbon fibers (High strength)	1.95	250	2.7	8.0
Kevlar fibers	1.45	125	3.2	12
Silicon carbide (mono- filament)	3.00	410	8.6	140
Silicon carbide (Nicalon)	2.50	180	5.9	14
Alumina (Saffil)	2.80	100	1.0	3

Reproduced with permission from Materials Science and Engineerings, An Introduction, 6th ed., W.D. Callister, Jr., John Wiley, New York, 2003.

grades at more or less regular numerical intervals. Table 7.31 illustrate some of the above features.

Steel-bonded titanium carbides are another type of cermet which can be machined and are corrosion, oxidation and wear resistant. In addition, they are light in weight and have high elastic modulus and vibration damping. One of the attractive properties of steel-bonded TiC is that it is heat treatable. These cermets are extremely stable and distort very little during

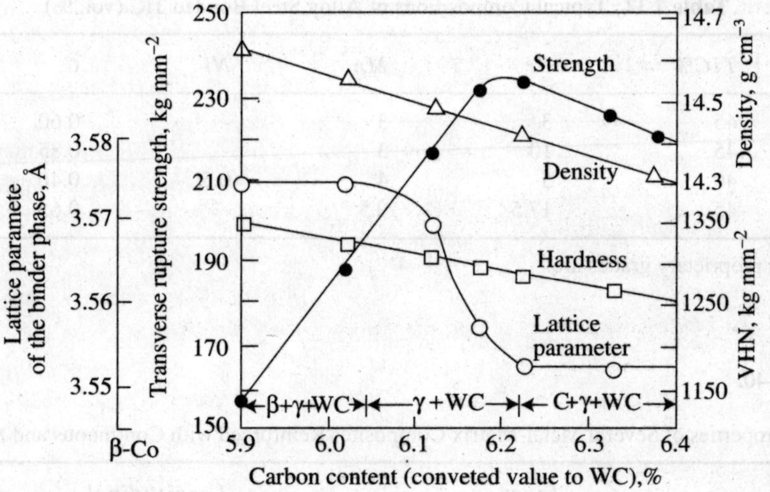

Fig. 7.16. Effect of the carbon content on density, lattice parameters, Vickers hardness and transverse rupture strength of WC-10 Co cemented carbide.

Table 7.31. Tungsten Carbide-based Hard Metal Groups according to Application.

Symbol	*Application categories*	*Range of composition (wt %)*	*Hardness HV*	*Bend strength MPa*
K	Metals with short chips (preferably cast iron) some non-ferrous alloys, plastics, wood	WC-Co Co 4-12% 0-3% TiC 0-4%(Ta,Nb)C	1300-1800	1200-2200
P	Metals with long chips (steel, steel castings, non-ferrous metals)	WC-TiC-Co WC-TiC-(Ta,Nb)C-Co Co 5-14% TiC-(Ta,Nb)C upto >50%	1300-1700	800-1900
M	Multi-purpose applications	WC-TiC-Co WC-TiC-(Ta,Nb)C-Co Co 6-15% TiC-(Ta,Nb)C 6-12%	1300-1700	1350-2100

heat treatment. This eliminates expensive machining after hardening. Among the steel-bonded titanium carbides, Ferro-TiC grades are very common, for which various grades have been developed (Table 7.32). Hardening and tempering procedures vary with the matrix alloy and these can be classified into oil-, or air- hardening types. Age-hardening type is also included.

In the group of continuous fiber reinforced metal-matrix composites aluminium, titanium and super-alloy based matrices are very common. The major applications are as structural parts in aerospace industry. Designs incorporating these composites permit higher operating temperatures and better efficiencies for turbine engines. It is worthwhile to mention that metal

Table 7.32. Typical Compositions of Alloy Steel Bonded TiC (vol. %).

Type[a]	*TiC%*	*Cr*	*Mn*	*Ni*	*C*	*Fe*
1	45	3	3	-	0.60	Balance
2	45	10	3	-	0.85	Balance
3	40	5	4	0.5	0.40	Balance
4	45	17.5	0.5	-	0.65	Balance

[a] Corresponding proprietary grades are:
1 Ferro-TiC C
2 Ferro-TiC CM
3 Ferro-TiC SK
4 Ferro-TiC CS-40.

Table 7.33. Properties of Several Metal-Matrix Composites Reinforced with Continuous and Aligned Fibers.

Fiber	*Matrix*	*Fiber Content (vol.%)*	*Density (g/cm^3)*	*Longitudinal tensile modulus (GPa)*	*Longitudinal tensile strength (MPa)*
Carbon	6061 Al	41	2.44	320	620
Boron	6061 Al	48	-	207	1515
SiC	6061 Al	50	2.93	230	1480
Alumina	380.0 Al	24	-	120	340
Carbon	AZ31 Mg	38	1.83	300	510
Borsic	Ti	45	3.68	220	1270

Source: Adapted from J.W. Weeton, D.M. Peters and K.L. Thomas, Engineers' Guide to Composite Materials, ASM International, Materials Park, OH, 1987.

matrix composites are much more expensive than the polymer matrix composites. These composites operate at higher temperature, are non-flammable and offer greater resistance to degradation by organic fluids. Processing of critical MMC's are generally done through powder metallurgy route including hot pressing. Table 7.33 gives density and mechanical properties data of continuous fiber reinforced MMCs titanium. One of the major limitations of MMCs is rather poor machinability, which makes the machining process rather costly.

7.18 CERAMIC MATRIX COMPOSITES

Ceramic matrix composites can be classified in ceramic-metal, ceramic-ceramic and ceramic-polymer composites, the reinforcement being either in particulate or fiber form. Most of the ceramic matrix composites are based on a limited range of high covalency or polyvalent ionic phases, e.g., α and β Si_3N_4, SiC, ZrO_2, Al_2O_3. For particulate composites, powder metallurgy processing is a convenient method. The directed molten-metal oxidation (DIMOX) is now commercially established (the 'Lanxide' process), in which Al_2O_3 in a contiguous polycrystalline form, contains residual intercrystalline aluminium melt. The net-shape low cost fabrication process suffers from the fact that application at high temperatures is not possible.

Table 7.34. Room Temperature Fracture Strengths and Fracture Toughnesses for Various SiC Whisker Contents in Al_2O_3.

Whisker content (vol. %)	*Fracture strength (MPa)*	*Fracture toughness (MPa $\sqrt{m}$)*
0	-	4.5
10	455 ± 55	7.1
20	655 ± 135	7.5 - 9.0
40	850 ± 130	6.0

Reproduced from Engineered Materials Handbook, Vol. 1, Composites, C.A. Dostal (Sr. Editor), A.S.M. International, Materials Park, Ohio, 1987.

Another variation of reactive forming is via sintering of mixed metallic/ceramic powders or via pressure induced molten metal penetration of porous ceramic preforms. A reaction like the following is possible:

$$3TiO_2 + 3Al \rightarrow 2Al_2O_3 + 3TiAl_3$$

In case the reaction is incomplete, it again poses a high temperature application limit. Although the process used cheap precursors and short cycle times, the properties so far produced do not exceed that of a good engineering ceramic ($K_c \sim 9$ MPa.m$^{1/2}$; MOR ~ 600 MPa).

In case of ceramic-ceramic composites, mention has already been made of stabilized zirconia and the mechanism of strengthening described in Chapter 5 (Section 5.15). Similarly mullite-zirconia composites has been developed. An increase in fracture toughness of the mullite-ZrO_2 composites (prepared by sol-gel method) with increase in total ZrO_2 content is attributed to microcrack toughening due to spontaneous phase transformation (tetragonal → monoclinic) of ZrO_2. The fracture toughness value of the 15 vol. per cent ZrO_2 composition is 4.9 MPa.m$^{1/2}$, whereas the flexural strength is 520 MPa. Beyond 15 vol. per cent ZrO_2, the values decrease due to both enhanced microstructures formed by the $t \rightarrow m$ transformation on cooling and the release of strain caused by thermal expansion mismatch between the mullite and ZrO_2.

Figure 7.17 shows the generalized picture of fracture toughness variation of oxide and monoxide based ceramics containing different volume fractions of ZrO_2. The number on top of the shaded bars indicates the volume per cent of ZrO_2 in the composite.

The development of CMC utilizing whisker ceramic reinforcements has been much in progress. The ceramic whiskers inhibit crack propagation (Section 5.15). Table 7.34 shows how the mechanical properties get improved by increasing the whisker content of SiC in Al_2O_3 matrix. These CMCs exhibit improved creep resistance, and resistance to thermal shock.

However, there is a negative aspect of using whiskers as they can become airborne during manufacture or disposal providing an opportunity for inhalation by workers or the public. Because of this health hazard, composites containing such materials are now deemphasized.

SiC-SiC (continuous fiber) composites have been formed by chemical vapour infiltration. At 1000 °C, they have a strength of 280 MPa, comparable to that of most steels. SiC monofilament-reinforced silicon nitride (Textron) has shown an ultimate tensile strength of 310 MPa at 1200 °C.

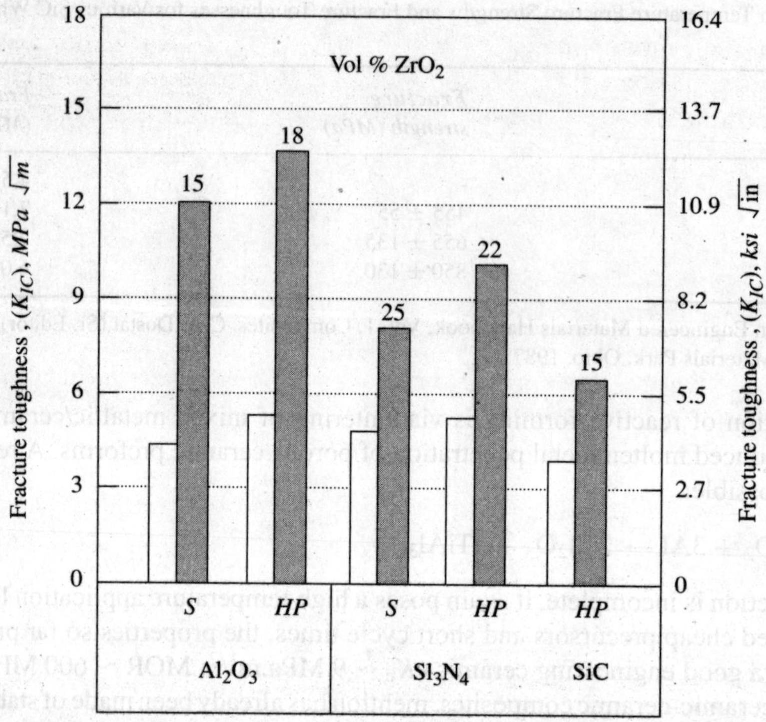

Fig. 7.17. Fracture toughness of alumina, silicon nitride and silicon carbide. Numbers denote volume % zirconia (S: sintered, HP: hot pressed).

One of the most useful development in CMC has been of carbon-carbon composites, in which amorphous carbon matrix is reinforced with continuous Textron fibers. The fibers are based on rayon, pitch or polyacrylonitrile (PAN) and increase in cost in the order listed. Such composites maintain their strength at temperatures as high as 2500 °C. One of the most important applications of C-C composites is in the nose cone and leading edge components of the space shuttle. These are also used in aircraft brakes, where their high thermal conductivity and light weight are advantageous. Other applications are turbine wheels as hypersonic aircraft, light weight housings for satellite solar cells. The only disadvantage of such composites is that they oxidize readily and hence require protective coating.

Crystalline ceramic fibers, have also been used to reinforce amorphous glasses. Borosilicate glass, reinforced with SiC fibers, has a strength of 830 MPa, four times that of the glass alone.

7.19 POLYMER MATRIX COMPOSITES

Polymer matrix composites may be classified into polymer-polymer, polymer-metal and polymer-glass composites.

A lot of development has been made in metal powder filled liquid polymers in order to prepare castable or paste like metal/polymer composites. While moulding, compounds may achieve a certain degree of fluidity at high temperature and high pressure, casting techniques are generally associated with room temperature and they produce compounds, which are pourable because of the lower viscosity.

The polymers available for pouring are low molecular weight thermosetting resins rather than thermoplastics. Types of thermosetting liquid polymers, which have been used, with metal powders are classified in two types.

1. Polymers which liberate water or volatile during curing as a consequence of condensation reactions. Example are: Phenol-formaldehyde resins, Urea-formaldehyde resins, Melamine-formaldehyde resins, Furfuryl alcohol systems, Glyceryl-phthalate systems.
2. Thermosetting polymers which are liquid and do not liberate volatiles during curing. Examples are: Epoxy resins, Polyester resins and Polyurathane resins.

Finely divided inorganic fillers in addition to metallic fillers, offer the advantage of reducing settling tendencies of the denser metallic fillers. Finely divided silica fillers such as "cabosil" or "bentonite" are used to aid the suspension of metallic particles. When casting of metal powder filled polymers are made, there is more separation and settling of the dense metallic particles when the viscosity of the pourable composite mixture is low.

Some commercial resin systems featuring a number of metallic powders dispersed in a liquid polymer (epoxy) are given in Table 7.35.

The metal components in the composites alter the original attributes of the composites, e.g.,

- Attenuation of electromagnetic radiation,
- Reduction in radiation from nuclear sources
- Reduction of electrical volume sensitivity and surface sensitivity.

The polymer components on the other hand contribute to the following:

- Increase in abrasion and weathering resistance,
- Electrical insulative protection,
- Serviceability of metal products is enhanced due to the thermal qualities.

Other group of polymer matrix composites (PMC) is fiber reinforced in polymer resin matrix. The common reinforcements are glass, carbon and aramid fibers. The largest quantity of PMC produced is that of glass fiber-reinforced polymer (GFRP) composites. The most common glass is of E-glass. The fiber diameters normally range between 3 and 20 μm and they can be either continuous or discontinuous. The surface characters of glass fibers are very important, as even small quantity of surface flaws can adversely affect the mechanical properties. Another limitation is the glass surfaces that have been exposed to the normal atmosphere for even short period of time. The freshly drawn glass fibers are, therefore, coated. The coating is removed just before the composite fabrication and replaced with a coupling agent, which promotes better bonding between the fiber and matrix. The limitations are not having high stiffness and limit of service temperature below 200 °C. The applications are in automotive and marine bodies, plastic pipes, storage containers etc.

Table 7.35. Typical Physical Properties of Metal Filled Epoxy Compounds (7 days room temperature).

	Plastic steel putty	*Aluminium putty*	*Bronze putty*	*Stainless steel putty*
Mix ratio. wt	9.0 : 1	9.0 : 1	9.0 : 1	9.0 : 1
Resin hardener vol.	2.5 : 1	4.0 : 1	3.0 : 1	3.0 : 1
Specific volume m^3/kg (10^{-4})	4.3	6.3	4.5	4.5
Operating temperature maximum °C	121	121	121	121
Compressive strength, MPa ASTM D 695	57	58	59	61
Cured hardness, Shore D ASTM D 1706	85	85	85	85
Flexural strength MPa ASTM D 790	38	46	42	43
Cure shrinkage m/m ASTM D 2566	.0006	.0008	.0010	.0010
Colour	Dark grey	Aluminium	Bronze	Grey

Reproduced from Engineered Materials Handbook, Vol. 1, Composites, C.A. Dostal (Sr. Editor), A.S.M. International, Materials Park, Ohio, 1987.

Laminate polymer matrix composites may have between 4 to 40 layers and the fiber orientation changes from layer to layer in a regular manner though the thickness of the laminate, e.g., a 0/90 ° stacking sequence results in a *cross ply* composite. Figure 7.18 illustrates Young's modulus/tensile strength relationship of cross laminated glass reinforced epoxy in different directions.

Carbon fiber-reinforced polymer is another group of PMC, where specified engineering properties are achieved because of physical and mechanical properties of the fibers. While describing C-C composites mention has already been made of different organic precursors for producing carbon fiber. Carbon fiber diameters are usually in the range of 4-10 μm. Like GFRP, in this case too continuous or noncontinuous (chopped) fibers are used. The fibers are normally coated with a protective epoxy to improve the adhesion with the polymer matrix. Such composites have applications in sports and recreational equipments, filament-wound rocket motor cases, pressure vessels and aircraft structural components.

Aramid fiber-reinforced polymer composites are the third category of materials, which have even a better tensile longitudinal strength on the carbon fiber reinforced ones. They are however weak in compression. Aramids are thermoplastics, but are resistant to combustion and stable in the temperature range -200 to 200 °C. Common matrix materials for this group of PMC are epoxies and polyesters. Common applications of such composites are in ballistic products, sporting goods, tyres and in pressure vessels. The material is suited as a replacement for asbestos in automotive brake and clutch linings. It is not out of place to caution about health hazard aspects of the fibers. Carbon fibers can irritate skin and aramid fibers can cause fibrosis in the lungs.

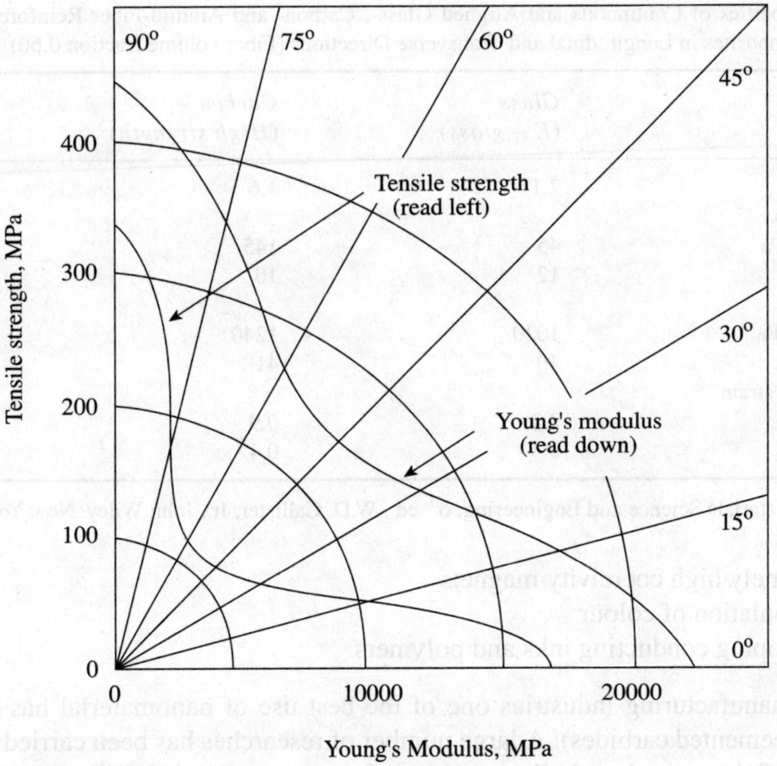

Fig. 7.18. Young's modulus and tensile strength of cross laminated, glass reinforced epoxy in different directions.

Specific gravity and tensile mechanical properties data of the continuous and aligned PMCs containing the above three types of fibers are given in Table 7.36. Kevlar is one of the trade name for aramid, which have different varieties e.g., Kevlar 29, 49 and 149.

NANOMATERIALS

A mention of nanomaterials has been made in Chapter 1. Nanomaterials contain a controlled morphology with at least one nanoscale dimension. If we go on reducing the size of a bulk material, the percentage of atoms at the surface increases. This is one of the main causes of the catalytic properties of nanocrystalline particles. With size reduction, one may achieve quantum wells (ultra thin quasi two dimensional structure), quantum wire (ultrafine one dimensional structure) and quantum dots (zero dimensional structure). The transition from micro to nano-scale has great bearing on most of the physical properties. Some of the special features of materials produced out of nanoparticles are, for example:

- Very high ductility
- Very high hardness, say 4 to 5 times more than the conventional material
- Transparent ceramics achievable

Table 7.36. Properties of Continuous and Aligned Glass-, Carbon- and Aramid-Fiber Reinforced Epoxy-Matrix Composites in Longitudinal and Transverse Directions (Fiber volume fraction 0.60).

Property	*Glass (E – glass)*	*Carbon (High strength)*	*Aramid (Kevlar49)*
Specific gravity	2.1	1.6	1.4
Tensile modulus			
Longitudinal GPa	45	145	76
Transverse GPa	12	10	5.5
Tensile strength			
Longitudinal MPa	1020	1240	1380
Transverse MPa	40	41	30
Ultimate tensile strain			
Longitudinal	2.3	0.9	1.8
Transverse	0.4	0.4	0.5

Adapted from Materials Science and Engineering, 6th ed., W.D. Callister, Jr., John Wiley, New York, 2003, p. 549.

- Extremely high coercivity magnets
- Manipulation of colour
- Developing conducting inks and polymers.

Among the manufacturing industries one of the best use of nanomaterial has been as nano-hard metals (cemented carbides). A large number of researches has been carried out at Rutgers University, U.S.A. to produce bulk quantities of nano-structured WC/Co powders. A process called spray conversion was developed, which allows tungsten and cobalt to be mixed at molecular level. The precursor powder is reduced and carburized in a fluidized bed reactor to yield nanophase cobalt/tungsten carbide powder.

Nano-technology is also growing in importance with very small-scale devices performing vital functions in automobile's operation. A good example of a successful application has been is the use of solid-state microchip mounted accelerometers to activate airbags, replacing the inertial ball and spring mechanical device.

One of the biggest applications of nanomaterial in foreseeable future is carbon nanotubes, particularly in the field of energy storage. The main challenge is to make nanotubes of controlled structure, in sufficient quantity and at affordable price.

We have described the structure of Fullerenes in Chapter 2. Nanotubes of carbon can be obtained in single wall or multiple walls. Single walled tube is of one atom in thickness and around ten atoms of diameter. Because of freedom from any defect, electrons propagate fast without scattering by the atoms. This, therefore, offers the advantage of zero thermal loss. These tubes are very flexible. For a nanotube with a diameter smaller than 8 nm, the Young's modulus is 1200 MPa slightly higher than for diamond and about six times the value for carbon steel. This property could be measured at atomic force microscope. The thermal conductivity is better to that of the best thermal conducting metals i.e., silver, copper.

The preparation of carbon nanotubes is generally through either direct current discharge or through chemical vapour deposition. Laser ablation of carbon target is also another method for preparation. The CVD process is achieved by decomposition of acetylene, ethylene and propylene at a temperature of 650-800 °C in the presence of transition metal catalysts. The separation of the nanotubes from the catalyst particles is carried out by a hydrofluoric acid treatment. Any residual amorphous carbon is removed by air oxidation.

With all the excitements described above, a caution towards toxicity against the use of nanomaterial is called for, although this has not yet been established conclusively.

QUESTIONS AND PROBLEMS

Carbon and Low Alloy Steels

7.1 What are the differences between a high strength low alloy steel and a carbon steel?

7.2 The ancient arch bridges made out of stone are now replaced by truss and suspension bridges built out of steels. Describe how the material plays the role in the configuration of the bridge.

7.3 Highlight the role of boron addition in steel from the viewpoint of (1) hardenability, (2) austenite grain size, (3) effect on M_s temperature and (4) concentration for maximum effectivity.

7.4 What are the special advantages of nitriding over carburizing and cyaniding?

7.5 British steel, Workington has been a major producer of pearlitic steel rails (composition 0.86 C, 0.86 Mn) in U.K. In earlier technology, the hot rails passed from the finishing stand of the mill at about 1000 °C, followed by normal air cooling (Hardness 290 Brinell). In 1985 the plant decided to change the technology so as to produce long life (4 to 5 times over conventional) rails of same composition but of hardness 400 Brinell to enable them to carry upto 150 million gross tons of traffic per year. The operation must be continuous and in-line. Suggest the proper operation and the resultant microstructure of the steel.

7.6 The substitution of HSLA steels for low carbon grades to achieve weight saving in automobiles is based on their higher yield strengths:

(a) What implication does such substitution have for the elastic stiffness of components?

(b) In view of corrosion on cars, what implication does such substitution have for component life? (Hint: The corrosion resistance of HSLA steels is almost identical to that of low carbon steels).

7.7 Explain why inclusion shape control is of much greater importance in high strength steels than in low carbon steels.

7.8 Steel ball bearings are assembled from through-hardened elements while roller bearings use surface hardened parts. Which of these will have the greatest shock resistance and why?

7.9 What mechanisms are responsible for the strength of high strength low alloy structural steels?

7.10 In developing high strength-low alloy steels the controls on composition, rolling, cooling and purity are ought to be stringent. Describe the practical challenges when these factors are combined in order to maximize the end properties.

7.11 Why dual-phase HSLA steels are categorized in a separate class altogether? Can they be satisfactorily classified as composites?

High Alloy Steels

7.12 (a) Name five major kinds of tool and die steels.

(b) What properties are desirable in tool and die steels?

(c) What are the effects of varying the carbon content of tool steels?

(d) What are the effects of the carbide formers on tool steels?

(e) What is the effect of austenite grain size on tool steels?

(f) Why is cobalt used as an alloying element in high speed tool steels?

7.13 (a) What is the difference in compositions of ferritic, austenitic and martensitic stainless steels?

(b) What is an important disadvantage of the ferritic stainless steels with respect to their fabrication?

(c) What causes inter granular corrosion in improperly welded austenitic stainless steels?

(d) What remedial steps may be taken to prevent intergranular corrosion in austenitic stainless steels?

7.14 You want to make a deep drawn vessel similar to a sink. What grade of stainless steel would you use? Why?

7.15 Explain the significance of sigma phase on the properties of stainless steels.

7.16 A carbon steel component operates satisfactorily at 500 °C under a stress of 50 MPa. If 304 stainless steel was substituted, upto what level could the temperature be raised?

7.17 For economic reasons a Fe-18Ni maraging steel is planned to be substituted by Fe-12Ni-6Mn system. Such a change is accompanied by a fall in the toughness. What treatments would you prefer to enhance the toughness from the view point of following factors ?

(a) Type of treatment thermal cycling or thermo-mechanical treatment, mainly from economic view point
(b) Prior austenite grain size
(c) Retained austenite in the microstructure
(d) Selection of temperature range for mechanical working
(e) Need for other minor additives and their role

[Hint: The grain refinement is an excellent tool to enhance toughness]

Cast Irons

7.18 Why are steel or cast iron valves usually specified for ammonia refrigeration system?

7.19 What factors determine whether a cast iron is grey or white?

7.20 What is the difference between ductile iron and malleable iron?

7.21 A malleable cast iron containing 3% C was initially solidified as austenite and carbide. It was then reheated to decompose the carbide. What is the maximum volume fraction graphite that may form at 800 °C? Assume for ferrite and austenite same density of 7.6 g/cm^3 at this temperature. The density of graphite is 2.0 g/cm^3.

7.22 Which types of cast irons would be preferred for (a) a sewage pipe, (b) the crankshaft of an integral combustion engine, (c) grinding mill parts, (d) a manhole cover, (e) a machine tool bed?

7.23 Briefly describe a thermal process to produce a cast iron consisting a martensite matrix that contains (a) graphite flakes, (b) graphite nodules and (c) graphite rosettes. Which of these cast iron would be impractical? Why?

Non-ferrous Alloys

7.24 Is tin-bronze as good as copper as an electric conductor? Explain.

7.25 Explain the significance of the number (a) 1035 and (b) 332.0 for aluminium or its alloy designation.

7.26 Given a magnesium alloy designation as AZ61A-F, indicate the meaning of each of the letters and numbers for this alloy.

7.27 Write the designation number for the Mg-alloy that has a composition of 5.5% Zn and 0.45% Zr and which has been partially solution treated and artificially aged. Assume this is to be the second Mg-alloy developed of this type.

7.28 Copper has a melting point of 1083 °C and lead has a melting point of 327 °C. What is the non-dimensional homologous temperature for each of these materials at 100 °C?

7.29 Explain why the fatigue strength/yield strength ratio for copper alloys varies so widely?

7.30 (a) Is beryllium used for conventional structural parts? Why?
(b) What are two important uses of beryllium?
(c) Describe the properties of beryllium copper.

7.31 Why is copper used in the Babbitt bearing alloys?

7.32 Study the Al rich side phase diagram of binary Al-Cu and outline the complete program for obtaining a maximum hardness in an alloy of Al-4 Cu if the ageing temperature is 403 K. When the maximum hardness is obtained, what would be the nature of the precipitate? Explain.

7.33 The supersonic passenger aircraft Concord, which is no longer operational because of financial constraints, is supposed to be an engineering marvel. The skin of the plane is made out of age hardening aluminium alloy-Hinduminium RR58, first developed by High Duty Alloys, Ltd., U.K. This alloy is based on Al-2.5 Cu-1.5 Mg composition and contains 0.2% silicon. Referring Handbooks, justify the selection of RR58 over

Al-Zn-Mg alloy from the point of (a) crack growth propensity and (b) elevated temperature mechanical properties.

7.34 Why in developing Fe and Co-base superalloys, we consult Fe-W-C and Co-W-C ternary phase diagrams, but not Ni-W-C in case of Ni-base superalloys?

7.35 Which strengthening mechanism(s) is(are) applicable for Ni-based superalloy MA754 with 0.6 wt % Y_2O_3, 0.5 wt % Ti and 0.3 wt % Al?

7.36 Titanium alloys are being considered for automotive engine valves. What advantages would the titanium provide? Suggest alloys for both the inlet and outlet valves.

7.37 Describe thermal treatments of Ti-6 Al-4 V alloy to get the following microstructures:

(a) equiaxed α and small amount of intergranular β
(b) equiaxed and acicular α and a small amount of intergranular β
(c) equiaxed α in an acicular α matrix
(d) small amount of equiaxed α in an acicular α matrix
(e) plate like acicular α and some α at prior β grain boundaries.
(Note: acicular α denotes transformed β).

7.38 What are the advantages of thoriated tungsten in comparison to either tungsten or thorium alone for hot cathode applications?

Glass and Ceramics

7.39 What is the highest melting glass?

7.40 Give an example where glass is used as a structural material. Explain why it is safe?

7.41 What steps would be followed, if you intend to replace laminate glass with other glass-based materials for enhancement of properties?

7.42 Alumina is one of the hybrid circuit substrates in electronic devices. Invariably a glassy phase pertaining to $SiO_2 - Al_2O_3 - MgO$ ternary system is desirable in it. Could you suggest some reasons? [Hint: Distortion in the substrate not permitted.]

7.43 An electrical porcelain is made into an insulator for 50 w/o clay [$Al_2Si_2O_5(OH)_4$], 40 w/o potassium feldspar [$K_2Al_2Si_6O_{10}$] and 10 w/o silica flour (SiO_2). What is its fired composition? Solve by computing as well as graphically.

7.44 A fireclay is a relatively pure clay with a composition of $Al_2Si_2O_5(OH)_4$. When it is used to produce refractory brick it dissociates according to $Al_2Si_2O_5(OH)_4 \rightarrow Al_2O_3 + 2SiO_2 + 2H_2O\uparrow$. Based on Al_2O_3–SiO_2 phase diagram and assuming equilibrium, what phases will form at 1500 °C? What is the maximum weight per cent of each? What are the crystal structures of these resultant phases?

7.45 Why chemically synthesized magnesium-aluminate spinel is used exclusively as refractory in molten salt electrolysis cell for producing magnesium?

7.46 In PLZT ceramic for piezoelectric and electrooptic applications, is the amount of lanthanum same? If not, what are the values?

7.47 In a heterogeneous multiphase refractory, what are the microstructural features?

Polymers

7.48 What is the ratio of ethylene mers to vinyl chloride mers in a copolymer of the two in which there is 30 wt % chlorine?

7.49 What fraction of butadiene (C_4H_6) is cross-linked if the product contains 18.5% sulphur?

7.50 How many kilograms of sulphur are required per 100 kg of final rubber produced to completely cross-link butadiene (C_4H_6) rubber?

7.51 Calculate the quantity of sulphur required to fully vulcanize 100 kg of raw polyisoprene and the percentage of sulphur in the final product.

7.52 Carbon black is often used as a particulate filler in both thermoplastic and thermosetting polymers. What advantages in various properties do you expect in the composites?

7.53 For each of the following applications, suggestions are given from the polymers. Select the most appropriate one in each case
(a) The housing for a computer: polyethylene, polypropylene, ABS, PVC
(b) A self-lubricating gear that is not exposed to moisture: polyethylene, ABC, Nylon 6.6, polycarbonate
(c) An electrical insulator at room temperature: polyethylene, PVC, ABC, PET
(d) A gasket exposed to fuels and oils: natural rubber, SBR, NBR.
Give reasons.

7.54 Copier toners are made up of random copolymers of polysterene ($T_g = 105\,°C$) and polybutadiene ($T_g = -60\,°C$). In case you wish to develop a toner material with T_g of 60 °C, what shall be its composition? (Note: use rule of mixture).

7.55 Why PMMA polymer cannot be used in circuit boards?

7.56 Why is that thermosetting plastics cannot be remelted?

7.57 What rubber would you use for a punch press die spring? Explain your reasons.

7.58 A polymer is said to be plasticized if it is made more pliable. Copolymerization can serve this purpose. Alternatively, additives, called plasticizers, can be used. A plasticizer is a lower molecular weight substance added to the polymer, usually a nonvolatile solvent that partially remains in the system in its ultimate use:
(a) What effect should a plasticizer have on T_g? Why?
(b) How will a plasticizer affect crystallinity? Why?
(c) Can a network polymer be plasticized?

7.59 Certain long-chain polymers, such as isotactic polypropylene assume a helical chain conformation because of geometric hindrance of the side groups. Is crystallinity favoured or disfavoured because of this? Why?

7.60 Can a fully cured Bakelite be ground up and reused? Explain.

7.61 (a) Give the structural formula for vinyl chloride.
(b) Show how polyvinyl chloride can be prepared by chain polymerization.
(c) Schematically show how the monomer concentration varies with time, relative to completion of the chain polymerization process.

7.62 Sketch the chain structure of Kevlar. What is the molecular weight of a mer? Calculate the degree of polymerization for a polymer with a total molecular weight of 140,000 g/mol.

7.63 What is the short form of polyetherethereketone? Is it thermoplastic or thermosetting? What is the value of its T_g in Centigrade?

Composites

7.64 Why is glass the most used reinforcement for composites?

7.65 A refrigerator door needs a rubber gasket for stopping air flow and a magnetic latch. Develop a technique to solve both these purposes.

7.66 Explain why a mixture of sand and gravel is used as the aggregate for concrete.

7.67 Why the coefficient of thermal expansion data for both the matrix and the reinforcement in CMC are very important?

REFERENCES

Engineering Materials, Properties and Selection, 6th ed., K.G. Budinski and M.K. Budinski, Prentice Hall Int., N.J., 1999.
Properties of Engineering Materials, 2nd ed., R.A. Higgins, Industrial Press, N.Y., 1994.
Engineering Materials 2, M.F. Ashby and D.R.H. Jones, Pergamon Press, Oxford, 1986.
Electrical Engineering Materials, A.J. Dekker, Prentice Hall, N.J., 1959.
Plastics Materials: Properties and Applications, A.W. Birley and N.J. Scott, Leonard Hill, Glasgow, 1982.
Industrial Materials Science and Engineering, ed. L.E. Murr, Marcel Dekker Inc., New York, 1984.
Physical Metallurgy and the Design of Steels, F.B. Pickering, Applied Science Publishers, London, 1978.

Steels, Metallurgy and Applications, 3rd ed.,D.T. Llewellyn and R.C. Hudd,Butterworth Heinemann, Oxford, 1998.
Light Alloys, 2nd ed., I.J. Polmear, Edward Arnold, London, 1989.
Industrial and Engineering Materials, H.R. Clauser, McGraw Hill, New York, 1975.
Metals, Ceramics and Polymers, O.H. Wyatt and D. Dew-Hughes, Cambridge University Press, Cambridge, 1974.
The Principles of Materials Selection for Engineering Design, P.L. Mangonon, Prentice Hall Inc., N.Y., 1999.
Composite Materials, F.L. Matthews and R.D. Rawlings, Chapman & Hall, London, 1994.
Engineering Materials, M.F. Ashby and D.R.H. Jones, Pergamon Press, Oxford, 1980.
Handbooks of Industrial Refractories Technology, S.C. Corniglia and G.L. Barna, Noyes Publications, Westwood, N.J., 1992.
Electroceramics, 2nd Ed.,A.J.Moulson and J.H.Herbert, John Wiley & Sons, Inc.,Hoboken,NJ, 2003.

8

MATERIALS FORMING

In order to produce useful products materials need to be formed or shaped. The forming processes may be widely classified into

- Solidification Forming
- Deformation Forming
- Powder Metallurgy Forming.

In the first type of forming, liquid to solid transformation is associated, while the latter, two types pertain mainly to solid state forming although sometimes minor volume fraction of melt may be present during processing. In addition to above three processes, solid free form fabrication has come in vogue, although in a limited sense. The process is entirely based on CAD (Computer aided design).

The following sections in this chapter shall deal with mainly on the technological aspects of above processes.

SOLIDIFICATION FORMING

8.1 SCIENTIFIC BASICS OF METAL CASTING

Metal casting basics are involved around following features:

- Structural evolution
- Heat transfer
- Melt flow
- Stress analysis.

8.1.1 Structural Evolution

Microstructural changes during solidification of pure metals and solid solutions have been described in Chapter 3. Also the role of impurities of the resultant cast microstructures has been highlighted. When a molten metal or alloy, say 70/30 brass or steel is cast in a metal mould at room temperature, the process is known as chill casting (Figure 8.1). Because of chilling effect, at the mould wall, the casting develops a skin constituting of very fine equiaxed grains. The grains grow in the direction opposite to the direction of heat transfer from the mould. Grains that have a favorable orientation called *Columnar grains* grow preferentially.

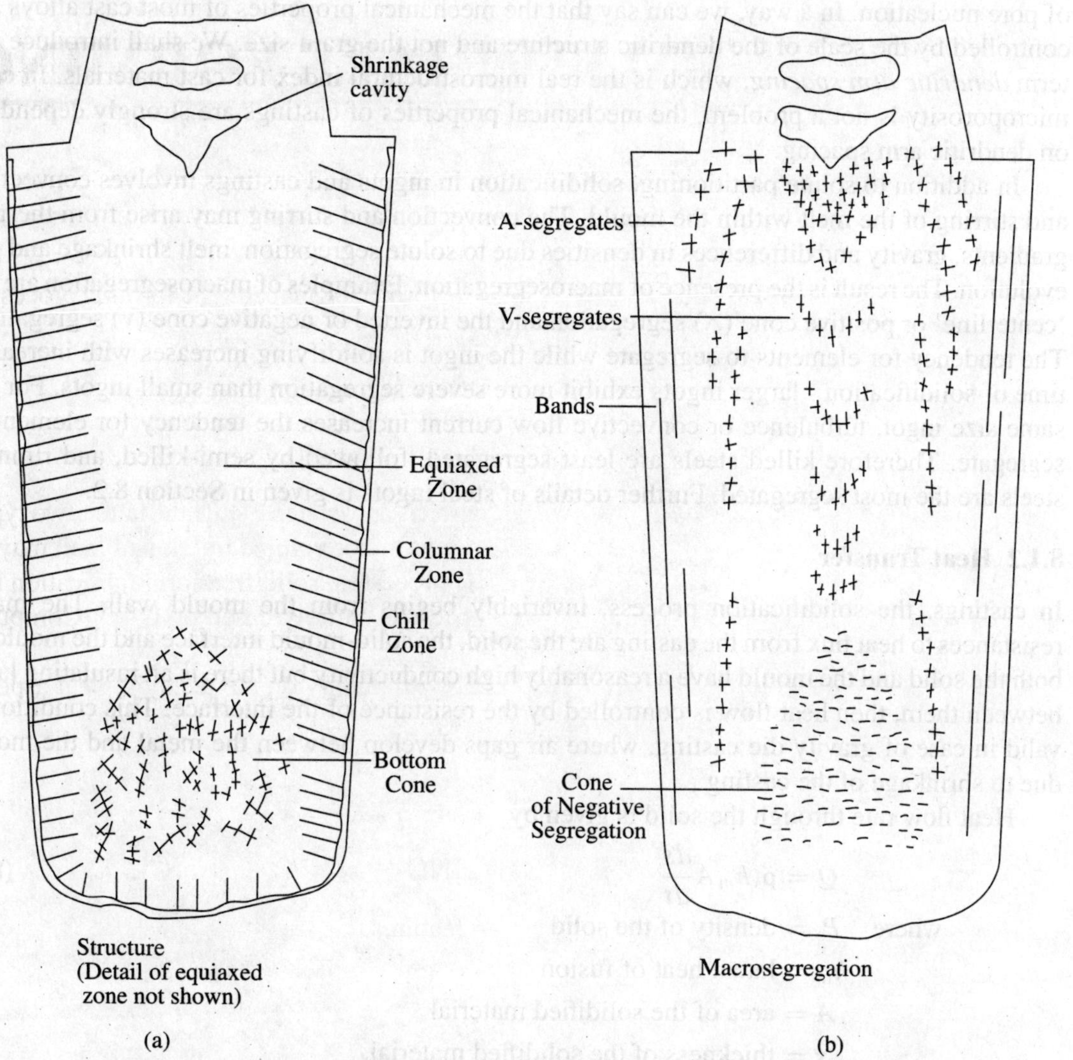

Fig. 8.1. Schematic representation of (a) structure and (b) segregation pattern in a killed, big end up steel ingot.

In single phase alloys nonequilibrium concentration gradients lead to *constitutional undercooling* which encourages the growth of projection from the grain surface. The prior conditions for such undercooling are:

- The solute must depress the freezing temperature of the melt.
- The actual temperature in the liquid ahead of the interface must be less than the equilibrium temperature.

In order to achieve planar growth in an alloy during solidification, there has to be a steep temperature gradient in the melt combined with a slow interface velocity.

Although, small grain size microstructure is advantageous in wrought alloys, the same is not true in case of cast ones. This is due to the fact that in the latter, there is more probability of pore nucleation. In a way, we can say that the mechanical properties of most cast alloys are controlled by the scale of the dendritic structure and not the grain size. We shall introduce the term *dendritic arm spacing*, which is the real microstructural index for cast materials. In case microporosity is not a problem, the mechanical properties of castings are strongly dependent on dendritic arm spacing.

In addition to solute partitioning, solidification in ingots and castings involves convection and stirring of the melt within the mould. The convection and stirring may arise from thermal gradients, gravity and differences in densities due to solute segregation, melt shrinkage and gas evolution. The result is the presence of macrosegregation. Examples of macrosegregation are the 'centerline' or positive cone (∧) segregation and the inverted or negative cone (v) segregation. The tendency for elements to segregate while the ingot is solidifying increases with increased time of solidification - larger ingots exhibit more severe segregation than small ingots. For the same size ingot, turbulence or convective flow current increases the tendency for element to segregate. Therefore killed steels are least segregated, followed by semi-killed, and rimmed steels are the most segregated. Further details of steel ingots is given in Section 8.2.

8.1.2 Heat Transfer

In castings, the solidification process, invariably begins from the mould wall. The major resistances to heat flux from the casting are the solid, the solid-mould interface and the mould. If both the solid and the mould have a reasonably high conductivity but there is an insulating layer between them, then heat flow is controlled by the resistance of the interface. This condition is valid in case of gravity die casting, where air gaps develop between the metal and the mould due to shrinkage of the casting.

Heat flow rate through the solid is given by

$$Q = \rho_s h_m A \frac{ds}{dt} \tag{8.1}$$

where P_s = density of the solid

h_m = latent heat of fusion

A = area of the solidified material

s = thickness of the solidified material.

'Q' can also be represented as

$$Q = KA(T_m - T_o) \tag{8.2}$$

where K = heat transfer coefficient

A = interface area

T_m = temperature of solid

T_o = temperature of mould.

Equating equation (8.1) and (8.2) and integrating we get

$$S = \frac{K(T_m - T_o)}{\rho_s h_m} t \tag{8.3}$$

Equation (8.3) can be generalized in term of volume to area ratio of casting and freezing time t_f as

$$t_f = \frac{\rho_s h_m}{K(T_m - T_o)} \left(\frac{V}{A}\right) \tag{8.4}$$

where V is the volume concerned.

The ratio (V/A) is known as the *modulus of a casting*. From technological point of view, it can be assumed that castings having the same modulus will have the same solidification time.

In case of sand casting and investment casting, the molten metal is a better conductor of heat than the mould. In this case, the main resistance to heat flow is the mould itself. Here the calculation is more complex, because the temperature and heat flux both vary across the thickness of the mould as well as time.

In such a case the thermal relationship is expressed as *Chvorinov's rule*, which is as

$$t_f = C\left(\frac{V}{A}\right)^2 \tag{8.5}$$

where C is a constant for a given mould and metal.

In case of pressure die casting for metals or injection moulding of polymers, the situation is at variance with what has been described so far. Here the solidification rate is controlled by heat transfer through the solid.

In case, the section of casting is complex, the success of a casting depends on the fact whether there is premature solidification of thinner sections. To solve this the introduction of riser (feeder head) is inevitable. They are made with a high V/A ratio, so that they solidify last and feed enough liquid to heavy sections of the casting to make up for shrinkage before and during solidification. The risers are, therefore, placed between runners and casting so that they are filled last and contain the hottest metal. Risers may be open to the atmosphere and exothermic compounds may be placed on them.

Another aspect is the variation in porosity level in different section of the casting. Porosity in a thin section can be avoided by initiating solidification in that very section and moving the solidification front toward thicker sections. In expendable moulds, this can be aided by placing chills (metal inserts) into refractory mould at points where maximum cooling is desired. In permanent moulds localized cooling is achieved by placement of cooling fins or pairs on external surfaces, or even by air or water cooling passages in the mould.

Heat extraction and mode of solidification, apart from affecting the soundness of the casting also influence the microstructure. This has been highlighted in earlier part of this section. Directional heat extraction offers directional solidification. A special example is directional solidification of eutectic alloy, in which the fibers of a hard, strong phase (e.g., intermetallic compound) becomes oriented in the direction of loading, thus giving rise to a desired composite material.

8.1.3 Melt Flow

The melt flow behaviour and the surface energy of the melt both affect its ability to fill the mould. Fluidity of a melt is an empirical measurement which depends not only on the material's viscosity but also on how quickly it solidifies and how it interacts with the surface it is flowing across. A spiral fluidity test is carried out, in which a molten metal is poured down a channel, where solidification begins at the entrance and ends till the flow is stopped. The length is the measure of fluidity. Mathematically, fluidity is expressed by following equation

$$Fluidity = \frac{\rho_s h_m r v}{2K(T_m - T_o)} \tag{8.6}$$

where ρ_s is solid density, h_m = latent heat of fusion, r = radius of spiral channel, v = melt flow velocity at die entrance, K = heat transfer coefficient, T_m = solidification temperature of the melt, and T_o = ambient temperature.

When filling a closed mould, it is desirable to maintain laminar flow. In case of any turbulence, mould erosion is the common feature. All impurities are possible sites for pore nucleation. This is the reason why castings having turbulent melt flow during feeding are often very porous and, hence, of poor quality.

An interesting example to show the effect of turbulence is the case of high pressure die casting. Here a lot of air gets introduced, which results in very poor quality castings. Improvement in quality can be achieved, if the melt injection is done in several stages. Improved design of gating system, multistage filling and the use of relatively low pressure, low speed upward mould filling are some of the solutions to prevent defects in the castings.

Fluidity of alloys varies with compositions. If one correlates the behaviour with the type of phase diagrams, it is noticed that maximum fluidity occurs at the eutectic composition and also for the pure metals and intermetallics. Primary dendritic crystals reduces fluidity of alloys and mushy stage of solidification hampers the flow of melt. Fluidity of metals and alloys can be manipulated by small additions of alloying elements and by altering the degree of superheat. The fluidity of steel is improved when they are deoxidized by silicon and aluminium and also by alloying with copper, silicon, manganese and nickel. Sulphur on the other hand decreases fluidity.

8.1.4 Stress Behaviour

During solidification of the metal, shrinkage is accompanied, which might introduce shrinkage stresses if the casting is not designed properly. Secondly when pouring into a mould, the molten metal tends to rise the upper half of the mould. To avoid this, half moulds are clamped with each other with the help of bolts, clamps etc. or loads are placed on the assembled mould.

The pressure on molten metal (P) on the upper half mould can be estimated according to the expression:

$$P = AH\gamma_m + V(\gamma_m - \gamma_c) \tag{8.7}$$

where H = hydrostatic pressure of metal
A = area of horizontal projection of the upper surface of mould cavity
V = volume of cores
γ_c = specific gravity of core
γ_m = specific gravity of liquid metal.

Stress levels in the casting should be kept consistent with the strength of the metal/alloy. In case the material is not ductile e.g., cast irons, the compressive strength is higher than the tensile strength. In many metals the as cast structure is likely to contain porosity, inclusions and unfavourably located second phases. These factors make the tensile or fatigue strength of cast material lower than that for wrought material.

Location of drilled holes should be strengthened by bosses, preferably shaped so as to make the drill enter perpendicular to the cast surface.

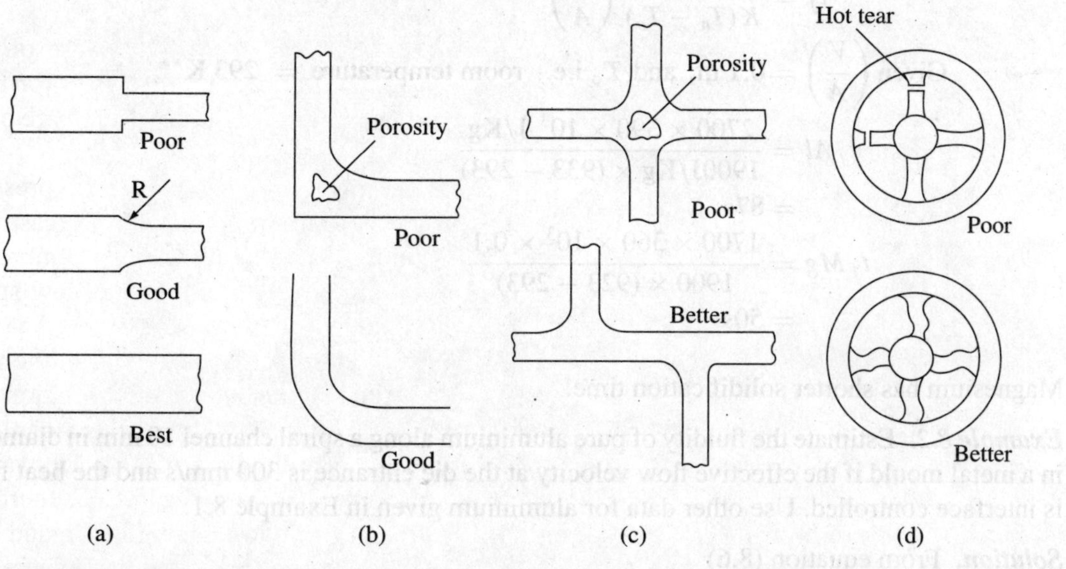

Fig. 8.2. Design features to minimize some casting problems due to (a) wall thickness variation, (b) hot spot in corner, (c) hot spot at cross ribs, and (d) hot tearing.

Some of the preferred geometrical configurations of the casting taking into account the stress distribution are illustrated in Figure 8.2. In Figure 8.2(a) the wall thickness transition is accompanied by generous radii. Small radii or sharp corners act as stress in the finished casting, create turbulence during pouring and prevent proper feeding during solidification.

Figures 8.2(b) and (c) illustrate the cases with localized heavy cross sections which can create hot spots, resulting into shrinkage cavities. The problem is avoided by applying a radius to the outer surface. Figure 8.2(d) shows how the shapes of the casting should allow deformation without moving large mould masses. For example a curved spoke can straighten somewhat by displacing little sand in the mould, and this accommodates the required shortening during or after solidification.

Example 8.1. For a die casting of modulus 0.1 m, where heat transfer is limited by the air gap at the metal mould interface which has a heat transfer coefficient $K = 1.9\,\text{kJ m}^{-2}\,\text{K}^{-1}\,\text{s}^{-1}$, which would possess shorter solidification time, aluminium or magnesium?
Given:

	Al	Mg
Density, kg m^{-3}	2700	1700
Melting point, K	933	923
Heat of fusion, kJ/kg	390	360

Solution. Since gravity die casting is interface controlled, from equation (8.4)

$$t_f = \frac{\rho_s h_m}{K(T_n - T_o)}\left(\frac{V}{A}\right)$$

$$\text{Given}\left(\frac{V}{A}\right) = 0.1\text{ m, and } T_o \text{ i.e., room temperature} = 293\text{ K}$$

$$t_f Al = \frac{2700 \times 390 \times 10^3\text{ J/Kg}}{1900\text{J/Kg} \times (933 - 293)}$$

$$= 87\text{s}$$

$$t_f Mg = \frac{1700 \times 360 \times 10^3 \times 0.1}{1900 \times (923 - 293)}$$

$$= 50\text{s}$$

Magnesium has shorter solidification time.

Example 8.2. Estimate the fluidity of pure aluminium along a spiral channel 10 mm in diameter in a metal mould if the effective flow velocity at the die entrance is 300 mm/s and the heat flow is interface controlled. Use other data for aluminium given in Example 8.1.

Solution. From equation (8.6)

$$\text{Fluidity} = \frac{\rho_s h_m r v}{2K(T_m - T_o)}$$

$$= \frac{2700 \times 390 \times 10^3 \times 5 \times 10^{-3} \times 0.3}{2 \times 1.9 \times 10^3 \times (933 - 293)}$$

$$= 0.65\text{ m.}$$

8.2 INGOT AND CONTINUOUS CASTING

Ingots are cast shapes produced with a cross section that is suitable for rolling, extrusion and forging operations. They are commonly used in steel and copper industries. In steel ingot casting, melt is poured into a top or bottom of cast iron moulds. The moulds are often provided with 'hot tops', combining insulation boards and exothermic compound to greatly reduce the depth of the shrinkage cavity formed during the solidification of ingot. Steel ingots can be classified into four types according to the deoxidation practice used or, alternately, by the amount of gas evolved during solidification. *Killed* steel ingots are fully deoxidized so that there is only a slight or practically no evolution of gas during solidification. These ingots are normally cast in hot topped, big end up moulds to reduce the depth of the shrinkage cavity (Figure 8.1). All steels having more than 0.3% C are killed. Alloy steels, forging steels and carburizing grades of steels are usually killed. *Semikilled* steels are partially deoxidized. An ingot skin of considerable thickness is formed prior to the beginning of gas evolution. A correctly deoxidized semikilled steel ingot does not have a pipe, but does have well scattered large blowholes in the top-center half of the ingot. Semikilled steels, generally, have a carbon content in the range of 0.15 to 0.3%. *Rimmed steel* ingots are characterized by only a small amount of deoxidation and a greater degree of gas evolution during solidification in the mould and a marked difference in chemical composition across the section and top to bottom of the ingot. This results in the formation of an outer ingot skin or rim of relatively pure iron and an inner core of the ingot with higher segregation/concentration of alloying elements, especially carbon, nitrogen, sulphur and phosphorus, having lower melting temperature. Most low carbon steels (<0.25%) are rimming steels.

Continuous casting is relatively modern method of production of steel/nonferrous products such as billets, blooms and slabs. The process offers increased productivity and yield, reduced energy and manpower, improved product quality, and consistency of quality over the traditional ingot casting process. Figure 8.3 shows the schematic representation of the continuous casting process for steel and aluminium. In case of direct chill (DC) casting, a process common for nonferrous metals and alloys, billets for extrusion and ingots for rolling are produced. Highly automated DC casting is equipped with a programmable computer to control all or some of the parameters such as cast length, casting speed, cooling water, metal temperature, metal flux and mould lubrication.

8.3 SINGLE CRYSTAL GROWING AND ZONE REFINING

Silicon is the most important material of the semiconductor industry. Specifications on semiconductor grade silicon typically require the removal of impurities like aluminium, iron and other metallic impurities from the metallurgical grade silicon to below 10^{-3} ppm, that is, a reduction in the levels of impurities by five or six orders of magnitude. This is done by Siemens process, whereby volatile chlorides of silicon are formed and then purified by distillation.

The polycrystalline silicon rods produced in this way are then melted and grown into single crystal ingots, usually in the Czochralski apparatus shown in Figure 8.4. The single 'seed' crystal oriented in the desired direction is mounted on a rotating shaft and contacted with the melt.

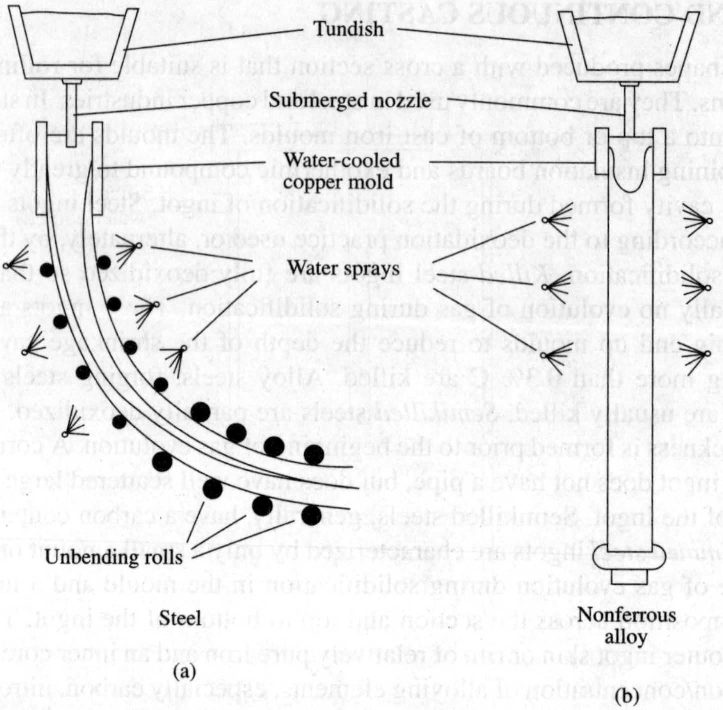

Fig. 8.3. Schematic representation of the continuous casting process of (a) steel and (b) aluminium.

The rotating crystal is then slowly withdrawn while precise control of temperature within the apparatus is maintained so that solidification occurs on the seed crystal and a long cylindrical crystal is produced. The crystal is then sawn into wafers, which are the raw materials for the semiconductor industry.

Another method to produce semiconductor grade material is by *zone refining*. In many cases it is desirable to eliminate impurities found by contact between the molten material and the crucible. One way of achieving this is by floating zone refining. Here, the rod of material to be refined is positioned vertically and the dimensions of the liquid pool made sufficiently small that surface tension serves to keep the pool in place between the two solid sections. A number of passes might be required for refining. Finally, the end of the solid rich in impurities can be cut off and rejected. Homogenization of the pure portion can be achieved either by melting and casting or by *zone leveling*. The latter technique is similar to zone refining except that the heater is passed backward and forward rather than only in one direction. The effect is to reduce the macrosegregation.

8.4 METAL CASTING TECHNOLOGY

Two forms of moulds are available for industrial casting: open and closed. The open mould is already dealt in the Section 8.2. In the *closed mould*, a passage way, called the gating system is provided to permit the molten metal to flow from outside the mould

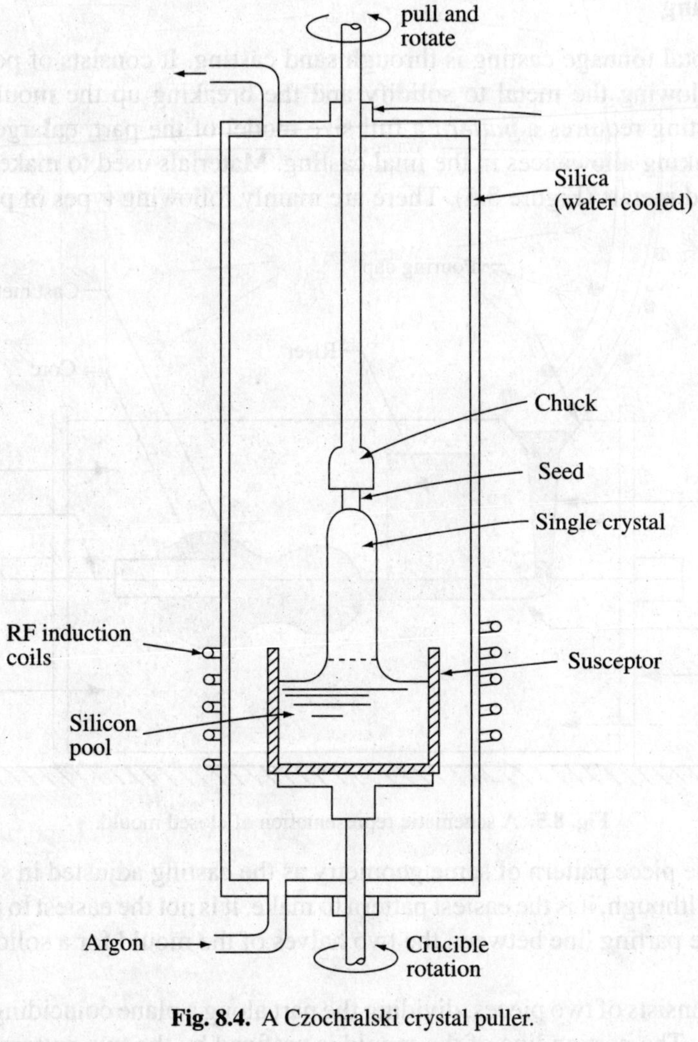

Fig. 8.4. A Czochralski crystal puller.

into the cavity. Once the casting is cooled sufficiently, it is removed from the mould. Depending on the casting material and metal used, further processing may be required. This may include trimming, the excess metal from the actual cast part, cleaning the surface, inspecting the product and heat treatment, if desired. In addition, machining (Chapter 9) may be required to achieve closer tolerances on certain part features and to remove the cast surface and its associated microstructure.

Another way to classify metal casting process is, whether the mould is expendable or permanent. In the former, the mould must be sacrificed in order to remove the cast part. In permanent mould casting, the mould is fabricated out of metal and can be used many times to make many castings.

8.4.1 Sand Casting

Majority of the total tonnage casting is through sand casting. It consists of pouring metal into a sand mould, allowing the metal to solidify and the breaking up the mould to remove the casting. Sand casting requires a *pattern* a full size model of the part, enlarged to account for shrinkage and making allowances in the final casting. Materials used to make patterns include wood, plastics and metals (Figure 8.5). There are mainly following types of pattern:

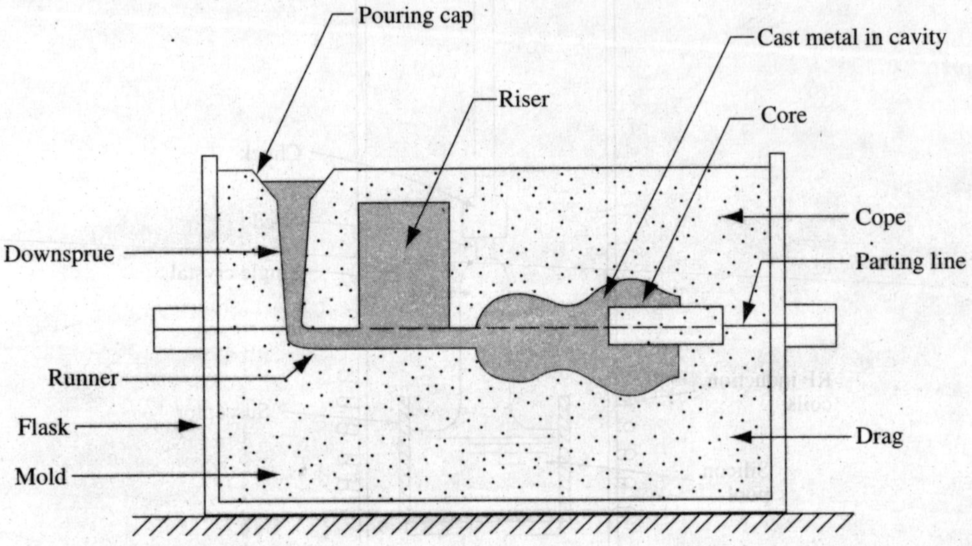

Fig. 8.5. A schematic representation of closed mould.

Solid pattern: One piece pattern of same geometry as the casting adjusted in size for shrinkage and machining. Although, it is the easiest pattern to make, it is not the easiest to use. Determining the location of the parting line between the two halves of the mould for a solid pattern can be a problem.

Split pattern: It consists of two pieces, dividing the part along a plane coinciding with the parting line of the mould. The parting line of the mould is prefixed by the two pattern halves.

Match plate pattern: In this case, the two pieces of the split pattern are attached to opposite sides of the wood or metal plate. Holes in the plate allow the top and bottom sections of the mould to be aligned.

Cope-and-drag pattern: This type of pattern is similar to match-plate patterns except that split pattern halves are attached to separate plates, so that the cope and drag section of the mould can be fabricated independently.

So far we have talked about solid casting. In case the casting has internal surfaces, a *core* is required, which is a full scale model of the interior surfaces of the part. It is inserted into the mould cavity prior to pouring and is usually made of sand, compacted into the desired shape. Depending on the geometry of the part, the core may or may not require supports to hold it in position in the mould cavity during pouring. These supports, called *chaplets* are made of a metal with higher melting point than the casting metal.

Foundry sands are silica or silica mixed with other minerals. The sand must have good refractoriness, and proper grain size, grain size distribution in the mixture and grain shape. The significance of powder character shall be dealt in powder metallurgy forming later in this chapter. In making the mould, the sand grains are held together by a mixture of water and binding clay. A typical mixture, by volume, is 90% sand, 3% water and 7% clay. Other bonding agents can be used in place of clay, including organic resins (e.g., phenolic resins) and inorganic binders (e.g., sodium silicate and phosphate). Sometimes special additions are used in the mixture to enhance properties such as strength and/or permeability of the mould.

The major factors in selecting the sand mould are strength, permeability, thermal stability, collapsibility and reusability. Some factors are not complementary. For example a mould with greater strength is less collapsible. Sand moulds are often classified as green-sand, dry-sand or skin-dried moulds. Green-sand mould contains moisture, which provides enough strength etc. and are the least expensive of the moulds. However, there is disadvantage of moisture too, as they sometimes, cause casting defects. A *dry-sand mould* is made using dry binders rather than clay and the mould is baked in an oven at about 200-320 °C. Such a treatment imparts strength and hardness. It also provides better dimensional control in the cast product, compared to green-sand moulding. The process is more expensive and production rate is reduced because of drying operation. In a *skin-dried mould* the advantages of a dry-sand mould is partially achieved by drying the surface of green-sand mould to a depth of 10 to 25 mm at the mould cavity surface, using torches, heating lamps etc.

The baking operation of the mould is many times dispensed with by using chemical-bonded moulds, which include furan resins (consisting of furfural alchohol, urea and formaldehyde), phenols and alkyd oils.

Feeding of molten metal: After the mould is prepared the melt with certain superheat is to be fed into the mould. Freezing range of alloys has been mentioned in the beginning. The first stage of feeding is *mass feeding*. In short freezing range alloys, such feeding is witnessed. In case the alloy has long freezing range, the formation of dendrites tends to restrict the flow of melt. This stage is known as *interdendritic feeding*. At this stage shrinkage porosity is produced. Insufficient interdendritic feeding can also lead to localized strains which result in macroporosity known as *solidification cracking* or *hot tearing*. The final stage of feeding is known as *solid feeding*, in which liquid regions become isolated from the feed by the intervening solid. In this stage, rather large stress develops in the liquid, which may cause plastic deformation of the surrounding solid shell. Dimensional errors resulting from solid feeding are usually minimal.

8.4.2 Shell-Mould Casting

In this type of process, the mould is a thin shell (~10 mm) made of sand and held together by a thermosetting resin binder. The advantages are smooth surface and good dimensional accuracy. The major disadvantage is cost, which makes difficult to justify for small quantities of parts. The moulding process can be mechanized for mass production. Some of the examples of parts made by this process include gears, valve bodies, bushings and camshafts.

8.4.3 Expanded Polystyrene Process

This process uses a mould of sand packed around a polystyrene foam pattern, which vaporizes when the molten mould is poured into the mould. The polystyrene pattern includes the sprue, risers and gating system, and it may also contain internal cores.

8.4.4 Investment Casting

This is also one of the expendable core type casting process. A pattern of wax is coated with a refractory material to make the mould, after which the wax is melted away prior to pouring the molten metal. This method gives casting of high accuracy and intricate detail. The process is also known as *lost-wax process*, because the wax pattern is lost from the mould prior to casting. Figure 8.6 illustrates the complete cycle of operation.

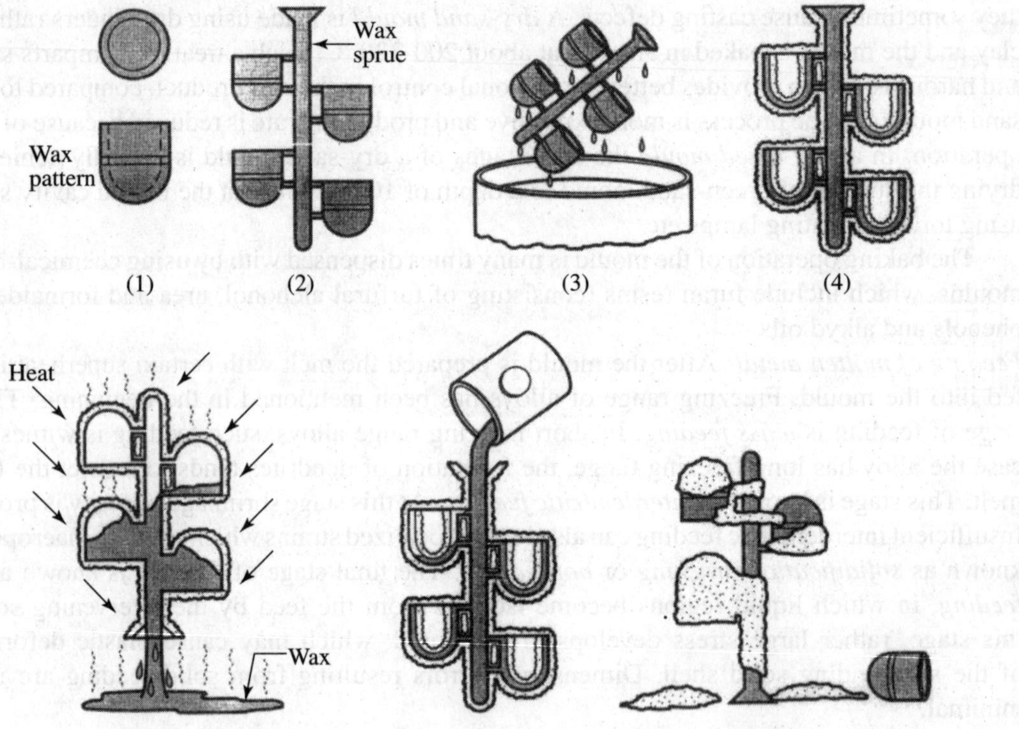

Fig. 8.6. Stages in investment casting (for details refer the text) (Reproduced with permission from Fundamentals of Modern Manufacturing, M.P. Groover, 2nd ed., John Wiley & Sons, Inc.,Hoboken, NJ, 2002, p. 223).

The steps involved are (1) wax pattern production, (2) attachment of several patterns to a sprue, (3) coating of the pattern with a thin layer of refractory slurry, (4) formation of full mould by covering it with refractory to make it rigid, (5) heating of the mould to melt the wax, (6) pretreating of mould and pouring of liquid metal and finally (7) breaking of the mould from the finished casting. Here the parts are separated from the sprue.

8.4.5 Permanent Mould Casting

Permanent-mould casting uses a metal mould constructed of two sections that are designed for easy, precise opening and closing. These moulds are commonly made of steel or cast iron. The cavity, with gating system included, is machined into the two halves to provide accurate dimensions and good surface finish. Metals commonly cast in permanent molds include aluminium, magnesium, copper-base alloys, and cast iron.

Cores can be used in permanent moulds to form interior surfaces in the cast product. The cores can be made of metal, but either their shape must allow for removal from the casting, or they must be mechanically collapsible to permit removal.

Steps in the basic permanent-mould-casting process are described in Figure 8.7. In preparation for casting, the mould is first preheated and one or more coatings are sprayed on the cavity. Preheating facilitates metal flow through the gating system and into the cavity. The coatings aid heat dissipation and lubricate the mould surfaces for easier separation of the cast product. After pouring, as soon as the metal solidifies, the mould is opened and the casting is removed.

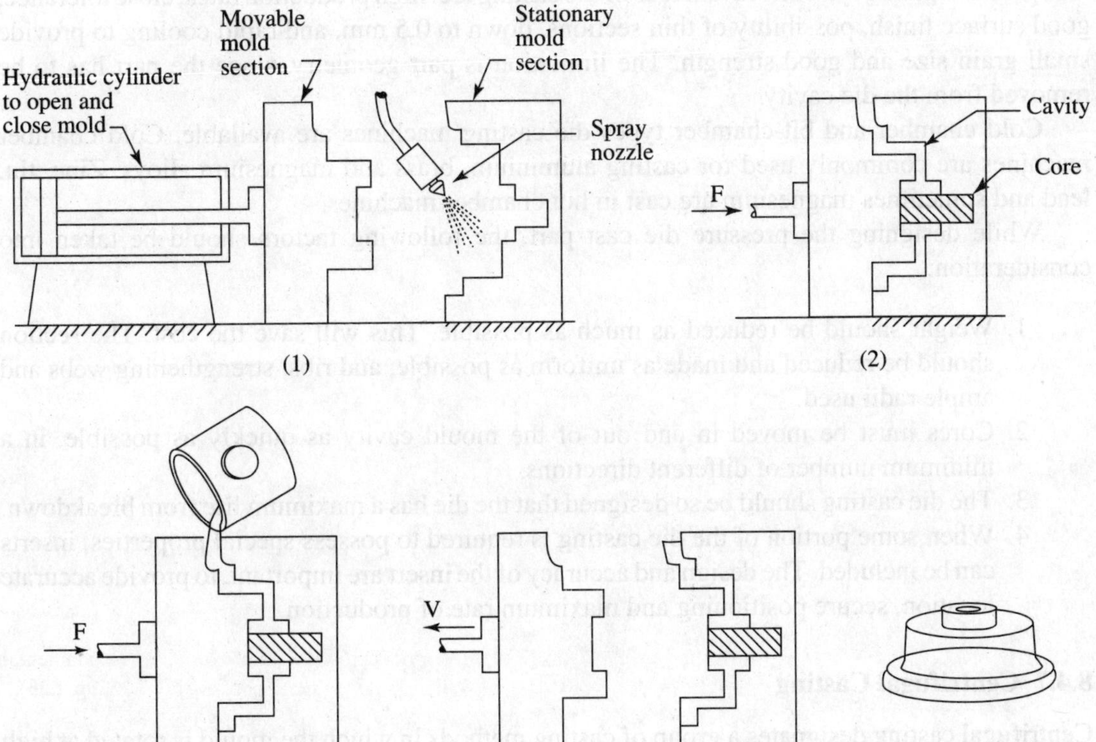

Fig. 8.7. Stages in permanent mould casting (a) mould is preheated and coated; (2) cones are inserted and mould is closed; (3) molten metal is poured into the mould; (4) mould is opened, and (5) finished part.

Advantages of permanent-mould casting include good surface finish and close dimensional control. In addition, more rapid solidification caused by the metal mould results in a finer grain structure, so stronger castings are produced. The process is generally limited to metals of lower melting points. Other limitations include simple part geometries compared to sand casting and the expense of the mould. Because mould cost is substantial, the process is best suited to high-volume production and can be automated accordingly. Typical parts include automotive pistons, pump bodies and certain castings for aircraft and missiles.

8.4.6 Die Casting

It is a form of permanent-mould casting in which the molten metal is injected into the mould cavity under high pressure (7 to 350 MPa). The pressure is maintained during solidification, after which the mould is opened and the part is removed. Mould in this type of operation is called *die*. These are made of tool steel or maraging steel. Tungsten and molybdenum are used in case of die-cast steel and cast iron. Dies can be single cavity or multiple cavity. Venting holes or passage ways are built into the dies at the parting line to evacuate the air and gases in the cavity. Formation of flash is common, which must be trimmed from the casting along with the sprue and gating system. Advantages of die casting are: high production rates, close tolerance, good surface finish, possibility of thin sections, down to 0.5 mm, and rapid cooling to provide small grain size and good strength. The limitation is part geometry, since the part has to be removed from the die cavity.

Cold chamber and bit-chamber types die casting machines are available. Cold chamber machines are commonly used for casting aluminium, brass and magnesium alloys. Zinc, tin, lead and sometimes magnesium are cast in hot chamber machines.

While designing the pressure die cast part, the following factors should be taken into consideration:

1. Weight should be reduced as much as possible. This will save the cost. The section should be reduced and made as uniform as possible, and ribs, strengthening webs and ample radii used.
2. Cores must be moved in and out of the mould cavity as quickly as possible, in a minimum number of different directions.
3. The die casting should be so designed that the die has a maximum life from breakdown.
4. When some portion of the die casting is required to possess special properties, inserts can be included. The design and accuracy of the insert are important, to provide accurate location, secure positioning and maximum rate of production.

8.4.7 Centrifugal Casting

Centrifugal casting designates a group of casting methods in which the mould is rotated at high speed so that centrifugal force distributes the molten metal to the outer region of the die cavity. The group includes true centrifugal casting, semicentrifugal casting and centrifuse casting. In true centrifugal casting, molten metal is poured into a rotating mould to produce a tubular part. The casting can be either horizontal or vertical. The molten metal is poured into a horizontal rotating mould at one end. The outside shape of the casting can be round, octagonal, hexagonal etc. However, the inside shape of the casting is round due to the radially symmetric forces at work. It must be borne in mind that the inside perfect round is a theoretical one.

G-factor (GF) is the ratio of centrifugal force divided by weight i.e.,

$$GF = \frac{F}{W} = \frac{mv^2}{R\,mg} = \frac{v^2}{RG} \tag{8.8}$$

where m = mass, v = velocity, R = inside radius of the mould, g = acceleration of gravity (9.8 m/s^2). Velocity v can be expressed as $\frac{2\pi RN}{60}$ *or* $\frac{\pi RN}{30}$, where N = rotational speed in RPM. Substituting the values in equation (8.8), we get

$$N = \frac{30}{\pi}\sqrt{\frac{2gGF}{D}} \tag{8.9}$$

where D = inside diameter of the mould.

The GF values in horizontal metal casting is in the range 60-80. It also depends on the metal being cast.

8.4.8 Squeeze Casting

The process involves solidification of the molten metal under high pressure and in a way it can be considered as a combination of casting and forging. Parts can be made to near-net-shape, with complex shapes and fine surface detail, from both ferrous and nonferrous alloys. The pressures required in squeeze casting are typically higher than those used in pressure die casting, but lower than those for hot or cold forging. Typical product made out of this process is automotive wheels.

8.4.9 Semi-solid Solidification Processing

The processing temperature is above of solidus temperature of the alloy, and the alloy exhibits *thixotropic* behaviour, i.e., its viscosity decreases when agitated. The process is also called as thixoforging or thixocasting. This technology combines the casting and forging operations. This method is very common for making cast metal matrix composites.

8.5 POLYMER SOLIDIFICATION FORMING

In polymer, casting involves pouring of liquid resin into a mould, using gravity to fill the cavity and allowing the polymer to harden. Both thermoplastics and thermosets are cast. Examples of the former include acrylics, polystyrene, polyamides and vinyls. Conversion of the liquid resin into a hardened thermoplastic can be achieved by following ways:

- Heating the thermoplastic resin to a highly fluid state so that it readily pours and fills the mould cavity and then permitting it to cool and solidify in the mould.
- Using a low molecular weight copolymer or monomer and polymerizing in the mould to form a high molecular weight thermoplastic.

- Pouring a plastisol (a liquid suspension of fine particles of a thermoplastic resin such as PVC in a plasticizer) into a heated mould so that it gels and solidifies.

Thermosetting polymers used for casting include polyurethene, unsaturated polyesters, phenolics and epoxies. The process involves pouring the liquid ingradients that form the thermoset into a mould so that polymerization and cross-linking occur. Heat and/or catalysts may be required depending on the resin system. The reactions must be sufficiently slow to allow mould pouring to be completed.

8.5.1 Injection Moulding

Injection moulding is a process in which a polymer is heated to a highly plastic state and forced to flow under high pressure into a mould cavity where it solidifies. The process produces discrete components that are almost always net shape. The product cycle time is typically in the range 10 to 30 seconds, although cycles of one minute or longer are not uncommon. The mould may contain more than one cavity, so that multiple parts can be produced each cycle. Part size can range from about 50 g upto about 25 kg, the upper limit representing products such as refrigerator doors and automobile bumpers. Injection moulding is the most widely used moulding process for thermoplastics. Some thermosets and elastomers can also be injection moulded with modification in equipment and operating parameters to allow for cross-linking of these materials. Figure 8.8 shows the schematic of such a machine, which consists essentially of two parts: (a) the plastic injection part and (b) the mould clamping part. The functions of the injection part are to melt and homogenize the polymer, and then inject it into the mould cavity. The unit is much like an extruder. The functions of the clamping part are:

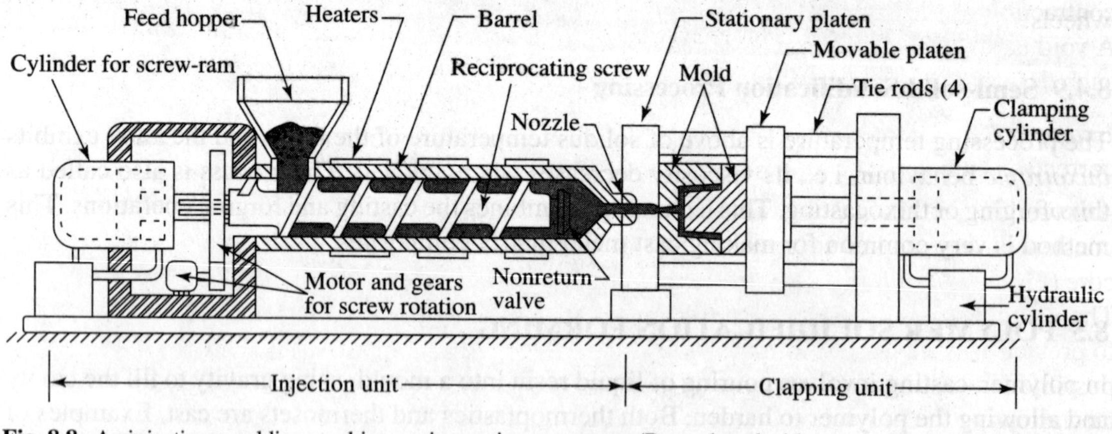

Fig. 8.8. An injection moulding machine, reciprocating screw type (Reproduced with permission from Fundamentals of Modern Manufacturing, M.P. Groover, 2nd ed., John Wiley & Sons, Inc., Hoboken, NJ, 2002, p. 276).

- hold the two halves of the mould in proper alignment with each other
- keep the mould closed during injection by applying a clamping force sufficient to resist the injection force
- open and close the mould at the appropriate times in the moulding cycle.

In summary, the mould consists of (i) one or more cavities that determine part geometry, (ii) distribution channels, through which the polymer melt flows to the cavities, (iii) an ejection system for part removal, (iv) a cooling system, and (v) vents to permit evacuation of air from the cavities.

Since polymers have high coefficients of thermal expansion, shrinkage is invariably associated during cooling of the plastic. To compensate for shrinkage, the dimensions of the mould cavity must be made longer than the specified part dimensions. The expression used is

$$D_c = D_p + D_p S + D_p S^2 \tag{8.10}$$

where D_c = dimension of cavity, D_p = moulded part dimension and S = shrinkage of the particular polymer (Table 8.1).

Table 8.1. Typical Values of Shrinkage for Moldings of Selected Thermoplastics.

Plastic	*Shrinkage*
ABS	0.006
Nylon-6.6	0.020
Polycarbonate	0.007
Polyethylene	0.025
Polystyrene	0.004
PVC	0.005

The common defects in injection moulding are: short shots, flashing, sink marks and voids and weld lines. The sink mark occurs when the outer surface on the moulding solidifies but, contraction of the internal material causes the sink to be depressed below its intended profile. A void is caused by the same basic reason. A change in the design of the part is the solution for such defects. An example shall illustrate this feature.

Example 8.3. Describe some common faults in polymer injection moulding and suggest the corrective measures.

Answer. Figure Ex. 8.3 shows some of the common moulding defects. Injection moulds of constant thin cross-sectional thickness is necessary to avoid sink marks and voids within the core (Figure a). This will also reduce distortion and result in more accurate tolerances for a part. The wall thickness of ribs and bosses must be kept to less than 2/3 of the main wall thickness to prevent sink marks appearing on the surface (Figure b). If thick ribs are necessary it may be possible to hide the sink mark by surface features (Figure c and d). Flat plates are susceptible to warping, so additional ribbing or corrugating the surface will produce a much more rigid structure (Figure e).

8.5.2 Extrusion

Although extrusion is a compression process (deformation forming), in case of polymer we prefer to include in the section of solidification, as thermoplastic or elastomer melts are involved. The items produced by this method are tubing, pipes, hose, structural shapes, sheet and film,

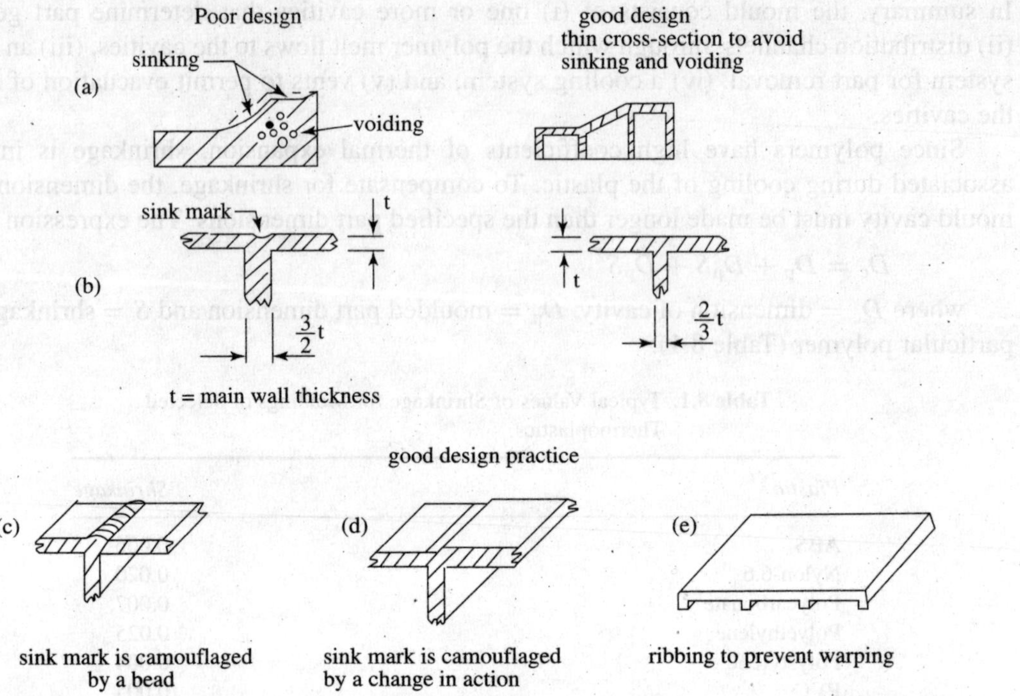

Figure Ex. 8.3

continuous filaments, and coated electrical wire and cable. The process is a continuous one and the extruded product (extrudate) is subsequently cut into desired lengths.

In polymer extrusion, feed stock in pellet or powder form is fed into an extrusion barrel, where it is heated and melted and forced to flow through a die opening by means of a rotating screw. The two main components of the extruder are the barrel and the screw. The die is not a component of the extruder. It is a special tool that must be fabricated for the particular profile to be produced. The internal diameter of the extruder barrel typically ranges from 25 to 150 mm. The length to diameter ratio (L/D) of the barrel is usually between 10 and 30. The length ratio are used for thermoplastic materials, while lower L/D ratios are for elastomers.

The design aspect of screw is important in polymer extrusion. The different sections of screw are feed section, compression section and metering section. The typical screw geometry is shown in Figure 8.9. The screw consists of spiraled 'flights' (threads) with channels between them through which the polymer melt is removed. Some of the designations are

w_c = width of channel

d_c = depth of channel

D = barrel diameter

w_f = width of flight land

A = flight angle (helix angle of screw)

p = pitch.

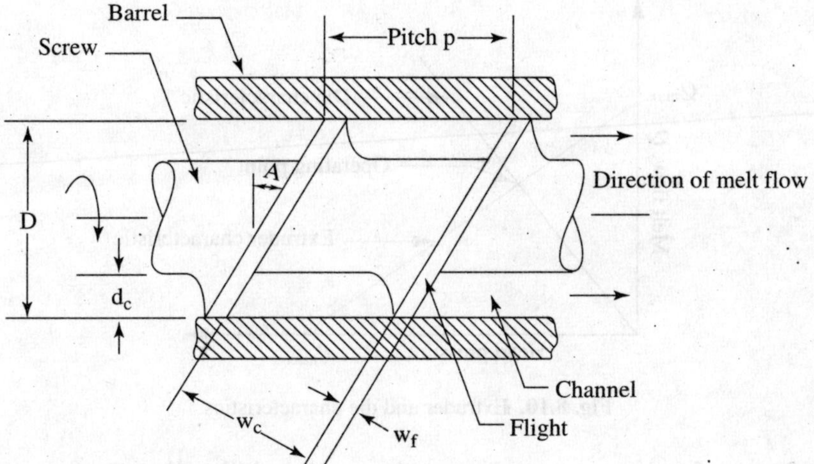

Fig. 8.9. Geometric representation of an extruder screw inside the barrel.

The relationship between these parameters is

$$\tan A = \frac{p}{\pi D} \tag{8.11}$$

In practice, we might have more than one screw e.g., in case of twin screw extruder, the screws are parallel and side-by-side inside the barrel. The twin-screw extruder is suited for rigid PVC, and for materials that require greater mixing.

The flow rate of polymer melt in the extruder Q_x is expressed as

$$Q_x = Q_d - Q_b$$

$$= 0.5\pi^2 D^2 N d_c \sin A \cos A - \frac{P\pi D d_c^3 \sin^2 A}{12\eta L} \tag{8.12}$$

where Q_d = drag flow rate

Q_b = back pressure flow

N = screw rotational speed, rev. per second

L = length of barrel

η = viscosity of the polymer melt

P = head pressure in the barrel.

The parameters in equation (8.12) can be divided into (a) design parameters and (b) operating parameters. The design parameters are those that define the geometry of the screw and barrel, i.e., D, d_c and A. The operating parameters are those that can be changed during the process to affect output flow, i.e., N, P and η. The relationship between Q_x and P is called *die characteristic* (Figure 8.10). The point at which die characteristic intersects *extruder characteristic*, we get the operating point for the extrusion process.

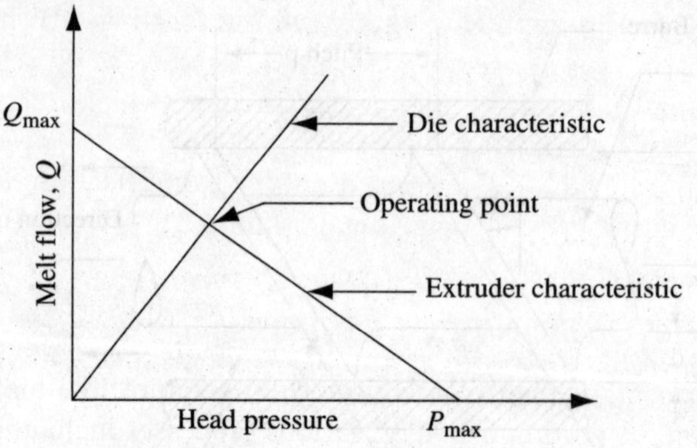

Fig. 8.10. Extruder and die characteristics.

Another forms of polymer extrusion processes are *chill roll extrusion* and *blown film extrusion* processes. The productivity is high in case of the former. The latter process is extensively used for making thin polyethylene film for packaging. This process combines extrusion and blowing to produce a tube of thin film.

Example 8.4. An extruder barrel has a diameter of 70 mm. The screw rotates at $N = 1$ rev/s. Channel depth = 6.0 mm, flight angle = 20. Head pressure at the end of the barrel = 7.0×10^6 Pa. If the barrel length is 2.0 m and the viscosity of the polymer melt is 100 Pas., calculate the volume flow rate of the plastic in the barrel.

Solution. From equation (8.12),

$$Q_d = 0.5\pi^2(70 \times 10^{-3})^2(1)(6 \times 10^{-3})(\sin 20)(\cos 20)$$

$$= 4.663 \times 10^{-5}\ \text{m}^3/\text{s}$$

$$Q_b = \frac{\pi(7 \times 10^6)(70 \times 10^{-3})(6 \times 10^{-3})^3(\sin 20)^2}{12(100)(2.0)}$$

$$= 1.621 \times 10^{-5}\ \text{m}^3/\text{s}$$

Therefore $Q_x = Q_d - Q_b$

$$= 4.663 \times 10^{-5} - 1.621 \times 10^{-5}$$

$$= 3.042 \times 10^{-5}\ \text{m}^3/\text{s}$$

8.5.3 Forming of Polymer Matrix Composites

Methods described for polymers are equally applicable for polymer matrix composites e.g., compression moulding, transfer moulding, injection moulding processes. One special form of forming for making glass fiber reinforced polymer (GFRP) is known as *pultrusion process*. The process is similar to extrusion, as it produces continuous straight sections of constant cross-section. In this process the continuous fiber rovings are dipped into a resin bath and pulled through a shaping die, where the impregnated resin cures. The process consists of

(a) filament feeding, (b) resin impregnation, (c) pre-die forming, (d) shaping and curing; and (e) pulling and cutting. E-glass fiber is the most widely used reinforcing material. Common resin used in poltrusion are polyesters, epoxies and silicones, which are thermosetting polymers. Thermoplastics have also been tried in many cases.

Centrifugal casting of PMC is also carried out for cylindrical products such as pipes and tanks. The process is similar to metal centrifugal casting.

8.6 GLASS SOLIDIFICATION PROCESSING

We know that glass does not melt at a constant temperature like metal and it is a viscous melt. Forming of glasses is accomplished by various processes including casting, pressing and blowing and rolling. In case the molten glass is sufficiently fluid, it can be poured into a mould. Relatively bigger objects, like astronomical lenses and mirrors are made by this method. These pieces must be cooled very slowly to avoid internal stresses and possible cracking. As molten glass is relatively viscous at normal working temperature, it becomes difficult for it to flow through small orifices. Smaller lenses are, therefore, made by pressing method. Molten glass can be spinned similar to centrifugal casting of metals. This is commonly known by the term *glass working*. Pressed glasses are extensively used in more quantities and a high level of automation has been, therefore, introduced. Press-and-blow method is applied for production of wide-mouth containers. Smaller-mouthed bottles are produced by blow-and-blow method. Reheating is sometimes required between blowing steps.

Flat glass plates can be produced by rolling, but the most significant method is *float process* (Figure 8.11). The glass flows directly from the melting furnace onto the surface of a molten tin bath. After moving into a cooler region of the bath, the glass hardens and travels through an annealing furnace, after which it is cut to size.

Glass tubing is produced by a drawing process. Molten glass flows around a rotating hollow mandrel through which air is blown while the glass is being drawn. The air temperature and its

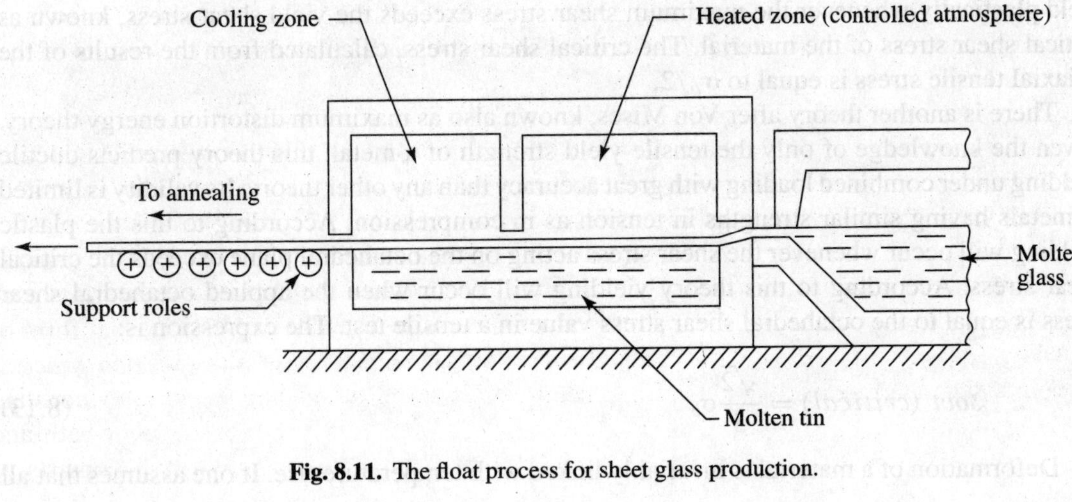

Fig. 8.11. The float process for sheet glass production.

volumetric flow rate, as well as the drawing velocity, determine the diameter and wall thickness of the tubular cross section.

DEFORMATION FORMING PROCESSES

One way of classification is based on the temperature of working i.e., hot working or cold working. This depends on the recrystallization temperature of the metal concerned. The recrystallization phenomenon has been described in Chapter 3 (Section 3.12).

Another method of classification is based on the stress generated in the work piece during deformation. These can be elaborated depending on the type of stressing which are as follows (Figure 8.12):

a) *Indirect compression:* Examples are wire drawing and deep drawing processes.

b) *Direct compression:* Forging, rolling and extrusion fall into this category.

c) *Biaxial tension:* In this case an applied biaxial tensile stress system introduces a compressive stress on a mutually perpendicular plane. Stretch forming operations pertains to this type of system.

8.7 PLASTIC YIELDING AND DEFORMATION PROCESSING

The determination of the state of stress is essential for describing the plastic deformation in metals. The state of stress at a part can be determined by computing the stresses acting on certain conventionally oriented planes passing through it. Stresses acting on any other planes can be determined by means of simple standardized analytical or graphical method. The concept of yielding in metals has been given in Chapter 5. A number of theories were propounded to offer a mathematical relationship between the magnitude of the applied triaxial stress system and the yield stress. According to Tresca theory, a material subjected to any combination of stress will yield plastically whenever the maximum shear stress exceeds the yield shear stress, known as critical shear stress of the material. The critical shear stress, calculated from the results of the uniaxial tensile stress is equal to $\sigma_o/2$.

There is another theory after Von Mises, known also as maximum distortion energy theory. Given the knowledge of only the tensile yield strength of a metal, this theory predicts ductile yielding under combined loading with great accuracy than any other theory. Its validity is limited to metals having similar strengths in tension as in compression. According to this the plastic yielding will occur whenever the shear stress acting on the octahedral plane exceeds the critical shear stress. According to this theory yielding will occur when the applied octahedral shear stress is equal to the octahedral shear stress value in a tensile test. The expression is:

$$\mathfrak{Z}oct\ (critical) = \frac{\sqrt{2}}{3}\sigma_o \tag{8.13}$$

Deformation of a material is invariably linked with temperature rise. If one assumes that all

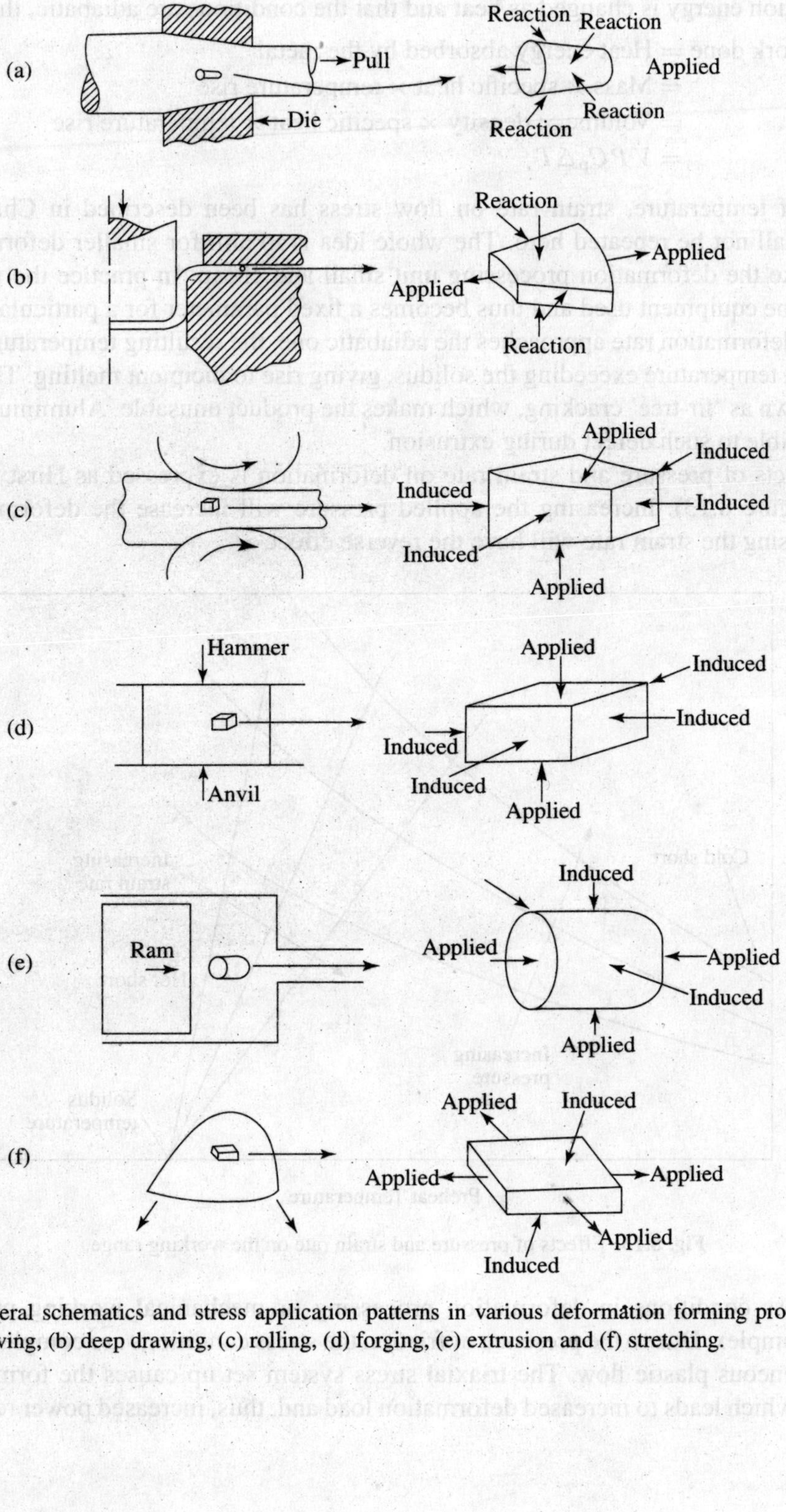

Fig. 8.12. General schematics and stress application patterns in various deformation forming processes (a) wire drawing, (b) deep drawing, (c) rolling, (d) forging, (e) extrusion and (f) stretching.

the deformation energy is changed as heat and that the conditions are adiabatic, the

$$\begin{aligned}\text{Work done} &= \text{Heat energy absorbed by the metal}\\ &= \text{Mass} \times \text{specific heat} \times \text{temperature rise}\\ &= \text{Volume} \times \text{density} \times \text{specific heat} \times \text{temperature rise}\\ &= VPC_p\Delta T \qquad (8.14)\end{aligned}$$

Effect of temperature, strain rate on flow stress has been described in Chapter 5 and, therefore, shall not be repeated here. The whole idea is to aim for smaller deformation load, so as to make the deformation processing unit small and cheap. In practice the strain rate is decided by the equipment used and thus becomes a fixed parameter for a particular machine.

In case deformation rate approaches the adiabatic one, the resulting temperature rise might lead to metal temperature exceeding the solidus, giving rise to incipient melting. This produces a defect known as 'fir-tree' cracking, which makes the product unusable. Aluminium alloys are very susceptible to such defect during extrusion.

The effects of pressure and strain rate on deformation is expressed as Hirst and Ursell's diagram (Figure 8.13). Increasing the applied pressure will increase the deformation range, while increasing the strain rate will have the reverse effect.

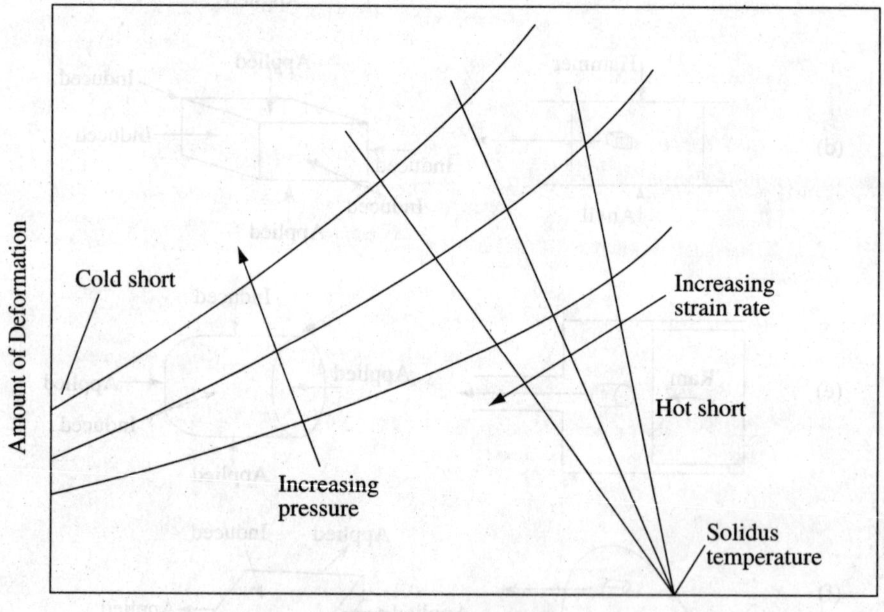

Fig. 8.13. Effects of pressure and strain rate on the working range.

The stress conditions in deformation processing or mechanical working processes are invariably complex. Due to the presence of friction the stress conditions are complex. This leads to inhomogeneous plastic flow. The triaxial stress system set up causes the formation of the *friction hill* which leads to increased deformation load and, thus, increased power requirements.

It is worth mentioning that a very thin work piece can cause extremely high working loads, and, thus, require extra power.

8.8 FORGING

There are two types of forging process: *impact forging* and *press forging*. In the former, the load is applied by impact and deformation takes place over a very short time. Press forging on the other hand, involves the gradual build upto pressure to cause the metal to yield. The time of application is relatively long. Forging can be done at lower temperature as well called cold forging. But, over 90% of forging operations are hot forging. The impact forging can be classified into drop forging and upset forging. The drop forging can be again subdivided into open drop forging and close-die drop forging. In the former the main aim is to reduce the thickness of the work piece in a stepwise sequence from end to end. Several passes may be required to complete the job and edging is generally carried out to control the width. The relative amounts of elongation and spread of the work piece depend on the ratio of the tool length to the metal width called the *bite ratio*.

In case of *closed die drop forging*, the machined die fits into the anvil and the top. The dies have a series of grooves and depressions cut into them and the work piece is passed in sequence through a shaping series. Care should be taken of the fact that the metal does not become greatly chilled before the last shaping operation. The excess metal appears as a *flash* at each stage, which comes out of the product as the parting line. This flash is trimmed away. This means that each size and shape of forging requires a separate set of forging and trimming dies, which makes this process a costly one.

Originally, upset forging was initiated to form head on bolts. But, now the process covers a wide range of forgings. It is essentially a double acting press with horizontal motions rather than vertical. In complex forges, many different dies and heading tools may be required, similar to the die drop forging process.

When the major dimension of the work piece is more than three times the smaller dimension, there is more chances of buckling of the work piece. Another factor in drop forging should be considered is the friction. In case there is friction between the press tool and the work piece, barreling is a common phenomena.

During forging, the work done during compression, dw, can be expressed as

$$dw = \sigma_o A dh = \sigma_o \frac{V}{h} dh \qquad (8.15)$$

where σ_o = yield stress of the metal
A = cross-section area
dh = amount of compression
V = volume of bar.

Over complete forging cycle the total work done can be obtained by integrating equation (8.15)

$$w = \sigma_o V \log_e \frac{h_1}{h_0} \tag{8.16}$$

but $\log_e \frac{h_1}{h_o} = \epsilon_1$ i.e., the natural strain.

In press forging, one uses hydraulic power to actuate the process. This is in contrast to impact forging, where mechanical presses are used. Press forges are rather bigger in size and they do not have load limitation. Forgings upto 100 tonnes weight are easily handled by this type of forging operation.

In case we wish to forge the work piece by a succession of small deformation, schematic is shown in Figure 8.14. In this case plane strain is involved rather than homogeneous deformation. The force required to produce the deformation in each bite is equal to sbw, where s = plane strain yield stress, b = tool breadth and w = work piece width. The work done per bite is, therefore, equal to

$$w = sbw\,\Delta h \tag{8.17}$$

where Δh = reduction in height and is known as draft.

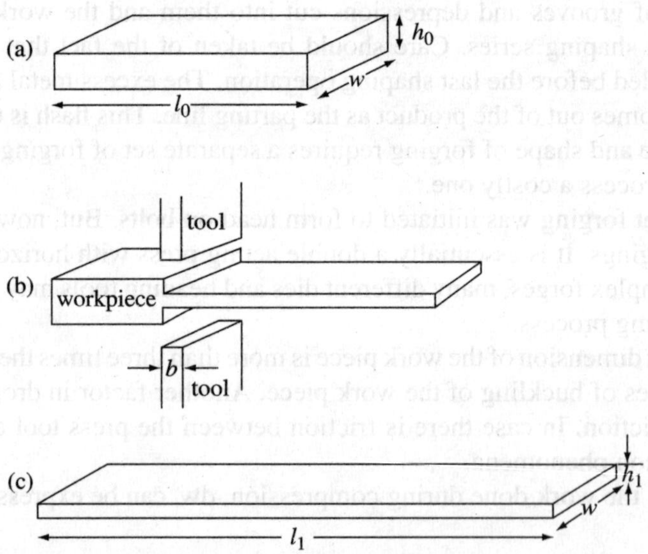

Fig. 8.14. Forging a bar by successive small deformations: (a) the initial shape of the bar; (b) the shape at some intermediate stage of working; (c) the final shape

Roll Forging Roll forging is a deformation processing method to reduce the cross section of a cylindrical or rectangular work piece by passing it through a set of opposing rolls that have grooves matching the desired shape of the final part. The process is classified in forging category, even though it utilizes rolls. The rolls do not turn continuously, but rotate through only a portion of one revolution. Roll forging is sometimes used for axles, blanks for airplane propellers, crowbars, knife blades, chisels, tapered tubing, and end of leaf springs. The products made out of such operation are generally stronger compared to competing processes.

Swaging and Radial Forging In swaging, the process is accomplished by means of rotating dies that hammer a work piece radially inward to taper it. In radial forging, the dies do not rotate around the work piece; instead the work is rotated as it feeds into the hammering dies.

Orbital Forging In this process, deformation occurs by means of a cone-shaped upper die that is simultaneously rolled and pressed into the work piece. The work is supported on a lower die, which has a cavity into which the work is compressed. Such type of forging results in a substantial reduction in press load required to accomplish deformation of the work.

Example 8.5. A mild steel ingot of circular cross-section has an initial diameter of 1 m and length 3 m. It is open drop-forged to a length of 2 m under frictionless condition. Calculate the total work done in carrying out the operation. Assume the flow stress of steel as 350 MPa.

Solution. Since there is no volume change after forging, we can get the final forged cross-section area A, from the expression

$$\pi \times \frac{1^2}{4} \times 3 = A \times 2$$

$$\text{or} \quad A = 1.178 \text{ m}^2$$

From equation (8.16), the total work done

$$W = \sigma_o V \log_e \frac{h_1}{h_0}$$

$$(350 \times 10^6) \times (\pi \times \frac{1^2}{4} \times 3) \times \log_e \left(\frac{3}{2}\right) \text{ (joules)}$$

$$= (350 \times 10^6) \times (0.589) \times (0.405) \text{ MJ}$$

$$= 83586632.04 \text{ J}$$

$$= 83.59 \text{ MJ}$$

Example 8.6. In a tabular form show various process parameters which might affect the geometrical accuracy and surface quality of a forged steel part.

Answer. *See* Table Ex. 8.6

8.9 ROLLING

Rolling mills are classified in various categories, which are highlighted in Figure 8.15. The two-high rolling mill is the simplest of all. A modified version of this is two-high reversing rolling mill, where the metal can be rolled from both directions. The next category of mill is three-high rolling mill, which combines both the two high reversing mill and non-reversing mills. These mills require elevating tables on both sides of the rolls. Unlike two-high reversing mill, the roll gap on the three-high mill cannot be adjusted between passes. All the above mentioned mills

Table Ex. 8.6.

Line:	*Machine*
	Automation
	Heating
	Cooling
Billet:	Starting material
	Material form
	Manufacturing process
	Pre-treatment
Process:	Processing sequence
	No. of steps
	Temperature
	Intermediate treatment
	Lubrication and cooling
	Clearance for part transport
Die:	Precision
	Wear
Human element:	Die designer, die manufacturer, operating staff, training and instruction

are used for primary rolling, where a drastic change in the cross-section is the aim. Four-high rolling mill is a special case of two-high mill. The middle low diameter rolls cause the lowering of rolling load. The rolling load is given by the equation

$$\text{Rolling load} = \sigma_o\, w\sqrt{R\,\Delta h}$$

where σ_o = Yield strength of the metal

w = Width of metal work piece

R = Radius of the roll

Δh = Reduction in thickness i.e., $h_1 - h_2$ also known as draft.

Draft is sometimes expressed as a function of starting stock thickness, called the *reduction* i.e.,

$$\text{Reduction} = \frac{\text{Draft}}{\text{Original thickness of the metal}}$$

In addition to thickness reduction, rolling usually increases work width. This is known as *spreading*. It tends to be most pronounced with low width thickness ratios and low coefficient of friction.

The rolls contact the work along a contact arc, which is defined by the angle θ. Let v_r be the rotational speed of the roll. This velocity is greater than the entering speed of the work v_o and less than the existing speed v_f. Since the metal flow is continuous, there is a gradual change in

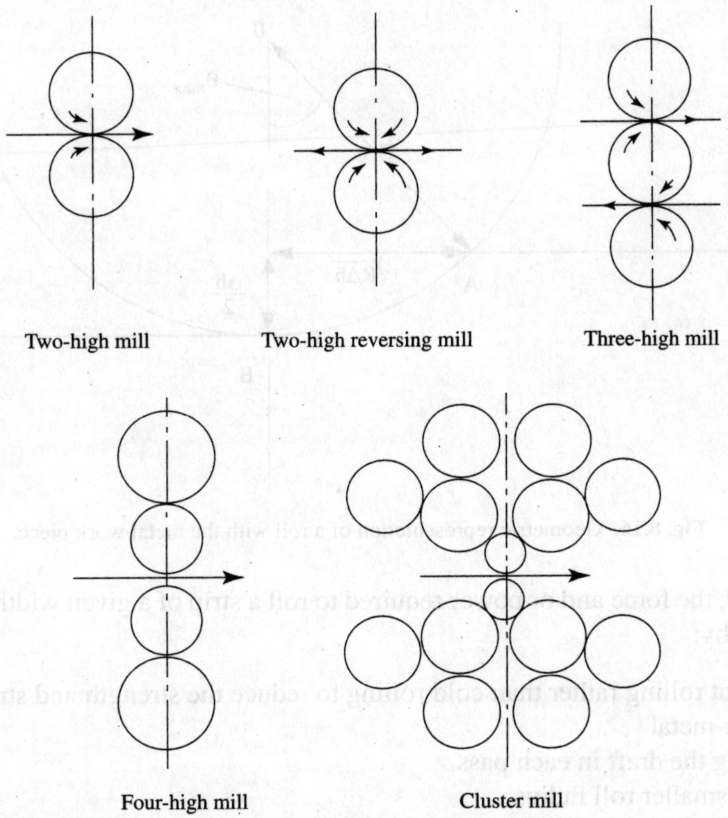

Fig. 8.15. Classification of rolling mills.

work velocity between the rolls. However, there is one point along the arc where work velocity equals roll velocity. This is called the *no-slip point* or *neutral point*. In either side of this part, slipping and friction occur between the roll and work.

The maximum angle of bite or the friction angle θ_{max} can be related to the coefficient of friction as

$$\mu = \tan \theta_{max} \tag{8.18}$$

From Figure 8.16, it is obvious that the maximum draft possible depends on two factors, the coefficient of friction μ and the roll radius R

$$\mu = \tan \theta_{max} = \frac{\sqrt{R\Delta h}}{R - \left(\frac{\Delta h}{2}\right)} \approx \sqrt{\frac{\Delta h}{R}} \tag{8.19}$$

Therefore maximum draft $\Delta h_{max} \approx \mu^2 R$.

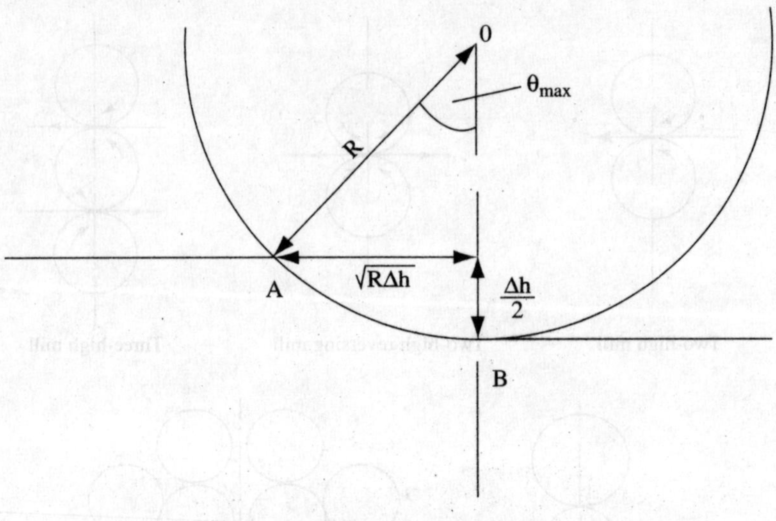

Fig. 8.16. Geometric representation of a roll with the metal work piece.

In summary, the force and or power required to roll a strip of a given width and work metal can be reduced by:

- Using hot rolling rather than cold rolling to reduce the strength and strain hardening of the work metal
- Reducing the draft in each pass
- Using a smaller roll radius
- Using a lower rolling speed.

Example 8.7. Determine the maximum reduction possible on a piece of steel 200 mm thick during cold rolling when $\mu = 0.1$ on a rolling mill with a roll radius of 500 mm.

Solution. From equation (8.18)

$$\mu = \tan \theta_{max}$$

$$\theta_{max} = \tan^{-1} 0.1 = 6°$$

From Figure 8.16

$$\sin \theta_{max} = \frac{\sqrt{R\Delta h}}{R}$$

$$\text{or} \quad 0.1045 = \frac{\sqrt{500\Delta h}}{500}$$

$$\Delta h = 5.46 \text{ mm}$$

Therefore maximum draft in rolling

$$= 5.46/200 \times 100$$

$$= 2.7\%$$

8.10 EXTRUSION

Extrusion process is of two types: direct or indirect extrusion. Direct extrusion is also known as *forward extrusion* (Figure 8.17a). A metal billet is loaded into a container and a ram compresses the material, forcing it to flow through one or more operations in a die at the opposite end of the container. The starting billet in a direct extrusion is usually round in cross-section, but the final shape is dependent on the shape of the die opening. The biggest problem in extrusion is the friction that exists between the work surface and the walls of the container. This causes a considerable increase in the ram force required for extrusion.

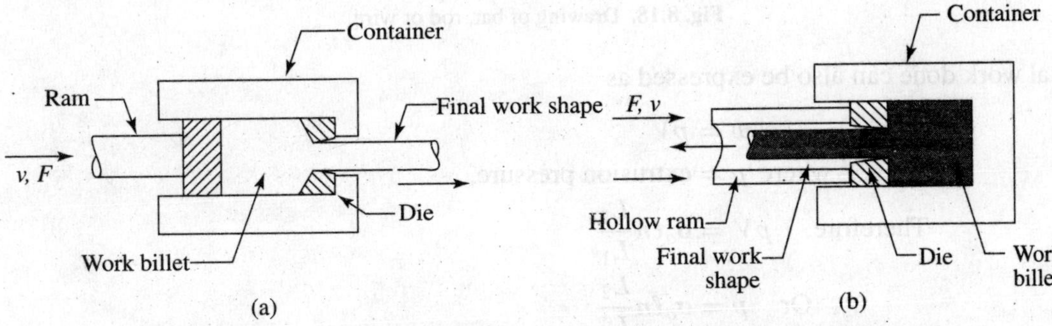

Fig. 8.17. Direct extrusion (a) and indirect extrusion (b) to produce solid cross-section.

In indirect extrusion, also called backward or reverse extrusion, the die is mounted to the ram rather than at the opposite end of the container (Figure 8.17b). As the billet is not forced to move relative to the container, there is no friction at the container walls, and the ram force is, therefore, lower than in direct extrusion.

As described in the introductory part, extrusion can also be classified as hot or cold extrusion, depending on the working temperature. During hot extrusion, any oxide formation on the billet surface adversely affects the extrusion process because it aggravates friction problem. One of the typical example of cold extrusion is *impact extrusion*, which has a relatively high speed. The process is a discrete one.

Extrusion Load: The work done dw in extruding a bar of length L and cross-section A by an amount $d\ell$ can be given as

$$dw = \sigma_o A\, d\ell \tag{8.20}$$

where σ_o = yield stress.

Since volume $V = AL$

$$dw = \sigma_o V \frac{d\ell}{L}$$

Therefore total work done $$w = \int_{L_1}^{L_2} dw = \sigma_o V \int_{L_1}^{L_2} \frac{d\ell}{L}$$

$$= \sigma_o V \ell n \frac{L_2}{L_1}$$

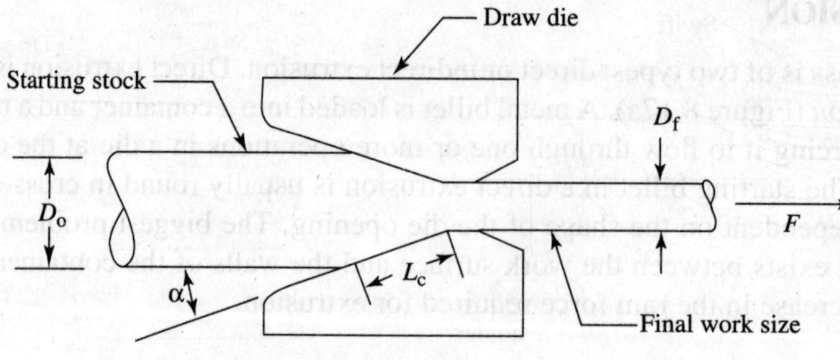

Fig. 8.18. Drawing of bar, rod or wire.

Total work done can also be expressed as

$$w = pV$$

where p = extrusion pressure.

Therefore, $$pV = \sigma_o \ell n \frac{L_2}{L_1}$$

Or $$p = \sigma_o \ell n \frac{L_2}{L_1}$$

The extrusion ratio $r = \dfrac{A_1}{A_2} = \dfrac{L_2}{L}$ (V is constant).

Hence $$p = \sigma_o \ell n\, r \tag{8.21}$$

The relationship expresses the ideal or theoretical pressure required for extrusion. This does not take into account the effect of friction or redundant work. These factors make the extrusion pattern rather inhomogeneous. In order to get the actual pressure, one has to place a multiplying factor α in the equation. Thus

$$p' = \alpha \sigma_o \ell n\, r \tag{8.22}$$

8.11 BAR AND WIRE DRAWING

Drawing is a process in which the cross-section of a bar or wire is reduced by pulling it through a die opening (Figure 8.18). The difference in bar drawing and wire drawing lies in the diameter of the stock. In the former, it is relatively large. Bar drawing is achieved as a single-draft operation, while in case of wire drawing the operation is through multiple dies, making the operation a continuous one.

The drawing equipment consists of three parts — a swift or captan to hold the coil of rod ready for drawing, the die which executes the actual reduction and the drawing block which supplies the load for reduction. These three parts are shown in Figure 8.19. The die consists of two parts, the casing and the nib. The nib contained inside the casing is made of tungsten carbide for large dies and industrial diamond for small. The hardness is essential, since it is the part where the actual reduction is carried out.

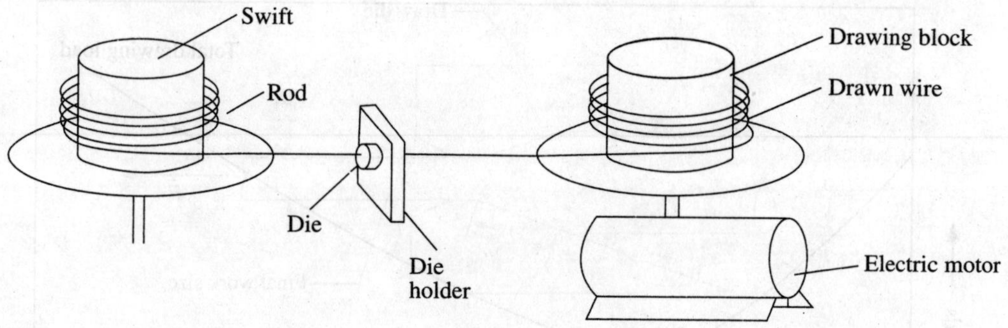

Fig. 8.19. Set up of a metal wire drawing assembly.

The die angle is an important parameter in wire drawing. This is the angle, that the approach makes to the center line of the die (Figure 8.18). The die angle controls to a large degree the drawing load. The total drawing load can be divided into three parts:

(a) Ideal load: It is required to deform the metal ideal or homogeneously. Its value can be obtained similar to equation (8.21). The ideal load is independent of working and in case of wire drawing it is independent of α.

(b) Friction load: This is required to overcome external friction. This is dependent on α since the area of contact decreases as the die angle is increased.

(c) Redundant work load: In wire drawing the redundant work load is increased with the die angle.

Figure 8.20 illustrates the sum of all these contributions with respect to die angle.

Example 8.8. Show that the maximum possible reduction per pass during wire drawing is 63% assuming no external friction or redundant work.

Solution. The limit of reduction during wire drawing reaches when the drawing load is equal to the breaking load of the drawn wire in tension.

If σ_o = yield stress of the drawn metal

σ = ultimate tensile stress of the drawn metal.

Then $\sigma_o = \sigma \ell n\, r$ (1)

where r = reduction ratio $= \frac{A_1}{A_2}$ (*see* Equation 8.21).

Thus $r = \frac{1}{1-R}$, where R = reduction per pass.

Substituting this value in equation (1),

$$\sigma_o = \sigma \ell n \left[\frac{1}{1-R} \right]$$

At the points of breakage $\sigma_o = \sigma$, therefore

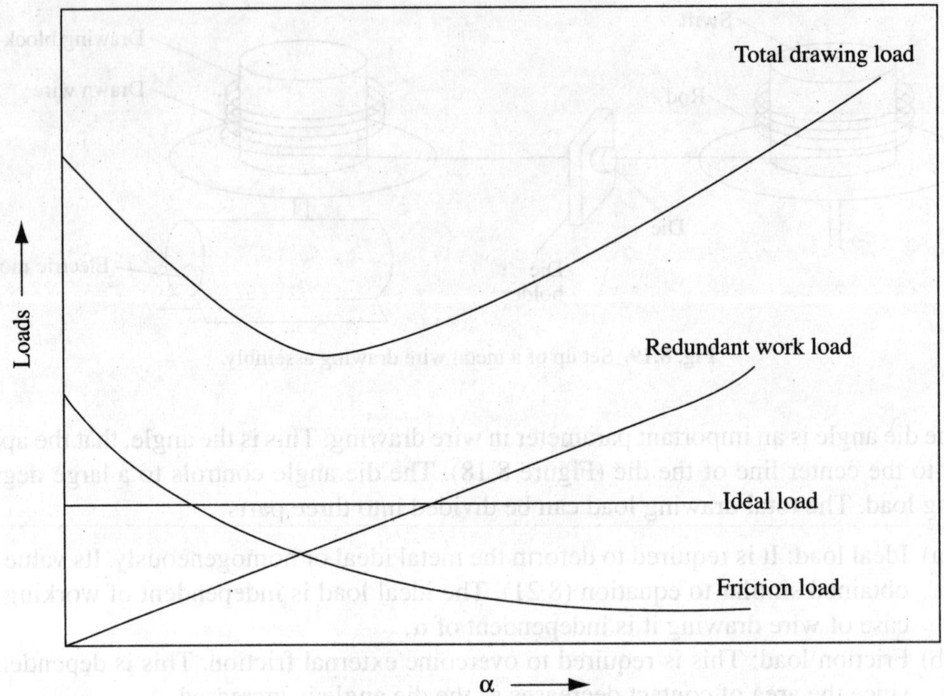

Fig. 8.20. Variation of total wire during load with respect to die angle (α)

$$\frac{\sigma_o}{\sigma} = 1 = \ell n \left[\frac{1}{1 - R_{\max}} \right]$$

where $R_{\max}$ = maximum possible reduction per pass.

Thus, $R_{\max} = 0.63$.

Example 8.9. Describe the role of segregations in steel wire drawing.

Answer. Steel wires in the form of filaments are used as reinforcement of tyres, surgical purposes, bridge cables, overload transmission cables etc. Strength level of upto 56 Pa can be obtained in steel filaments. Conventional galvanized steel wire for bridge cable has a tensile strength of 1.6 GPa. The behaviour of such filaments in torsion and fatigue is important. Control of inclusions is extremely important. C, Mn and Si are required as alloying elements in steels, but their microsegregation must be controlled. In addition, P and S must be eliminated. The inclusions typically consist of carbides as well as oxides e.g., alumina, mullite, spiral and calcium hexa-aluminate. If the inclusions are of nondeformable type, decohesion of the matrix/inclusions interface takes place. In the wire, the inclusions tend to segregate to the centre of the wire. Steel wires can be very strong as compared to the bulk steel, although its modulus does not change very much.

A homogeneous steel microstructure is very essential for a successful wire drawing. For example, any untransformed austenite in the central zone will transform to martensite, because

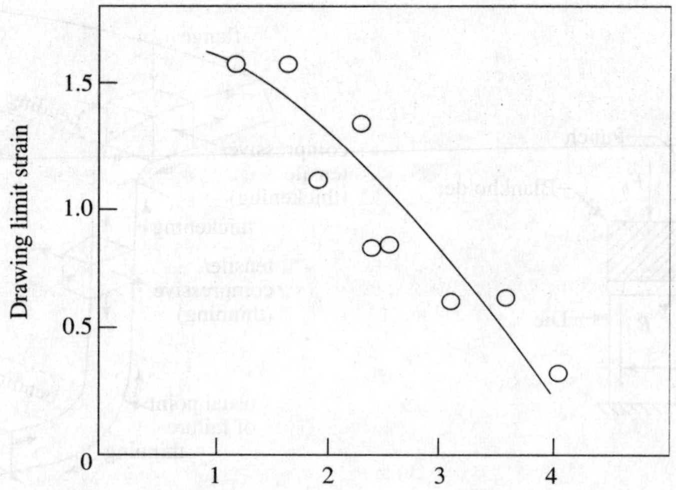

Figure Ex. 8.9

of the presence of manganese there. A reduction in inclusion size improves drawability. Figure Ex. 8.9 shows the drawing limit strain as a function of the centre size segregation in the starting steel billet. This confirms the need for clean steels. The pouring temperature of the molten steel must not be high and it should be preferably electromagnetically stirred.

8.12 DEEP DRAWING OF METAL SHEET

Deep drawing is a compression-tension forming process. Rigid tooling involves draw punches, a blank holder and a female die (Figure 8.21). The blank is generally pulled over the draw punch into the die, the blank holder prevents any wrinkling taking place in the flange. In pure deep drawing, the thickness of the sheet metal remains unchanged. In deep drawing, a cup is produced from the flat blank. This cup can be further processed, for example by an additional drawing, reverse drawing, ironing or extrusion. The work piece during deep drawing is subjected to radial tensile force and tangential compressive force.

The problem of calculating the drawing load is a complicated one, since there are three contributions involved. First is the ideal deformation load, which is the function of the degree of deformation carried out, i.e., *drawing ratio*. On the load-punch movement diagram (Figure 8.22), this contribution is constant. The friction contribution depends upon the surface area of contact between the metal and the punch and die. As the punch moves, the area of contact decreases and hence the friction load will fall. The third contribution is the ironing load, which occurs late in the cycle and increases as drawing progresses. The total load is, therefore, the sum of these contributions (Figure 8.22). Thus the punch load consists of two peaks, one early and the other late in the draw cycle. The position of these peaks and their relative magnitudes depends on many factors such as drawing ratio, blank-holder clearance and lubrication condition.

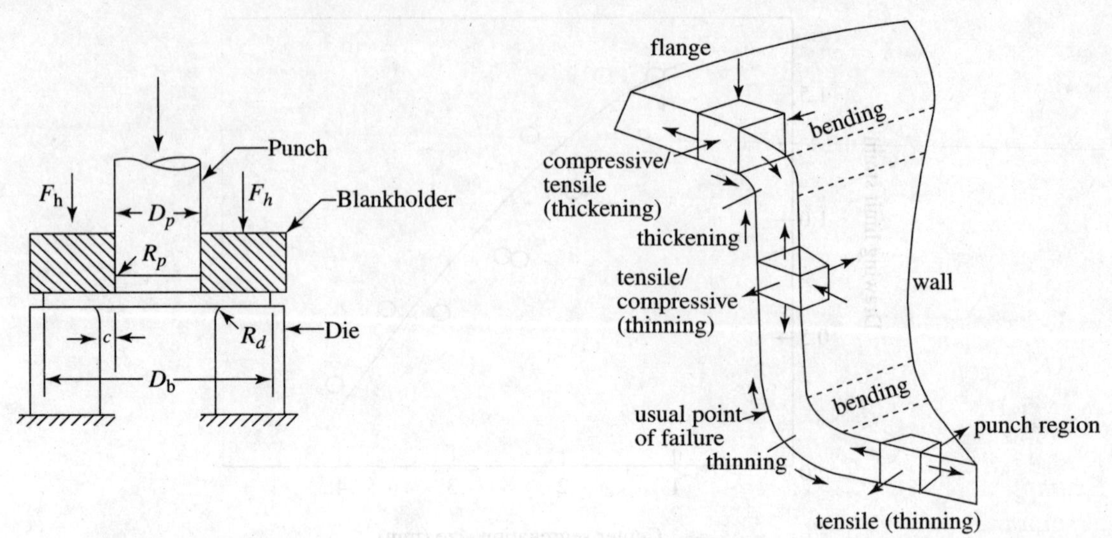

Fig. 8.21. (Left) Set up for deep drawing. (Right) Stress acting on different elements of metal during the deep drawing of a cup.

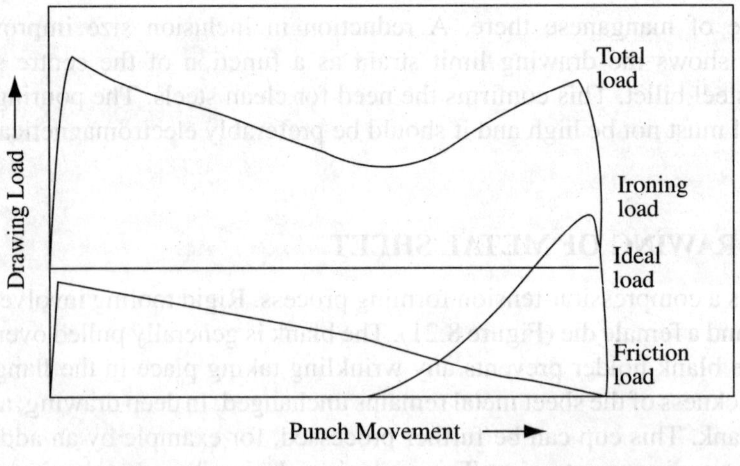

Fig. 8.22. Variation of total deep drawing load with respect to punch movement.

Now the question comes up, how to calculate the diameter of the blank metal sheet from the final cup diameter and height data. It must be large enough to supply support metal to complete the cup, but at the same time, if there is too much material, unnecessary waste will result. Because the volume of the final product is the same as that of the starting sheet metal blank, the blank diameter can be calculated by setting the initial blank volume equal to the final volume of the product. It is assumed that negligible thinning of the part wall occurs. In case of simple hollow cylinder the blank sheet diameter (D) is given by the expression

$$D = \sqrt{d^2 + 4d\,h} \tag{8.23}$$

where d = cup diameter

$h =$ cup height.

The maximum punch force is given by the expression

$$F = 2\pi\, r_i t_o \left\{ \sigma_f \ell n \left(\frac{r_o}{r_i} \right) + \frac{\mu B}{\pi r_o t_o} \right\} \exp\left(\frac{\mu \pi}{2} \right) \tag{8.24}$$

where $\mu =$ coefficient of friction

$\sigma_f =$ flow stress of the metal

$B =$ blank holder force

$t_o =$ sheet thickness

$r_o =$ blank radius

$r_i =$ cup radius.

Example 8.10. In a 1 mm thick mild steel sheet, a hollow cylindrical part with a diameter of 50 mm and height of 90 mm is to be drawn. Calculate the sheet metal blank diameter and the total draw ratio. In case there is limiting draw ratio of 2.0 for the material, what possible solutions do you propose? Also calculate the maximum punch force.

Solution.

$$D = \sqrt{d^2 + 4dh}$$

$$= \sqrt{50^2 + 4 \times 50 \times 90}$$

$$= \sqrt{2500 + 18000}$$

$$= \sqrt{20500}$$

$$= 143 \text{ mm}$$

The total draw ratio

$$\text{DR} = \frac{D}{d} = \frac{143}{50} = 2.86$$

Since the ratio is greater than 2, we can do the following:

(a) Drawing in more than one step.

(b) An intermediate annealing may be introduced.

8.13 STRETCH FORMING

This is a cold working process and not used very extensively. In this process the sheets are drawn over shaped formers to the extent that they deform plastically and assume the required profile. The work piece is gripped by one or more jaws in each end and then stretched and bent over a positive die containing the desired form. Stretch forming is widely used in the aircraft industries to economically produce large sheet metal parts in rather low quantities.

8.14 FORMABILITY OR WORKABILITY OF METALS AND ALLOYS

Workability signifies the relative case with which a metal can be formed through plastic deformation. A complete description of workability is specified by its flow stress dependence on processing variables (e.g., strain, strain rate, preheat temperature and die temperature), its failure behaviour and phase transformations that characterize the alloy system. A number of tests are used to evaluate the workability of a metal. Some of them are tension, torsion, compression and bend. Out of these the compression test is most common. There are other specialized tests like in the plane strain compression test, the partial-width indentation test and ring compression test. Workability analysis is carried out using the fracture limit line.

The formability of a metal is not a fixed quantity. It depends on the mean hydrostatic pressure exerted during the forming operation. For example, a material can have a low formability for one type of forming operation where the mean hydrostatic pressure is low. However, if a different forming process is employed in which the mean hydrostatic pressure is higher, the same material can be formed without any problem. Even a ceramic can be plastically deformed if the mean hydrostatic pressure exerted is sufficiently large.

Finite element analysis is a computer based analytical technique, which allows numerical solutions to be obtained by creating a discrete or finite number of individual nodes and elements. It is powerful analysis tool, which is flexible for solving problems with irregular shapes, variable material properties and irregular boundary conditions. Common problem solved by FEA include insufficient die filling, poor shape control, poor flow of metal, cracks and voids that lead to fracture and inadequate properties from microstructural variations.

8.14.1 Metallurgical Factors Influencing Workability

Workability greatly depends on the microstructure of the material i.e., grain structure and size. In conventionally cast ingot structure, workability is lower, since the grain size is relatively large. In addition, the segregation of impurities in cast structure also affects workability, because of inhomogeneous distribution of the chemical elements. Hot shortness is very common in working of cast products. This can be prevented by using a sufficiently low deformation rate that allows the heat developed by deformation to be dissipated by the tooling, or by using lower working temperature or by subjecting the work piece to a homogenization heat treatment prior to hot working.

The type of phase equilibria in an alloy system is vital in understanding the response of say temperature on hot workability. Figure 8.23 illustrates hot forgeability response with temperature for various types of alloy systems of technological significance. The forgeability of precipitation hardening aluminium alloys of different compositions with temperature is another interesting example.

The texture of the polycrystalline metal or alloy is another factor, which influences workability. The anisotropy in mechanical properties is greatest between the longitudinal and transverse directions. A typical manifestation of the effect of preferred orientation on aluminium alloy and steel sheets is in the form of '*earing*' (Figure 8.24). The development of such structures is conveniently characterized by the measurement of *plastic strain ratio* or *plastic anisotropy*

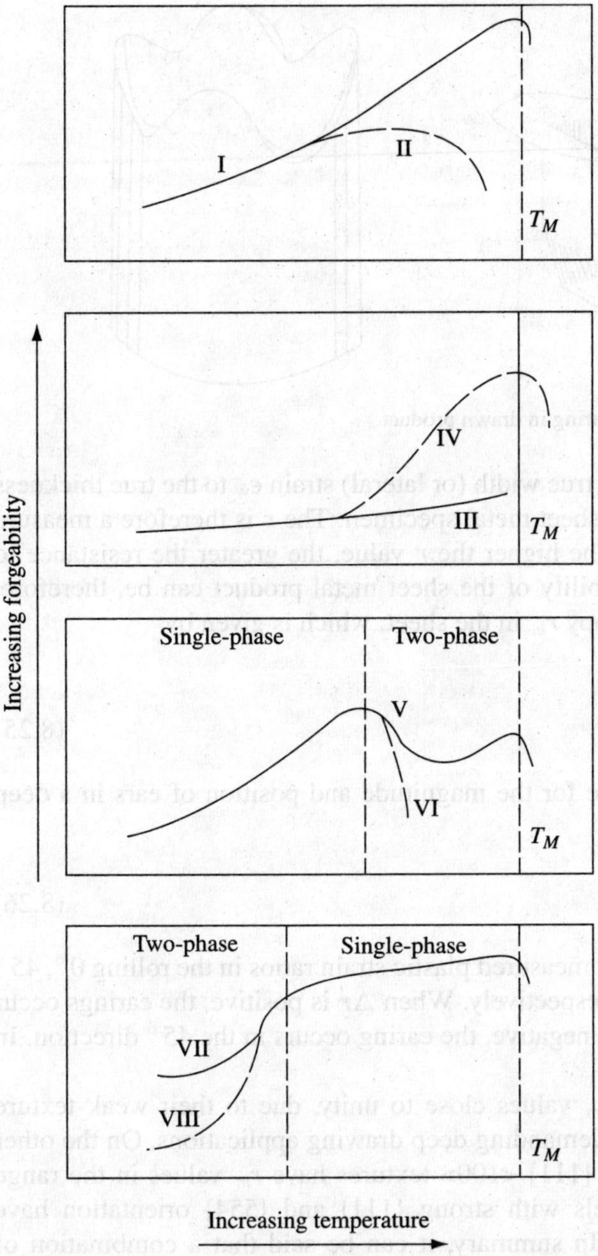

I. Pure metals and single phase alloys
Aluminum alloys
Tantalum alloys
Niobium alloys

II. Pure metals and single phase alloys exhibiting rapid grain growth
Beryllium
Magnesium alloys
Tungsten alloys
All-beta titanium alloys

III. Alloys containing elements that form insoluble compunds
Resulfurized steel
Stailness steel containing selenium

IV. Alloys containing elements that form soluble compunds
Molybdenum alloys containing oxides
Stailness steel containing soluble carbides or nitrides

V. Alloys forming ductile second phase on heating
High-chromium stainless steel

VI. Alloys forming low-melting second phase on heating
Iron containing sulfur
Magnesium alloys containing zinc

VII. Alloys forming ductile second phase on cooling
Carbon and low-alloy steels
Alpha-beta and alpha-titanium alloys

VIII. Alloys forming brittle second phase on cooling
Superalloys
Precipitation-hardenable stainless steels

Fig. 8.23. Typical workability behaviour exhibited by different alloy systems. T_m: absolute melting temperature. (Reproduced with permission from Handbook of Workability and Process Design, Ed. G.E. Dieter, H.A. Kuhn and S.L. Semiatin, ASM International, Ohio, 2003, p. 31).

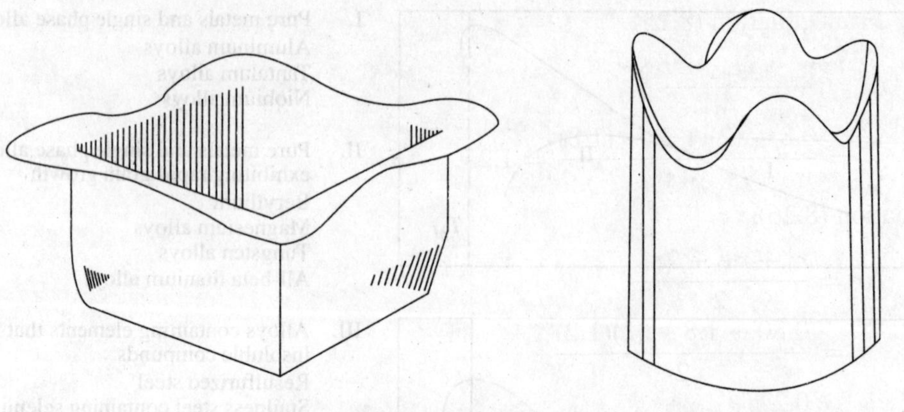

Fig. 8.24. Earing in drawn product.

factor r, which is defined as the ratio of the true width (or lateral) strain ϵ_w to the true thickness strain ϵ_t in a tensile test carried out on the sheet metal specimen. The r is therefore a measure of capacity of a sheet to resist thinning. The higher the r value, the greater the resistance to thinning during deep drawing. The drawability of the sheet metal product can be, therefore, predicted from the average normal anisotropy r_m in the sheet, which is given by

$$r_m = \frac{r_o + 2r_{45} + r_{90}}{4} \tag{8.25}$$

The planar anisotropy which is responsible for the magnitude and position of ears in a deep drawn cylindrical cup is given by

$$\Delta r = \frac{r_o + r_{90} - 2r_{45}}{2} \tag{8.26}$$

where r_o, r_{45} and r_{90} are the experimentally measured plastic strain ratios in the rolling 0°, 45° and transverse 90° direction in the sheet, respectively. When Δr is positive, the earings occur in the 0° and 90° directions. In case Δr is negative, the earing occurs in the 45° direction. In absence of any earing, $\Delta r = 0$.

Rimmed and semikilled steels have r_m values close to unity, due to their weak texture and are, therefore, not suited for the most demanding deep drawing applications. On the other hand, aluminium-killed steels with strong [111] <100> textures have r_m values in the range 1.5 to 1.8. The Ti- or Nb-stabilized steels with strong {111} and {554} orientation have r_m values approaching or exceeding 2.0. In summary, it can be said that a combination of high r_m value and low Δr value gives optimum drawability.

Example 8.11. A given metal sheet has r values of 0.8, 1.2 and 1.8 for the 0°, 45 and 90° direction to rolling respectively. Will ears be formed after deep drawing? In case the sheet blank is of 100 mm diameter, what shall be the smallest cup diameter drawn successfully?

Solution. From equation (8.25)

$$r_m = \frac{r_o + 2r_{45} + r_{90}}{4}$$

$$= \frac{0.8 + (2)(1.2) + (1.8)}{4} = 1.25$$

From equation (8.26),

$$\Delta r = \frac{r_o + r_{90} - 2r_{45}}{2}$$

$$= \frac{0.80 + 1.8 - (2)(1.2)}{2}$$

$$= \frac{0.2}{2} = 0.1$$

Since $\Delta r > 0$, ears will form.

For estimating the cup diameter, refer Handbook for a plot of limiting drawing ratio (LDR) vs. r_m. For $r_m = 1.25$, the LDR value is ~2.3. Therefore the smallest cup diameter is 43.5 mm.

Example 8.12. Show in a tabular form various defects in deep drawn metal sheet, their possible causes and remedies.

Answer. See Table Ex. 8.12.

8.15 THERMOMECHANICAL FORMING

Thermomechanical treatment or processing involves a combination of heat treatment, plastic deformation and phase transformation within a single hot deformation process (e.g., rolling, forging, extrusion) to control the structure, morphology and grain size of parent and product phases. The process is very common in steels and aluminium alloys. Detailed description of phase transformation involved during heat treatment has been made in Chapter 3 as well as in Chapter 7.

Thermomechanical forming of steels can be classified into three categories (Figure 8.25):

(a) Deformation completed prior to the transformation of austenite (Figure 8.25a, b and c).
(b) Deformation occurs during the transformation of austenite (Figure 8.25d and e).
(c) Deformation occurs after the transformation of austenite (Figure 8.25f).

The details of some of the processes, have been incorporated as Case Studies in Chapter 12.

In case of aluminium alloys a combination of hot deformation and recrystallization leads to desired microstructural evaluation. In such alloys, the amount of stored energy is reduced by dynamic recovery. The process is strongly dependent on alloy composition, temperature, strain rate, strain and interpass time, particularly in multistand hot rolling lines with thin gauges, relatively high temperature and high rolling speed. A processing sequence used for grain

Table Ex. 8.12.

	Defect	*Possible cause*	*Prevention, remedy*
(a)	drawing grooves	tool wear	replace draw punch/female die, improve lubrication
(b)	tear marks on the base of the drawn part	drawing ratio set too high/ drawing gap too narrow/ curvature on draw punch or female die too small/ blank holder force too great/ drawing speed too great	use a sheet metal with greater deep drawing capability or employ multi-stage drawing process/extend drawing gap/ check sheet metal thickness/ increase die radii/reduce blank holder force/reduce forming speed
(c)	vertical tear marks on upper edge	insufficient material in the defective area/drawing gap too wide or drawing edge curvature too great	replace the female die
(d)	on rectangular parts, the corners are higher than the side walls	too much material at the corners	change the blank shape/ recess material in the affected areas/trim the edge afterwards
(e)	vertical creases in the upper area of the vessel wall, also in conjunction with vertical cracks	blank holder force too low/ drawing gap too great/curvature at the female die too great	increase blank holder force/ renew female die
(f)	crease formation at the flange	blank holder pressure too low	increase blank holder pressure
(g)	other non-specific defects/faulty or irregularly formed drawn parts	incorrect blank shape/ asymmetrical material location/ unsuitable sheet metal/ incorrect lubricant	correct blank shape/check die and stops/use suitable sheet metals and lubricants

refinement of 7075 Aluminium alloy is schematically shown in Figure 8.26. This consists of (1) solution treatment in a molten salt bath to process a standard starting material by dissolving soluble precipitates, leaving dispersoid particles undissolved; (2) overaging to precipitate a high density of uniform and large second-phase particles with 0.5 to 1μ m in diameter to act as nucleation site for recrystallizing grains, (3) warm-working below the recrystallization temperature to -85% reduction to introduce defects into the alloy and produce local intense deformation zones around the hard, large precipitate particles serving as nucleation sites during recrystallization; and (4) rapid heating of the alloy above the recrystallization temperature and solution treatment in order to dissolve large precipitates leaving the fine dispersoid particles in the fine recrystallized grained matrix by effectively pinning grains and thus preventing grain growth.

Deformation temperature and metal phase	Name	Lay-out of the treatment	Treatment suitable to
Before transformation	(a) HTMT	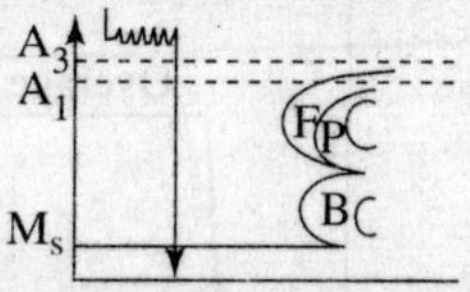	high or low alloy steels
Beyond A_3 stable austenite	(b) controlled rolling	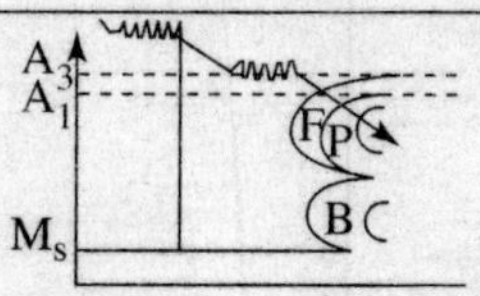	structural steels
Below A_3 metastable austenite	(c) LTMT ausforming	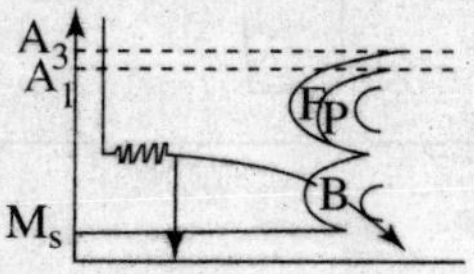	high or low alloy steels
During transformation between A_1 and M_s austenite and ferrite pearlite or pearlite	(d) isoforming	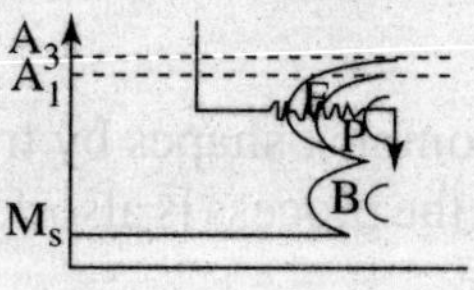	pearlite-forming steels
near M_s austenite and martensite	(e) * treatment of TRIP steels ** zerolling	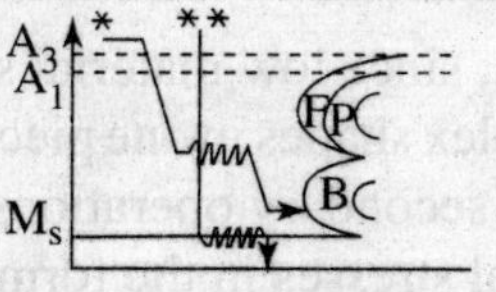	Stainless or semi-stainless steels
below M_s austenite	(f) marstraining	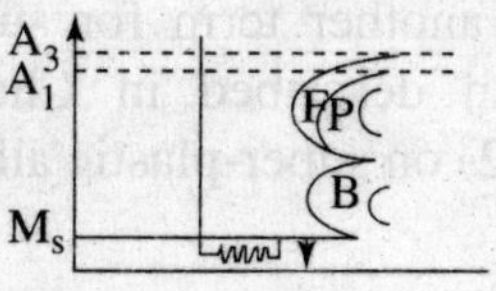	martensitic steels

Fig. 8.25. Classification of thermo-mechanical treatment of steel.

8.16 SUPER-PLASTIC FORMING

Super-plastic forming can be achieved by bulk deformation processes, such as compression moulding, closed-die forming, and extrusion. Sheet forming of alloys can also be done using operations such as thermoforming, vacuum forming and blow moulding. Super-plastic behaviour of fine grained alloys has been highlighted in Chapter 5. Super-plastic alloys of Zn-Al and Ti-

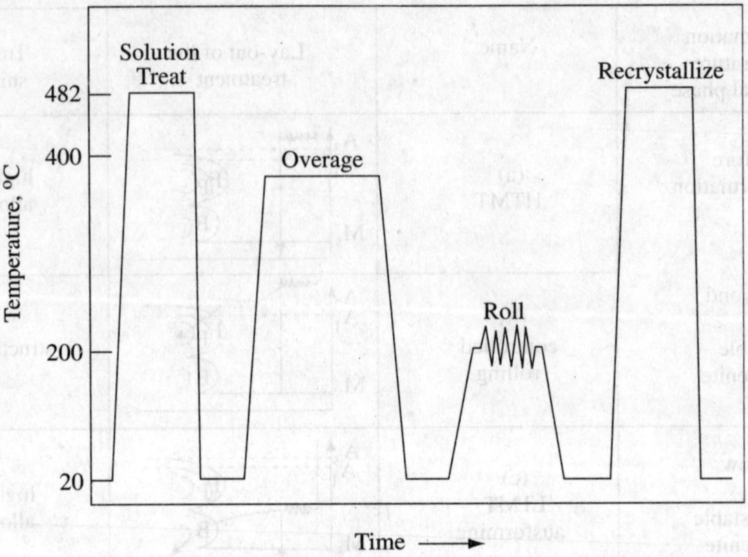

Fig. 8.26. Schematic diagram showing the four step thermo-mechanical treatment developed for grain refinement in 7075 Al alloy.

based can be formed into complex shapes by traditional metal working or polymer processing techniques. In press forging the process is also called as isothermal (hot die) forging. Advantages of super-plastic forming are

- Lower tooling cost, since low material strength at forming temperature.
- Formation of complex shapes in one piece, with fine details, close dimensional tolerances and elimination of secondary operations.
- Little or no residual stresses in the formed parts.

Diffusion bonding is another term for super-plastic operation to fabricate sheet metal structures. This has been described in Chapter 10 (Materials Joining). A case study (Section 12.12, Chapter 12) on super-plastic alloys may be referred by the readers.

8.17 THERMOFORMING OF POLYMERS

The thermoforming of polymer is a combination of drawing and stretching operators, similar to sheet metal forming. In this case thermoplastic sheet or film is formed over a mould with the application of heat and pressure. The sheet is heated in an oven to the softening point and placed over a mould with subsequent vacuum application. Because of the low strength at elevated temperature, the pressure differential caused by the vacuum is, generally, sufficient for forming. Other methods like air pressure or mechanical means is also common. The pressure thermoforming is also known as *blow forming*.

One of the problem in thermoforming is the thinning of the plastic sheet. Unless the contour of the mould is very shallow, there will be significant thinning of sheet as it is stretched to conform to the mould contour. It is obvious that only thermoplastics can be thermoformed, since extruded sheets of thermosetting or elastomeric polymers have already been cross-linked and cannot be softened by reheating. Mass production of thermoforming operations are carried out in the packaging industry. Relatively large parts produced from thicker sheet are also common by thermoforming. Some of the examples are covers for business machines, boat hulls, shower stalls, diffusers of lights, advertising displays and signs, bath tubs and certain toys.

Moulds for thermoforming are usually made of aluminium, since high strength is not a requirement. Quality consideration include tears, nonuniform wall thickness, improperly filled moulds, and poor part surface details.

8.18 PROCESS DESIGN IN DEFORMATION PROCESSING

The purpose of the process design is to achieve:

- Proper part shape and formal geometry
- Absence of any defect
- Adequate properties through proper tailoring of microstructure
- Maximum productivity/minimum energy consumption.

Each of the above objectives requires different types of analytical tools and models.

The above mentioned objectives in designing of a bulk forming process can be elaborated as:

- Material Flow: Flow stress behaviour can be modeled by constitutive equations that describe the relationship between stress and strain during plastic deformation.
- Fracture Behaviour: The effects of temperature, stress state and strain rate on fracture of material must be considered. Workability of materials comes in this category, which has already been described in Section 8.13.
- Microstructure: Thermal behaviour and associated analytical phenomena comes in this category. Finite element modeling is often used to accurately predict the thermomechanical history of the material.
- Available Equipment and Economics: This feature includes factors of economic productivity, efficiency, tool wear and scrap loss.

The interlinkage of processes in a deformation forming, say forging, is illustrated in Figure 8.27. This amply justifies the vital role of materials engineering and mechanics.

POWDER METALLURGY (P/M) FORMING

All powder metallurgy forming routes involve the production of powder compact (shaped), which is subsequently sintered to a solid component. Both metallic and ceramic powders or their composites are used in P/M forming. Figure 8.28 illustrates various steps involved in P/M forming.

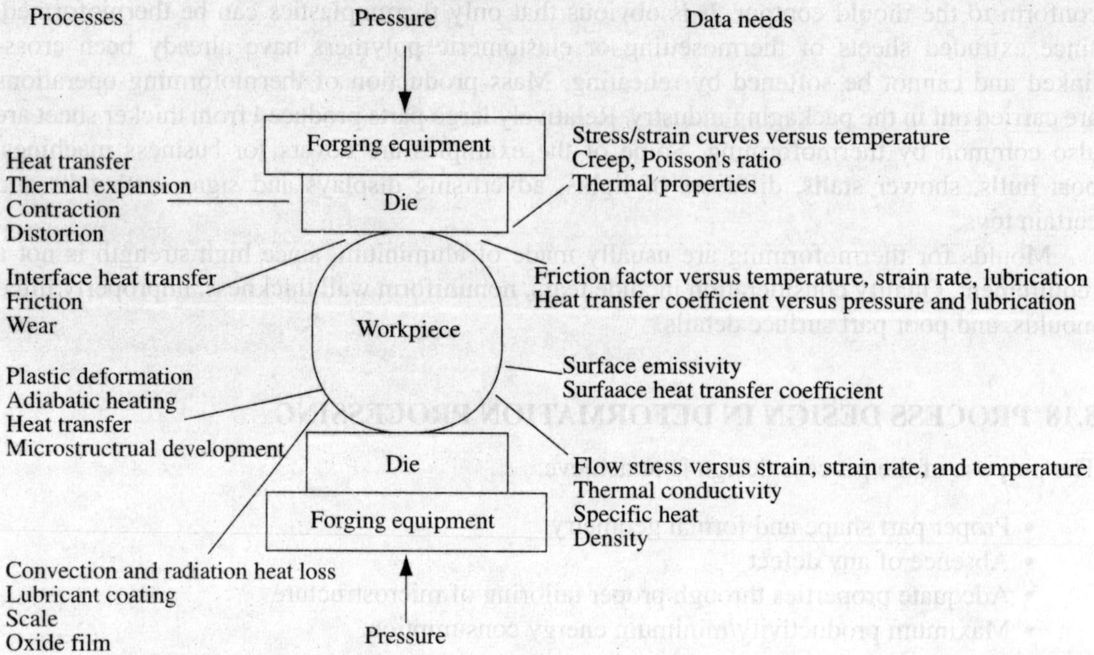

Fig. 8.27. Typical physical phenomena occurring during a forging operation. (Reproduced with permission from Handbook of Workability and Process Design, Ed. G.E. Dieter, H.A. Kuhn and S.L. Semiatin, ASM International, Ohio, 2003, p. 10).

The P/M process is many times more competitive than other fabrication methods like casting, stamping or machining. P/M is the choice when requirement for strength, wear resistance or high operating temperatures exceed the capabilities of die casting alloys. P/M offers greater precision, eliminating most or all of the finish machining operations required for castings. It avoids casting defects such as blow holes, shrinkage or inclusions. It should, however, be borne in mind that P/M forming is economical only when production rates are higher, since the tooling cost is quite appreciable. Apart from conventional powder compaction/sintering process, a number of related consolidation techniques, e.g., powder rolling/forging/extrusion, hot isostatic pressing and powder injection moulding have emerged.

Some of major applications of P/M forming are in the following group of materials:

- Refractory metals: As melting and casting of refractory metals are difficult, P/M is the most convenient route.
- Composite materials: MMC and CMC can be formed advantageously by P/M forming route. Examples are electrical contact materials, cemented carbides, friction materials, grinding wheels, oxide dispersion strengthened materials.
- Porous materials: Filters and self lubricating porous bearings are examples in this category.
- Structural parts: This category encompases the largest share of P/M formed products.

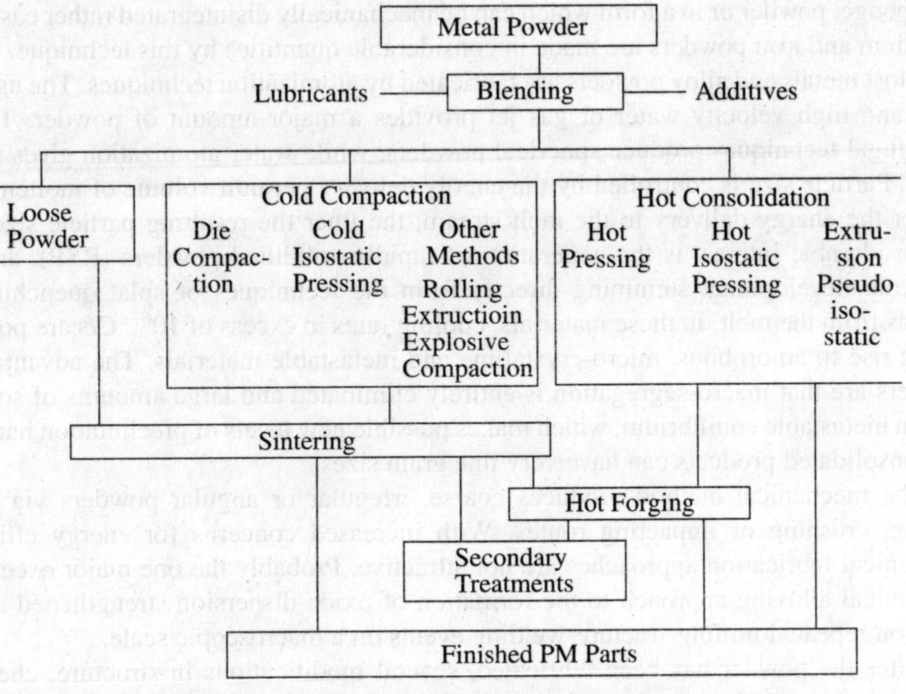

Fig. 8.28. Basic steps of the powder metallurgy process.

In general such parts do not have mechanical properties superior to those of equivalent parts made from wrought materials, but they suffice for the required duty. This is due to the fact that conventional pressed/sintered products do have some residual porosity. The main advantage is the production of net shape complicated engineering parts by P/M route. However, other full dense consolidation processes have properties equal to or even superior to those of parts made by casting and working route.

More details on such products with their relevant alloy composition have already been made in Chapter 7. Chapter 12 includes some typical case studies.

8.19 METAL POWDER PRODUCTION AND TREATMENT

The major production methods of metal powders can be classified as:

- Chemical reaction and decompositions
- Electrolytic deposition
- Atomization of molten metals
- Mechanical processing of solid materials.

The classic example of the first method is the reduction of iron oxide, producing a fine sponge-like iron powder. Fine powders of nickel and iron are produced by decomposing their carbonyls. A number of metals can be made to precipitate on the cathode of an electrolytic cell as a sponge, powder or in a form which can be mechanically disintegrated rather easily. Copper, beryllium and iron powders are made in considerable quantities by this technique.

Most metals and alloy powders are fabricated by atomization techniques. The use of molten alloy and high velocity water or gas jet provides a major amount of powder. The gas and centrifugal techniques produce spherical powders, while water atomization gives an irregular shape. Particle size is controlled by the energy delivery per unit volume of molten metal. The greater the energy delivery to the melt stream, the finer the resulting particle size. One area of considerable interest is the generation of rapidly solidified powders (RSP), an American industrial development stemming directly from the techniques of splat quenching of alloy ribbons from the melt. In these materials, cooling rates in excess of 10^6 °C/s are possible; thus giving rise to amorphous, micro-crystalline and metastable materials. The advantage of such powders are that macro-segregation is entirely eliminated and large amounts of solute can be held in metastable equilibrium, which makes possible new levels of precipitation hardening and the consolidated products can have very fine grain sizes.

The mechanical method produces coarse, irregular or angular powders via machining, milling, crushing or impacting routes. With increased concerns for energy efficiency, the mechanical fabrication approaches are not attractive. Probably the one major exception is the mechanical alloying approach to the formation of oxide dispersion strengthened alloys. This relies on repeated milling-fracture-welding events on a macroscopic scale.

After the powder has been fabricated, several modifications in structure, chemistry and size might be necessary for best compaction. This includes removal of oxides and inclusions, mixing, addition of lubricants and other sintering aids. Care must be exercised at this point to avoid segregation of components or particle sizes. Control in each handling step is crucial to successful powder processing.

Example 8.13. What are the factors, which shall favour fine particle size metal powders by atomization process?

Answer. Fine particle sizes are favoured by:

(a) Low metal viscosity
(b) Low metal surface tension
(c) Superheated metal
(d) Small nozzle diameter, i.e., low metal feed rate
(e) High atomizing pressure
(f) High atomizing agent volume
(g) High atomizing agent velocity
(h) High atomizing agent viscosity
(i) Short metal stream (F)
(j) Short jet length (E)
(k) Optimum apex angle (α^o).

For various parameters refer Figure Ex. 8.13.

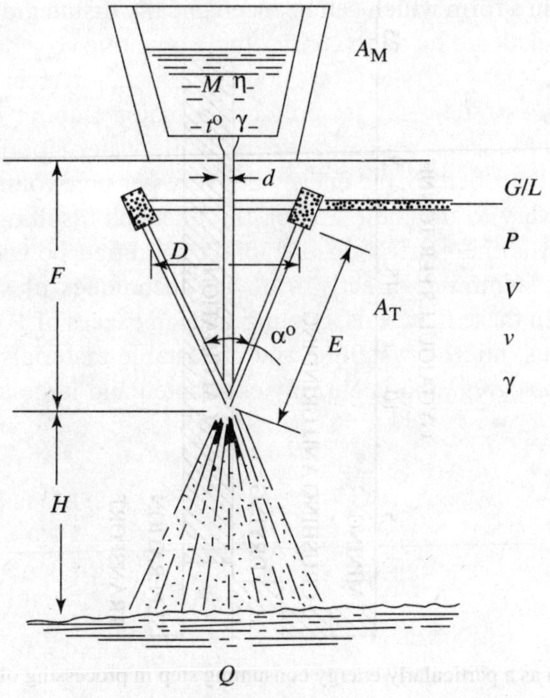

Figure Ex. 8.13

8.20 CERAMIC POWDER PRODUCTION

Ceramics raw materials are usually classified according to their function in ceramic manufacture and then according to their basic properties. One simplest method is to divide into plastic and non-plastic raw materials. Plastic ceramic raw materials involve any clay substances that mixed with enough water reveals plasticity. The non-plastic category involves minerals, rocks and synthetic chemicals that when mixed with water, do not exhibit plasticity. Some of them need not be blended at all with a plastic raw material and compose nearly 100% of the ceramic body. Some specific non-plastic raw materials are used for sintering, fluxing and melting while others are applied for enhancing the refractoriness. Some non-plastic raw materials contribute to high porosity, lower bulk density, insulating capacity, while others to chemical or mechanical resistance in different environments.

The requirements regarding shape, size and size distribution of ceramic particles or reactivity can be varied on a wide range by proper choice of materials and conditions for synthesis. One of the biggest limitations in synthetic ceramic powders is that they are more expensive than natural materials. Out of various ceramic powders processing methods, calcination is the most energy intensive process (Figure 8.29).

Advanced ceramic powders, both pertaining to oxide and nonoxide categories, are prepared by chemical methods. These are sub-classified as (a) solid state reaction, (b) liquid solutions, and

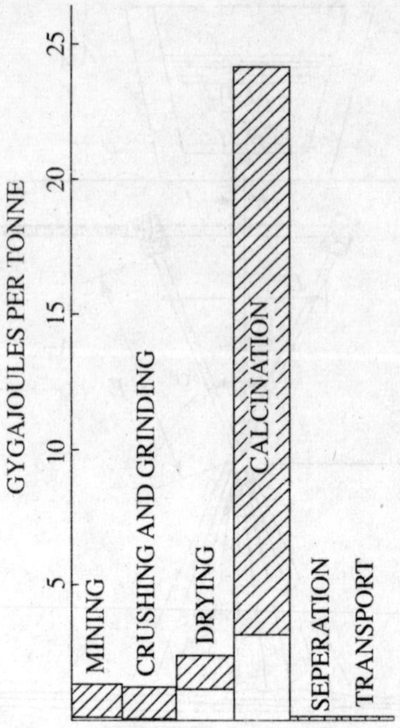

Fig. 8.29. Calcination as a particularly energy consuming step in processing of ceramic raw materials.

(c) vapour phase reaction. In the first category examples are production of powders of simple oxides from carbonates, hydroxides, nitrates, sulphates, oxalates, alkoxides and other metal salts. Chemical reactions between solid state materials, usually in the form of mixed powders, are common for the production of powders of complex oxides such as titanates, ferrites and silicates. Liquid solution preparation involves the dissolution of soluble salts, usually in water or dissolution of metals in acid solutions. The *sol-gel process* is another important method in which sols, or suspensions of colloidal particles, are formed by controlled precipitation from aqueous solutions. In some cases a stable sol is made by peptizing a slurry using pH control. Gelling to a semi rigid body is carried out by evaporating the water or changing the pH. Another category of solution preparation is *precipitation* or *co-precipitation* in which the desired cations are precipitated from the solutions as an insoluble salt using anions such as hydroxide, carbonate or oxalate. Vapour-phase techniques for preparation of ceramic powders rely on the nucleation of particles from the vapour phase. This is well suited to the preparation of powders which are much less aggregated than those prepared by solution techniques. It is considerably more difficult to make homogeneous powders of complex compositions by vapour-phase techniques.

As materials engineers, we are not so much concerned with the synthesis aspects of ceramics. Rather we are more involved in optimizing the application of synthetic raw materials, owing to their specific properties or higher cost. The calcined ceramic materials include metakaolin, calcined brick clays, fireclay and fire brick, calcined bauxite, sea water magnesia, dolomite, calcined bone and wollastenite etc. In the cateogy reaction synthesized raw materials, examples are cristabolite, mullite, spinel, ferrites, titanates, zirconates and related compounds, cordierite, silicon nitride etc. The detailed description of their preparation is beyond the scope of this book.

8.21 METAL/CERAMIC POWDER CHARACTERIZATION

The powder production method influences particle chemistry and structure, apart from the precise nature of particle size distribution. These properties also influence the behaviour of the powder during compaction and sintering and finally the composition, structure and properties of the sintered material. The following are the major characteristics:

Chemical compositions and structure: It is necessary to know whether the impurity elements are present in the elemental form or in a combined form. For example, in reduced iron powders silicon is present as impurity in the form of silica. Other ceramic and inert compounds present in metal powders are reported in terms of an acid insoluble figure. The effect of impurity elements on the hardness of particles and the degree of chemical reactivity during sintering will differ widely, depending on the actual form they are in.

The microstructure of the crystalline powder has a significant influence on the behaviour of powder during compaction and sintering and the properties of the final product. Fine grain size is always desirable, as it improves the mechanical properties apart from the sinterability and the uniformity of dimensional changes.

Particle size and shape: The shape of the powder is characterized by the dimensionally of the particle and its contour surface. Figure 8.30 shows the shapes along with the processing route adopted to get such shapes. In a real powder mass, all the particles will not have the same exact size, even though the shape may be essentially the same. Thus particle size distribution is an important characterizing feature.

Particle surface topography: The nature of the particle surface is an important characteristic. The smoothness or roughness of the surface is a function of powder production method. For example, atomized powder has smooth surface, while reduced metal powder possesses a rough surface. The exact nature of surface topography influences the friction forces between particles. These are important in case of bulk movements of particles and sintering. Chemical reactivity of the powder also increases with increasing surface roughness, especially the irregular type.

Surface area: The actual amount of surface area per unit mass of powder is known as specific surface area. For a very irregular shaped particle with a high degree of roughness, the specific surface area can be very high. This affects sinterability.

Apparent and tap density: The apparent density of a powder refers to the mass of unit volume of loose powder, expressed in g cm^{-3}. Very often a mass of loose powder is mechanically vibrated or tapped. The density of the loose powder increases due to such a treatment and is always higher than the apparent density.

Flow rate: The standard method for estimating powder flow rate is by the Hall flowmeter, where the time necessary for 50 g of powder to flow through a prescribed small orifice is measured. Very fine powders do not flow through the small orifice and the powder is known as non-free flowing one. For a given powder, the high the apparent density, the lower the flow time.

Compressibility: Compressibility is a measure to which a powder will compress or densify upon application of external pressure. It is reported as the density in g cm^{-3}, rounded to the nearest 0.01 g cm^{-3}, at a specified compaction pressure. Compressibility of the powder is influenced by factors like internal hardness of the concerned material particles, particle shape, internal porosity, particle size distribution, presence of non-metallics in metal powders, addition

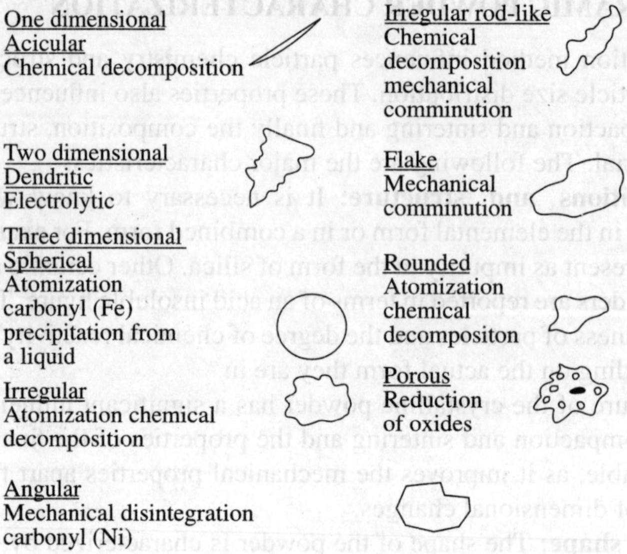

Fig. 8.30. System of particles shape characterization.

of alloying elements or solid lubricants. Alternatively, compressibility is also indicated in terms of densification parameters, such as

$$Densification\ Parameter = \frac{Green\ density - Apparent\ density}{Theoretical\ density - Apparent\ density}.$$

Table 8.2 shows the compaction pressure data of some major metallic, ceramic and cermet powders.

Green strength: Green strength is the mechanical strength of a green, i.e., unsintered powder compact. This characteristic is very important as it determines the ability of a green compact to maintain its size and shape during handling prior to sintering. The standard green strength is a transverse bend test of a 12.7 by 31.7 mm rectangular specimen of 6.35 mm (0.25 inch) thickness.

Table 8.3 summarizes various characteristics of metal powder along with their routes for production.

8.22 POWDER COMPACTION

The compaction of metal powders has the following major functions:

- to consolidate the powder into desired shape
- to impart, to as high a degree as possible, the desired final dimensions with due consideration to any dimensional changes resulting from sintering
- to impart the desired level and type of porosity
- to impart adequate strength for subsequent handling.

Table 8.2. Compacting Pressures for Various Metal and Ceramic Powders.

Material	*Pressure MPa*
Aluminum	70-275
Brass	400-700
Bronze	200-275
Iron	350-800
Tantalum	70-140
Tungsten	70-140
Aluminum oxide	110-140
Carbon	140-165
Cemented carbides	140-400
Ferrites	110-165

Table 8.3. Some Characteristics of Metal Powders Made by Various Commercial Methods.

		Particle characteristics				
Method of production	*Typical purity (est.)*	*Shape*	*Meshes available*	*Compressibility*	*Apparent density*	*Green strength*
Atomization	High 99.5+	Irregular to smooth, rounded dense particles	Coarse shot to 325 mesh	Low to high	Generally high	Generally low
Gaseous reduction of oxides	Medium 98.5 to 99.0+	Irregular, spongy	Usually 100 mesh and finer	Medium	Low to medium	High to medium
Gaseous reduction of solutions	High 99.2 to 99.8	Irregular, spongy	Usually 100 mesh and finer	Medium	Low to medium	High
Reduction with carbon	Medium 98.5 to 99.0+	Irregular, spongy	Most meshes from 8 down	Medium	Medium	Medium to high
Electrolytic	High +99.5+	Irregular, flaky to dense	All mesh sizes	High	Medium to high	Medium
Carbonyl decomposition	High 99.5+	Spherical	Usually in low micron ranges	Medium	Medium to high	Low
Grinding	Medium 99.+	Flaky and dense	All mesh sizes	Medium	Medium to high	Low

Several approaches exist for achieving these goals. In general the techniques can be categorized as: (a) discontinuous vs. continuous process, (b) pressure-high vs. low, (c) compaction velocity-high vs. low; (d) temperature-room to elevated temperatures; (e) uniaxial vs. hydrostatic pressures.

Die compaction represents the most widely used method and is considered as the 'conventional' technique. This involves rigid dies and special mechanical or hydraulic presses. Densities of upto 90% of full solid density can be achieved following the compaction cycle, the duration of which may be of the order of just a few seconds for very small parts. Mechanical presses are now available in capacities upto 600 tons force or several thousand tons force for hydraulic press; these capacities are determined by the size of the part that can be pressed to a given 'green' density.

Complex green parts are generally compacted through *withdrawal tooling*. In this case, the lower punch does not move during compacting cycle. After the upper punch has entered the die cavity, both upper punch and die plate move downwards. After the compact has been pressed, the upper punch moves up, but the die plate and lower coupler move further down until the top of the die plate is flush with the lower punch (Figure 8.31). The compact is ejected and can be moved out of the way by the loading shoe. Die plate and lower coupler then move back into the filling position and the cycle repeats.

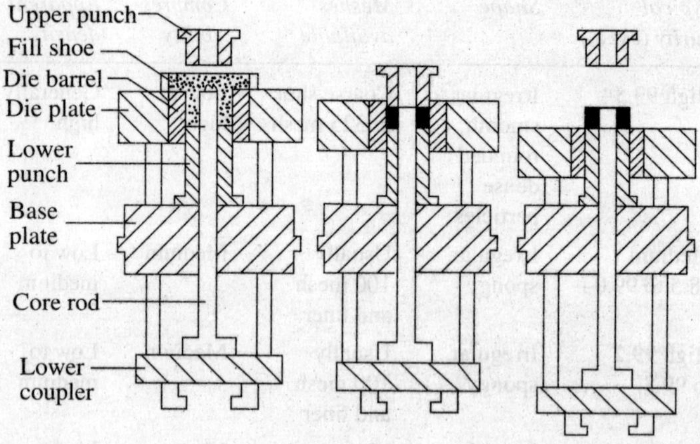

Fig. 8.31. Sequences of operations for pressing powder with the withdrawal system.

The major advantage of withdrawal system of tooling is that the lower punches are relatively short and are well supported during compaction and ejection. When there are multiple lower punches, as many of them as possible rest directly on the base plate. Withdrawal tooling can be built for very complex parts. On the other hand, in the tooling system with ejection by the lower punches the motions of the punches are built into the multiple action presses. In many cases no tool holders are required.

P/M Part Classification: P/M parts usually are classified by evaluating the complexity of part design on a range of I through IV. They are signified by:

Class I parts: Single level components with the compacting force applied from one direction only. In this case the part thickness is generally limited to a maximum of 6-7 mm.

Class II parts: Single level components with the force applied from two directions.

Class III parts: Two level components pressed with forces from two directions.

Class IV parts: Multilevel parts pressed with forces from two directions.

Figures 8.32 to 8.35 illustrate the above classes.

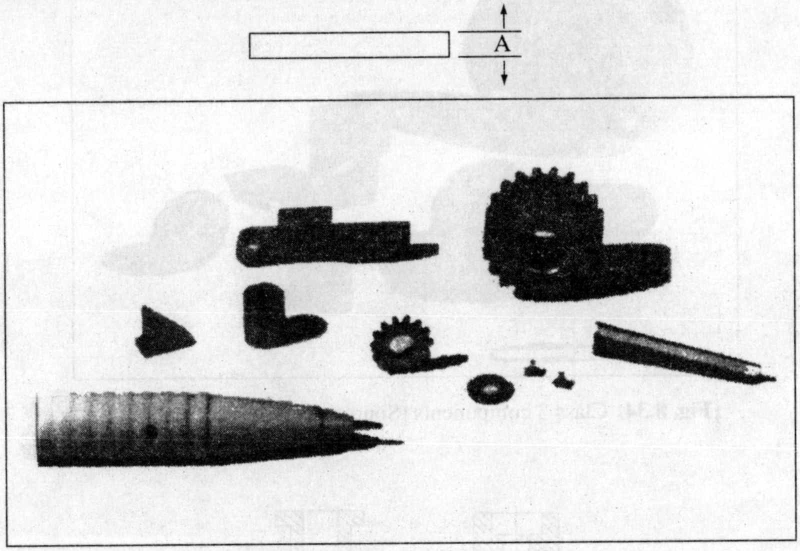

Fig. 8.32. Class 1 components (Reproduced with permission from the Powder Metallurgy Design Manual, 1995 edition, Metal Powder Industries Federation, Princeton, N.J., U.S.A., 1995).

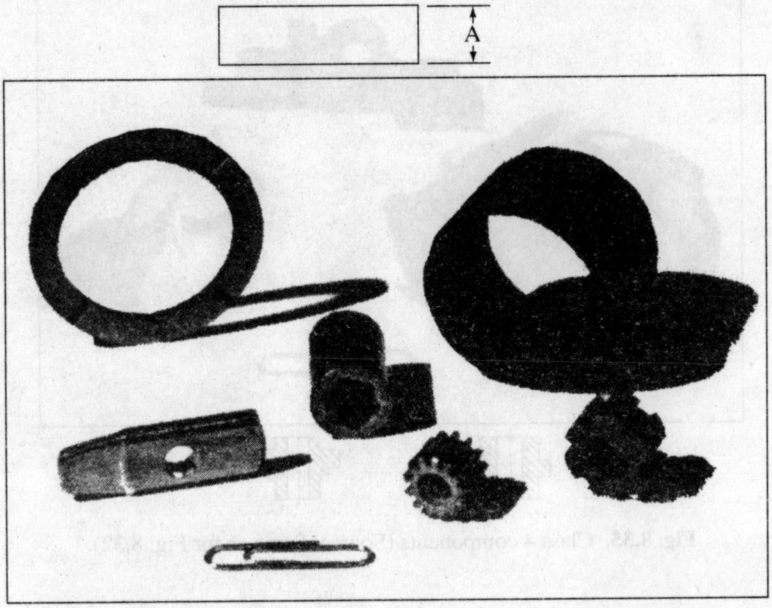

Fig. 8.33. Class 2 components (Source: Same as for Fig. 8.32).

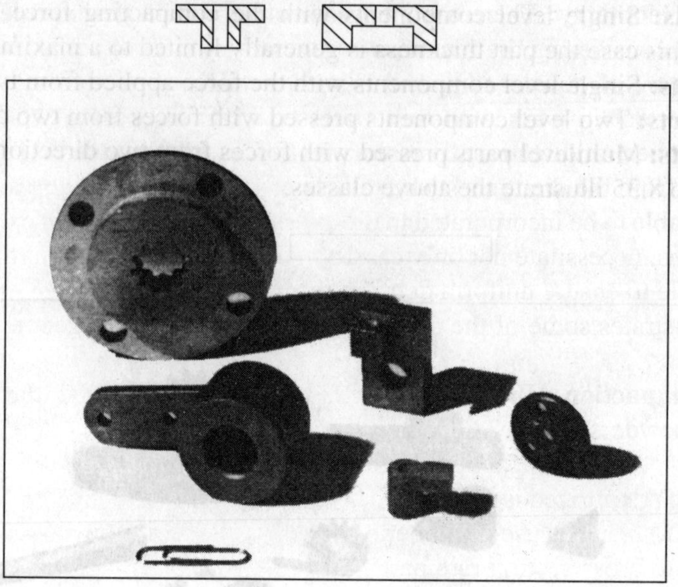

Fig. 8.34. Class 3 components (Source: Same as for Fig. 8.32).

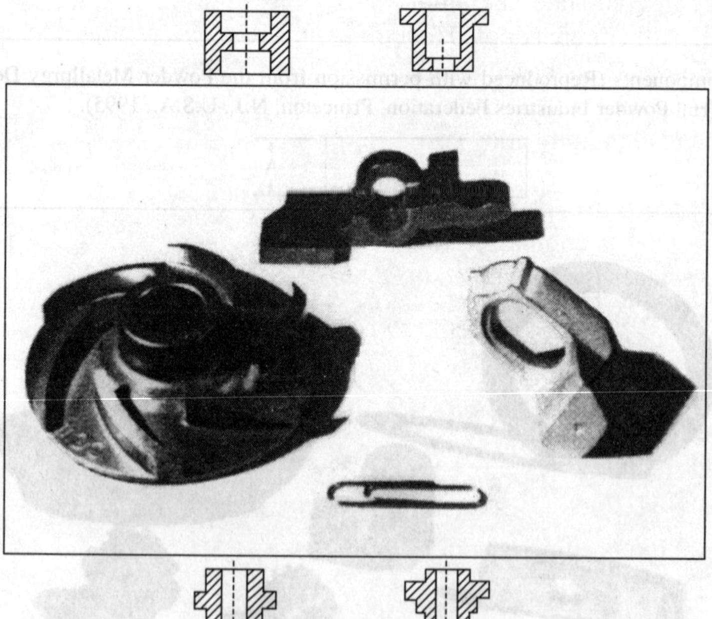

Fig. 8.35. Class 4 components (Source: Same as for Fig. 8.32).

Selection of Part Geometry: There are a number of important design rules, that should be followed when first considering the manufacturing of parts by P/M. The design must be such that the proper amount of powder is placed in the die close to its final mould location. This is particularly important in parts with thin walls and variable cross sections. Furthermore, since pressure is applied unidirectionally in die pressing, it is important that the design permits ejection of the part from the die; thus such features as undercuts, reentrant angles, cross holes and threads are not able to be incorporated in die-pressed parts. Very complex parts, with multi-height configurations, necessitate not only careful tool design, but different tool movements in the compacting cycle to ensure uniform density throughout the part.

Figure 8.36 illustrates some of the design aspects for proper part geometry in powder die compaction.

Other Cold Compaction Processes: So far we have described the die compaction of metal/ceramic powders. There are other cold compaction processes, like cold isostatic compaction and powder roll compaction. In cold isostatic compaction, a flexible mould is filled with the powder and pressurized isostatically using a liquid such as oil or water. The pressure is generally below 350 MPa. The use of rubber mould provides the possibility of complex shaped parts. Large length to cross section rate parts and high uniform green density are possible by this method.

Powder roll compaction is common for ductile metallic powders. The mechanics is similar to what has been described in 'deformation forming' section. Unlike wrought products, powder rolling is done only in cold state i.e., ambient temperature. Sintering invariably follows roll compaction. Special methods are adopted for controlling the powder feed within the compacting rolls. Speciality strip materials based on nickel and cobalt alloys for electronic and magnetic applications, porous powder rolled strip, in particular stainless steel strip for filters and nickel strip for electrodes and aluminum alloy strip for bimetal bearing applications are very common through this P/M route.

Example 8.14. A reduced iron powder has an apparent density of 3 g cm^{-3} and a tap density of 4 g cm^{-3}. The powder is compacted to a green density of 6 g cm^{-3} at 300 MPa (uniaxial compaction). If the powder is compressed to 10 mm, what will be the initial powder fill depth?

Solution. We know that

$$Apparent\ density = \frac{Mass\ of\ loose\ powder}{Volume\ of\ loose\ powder}$$

$$Green\ density = \frac{Mass\ of\ green\ compact}{Volume\ of\ green\ compact}$$

Since cross section of die opening is uniform

$$\frac{Apparent\ density}{Green\ density} = \frac{Height\ of\ green\ compact}{Height\ of\ loose\ powder}$$

$$\text{Height of loose powder} = \frac{6}{3} \times 10$$

$$= 20 \text{ mm.}$$

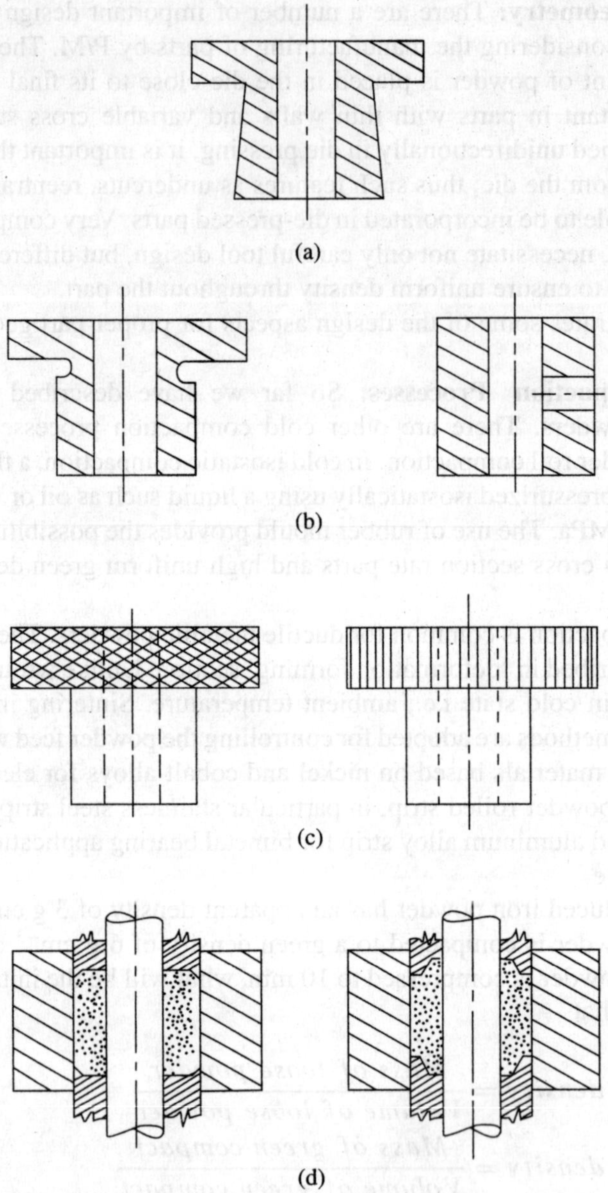

Fig. 8.36. **Design aspects for P/M components : (a) reverse taper should be avoided,(b) undercuts and holes at right angle to pressing direction should be avoided,(c) diamond knurls should be replaced by straight serrations,(d) part shapes that require feather edges and similar features should be avoided..**

The tap density value is irrelevant in solving this problem.

Example 8.15. The designs of P/M parts shown in Figure Ex. 8.15a are improper. Give the correct version, justifying your answer.

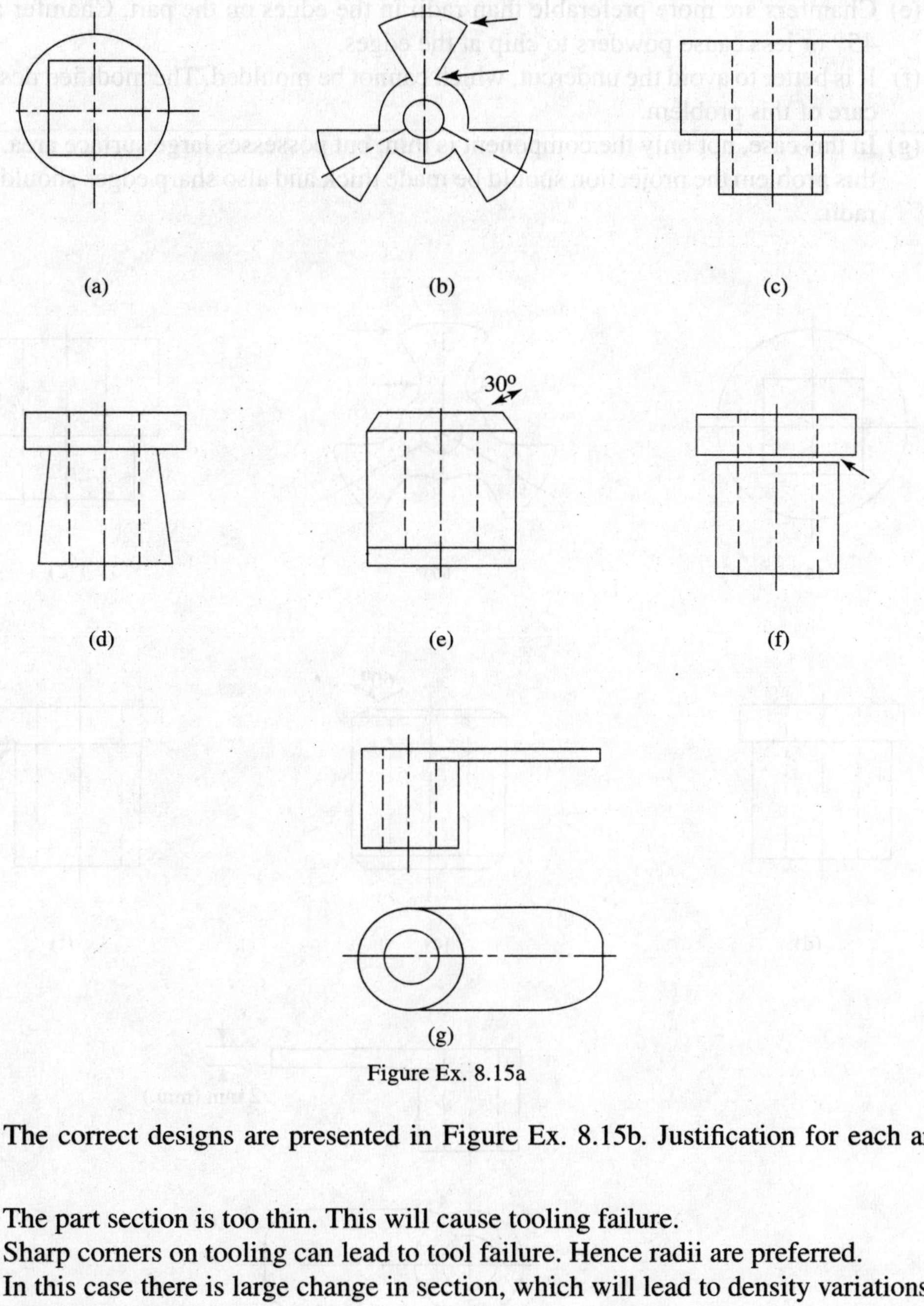

Figure Ex. 8.15a

Answer. The correct designs are presented in Figure Ex. 8.15b. Justification for each are as follows:

(a) The part section is too thin. This will cause tooling failure.
(b) Sharp corners on tooling can lead to tool failure. Hence radii are preferred.
(c) In this case there is large change in section, which will lead to density variations and hence part weakness.
(d) In this case the re-entrant features have to be redesigned, otherwise the part cannot be ejected out from the die during the pressing cycle.

(e) Chamfers are more preferable than radii in the edges on the part. Chamfer angles of 45 ° or less cause powders to chip at the edges.
(f) It is better to avoid the undercut, which cannot be moulded. The modified design takes care of this problem.
(g) In this case, not only the component is thin, but possesses large surface area. To solve this problem the projection should be made thick and also sharp edges should be given radii.

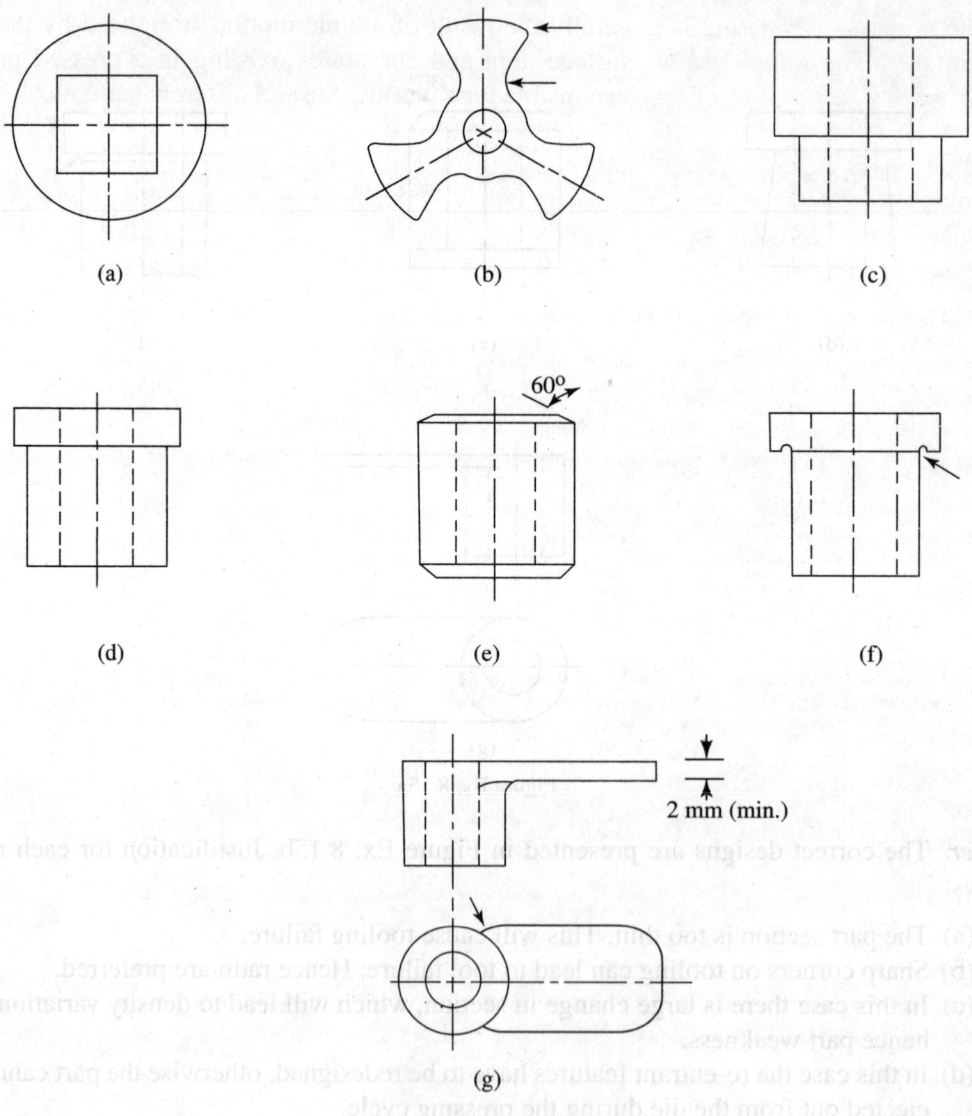

Figure Ex. 8.15b

8.23 SINTERING AND POST-SINTERING TREATMENTS

In this step, heating of the 'green' compacts is carried out in a protective atmosphere within a furnace at a temperature below the melting point of the base metal. This temperature is generally 75% of the absolute melting temperature. The process leads to a decrease in the surface area, an increase in compact strength and in many cases a shrinkage in the compact. With prolonged high temperature sintering there will be a decrease in the number of pores, the pore shape will become smooth and grain growth can be expected. It is possible that some melting or liquid phase sintering of other powder additives may take place during sintering, an example being sintering ferrous powders mixed with copper.

The process of sintering is generally the result of atomic motion stimulated by the high temperatures. The initial strains, surface area and curvatures existing in a pressed powder compact drive the atomic motions responsible for sintering. Several different patterns of atomic motion can contribute to the effect including evaporation and condensation, volume diffusion grain boundary and surface diffusion and plastic flow. In most cases the sintering kinetics are determined by several parameters including pressed density, material, particle size, sintering atmosphere, temperature and even the degree of sintering. As material transport takes place, the geometric progression can be divided into a number of stages representative of driving forces. These are:

- initial bonding among particles
- neck growth
- pore channel closure
- pore rounding
- pore shrinkage
- pore coarsening.

To enhance the overall densification process, it is common to impose an external force through pressure or to provide a highly active kinetic path through the addition of a second phase. Hot pressing techniques use superimposed temperature and pressure effects to give rapid, controlled densification. The form of hot pressing can be manifested as either uniaxial or hydrostatic forces. In recent years, hot isostatic pressure has been the most rapidly expanding commercial form of pressure enhanced sintering.

Presence of a liquid phase during all or part of the sintering cycle of material also represents an enhanced sintering method. There are two variations of the process:

(a) normal liquid phase sintering for which the formation of the liquid phase is associated with one or more components contained in the original green compact
(b) infiltration of the original green compact with a liquid formed outside the compact during the very early period of sintering. Although simultaneous infiltration and sintering appears to be dominant, infiltration of a previously sintered material is also practiced.

One of the main types of sintering furnace commonly used throughout the whole operation is that of the continuous mesh belt design (Figure 8.37). Various attempts have been made to

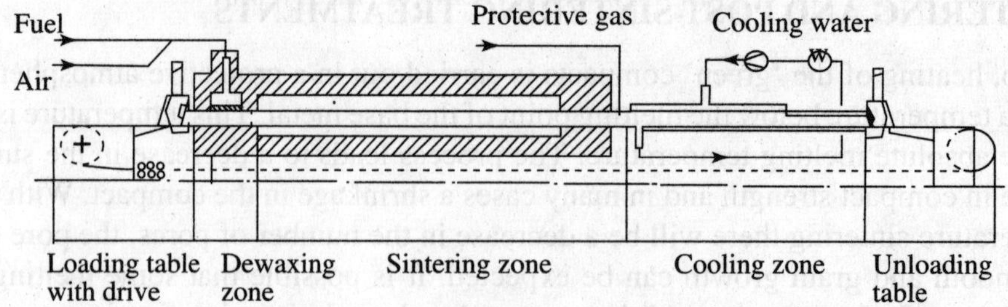

Fig. 8.37. Schematic of a conveyor belt continuous sintering furnace.

improve the throughput of these furnaces, partly by more advanced and heavier loading systems and also by having a shorter and more efficient burn-off stage.

Sintered compacts in most cases can be subjected to treatments analogous to wrought metals. The compact can be adjusted for surface finish or final dimensions using coining or sizing operations. Another operation may be coating. The pore structure must be sealed before coating in order to prevent intrusions of the fluid. Many times heat treatment is desirable to improve the mechanical properties. The presence of pores requires a protective atmosphere. It is generally effective with pre-alloyed powders. Common heat treatments include normalizing, aging, carburizing and carbonitriding. These treatments are most effective, when the sintered density exceeds 90% of the theoretical one. Impregnation treatments are used with self lubricating bearings to provide lubrication during services.

Other post sintering treatments for powder metallurgy compacts include machining, welding, brazing and surface finishing. These operations are generally dependent on sintered density.

Stages of Liquid Phase Sintering: During liquid phase sintering three stages 'rearrangement' or 'liquid flow', 'accommodation' or 'dissolution and reprecipitation' and 'coalescence' or 'solid phase bonding' take place. These stages follow in the approximate order of their occurrence, but there may be significant overlapping for any specific system (Figure 8.38).

In the first stage the melt must wet the solid substrate, which is accompanied with rapid initial densification due to the capillary force excited by the wetting liquid on the solid particles. The porosity is drastically reduced, as the system minimizes its surface energy. During, subsequent particle rearrangement, the compacts respond as a viscous solid to the capillary action. This process depends on the amount of liquid, particle size and solubility of the solid in the melt. Usually finer particles give better rearrangement. In the second stage of liquid phase sintering the solubility and diffusivity effects become dominant. In this stage microstructural change i.e., coarsening occurs. This is also known as *'Ostwald ripening'*. Solution-reprecipitation contributes to grain coarsening and to densification, but in latter the degree is not that significant as in the first stage, i.e., rearrangement. One of the best examples for this stage is liquid phase sintering of W-Ni-Fe heavy alloys, where the tungsten grains of many times larger than the original particle size are observed. In case the second stage is not effective, the melt will penetrate along the interparticle interfaces and cause particle separation. This will contribute to swelling. A small dihedral angle (for details refer Chapter 2), inhibits coalescence of

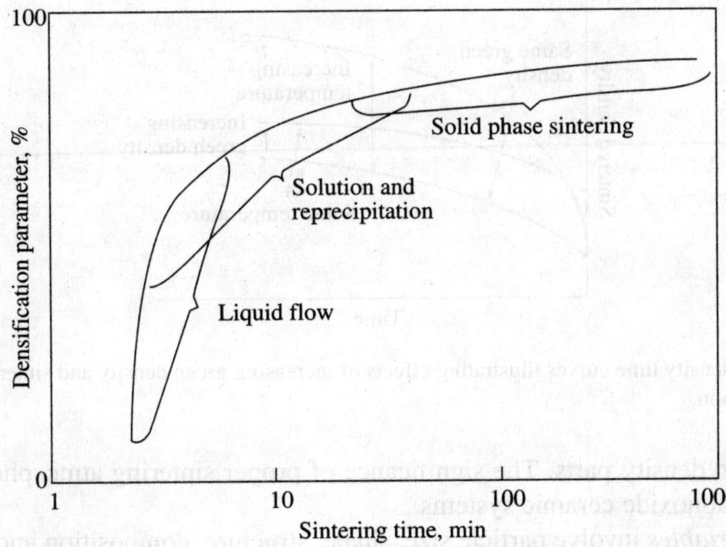

Fig. 8.38. A schematic plot of the densification of a powder compact during liquid phase sintering, showing the three stages.

neighbouring particles. The last stage is referred to as solid state controlled sintering. Densification is slow in this case because of the existence of solid skeleton. This is observed in case of *incipient liquid phase sintering* systems of Cu-Sn premix bronze, where elemental tin dissolves into the copper matrix during sintering (temperature > M.P. of tin), leaving the solid bronze skeleton at the end. On the other hand, there are systems e.g., heavy alloy (W-Ni-Fe) or cemented tungsten carbides (W-Co), where *persistent liquid phase sintering* occurs. In this case, the last portion of the melt, solidifies purely as described in the section on 'solidification processing'. All solidification defects e.g., porosity, blowholes etc. are also present in this type of liquid phase sintering.

A special variant of liquid phase sintering is *super-solidus liquid-phase sintering*, where the starting powder is invariably pre-alloyed. The green part is heated to a temperature somewhere above the solidus temperature, where the individual particles become mushy and densify by the first stage of liquid phase sintering. The process, is useful for systems having large melting ranges and makes high densification possible even with large particle sizes. One of the typical applications of this process is in sintering of high speed steels and superalloys.

Variables during Sintering: Various factors affecting sintering may be classified under following categories:

- Process variables
- Material variables
- Factors affecting dimensional change of part
- Factors affecting microstructural changes.

Process variables involve sintering temperature, time and atmosphere. Increase in sintering temperature/time promotes sintering. Figure 8.39 illustrates the effect of increasing temperature

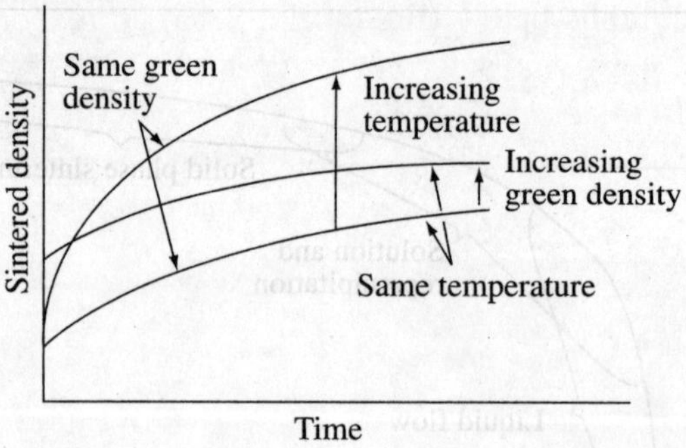

Fig. 8.39. Sintered density time curves illustrating effects of increasing green density and sintering temperature on densification.

on constant green density parts. The significance of proper sintering atmosphere is important for metallic and nonoxide ceramic systems.

Material variables involve particle size, shape, structure, composition and green density.

Dimensional changes resulting from sintering may be associated with entrapped gases, chemical reaction during sintering, alloying and inhomogeneity in green density of the P/M part.

Microstructural changes may be associated with the state of the powder, phase transformation in the material during sintering, pore rounding during liquid phase sintering etc. Any exaggerated grain growth must be avoided by proper microstructural control of the material.

Sintering Atmospheres: Sintering atmosphere can be categorized in various categories, like oxidizing, reducing and neutral. Hydrogen, reformed hydrocarbon gases, cracked ammonia and nitrogen based atmospheres belong to reducing type. Argon and helium are inert in nature, while vacuum is neither oxidizing nor reducing type. Dew point of the gas is an important index to indicate the non-oxidizing capability of the gas or gas mixtures. In sintering of steel parts, carbon potential control of the furnace is important, which shall prevent the steel from either carburizing or decarburizing. Either dew points or carbon dioxide analysis of the atmosphere can be used as a measure of its carbon potential.

Reformed hydrocarbon gases are classified into categories - *exothermic* and *endothermic*. In case of former, the fuel gas and air are mixed in such a ratio, that incomplete combustion takes place. The gas is named exo, since the heat generated in the combustion chamber is sufficient to support the reaction. Endothermic gas is produced at a lower air : gas ratio, and heat must be supplied to support the reaction. It is richer than exogas in CO, H_2 and CH_4. This gas is not only strongly reducing, but is also carburizing. Nitrogen based atmospheres are produced by careful mixing of predominantly nitrogen with hydrogen, and for the sintering of steel, a hydrocarbon gas in predetermined proportions. These, though possibly more expensive than exo or endogas, have the advantage of cleanliness, more reliable adherence to the specified composition, and inherently low water vapour content.

Table 8.4 shows the compositions of some of the principal furnace atmospheres according to the classification of the American Gas Association.

Example 8.16. Mo-Re products so useful for extreme harsh service environments, are fabricated via P/M route:

(i) The photomicrographs below (Figure Ex. 8.16) correspond to sintered pellets after 1650°C for 110 hours sintering in hydrogen of Mo-41 wt % Re alloy prepared from (a) blended and (b) composites powders. Indicate the type of original powders. In case the sintering time is lower say 50 hours, what difference in microstructure would be there?

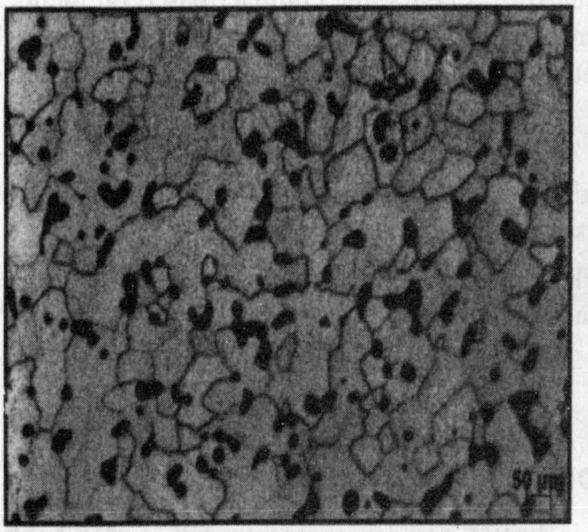

(a)

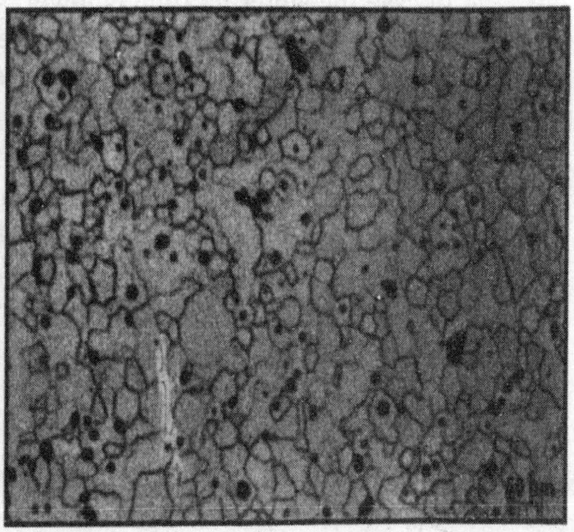

(b)

Figure Ex. 8.16

(ii) The Table below gives the mechanical properties and density data of Mo-41 wt % Re alloy sintered at 1650 °C for 61 hours. The initial powders were in blend form and composite powder respectively. Further the alloy was cold worked in sheet form. Indicate the type of material/product. (Int J. of P/M, Sept. 2003, p. 9; Reproduced with permission from MPIF, Princeton, USA).

Condition	UTS, MPa	0.2% Y.S, MPa	% El.	Density, % Theo.	Hardness, HRA
a	765	668	13	93.2	58
b	948	754	19	100	65
c	910	834	14	96.3	63

(iii) Suggest typical applications of this alloy.

Table 8.4. Composition of Common Sintering Atmospheres.

Constituent	*Endothermic*	*Exothermic*	*Dissociated Ammonia*	*Nitrogen based*
% Nitrogen	39	70-98	25	75 to 97
% hydrogen	39	2-20	75	20-2
% carbon monoxide	21	2-10	-	-
% carbon dioxide	0.2	1-6	-	-
ppm oxygen	10-150	10-150	10-35	5
dew point, °C	-16 to -10	-25 to -45	-30 to -50	-50 to -75

Answer.

(i) The microstructure (a) exhibits interconnected and triple point porosity and pertains to blended powder compact. The microstructure (b) pertains to composite powder, since the majority of pores are within the grains.

In case the sintering time is lower, say for 50 hours, the compact based on blended powder would have large size interconnected porosity and grains might be still distinct. The composite powder based compact would show limited inter-connected porosity.

(ii) (a) is based on blend powder, while (b) must correspond to cold worked condition, as in this case there is no porosity and the part is fully dense. Naturally, the condition (c) would be related for the compacts based on composite powder.

(iii) Typical applications may be heat sinks, heating elements, thermocouple sheathings, vacuum furnace components and in aerospace.

8.24 HOT CONSOLIDATION OF POWDERS

Hot consolidation is very common for ceramic brittle powder compacts, but is also useful to get fully dense components based on metal powders. This encompasses many diverse operations, including uniaxial hot pressing and pressure sintering, hot isostatic pressing, hot extrusion and hot forging. The basic principles of some of the processes are similar to those highlighted in earlier sections for bulk solid deformation forming. In hot isostatic process (HIP) the pressure medium used is an inert gas. Heat is applied by an internal furnace in the pressure vessel. In most cases, the powder or work piece is encapsulated in a gas tight material, which can withstand the pressure and temperature and, which, does not react with the powder. Typical temperatures are 1000-1750°C and typical pressures 100-320 MPa. HIP has many applications, but in the context of powder metallurgy, it has two major applications, namely the densification of cemented carbides and consolidation of superalloys. A variation of HIP is containerless HIP'ing, where a higher throughput of as much as 50-90% can be achieved as compared to containerized parts. The process is known as *Sinter HIP* or *Overpressure sintering*.

Powder forging can be broadly classified into two classes - (i) conventional powder forging and (ii) perform powder forging. The conventional forging, as applied to original compacting of loose powder consists of 'canning' the powder in some type of metal container. Once the powder is within such an enclosure, it can be evacuated, and eventually the assembly is treated in a conventional forging press. After forging, the can material may be removed by chemical or mechanical means. The resulting compact can have a very high density and may require no sintering if the powder was heated prior to the forging step. Powder perform forging, on the other hand, is a combination of powder metallurgy and forging. Powders used for perform forging are essentially those used for conventional P/M process. There are two variations in perform forging:

(a) Compression of a preform very similar in cross section to the final part. This is known as *hot repressing* (Figure 8.40a).

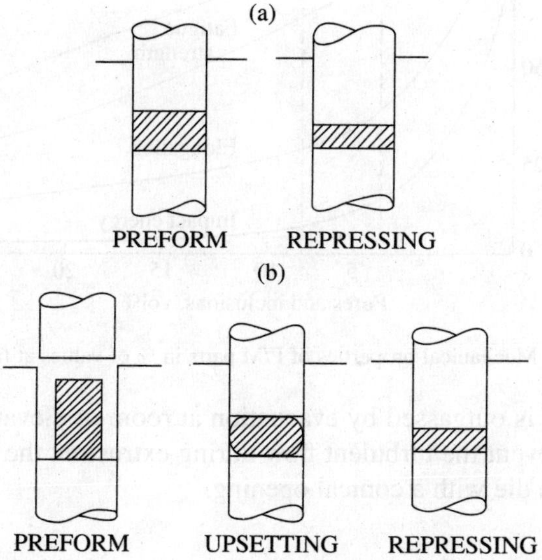

Fig. 8.40. (a) Schematic diagram of hot repressing; (b) Forming in a confined die showing upsetting and repressing modes.

(b) Compression of a perform of relatively simple shape into a part having a much different shape (Figure 8.40b). This is similar to true forging. In the first case, densification occurs with little or no lateral flow, while in the second case, densification and shape change are achieved simultaneously through a large degree of plastic deformation and lateral flow.

Figure 8.41 illustrates the effect of porosity (i.e., density) on the mechanical properties of conventional P/M and powder forged materials. Any residual porosity results in marked decrease in tensile and fatigue properties as well as impact strength. In general, higher forging temperature and less perform chilling enhance the densification of the forged part and consequently increase mechanical properties. Dynamic mechanical properties such as fatigue and impact strength get improved with increasing lateral flow in the perform during forging, because of the extensive particle shearing caused by lateral flow.

In P/M hot extrusion, three basic variations are available (Figure 8.42). In the first method loose powder is placed into a heated extrusion container and extruded directly through the die. In the second method, the powder is cold compacted and then hot extruded. In the third method, the metal powders are placed into a metallic capsule or 'can', heated and extruded with the can.

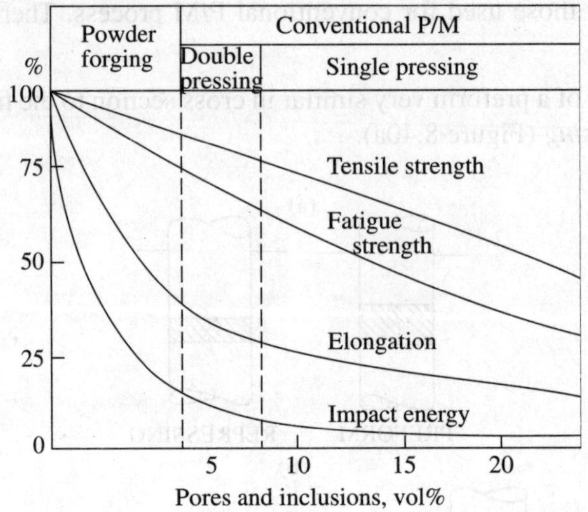

Fig. 8.41. Mechanical properties of P/M parts in % of values at full density.

Prior to heating, the can is outgassed by evacuation at room or elevated temperature and then can be sealed off. To prevent the turbulent flow during extrusion, the end of the can is conical and fits into an extrusion die with a conical opening.

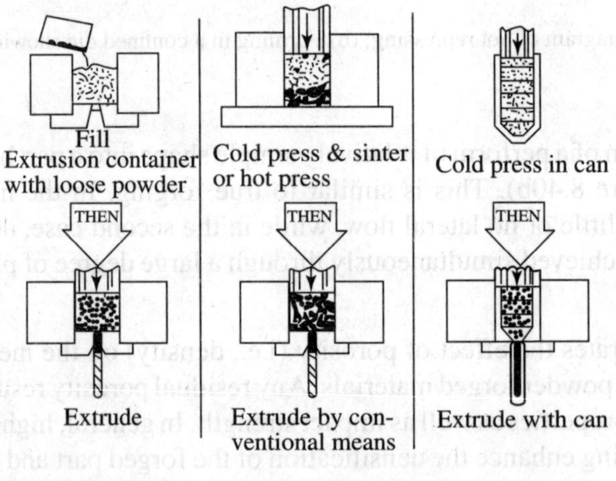

Fig. 8.42. Three variants of hot extrusion of metal powders.

8.25 SPRAY FORMING

Spray forming is not strictly a powder metallurgical process. It involves the atomization of molten metal, but instead of being allowed to solidify as powder, the spray is collected on a substrate to form billets for subsequent rolling or forging. The success of spray forming lies in the fact that each deposited atomized particle from the melt should splat at its point of impact on the substrate surface and stay exactly in its original position. The liquid particles flatten on the substrate and then cumulatively build up a substantial layer. The splats cool extremely rapidly at speeds upto 10^6 °C s^{-1}, because of the thinness of the splat and the excellent contact with the cool substrate. Typical splats are 10-20 μm thick and 500-1000 μm in length. Surface preparation of the substrate is very crucial; it should not be very rough or very smooth. In the former, mechanical interlocking with substrate may result, while in the latter porous non-adherent coatings result. Although all commercially used metals can form undesirable oxides, the greatest difficulty arises when the melt contains metals like aluminum, chromium or titanium. The oxides prevent the welding of one splat with another during spray deposition. The design of mould cavity is done in such a way that, the perform can be readily separated from it, prior to hot rolling/forging operation.

The spray forming process is used in producing a wide variety of perform shapes and sizes. Typical perform shapes are tubes, rings, cylinders, discs or simple billet. Typical spray deposition rates range from 10 to 50 kg/min. Size is dictated by economics, with the melt facility, atomizer and inert gas chamber size for a particular product line.

8.26 POWDER INJECTION MOULDING

Powder injection moulding (PIM) is similar to polymer injection moulding, with the difference that, the polymer is filled with dispersed metallic or ceramic powders. The technology permits production of stronger, more uniform and more complex P/M parts. The process is, however, more costly than conventional P/M techniques. The process steps involved in PIM are (Figure 8.43):

- Selection and production of metal/ceramic powder
- Mixing
- Moulding
- Debinding
- Sintering.

Typical powder sizes used for PIM are in the range 0.5-20 μm, in contrast to the relatively coarse free-flowing powders having particle sizes greater than 40μm used in conventional P/M. Since spherical particles are preferred, gas atomized or carbonyl (Fe or Ni) powders are selected. The thermoplastic binder system used must be stable and repeatable. Because the green part undergoes considerable shrinkage, the volume ratio of binder to powder must be held within a tight tolerance. Mixing is performed at an intermediate temperature, where shearing is predominant. The feedstock mass is then extruded and cut into pellets for ease in feeding into the moulding machine. Moulding process consists of heating the feedstock pellets to a sufficiently high temperature, so that they are melted, then forcing this melt into a cavity where, it cools and assumes the compact shape. The moulding parameters are highly dependent on particle characteristics, binder formulation, feedstock viscosity, tool design and machine operating condition. The probability of defects and the presence of distortion in sintering is highly dependent on the moulding step. Among all the steps, the debinding is very crucial,

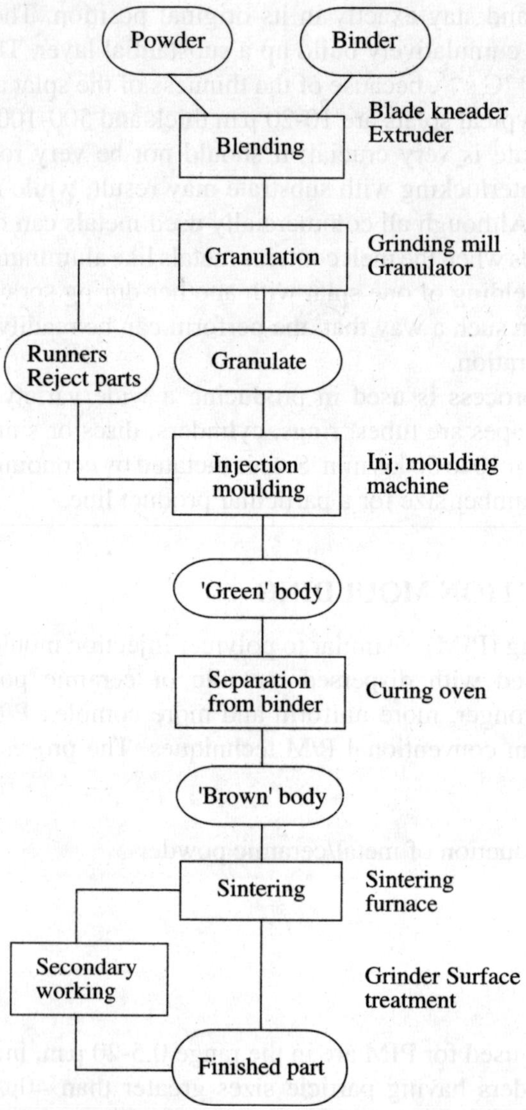

Fig. 8.43. Flow chart of powder injection moulding of metal/ceramic powders. (Reproduced with permission from Introduction to Powder Metallurgy, G. Dowson, European Powder Metallurgy Association, Shrewsbury, U.K.).

which is done thermally. Failure to remove the binder before sintering results in compact cracking. Thermal debinding is carried out in either vacuum or an atmosphere of air, hydrogen, nitrogen-hydrogen, hydrogen-argon or nitrogen-hydrogen-water. Debinding in air is cheap and efficient but will oxidise a metallic part and create subsequent problems in sintering. The final operation in PIM i.e., sintering is similar to conventional P/M sintering. For removal of residual porosity, recourse of HIP or infiltration may be taken.

The main field of application of PIM products are in fire arms, business machines and printers, hand tools, aircraft, automotive, ordnance, medical and dental, cameras and controls. The process is finding acceptance as a part by part basis, and is expected to penetrate almost all market segments in competition with investment castings.

SOLID FREEFORM FABRICATION

This method of shaping comes within the ambit of rapid prototyping, which is based on computer-aided design (CAD). Rapid prototyping allows the designer to complete a design iteration loop significantly faster than, was traditionally possible by reducing the time and cost required to produce a prototype. The manufacturing processes fall into four categories:

- substractive processes
- forming processes
- additive processes
- hybrid processes.

Most traditional manufacturing processes fall into one of the first two categories, while the most rapid prototyping processes fall into one of the last two categories. Figure 8.44 illustrates these categories. For additive processes, parts are produced by adding material to create a part. Some techniques are considered hybrid, as they require an initial substractive process to cut each layer from a sheet and then additively bond the layers together to create the part. Many people refer these processes as three-dimensional printing or desk top manufacturing. The original rapid prototyping process; known as *stereolithography*, used a liquid vat of photopolymer resin. Where the liquid polymer is exposed to a laser light, it solidifies. Modifications of this use a bed of metal or ceramic powder coated with polymer binder. Other rapid prototyping processes form the solid by extruding polymer from a small orifice or build up the solid by depositing droplets of polymer binder, like from an inkjet printer head; into a bed of powder.

Selective laser sintering is another process of free forming, which is based on the sintering of powders selectively into an individual object. A thin layer of powder is first deposited into the part build cylinder. A laser beam guided by a process-control computer using instructions generated by the 3-D CAD program of the desired part, is then focused on that layer, tracing and sintering a particular cross-section into a solid mass. The cycle of deposition is repeated until the entire three-dimensional part is produced. The loose particles are then shaken off and the part is removed. It is most common to use polymer, because of the smaller, less expensive, less complicated lasers required for sintering. With ceramics and metals, it is common to laser sinter only a polymer binder followed by actual sintering of the green shape in a furnace. Part infiltration e.g., porous steel by molten copper can be done similar to any powder metallurgy product. As an example the dies used in electro-discharge machining are made by such a process.

Since the process described above is rather slow, the solid base curing process has been devised, in which the entire slices of a part are manufactured at one time. A large throughput is possible by this method.

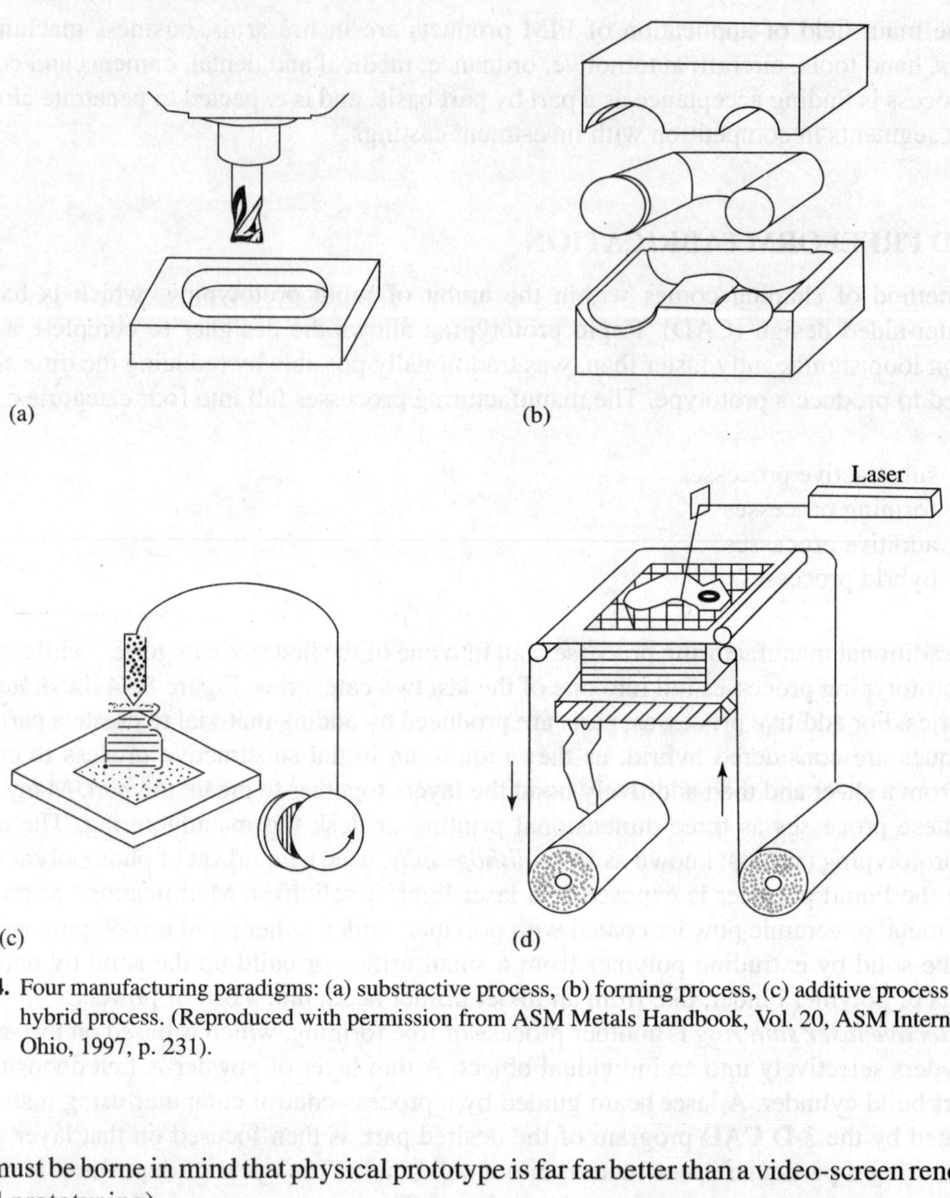

Fig. 8.44. Four manufacturing paradigms: (a) substractive process, (b) forming process, (c) additive process and (d) hybrid process. (Reproduced with permission from ASM Metals Handbook, Vol. 20, ASM International, Ohio, 1997, p. 231).

It must be borne in mind that physical prototype is far far better than a video-screen rendering (virtual prototyping).

The technology described above is already beginning to displace conventional manufacturing in some areas, like specialized parts for jets. At 'Demand Manufacturing', a Boeing subsidiary, 60 complex tubes for the environmental systems inside fight jets are printed out by such method. Such type of processing works best for relatively low volume fabrication of high cost items. Siemens Hearing Instrument is manufacturing custom-filled hearing aid shells. The advantage of printing parts is to compress manufacturing schedules by 50%.

QUESTIONS AND PROBLEMS

Solidification Processing

8.1 Compare the modes of solidification in sand moulds of (a) a pure metal and (b) an alloy of wide freezing range. Indicate the structures you would expect to find in such cases, and comment briefly on those aspects of the solidification process and resultant structures that are of special importance in materials engineering.

8.2 What kind of cast microstructures are possible during constitutional undercooling? Which of the following alloys will typically not show constitutional supercooling: Cu-12.6 Pb, Ni-12.6 Cu and Al-12.6 Si? Justify your answer.

8.3 Write a brief account of the possible causes of blowholes in cast metals and how they may be obviated.

8.4 Give an account of the variables which determine the incidence of shrinkage defects in castings. How may these defects be minimized?

8.5 What are the causes and effects of gas evolution associated with the solidification of metals? Give examples where removal of gases plays an important part in the production of a sound casting, and discuss the means adopted for such removal.

8.6 How may primary silicon be avoided in an aluminium-12 w/o silicon alloy? What is this method called?

8.7 (i) How can poor ductility in casting be explained? (ii) Rank the casting processes with respect to surface qualities of the castings. (iii) Which casting processes are best able to reproduce mould details and thin section? (iv) Which casting process is poorest with respect to dimensional control?

8.8 Compare the strength, ductility, wear resistance, creep strength and fatigue strength of zinc-base die castings with that of aluminium.

8.9 Make a guess at how you think the following variables would affect the dendrite zone length, and then explain the reason for your guess:
(a) an increase of alloy composition.
(b) a decrease of thermal conductivity of the mold metal.
(c) an increase of the latent heat of the alloy.
(d) an increase of convective mixing within the liquid ahead of the interface.

8.10 6XXX series Al-alloys used for extrusion are invariably cast by Direct Chill billet casting method (DC). Describe the controls at the following stages: (a) Furnace charge, (b) Holding furnace, (c) In-line metal treatment during laundering, (d) Casting and (e) Homogenization.

8.11 The following three products (Figure P.8.11) obtained from pressure die casting are designed improperly. Give the correct design justifying your answer.

8.12 Draw a graph showing relationship between relative cost per unit component and production volume pertaining to following casting processes: (a) Sand casting, (b) Permanent mould casting, (c) Low pressure casting, and (d) Die casting (Refer Section 12.8 of Chapter 12).

8.13 What are various liquid state processes for producing MMC? What are the limitations?

8.14 Assuming fully efficient operation, how many passes by an r-f induction coil are required to refine a rod of 90 Si-10 Al materials to "8-nines" purity, that is, 99.999999% Si at the starting end of the rod? (Hint: Near the melting temperature of silicon, the ratio C_S/C_L, of aluminium contents in the solid and in the liquid is ~0.001.)

8.15 Silicon is extensively used in solar cells. The growing of silicon crystal ingot and its further slicing into wafers is a costly affair. An alternative exists in forming a silicon sheet or ribbon directly from the melt. Describe the possible steps in such processing and the precautions required.

8.16 Give reason why the annealing temperature of any glass is defined to give it a viscosity of 10^{13} poise.

8.17 The molecules within a polymer may be spatially oriented or not, depending upon the fabrication process. Suppose a polymer is either stretched, or drawn through a die, above its glass transition temperature.
(a) Compare the molecular configuration with that of the same polymer solidified from the melt in a container.
(b) Why might stretching or drawing promote crystallization? Would this occur for all long-chain polymers? Explain.

8.18 During injection moulding in which case the shrinkage shall be felt more (a) crystalline polymer, (b) amorphous polymer?

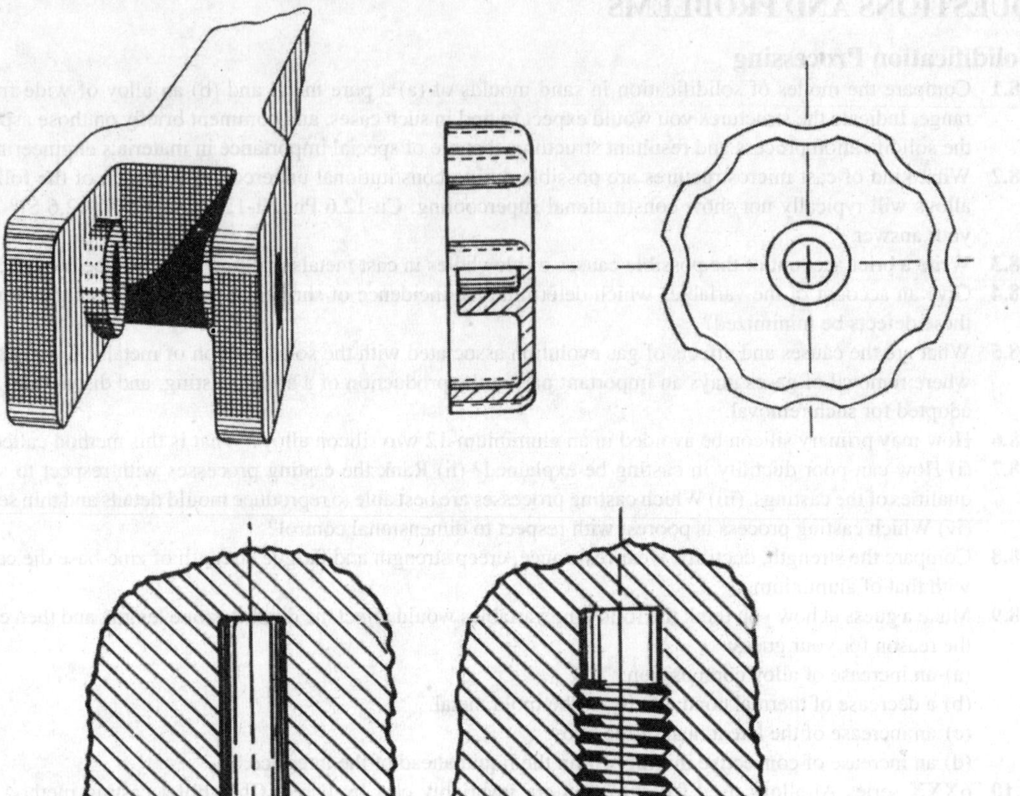

Figure P. 8.11

8.19 Identify the length (short or long) in the compression zone during extrusion of the following polymer:
(a) nylon
(b) polyvinyl chloride.
Give reasons.

8.20 Why processing of rubber must be done at lower temperatures than for thermoplastic polymers?

8.21 Can a thermoset be melted so that it can be reshaped?

Deformation Processing

8.22 Why are the different forms of passes used on grooved rolls? Describe with the aid of careful sketches and with the necessary relevant detail, the types used. What passes and reduction would you use to roll a 100 mm by 100 mm billet to a 20 mm round bar?

8.23 Explain why in case of primary rolling, the roll diameters are large and their surface rough or ragged.

8.24 What are the advantages and limitations of extrusion over other shaping processes such as rolling and drawing? Discuss the characteristics of extruded material and trace them back to conditions in the extrusion process.

8.25 Give a brief description of the method employed in extruding (i) lead, (ii) brass. What is the usual composition of the brass used for this purpose? Comment upon the difficulties likely to accrue in these extrusion processes.

8.26 Explain how in wire drawing there is an optimum die angle to carry out the operation at a minimum load.

8.27 Calculate the required force for drawing a cup from steel having a flow stress of 400 MPa

Given: Blank dia = 100 mm,
cup dia = 50 mm,
Blank holder force = 30 kN,
thickness of sheet = 1.2 mm,
Coefficient of friction = 0.1.

What will be the height of drawn cup?

8.28 In drawing complex body panel for a passenger car, stretch forming and deep drawing may be involved simultaneously. What is the role of blank hold in both of these operations.

8.29 What is meant by preferred orientation? State an application for cold worked alloy in which preferred orientation is useful and one in which it is harmful.

8.30 Why high temperature is required during super-plastic forming? Justify why all practical super-plastic alloys are two phase in nature? Could you further throw light on the morphology of these phases?

8.31 Draw hot deformability/temperature plots in the following cases:
(a) Pure metals and single phase alloys
(b) Pure metals and single phase alloys exhibiting rapid grain growth
(c) Alloys containing elements that form insoluble compounds
(d) Alloy containing elements that form soluble compounds
(e) Alloys forming ductile second phase on heating
(f) Alloys forming low melting second phase on heating
(g) Alloys forming ductile second phase on cooling
(h) Alloys forming brittle second phase on cooling.

8.32 Mention in which category in the previous questions, the following alloys correspond:
(i) Precipitation-hardenable stainless steels
(ii) High-chromium stainless steels
(iii) All-beta titanium alloys
(iv) Cu-Ni alloy
(v) Resulphurized steel
(vi) Mo-alloy containing oxides
(vii) Iron containing sulphur
(viii) Carbon and low alloy steels.

8.33 Table below summarizes the properties and basic principle involved in developing magnetic materials. What shall be their implications during processing?

Property	Principle
(a) high B_r	many randomly oriented easy axes, aligned easy axes
(b) high H_c	hard to reverse single domains, impaired reverse domain nucleation
(c) low H_c	special low anisotropy

8.34 Although thermomechanical processing of steels enhances the strength, the method could not be adopted by the industry to a greater extent. What are the reasons?

8.35 During ausforming unstable austenite is subjected to thermomechanical treatment. Enumerate the changes in the microstructure of the martensite obtained after such a treatment.

8.36 How would you make a pipe from a thermoplastic composite?

Powder Metallurgy Forming

8.37 Why should the flowability of a powder be as good as possible for powder forming processes?

8.38 What do you understand by diffusion alloying of powders?

8.39 Atomization is a common production method for spherical metal/alloy powder. What are the features which shall favour sphericity of the particles?

8.40 A 10 mm dia. final dimension powder compact is needed with a final density of 85% which is to be formed from an 82% dense pressing. What die diameter is needed for this process?

8.41 Three powders are die pressed using a compaction pressure of 450 MPa. One of the three is -325 mesh Cu, the second is -325 mesh superalloy and the third is -100 mesh copper. Which powder gives highest green strength? Why?

8.42 A P/M part shown in Figure P.8.42 is to be made out of iron -10% copper powder premix at a compaction pressure of 520 MPa. Determine

(a) the required press tonnage to compact the part,

(b) the final weight of the part if the porosity is 10%. Assume shrinkage during sintering to be neglected. (All dimensions are in mm).

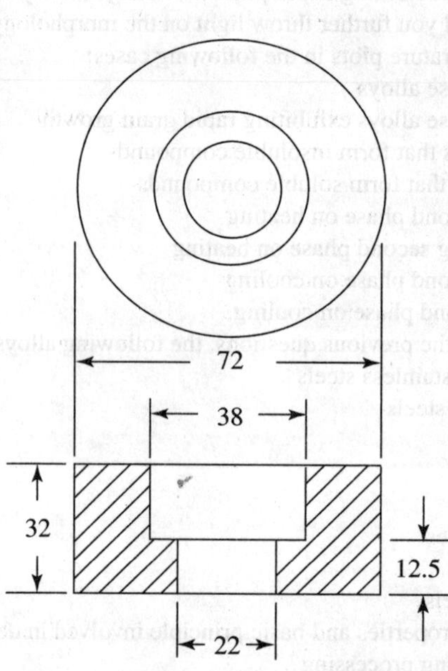

Figure P. 8.42

8.43 An iron powder compact is sintered for 1 h at 890 °C and is found to have superior mechanical properties to the same material sintered at 930 °C, with all variables remaining the same. Explain the reason.

8.44 A green compact of 68% density is to be sintered to 87% density. What is the densification parameter and estimate the linear shrinkage?

8.45 A Cu-5 Sn bronze is compacted to 90% of theoretical density. When sintered at 200 °C for 2 h the compact shrinks and densifies to 95% of its theoretical density. However, when sintered at 300 °C for 2 h, the compact had only 82% of the theoretical density. Explain this trend. Calculate the densification parameter for the compacts sintered at 300 °C.

8.46 At 1400 °C, W-1 Cu alloy undergoes liquid phase sintering, whereas W-1 Ni alloy is solid-state sintered. The densification of latter alloy was found to be greater than that of former. Why? (Hint: Refer W-Cu, W-Ni phase diagrams).

8.47 Wear resistant sintered steels are often subjected to case carburizing or nitriding. Such treatment has a disadvantage of substantial dimensional changes. For proper tolerance, a subsequent grinding operation becomes necessary. What other alternative do you have in mind to avoid the heat treatment? [Hint: Manganese surface alloying.]

8.48 In which case the vacuum sintering is applied?

8.49 The mechanical properties of HIP consolidated material have smaller sample-to-sample variation than the wrought or cast counterparts. Give reasons.

8.50 Why must a low porosity in permanent magnet processed through sintering route be provided?

8.51 A P/M Ni-base superalloy aerospace component, which was hot isostatically pressed, was found to contain following defects:

(a) ceramic inclusions
(b) metallic inclusions
(c) voids and pores
(d) prior particle boundary contamination.

Enumerate methods to minimize the above defects.

8.52 A batch size of 100000 of spark plug insulator is to be manufactured in a plant. The weight of plug is about 0.05 kg and its minimum section thickness is 1.2 mm. The mean precision has to be high ($< \pm 0.2$ mm). What possible methods are available in principle to manufacture such a part. Which one; would you finally select? Justify. (Adapted from M.F. Ashby, Selection of Materials in Design, 2nd ed., Butterworth-Heinemann, Oxford, 1999, p. 298).

8.53 Circular wheels containing abrasive grains (e.g., alumina, emery, or diamond) frequently are used to cut metals. The abrasive grains are bonded together with bronze or by a polymeric or vitreous material. Describe how such cut off wheels could be made.

8.54 Why Cu-graphite composite is mostly produced by powder metallurgy?

8.55 Sintering of Al-powder compact is strongly influenced by oxide layer. Indicate a solution for this?

8.56 What are the main sintering atmospheres for sintering of tungsten carbide based cemented carbides? In what ways the end properties differ?

8.57 During conventional drying of monolithic refractories, explosive spalling is sometimes witnessed. Could you suggest some additive to take care of this problem? Give reasons.

REFERENCES

General:

Processes and Design for Manufacturing, 2nd ed., S.D. Wakil, P.W.S. Publishing Co., Boston, 1998.

Manufacturing Processes for Engineering Materials, 4th ed., S. Kalpakjian, S.R. Schmid, Pearson Education (Singapore) Ltd., Singapore, 2003.

Manufacturing Processes and Systems, 9th ed., P.F. Ostwald and J. Munoz, John Wiley & Sons, New York, 1997.

Manufacturing with Materials, (Ed.) L. Edwards and M. Endean, Butterworth-Heinemann, Oxford, 1990.

Solidification Forming:

Foundry Technology, P.R. Bealey, John Wiley, New York, 1972.

Solidification Processing, M.C. Flemings, McGraw Hill, New York, 1974.

Principles of Metal Casting, 2nd ed., R.W. Heine, C.R. Loper Jr. and C. Rosenthal, McGraw Hill, New York, 1967.

Castings, J. Campbell, Butterworth-Heinemann, Oxford, 1991.

Casting Design, R. Johns, American Foundry Men's Society, 1987.

Semi-solid Processing. K.P. Yong, Chapman & Hall, London, 1997.

Deformation Forming:

Mechanical Metallurgy, G.E. Dieter, McGraw Hill, N.Y., 1988.

Metal Forming Handbook, Springer, Berlin, 1998.

Introduction to Manufacturing Processes, 2nd ed., J.A. Schey, McGraw Hill Book Co., New York, 1987.

Mechanical Working of Metals, J.N. Harris, Pergamon Press, Oxford, 1983.
Fundamentals of Modern Manufacturing, 2nd ed., M.P. Groover, John Wiley & Sons, New York, 2002.
Mechanics of Sheet Metal Forming, 2nd ed., Z. Marciniak, J.L. Puncan and S.J. Hu, Butterworth-Heinemann, Oxford, 2002.

Powder Metallurgy Forming:

The Production and Processes of Inorganic Materials, J.W. Evans and L.C. De Jonghe, TMS, Warrendale, 2002.
Powder Metallurgy: Principles and Applications, F.V. Lenel, Metal Powder Industries Federation, Princeton, 1980.
Powder Metallurgy Technology, G.S. Upadhyaya, Cambridge International Science Publishing, Cambridge, 1997.
Powder Metallurgy: Principles and Applications, European Powder Metallurgy Association, Schrewsbury, U.K., 1997.
Powder Metallurgy Science, 2nd ed., R.M. German, Metal Powder Industries Federation, Princeton, 1999.
Introduction to Powder Metallurgy, F.Thummler and R.Oberacker, The Institute of Materials, London,1994.
Handbook of Powder Metallurgy, 2nd ed., H.H. Hausner and M.K. Mal, Chemical Pub. Co. Inc., New York, 1982.
Powder Metallurgy Design Manual, 2nd ed., Metal Powder Industries Federation, Princeton, 1995.
Powder Injection Moulding, R.M. German, Metal Powder Industries Federation, Princeton, 1990.
Principles of Ceramic Processing 2nd ed., J.S. Reed, John Wiley & Sons, New York, 1995.
Ceramic Fabrication Technology, R.W. Rice, Marcel Dekker, New York, 2003.

9

MATERIALS REMOVAL

Material removal becomes necessary because of imparting a precise shape to an engineering part. Many authors, for conveying this, use another term 'machining'. There is another aspect of material removal, which is unintentional. This occurs because of corrosion and wear. Corrosion has been dealt in detail in Chapter 6. The current chapter would also include the topic of wear.

9.1 MACHINING

There is a general misconception that machining is solely important for mechanical products. Even in electronic and optical products, the significance of machining to different degrees of tolerances is of no less importance (Table 9.1).

The advantages of machining operations are:

- Closure dimensional accuracy.
- Processing of parts with external and internal geometric features (sharp corner, flatness etc.), which can be generated by forming processes.
- Special surface texture.

However, machining has limitations too. They are:

- Scrap material generation.
- More time consuming than other forming processes.
- Limited surface integrity of the product.

Machining is an essential processing on cast and wrought products. Many times it is not necessary for precision sintered parts, processed through powder metallurgy.

Each machining operation produces a characteristic geometry due to two factors: (a) the relative motions between the tool and the work piece and (b) the shape of the cutting tool. These operations are classified as generating and forming operations.

Table 9.1. Examples of Normal and Precision Machined Products

(Reprinted from Metal Cutting, 4th ed., E.M. Trent and P.K. Wright, Butterworth-Heinemann, Oxford, 2000, with permission from Elsevier).

	Tolerance Band	Mechanical	Electronic	Optical
Normal machining	200 μm	Normal domestic appliances, automotive fittings etc.	General purpose electrical parts, e.g., switches, motors and connectors	Camera, telescope, binocular bodies
	50μm	General purpose mechanical parts for typewriters, engines etc.	Transistors, diodes, magnetic heads for tape recorders	Camera shutters, lens holders for cameras and microscope
	5μm	Mechanical watch parts, machine tool bearings, gears, ballscrews, rotary compressor parts	Electrical relays, condensors, silicon wafers, TV color masks	Lenses, prism, optical fiber and connectors (multimode)
Precision machining	0.5μm	Ball and roller bearings, precision drawn wire, hydraulic servo- valves, aerostatic gyro bearings	Magnetic scales, CCD, quartz oscillators, magnetic memory bubbles, magnetron, IC line width, thin film pressure transducers, thermal printer heads, thin film head discs	Precision lenses, optical scales, IC exposure masks (photo, X - ray), laser polygon mirrors, X - ray mirrors, elastic deflection mirrors, monomode optical fiber and connectors
	0.05μm	Gauge blocks, diamond indentor top radius microtome cutting edge ra, ultra-precision X - Y tables	IC memories, electronic video discs, LSI	Optical flats, precision Fresnel lenses optical diffraction gratings, optical video discs
Ultra - precision machining	0.005 μm		VLSI, super-lattic thin films	Ultra-precision diffraction gratings

Notes:
CCD - charge couple device
IC - integrated circuit
LSI - large scale integration
VLSI - very large scale integration.

The machining can be subclassified as conventional machining, abrasive machining and non-conventional machining. The conventional includes operations like turning, drilling, milling and other operations e.g., shaping, planing, broaching and sawing. In abrasive machining the material removal results because of the action of hard, abrasive particles. This includes processes like grinding, honing, lapping and super-finishing. In conventional machining, we use a variety of energy forms other than a sharp cutting tool or abrasive particles to remove material. The energy forms include mechanical, electrochemical, thermal and chemical.

There are three types of major machining operations, namely turning, drilling and milling. In turning, a cutting tool with a single cutting edge is used to remove material from a rotating work piece to form a cylindrical shape (Figure 9.1a). The speed motion in turning is provided by the rotating work piece and the feed motion is achieved by the cutting tool moving slowly in a direction parallel to the axis of rotation of the work piece. Drilling is used to create a round hole. It is commonly achieved by a rotating tool that has two cutting edges. The tool is fed in a direction parallel to its axis of rotation into the work piece to form the round hole (Figure 9.1b). In milling, a rotating tool with multiple cutting edges is moved slowly relative to the material to generate a plane or straight surface. The direction of the feed motion is perpendicular to the tool's axis of rotation. The speed motion is provided by the rotating milling cutter. The most basic form of milling are peripheral milling and face milling as shown in Figure 9.1c.

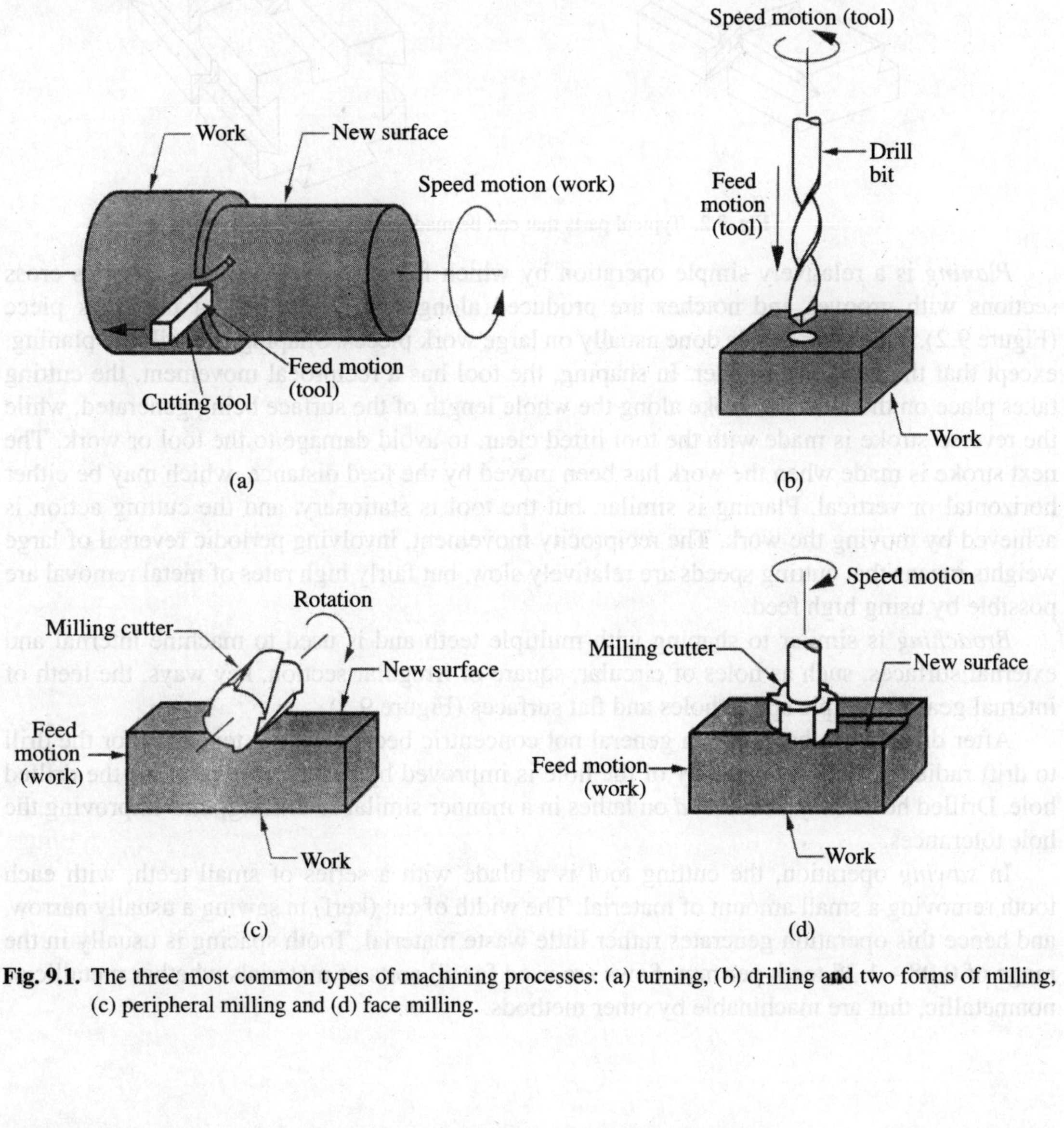

Fig. 9.1. The three most common types of machining processes: (a) turning, (b) drilling and two forms of milling, (c) peripheral milling and (d) face milling.

A number of different shapes of milling cutters are available for various applications. Single toothed cutters are possible, but typical milling cutters have a number of teeth (cutting edges) which may vary from three to over one hundred. The new surface is generated as each tooth cuts away an arc shaped segment, the thickness of which is known as feed or tooth load. Feeds are usually light, not often greater than 0.25 mm per tooth. Because of the large number of teeth, the rate of metal removal is rather high. The feed often varies through the cutting part of the cycle. Milling is also used for the production of curved shapes, while end mills are employed in the production of hollow shapes such as die cavities. The end mill can be used to machine a feature such as a rectangular pocket with vertical walls.

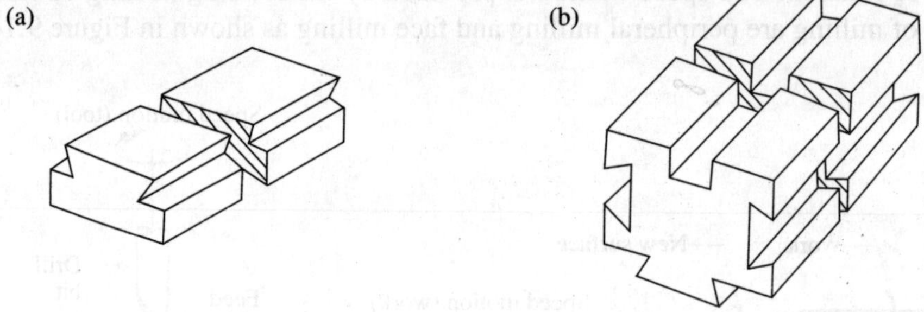

Fig. 9.2. Typical parts that can be made on a planer.

Planing is a relatively simple operation by which flat surfaces, as well as various cross sections with grooves and notches are produced along with the length of the work piece (Figure 9.2). This operation is done usually on large work pieces. Shaping is similar to planing, except that the parts are smaller. In shaping, the tool has a reciprocal movement, the cutting takes place on the forward stroke along the whole length of the surface being generated, while the reverse stroke is made with the tool lifted clear, to avoid damage to the tool or work. The next stroke is made when the work has been moved by the feed distance, which may be either horizontal or vertical. Planing is similar, but the tool is stationary, and the cutting action is achieved by moving the work. The reciprocity movement, involving periodic reversal of large weights means that cutting speeds are relatively slow, but fairly high rates of metal removal are possible by using high feed.

Broaching is similar to shaping with multiple teeth and is used to machine internal and external surfaces, such as holes of circular, square or irregular section, key ways, the teeth of internal gears, multiple spline holes and flat surfaces (Figure 9.3).

After drilling the holes are in general not concentric because of the tendency for the drill to drift radially. The concentricity of the hole is improved by subsequently *boring* the drilled hole. Drilled holes may be *reamed* on lathes in a manner similar to drilling, thus improving the hole tolerances.

In *sawing* operation, the cutting tool is a blade with a series of small teeth, with each tooth removing a small amount of material. The width of cut (kerf) in sawing a usually narrow, and hence this operation generates rather little waste material. Tooth spacing is usually in the range of 0.08 – 1.25 teeth per mm. Saws are used for all sorts of materials whether metallic or nonmetallic, that are machinable by other methods.

(a)

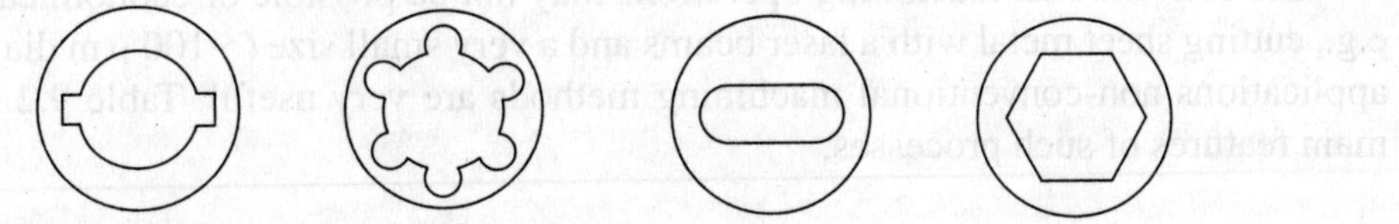

(b)

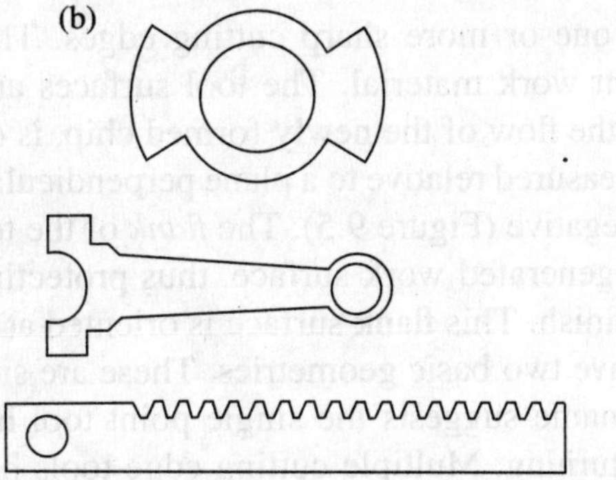

Fig. 9.3. Typical parts made by (a) internal broaching, (b) surface broaching. Heavy lines indicate broached surfaces.

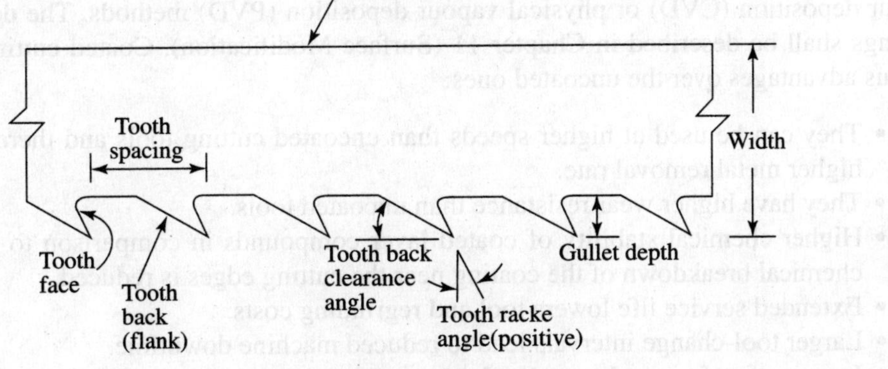

Fig. 9.4. Terminology for saw teeth.

Figure 9.4 shows the portion of the saw indicating various saw teeth terminology. The saws have straight blades, circular blades and belong to either reciprocating or continuous motion (hand saws or band saws).

Saw blades are made from carbon and high speed steels, steel blades with tips of cemented carbides, or high speed steel. Cutting speed in sawing usually ranges upto 1.5 m/s with lower speeds for high strength metals.

The conventional machining operations may not be possible or economical for some jobs, e.g., cutting sheet metal with a laser beams and a very small size (~100 μm dia.) gear. For such applications non-conventional machining methods are very useful. Table 9.2 summarizes the main features of such processes.

9.2 CONVENTIONAL METAL CUTTING

9.2.1 The Cutting Tool

A cutting tool has one or more sharp cutting edges. The cutting edge serves to separate a chip from the parent work material. The tool surfaces are rake face and the flank. The *rake face*, which directs the flow of the newly formed chip, is oriented at a certain angle, called the *rake angle* α. It is measured relative to a plane perpendicular to the work surface. The rake angle can be positive or negative (Figure 9.5). The *flank* of the tool provides a clearance between the tool and the newly generated work surface, thus protecting the surface from abrasion, which would degrade the finish. This flank surface is oriented at an angle known as *relief angle*.

Cutting tools have two basic geometries. These are single point tools and multiple cutting edge tools. As the name suggests the single point tool has one cutting edge and is used for operations such as turning. Multiple cutting edge tools have more than one cutting edge and generally achieve their motion relative to the work piece by rotating. Drilling and milling use rotating multiple cutting-edge tools.

Coated cutting tools are finding more and more applications, when a hard ceramic phase e.g., TiC, TiN, Al_2O_3, diamond is deposited in single or multiple layers preferably by chemical vapour deposition (CVD) or physical vapour deposition (PVD) methods. The details of such coatings shall be described in Chapter 11 (Surface Modification). Coated cutting tools offer various advantages over the uncoated ones:

- They can be used at higher speeds than uncoated cutting tools and thereby provide a higher metal removal rate.
- They have higher wear resistance than uncoated tools.
- Higher chemical stability of coated layer compounds in comparison to WC, and the chemical breakdown of the coating near the cutting edges is reduced.
- Extended service life lowers tool and regrinding costs.
- Larger tool-change intervals lead to reduced machine downtime.
- Improved surface quality makes further operations easier or even superfluous and reduces manufacturing.

Tool designers and machinists optimize the shapes of tools to give long tool life at high cutting speed.

There are various ways in which the cutting edge of the tool is held in position. High speed steel cutting tool is generally ground from a solid shank and imparted the proper geometry. As cemented carbide and ceramic tools are costly as well as relatively brittle, they are used as inserts, either in brazed on a shank or mechanically clamped condition to a tool holder. The shank is made of tool steel for strength and toughness. Mechanical clamping is used for cemented carbides, ceramics and other hard materials. The major advantage of the mechanically

Table 9.2. General Non-conventional Machining Processes.

Process	*Characteristics*	*Process parameters and typical material removal rate or cutting speed*
Chemical machining (CM)	Shallow removal (upto 12 mm) on large flat or curved surfaces; blanking of thin sheets; low tooling and equipment cost; suitable for low production runs.	0.0025-0.1mm/min.
Electrochemical machining (ECM)	Complex shapes with deep cavities; highest rate of material removal among nontraditional processes; expensive tooling and equipment; high power consumption; medium to high production quantity.	V: 5-25 dc; A: 1.5-8 A/mm^2 ;2.5 - 12 mm/min. depending on current density.
Electrochemical grinding (ECG)	Cutting off and sharpening hard materials, such as tungsten-carbide tools; also used as a honing process; higher removal rate than grinding	A: 1 -3 A /,mm^2; Typically 25 mm^3 /s per 1000 A.
Electrical-discharge machining (EDM)	Shaping and cutting complex parts made of hard materials; some surface damage may result; also used as a grinding and cutting process; expensive tooling and equipment.	V: 50-380; A: 0.1-500; Typically 300 mm^3/min
Wire EDM	Contour cutting of flat or curved surfaces; expensive equipment.	Varies with material and thickness.
Laser-beam machining (LBM)	Cutting and holemaking on thin materials; heat-affected zone; does not require a vacuum; expensive equipment; consumes much energy.	0.50-7.5 m/min.
Electron-beam machining EBM	Cutting and holemaking on thin materials; very small holes and slots; heat-affected zone; requires a vacuum; expensive equipment.	1-2 mm^3/min.
Water-jet machining (WJM)	Cutting all types of non-metallic materials to 25 mm and greater in thickness; suitable for contour cutting of flexible materials; no thermal damage; noisy.	Varies considerably with material.
Abrasive water-jet machining (AWJM)	Single or multilayer cutting of metallic and nonmetallic materials.	Upto 7.5 m/min.
Abrasive-jet machining (AJM)	Cutting, slotting, deburring, deflashing, etching, and cleaning of metallic and non-metallic materials; manually controlled; tends to round off sharp edges; hazardous.	Varies considerably with material.

Adapted from Manufacturing Engineering and Technology, 4th ed., S. Kalpakjian and S.R. Schmid, Pearson Education (Singapore) Pvt. Ltd., Singapore, 2001.

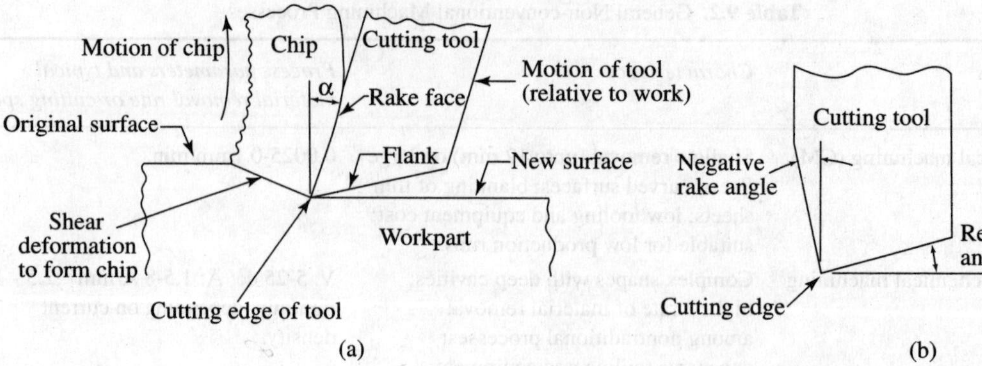

Fig. 9.5. (a) cross-sectional view of the machining process, (b) tool with negative rake angle.

clamped insert is that each insert contains multiple cutting edges. When an edge wears out, the insert is unclamped, indexed (rotated in the tool holder) to the next edge, and reclamped in the tool holder. Cutting tool inserts are widely used in machining because they are economical and adaptable in many different types of machining operations. They are available in a variety of shapes and sizes for the variety of cutting conditions. Inserts with larger point angles are inherently stronger and less likely to chip or break during cutting, but they require more power, and there is a greater probability of vibration. In most negative rake angle applications, both sides of the insert can be used. Edge preparation of tool insert is generally done to increase the strength of the cutting edge by providing more gradual transition between the clearance edge and the rake face of the tool.

9.2.2 Chip Formation during Machining

In actual chip formation, four basic types of chip may result. These are:

- **Dicontinuous chip:** When relatively brittle material e.g., cast irons are machined at low cutting speeds, such type of separate segments results (Figure 9.6a).
- **Continuous chip:** When ductile work materials are cut at high speeds and relatively small feeds, long continuous chips are formed (Figure 9.6b). Long continuous chips can cause problems with regard to chip disposal and/or tangling about the tool. To solve these problems, turning tools are often equipped with *chip breakers*.
- **Continuous chip with build-up-edge:** This occurs when ductile materials are machined at low to medium cutting speeds. Friction tends to cause portion of the work metal to adhere to the rake face of the tool near the cutting edge. This formation is called a *built-up-edge* (Figure 9.6c).
- **Serrated chips:** This is also known as shear-localized or segmented chips. These chips are semi-continuous. Such types of chip are generally associated with difficult-to-machine metals such as titanium alloys, nickel base superalloy, and austenitic stainless steels, when they are machined at higher cutting speeds. The appearance of chip is a saw-tooth appearance that is produced by a cyclic chip formation of alternating high shear strain followed by low shear strain (Figure 9.6d).

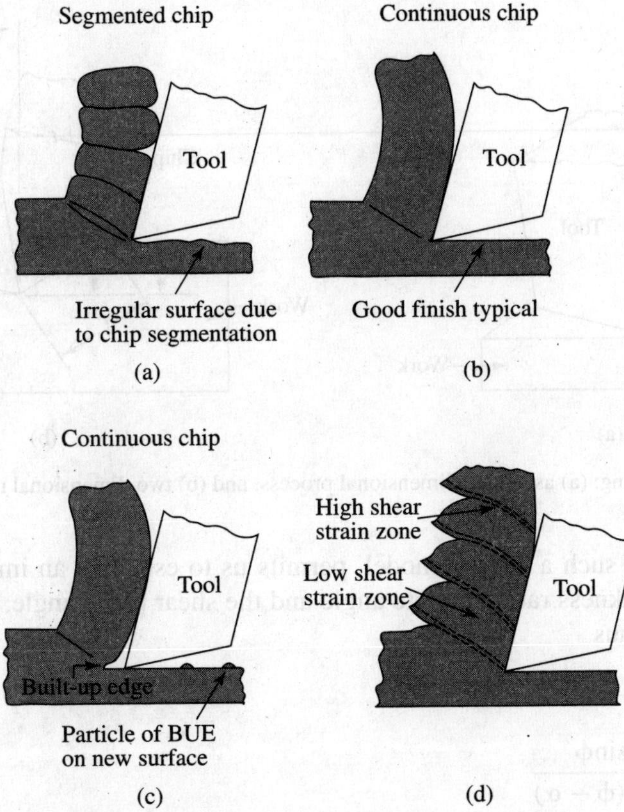

Fig. 9.6. Four types of chip formation in metal cutting (a) discontinuous, (b) continuous, (c) continuous with built up edge and (d) serrated.

The geometry of most practical machining operations is complex. A simplified model of machining is known as *orthogonal cutting model*. As a matter of fact actual machining is three dimensional, but in orthogonal model we consider only two dimensions. In this type of model, the tool is assumed wedge-shaped, in which the cutting edge is perpendicular to the direction of cutting speed. As the tool is forced into the material, the chip is formed by shear deformation along a plane called the *shear plane*, which is oriented at an angle ϕ with the work surface. Along the shear plane, where the bulk of the mechanical energy is consumed in machining, the material is plastically deformed (Figure 9.7).

During cutting, the cutting edge of the tool is positioned with a certain distance below the original work surface. This corresponds to the thickness of the chip prior to chip formation, t_o. As the chip is formed along the shear plane, its thickness is increased to t_c is called the *chip thickness ratio*, γ, i..e.

$$\gamma = \frac{t_o}{t_c} \tag{9.1}$$

This ratio is always less than unity.

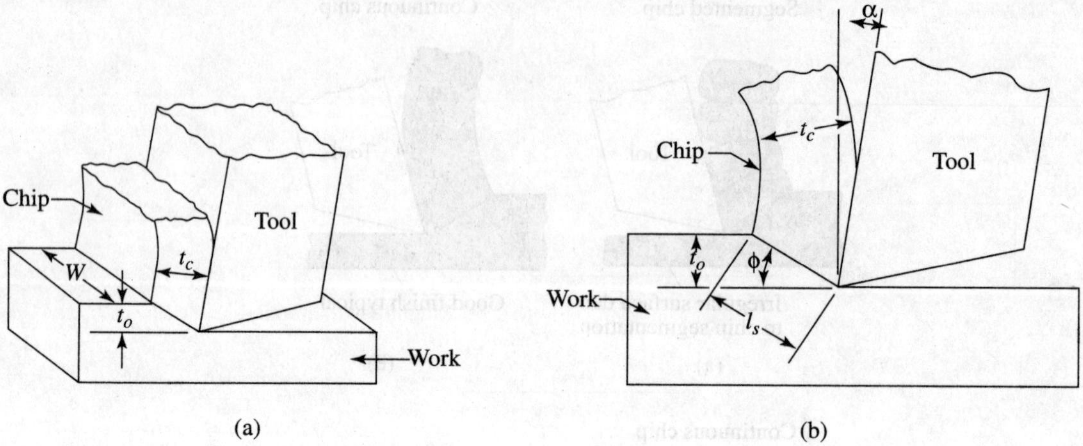

Fig. 9.7. Orthogonal cutting: (a) as a three dimensional process, and (b) two dimensional representation in the side view.

The geometry of such a cutting model, permits us to establish an important relationship between the chip thickness ratio, the rate angle and the shear plane angle. Let ℓ_s be the length of the shear plane. Thus

$$\gamma = \frac{t_o}{t_c}$$

$$= \frac{\ell_s \sin\phi}{\ell_s \cos(\phi - \alpha)}$$

$$= \frac{\sin\phi}{\cos(\phi - \alpha)}$$

Therefore, after rearranging we get

$$\tan\phi = \frac{r\cos\alpha}{1 - r\sin\alpha} \tag{9.2}$$

Shear deformation during cutting can be approximated by a series of parallel plates sliding against one another to form the chip (Figure 9.8). Each plate would experience the shear strain (γ) as shown in Figure 9.8b. From Figure 9.8c, it is evident that

$$v = \frac{AC}{BD} = \frac{AD + DC}{BD}$$

$$= \tan(\phi - \alpha) + \cot\phi \tag{9.3}$$

Let ν is the cutting speed and nu_c the chip velocity. Then chip thickness ratio r using Figure 9.8 can be equally expressed as

$$\gamma = \frac{\nu_c}{\nu} = \frac{\sin\phi}{\cos(\phi - \alpha)} \tag{9.4}$$

In reality, it may be said that in actual machining process, the shear deformation does not occur along a plane, but within a zone. In addition, there is secondary shear resulting from

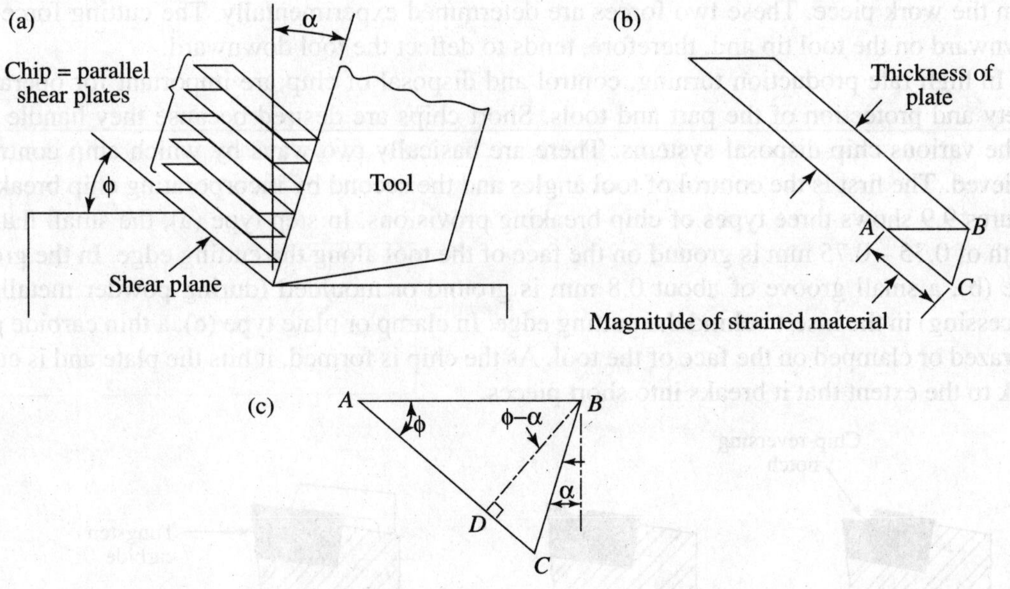

Fig. 9.8. Shear strain during chip formation: (a) series of parallel plates sliding relative to each other; (b) details on an isolated plate; (c) shear strain triangle.

friction between the chip and the tool as the chip slides along the rake face of the tool. Its effect increases with increased friction between the tool and chip. Another contributory factor is the type of material being machined and the cutting conditions of the operation.

The cutting conditions of any turning process are defined by speed (v), feed (f) and depth of cut (d). The material removal rate for the process is expressed as:

$$\text{MRR} = vfd \tag{9.5}$$

Here MRR unit is mm³/s, where v is in mm/s, f is in mm and d is mm respectively.

While machining a cylindrical rod of dia D_o, into a final diameter D_f, the MRR in mm³/min can be expressed as:

$$\text{MKR} = \pi D_{\text{av.}}\, dfN \tag{9.6}$$

where $D_{\text{av.}}$ = Average diameter = $\dfrac{D_o + D_f}{2}$ in mm

N = Rotation speed of work piece in rpm.

Similarly the cutting time for work piece of length ℓ can be calculated using the expression

$$t = \frac{\ell}{fN} \tag{9.7}$$

Here ℓ is in mm, f is in mm/rev and N is rpm.

This time is actual cutting time and does not include approach and retraction times.

The next question now arises about the cutting force. Apart from this, there are two more forces i.e., feed force in the direction of feed and radial force, which tends to push the tool away from the work piece. These two forces are determined experimentally. The cutting force acts downward on the tool tip and, therefore, tends to deflect the tool downward.

In high rate production turning, control and disposal of chip are important for operator's safety and protection of the part and tools. Short chips are desired because they handle well in the various chip disposal systems. There are basically two ways by which chip control is achieved. The first is the control of tool angles and the second by incorporating chip breakage. Figures 9.9 shows three types of chip breaking provisions. In step type (a), the small flat to a depth of 0.35 – 0.75 mm is ground on the face of the tool along the cutting edge. In the groove type (b), a small groove of about 0.8 mm is ground or moulded (during powder metallurgy processing) in the insert behind the cutting edge. In clamp or plate type (c), a thin carbide plate is brazed or clamped on the face of the tool. As the chip is formed, it hits the plate and is curled back to the extent that it breaks into short pieces.

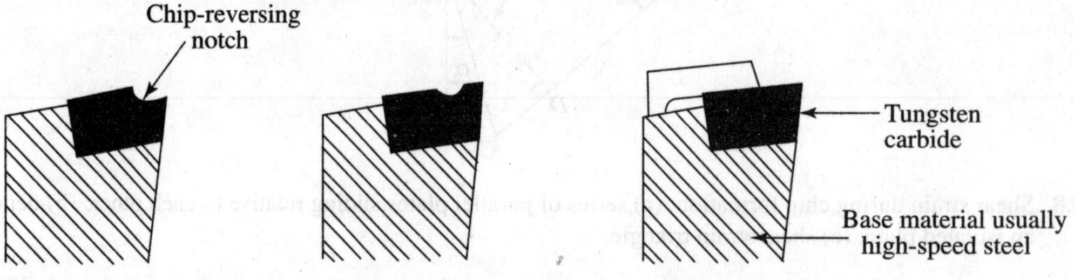

Fig. 9.9. Typical chip breakers used on single point tools.

As far as heat generation during turning operation is concerned, there are basically three zones of heat generation (Figure 9.10). As the shear angle is increased, the per cent heat generated in the shear plane A will decrease, since the plastic flow of the metal takes place over a shorter distance. The shear angle can be increased by applying a coolant and reducing the friction between the chip and the tool as well as by properly grinding the tool. Of all the cutting variables, cutting speed has the most effect on temperature.

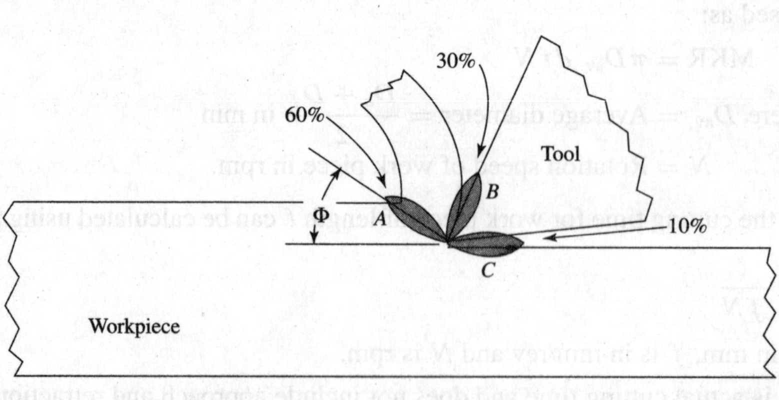

Fig. 9.10. Sources of heat in three zones: A, shear plane, B, friction plane and C, surface plane.

Example 9.1. In machining operation, approaching to orthogonal cutting, the cutting tool has a rake angle of 12 °. The chip thickness before the cut is 0.5 mm and after the cut the value is 1.125 mm. Calculate the shear plane angle and the shear strain in the cutting process.

Solution. The chip thickness ratio according to equation (9.1)

$$\gamma = \frac{0.5}{1.125} = 0.444$$

The shear plane angle is given by equation (9.2)

$$\tan\phi = \frac{0.444 \cos 12^\circ}{1 - 0.444 \sin 12^\circ}$$
$$= 0.4785$$
$$\therefore \quad \phi = 25.56^\circ$$

The shear strain can be calculated using the equation (9.3)

$$\gamma = \tan(25.56\text{–}12) + \cot 25.56$$
$$= 0.241 + 2.09$$
$$= 2.332$$

This infers that the shear strain in metal cutting is rather large.

Example 9.2. How much time shall be required to rough machine a 1.0 mm long aluminium alloy round bar with a 7.5 mm diameter if the feed is 2 mm/rev by cemented carbide and high speed steel tools? Given the maximum allowed cutting speed in case of cemented carbide as 7 m/s.

Solution. For cemented carbide the maximum allowed cutting speed is 7 m/s. Therefore in case of high speed steel tool, the value shall be less. Let us assume that for HSS, the value is 4 m/s.

For high speed steel –

$$V = N\pi D, \quad N = \frac{V}{\pi D} = \frac{4}{\pi(0.075)}$$
$$= 16.97 \text{ rev/s}$$
$$= 1018 \text{ rpm.}$$

For cemented carbide –

Using above expression

$$V = 1782 \text{ rpm}$$

Given the feed = 2 mm/rev.

Time to perform one pass in case of high speed steel is given by

$$t = \frac{L}{fN} = \frac{1.0}{(0.002)(1018)} = 0.49 \text{ min}$$
$$= 29 \text{ s}$$

Similarly, for carbide tool the value is

$$t = \frac{1.0}{0.002 \times 1782}$$
$$= 0.28 \text{ min.}$$
$$= 16.8 \text{ s}$$

Example 9.3. A 200 mm long, 75 mm diameter stainless steel rod is being reduced in diameter to 65 mm by turning operation. The spindle rotates at 400 rpm and the tool is traveling at an axial velocity of 250 mm/min. Calculate the cutting speed, material removal rate, time of cut, power required and cutting force. Given approximate specific energy required for stainless steel 3.0 – 5.2 Ws/mm^3.

Solution.

$$\text{Spindle speed} = 400 \text{ rpm} = 41.89 \text{ rad/s}$$
$$\text{Depth of cut } d = \frac{75 - 65}{2} = 5 \text{ mm}$$
$$\text{Feed } f = \frac{250 \text{ mm/min}}{400 \text{ rpm}} = 0.625 \text{ mm/rev.}$$

Material removal rate (MRR) can be calculated using equation (9.6)

$$\text{MRR} = \pi \, D_{av} \, d \, f \, N$$
$$= \pi(70 \text{ mm}) \, (5 \text{ mm}) \, (0.625 \text{ mm/rev}) \, (400 \text{ rpm})$$
$$= 4580 \text{ mm}^3\text{/s}$$

The actual time to cut

$$t = \frac{\ell}{fN} = \frac{200 \text{ mm}}{250 \text{ mm/min}}$$
$$= 0.8 \text{ min} = 48 \text{ s}$$

The approximate energy required is 3.0 – 5.2 W.s/mm^3. So we can assume an average value of 4.0 Ws/mm^3. The power (P) required is

$$P = (4.0 \text{ W.s/mm}^3)(4580 \text{ mm}^3\text{/s})$$
$$= 18320 \text{ kW}$$

Let T = torque and w = rotational speed

Then $P = T\,w$

$$\text{Or } T = \frac{P}{W} = \frac{18320}{41.89}$$
$$= 437.3 \text{ Nm}$$

The cutting force F_c can be obtained after dividing torque by average work piece radius. Thus

$$F_c = \frac{T}{D_{av}/2} = \frac{437.3 \text{ Nm}}{0.035 \text{ m}}$$
$$= 12495 \text{ N}$$

9.2.3 Cutting Tool Materials

Cutting tool materials must possess following characteristics:

- Room temperature hardness
- Resistance to thermal softening
- Toughness
- Stiffness
- Wear resistance
- Resistance to chemical reaction
- High thermal conductivity
- Low thermal expansion
- Dimensional stability
- Machinability and grindability
- Cost, both material and processing.

There is no single tool material which is best with regard to all of these considerations. The relative importance of each item will shift with the nature of the product machined, the volume of production, the type of machining operation (intermittent or continuous cut, roughing or finishing, high or low speed etc.), the tool design details (cutting and clearance angles, method of holding, rigidity etc.), the general condition of machine tool and the physical characteristics of work material.

The range of tool materials available include steels, non-ferrous alloys containing chromium, cobalt and tungsten, cemented carbides, cermets, ceramics and super-hard materials. In some cases the tool material might have the required properties intrinsically e.g., diamond (very hard material) or conferred on it by virtue of heat treatment or some other processing e.g., coatings. In Chapter 7, we have described these materials in detail under their respective categories. Some case studies are also included in Chapter 12 (Materials Design).

Figure 9.11 lists the cutting tool materials indicating the ranges of hardness, strength and temperature stabilities. Also illustrated is the influence of cutting conditions on properties of tool materials (Figure 9.12). Figure 9.13 gives a broad range of cutting speeds and feeds applicable for various types of tool materials.

Various factors affecting the selection of cutting tool material for a specific application may be summarized as:

- Hardness and condition of the work piece material.
- Operations to be performed-optimum tool selection may reduce the number of operations required.
- Amount of stock to be removed.
- Accuracy and finish requirements.
- Type, capability and condition of the machine tool to be used.
- Rigidity of the tool and work piece.
- Production requirements influencing the speeds and feeds selected.
- Operating conditions such as cutting forces and temperatures.

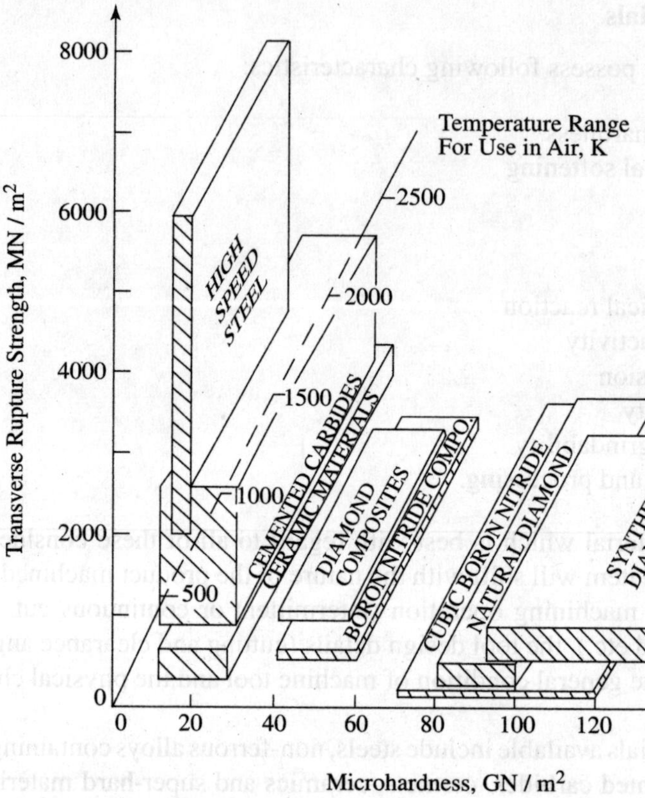

Fig. 9.11. **Schematic diagram of the property range of cutting tool materials. The hatched areas show room temperature properties, the temperature axis gives the maximum range of application in air disregarding temperature dependence of properties.**

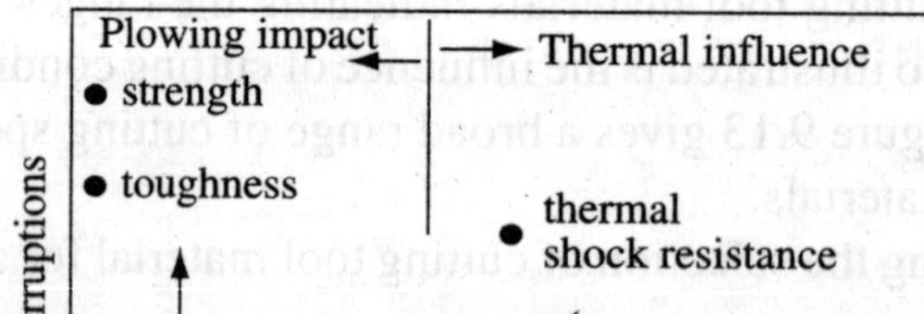

Fig. 9.12. **Influence of cutting condition on properties of tool materials.**

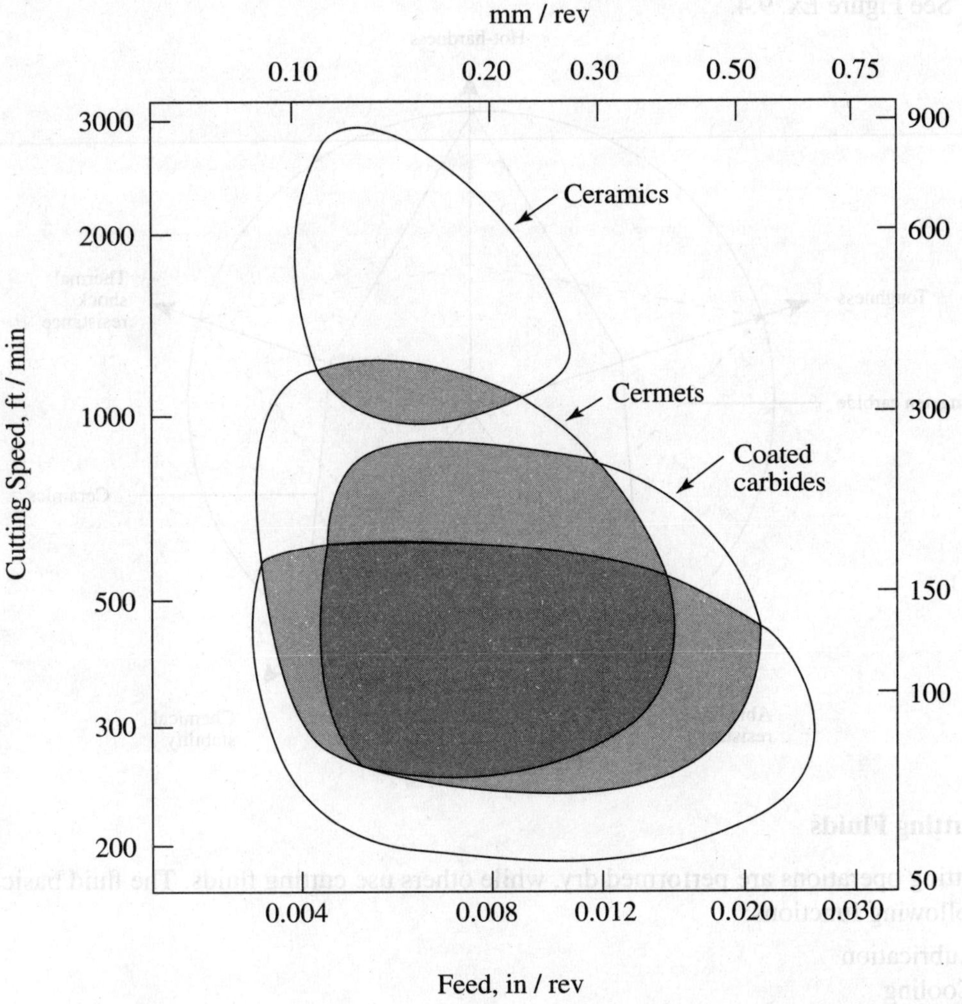

Fig. 9.13. The range of applicable cutting speeds and feeds for a number of tool materials.

- **Tool cost per part machined, including initial tool cost, grinding cost, tool life, frequency of regrinding or replacement, and labour cost-the most economical tool is not necessarily the one providing the longest life, or the one having the lowest initial cost.**

Example 9.4. **Show graphically how cemented tungsten carbide and ceramic cutting tools differ in following properties:**

(a) Hot hardness
(b) Thermal shock resistance
(c) Chemical stability
(d) Abrasion resistance
(e) Toughness.

Answer. See Figure Ex. 9.4.

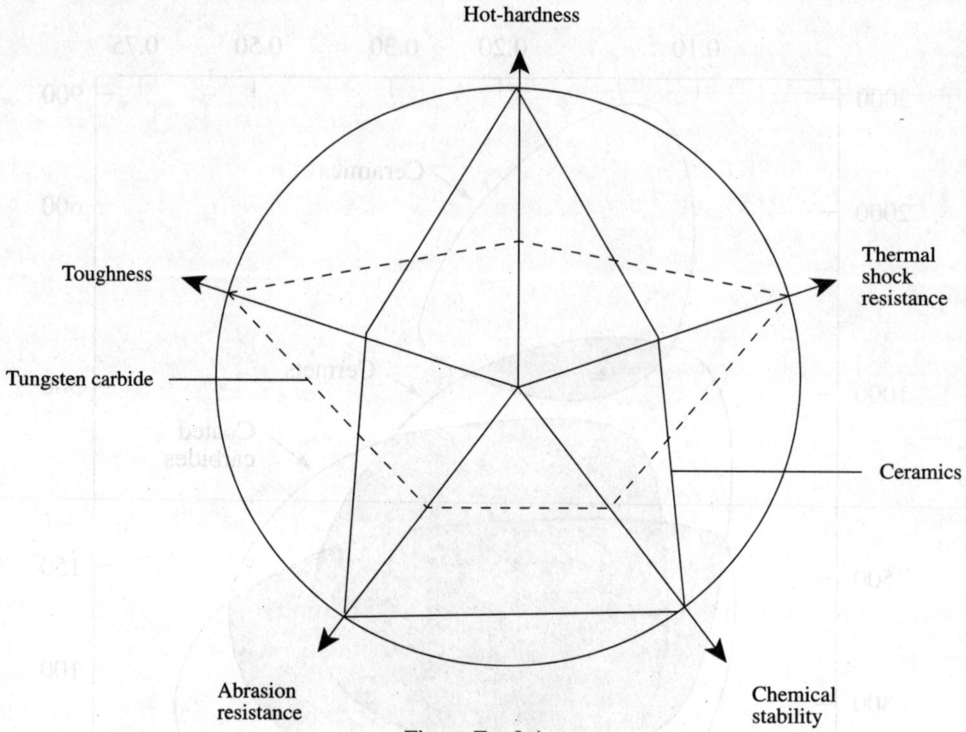

Figure Ex. 9.4

9.2.4 Cutting Fluids

Some cutting operations are performed dry, while others use cutting fluids. The fluid basically fulfills following functions:

- Lubrication
- Cooling
- Chip removal.

Cutting fluids fall into two main categories: cutting oils and water based fluids. The cutting oils are based on mineral oils with appropriate additives, and are used mostly at lower speeds and with high speed steel tooling. Water based fluids may be emulsions, semisynthetic fluids or synthetic fluids. The last one is also called chemical fluids, which contain no oil, only water soluble wetting agents, corrosion inhibitors and salts.

There is some overlap between the two types of cutting fluids. Coolants are formulated with ingredients that help reduce friction. Also, lubricants have thermal properties that, although not as good as those of water, act to remove heat from the cutting operation.

The cutting fluids may be applied in one of the following manners:

- Manual application.
- Flooding.
- Mist application.

Cutting fluids become contaminated over time with a variety of foreign substances such as oil, small chips, fungi and bacteria. In addition to causing odors and health hazards, contaminated cutting fluids do not perform their lubricating function as well. Alternative ways to deal with this problem is to:

- Replace the cutting fluid at regular and frequent intervals
- Use a filtration system
- Switch over to dry machining.

Because of the growing concern over pollution and associated legislation, fluid disposal has become costly. Of late, much research is devoted to dry machining and cutting tool manufacturers have developed certain grades of carbide and coated carbides for use in dry machining.

Example 9.5. During machining operation, if all of a sudden the cutting fluid flow is stopped, what shall be its effect on

(a) Friction at the tool-chip interface
(b) The shear angle
(c) Shear strain
(d) Thickness of chip
(e) Built up edge.

What shall be the effects of the resultant features?

Solution.

(a) Increases with the result that friction energy in the secondary zone increases.
(b) Increases, and hence total energy increases.
(c) Increases and hence surface finish gets adversely affected.
(d) Increases, and hence the temperature of the cutting zone increases. This promotes tool wear.
(e) The likelihood of formation, with the result that the dimensional tolerance difficult to maintain.

9.2.5 Cutting Tool Wear

Tool is an important aspect of machining operations. Wear depends on tool and workpiece materials, tool geometry, and type of cutting fluids. Figure 9.14 shows the wear behaviour of a cutting tool. The various regions of wear are identified as flank wear, crater wear, nose wear and chipping of the cutting edge. In the crater region (common in steel machining), the tool is subjected to substantial compressive and shear stresses, to temperatures which may reach 1000 °C or more, and to a very steep temperature gradients on the order of 1 °C/μm. The crator wear is also known as *diffusion wear*, in which the tool shape is changed by the diffusion of the tool surface by the chip material flowing over it. Clearly this depends upon the solubility of tool material in the work metal and the interface temperature obtained during cutting. The interface temperature is proportional to the cutting speed. Figure 9.15 indicates a direct relation between crater wear rate and tool-chip interface temperature. It is evident that a sharp increase in crater wear results after a certain temperature range has been reached.

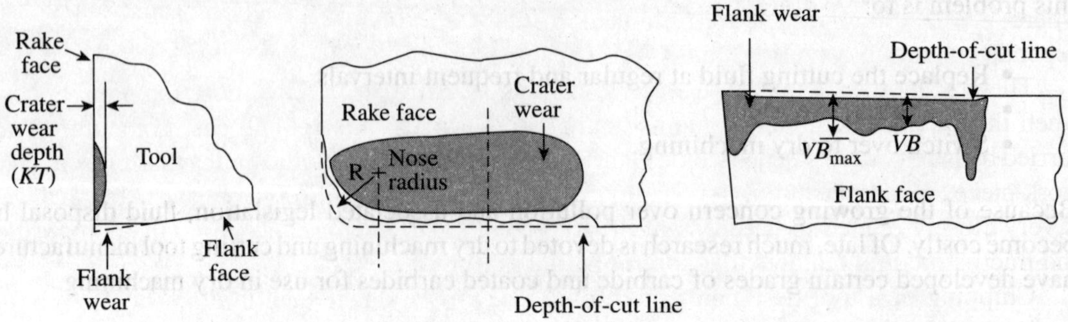

Fig. 9.14. Features of tool wear in a turning operation (V_B indicates average flank wear).

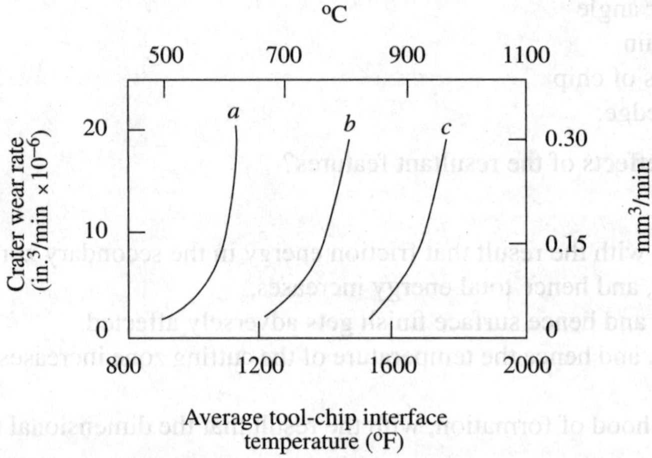

Fig. 9.15. Relation between crater-wear rate and average tool-chip interface temperature in turning: (a) high speed steel tool (b) C-1 carbide, (c) C-5 carbide.

Flank wear is synonym to *attrition wear*, in which the particles of microscopic size are detached from the tool surface and carried away in the stream of work metal. It is due to the cold welding effect which can occur readily in metal cutting, especially at low surface speed. Fresh, clean metal surfaces are formed continuously by the cutting action and, under the high pressure produced, intermittent welding between tip and chip is inevitable. The motion of the tool piece and chip continually breaks the junctions so formed and discrete particles of tool material are carried away. The wear rate in the category is associated with tool materials' grain size rather hardness. A finer grain cemented carbide with 16% Co and a hardness of 1265 HV can show better resistance to attrition wear than a coarser grain hard metal with 4.5% Co and a hardness of 1600 HV.

For flank wear Taylor established a relationship for cutting various steels as:

$$VT^n = C \qquad (9.8)$$

where V = cutting speed, T = time taken to develop a flank wear land, n = exponent that depends on cutting conditions and C is a constant.

There is another type of tool wear, known as *abrasive wear*. This most commonly occurs when the material being cut contains many hard particles. In steel and cast iron cutting, it is carried by hard particles e.g., carbides, nitrides, oxides and nonmetallic inclusions, present in the work piece. When machining Al-Si alloy, tool wear is almost entirely abrasive. Abrasive wear is very closely related to hardness of the tool metal, and, in consequence, to the composition and microstructural features that control hardness.

Chipping of a tool designates more of a tool failure rather than tool wear. Chipping by mechanical shock may occur in a region in the tool where a small crack or defect already exists. High positive rake angles also can cause chipping because of the small included angle of the tool tip. Thermal cracks, which are generally perpendicular to the cutting edge are caused by the thermal cycling of the tool in interrupted cutting. In case, the crater wear progresses towards the tool tip, chipping is promoted.

For various cutting operation the values of average wear land data for high speed steel and cemented carbide cutting tools are given in Table 9.3.

Table 9.3. Allowable Average Wear Land in Steels for Cutting Tools for Various Operations.

	Allowable Wear Land (mm)	
Operation	*High-Speed Steels*	*Carbides*
Turning	1.5	0.4
Face milling	1.5	0.4
End milling	0.3	0.3
Drilling	0.4	0.4
Reaming	0.15	0.15

Table 9.4 summarizes the advancing tool wear with cues to the machinist.

Example 9.6. Suggest the possible solutions for following problems noticed during metal machining using cemented carbide tool:

(a) Flank wear, (b) cratering, (c) chipping, (d) built-up-edge, (e) edge deformation, (f) thermal cracking and (g) fracture.

Solution.

(a) Reduce cutting speed and/or feed rate. Use a more abrasion-resistant grade of carbide.
(b) Use harder grade of alloyed carbide or coated insert. Reduce cutting speed. Apply cutting fluid.

(c) Hone cutting edge. Reduce deflection. Select stronger tool geometry or more shock-resistant grade of carbide.
(d) Increase cutting speed. Apply cutting fluid. Use positive rake angles to reduce cutting forces. Try a coated-carbide insert.
(e) Reduce-cutting speed and/or feed rate. Use carbide grade with higher hot hardness and more wear resistance. Apply cutting fluid.
(f) Reduce cutting speed. Apply cutting fluid correctly or discontinue its use. Use a more shock-resistant grade of carbide.
(g) Reduce feed rate and/or depth of cut. Use a stronger tool geometry and/or thicker insert. Use a softer, tougher, more shock-resistant grade of carbide.

Table 9.4. Advancing tool wear with cues to the machinist.

Characteristics	*Running-in Wear*	*Steady-State Wear*	*Rapid, Fatal Wear*
Visual Clues	• Repetitive radius of curvature of chip with smooth reflective underside • Part's surface finish smooth • No collar formation • Slight flank wear	• Chips begin to show changing curvature and duller underside • Surface finish on part poorer • Collar begins to form • More flank wear	• Chips irregular and broken with uneven serrations and dull underside • Dull surface finish • Large collar • Heavy flank wear
Aural Clues	• 'clean hiss' of serrations and 'constant trickel' of regular chip sections • no unusual 'spikes'	• Less regular chip section interrupt even 'trickel' • Some 'spikes' heard	• Chips obviously uneven • Spikes intense with squealing possible
Tactile	• Part surface feels smooth	• Some 'ears' on part surface	• Rough 'ears' on part
Machinist Thoughts	• Tool basically 'new' and working smoothly	• Expected deterioration occuring	• Must stop before damage occurs

9.2.6 Machinability

Machinability of a material is often defined in terms of the following features: (a) surface finish and integrity of the machined part, (b) tool life obtained, (c) force and power requirements, and (d) chip control. For good machinability, all the above criteria must be on higher side. Because of the complex nature of machining operation, the exact relationship to define quantitatively the machinability of a material is difficult to establish.

Machinability is naturally a function of alloy composition/its microstructure and mechanical properties. For example, the additives like lead and sulphur are used in steel in order to make it *free machining grade*. Lead is insoluble in iron, copper and aluminium and their alloys. Hence, during cutting, the lead particles are sheared and smeared over the tool chip interface. Because of their low shear strength, the lead particles act as a solid lubricant. Because of ecological problems, lead is not a preferred additive. Other examples of better machinability steels are resulphurized, rephosphorized and calcium deoxidized steels. Sulphur in steel forms manganese sulphide inclusions, which act as stress raisers in the primary shear zone. The chips produced, therefore, are small and break up easily. Elements like tellurium and selenium is also added along with sulphur in order to modify the inclusions. Phosphorus in iron is ferrite stabilizer and strengthens it with obvious hardening, and improvement in machinability. Oxide flakes of calcium alumino silicate improves machinability because of decrease in the tool-chip interface friction and wear. Because of these the tool is cool and hence less crater wear, at high cutting speed.

Table 9.5 gives the approximate ranges of recommended cutting speeds for turning operation for a number of ferrous and non-ferrous alloys.

Table 9.5. Approximate Ranges of Recommended Cutting Speeds for Turning Operations.

Workpiece Material	*Cutting Speed, m/min*
Aluminium alloys	200-1000
Cast iron, gray	60-900
Copper alloys	50-700
High-temperature alloys	20-400
Steels	50-500
Stainless steels	50-300
Thermoplastics and thermosets	90-240
Titanium alloys	10-100
Tungsten alloys	60-150

Note:

(a) The speeds given in this table are for carbides and ceramic cutting tools. Speeds for high speed-steel tools are lower than indicated. The higher ranges are for coated carbides and cermets. Speeds for diamond tools are significantly higher than any of the values indicated in the table.

(b) Depths of cut, d, are generally in the range of 0.5-12 mm.

(c) Feeds, f, are generally in the range of 0.15-1 mm/rev.

Earlier, we indicated that surface finish is also a measure of machinability. Surface finish influences the dimensional tolerances, which is the degree of deviation from ideal. Each machining operation or in that respect any manufacturing process has the capability of producing a part with a certain range of tolerance and surface finish. Figure 9.16 amply illustrates this aspect. Polymers are different from metals and ceramics in that they can processed to a very high surface smoothness, but tight tolerances are seldom possible because of internal stresses, left by moulding and creep at service temperatures.

Sometimes, poor machinability of a material, has given more prominence of its processing by a particular method. A typical example is titanium and its alloys. Their relatively poor machinability has given more impetus to their processing by powder metallurgy processing route

In brief, it can be said that machinability should be looked into a more bigger perspective of material, tool and machinery and not limited to only material of work job, which has been the common notion so far. Figure 9.17 illustrates the system approach encompassing all the above three factors. The combination of hardware and software is a major contemporary tendency, that must be fully exploited in understanding machinability.

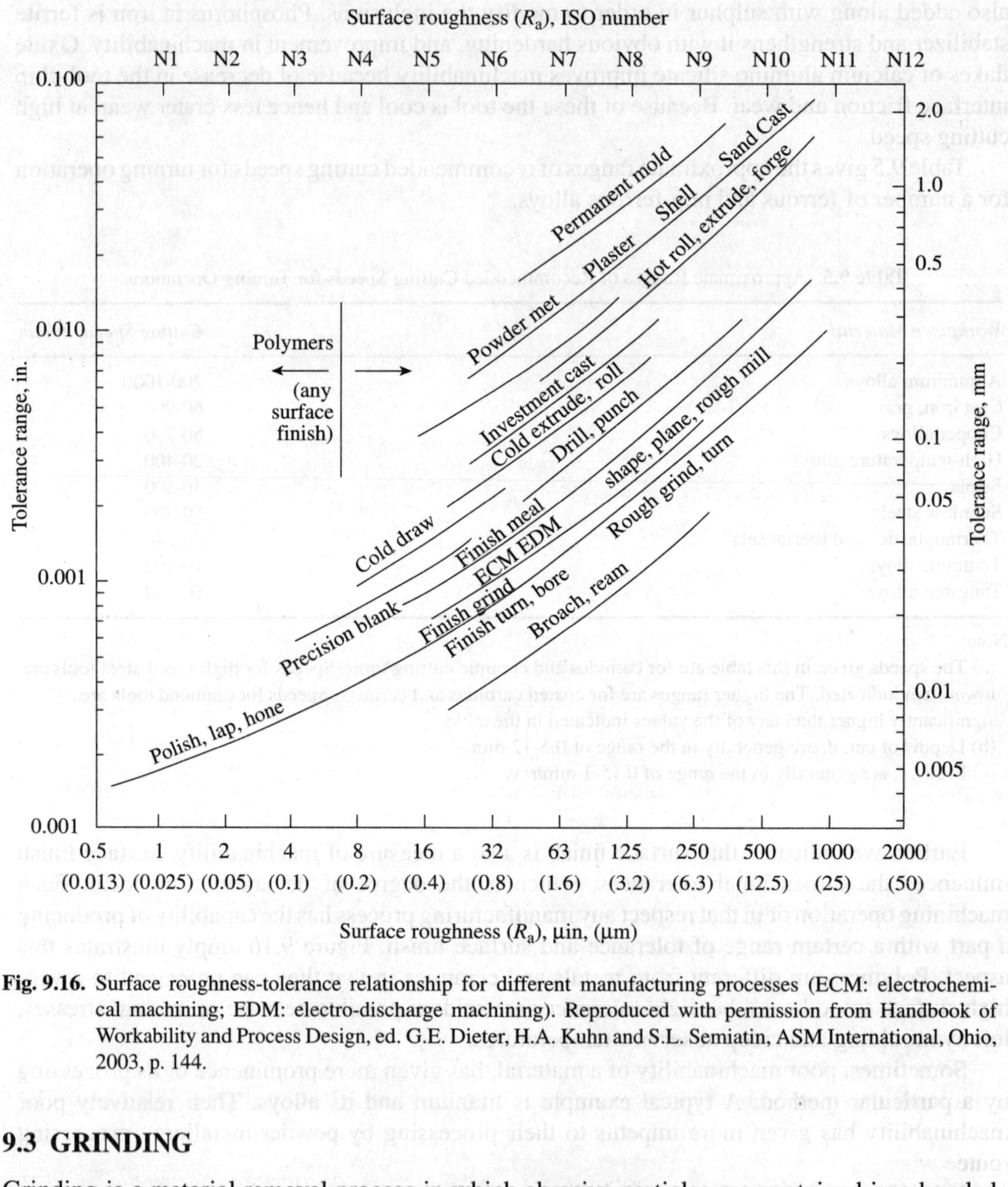

Fig. 9.16. Surface roughness-tolerance relationship for different manufacturing processes (ECM: electrochemical machining; EDM: electro-discharge machining). Reproduced with permission from Handbook of Workability and Process Design, ed. G.E. Dieter, H.A. Kuhn and S.L. Semiatin, ASM International, Ohio, 2003, p. 144.

9.3 GRINDING

Grinding is a material removal process in which abrasive particles are contained in a bonded grinding wheel that operates at very high surface speeds. Like milling, in this operation too peripheral and face grindings can be achieved. The rotating grinding wheel consists of many abrasive particles, which act as teeth. The work is fed relative to the wheel to accomplish material removal.

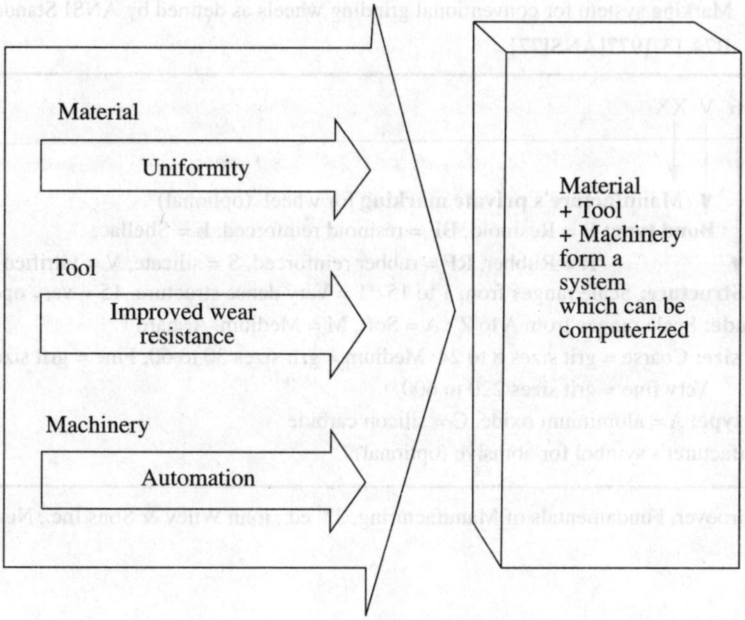

Fig. 9.17. System approach to machinability.

Grinding wheels have following parameters: abrasive material, grain size, bonding metal, wheel grade and wheel structure. The abrasive material must possess high hardness, wear resistance, toughness and friability. The last one is unique property, which reflects the ability of the abrasive metal to fracture when the cutting edge of the grain becomes dull, thereby exposing a new sharp edge. The common materials are aluminium oxide, silicon carbide, cubic boron nitride and diamond. The details of these ceramic materials are already included in Chapter 7.

The bonding material holds the abrasive grains and establish the shape and structural integrity of the grinding wheel. The major properties are strength, toughness, hardness and temperature resistance. The materials in this category are vitrified bond, silicate bond, rubber bond, resinoid bond, shellac bond and metallic bond. The vitrified bond is very common, which consists mainly of baked clay and ceramic materials. These bonds are relatively unaffected by water and oil, that might be used in grinding fluids. A typical case study for the metallic bond material is given in Chapter 12.

Wheel grade indicates the grinding wheel's bond strength in retaining the abrasive grinding/cutting operation. This is mainly dependent on the amount of bonding material present in the wheel structure. Soft grade wheels lose grains readily, while opposite is true for hard grade wheels. Soft wheels are used for low metal removal rate. It is interesting that soft wheels are useful for grinding hard work materials. Opposite is the case for hard wheels.

The American National Standards Institute (ANSI) has presented the specification, which uses numbers and letters to specify the parameters for the wheel. Table 9.6 illustrates an abbreviated version of the ANSI Standard for a conventional grinding wheel. For super hard material based grinding wheels, the standard is a slightly modified one.

Table 9.6. Marking system for conventional grinding wheels as defined by ANSI Standard B74.13-1977[ANSI77]

30 A 46 H 6 V XX

Manufacture's private marking for wheel (optional)

Bond type: B = Resinoid, BF = resinoid reinforced, E = Shellac, R = Rubber, RF = rubber reinforced, S = silicate, V = vitrified

Structure: Scale ranges from 1 to 15 : 1 = Very dense structure, 15 = very open structure.

Grade: Scale ranges from A to Z : A = Soft, M = Medium, A=hard

Grain size: Coarse = grit sizes 8 to 24, Medium = grit sizes 30 to 60, Fine = grit sizes 70 to 180 Very fine = grit sizes 220 to 600

Abrasive type: A = aluminium oxide, C = silicon carbide

Prefix: Manufacturer's symbol for abrasive (optional)

Source: M.P. Groover, Fundamentals of Manufacturing, 2nd ed., john Wiley & Sons Inc., New York, 2002,p.589.

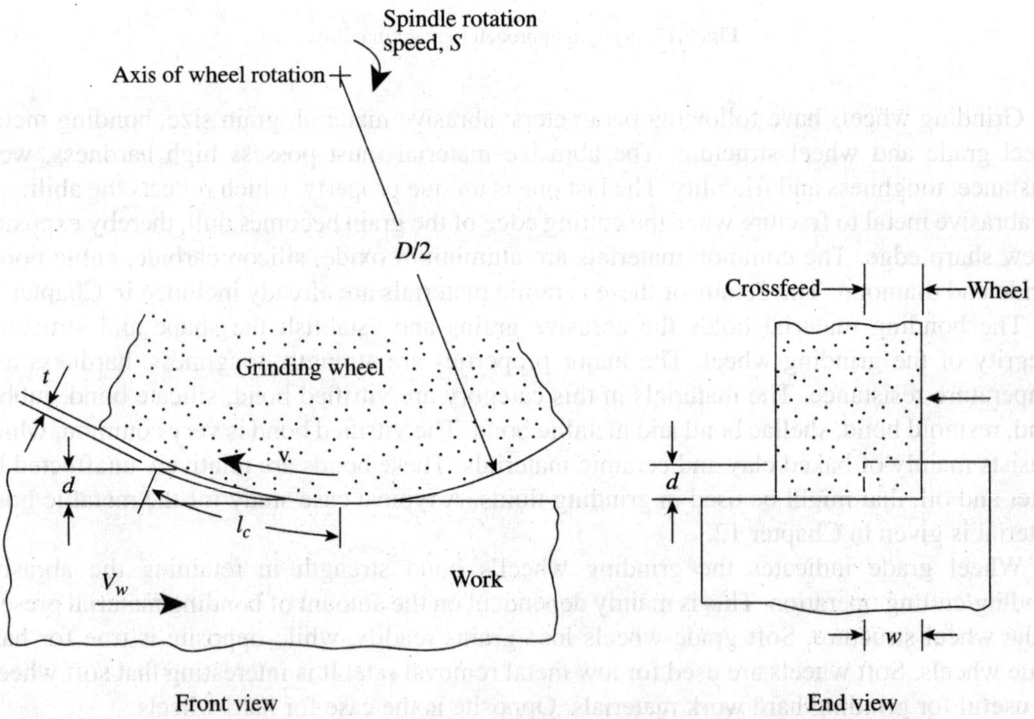

Fig. 9.18. The geometry of surface grinding, showing the cutting condition; cross-section of a single chip.

The peripheral speed of the grinding wheel V (m/min) is given as (Figure 9.18)

$$V = \pi D N \tag{9.9}$$

where D = wheel diameter (m)

N = spindle speed (rev/min).

Further, the material removal rate (MRR) is expressed as

$$\text{MRR} = V_w W d \tag{9.10}$$

where V_w = work speed

W = width of the grinding path

D = depth of cut (also called in feed).

The average length of the chip (ℓ_c) is given by

$$\ell_c = \sqrt{Dd} \tag{9.11}$$

where D = wheel diameter (mm)

d = depth of cut (mm).

The specific energy in grinding U (J/mm^3) is expressed as

$$U = \frac{F_c V}{V_w} \text{w.d} \tag{9.12}$$

where F_c = cutting force (N)

V = wheel speed (m/min)

V_w = work speed (mm/min)

w = width of cut (mm)

d = depth of cut (mm).

Similar to cutting tool wear, grinding tools do wear. Three mechanisms: grain fracture, attrition wear and bond fracture are the principal causes of wear. *Grinding ratio* of wheel is expressed as

$$\text{G.R.} = \frac{v_w}{v_g} \tag{9.13}$$

where v_w = Volume of work material removed

v_g = Corresponding volume of the grinding wheel worn out during the process.

Typical values of grinding ratios are between 95 and 125. This is about five orders of magnitude less than the analogous ratio in conventional machining.

After excessive grinding operation, the grinding wheels are subjected to *dressing*, which consists of breaking off the dulled grit on the outside periphery of the grinding wheel. This exposes fresh sharp grains. Furthermore, the clogged chips on the wheel are also removed. Although, dressing sharpens the wheel, it does not guarantee the shape of the wheel. Both are achieved by *truing* operation, which involves diamond pointed tool, that is fed across the wheel as it rotates.

Example 9.7. Explain why the specific energy in grinding operation is much greater than in conventional machining.

Answer. The reasons are following:

I. Size effect: The chip thickness in grinding is much smaller than for other machining operations, e.g., milling. Therefore, the energy required to remove each unit volume of metal is considerably higher than for conventional machining.
II. Negative rake angles: The individual hard ceramic grains have negative rake angles. This results in low values of shear plane angle and high shear strains. This results in high energy requirement.
III. Ineffective grit engagement: Not all the individual grits are engaged in actual cutting. Some grains, thus, may not be projected far enough in to the work surface to accomplish grinding. It may be mentioned that grains have three types of actions: cutting, plowing and rubbing. The first action causes material removal.

Example 9.8. In a surface grinding operation the following data are available:

Wheel dia = 200 mm
Depth of cut = 0.05 mm
Width of cut = 2.54 cms
Feed rate = 1524 mm/min

Determine

(a) The length of undeformed chip.
(b) Material removal rate.

Solution. From equation (9.11)

(a)

$$\ell = \sqrt{Dd} = \sqrt{200 \times 0.05} = 3.2 \text{ mm.}$$

(b) From equation (9.10)

$$\text{MRR} = V_w W d = (60 \times 25.4)(25.4)(0.05) = 1935 \text{ mm}^3/\text{min.}$$

9.4 NON-CONVENTIONAL MACHINING

Under this category, we shall take up only two types of machining viz. electro-chemical machining and electro-discharge machining typical of electro-chemical and thermal energy processes respectively.

Electro-chemical Machining (ECM)

In electro-chemical machining metal is removed from an electrically conductive work piece by anodic dissolution. This is in essence is a depleting operation. Such type of machining is used in applications where the work metal is very hard or difficult-to-machine, or where the work point geometry is difficult to accomplish by conventional machining methods. The operation is not affected by the hardness of the work piece, since the metal removal is not mechanical. Typical ECM applications include die sinking, multiple hole drilling, non-round hole making and deburring.

Material removal rate during electro-chemical machining is determined by Faraday's first law. The feed rate is ECM process, f_r , is expressed as

$$f_r = \frac{CI}{A} \tag{9.14}$$

where A = Frontal area of the electrode, mm^2. This is the projected area of the tool in direction of the feed in to the work.

I = Current, in amperes

C = Specific removal rate, mm^3/amp.s

Table 9.7 gives values of specific removal rate, C, for various work materials.

Example 9.9. In electro-chemical machining of Ti-alloy block, determine the machining time to drill a 20 mm deep hole when current density is 6 A/mm^2. Assume the efficiency as 90%. For 'C' (specific removal rate) value refer Table 9.7.

Solution.

$$C = 2.73 \times 10^{-2}\ mm^3/amp.s$$

Table 9.7. Typical Values of Specific Removal Rate C for Selected Work Materials In Electrochemical Machining.

Work material[a]	*Specific Removal Rate C mm^3 /amp-s*
Aluminium (3)	3.44×10^{-2}
Copper (1)	7.35×10^{-2}
Iron (2)	3.69×10^{-2}
Nickel (2)	3.42×10^{-2}
Steels:	
Low alloy	3.0×10^{-2}
High alloy	2.73×10^{-2}
Stainless	2.46×10^{-2}
Titanium (4)	2.73×10^{-2}

[a]Most common valence given in parentheses ( ) - assumed in determining specific removal rate C. For different valence compute C by multiplying C by most common valence and dividing by actual valence. Adapted from Fundamentals of Modern Manufacturing 2nd ed., M.P. Groover, John Wiley & Sons Inc., New York, 2002.

$$\text{Feed rate} = \frac{CI\eta}{A_o}$$

$$= 2.73 \times 10^{-2} \times 6 \times 0.9$$

$$= 14.75 \times 10^{-2} \text{ mm/s}$$

$$\text{Machining time} = \frac{20}{14.75 \times 10^{2}}$$

$$= 1.35 \times 10^{2}$$

$$= 135 \text{ s}$$

Electric Discharge Machining (EDM)

In this type of machining, metal is removed by a series of discrete electric discharges that cause localized temperature high enough to melt or vaporize the metal. This process is carried out only on electrically conducting work material. Important parameters in the process are (a) discharge current and (b) frequency of discharges. With increase in either of these parameters, the metal removal rate increases. It must be remembered that high spark temperature also melts the tool, depending on its composition. Like electro-chemical machining, EDM process does not depend on the hardness of work-piece or tool material. Melting point of the work material is important in this type of machining. The metal removal rate, MRR (mm^3/s) during EDM is expressed as

$$\text{MRR} = \frac{KI}{T_m^{1.23}} \tag{9.15}$$

where K = Constant of proportionality, equal to 664 in SI units

I = Discharge current in amperes

T_m = Melting point of work material is °C.

The EDM process must take place in the presence of a dielectric fluid, which creates a path for each discharge as the fluid becomes ionized in the gap. The discharges are generated by a pulsating direct current power supply connected to the work and the tool.

Applications of EDM include both tool fabrication and parts production. For tool making, the material concerned is often difficult to machine by conventional methods. Hole drilling, where the axis is at an acute angle to the surface is common by EDM.

9.5 DESIGN ASPECTS IN MACHINING

Design aspects are described below under major machining process.

Turning

- Parts should be designed so that they can be fixtured and clamped in work holding devices with relative ease.
- The dimensional accuracy and surface finish specified should be as wide as permissible for the part to function properly.
- Sharp corners, tapers and major dimensional variations in the part should be avoided.

- The cutting tools should be able to travel across the work piece without hindrance.
- Materials should as much as possible, be selected for their machinability.

Drilling

- Design should allow holes to be drilled on flat surfaces and perpendicular to the drill motion.
- Interrupted hole surfaces should be avoided.
- Through holes are preferred over blind holes.
- Work piece should be designed so that all drilling can be done with a minimum of fixturing and without repositioning it.

Milling

Apart from the factors mentioned for turning, additional following aspects should be considered:

- Work piece should be rigid to avoid any distortion resulting from clamping and cutting force.
- Chamfers should be used instead of radii because of the difficulty of proper machining various intersecting surfaces.
- Internal cavities and pockets with sharp corners should be avoided.
- Standard milling cutter should be used as far as possible.

Grinding

Most of the design factors related to machining are valid in grinding process also. Some special features to be taken into account are:

- If high dimensional accuracy is required, interrupted surfaces such as holes and key ways should be avoided because of vibration problem.
- In cylindrical grinding, parts should be balanced and long.
- Design requiring accurate form grinding should be kept simple to avoid frequent wheel dressing.

Electro-Chemical Machining/Grinding

- Not suited for producing sharp square corners or flat bottoms, as electrolytes erode away sharp corners.
- Irregular cavities may not be possible to the desired shape and accuracy.
- If flat surfaces are to be produced, the electro-chemically ground surface should be narrower than the width of the grinding wheel.

Electro-Discharge Machining

- Parts should be designed so that the required electrodes can be shaped properly.
- Deep slot and narrow openings should be avoided.
- For economic production, the surface finish specified need not be too fine.

9.6 ECONOMICS OF MACHINING

In the beginning of this chapter, we have outlined the advantages and limitations of machining.

In the economics of machining, one has two major goals i.e., minimum cost per part and the maximum production rate. The total cost per piece (C_p) can be presented as:

$$C_p = C_m + C_s + C_\ell + C_t \tag{9.16}$$

where C_m = machining cost

C_s = setting up cost

C_ℓ = cost of loading, unloading and machine handling.

C_t = tooling cost, which includes tool changing regrinding and depreciation of the cutter.

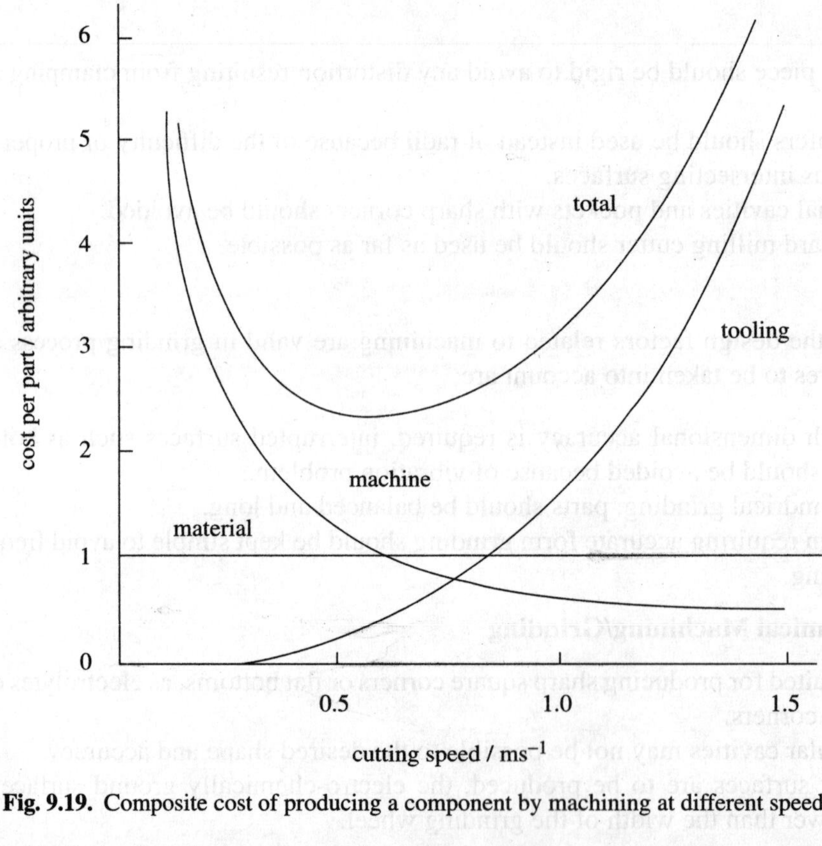

Fig. 9.19. Composite cost of producing a component by machining at different speeds.

Figure 9.19 shows a variation of machining and tooling costs with respect to cutting speed. The material cost, of course, is independent of this variable. It is evident that the total cost is always optimal at a critical cutting speed for a certain work piece material and the tool. The importance of accurate data is obvious as small changes in cutting speed can have a considerable effect on the minimum cost or time per piece.

9.7 FRICTION AND WEAR

Friction is defined as the resistance to relative motion between two bodies in contact, under a normal load. Friction plays an important role in metal working and manufacturing processes. Mechanical components often slide against another body. The normal force (N) exerts a force called interface pressure. The force required to move the body parallel to the surface is called shear force (F). The coefficient of friction is defined as:

$$\mu = \frac{F}{N} \tag{9.17}$$

In many applications it is necessary to minimize μ, either by the use of a lubricant, or by selecting materials that show inherently low friction or both. Materials pairs that show low adhesion usually – but not always – also give low friction.

Wear is defined as deterioration of surface due to use. No machine element is immune from wear. Scuffing of piston in internal combustion engines, pitting in power transmission gears, frettage in press – fitted assemblies and cavitation corrosion in cylinder liners are some of the examples. Wear is affected by a variety of conditions, such as the type and mode of loading, speed, quantity and type of lubricant, temperature, hardness, surface finish, presence of foreign materials and the chemical nature of the environment. Sometimes, the wear may be a combination of one or more elementary forms. The most common types of wear are abrasive wear and adhesive wear. In the former type, hard particles in one of the sliding elements (or in the form of hard third particles) cut grooves in the softer materials in a manner similar to that involved in fine grinding. In the case of adhesive wear, the high points (asperities) on material surfaces come close enough to establish strong bonds. If a weak point in the bulk metal happens to lie in the vicinity of a contacting asperity, a wear particle is generated.

It is found empirically, that for both abrasive and adhesive wear, the wear volume (V) is the function of applied load (F), the sliding distance (L), and the hardness of the softer of the two sliding elements (H)). This is expressed as

$$V = K\frac{LF}{H} \tag{9.18}$$

where K is constant depending on the materials in sliding contact and the sliding conditions. This non-dimensional quantity (K) is called the *wear coefficient*. Table 9.8 gives 'K' values for a number of representative adhesive wear situations for dry surfaces sliding in air. The value of 'K' for dry abrasive wear in air will be considerably higher than those in Table 9.8. Some of the examples of adhesion wear are galling, scuffing, scoring and seizing. Depending upon severity, abrasive wear may be of the gouging and scratching form. Other types of material wear e.g., pitting, fretting, cavitation and corrosive wear are described in other relevant chapters.

Erosion and impact wears are other forms of wear. In the former, the loose abrasive particles abrade a surface, while in the latter, the reason is the impact over the surface. Deburring by vibratory finishing and tumbling is an example of impact wear.

Wear design involves four elements: system analysis, modeling, data gathering and verification. The most important of these is system analysis, which includes the determination of how much wear can be allowed and the establishment of a failure criterion. In general, it consists of all those elements needed to implement the modeling and data gathering elements.

Table 9.8. Values of Wear Coefficient for Several Dry Sliding Pairs of Metals.

Materials		
Wear Surface	*Mating Surface*	*K for Wear Surface*
BB steel	BB steel	10^{-10}
Mild steel	Mild steel	2×10^{-3}
Tool steel	Mild steel	10^{-5}
Cast alloy (Co) steel	Tool steel	2×10^{-5}
Polytetrafluoroethylene	Tool steel	4×10^{-8}
Copper	Mild steel	30×10^{-4}
Stainless steel	Mild steel	2×10^{-5}

BB = Hard ball bearing steel.

Adapted from M.C. Shaw, Engineering Problem Solving, William Andrew Publishing, Norwich, N.Y., 2001.

For selecting materials against wear, one must consider primary criteria (chemical, mechanical and thermal stability) as well as secondary criterion e.g., nominal contact stresses. Earlier, we have highlighted the role of hardness in wear control. But other properties of the materials must be included, since in real situation one does consider different types or classes of materials. Because of the many factors associated with wear behaviour, different types of materials are used for different wear situations. Table 9.9 summarizes typical wear applications for selected engineering materials.

Rabinowicz has correlated wear behaviour of alloys from the point of view of material compatibility as reflected by their phase diagram. This was done on only binary systems, but is handy qualitative tool to understand wear. A material rubbing with the same material or an alloy system components (A-B) forming a wide range of solid solution rubbing together is considered as *compatible*. However in case the solid solubility is too limited (between 0.1 to 1%), the system is known as *partially compatible*. In case the alloy system does not form miscible liquid phase, e.g., Cu-Pb, the system is known as *incompatible*. It must be, however, remembered that metallurgically compatible systems are frictionally incompatible. There are certain exceptions, as other factors like intrinsic alloy hardness also comes into factor.

Ashby generated wear mechanism maps for a particular set of contacting materials, in order to identify the ranges of normal pressure and velocity that result in a particular wear mechanism. Figure 9.20 shows the general structure of such a map when the area of the contacting wear surface remains constant. Temperature is not a mapped variable. The dashed lines in figure marks possible boundaries that could be used to define a safe operating zone.

9.7.1 Design Against Wear

We have described in earlier section adhesive and abrasive wears. Many times a combination of adhesion and abrasion may exist, e.g., plain bearing and cylinder liners. To design against wear, it is desirable that the mating metals do not have mutual solubility. For example lead, tin and indium have very limited solubility in iron and a combination of these elements with

Table 9.9. Typical wear applications for selected engineering materials.

Materials	*Sliding*			*Rolling*								
	Unlubricated wear	*Lubricated wear*	*Abrasion*	*Unlubricated wear*	*Lubricated wear*	*Abrasion*	*Impact wear*	*Three body abrasion*	*Fluid erosion*	*Cavitation*	*Drop erosion*	*Particle erosion*
Structural alloys												
Surface treatments	X	X	X			X		X				
Hard surfacing	X	X	X			X	X	X			X	X
Soft coatings	X	X										X
Alloy steels		X	X	X	X	X	X	X				X
Tool steels	X	X	X	X	X	X	X	X		X		
Stainless steels												
Precipitation hardened			X	X				X				
Martensitic		X	X	X	X			X	X			
Cast irons												
Graphitic	X	X	X	X		X		X				
White	X		X			X		X				
High temperature alloys												
Refractory materials	X	X					X	X				
Superalloys		X		X						X	X	X
Copper-base alloys												
Bronze		X									X	
Beryllium copper		X					X					
Soft bearing alloys (babbitts)		X										
Carbides	X	X	X	X				X			X	X
Ceramics	X		X					X	X		X	X
Polymers												
Thermosets	X											
Thermoplastics	X	X									X	
Elastomers				X	X		X	X		X	X	X
Carbons	X											
Lubricating composites	X	X										

Source: ASM Handbook, v. 20, Materials Selection and Design, ASM International, Materials Park, OH, 1997, p. 607

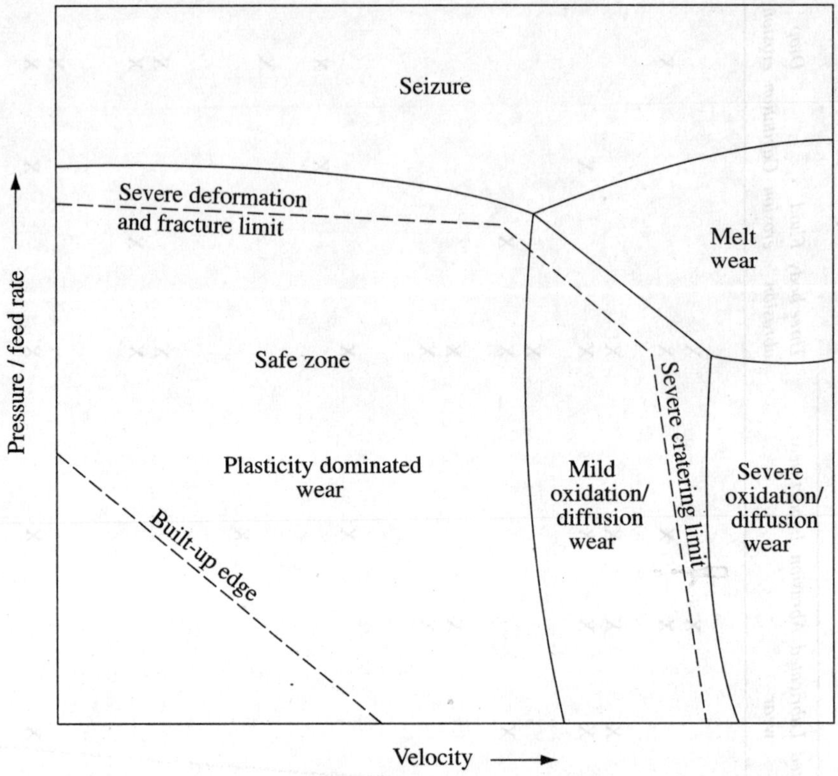

Fig. 9.20. The wear mechanism map and safe operating region for cutting tools.

iron provides a good bearing surface. On the other hand, very poor score resistance is observed with a combination of iron and nickel, which have a rather large mutual solid solubility. Both the adhesive and abrasive wear confirm that the harder a component is the more resistant it is to wear. However, in case the material is too hard, then the problem arises because of its brittle failure. The solution, therefore, lies that the component be tough. A tough core and hard surface can be achieved by heat and other surface treatments.

There are three ways by which the wear may be minimized:

(a) **Protective layer:** Such layer can be obtained by either deposition methods or diffusion treatment like carburizing/nitriding/carbonitriding. The details are described in Chapter 11.

(b) **Principle of conversion:** This method allows wear of one part of the system to offer protection to more important components. An example is the use of cast iron piston rings, which, if allowed to wear rapidly, should prevent scuffing of the cylinder liner.

(c) **Principle of diversion:** The method allows the less costly component to wear so as to protect the more expensive component. For example white metal containing tin, lead, copper and antimony over the journal bearings protects the harder bronze bearing surface beneath them.

QUESTIONS AND PROBLEMS

Machining

9.1 Distinguish between machining and other manufacturing processes.

9.2 Describe the three types of metal chips and list the conditions which promote the formation of each.

9.3 What are the effects of material properties and process variables on serrated chip formation?

9.4 Comment on the statement: To increase the metal removal rate, an increase in feed is preferred over an increase in speed.

9.5 For a rake angle 8 °, undeformed chip thickness 0.3 mm, actual chip (curled type) thickness of 0.48 mm and cutting speed of 0.6 m/s, calculate (a) chip thickness ratio, (b) shear angle, (c) chip velocity. In case the shape of chip changes from curled to straight, what shall be the effect on chip thickness ratio and velocity?

9.6 Explain how the minimum power consumption during metal cutting occurs where shear plane angle (ϕ) is 45 °.

9.7 Maximum current density and material removal rate by electro-chemical machining are 8 A/mm^2 and 12 mm/min. Estimate the maximum current and time required to machine a 20 mm diameter and 50 mm deep hole.

9.8 Why all room temperature mechanical properties data of materials are rather irrelevant for developing cutting tools?

9.9 What are the reasons for so many different types of cutting tool materials? Justify with suitable examples.

9.10 What are the major alloying additions in a high speed steel? Explain why they are used?

9.11 Solid cemented carbide hubs can be used at very high cutting speeds. However they are expensive and they have a low toughness. Powder Metallurgy high speed steels, on the other hand, have excellent toughness, but they have limitation due to reduced life due to high temperature operation. What further improvements would you suggest for making the HSS applicable?

9.12 Suggest the group of cemented carbide/cermet for following applications:
(a) Grey cast iron machining
(b) Mining application
(c) Cutting of steels
(d) Fine machining with high cutting speeds.

9.13 What are the most important factors influencing the properties of cemented carbide cutting tools?

9.14 Distinguish between the TiC, TiN and Al_2O_3 coated cemented carbide tool from the machining point of view.

9.15 What are super hard materials and where are their areas of application?

9.16 What are the effects of tool wear on the work piece and on the machining operation?

9.17 Identify possible causes for the following troubles noticed during metal machining using ceramic tool: (a) Edge chipping, (b) Tool cracking, (c) Chatter, (d) Torn finish, (e) Rapid wear.

9.18 Describe the functions of cutting fluids. Under what conditions would you discourage the use of cutting fluids?

9.19 Why is the machinability of a material difficult to assess?

9.20 What is the effect of residual stresses on machinability?

9.21 Explain why the grinding ratios of grinding wheels are less than the analogous ratio in metal cutting.

9.22 What types of bonding materials shall be used in grinding wheels for following operations?:
(a) grinding of cutting tools
(b) cut off wheels
(c) rough grinding wheels
(d) good finish grinding
(e) super hard abrasive wheels.

9.23 What are the reasons by which a machinist is more inclined to finish the job by machining rather than grinding?

9.24 Explain what shall be the changes in machining costs if one switches from manual to CNC machine. Features like batch size and part complexibility should be considered.

Friction and Wear

9.25 If hard ball bearing steel rubs against identical hard ball bearing steel in the absence of a lubricant, would you expect the adhesive wear rate to increase or decrease with
(a) reduction of hardness of one of the items
(b) decrease in the applied load
(c) introduction of a lubricant?

9.26 Why garnets are used in bubble devices for recording-head? Why ferrite-particle tapes have deleterious effect on permalloy heads?

9.27 Consult the binary phase diagram of following systems from the Handbooks: CaO-NiO, Ag-Cu; Cr_2O_3-SiO_2, Al-Sn. What prediction of tribological behaviour can be made when these pairs are rubbing against each other in unlubricated condition. If the surface energy values for Al_2O_3 and ZrO_2 are 740 and 530 ergs/cm^2 respectively and their hardness 2150 and 1150 kg/mm^2, what conclusion can you draw when each of them rub on the same species?

9.28 Figure P. 9.28 represents three recording head core designs. Out of these the head (c) is ideally suited. Give reasons. [Hint: Both ferrite and permalloy are soft magnetic. Consider both magnetic and wear behaviour]. Adapted from Electronic Materials, 2nd ed., N. Braithwaite and G. Weaver, Butterworth-Heinemann, London, 1998, p. 128.

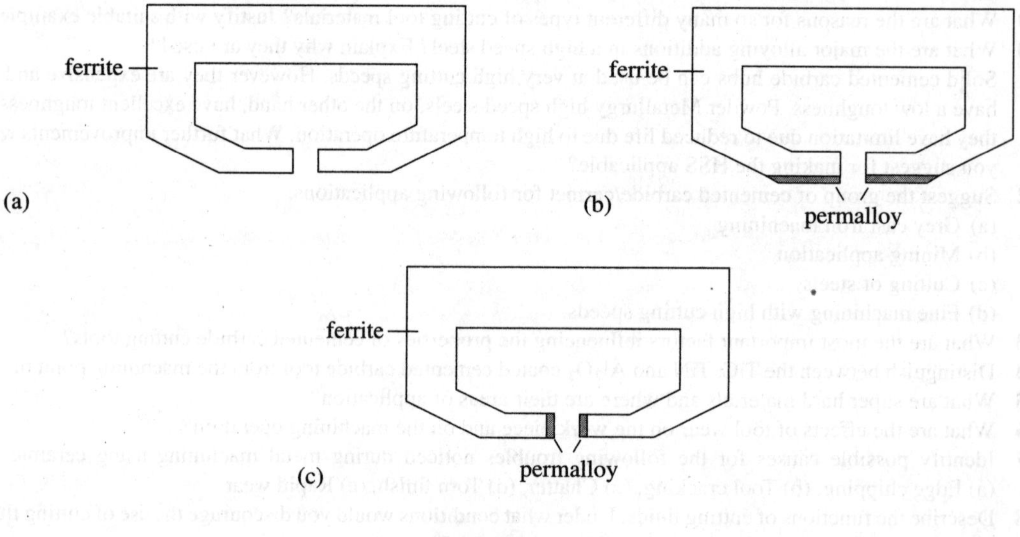

Figure P. 9.28

REFERENCES

Metal Cutting, 4th ed., E.M. Trent and P.K. Wright, Butterworth-Heinemann, Boston, 2000.
Friction and Wear of Materials, 2nd ed., E. Rabinowicz, John Wiley & Sons, Inc., New York, 1995.
Cemented Tungsten Carbides: Production, Properties and Testing, G.S. Upadhyaya, Andrew William Publishing, Norwich, N.Y., 1998.
Machining with Abrasives, R.L. McKee, Von Nostrand Reinhold, New York, 1983.
Introduction to Manufacturing Processes, 2nd ed., J.A. Schey, McGraw Hill, New York, 1987.

10

MATERIALS JOINING

Joining covers a wide range of processes, which are important in product manufacturing. There is another term 'fastening', which is more pertinent in connection with mechanical assembly where variety of fasteners both screws and rivets are used. The bond in the latter is not metallurgical one. However, many authors have erroneously grouped fastening in joining. The need for joining arises under following circumstances:

- Impossible or uneconomical to manufacture the product as a single piece,
- Transportation of big size products, say gas pipelines, to the field eventually followed by joining.

Joining processes are welding, brazing, soldering and adhesive bonding. Welding occurs at elevated temperature either by fusion or solid state welding methods. Brazing and soldering use filler metals and involve lower temperatures than used in welding.

10.1 WELDING PROCESSES

American Welding Society (AWS) has given letter designations to various welding processes. These are grouped in Table 10.1. In fusion welding, heat sources for the gas, arc and high energy beam welding processes are a gas flame, an electric arc and a high energy beam respectively. As the power density of the heat source increases, the heat input to the work piece that is required for welding decreases. If one increases the power density of the heat source, the weld penetration is deeper, welding speed could be increased and weld quality is better.

Before going in details of various welding processes, it is worthwhile to mention about weld joint designs. Figure 10.1 shows the basic designs in fusion welding. The surface of the weld is called the *face,* the two junctions between the face and the work piece surface are called the *toes* and the portion of weld beyond the work piece surface is called the *reinforcement*.

Table 10.1. American Welding Society Letter Designation for Various Welding Processes.

Group	*Welding*	*AWS Letter Designation*
Arc welding	Carbon arc	CAW
	Electrogas welding	EGW
	Flux cored arc welding	FCAW
	Gas metal arc welding	GMAW
	-pulsed arc	GMAW-P
	- short-circuiting arc	GMAW-S
	Gas tungsten arc	GTAW
	-pulsed arc	GTAW-P
	Plasma arc welding	PAW
	Shielded metal arc welding	SMAW
	Stud arc welding	SW
	Submerged arc welding	SAW
Oxyfuel gas welding	Oxyfuel welding	OFW
	Air acetylene welding	AAW
	Oxyacetylene welding	OAW
	Oxyhydrogen welding	OHW
	Pressure gas welding	PGW
Resistance welding	Flash welding	FW
	Projection welding	PW
	Resistance seam welding	RSEW
	Resistance spot welding	RSW
Other welding processes	Electron beam welding	EBW
	Electroslag welding	ESW
	Flow welding	FLOW
	Induction welding	IW
	Laser beam welding	LBW
	Percussion welding	PEW
	Thermit welding	TW
Solid state welding	Cold welding	CW
	Diffusion welding	DFW
	Explosion welding	EXW
	Forge welding	FOW
	Friction welding	FRW
	Hot pressure welding	HPW
	Roll welding	ROW
	Ultrasonic welding	USW

10.1.1 Gas Welding

In gas welding the heat is generated by the reaction between the fuel gas and oxygen. Oxyacetylene welding (OAW) is the most commonly used gas welding process because of its high flame temperature (Figure 10.2a). A flux over the filler rod is used to deoxidize and cleanse the weld metal. Figure 10.3 shows three different types of flames: neutral, reducing and oxidizing. In neutral flame the oxygen and acetylene are mixed in equal amounts and burned at the tip of the

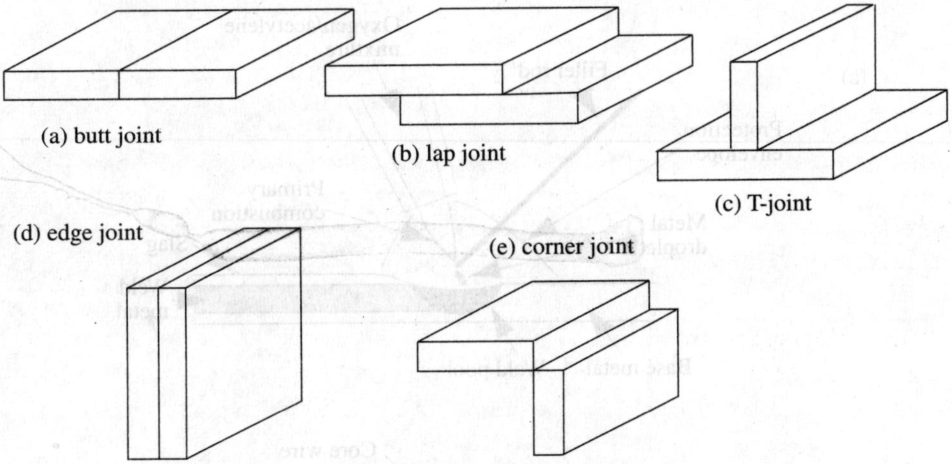

Fig. 10.1. Five basic types of weld joint designs.

welding torch. In the inner cone, primary combustion takes place. In addition, there is secondary combustion too, where the products of primary combustion CO and H_2 react with oxygen from the surrounding air to form CO_2 and H_2O. The area, where the secondary combustion takes place is known as outer envelop. In reducing flame, an excess of fuel i.e., acetylene is used. Such a flame is desirable for welding aluminium alloys, which has high activity towards oxidation. This is also useful in case of high C-steel welding to prevent the risk of decarburization. The acetylene feather is of greenish colour and is between the inert cone and the outer envelop. In case of oxidizing flame, there is excess of oxygen. This flame is characterized by a short white inner cone. Such a flame is desirable while welding brass, so that the resultant copper oxide over the weld pool, prevents dezincification of the alloy.

10.1.2 Arc Welding

Various types of arc welding are categorized in Table 10.1. The shielded metal arc welding (SMAW) is one of the most widely used processes (Figure 10.2b). The core of the covered electrode, known as core wire, conducts the electric current and provides filler metal for the joint. The lighter molten flux is on the surface of the weldment and is removed. The functions of electrode covering are protection, deoxidation, arc stabilization and alloying addition. As the welding equipment is simple, it is often used for maintenance, repair and field construction. The gas shielding is not, however, applicable for reactive metals like aluminium and titanium.

Gas-tungsten arc welding (GTAW) is another process, which is also known as TIG (tungsten-inert gas) welding (Figure 10.2c). The tungsten electrode is usually in contact with a water-cooled copper tube, called the contact tube. Protection against air is much better in GTAW than in SMAW because of the use of argon or helium as protective gas. In case the thickness of the material is more, a filler rod is required. Since the deposition rate in GTAW is rather low, care must be exercised against tungsten inclusion in the weld metal because of extremely high temperature. By using preheated filler metals the deposition rate can be increased.

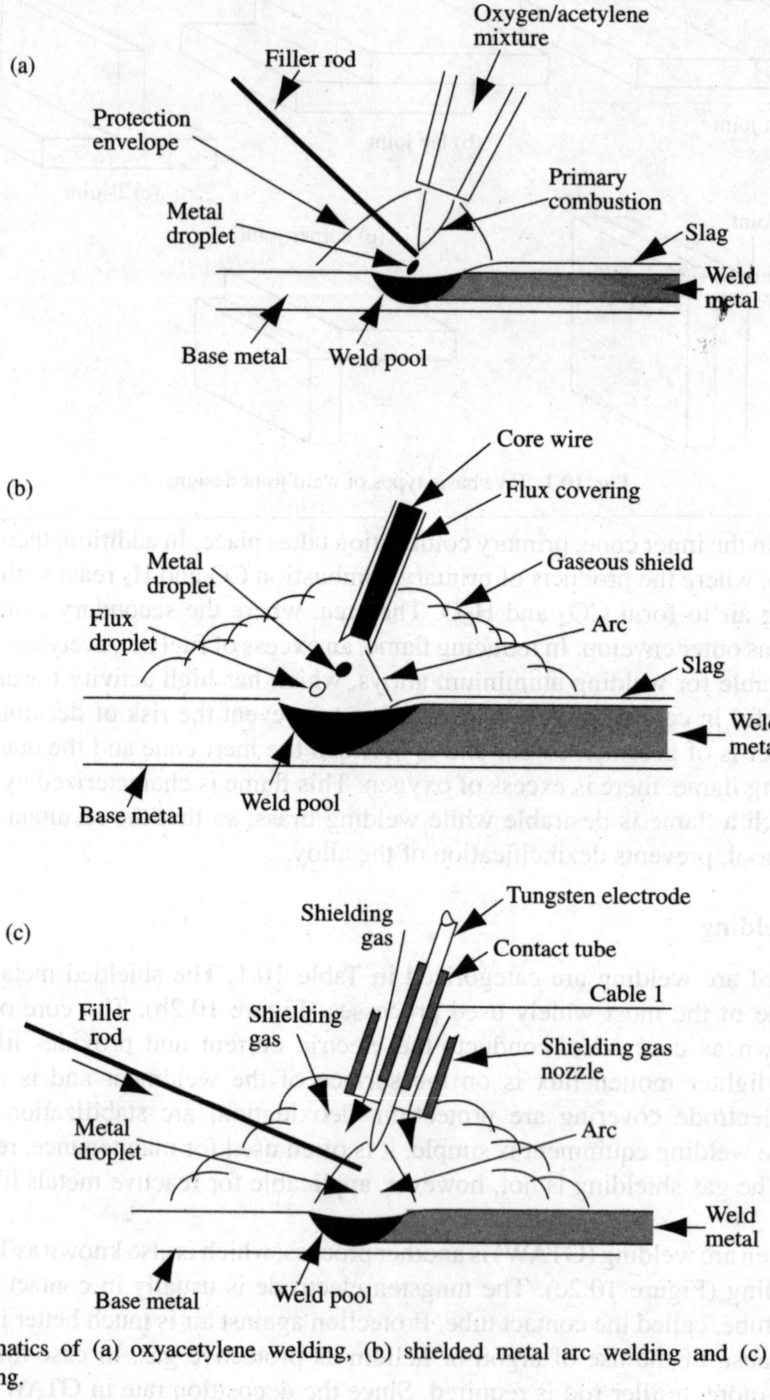

Fig. 10.2. Schematics of (a) oxyacetylene welding, (b) shielded metal arc welding and (c) gas-tungsten arc welding.

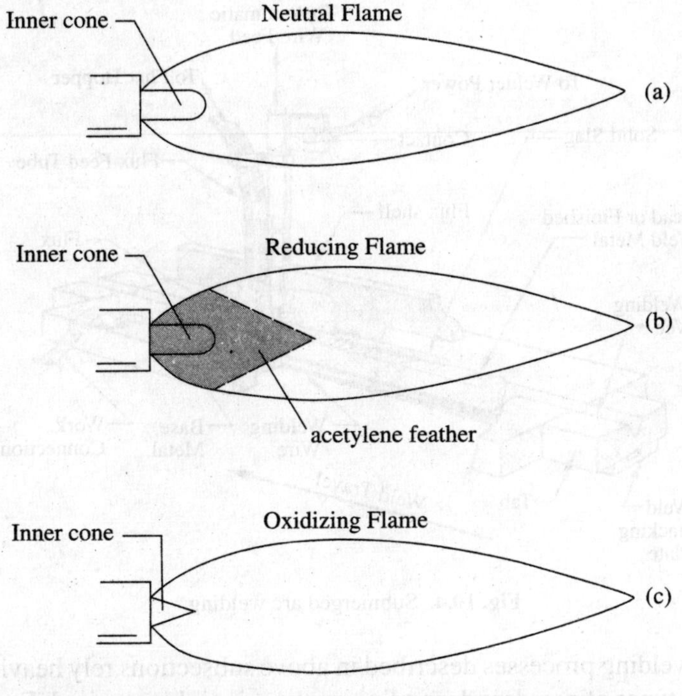

Fig. 10.3. Different types of flames in oxyacetylene welding.

In many cases, the tungsten electrodes have additives of rare earth oxides (0.25-2%), which are used for improving arc starting and stability. They also improve burn off. These electrodes are used for DC operation.

Gas-metal arc welding (GMAW), is also known as metal-inert gas (MIG) welding process. As non-inert gas like CO_2 is also used, the former designation is more scientific. The arc energy is found to be less uniformly dispersed in an argon arc than in a helium arc, because of the lower thermal conductivity of the latter.

In case of submerged arc welding (SAW), the metal and the arc are shielded by a molten slag and granular flux (Figure 10.4). In this case the arc is not visible and hence no shielding gas is required. The refining action of the slag gives rise to clean welds. For increasing the deposition rate, multi-electrodes can be used.

10.1.3 Electroslag Welding

In electroslag welding (ESW) the molten slag protects the weld metal. However, strictly speaking, this is not an arc welding process, because arc exists only during the initiation period of the process. After the flux melting, the resistance heating generated by the electric current passing through the slag keeps it molten. In this process the deposition rate is very high with no distortion. Because of very high heat input, the weld quality can be rather poor, resulting in low toughness.

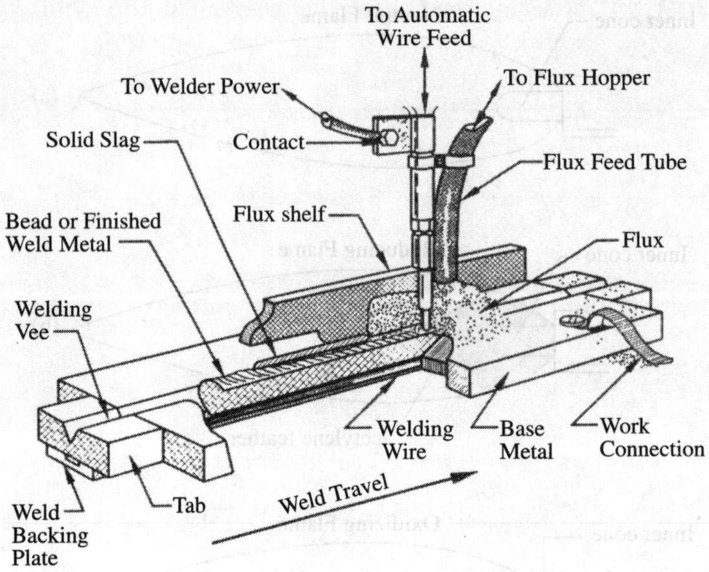

Fig. 10.4. Submerged arc welding.

All the fusion welding processes described in above subsections rely heavily on flux coating comprising various types of powders depending on the principles required. For example, the use of high rutile (TiO_2) fluxes improves arc properties, while basic fluxes provide a cleaner weld deposit with lower inclusion content than a rutile flux. Table 10.2 summarizes these coatings.

Table 10.2. Some Electrode Coating Types.

Type	*Formulation*	*Comments*
Cellulosic	20-60% rutile, 10-20% cellulose, 15-30% quartz, 0-15% carbonates, 5-10% ferromanganese	Cellulose promotes gas shielding in the arc region. Hydrogen increases heat at weld. Hydrogen content high (30-100 ppm). Deep penetration, fast cooling Weld
Rutile	e.g., 40-60% rutile, 15-20% Quartz, 0-15% carbonates, 10-14% ferromanganese, 0-5% organics	Slags mainly for slag shielding. Hydrogen content fairly high (15-30 ppm). High inclusion content in weld deposit
Acid	Iron ore-manganese ore, quartz, Complex silicates, carbonates,	Fairly high hydrogen content. High slag content in weld metal
Basic	20-50% calcium carbonate, 20-40% fluorspar, 0-5% quartz, 0-10% rutile, 5-10% ferro-alloys	Fairly low hydrogen levels ($\leq$10 ppm), hence commonly used in welding low alloy constructional steels. Electrodes should be kept dry. Low inclusion content in weld deposit

Example 10.1. In GMAW often helium or carbon dioxide is added to argon shielding gas. Explain why?

Solution. Figure Ex. 10.1 shows the relationship between melt drop radius vs welding current for steel electrodes. In pure argon, it is evident that excessive taper of electrode tip occurs. By addition of He or CO_2 to the argon shielding gas, the anode spot is constricted and does not climb the cylindrical side walls of the electrodes as easily.

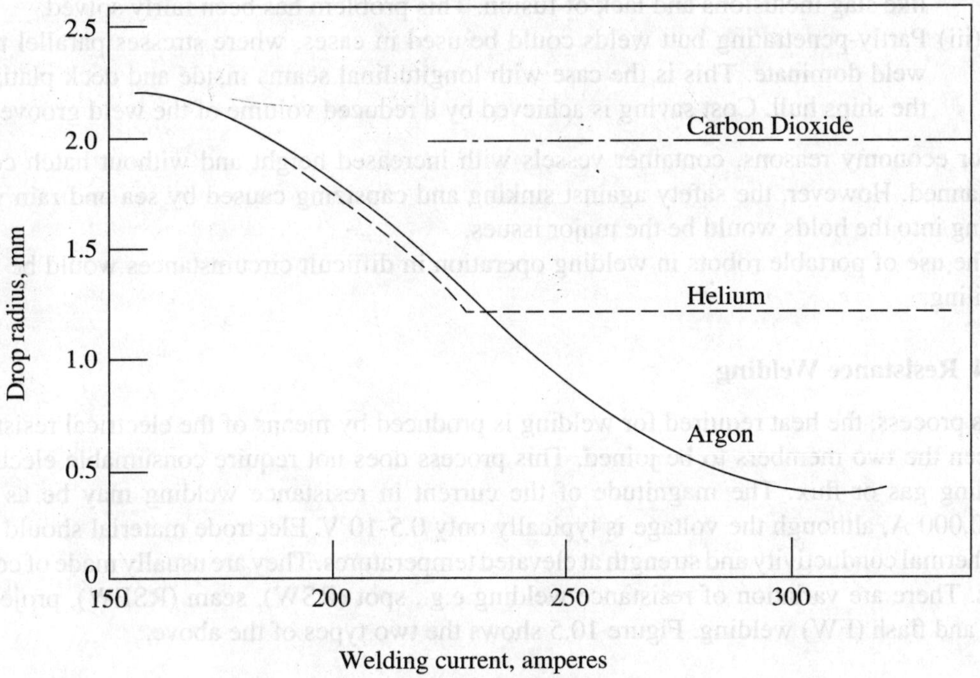

Figure Ex. 10.1

Example 10.2. What is the difference in material requirements for a heavy cargo vessel and high speed small crafts (surface effect ships)? What effects will it pose on joining processes?

Answer. Low carbon steels with yield strengths up to 355 N/mm^2 are generally used for heavy cargo vessels. Alloyed steels are also used, where low temperature application or chemical resistance are required. In case of high speed crafts, the speed ranges between 30 and 100 knots. Hence weight saving is essential. Materials used for such crafts include high and ultra high strength steels, aluminium alloys and glass reinforced plastics. Other light materials as titanium alloys, carbon reinforced plastics are attractive materials, although these are costly ones. Increased notch sensitivity of high strength materials becomes essential. The demand on weld quality becomes high. Sandwich elements are increasingly used as building elements for fast passenger vessels. However, their acoustical properties can be very poor. The need is to predict the acoustical characteristics from material parameter, plate geometry and frequency range.

Some of the special features in welded joints are:

(i) Fillet welds between plate and stiffener are often oversized due to production requirements.
(ii) Downward welding of vertical fillet welds, although much economical against conventional upward welding, is not recommended because of danger of defects like slag inclusions and lack of fusion. This problem has been fairly solved.
(iii) Partly-penetrating butt welds could be used in cases, where stresses parallel to the weld dominate. This is the case with longitudinal seams inside and deck plating of the ships hull. Cost saving is achieved by a reduced volume of the weld groove.

For economy reasons, container vessels with increased height and without hatch covers are planned. However, the safety against sinking and capsizing caused by sea and rain water entering into the holds would be the major issues.

The use of portable robots in welding operation in difficult circumstances would be more prevailing.

10.1.4 Resistance Welding

In this process, the heat required for welding is produced by means of the electrical resistance between the two members to be joined. This process does not require consumable electrode, shielding gas or flux. The magnitude of the current in resistance welding may be as high as 100,000 A, although the voltage is typically only 0.5-10 V. Electrode material should have high thermal conductivity and strength at elevated temperatures. They are usually made of copper alloys. There are variation of resistance welding e.g., spot (RSW), seam (RSEW), projection (PW) and flash (FW) welding. Figure 10.5 shows the two types of the above.

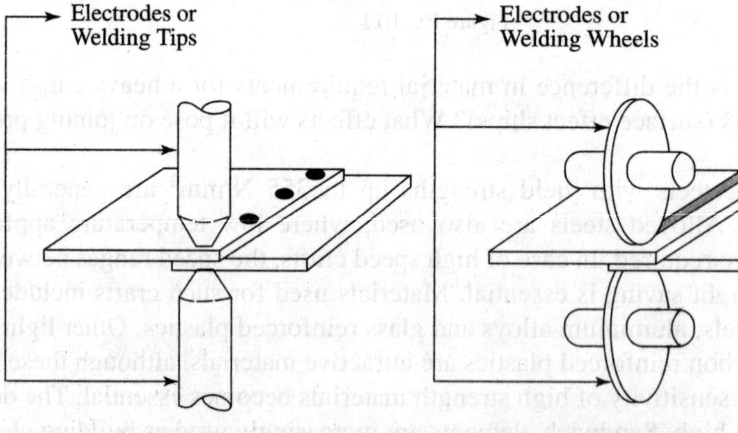

Fig. 10.5. Spot welding (left) and seam welding (right).

Flash welding (FW) is very common for rail steels (for details on composition etc. refer Case Study 12.3 in Chapter 12). The joining process is a variant of electrical resistance welding. It involves a fusion stage followed by a forging stage during which a substantial amount of deformation occurs just as in friction welding. The deformed metal must be stripped from around the weld and the joint shaped into a roughly correct profile in a forming press before being ground down to the final dimensions. One must be careful that the transformation is complete, and only after that the cooling rate be increased by cold air spraying. The reason for this rapid cooling is due to the fact that, the dimensional accuracy is better if the forming is carried out when the rail is cold.

10.1.5 Thermit Welding (TW)

When a metal oxide of low free energy of formation is brought into intimate contact with a metal of higher free energy of oxide formation, the metal oxide is reduced in an exothermic reaction. *Thermit powder* (a registered trade mark of Th. Goldschmidt AG, Essen, Germany) is a mixture of a metal and an oxide, e.g., aluminium and iron oxide. The reaction, initiated with a special ignition powder, liberates iron

$$3Fe_3O_4 + 8Al \rightarrow 4Al_2O_3(\text{slag}) + 9Fe \qquad (10.1)$$

Iron alloy pellets are added to reduce the reaction temperature to about 2500 °C.

The process finds application for joining heavy sections such as rails or reinforcing rods in the field. Large electrical bus bars are similarly welded, using aluminium and Cu_2O.

10.1.6 Solid State Welding

Various solid state welding processes are enumerated in Table 10.1. Among all the methods, friction welding (FRW) is very common. The two parts to be joined are secured in a lathe-like machine. One of the parts remains stationary and the other part rotates, when the rotating part is brought up to speed, the surfaces of the parts to be welded are squeezed together. Friction generates heating and eventual incipient melting. When the heating is achieved additional force is applied along the rotating axis (Figure 10.6). The process is very quick and used for both ferrous and nonferrous systems. Dissimilar metals may also be welded. One of the typical example of this type of joining is to secure larger parts on to the ends of shafts in most cost effective manner.

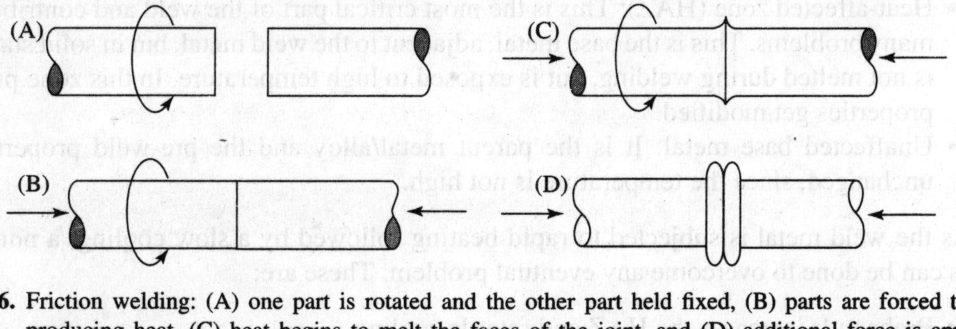

Fig. 10.6. Friction welding: (A) one part is rotated and the other part held fixed, (B) parts are forced together producing heat, (C) heat begins to melt the faces of the joint, and (D) additional force is applied to complete the weld.

One of the variations in friction welding is *friction stir welding*, which is very common for aerospace alloys. In friction stir welding, unlike friction welding, a third body is rubbed against the surfaces to be joined. The probe at the tip of the rotating tool forces heating and mixing or stirring of the material in the joint. The thickness of the welded material can be as little as 1 mm and as much as 30 mm. The weld quality is high with minimal pores and with uniform material structure as compared to fusion welding. For welding aluminium alloys, the tool is usually made of tool steel. In case of aluminium alloys three zones: thermally affected zone, thermomechanically affected zone and dynamically recrystallized zone, are noticed. Reversion of precipitation, overageing and solutionizing occur during welding, from near the base metal to the weld centre line.

Another solid state joining process is known as diffusion bonding or *diffusion welding* (DFW), in which the joint strength results primarily from diffusion and to a lesser extent, some plastic deformation of faying surfaces. The process requires temperatures of about 0.5 T_M (T_M is the melting point of metal on the absolute scale) in order to have a sufficiently high diffusivity between the parts to be joined. The strength of the bonded interface depends on pressure, temperature, time of contact and the cleanliness of the faying surfaces. The combination of diffusion bonding with superplastic forming (refer Chapter 8) has contributed a lot in aircraft and aerospace applications. This leads to improved productivity by eliminating mechanical fasteners, reducing the number of parts required, producing parts with good dimensional accuracy and low residual stresses. Some of the typical alloys subjected to diffusion bonding are Ti-6 Al-4 V and 7475-T6 Al-alloy structures.

10.2 WELDABILITY OF METALS AND ALLOYS

A materials engineer engaged in welding must be well conversant with metallurgy right from melting, solidification, gas-metal reaction, slag-metal reaction, surface phenomena to solid state reactions including heat treatment. Full details of solidification and phase transformation have already been dealt in earlier chapters.

The weld may be classified in three parts according to its level of heat exposure during welding. These are (Figure 10.7):

- Solidified weld metal: This is constituted of base and filler metals.
- Heat-affected zone (HAZ): This is the most critical part of the weld and contributes for many problems. This is the base metal, adjacent to the weld metal, but in solid state. This is not melted during welding, but is exposed to high temperature. In this zone pre-weld properties get modified.
- Unaffected base metal: It is the parent metal/alloy and the pre-weld properties are unchanged, since the temperature is not high.

As the weld metal is subjected to rapid heating followed by a slow cooling, a number of things can be done to overcome any eventual problem. These are:

- Preheat: It increases the HAZ and retards cooling.
- Base metal plate thickness: Increase in thickness reduces HAZ, because of mass effect.

- Thermal conductivity of base metal: High diffusivity metals like copper and aluminium permit heat to flow more rapidly than iron and nickel. In fact, copper has such a high diffusivity, that a low energy GTAW arc can be struck without melting the base metal.

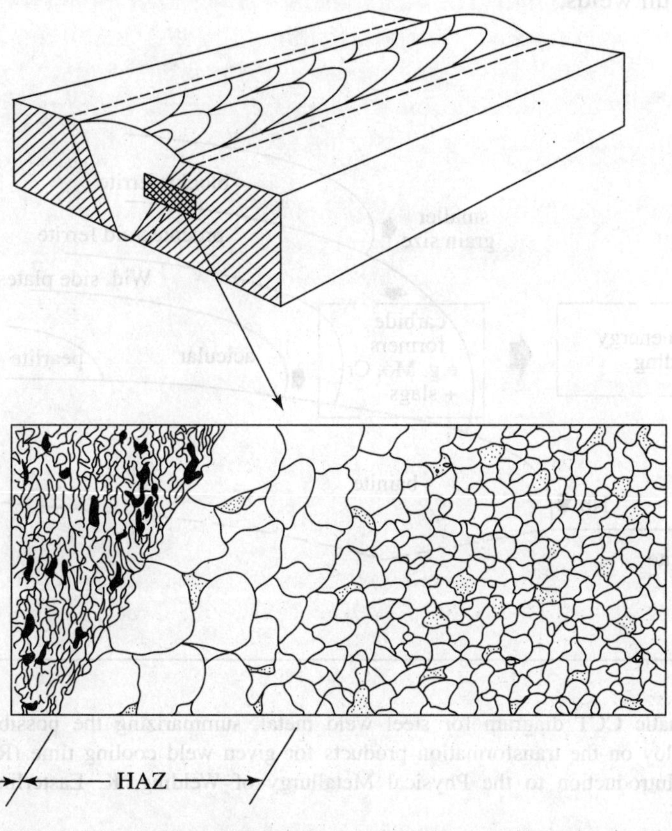

Fig. 10.7. Weld showing weld metal, heat-affected zone and base metal.(Reproduced with permission from Welding Essentials, W. Galvery and F.Marlow, Industrial Press, N.Y., 2001, p.299).

10.2.1 Weldability of C- and Low Alloy Steels

As steel is the most important engineering structural material, a more detailed description on the welding aspect is called for. A number of factors affect the welding of steels. Some of these are:

- The welding process itself which determines the weld pool size and geometry.
- The final composition of the melt as influenced by the filler wire, the base metal, fluxes, gases, moisture in the air, etc., and its effect on constitutional supercooling and segregation.
- The speed of welding and its effect on solidification speeds, crystal morphology and segregation.

- The weld thermal cycle and its influence on microstructural coarseness and type of transformation product produced during cooling.
- The effect of weld metal composition, particularly from dilution in high energy welding of microalloyed steels, on precipitation reactions, and especially during reheating or in multi-run welds.

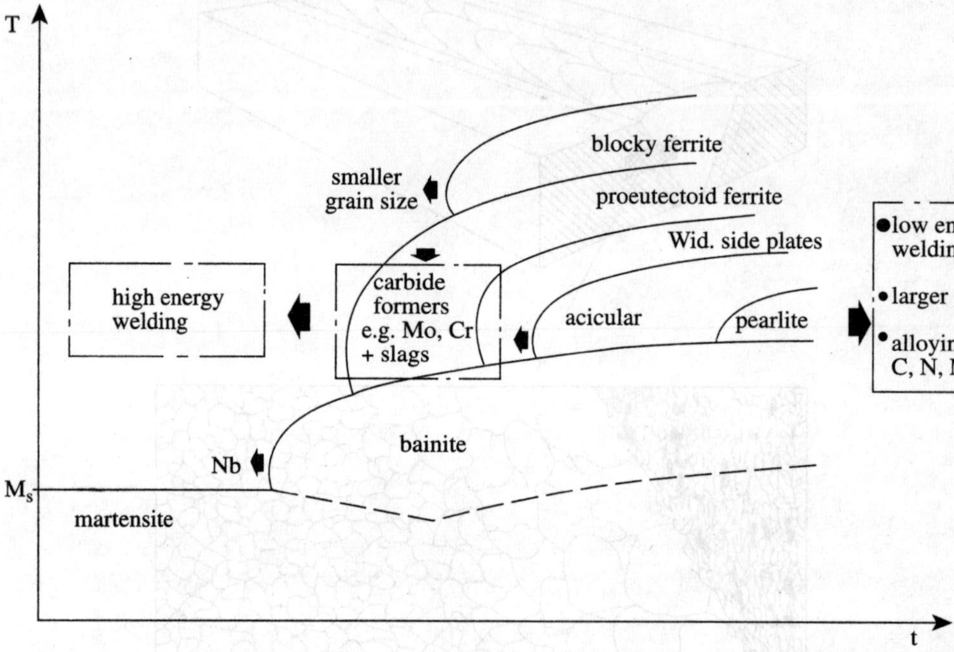

Fig. 10.8. Schematic CCT diagram for steel weld metal, summarizing the possible effect of microstructure and alloy on the transformation products for given weld cooling time (Reproduced with permission from Introduction to the Physical Metallurgy of Welding, K. Easterling, Elsevier, Oxford, 1983, p. 99).

To know the microstructure of the weld metal in steels, it is necessary to realize the influence of various factors cited above on the schematic CCT diagram. Figure 10.8 illustrates these features. Arrows pointing to left and right of the diagram denote shift of transformation to shorter times and vice versa. Strong carbide or nitride forming elements (e.g., Mo, Cr, Nb, V, Ti, Al), tend to suppress the blocky and proeutectoid ferrite, but not acicular ferrite or bainite. Niobium in particular tends to promote bainite formation. Slag inclusions also tend to promote the nucleation of acicular ferrite.

The HAZ in steel can be divided into a number of sub-zones. Figure 10.9 amply illustrates this feature. Each sub-zone refers to different type of microstructure and the associated mechanical properties. The thermal history plays a big role in width of each sub-zone. The changes in the HAZ is also dependent upon the prior thermal and mechanical history of the material. For example, the recrystallization behaviour during the heating cycle is affected by whether the original material was in cold rolled or annealed condition prior to welding. The onset and extent of grain growth zone is dependent on the presence of precipitates, if any, and their solubility at high temperatures.

Martensite obtained in HAZ is prone to hydrogen induced cracking (HIC). Such failures are commonly called as *underbead cracking*. To prevent HIC, careful selection of base- and weld-steel hardenability, cooling rate and post-weld heat treatment are required so as to keep

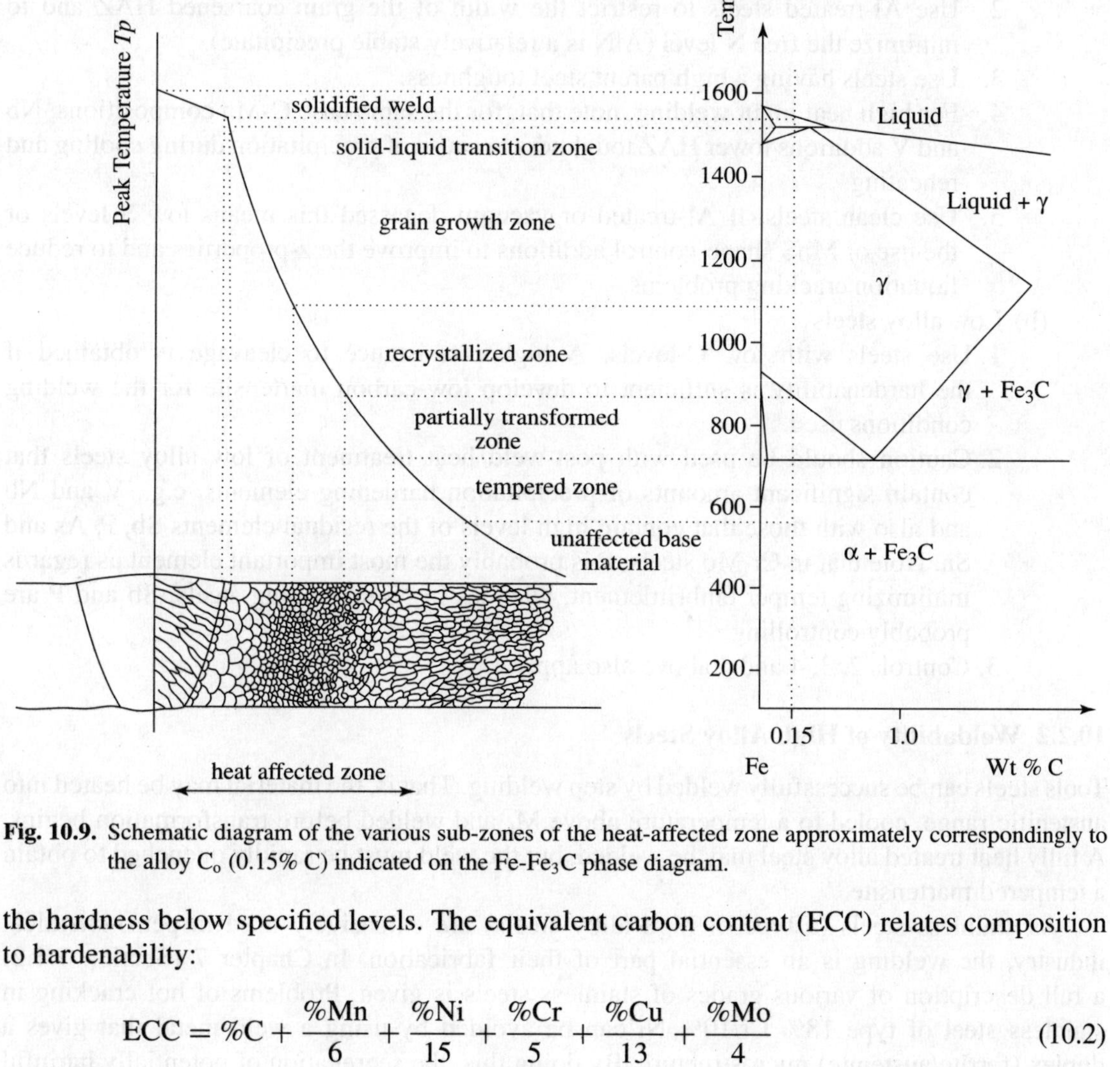

Fig. 10.9. Schematic diagram of the various sub-zones of the heat-affected zone approximately correspondingly to the alloy C_o (0.15% C) indicated on the Fe-Fe_3C phase diagram.

the hardness below specified levels. The equivalent carbon content (ECC) relates composition to hardenability:

$$ECC = \%C + \frac{\%Mn}{6} + \frac{\%Ni}{15} + \frac{\%Cr}{5} + \frac{\%Cu}{13} + \frac{\%Mo}{4} \quad (10.2)$$

The general rule is that HIC is unlikely if ECC < 0.4%. If the ECC is between 0.4 and 0.65%, preheating the weld is recommended to increase cooling time and permit more transformation of martensite to ferrite. If ECC > 0.65, post-weld heat treatment is required.

Example 10.3. Summarize the factors related to the selection of steels for obtaining optimum weldability and toughness under following categories: (a) C-Mn and microalloyed C-Mn steels, with C > 0.1 wt %, (b) low alloy steels.

Answer.

(a) C-Mn and microalloyed C-Mn steels with C > 0.1 wt %

1. Use steels with low levels of C and low carbon equivalent if possible.
2. Use Al-treated steels to restrict the width of the grain-coarsened HAZ and to minimize the free N level (AlN is a relatively stable precipitate).
3. Use steels having a high parent steel toughness.
4. For high heat input welding, note that, for the same base C-Mn compositions, Nb and V additions lower HAZ toughness because of precipitation during cooling and reheating.
5. Use clean steels. If Al-treated or vacuum degassed this means low S levels or the use of MnS shape-control additions to improve the z-properties and to reduce liquation cracking problems.

(b) Low alloy steels

1. Use steels with low C-levels. A higher resistance to cleavage is obtained if the hardenability is sufficient to develop low-carbon martensite for the welding conditions used.
2. Caution should be used with post-weld heat treatment of low alloy steels that contain significant amounts of precipitation hardening elements, e.g., V and Nb and also with those that contain high levels of the residual elements Sb, P, As and Sn. Note that in Cr-Mo steels, P is probably the most important element as regards minimizing temper embrittlement, whereas in Ni-containing steels, Sb and P are probably controlling.
3. Controls 2, 3, 4 and 5 above also apply.

10.2.2 Weldability of High Alloy Steels

Tools steels can be successfully welded by step welding. That is, the material may be heated into austenitic range, cooled to a temperature above M_s and welded before transformation begins. A fully heat treated alloy steel may be welded, but the weld must be rapidly quenched to obtain a tempered martensite.

As stainless steels (a type of high alloy steels) are extensively used in petro-chemical industry, the welding is an essential part of their fabrication. In Chapter 7 (Section 7.1.2) a full description of various grades of stainless steels is given. Problems of hot cracking in stainless steel of type 18% Cr/10% Ni can be avoided by using a weld metal that gives a duplex (ferrite/austenite) microstructure. By doing this, the segregation of potentially harmful impurity e.g., P and S can be reduced, since the dispersion of phases is fine and the phase boundary area is increased. The amount of ferrite (δ-ferrite 6-10 vol. %) that remains at ambient temperature due to fast cooling rates of welds in stainless steels is very important. This can be predicted with the help of *Schaeffler diagram* (Figure 7.2), which plots volume fraction of phases on the basis of Ni- and Cr-equivalents. If nitrogen is present in the weld metal, which is strong austenite stabilizer, a *De Long diagram* has to be used instead to estimate the residual ferrite (Figure 10.10). Any phosphorus impurity in the weld metal is contained in the δ-ferrite. Table 10.3 highlights the welding problems associated with stainless steels.

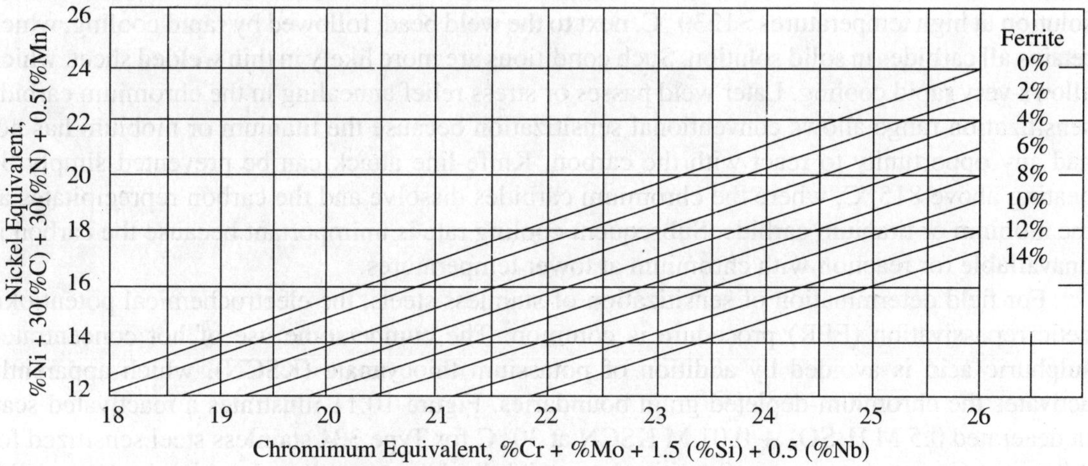

Fig. 10.10. A DeLong diagram for stainless steels.

Table 10.3. Welding of Stainless Steels.

Steel	*Typical Composition*	*Welding problem*	*Practical solution*
Austenitic stainless steels	18% Cr-8% Ni 25% Cr-20% Ni 25% Cr-12% Ni 15% Cr-35% Ni	Solidification cracking in weld metal	Maintain 4-10% ferrite phase in weld metal. Keep $C/\sqrt{Si} > 0.22$ in fully-austenitic weld metals.
		Weld decay	See text.
	18% Cr-13% Ni 1% Nb	Reheat (stress relief) cracking in the HAZ	Rapid heating through the critical temperature range. Use of electrodes having lower hot strength. Grind welds to remove stress concentration effects.
Martensitic stainless steel	12-16% Cr- 0.3% C (max)	Hydrogen-induced cracking in the HAZ	Preheat 200-400°C and post heat at 750°C low-hydrogen electrodes or austenitic stainless steel electrodes.
Ferritic stainless steel	16-30% Cr- 0.1% C	Low toughness in grain-coarsened HAZ	Preheat 200°C.

Weld decay in stainless steels: Intergranular corrosion with the heat affected zone of welded stainless steels is known as *weld decay*. The fundamental aspects have been described in

Chapter 6 (Section 6.8.6). Intergranular corrosion is related to the intergranular stress corrosion cracking of sensitized stainless steels. *Knife-line attack* is a highly localized form of intergranular corrosion that occurs only a few grain diameters immediately adjacent to the weld bead in stabilized Type 321 and 347 austenitic stainless steels. In Chapter 7, full details on stainless steels have been made. In knife-line attack, titanium or niobium carbides dissolve in solid solution at high temperatures >1230 °C, next to the weld bead, followed by rapid cooling, which retains all carbides in solid solution. Such conditions are more likely in thin welded sheet, which allows very rapid cooling. Later weld passes or stress relief annealing in the chromium carbide sensitization range allows conventional sensitization because the titanium or niobium has not had any opportunity to react with the carbon. Knife-line attack can be prevented simply by heating above 815 °C, where the chromium carbides dissolve and the carbon reprecipitates as the niobium or titanium carbide. Subsequent cooling rate is unimportant because the carbon is unavailable for reaction with chromium at lower temperatures.

For field determination of sensitization of stainless steels, the electrochemical potentiokinetic repassivation (EPR) procedure is common. The cumbersome use of hot concentrated sulphuric acid is avoided by addition of potassium thiocyanate (KSCN), which apparently activates the chromium-depleted grain boundaries. Figure 10.11 illustrates a reactivated scan in deaerated 0.5 M H_2SO_4 + 0.01 M KSCN at 30 °C for Type 304 stainless steel sensitized for 4 hrs at various temperatures. Sensitized material displays current peak which increases with degree of sensitization. A polished controlled surface area of stainless steel is passivated at +0.200 V vs saturated calomel electrode (SCE) for 2 minutes. Potential is then scanned in the

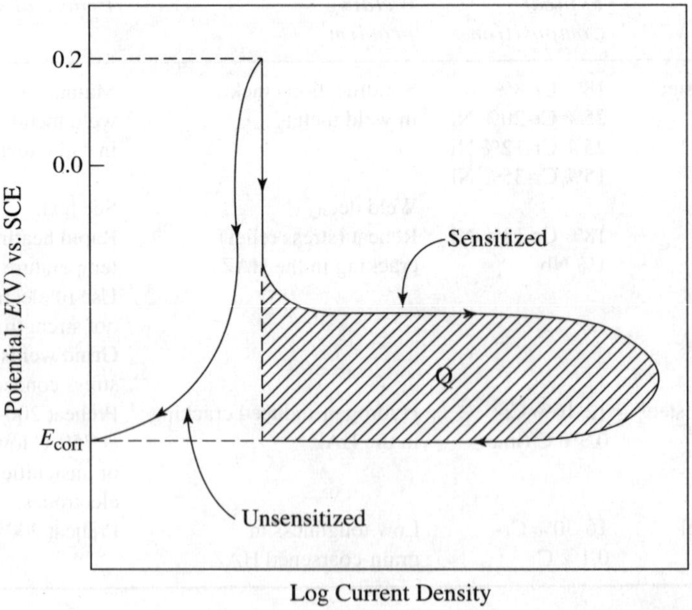

Fig. 10.11. EPR procedure for determining the sensitization to intergranular corrosion in an austenitic stainless steel.

active direction at a rate of 6 V/hr, down to the corrosion potential, where no further current is indicated. The charge, Q, the shaded area under the reactivation peak is measured automatically with coulometer in the polarization circuit. Q (Coulombs), a measure of sensitization is normalized by the grain boundary area GBA (cm^2) to a sensitization number, P_a, thus

$$P_o(C/cm^2) = \frac{Q}{\text{GBA}} \tag{10.3}$$

GBA is calculated from

$$\text{GBA} = A_s[5.095 \times 10^{-3} \exp(0.347x)] \tag{10.4}$$

where A_s is the sample area in cm^2 and x is the ASTM grain size at 100 X magnification. Various sensitization numbers for stainless steels are grouped in Table 10.4.

Table 10.4. Various Degrees of Sensitization Values (P_a) in Stainless steels.

P_a Value	*General Interpretation*
<2	Unsensitized microstructure, no pitting
2-5	Slightly sensitized microstructure; Pitting and limited intergranular attack
5-15	Sensitized microstructure; Pitting and attack of entire grain boundaries
>15	Heavily sensitized; Strong grain boundary attack

A filler alloy for austenitic stainless steel weldments is usually at different composition to maximize resistance to weld decay. For example, Type 308, usually specified as the filler alloy for Type 304 stainless steel, has somewhat higher chromium and nickel, which confers higher corrosion resistance to the weld metal. Further, high chromium in Type 308 causes weldments to freeze in a duplex, austenite plus ferrite structure. This has the following advantages:

- Prevention of microcracking at interdendritic boundaries.
- Resistance to intergranular corrosion, because carbides segregate at the dispersed austenite-ferrite boundaries and cannot easily produce chromium depletion at the austenite grain boundaries.
- Prevention of stress corrosion cracks propagation.

Example 10.4. Explain why ferritic stainless steels are more difficult to weld? Comment on its filler alloys.

Answer.

1. They are susceptible to grain growth and coarsening at welding temperatures, which may result to losses of low temperature toughness.
2. Interstitial contamination from carbon, nitrogen, oxygen and hydrogen sensitize the alloy to intergranular corrosion. Therefore, extreme cleanliness and care with gas-shielded tungsten arc processes are recommended.
3. Variations within specified composition limits often result in the formation of austenite, which embrittles the weldment when transforming to martensite.

4. Postweld annealing, which transforms the martensite to more ductile ferrite, is often required, but is expensive.

The filler alloys for ferritic stainless steel have the advantage of similar thermal expansion and corrosion properties. However, austenitic filler alloys (Type 308 and 309) are often used with the ferritic stainless steels to improve weldment ductility. It must also be noted that dissimilar metal galvanic cells may form in some environments.

10.2.3 Weldability of Cast Irons

The weldability of cast iron varies greatly. A high nickel filler is frequently used to stabilize the graphite form. Preheating and slow cooling are also useful. In welding grey iron the welding rod is enriched in silicon; while for spheroidal graphite formation magnesium is incorporated in the rod for welding nodular cast iron. Malleable iron reverts to brittle white iron, thus reducing the weld toughness. To remove this problem, the weldment is heat treated. Otherwise the joint can be prepared by brazing instead of welding.

10.2.4 Weldability of Nonferrous Metals and Alloys

Aluminium and magnesium bear similarity from the welding aspect. Most alloys are readily weldable, particularly with an inert gas envelop. Otherwise the oxide film must be removed with a powerful flux, which in turn must be washed off after welding to prevent corrosion. Moisture must be kept out, because it reacts to give an oxide and hydrogen, which embrittles the joint by causing porosity. Adequate precaution against overheating must be exercised. Preheating is desirable because of high thermal expansion coefficient. Alloys amenable to heat treatment are often heat treated after welding. If this is not possible, a different filler is used, e.g., Al-Si for aluminium alloys.

Deoxidized copper is readily welded, especially if the filler contains phosphorus. Tough-pitch copper cannot be welded because its oxygen content (~0.15%) reacts with hydrogen and carbon monoxide to form water and carbon dioxide. Both of these cause embrittlement due to porosity formation. In welding brasses, dezincification is a problem. This is mitigated either by selecting filler rich in zinc or Al/Si addition to preferentially form an oxide, which will reduce vaporization. Tin bronze suffers from hot shortness because of a very wide solidification range. Two remedies are available: addition of P in the filler rod or post heating operation so as to dissolve brittle nonequilibrium intermediate phase. In case of aluminium bronze, like aluminium, the oxide formed must be properly fluxed.

Nickel and its solid solution are easily weldable. Care must be practiced in case of superalloys where some low ductility temperature range exists. All nickel alloys are very sensitive to sulphur impurity.

Reactive metals like titanium and zirconium require inert gas to prevent against oxidation. In case of refractory metals e.g., W, Mo, Nb, Ta, electron beam welding method is followed, since their oxides are rather volatile.

Example 10.5. Where can the 4000 (Al-Si) and 5000 (Al-Mg) series Al-alloy welding rods be used? Justify your answer.

Answer. 4000 series shall be used for welding castings as it has more fluidity and lower melting point. It is not suited to weld Al-Mg alloys; as it forms Mg_2Si, which decreases ductility and increases crack sensitivity. 5000 series is useful for welding wrought structural aluminium alloys such as 5XXX and 6XXX. The limitation is that the latter type filler is not suited for service temperature exceeding 65 °C. The formation of Al_2Mg at elevated temperatures at the grain boundaries makes the alloys prone to stress corrosion.

10.3 WELDMENT CRACKING AND POROSITY

Cracking in a weldment is very common, unless the welder is knowledgeable. Mainly these are of two types:

- Those attributed to the welding process itself.
- Those occurring during service.

The latter is mainly concerned with external factors e.g., environment, vibration or thermal cycling. These factors have been dealt in generality in Chapters 5 and 6. In this section we are mostly associated with cracking associated with the welding process.

Some of the attributes for weldment cracking are:

- Hydrogen attack
- Gas porosity
- Solidification cracking
- Liquation cracking, HAZ burning or hot tearing
- Lamellar tearing
- Cold cracking or hydrogen cracking
- Reheat cracking.

Many times the cracks are categorized with respect to the position to the weld. Figure 10.12 illustrates major types of cracks.

The major causes and remedies for weld metal and heat affected zone (HAZ) cracking are summarized in Table 10.5.

In case of carbon or C-Mn steels welding conditions can be selected to avoid the cooling rates at which martensite is formed. But in low alloy steels, since hardenability is high, this cannot be avoided. HAZ is highly susceptible to hydrogen. One of the method to minimize this is to preheat the steel, so that hydrogen is diffused out of the weld.

The reheating of weldment is the final operation to release any residual stresses, which are elastic in nature. These are converted into plastic strains. Reheat cracking often occurs in HAZ of the steel. Cracking is more likely in very heavy section and where there is a stress concentration such as an unused root or the toe of large fillet weld. From the microstructural point of view, the precipitation of carbides of secondary hardening elements during stress relief is the main cause of such cracking. Steel composition plays a role in stress relief cracking. A typical formula applicable is

$$P_{Sr} = Cr + Cu + 2Mo + 10V + 7Nb + Ti - 2 \qquad (10.5)$$

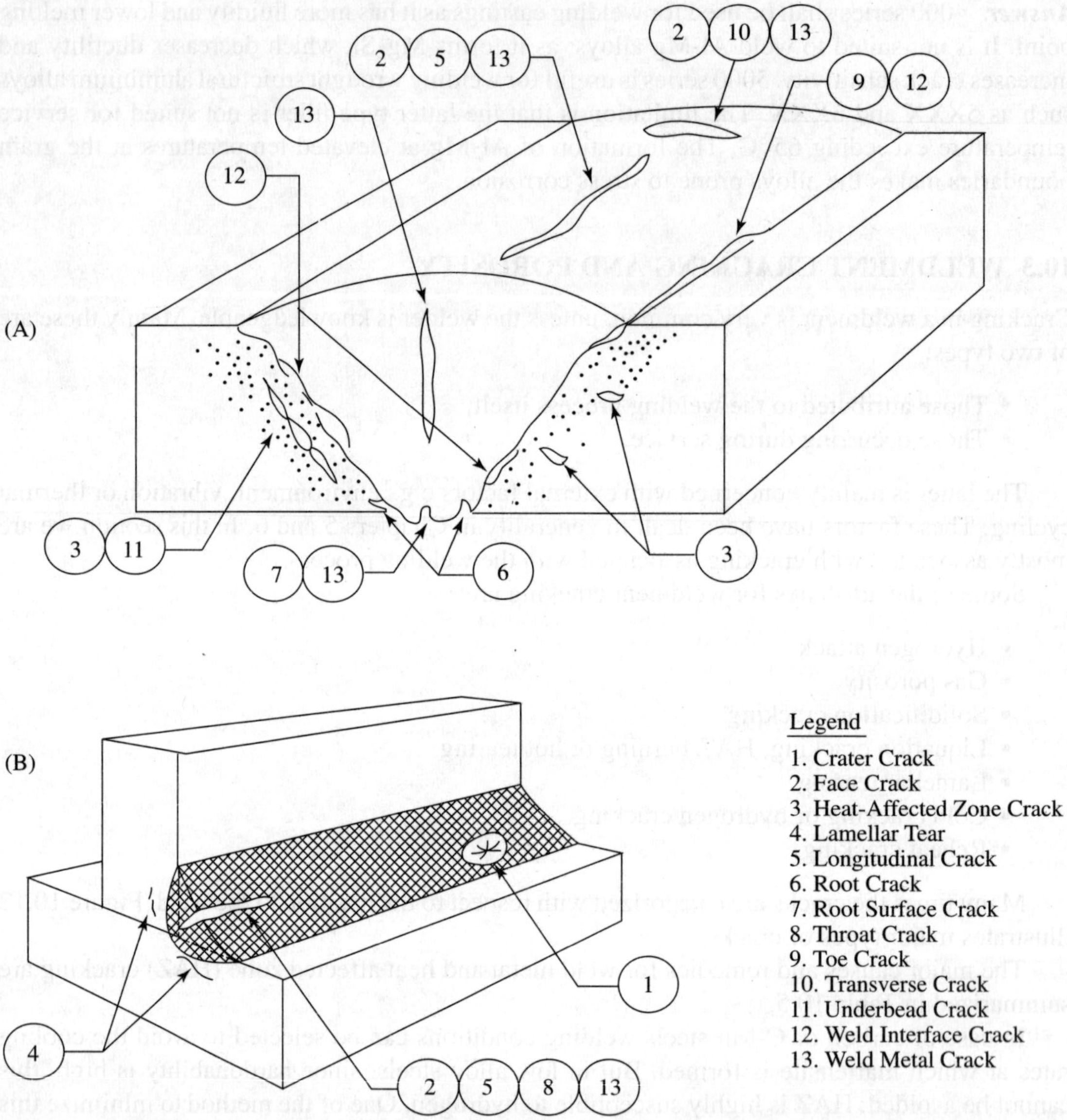

Fig. 10.12. Cracks in and around a weld (Reproduced with permission from Welding Essentials. W. Galvery and F. Marlow, Industrial Press, N.Y., 2001, p. 260).

When $P_{Sr} \geq 0$, stress relief cracks may occur. This formula is not applicable to low carbon (<0.1%) or high chromium (>1.5%).

Porosity in the weldment depends on three consecutive factors- the time allowed for gas absorption at high temperatures, the time required for bubble nucleation and growth and the ability of the gas bubbles to rise through the weld pool and break free at the weld surface. For these reasons, changes in prewelding parameters might have different effects on different

Table 10.5. Common Causes and Remedies of Cracking.

Causes	*Remedies*
Weld Metal Cracking	
Highly rigid joint	Use preheat. Relieve residual stress mechanically Minimize shrinkage stresses using backstep welding sequence.
Excessive dilution	Change welding current and travel speed.
Defective electrodes	Change to new electrodes. Bake electrodes to remove moisture.
Poor fit-up	Reduce root opening; build up the edges with weld metal.
Small weld bead	Increase electrode size, raise welding current; reduce travel speed.
Angular distortion	Change to balanced welding on both sides of joint.
Crater cracking	Fill crater before extinguishing arc; use a welding current decay device when terminating weld bead.
Heat-Affected Zone Cracking	
Hydrogen in welding atmosphere	Use low-hydrogen welding process; preheat and hold after welding or postweld heat treat immediately.
Hot cracking	Use low heat input, deposit thin layers; change base metal.
Low ductility	Use preheat; anneal the base metal.
High residual stress	Redesign the weldment, change the welding sequence; apply intermediate stress-relief heat treatment.
High hardenability	Preheat; increase heat input.
Brittle phases in microstructure	Solution heat treat prior to welding.

base metals. For example, in aluminium and steel welding, raising the welding speed, promotes porosity, as the generated gas bubbles are more easily trapped. In titanium, the opposite is true. Type of welding process also plays its role. For example if low heat input, low hydrogen manual metal-arc practice is specified, starting porosity may be problematic. This is due to the fact that the small quantities of shielding gas evolved from the electrode coating may be insufficient to exclude air. Table 10.6 gives a listing of major causes and preventions of porosity in metals.

10.4 WELD DESIGN

After understanding the basic science and technology behind welding, described in earlier sections, the following general guidelines for welding may be appreciated:

- Product design should minimize the number of welds, as welding can be costly when it is not automated.
- The weld location should be selected to avoid excessive stresses or stress concentrations in the welded structure, as well as for appearance.
- The weld location should be selected so as not to interfere with subsequent processing of the part or with its intended use and appearance.

Table 10.6. Cause and Prevention of Porosity.

Metal	*Absorbed gas causing porosity*	*Chemical reaction in weld pool*	*Prevention typical deoxidant*
Mild steels and low-alloy steels	oxygen and nitrogen	$C + [O] = CO_{(gas)}$	Ti, Al, Si, Mn
Stainless steels			Ti, Si
Nickel +			Ti
Copper and copper alloys	hydrogen	$[2H] + [O] = H_2O_{(gas)}$	P
	oxygen		Ti
	nitrogen		B
Aluminium +	hydrogen		
Titanium +	hydrogen oxygen argon		
Magnesium +	hydrogen		
Niobium +	hydrogen		

+ Pure shielding gas important.

- Parts should fit properly before welding; the method employed to produce the edges (sawing, machining, shearing and flame cutting) can affect weld quality.
- Modification of the design may avoid the need for edge preparation.
- Weld-bead size should be kept to a minimum to conserve weld metal.
- Mating surfaces for some processes may require uniform cross-sections at the joint.

Weld Distortion a common feature in fusion welding must be avoided. To achieve this various methods may be adopted. Joint preparation angle and the root pass should be minimized (Figure 10.13a). Balancing welding by using a double-V joint in preference to a single-V joint can reduce angular distortion (Figure 10.13b). Placing welds about the neutral axis also reduces distortion (Figure 10.13c). Other methods to avoid distortions are presetting, elastic pre-springing and pre/post-weld heating.

10.5 BRAZING AND SOLDERING

Brazing and soldering are materials joining processes that rely on the melting, flow and solidification of a filler metal to form leak-tight seals or strong structural bonds. The common feature in both the processes are: (a) only the filler metals and not the parts being joined are melted during processing and (b) the filler metals must wet the base metal and be drawn into or held in the joint by capillary attraction. According to American Welding Society a temperature of 450 °C is chosen as a boundary line. Above this temperature the process is known as brazing, while below it as soldering. This distinction is no doubt is rather arbitrary. There are two types of brazing processes: (a) brazing as described above, and (b) braze welding, in which the filler metal is deposited at the joint (Figure 10.14).

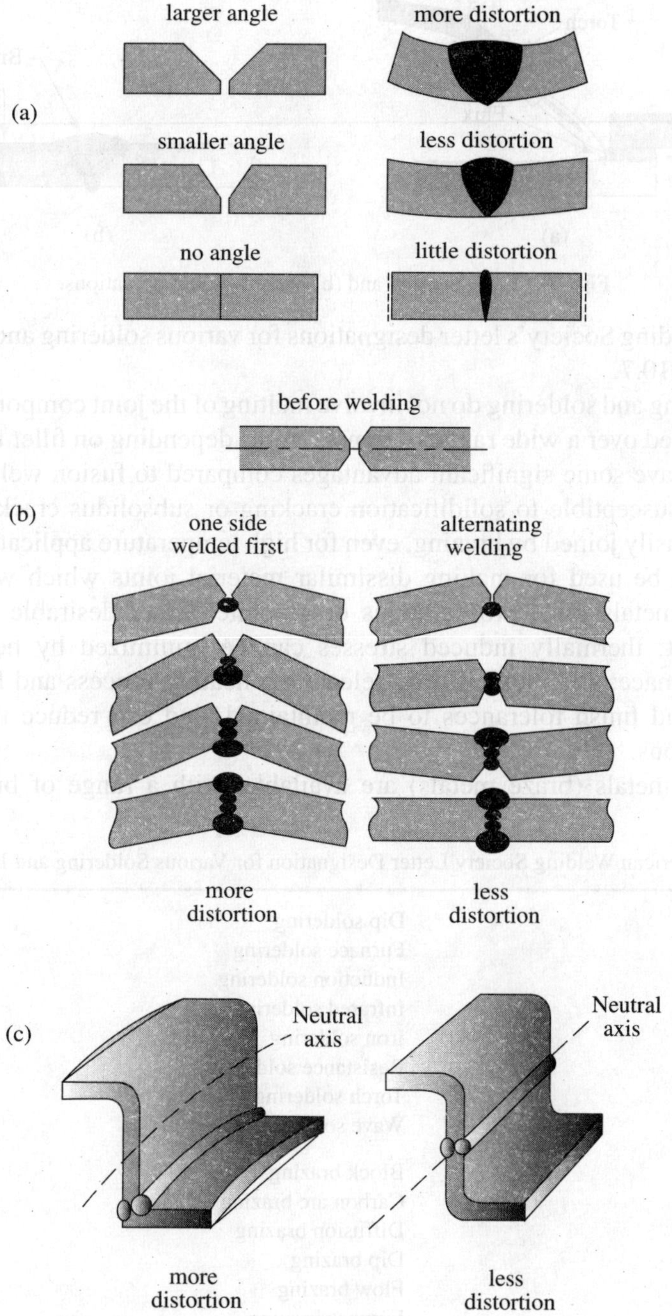

Fig. 10.13. Reducing distortion during welding: (a) by reducing volume of weld metal and by using single pass penetration welding, (b) by using double V joint and welding alternately on either side of joint, and (c) by placing weld around neutral axis. (Reproduced with permission from Welding Metallurgy, 2nd ed., S. Kou, John Wiley & Sons, Inc., N.Y., 2003, p. 129).

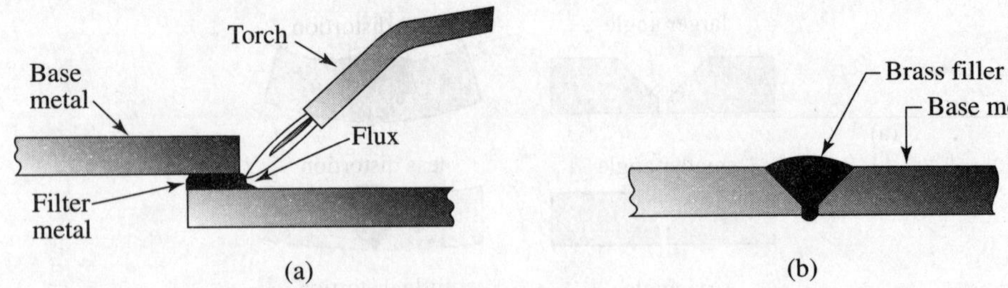

Fig. 10.14. (a) brazing and (b) braze welding operations.

American Welding Society's letter designations for various soldering and brazing processes are given in Table 10.7.

Because brazing and soldering do not involve melting of the joint component materials, they can be accomplished over a wide range of temperatures depending on filler metal composition. These processes have some significant advantages compared to fusion welding. For instance, alloys which are susceptible to solidification cracking or subsolidus cracking due to limited ductility may be easily joined by brazing, even for high temperature applications. Also, brazing and soldering can be used for making dissimilar material joints which would otherwise be impossible, e.g., metals joined to ceramics or graphite. Other desirable attributes of these processes are that: thermally induced stresses can be minimized by heating components uniformly in a furnace; and, flexibility in selecting a heating process and filler metal permits close assembly and finish tolerances to be maintained, and can reduce the need for costly secondary operations.

Several filler metals (braze metals) are available, with a range of brazing temperature

Table 10.7. American Welding Society Letter Designation for Various Soldering and Brazing Processes.

Soldering	Dip soldering	DS
	Furnace soldering	FS
	Induction soldering	IS
	Infrared soldering	IRS
	Iron soldering	INS
	Resistance soldering	RS
	Torch soldering	TS
	Wave soldering	WS
Brazing	Block brazing	BB
	Carbon arc brazing	CAB
	Diffusion brazing	DB
	Dip brazing	DB
	Flow brazing	FLB
	Furnace brazing	FB
	Induction brazing	IB
	Infrared brazing	IRB
	Resistance brazing	RB
	Torch brazing	TB

Table 10.8. Characteristics of Typical Solders and Brazing Filler Metals.

Material	*Composition*	*Melting range °C*	*Typical uses*
Solders:			
35-65 solder	35Sn 65Pb	183-245	Wiping solder
45-55solder	45Sn 55Pb	183-226	For automobile radiator cores; general purpose solder
60-40 solder	60Sn 40Pb	183-187	For use where temperature requirements are critical
Cd-Ag solder	95Cd 5Ag	337-390	High-temperature solder
Brazing filler metals			
B-CuZn-2	57Cu 42Zn 1Sn	887-900	General-purpose alloy for steel, copper alloys, and nickel alloys
B-Cu	99+Cu	1000	Furnace brazing alloy
B-CuP-2	93Cu7P	725 - 807	For self-fluxing brazing of copper alloys
B-AG-1*	45Ag 24Cd 16Zn 15Cu	608-618	Low-melting, free-flowing braze for general-purpose work
B-Ag-8*	72Ag 28Cu	780	For applications, such as vacuum tubes, in which volatile elements are harmful
B-AlSi-1	95Al 5Si	575-630	General-purpose alloy for brazing aluminium
B-Mg	89Mg 9A1 2Zn	410-600	General-purpose alloy for brazing magnesium
BNiCr	70Ni 16.5 Cr bal. mostly Fe	1010-1065	Heat-resistant alloy used for fabricating stainless steels or high-nickel alloys for jet engines

*Silver "solders."

(Table 10.8). The choice of filler metal and its composition is important in order to avoid embrittlement of joint, formation of brittle intermetallic compounds at the joint and galvanic corrosion in the joint. Braze metals unlike those for other welding operations, generally have significantly different compositions than those of the metals to be joined.

Another important ingredient during brazing is flux, which prevents oxidation and also removes any oxide films from the work piece surfaces. Brazing fluxes are generally made of borax, boric acid, borates, fluorides, and chlorides and are available as paste, slurry or powder.

There are various variations in brazing methods. Some of them are:

Torch Brazing,
Furnace Brazing,
Induction Brazing,
Resistance Brazing,
Dip Brazing,
Infrared Brazing,
Diffusion Brazing,
High Energy Beams Brazing.

Solders are usually tin-lead alloys in various proportions. Other alloys like Sn-Zn, Pb-Ag, Cd-Ag and Zn-Al are also used for better joint strength and special applications. Because of environmental issues lead is now-a-days avoided and tin based solders (96.5 Sn-3.5 Ag and 42 Sn-58 Bi) are more and more used.

Metals like copper and precious metals are easy to solder, while iron and nickel are difficult to solder. Aluminium and stainless steels are difficult to solder, because of their strong, thin oxide film (Al_2O_3, Cr_2O_3). Sometimes a preliminary plating of the material before actual soldering is desirable. Examples are cast iron, magnesium, titanium, graphite and ceramics.

Example 10.6. Generally cemented carbide cutting tools are brazed on steel shanks. A filler metal of typical composition 52 Cu-38 Mn-5.5 Ni is used. Answer the following:

1. What is the role of Ni in the filler metal?
2. A use of composite filler metal, consisting of a sandwich of two layers of a silver base filler metal and a copper shim is often recommended. Why?
3. Why the hydrogen sintered cemented carbides are poorly wetted by copper?
4. What are the other reasons for weak brazing?
5. Suggest the reasons why the right hand side joint design in Figure Ex. 10.6 is recommended over those given in left hand side.

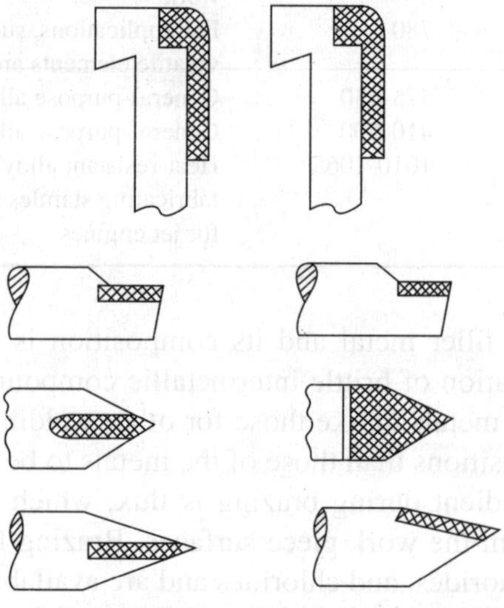

Figure Ex. 10.6

6. For mass production induction brazing is carried out. What limitation it has?
7. What composition/microstructure of cemented carbide are preferred for brazing?

Answer.

1. Increases bond strength.
2. The ductile copper layer diffuses during brazing to decrease the residual stress generated in the joint.
3. Alumina is often used as packing in the sintering boats. Cobalt binder in the cemented carbide gets contaminated by aluminium. Al during brazing gets oxidized into alumina, which prevents brazing.
4. Draining of cobalt from the interfaces and also of carbon causing η-phase formation.
5. The spreading of the filler metal more efficient in the right hand side joint designs (Figure Ex. 10.6).

6. Since the method is fast, the relief of internal stresses is problematic.
7. Coarse grained carbides with a relatively larger proportion of cobalt binder is preferred, as it minimizes brazing cracks.

(For further details about cemented carbides refer Chapters 7 and 9).

10.6 TRANSIENT LIQUID PHASE BONDING (TLP)

Many times the process is also known as transient liquid insert metal (TLIM) joining. The process is unique in the sense that the joint is nearly invisible and possesses strengths and other properties similar to the base metal. A thin interlayer material, such as a braze foil, is clamped at the faying surface and the entire assembly is brought to an elevated temperature. Bonding results in melting point depressants in the liquid interlayer to diffuse into the metal, which results in isothermal solidification. As the liquid phase is temporarily formed, the solidified bonding region consists of a primary solid solution similar to the composition of the base metal. Consequently, the excellent joint quality can be expected without any fusion welding defects and any brittle phases. The process, thus, consists of three stages, viz. a dissolution process of base metal, an isothermal solidification process and a homogenizing process.

TLP process has been applied in joining nickel-base super-alloys, titanium, boron-aluminium composites and semiconductor materials. The interlayer composition is selected to minimize bonding time, provide adequate wetting and avoid the formation of intemetallics.

The advantages or disadvantages of the TLP process is as follows:

Advantages

- Base metal properties at joint
- No interface remains after bonding
- Self homogenizing
- Low joint pressure required
- Intermetallic formation can be avoided
- Minimum surface preparation
- Large and complex shapes bonded simultaneously

Disadvantages

- Long bonding times (hours)
- Restricted to high temperatures, $T/T_m = 0.6$
- Fast diffusers required, preferably interstitial elements
- Rapid heat-up required
- Close fit-up required
- Post bonding heat treatment for age hardening alloys is required.

10.7 ADHESIVE BONDING

Adhesive bonding is gaining significance, of late, as it is used in large scale assembly of load bearing components in aircraft. A typical common example of adhesive bonding is that of several layers of wood and glue. Three basic types of adhesives are used: natural adhesives, inorganic adhesives, and synthetic organic adhesives. Out of these, synthetic organic adhesives are the most important, as load bearing structural parts are successfully bonded. Adhesives are available in various forms such as liquid, pastes, solutions, emulsions, powder, tape and films. When applied, adhesives are typically on the order of 0.1 mm thick.

Many of the organic adhesives can be combined to optimize their properties; examples are epoxy-silicone, nitrile-phenolic and epoxy-phenolic. The least expensive adhesives are epoxies and phenolics, followed by polyurethanes, acrylics, silicones and cyanoacrylates. Adhesives for high temperature applications up to about 260 °C, such as polymides and polybenzimidazoles are generally expensive ones.

Table 10.9 summarizes the characteristics of some chemically reactive structural organic adhesives. Surface preparation is very important in adhesive bonding, as joint strength depends greatly on the absence of dirt, dust, oil and various other contaminants. On the other hand, a porous or thin and strong oxide film may be desirable, particularly one with some surface roughness to improve adhesion. However, the roughness must not be too high, or else air may be trapped and joint strength becomes poor.

Table 10.9. Typical Properties and Characteristics of Chemically Reactive Structural Adhesives.

	Epoxy	*Polyurethane*	*Modified Acrylic*	*Cyanocrylate*	*Anaerobic*
Impact resistance	Poor	Excellent	Good	Poor	Fair
Tension-shear strength, MPa	15.4	15.4	25.9	18.9	17.5
Peel strength, N/m	<525	14,000	5250	<525	1750
Substrates bonded	Most	Most smooth, nonporous	Most smooth, nonporous	Most nonporous metals or plastics	Metals, glass, thermosets
Service temp. range,°C	-55 to 120	-160 to 80	-70 to 120	-55 to 80	-55 to 150
Heat cure or mixing required	Yes	Yes	No	No	No
Solvent resistance	Excellent	Good	Good	Good	Excellent
Moisture resistance	Excellent	Fair	Good	Poor	Good
Gap limitation, mm	None	None	0.75	0.25	0.60
Odor	Mild	Mild	Strong	Moderate	Mild
Toxicity	Moderate	Moderate	Moderate	Low	Low
Flammability	Low	Low	High	Low	Low

Source: Advanced Materials & Processes, July 1990, ASM International.

Design of adhesive bonding should ensure that joints are subjected to compressive, tensile or shear force, but not peeling or clearage. Figure 10.15 shows several joint designs for adhesive boning.

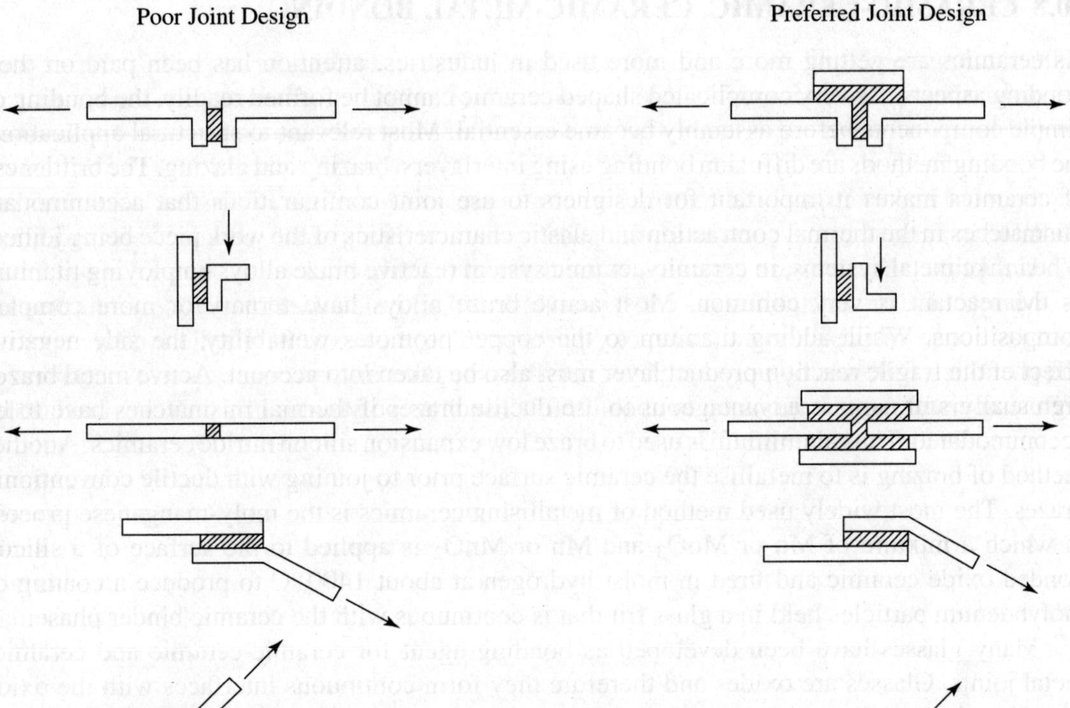

Fig. 10.15. Adhesive joint designs.

10.8 CERAMIC-CERAMIC, CERAMIC-METAL BONDING

As ceramics are getting more and more used in industries, attention has been paid on their bonding aspects. As very complicated shaped ceramic cannot be formed readily, the bonding of simple components before assembly became essential. Most relevant to structural applications the bonding methods are diffusion bonding using interlayers, brazing and glazing. The brittleness of ceramics makes it important for designers to use joint configurations that accommodate mismatches in the thermal contraction and elastic characteristics of the work piece being joined.

Unlike metal systems, in ceramic-ceramic system reactive braze alloys employing titanium as the reactant is very common. Most active braze alloys have ternary or more complex compositions. While adding titanium to the copper promotes wettability, the side negative affect of the fragile reaction product layer must also be taken into account. Active metal brazes are usually stiff, so it is advantageous to use ductile brazes if thermal mismatches have to be accommodated. Thus aluminium is used to braze low expansion silicon nitride ceramics. Another method of brazing is to metallise the ceramic surface prior to joining with ductile conventional brazes. The most widely used method of metallising ceramics is the moly-manganese process in which a mixture of Mo or MoO_3 and Mn or MnO_2 is applied to the surface of a silicon bonded oxide ceramic and fired in moist hydrogen at about 1400 °C to produce a coating of molybdenum particles held in a glass frit that is continuous with the ceramic binder phase.

Many glasses have been developed as bonding agent for ceramic-ceramic and ceramic-metal joints. Glasses are oxides and therefore they form continuous interfaces with the oxide ceramic substrate. In case, metals are to be wetted, it is preferred to oxidize their surfaces to get adherent film. In case of nonoxide ceramics, the joining is a bit more tedious. A ceramic could be coated with a metal whose outer surface was then oxidized. Another method is based on the development of glassy binder base ceramic. For example SiO_2-MgO-Al_2O_3, Y_2O_3-SiO_2-Al_2O_3 and Y_2O_3-SiO_2-Al_2O_3-N glasses have been developed to bond silicon nitride, doped with magnesia or yttria. A range of useful glazes have also been developed based on glass-ceramics.

To a design engineer the thermal expansion coefficient data of metals and ceramic type materials are essential. Table 10.10 summarizes those data. Table 10.11 summarizes the characteristics of joint and seal types for ceramics and glasses to metals or to themselves.

Example 10.7. Explain why brazing Cu for joining Al_2O_3 invariably contains some titanium? Further instead of Cu, many times Cu-Ag or Cu-Sn alloys is selected. Why?

Answer. Titanium promotes wetting and this type of brazing is known as *reactive brazing*. Presence of Ag, Sn or In in copper, decreases titanium solubility in copper and thus allows Al_2O_3 to get wetted at much lower titanium concentration. Sn or In in copper is more potent in decreasing the solubility of titanium as compared to Ag in Cu.

10.9 JOINING OF POLYMERS

Thermoplastics can be joined using external or internal heat source. On the other hand, thermosetting plastics can be joined either by mechanical means (fasteners) or by solvent bonding.

Table 10.10. Thermal Expansion Coefficient (α, ppm K^{-1}) of Metals, Their Alloys and Ceramic-Type Materials.

Metals	α	*Silicate glasses*	α	*Polycrystalline ceramics*	α
Zinc[a]	19-35				
Aluminium[a]	16-25			Ceramic type materials are not likely to be available in this range	
Copper	17-21				
Austenitic stainless steel	15-18				
Cast irons	11-14			Forsterite	11
Carbon and alloy steels	11-13	Potash-soda-lead	9-12	High-expansion glass-ceramics	b
Ferritic stainless steels	9-11				
Titanium[a]	8-11			Enstatite	8
Precious metals[a]	7-17	Soda-lime	8	Alumina	7
				Cemented carbide	5
Nickel (also Fe-Ni-Co and Ni-Fe alloys)	1-15	Low-expansion borosilicate	3	Silicon carbide	4
				Cordierite	3
				Silicon nitride	3
		High silica	1		
		Silica	0.5	Low-expansion glass-ceramics	0-1

[a]Including alloys of these metals.

b Glass-ceramics can, in theory, be 'tailored' to cover the expansion range from zero (or slightly negative) to about 13×10^{-6}/K. Check with manufacturers to see what is available from their production lines.

Table 10.11. Characteristics of Joint and Seal Types for Ceramics and Glasses to Metals or to Themselves.

Property	*Type of joint or seal*				
	Mechanical	*Cemented*	*Adhesive*	*Braze metal*	*Glass fusion*
Cost	Low-medium	Low	Low	Medium high	Low-medium
Strength (MN m^{-2})	Range 5-50	Typically 5	Range 5-30	Range 50-250	Range 10-20
Resistance to temperature and/or chemical attack	Poor-medium	Poor	Poor	Medium-good	Medium-good
Vacuum tightness	No	No	Hermetic (usually)	Yes	Yes

The fusion welding of metals and thermoplastics have many similarities. Like in metals, filler material of the same type of polymer may also be used. Inert shielding gas e.g., nitrogen is useful in preventing degradation at elevated temperature. Because of the low thermal conductivity of thermoplastics, a high rate of heating is avoided. Otherwise, the burning or charring of the surfaces is witnessed.

External heat sources for joining thermoplastics are: hot air or gases, heated tools and dies, radio-frequency or dielectric heating electrical resistance heating, and low power laser. The internal heat sources lead to ultrasonic, friction or orbital welding processes. Ultrasonic welding is useful for amorphous polymers, while friction welding (also known as spin welding or vibration welding) is useful for joining polymers with high degree of crystallinity.

Adhesive bonding (Section 10.7) is very common for polymers. Adhesive bonding of polyethylene, polypropylene and Teflon can be difficult, because these polymers have low surface energy. The surface, therefore, has to be treated chemically to improve bond strength.

10.10 JOINING OF SINTERED METALS AND ALLOYS

The description of sintering has been made in the section powder forming of Chapter 8. One method of joining sintered parts is equivalent to 'shrink' fit by using two materials with different growth during sintering. Powder Metallurgy parts can also be joined during the infiltration process by assembly the component parts and infiltrating the assembly during sintering. In sinter-welding, joining of two metal parts is done by inserting a suitable powder of the same metal between the parts to be joined, and then hot pressing the assembly in a suitable atmosphere. In case of activated sinter welding, the reactive metal hydride decomposition and hot pressing are combined.

There are no special technical problems in joining sintered to wrought steel parts as long as the joints are conveniently dimensioned and the operational parameters of the welding process are well selected. Best welding results for P/M steels are achieved when the component density is 6.8 g/cm^3 (88% of theoretical density) or higher and carbon content is held to the minimum practical for the application.

Sometimes the very presence of pores in sintered parts may pose problems during joining for example, brazing is unsuitable because the molten brazing alloying prefers to infiltrate the pores by capillary action. However, in case of necessity, the parts to be joined should first be infiltrated using an alloy with melting point equal to or higher than that of the brazing alloy.

QUESTIONS AND PROBLEMS

Welding Processes

10.1 What are the advantages and disadvantages of submerged arc welding?

10.2 What are the adverse effects of welding heat on metals?

10.3 What are the advantages and disadvantages of friction welding?

10.4 In Chapter 8, you have studied about solidification processing in connection with casting. How fusion-welding solidification is similar or dissimilar to casting solidification?

10.5 Why is diffusion bonding, when combined with superplastic forming of sheet metals, an attractive fabrication process?

10.6 The following weldment designs (Figure P.10.6) are faulty from the distortion point of view. Suggest the correct ones giving reasons.

10.7 Show by calculation that there is about 50% more weld metal in a fillet weld than in a butt.

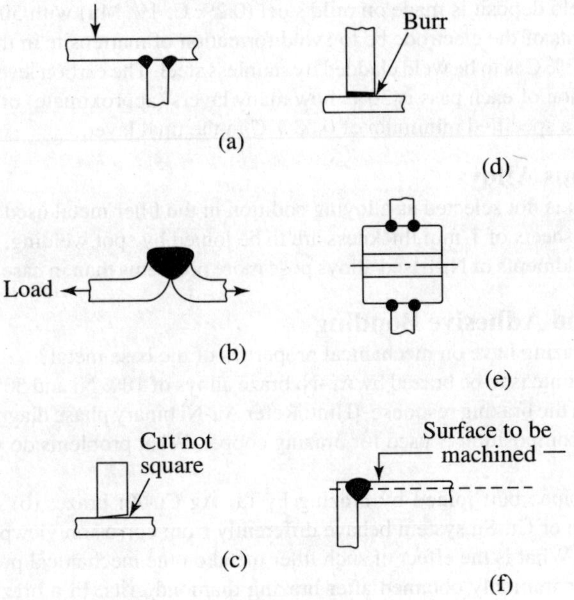

Figure P. 10.6

10.8 Which welding process would you select for the following:
(a) metal foil or thin sheet
(b) joining high speed tool tips to cheaper mild steel shank
(c) joining very thick steel plates say 75 cm thickness
(d) a process, which can weld all metals.

10.9 Would you expect a 0.2% C, 1% Cr, 1% Mo, 0.35% V steel to be sensitive to stress relief cracking?

Welding

10.10 Calculate the equivalent carbon content for steel of the following composition. Recommend pre- or post-weld heat treatments if necessary. Does this composition conform to the specifications for ASTM A588?
%C: 0.18, %Mn: 1.13, %Cr: 0.22, %Ni: 0.47, %Mo: 0.06, %Cu: 0.31

10.11 On a CCT diagram for a steel, how shall you select preheating temperature and time prior to welding?

10.12 What is the difference in the toughness variation of the heat affected zone in plain carbon and HSLA weld steels with respect to heat input?

10.13 Why a postweld heat treatment (stress relief) of a steel can cause a substantial decrease in toughness in the HAZ?

10.14 High silica fluxes have many desirable properties, such as high current capacity and excellent bead shape control, but they are recommended for single pass welds only. Why?

10.15 Draw a schematic presentation of the stress distribution along the thickness of a welded metal plate. At what location of the weldment, the stress is at peak and why? (Refer Case Study 12.6 in Chapter 12).

Welding of Stainless Steel

10.16 Use the De Long diagram to estimate the ferrite number of a stainless steel with composition: 13.3% Ni, 20.1% Cr, 0.03% C, 0.6% Si, 1% Mn, 0.01% N.

10.17 An accidental arc strike from a welding tool causes surface melting and refreezing at a spot on the surface. Would this likely cause intergranular corrosion problems in the spot if the alloy were Type 304 stainless steel? Type 347 stainless steel?

10.18 A stainless steel weld deposit is made on mild steel (0.2% C, 1% Mn) with 50% dilution. What should the Ni and Cr equivalents of the electrode be to avoid formation of martensite in the deposit?

10.19 A carbon steel of 0.2% C is to be weld cladded by stainless steel. The carbon level in the electrode is 0.02% C. Assuming the dilution of each pass is 30%, how many layers (approximate) of stainless steel weld metal is required to achieve a specified minimum of 0.03% C in the final layer.

Welding of Non-ferrous Alloys

10.20 Suggest why silicon is not selected as alloying addition in the filler metal used for joining Al-Mg alloy.

10.21 Two 5052 Al-alloy sheets of 1-mm thickness are to be joined by spot welding. Describe the procedure.

10.22 Discuss how the weldments of Ni-based alloys pose more problems than in case of austenitic stainless steels.

Brazing, Soldering and Adhesive Bonding

10.23 What effect does brazing have on mechanical properties of the base metal?

10.24 A nickel based substrate is to be brazed by Au-Ni braze alloys of 18% Ni and 50% Ni respectively. Comment on the differences in the brazing response. [Hint: Refer Au-Ni binary phase diagram from Metals Handbook].

10.25 Ag-28 Cu eutectic composition is used for brazing copper. What problems do you envisage during cooling the brazed joint?

10.26 Copper tap water pipes butt joined by brazing by (a) Ag-Cu-Zn braze, (b) Ag-Cu-P braze and (c) Tin rich solder in Ag-Sn or Cu-Sn system behave differently from corrosion viewpoint. Which system shall be preferred and why? What is the effect of such filler metal on the mechanical properties?

10.27 Diamond tools are commonly obtained after brazing diamond grits. In a brazing shop, two brazing alloy powders are in stock: (1) Cu-15 Ti-10 Sn and (2) Ni-12 Cr-4 Fe-2.5 B. The grits are to be brazed in vacuum for 5 minutes at 925 °C and 1000 °C respectively. Which brazing alloy would you select? In case the other alloy is unsuited could you suggest some treatment on the diamond grit?

10.28 For soldering copper Pb-Sn solder is used. Due to toxicity problem, if your management decides to replace lead by bismuth, what problems may be encountered? Can you suggest some way to minimize the problem?

10.29 Ag-96.5 Sn eutectic soldering alloy is used for silver coated substrate. Is the thickness of silver coating crucial for a prescribed soldering cycle? If so, why? [Refer: Binary Ag-Sn phase diagram].

10.30 In lead-tin solders tin is considerably costly. What cheaper metals can partially replace tin, without greatly widening the melting range of the solder. [Refer Metals Handbook for ternary phase diagram].

10.31 Tin based solders cannot be used for soldering to thick gold metallization. In-Pb solders are used for this purpose. Comment giving justification. Do you get an equilibrium microstructure in the latter case? [Hint: Refer In-Pb phase diagram from Metals Handbook].

10.32 What is the reason for the fact that Mo and W can be bonded with molten glass?

10.33 In a hybrid semiconductor device, what modification needs to be done on aluminium alloy pad before soldering by the Pb-Sn alloy?

10.34 Why is surface preparation important in adhesive bonding? List the limitations of adhesive bonding.

10.35 Why adhesive bonding of polyethylene, polypropylene and Teflon is difficult as compared to other polymer say polyvinyl chloride?

REFERENCES

Fundamentals of Welding Metallurgy, H. Granjon, Abington Publishing, Cambridge, 1991.
Welding Essentials, W. Galvary and F. Marlow, Industrial Press, New York, 2001.
Welding Metallurgy of Steels, Abington Publishing, Cambridge, 1986.
Welding Metallurgy of Stainless Steels, Abington Publishing, Cambridge, 1980.
Rational Welding Design, 2nd ed., T.G.F. Gray and J. Spence, Butterworth, London, 1982.
Advanced Joining Technologies, Ed. T.H. Norton, Chapman and Hall, London, 1990.
Introduction to the Physical Metallurgy of Welding, K. Easterling, Butterworth, London, 1983.

Welding Metallurgy, 2nd ed., K. Kou, John Wiley & Sons, Inc., Hoboken, N.J., 2003.

The Metal Science of Joining, Ed. M.J. Cieslack, J.H. Perepezko, S. Kang and M.E. Glicksman, TMS, Warrendale, 1992.

A Guide to Designing Weld, J. Hicks, Abington Publishing, Cambridge, 1989.

Principles of Soldering and Brazing, G. Humpton and D.M. Jacobson, ASM International, Ohio, 1993.

11

SURFACE MODIFICATION

Surface engineering covers a wide range of industrial sector activities. It encompasses techniques and processes capable of creating and/or modifying surfaces to provide enhanced performance such as wear, corrosion and fatigue resistance and biocompatibility. Various methods for surface modification and coating are categorized in Table 11.1. Surface engineering technologies span five orders of magnitudes in thickness and three orders of magnitudes in hardness. The thickness of the engineered surface can vary from several millimeters for weld overlaps to a few micrometers or even nanometers for physical vapour deposited or chemical vapor deposited coatings. In case of ion implantation the depth of surface modification is even less than 0.1 μm (Figure 11.1). Figure 11.2 shows the typical ranges of processing temperatures for surface engineering. In the following sections, they are described in some detail.

11.1 MECHANICAL TREATMENT

In this category common methods are shot penning, water-jet penning, laser penning, roller burnishing, explosive hardening, clad bonding and mechanical plating.

11.2 THERMAL TREATMENT

In this category of processes, we take recourse of phase transformation in material in the solid, which changes the overall chemistry of the alloy on the macro level. The methods are induction hardening, flame hardening, laser hardening and electron beam hardening. Among these the induction heating is the most widespread. Typically a medium carbon (0.3-0.6% C) steel is used for such treatment and the case depth is usually 1-10 mm.

Table 11.1. Surface modification and coating techniques.

Surface treatments	*Overlay coatings*
Thermal treatments	**Plating**
• induction hardening	• electroplating
• flame hardening	• mechanical plating
• laser hardening	
• electron beam	
Thermochemical treatments	**Weld cladding**
• carburising	• oxy-acetylene
• nitriding	• tungsten inert gas
• carbonitriding	• metal inert gas
• chemical - etching, oxidation	
Mechanical treatments	**Thermal spraying**
• shot-peening	• flame
• grinding	• detonation (D-) gun
• cold working	• plasma
Ion implantation	**Chemical vapour deposition (CVD)**
Laser glazing	**Physical vapour deposition (PVD)**
	• sputtering
	• evaporation

A full discussion on thermal treatment has been made in Chapter 3 and hence shall not be repeated here.

11.3 THERMOCHEMICAL TREATMENTS

In these methods the existing materials surface is modified, by changing the surface chemistry, but no external coating is adopted. Carburizing is carried out on steels (<0.45% C). There are a number of carbonitriding and nitrocarburizing processes (e.g., Tufftride or Sulfinuz), which are carried out on steels at comparatively low temperature (eg., 570 °C). With many of these types of processes a porous compound layer is often produced, which should be removed prior to in-service use. These pores can be sealed with a sealant to enhance the corrosion resistance.

Nitriding of a steel is very attractive process as it is a ferritic thermochemical treatment involving the introduction of atomic nitrogen into the ferrite phase in the temperature range 500–590 °C and consequently no phase transformation occurs on cooling to room temperature. Nitrogen can be incorporated in liquid (Tufftriding), gas (ammonia) on plasma (nitrogen-hydrogen) medium. In case of plasma nitriding (or called ion nitriding), N_2 and H_2 are fragmented to ions and radicals by the glow discharge plasma surrounding the part to be surface modified. The ions diffuse inside the surface because of high temperature. The part temperature is maintained by ion or neutral (argon) bombardment and by external heating. In case the reactor is small, external heating may not be required. The interesting question is the relative merits of plasma nitriding. Table 11.2 illustrates the distinctive features.

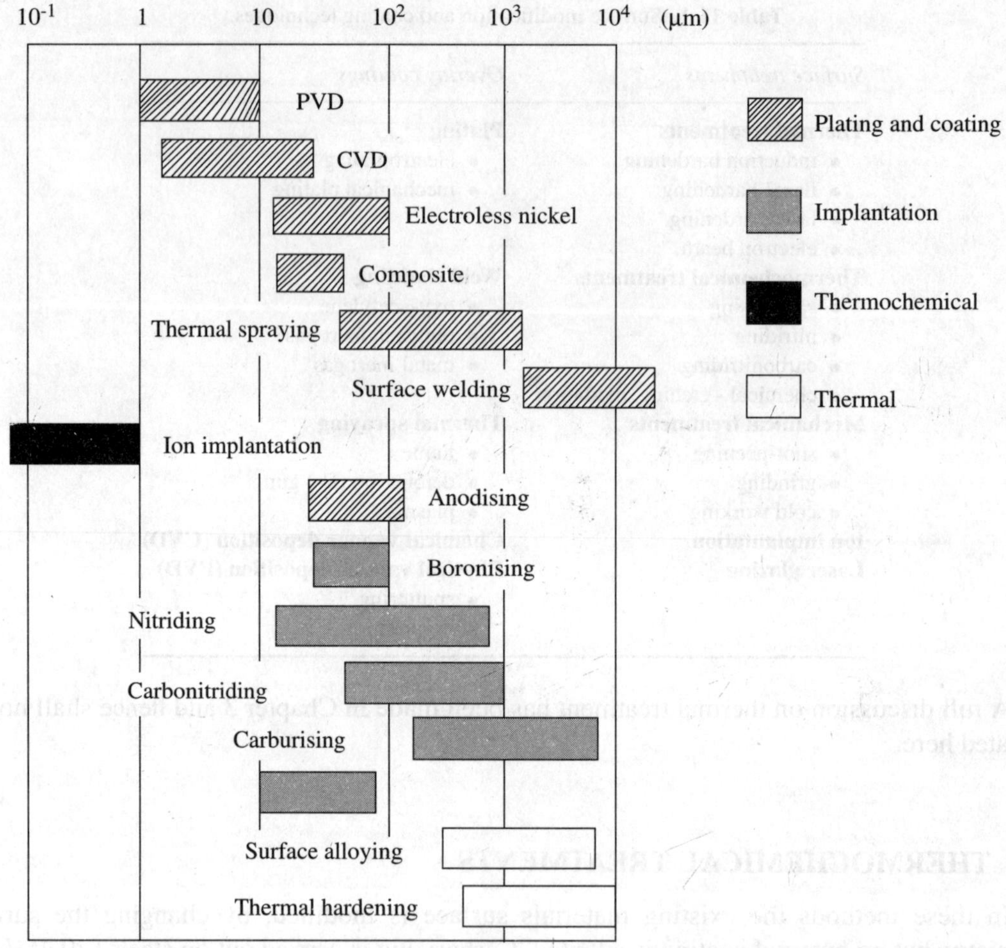

Fig. 11.1. Thickness of coatings or depths of surface treatments.

In ion implantation (not to be confused with ion plating), the ions are accelerated in a vacuum to such an extent that they penetrate the substance to a depth of a few micrometers. The process can be accurately controlled, and the surface can be masked to prevent ion implantation in unwanted locations. The process is particularly effective on metals such as aluminium, titanium, stainless steels, tool and die steels, cemented carbides and chromium. When used in semiconductor industries, the process is known as 'doping', meaning alloying with small amount of various elements.

11.4 ELECTRO AND ELECTROLESS PLATING

We have already described electrochemical properties of metals in chapter 6. Electroplating consists of:

- The metal ions from the anode are discharged using the potential energy from the external source of electricity.

- The metal ions combine with the ions in the solution and are deposited on the cathode.

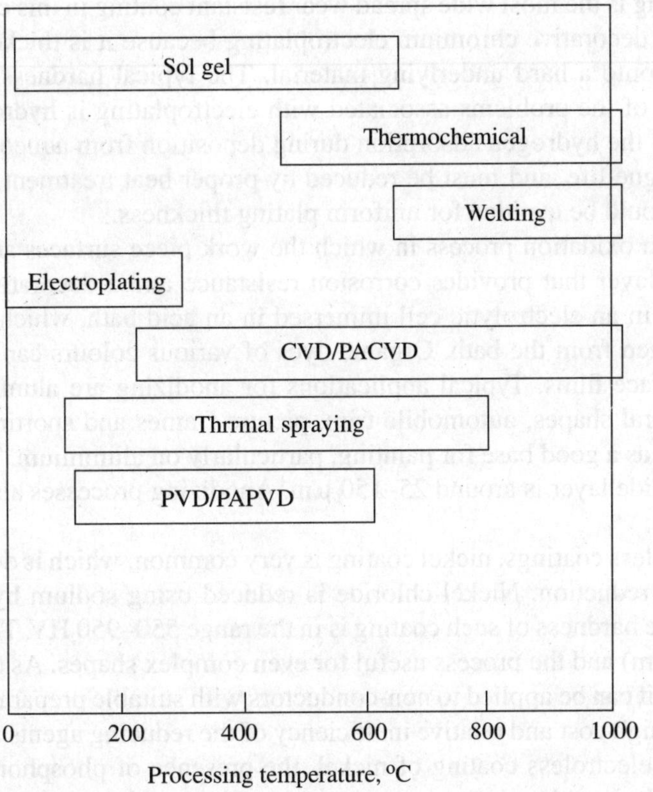

Fig. 11.2. Typical ranges for the processing temperatures for surface engineering.

Table 11.2. Comparison of plasma nitriding and gas nitriding.

Criterion	*Plasma Nitriding*	*Gas Nitriding*
Minimal distortion	Yes	Yes, but not in all cases
Treatment temperature	340-565°C	520-540°C
Corrosion protection	Yes	Less
Increase of surface roughness	Negligible	Yes
Very good polish capability	Yes	Not in all cases
Treatment duration at the same hardness depth	Low	Approx. 3 times as long
All ferrous alloys can be treated	Yes	Not all cases (no high alloy steels)
Covering of areas not to be nitrided	No problem	Difficult
Weldability	Yes	No
White layer on surface is free of pores	Yes	No
Thin white layer possible	Yes	Yes (with low hardness depth)

Chromium, nickel, cadmium, copper, zinc and tin are common plating materials. Chromium plating is done by plating the metal, first with copper, then with nickel and finally with chromium. Hard chrome plating is the most wide spread wear resistant coating in this category. This differs from conventional decorative chromium electroplating because it is thicker (10-500 μm) and usually deposited onto a hard underlying material. The typical hardness for such coating is 600–750 HV. One of the problems associated with electroplating is hydrogen embrittlement, which is caused by the hydrogen adsorption during deposition from aqueous solutions. This is harmful to the fatigue life, and must be reduced by proper heat treatment. Sharp external and internal corners should be avoided for uniform plating thickness.

Anodizing is an oxidation process in which the work piece surfaces are converted to hard and porous oxide layer that provides corrosion resistance and a decorative finish. The work piece is the anode in an electrolytic cell immersed in an acid bath, which results in chemical adsorption of oxygen from the bath. Organic dyes of various colours can be used to produce stable durable surface films. Typical applications for anodizing are aluminium furniture and utensils, architectural shapes, automobile trim, picture frames and sporting goods. Anodized surfaces also serve as a good base for painting, particularly on aluminium. The thickness of the hard aluminium oxide layer is around 25–150 μm. Anodizing processes also exist for zinc and titanium.

Among electroless coatings, nickel coating is very common, which is derived from an auto-catalytic chemical reduction. Nickel chloride is reduced using sodium hypophasphite as the reducing agent. The hardness of such coating is in the range 550–950 HV. The layer thickness is uniform (up to 5 mm) and the process useful for even complex shapes. As the process does not require any power, it can be applied to non-conductors with suitable preparation. A drawback of this process is the high cost and relative inefficiency of the reducing agents and the unavoidable solution losses. In electroless coating of nickel, the presence of phosphorus is very common because of the catalytic action due to hypophosphite. The reactions are

$$Ni^{2+} + H_2PO_2^- + H_2O \rightarrow Ni^o + H_2PO_3^- + 2H^+ \quad (11.1)$$

$$H_2PO_2^- + H \rightarrow P^o + OH^- + H_2O \quad (11.2)$$

$$H_2PO_2^- + H_2O \rightarrow H_2PO_3^- + H_2 \quad (11.3)$$

By continuing the initial bath conditions, the composition and properties of the coatings can be controlled. For example, baths at high pH ranges, favour the production of low phosphorus containing coatings.

Of late nickel/powder composites have been prepared by electroless method. The common powders are generally non-metallic like diamond, SiC, Al_2O_3 and PTFE.

11.5 VAPOUR DEPOSITION PROCESSES

The main methods in this group are chemical vapour deposition (CVD) and physical vapour deposition (PVD). Two important parameters of a coating process - process temperature and working pressure largely determine the production equipment and the properties of the end products. The temperature-pressure ranges of the most important vapour deposition technologies

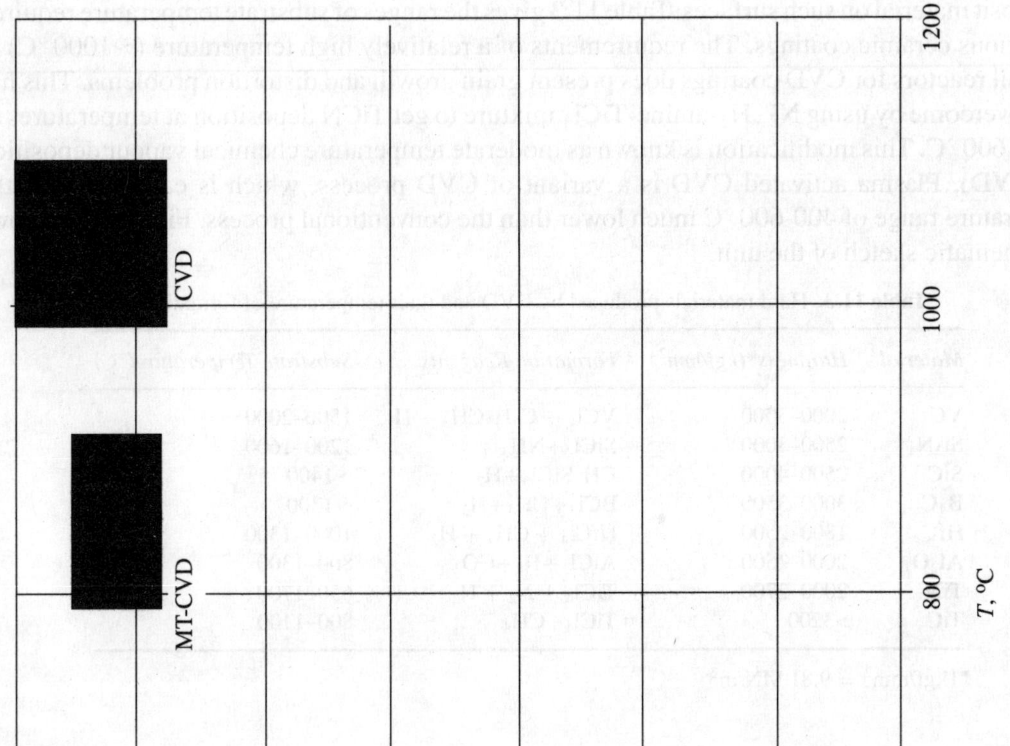

Fig. 11.3 Temperature/pressure relations for CVD and PVD processes.

are shown in Figure 11.3. The chemical vapour deposition (CVD) can be broadly defined as the process whereby a chemical reaction between gaseous reaction at or near heated surfaces is used to deposit material on such surfaces. Table 11.3 gives the ranges of substrate temperature required for various ceramic coatings. The requirements of a relatively high temperature (~1000 °C) in hot wall reactors for CVD coatings does present grain growth and distortion problems. This has been overcome by using N_2_H_2-amine-$TiCl_4$ mixture to get TiCN deposition at temperatures as low as 600 °C. This modification is known as moderate temperature chemical vapour deposition (MTCVD). Plasma activated CVD is a variant of CVD process, which is carried out in the temperature range of 400-600 °C much lower than the conventional process. Figure 11.4 shows the schematic sketch of the unit.

Table 11.3. Hard materials produced by CVD and their temperature of formation.

Material	*Hardness*(kgf/mm²)*	*Formation Reagents*	*Substrate Temperature(°C)*
VC	2000–3000	$VCl_4 + C_6H_5CH_3 + H_2$	1500–2000
Si_3N_4	2500–3000	$SiCl_4$+NH_3	1200–1600
SiC	2500–4000	CH_3SiCl_3+H_2	~1400
B_4C	3000–3500	BCl_3+CH_4+H_2	~1300
HfC	1800–2500	$HfCl_4 + CH_4 + H_2$	1000–1300
Al_2O_3	2000–2500	$AlCl_3$+H_2+CO_2	800–1300
TiN	2000–2700	$TiCl_4 + N_2 + H_2$	650–1700
TiC	>3200	$TiCl_4$+CH_4	800–1100

*1kgf/mm² = 9.81 MN/m²

In physical vapour deposition (PVD), the coating material is evaporated from a solid source under a partial vacuum. The most advanced ones involve the creation of a glow discharge or plasma. Other methods are sputtering, activated reactive evaporation and most recently, ion implantation. PVD coating on cutting tool inserts produces damage free interface, as a consequence of which the total tool life is increased.

Of late, diamond coating has attracted much attention. Coatings with predominantly sp^3 bonds are termed 'diamond coating' (DC), while the term diamond like coating (DLC) is applied to films of carbon exhibiting a mixture of sp^3 and sp^2 bonds.

The microstructure of the interface and the coating is dictated by the surface mobility of the adsorbed atoms. Low mobility leads to amorphous or small grain materials, while high mobility does the reverse. Bulk diffusion determines the crystallography of materials deposited at elevated temperature. Preferential crystallographic orientation results from the differential surface mobilities at low temperatures or recrystallizastion kinetics at high temperature.

It is now possible to produce coatings of novel composition and microstructure in multilayer/multicomponent format (Figure 11.5). It has been observed that a presence of a TiC precoat on cemented carbide leads to a more uniform growth of CVD alumina than on the bare substrate. The TiC probably acts as a nucleator. The TiC under layer may serve to reduce the lattice strain and mismatch at the Al_2O_3 interface, since the lattice parameters of Al_2O_3 and TiC are more closely matched than those of Al_2O_3 and WC.

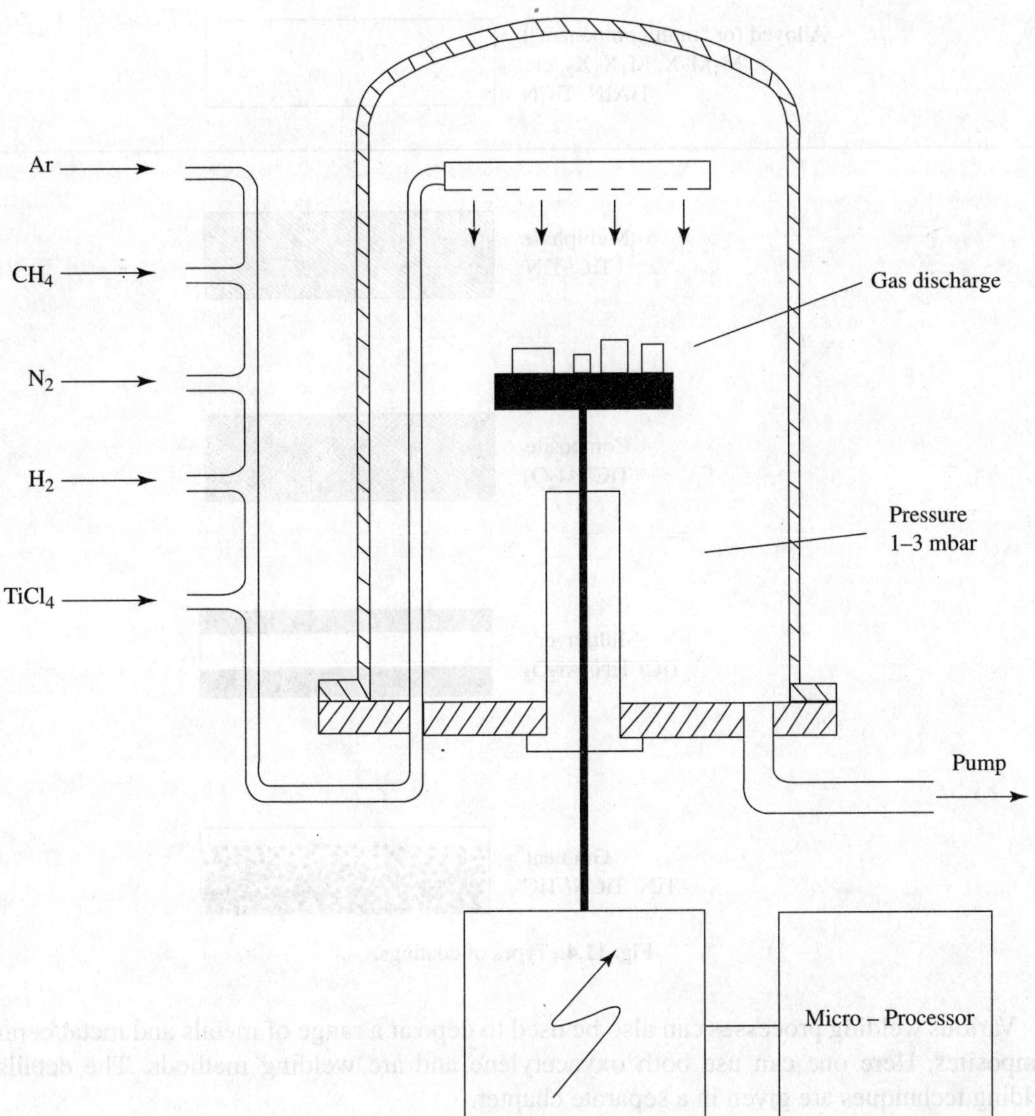

Fig. 11.3. Schematic set up for plasma CVD.

11.6 MOLTEN STATE DEPOSITION

The molten state deposition includes 'hard facing' method such as thermal and plasma spraying. A very wide range of materials can be deposited by this method. The thickness produced can be from a few micrometers to several millimeters. This type of deposition method offers an economical way of applying wear and corrosion resistant coatings. As the porosity can be greater than the acceptable limit, several processes such as detonation (D) gun and high velocity oxy fuel (HVOF) thermal spraying are developed.

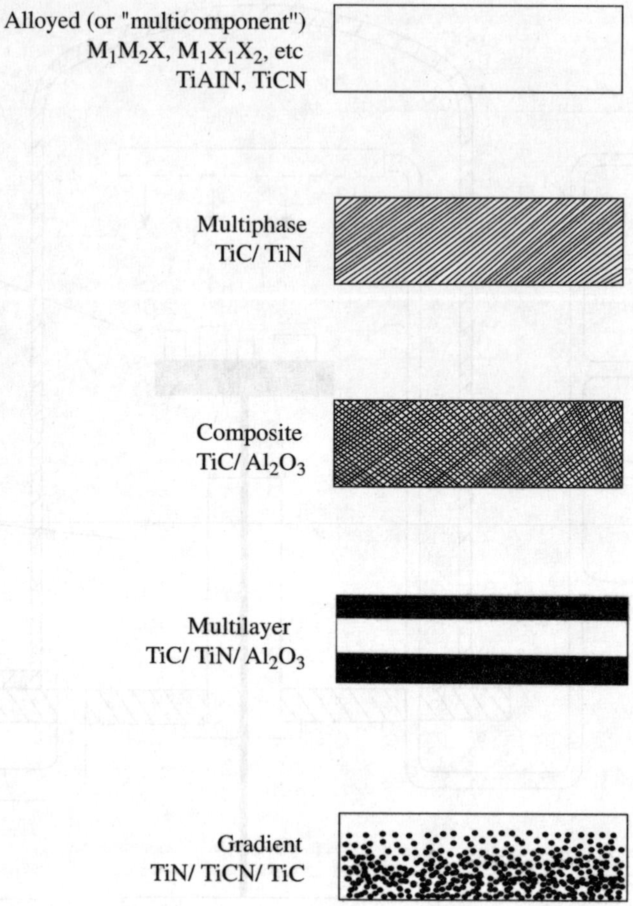

Fig. 11.4. Types of coatings.

Various welding processes can also be used to deposit a range of metals and metal/ceramic composites. Here one can use both oxyacetylene and arc welding methods. The details of welding techniques are given in a separate chapter.

Figure 11.6 illustrates the principle of thermal spraying. The spray material is in the form of a wire or powder, which is fed into a heat source. The near molten particle is rapidly accelerated to impact on to a prepared substrate. On collision, the particle deforms to form a splat, cools rapidly and adheres to the substrate. Subsequent particle build up gives rise to the coating. Although the term 'thermal' is prefixed with this type of coating, the thermal spraying is metallurgically cool process, as vary little heat is carried over to the base material. With due precautions even polymers can be successfully thermal sprayed. Figure 11.7 illustrates various modes of thermal spraying operations. Traditionally metals that could be easily drawn into wire were used as feed stock. The extensive use of powder mixtures has greatly expanded the choice of spray materials. The production of fully alloyed, smooth, spherical powders has made the more improved spray rate and consistency leading to high quality coatings.

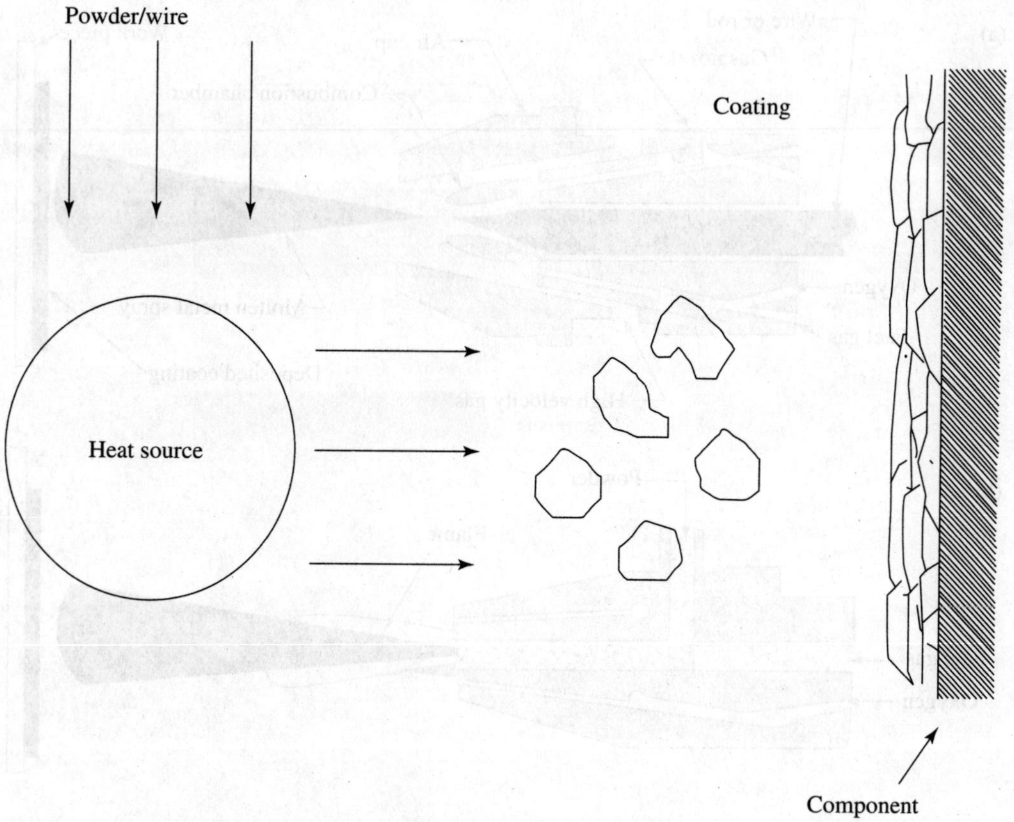

Fig. 11.5. Basic principles for thermal spraying.

Thermal sprayed coatings can offer numerous advantages and thus applications. Some of them are:

- Restoration of worn or mismachined dimensions
- Wear resistance
- Oxidation resistance
- Atmospheric corrosion resistance
- Resistance to chemical attack
- Electrical resistance and conduction
- Heat resistance and conduction
- Abradable coatings for clearance control
- Biomedical
- Repair and remanufacture
- Production of free-standing shapes of metals, ceramics or matrix composites

There are certain limitations of thermal sprayed coatings. Some of them are

- Porous by nature, thus posing corrosion problems
- Does not add to mechanical strength

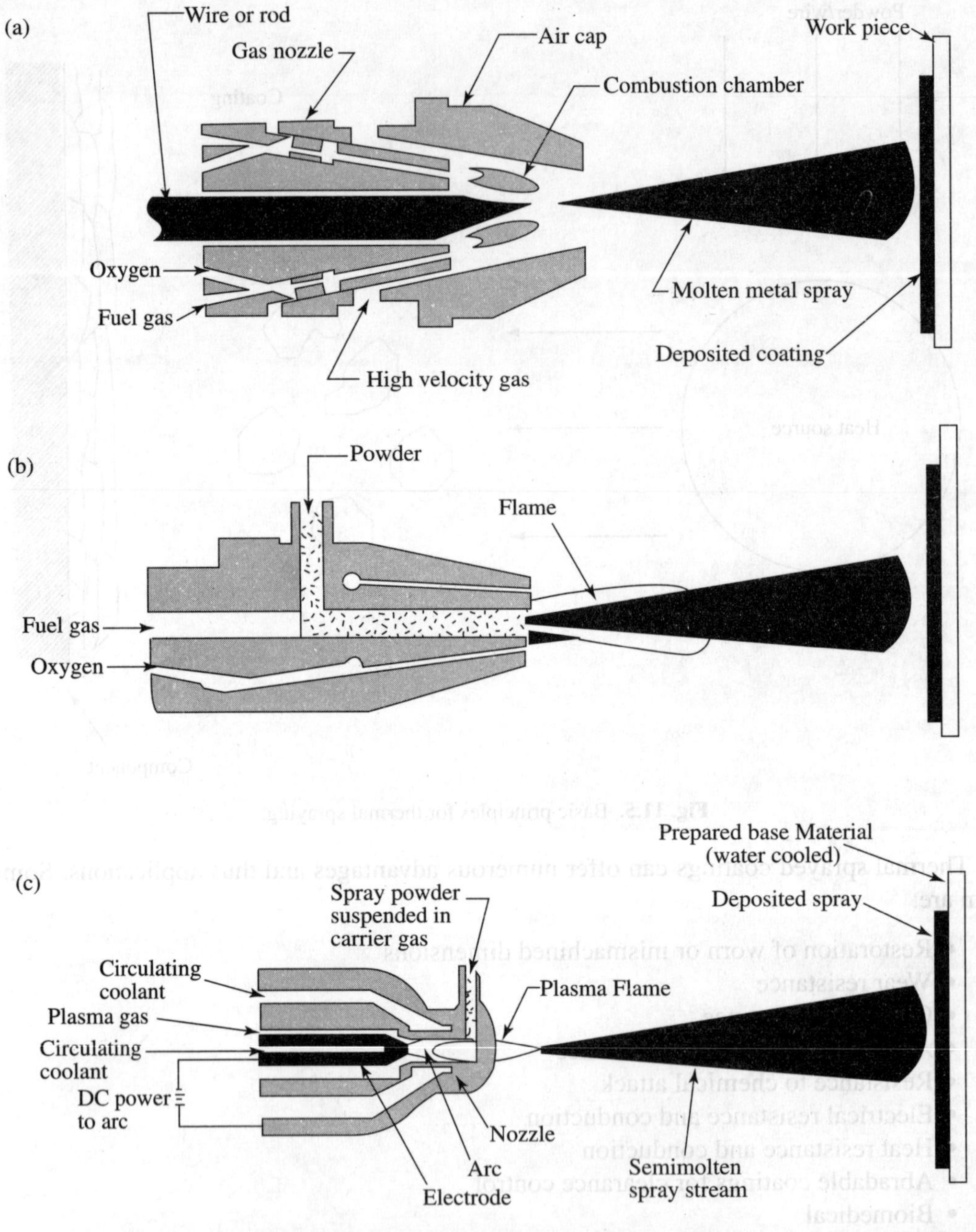

Fig. 11.6. Schematic illustrations of thermal spray operations: (a) thermal wire spray, (b) thermal metal powder spray, and (c) plasma spray. (Reproduced with permission from Manufacturing Engineering and Technology, S. Kalpakjian and S.R. Schmid, 4th ed., Pearson Education, Inc., Upper Saddle River, NJ, 2000, p. 907).

- Fatigue strength may be adversely affected
- Problems of edge damage
- Proper selection of the spray material. In case of high fusion temperature e.g., Ni-Cr-Si-B system, elevated temperature heating to 1050 °C becomes imperative for adequate diffusion process to take place.
- Any previous hard surfacing or surface treatment is to be removed prior to fresh thermal spraying.

11.7 PORCELAIN ENAMELING/CERAMIC COATING

Metals can be coated with a variety of glassy coatings to provide corrosion and electrical resistance and to provide protection at elevated temperatures. Glasses and enamels are glassy or partially crystalline coatings applied in the form of an aqueous slip. For optimum control of composition, the glass is often premelted and granulated in water. The resulting frit is comminuted and suspended in water with the addition of antiflocculants. To retain the slip on vertical surfaces, its properties are adjusted by the addition of clay or organic binders to impart pseudoplastic behaviour. The composition is so selected that the coating has a slightly lower thermal expansion coefficient than the part and will thus be put into compression on cooling. Lead reduces the firing temperatures but must be avoided in all food-related applications. Coatings range from transparent to completely opaque and a wide range of colors and special effects may be obtained.

Vitreous enamels are applied to low carbon steel, cast iron or aluminium products for improved corrosion resistance. Steels of very low carbon be selected, as iron carbide (cementite) reacts with enamels to form CO_2. Adhesion on cast iron is rather mechanical. On aluminium adhesion is aided by the surface oxide.

Glazes are also applied to ceramics in order to make them impermeable.

11.8 ECONOMICS OF VARIOUS SURFACE TECHNOLOGIES

Depending on the applications, each surface engineering method has different advantages. Table 11.4 summarises an economic comparison of different processes meant for tribological applications. The group titled as 'nitriding' relates to diffusion or implantation techniques which modify surface behaviour by the penetration of nitrogen into the surface. The other group is titled as 'coatings'. Keeping the base of cost for plasma nitriding as 100, relative values for other techniques are mentioned. The column 'fixed costs' refers mainly to investment and depreciation of a plant. The variable costs include the material, gases, energy etc. for the process and the labour costs. In the last column of the Table, the profitability of the surface treated work pieces, is considered.

In conclusion, it can be said that it is impossible to find a best surface coating process for all applications, because there are advantages and disadvantages in every case. To decide which process will be the best, for each application an optimization is ought to be made. Recently, the importance of ecological aspects has increased greatly. In most cases these questions are closely linked to the economic aspects. For example although the running cost of plasma assisted

Table 11.4. Economic comparisons of selected surface technologies for wear applications.

Process	*Fixed costs*	*Variable costs*	*Profitability*
Nitriding			
Salt bath	30	200	100
Gas	80	120	100
Plasma	100	100	100
Ion implantation	500	100	100
Coating			
Galvano-techniques	30	200	100
CVD	100	150	300
PVD	400	300	500
PACVD	200	150	500

(From R. Grum, Surface and Coating Technology, Vol. 60, 1993, 613-618).

processes is pretty low, the cost of use or production of toxic media from the point of view of ecology is pretty high.

11.9 CASE STUDY: PISTON RING

Piston ring is a moving metallic seal used in automotive engines. The roles of a piston ring are

- Sealing of hot gases
- Heat transferring media
- Lubricate film metering device
- Sliding support for piston

Generally there are three piston rings in a piston. They are:

- Top compression ring for sealing
- Second compression ring for sealing and oil scraping
- Oil control ring for oil film control

The material of ring is generally SG Iron, but of late martensitic stainless steel is very common, mainly developed in Japan.

The surface engineering of piston ring can be done by one of the following ways:

- Plasma spraying
- Hard chromium plating
- Gas nitriding

In case of plasma spraying molybdenum alloys are sprayed on steel or cast iron surfaces by plasma spraying process either in-lay or over-lay. The main advantages/disadvantages are:

Advantages

- Very high flame temperature-upto 10000 °C, even ceramics can be melted and sprayed.
- Scuff, wear and thermal resistant coatings.
- Good control of process parameters.

Disadvantages

- Bonding is only mechanical. Bond strength is less compared to metallurgical coatings (Diffusion type coatings)
- Differential thermal property
- Thermo-chemical degradation
- Cost

Chromium coating is produced by electro-deposition from chromic acid bath CrO_3. It provides hard wear resistant coating of minimum $800HV_{0,1}$ hardness. Piston rings can be selectively plated on ID, OD or all over the surface. The main advantages/ disadvantages of this process are:

Advantages

- Good wear and corrosion resistance
- Very hard deposit
- Resists galling and abrasive wear

Disadvantages

- Poor throwing power
- Poor plating efficiency
- Hardness drop at >400 °C
- High rejections during process
- Generation of chrome effluent
- Scuffing resistance poorer than plasma coating
- Fatigue strength of base metal is reduced about 50%

In case of gas nitriding, the process is carried out in gaseous ammonia or nitrogen. Nitriding of martensitic stainless steel requires special depassivation and activation techniques e.g., dry honing and activation using NH_4Cl. One essential feature in the steel is that it must contain nitro-alloying elements e.g., Cr, Al, Mo, V etc. The advantages/disadvantages of the process are:

Advantages

- High hardness
- Good wear resistance
- High fatigue strength

Disadvantages

- Formation of white layer (brittle layer) which needs to be removed.
- Longer nitriding period

Figure 11.8 shows the ring wear and bore wear data of some surface engineered piston rings. The summary of all these three surface engineering processes in relation to various properties of the ring is presented in Table 11.5.

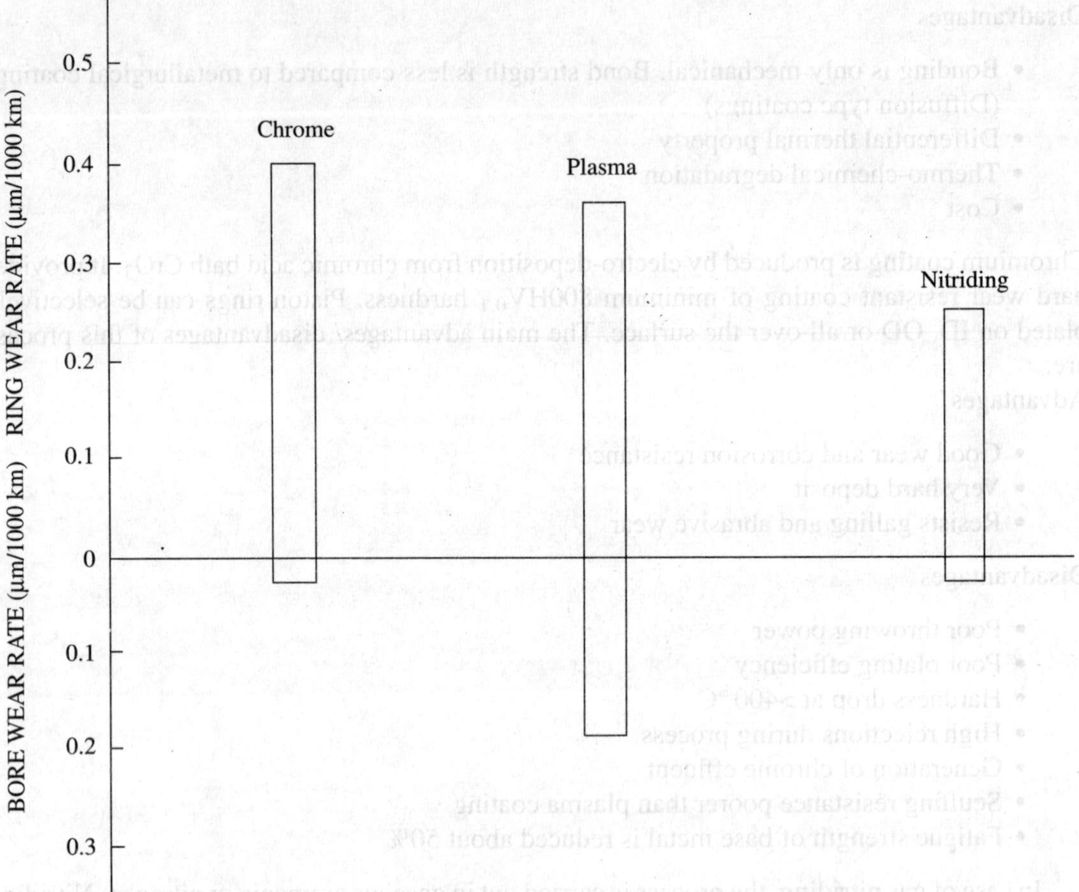

Fig. 11.7. Bore and ring wear rates of martensitic stainless steel piston rings for automobiles surface engineered by various methods.

QUESTIONS AND PROBLEMS

PVD/CVD

11.1 What is the difference between PVD and ion plating?

11.2 A Fe-0.2% C steel plate (hardness 400 VHN) is to be surface hardened to at least 1050 VHN. How could ion implantation processing be utilized to accomplish this task? Provide some experimental evidences to support the prospects of surface hardness in excess of 1050 VHN.

Table 11.5. Coating comparison of piston rings.

Property	*Gas Nitriding*	*Plasma Coating*	*Chrome Plating*
Wear Resistance	Best	Better	Good
Scuff Resistance	Good	Best	Better
Corrosion Resistance	Good	Better	Best
Fatigue Strength	Best	Better	Good

11.3 TiC phase in Binary Ti-C phase diagram exists with homogeneity range. Can you deposit a gradient CVD of this ceramic on cemented carbide (WC-Co)? Schematically draw a diagram between partial pressure of CH_4 and $TiCl_4$ vapours at a constant temperature say 1300K and pressure, say 1 atm.

11.4 A ceramic specimen, which is electrically insulating is to be examined in a scanning electron microscope. This necessitates a thin layer of gold on to the specimen. Which method would you follow and why?

11.5 Select proper inorganic coating by CVD process for following applications:
(a) Drawing punches and dies
(b) Coating on dressing sticks for grinding wheels
(c) Moulding tools and dies for glass filled plastics
(d) Pumps and valve parts for corrosive liquids
(e) Heating elements
(f) Abrasive slurry transport
(g) Oxidation resistant coating for C-C composites
(h) Cathode coating for aluminium production
(i) Coatings for neutron flux control in nuclear reactors
(j) Thermionic cathodes
(k) Target coatings for X-ray cathodes

Electroplating

11.6 What current density is required to plate 20 μm of chromium on to a surface in 20 minutes from a Cr^{2+} solution?

11.7 How long does it take to electroplate 225 g of nickel from a Ni^{2+} electrolyte with a current of 160 amperes?

11.8 An average current density of 100 A/m^2 is used to build up an anodized coating on aluminum. The Al_2O_3 is to be 1 μm thick. What time is required? Density of alumina is 3.8 g cm^{-3}.

11.9 Specify a plating to repair a worn cylinder in an automobile engine.

11.10 What makes electrodeposites hard?

11.11 Compare between plasma nitriding and chrome plated surface modification processes from the following view points (a) modified region; (b) sharpness of the cutting tool edge, (c) resistance against indentation, (d) fatigue strength, (e) environment friendliness and (f) cost effectiveness.

Other Methods

11.12 What is HVOF? How does it differ from plasma spray?

11.13 What is the thickness limit of hard facing?

11.14 List various engineering components, which can not function well in service, without surface modification.

REFERENCES

Surface Engineering: Case Book, (Ed.) J.S. Burrel-Gray and P.K. Datta, Woodhead Publishing Ltd., Cambridge, 1996.

Surface Engineering of Metals; Principles, Equipment, Technologies, T. Burakowski and T. Wiershon, CRC Press, Boca Raton, 1998.

Coatings and Coating Processes for Materials, Ed. J.H. Lindsay, ASM International, Ohio, 1998.

Surface Modification Technologies, Ed., T.S. Sundarshan, ASM International, Ohio, 1998.

12

MATERIALS DESIGN

12.1 INTRODUCTION

The materials available to the designer are numerous; several thousands exist. However, those used extensively for engineering applications are only in hundreds. The task of material's selection will be easier, if the designer has some knowledge of the different types of materials. Materials properties have been described in Chapters 4, 5 and 6, while fabricating aspects have been discussed in Chapter 8. Values of some of the properties can be obtained from Handbooks. In case, they are not available, experiments have to be made to get the data. The appropriate value of any property of a given material depends on its chemistry and the conditions e.g., temperature and environment under which it was determined. Sometimes in-service conditions are subject to change or fluctuation, perhaps in a random or unpredictable manner. To take care of such factors, safety factors are incorporated into the design.

Apart from material specification, two other factors are important. They are cost and availability. The cost is directly related to the quantity of material used, which itself depends upon the specification of property, size etc. Different quantities of alternative materials will be required to satisfy the same specification. Cost is a prime consideration but it must be determined in conjunction with other factors. In a way, the cost of a material is interlinked with its availability. The design engineer must investigate the availability of the materials intended for use in a design, before the final specification is prepared. A mass-produced item must be assured of stable, long-term supplies of the raw materials required. Other constraints may also emerge, which were not present during the design stage, but that may become significant during manufacture. These are often legal or political constraints resulting from environmental considerations or government-imposed trade embargoes.

The designer must decide whether to select a material and then specify suitable fabrication methods or to decide upon the fabrication process and then identify appropriate materials. Many times a compromise has to be made. The probability of success of a design product will be improved if processing methods and the associated production stages are considered in some detail during the preliminary and detailed design stages.

From the above description it is obvious that, any material design has two parts: qualitative and quantitative. Many times the qualitative component becomes very important, particularly in a complex situation, for which the exact quantitative analysis is still not available.

In short it can be said that the final decision will attempt to combine the technical 'best' at the lowest cost, with the minimum of problems due to outside influences. In the sections to follow in this chapter, numerous case studies are incorporated to give a feel of design aspects to the beginners in the field of materials engineering.

12.2 MATERIALS SELECTION CHARTS AND PERFORMANCE INDEX

In the introductory part of this chapter, we have described various factors related to materials design. Materials selection is synonym to materials design as a correct material selection would ultimately lead to a correct design. In the initial stage the choices are abundant, which narrows down to small subsets and then finally pinpointing the decision. The choice of material cannot be made independently of the choice of process by which the material is to be formed, joined, finished and otherwise treated. Ashby at the University of Cambridge has done seminal work in developing materials selection charts for mechanical design (Materials Selection in Mechanical Design, 2nd ed., M.F. Ashby, Butterworth, Heinemann, Oxford, 1999). A similar approach can be thought of for electrical design as well.

Performance of a component invariably depends on more than one property, say for example strength-to-weight ratio or stiffness-to-weight ratio for developing light weight design. Figure 12.1 illustrates the plot of modulus E against another property density on logarithmic scale. The range of the axes is chosen to include all materials from the lightest to the heaviest. As data for a given group of material cluster together, they can be enclosed in a proper envelop. In this chart more information can be added. We know that the speed of sound in a solid depends on the modulus E and the density ρ such that

$$v = \left(\frac{E}{\rho}\right)^{1/2} \tag{12.1}$$

$$\text{or} \quad \log E = \log \rho + 2 \log v \tag{12.2}$$

For a fixed value of v, this equation plots as a straight line of slope 1 on Figure 12.1. This allows us to add contours of constant wave velocity to the chart. These are the family of parallel diagonal lines, linking materials in which longitudinal waves travel with the same speed. All the charts allow additional fundamental relationship of this sort to be displayed. Ashby introduced another design optimizing parameters called **materials indices**, which are plots as contours on the charts. Ashby in his book included 18 primary mechanical and thermal properties, as listed in Table 12.1.

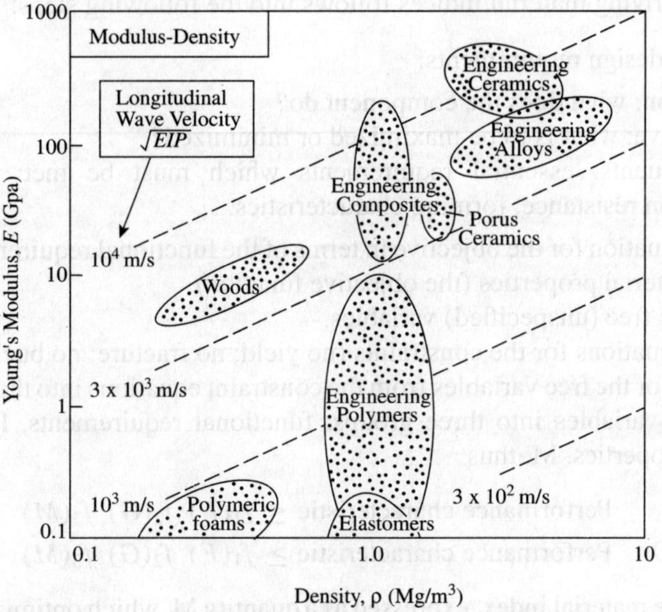

Fig. 12.1. Ashby materials properties chart between Young's modulus against the density (Reproduced with permission from Materials Selection in Mechanical Design, M.F. Ashby, 2nd ed., Butterworth-Heinemann, Oxford, 1999, p. 34).

Table 12.1. Design-Limiting Material Properties and Their Usual SI Units.

Class	*Property*	*Symbol and units*	
General	Cost	C_n	(\$/kg)
	Density	ρ	(kg/m^3)
Mechanical	Elastic moduli (Young's, shear, bulk)	E, G, K	(GPa)
	Strength (yield, ultimate, fracture)	σ_f	(MPa)
	Toughness	G_c	(kJ/m^2)
	Fracture toughness	K_{Ic}	(MPa m$^{1/2}$)
	Damping capacity	η	(-)
	Fatigue endurance limit	σ_e	(MPa)
Thermal	Thermal conductivity	λ	(W/mK)
	Thermal diffusivity	a	(m^2/s)
	Specific heat	C_p	(J/kg K)
	Melting point	T_m	(K)
	Glass temperature	T_g	(K)
	Thermal expansion coefficient	α	(°K^{-1})
	Thermal shock resistance	ΔT	(°K)
	Creep resistance	-	(-)
Wear	Archard wear constant	k_A	(MPa^{-1})
Corrosion/	Corrosion rate	K	(mm/year)
Oxidation	Parabolic rate constant	k_P	(m^2/s)

The method for deriving material indices follows into the following steps:

- Define the design requirements:
  (a) Function: what does the component do?
  (b) Objective: what is to be maximized or minimized?
  (c) Constituents: essential requirements which must be met: stiffness, strength, corrosion resistance, forming characteristics. . .
- Develop equation for the objective in terms of the functional requirements, the geometry and the material properties (the objective function).
- Identity the free (unspecified) variables.
- Develop equations for the constraints (no yield; no fracture; no buckling, etc.).
- Substitute for the free variables from the constraint equations into the objective function.
- Group the variables into three groups: functional requirements, F, geometry, G and material properties, M, thus

$$\text{Performance characteristic} \leq f_1(F)\ f_2(G)\ f_3(M)$$

$$\text{Or}\quad \text{Performance characteristic} \geq f_1(F)\ f_2(G)\ f_3(M)$$

- Read off the material index, expressed as a quantity M, which optimizes the performance characteristic.

The examples of material indices for various function in mechanical devices are given in Table 12.2.

Performance Index: After preliminary screening of possible materials, a still intensive work of comparing alternative solutions remains. Here the role of performance index comes into picture. The material performance index γ can be defined as

$$\gamma = \sum_{i=1}^{n} A_i \alpha_i \tag{12.3}$$

where A = scaled property and α = weighting factor.

Each property of the material is scaled such that its highest numerical value does not exceed 100. When evaluating a list of candidate materials, one property is considered at a time. The best value in the list is rated as 100 and the others are scaled proportionally. Introducing a scaling factor facilitates the conversion of normal material property value to scaled dimensionless values. The scaled property A can be defined as

$$\text{A} = \frac{\textit{Numerical value of the property} \times 100}{\textit{Maximum value in the list}} \tag{12.4}$$

For properties such as cost, corrosion or wear loss, weight gain in oxidation etc., a lower value is more desirable. In that case the lowest value is rated as 100 and B can be expressed as

$$\text{B} = \frac{\textit{Minimum value in the list} \times 100}{\textit{Numerical value of property}} \tag{12.5}$$

Table 12.2. Examples of Material Indices.

Function, objective and constraint	*Index*
Tie, minimum weight, stiffness prescribed	$\frac{E}{\rho}$
Beam, minimum weight, stiffness prescribed	$\frac{E^{1/2}}{\rho}$
Beam, minimum weight, strength prescribed	$\frac{\sigma_y^{2/3}}{\rho}$
Beam, minimum cost, stiffness prescribed	$\frac{E^{1/2}}{C_m \rho}$
Beam, minimum cost, strength prescribed	$\frac{\sigma_y^{2/3}}{C_m \rho}$
Column, minimum cost, buckling load prescribed	$\frac{E^{1/2}}{\rho}$
Spring, minimum weight for given energy storage	$\frac{\sigma_y^2}{E\rho}$
Thermal insulation, minimum cost, heat flux prescribed	$\frac{1}{\lambda C_m \rho}$
Electromagnet, maximum field, temperature rise prescribed	$kC_p\rho$

(ρ = density; E = Young's modulus; σ_y = elastic limit; σ_m = cost/kg; λ = thermal conductivity; k = electrical conductivity; C_p = specific heat).

Cost can be treated as one of the properties and given the appropriate weighting factor. A figure of merit, M, for the material can then be assigned as:

$$M = \gamma / C \quad (12.6)$$

where C = cost of material per unit weight.

When the function of the meterial is to bear load, it is more appropriate to use the cost of unit strength instead of the cost per unit weight or volume. Then the relation (12.6) is rewritten as

$$M = \gamma / C' \quad (12.7)$$

Here C' depends on the type of loading.

12.3 MATERIALS DESIGN FOR STRUCTURAL APPLICATIONS

12.3.1 Room and Low Temperature Service

Case Study 12.1: Materials for Fuel Efficient Automobiles: Automobile is a typical example, where numerous classes of engineering materials are used depending on the function of a

Table 12.3. Relationships Between System Functions and Material Properties for Automobiles.

Vehicle functions	*Mechanical*	*Thermal*	*Electrical*	*Magnetic*	*Chemical*	*Optical*
High Strength engine and chassis	*	*				
Vehicle weight reduction	*					
Ease of driving	*		*	*		*
Data transmission			*	*		*
Electronic systems			*	*		*
Improved engine efficiency	*	*				
Durability and ease of maintenance	*	*			*	
Exhaust gas purification		*			*	
Replacement energy sources	*	*	*	*	*	*
Reduced noise	*					
Improved safety	*					*

particular component. Table 12.3 shows such objectives and dominant vehicle and material functions.

One of the biggest problems faced in enhancing the use of automobiles is the emission control.

The emissions from automobiles can be reduced by

- Using cleaner fuels
- Improving engine design
- Streamlining the car design
- Reducing overall vehicle mass.

As a materials engineer, we shall mainly be interested with the last point i.e., reducing overall vehicle mass. A 10% reduction in mass leads to 5% fuel saving in automobile. One of the major items under this consideration is the body structure of the automobile. Steel has been a traditional material. But there are possible substitutes for that (Table 12.4). Figure 12.2 shows how the preference material is a function of the production volume of the automobile. For larger production, the zinc coated steel is more economical.

The main disadvantages of steel in autobody applications are:

- Heavier than alternative materials
- Corrosion if uncoated.

Both these problems have now been adequately addressed. High strength steels with a wide range of yield strength values now extending to 1200 MPa - are available. Although stiffness remains unchanged it is possible to offset the decreased torsional rigidity. Figure 12.3 shows the relationship of the strength and deformability of flat rolled steel products. At the lower end of the scale it is now possible to utilize very low carbon interstitial free 'IF' steel for some

Table 12.4. Requirements of Different Panels Comprising the B/W Structure.

Zone/Assembly	*Requirements*	*Materials choice*		
		Steels		*Possible* Alternatives (material/form)*
		Type	*YS, MPa*	
Main structure:				
• Front/rear longit mbrs	Impact resistance	HSS	300	DP600, AP
• 'A' post inner/outer	Rigidity, strength	HSS	300	AP, HT
• Cantrail	Rigidity, strength	HSS	260	AP, HT
• Main/rear floor	Moderate strength	HSS	180	AS
• Bodyside	Moderate strength, formability	HSS	180	AS, TWB
• Spare wheel well	Deep drawability	FS	140	SWS, SPA
• Wheelhouse, valance	Formability	FS	140	AS
'Bolt-on' assemblies				
Outer panels				
Door skins	'A' class surface Dent resistance	FS	140	BH180, AS, SPA PLA-RRIM,
Bonnet	'A' class surface Dent resistance	FS	140	BH180, AS, SPA PLA-SMC
Boot	'A' class surface Dent resistance	FS	140	BH180, AS, SPA PLA-SMC
Roof	'A' class surface Dent resistance	FS	140	BH180, HS
Inner panels				
Doors	Drawability	FS	140	TWB
Intrusion beams, rails	High impact strength	UHS	1200	AT, DP600

* Code:
HSS = high strength steel; UHS = ultra HS steel; FS = forming steel; AS = aluminium sheet; AP = aluminium profile; SPA = superplastic Al; HT = hydroformed tube; HS = hydroformed sheet; BH180 = Bake hardened steel; DPyyy = Dual phase steel; TWB = Tailor welded blank; PLA-xxx = Plastic-xxx type; RRIM = reinforced reaction injection moulded; SMC = sheet moulding compound; SWS = sandwich steel.

Reproduced with permission from Materials for Automobile Bodies, Geoff Davies, Elsevier/Butterworths-Heinemann, Amsterdam, 2003, p. 12.

parts requiring very high deep drawing properties. Regarding corrosion problem, a range of zinc coated steel in electrogalvanized, hot-dipped, alloyed and duplex form are now available. A somewhat reduced ductility is associated with hot-dip galvanized sheet due to the effect of the heat treatment cycle.

Bake-hardening steels derive their increase in strength from a strain ageing process that takes place as paint baking at around 180°C. Dual-phase and TRIP steels are being increasingly used to improve crash performance by utilizing improved energy absorption characteristics resulting from high strength and associated ductility level. A more detailed physical metallurgical aspects

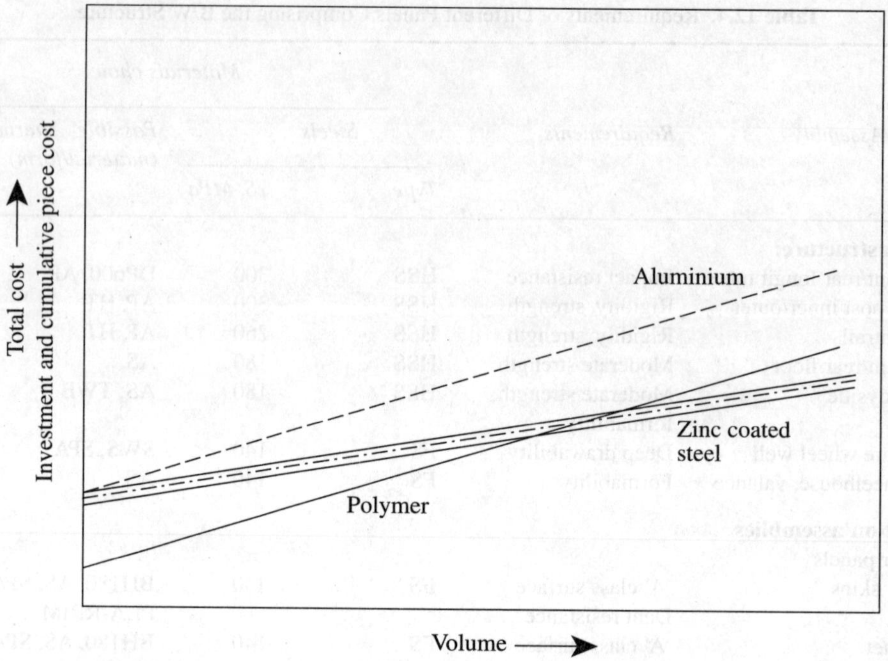

Fig. 12.2. General total cost for automotive skin materials with respect to production volume (Reproduced with permission from Materials for Automobile Bodies, Geoff Davies, Elsevier/Butterworth-Heinemann, Amsterdam, 2004).

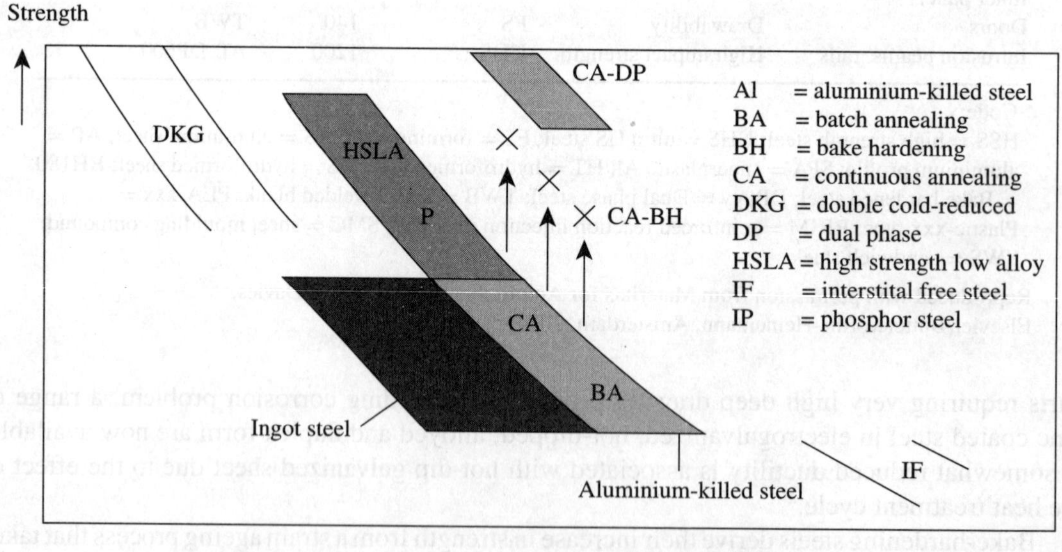

Fig. 12.3. Range of flat rolled steel Products (courtesy of Corus, U.K.).

of high strength steels are given in Case Study 12.2. Of late, duplex coating is very common, which consist of a zinc coating usually electrogalvanized with a superimposed organic coating. Resistance welding is the most common mode of joining in the assembly of steel automotive bodies.

Let us pick up three additional structural materials aluminium alloy, composite and magnesium, since they all possess densities lower than steel. The densities of these materials are 2.71, 1.55–1.93 and 1.74 g cm^{-3} respectively as compared to 7.85 g cm^{-3} for steel.

Aluminium Alloys: It is strong, but Young's modulus (9 GPa) is far lower than that of steel (200 GPa). This means that stiffening ribs need to be introduced. Remember the cost of Al-alloy is about four times, the steel cost. The addition of ribs would enhance the cost further. Aluminium alloys are used for luxury cars by some of the automobile makers, particularly in Germany.

5XXX series wrought Al-alloys has been used for panel production due to relatively low cost. Note that the cost is three times that of zinc coated steel compared to 6XXX alloy, which is five times costlier. The main problem with Al-alloy is that they are prone to stretcher strain markings or Luder bands. In spite of proper rolling and heat treatment solution these reappear on forming and can show through the paint finish unless reworked by abrasive discing. The 6XXX alloys have high yield strength than Al-Mg alloys and are heat treatable. They offer a significant degree of bake hardening at temperatures of about 200°C. Significant changes are required for the resistance welding of aluminium compared with steel. Higher current levels are required which necessitate larger welding transformers.

Composite: Young's modulus (31 GPa) rather low although strength has been good. Costly at higher volumes and gives problems in recycling. Advanced composites manufacturing is labour intensive. The basic method of polymer processing methods has been described in Chapter 8. Tailoring of fiber orientation and concentration can increase the strength in a particular region of a component.

Magnesium: Density is lower than that for aluminium, but the problem is poor fabricability because of hexagonal crystal structure. Moreover, it is costly and has corrosion problems. The full details of magnesium alloys are described in Chapter 7.

Hybrid material combinations, e.g., aluminium and steel, can cause serious galvanic corrosion problems resulting in panel perforation or fastener failure when incorrect selections are made. A metal with considerable weight saving potential is sandwich steel. This consists of two thin sheet outers encapsulating a thicker polypropylene central layer. However, such type of material is viable for components that are assembled into the body after the painting process, because of temperature limitations. High performance cars are using honeycomb materials, which provide ultralight structural sections. The potential application of such structures are for electric and alternatively fuelled cars.

In overall fuel economy, one must not singularly focus on weight of the car. There are other factors like power train effectiveness, aerodynamics, rolling resistance and auxiliary equipments. Table 12.5 shows how various technologies reduce fuel consumption. The role of power train (engine and transmission) is quite appreciable, where materials engineers can contribute quite a bit. The new advancements in fuel admission control optimization with promising solutions like the camshaft-free engine having all valves independently controlled, and in combustion optimization, with the expected developments concerning direct injection of

Table 12.5. Contribution of Various Technologies to Reduce Fuel Consumption.

% improvement due to	*Consumption savings*
Power train effectiveness	10%
Aerodynamism	2 – 3%
Rolling resistance	2 – 3%
Car weight	4 – 6%
Auxilliary equipment	1 – 2%

a perfectly homogeneous low-content mixture or through a variable engine displacement are expected soon.

Case Study 12.2: High Strength Steels for Automobiles: In Chapter 7, we have briefly described various grades of steels. High strength steels at low cost is the premier requirement for wide application of steels in automobile industry. Such a steel would contribute in reducing the weight of the automobile. If we look into the whole aspect historically, the picture emerges as follows:

Microalloyed steels: Here the role of precipitation strengthening is used, where microalloying additions e.g., niobium form very fine precipitate during the transformation process.

P-alloyed steels: Phosphorus is not used in structural steels, but in case of interstitial free steels (very low carbon and nitrogen), phosphorus addition up to 0.1% is advantageous. It enhances strength with little extra cost or sacrifice in formability. The mechanism of strengthening here is solid solution strengthening.

Bake hardening steels: Dent resistant bake hardening steel parts are produced by batch annealing, continuous annealing processes, including continuous annealing or galvanizing or aluminizing lines. Such steels are Al-killed, containing dissolved C and N levels of about 0.001%. These are characterized by their ability to exhibit accelerated (carbon and nitrogen) strain ageing, i.e., an increase in yield strength (~50 MPa) during paint baking on curing operation of press formed auto-body paint at 150-250°C for about 20 minutes. The extent of bake hardening effect increases with the decrease of the ferrite grain size. Good drawability is controlled by appropriate Al and N content, low coiling temperature, batch annealing heating rate, soak temperature and time.

Dual phase steels: As described in Chapter 7, dual phase steels are a mixture of ferrite and martensite. The strength of such steels is directly proportional to the martensite content. These steels offer great opportunity for weight reduction, ability to resist fatigue, good formability (high 'n' value), but low drawability (low R value). The schematic of thermomechanical path for the production of dual phase steel in hot strip mill is shown in Figure 12.4. Application of such steels include front and rear rails, crush cans, rocker reinforcement, pillar reinforcements, back panels, cross members, bumpers and door intrusion beams.

TRIP (transformation induced plasticity) steels: The microstructure consists of a continuous ferrite matrix with a dispersion of hard second phase martensite and/or bainite along with 5 to 20% of retained austenite. TRIP steels exhibit better ductility at a given strength level. Because of increased formability TRIP steels are used to produce more difficult and complicated parts enabling weight reduction at required strength levels. The schematic of thermomechanical aspect

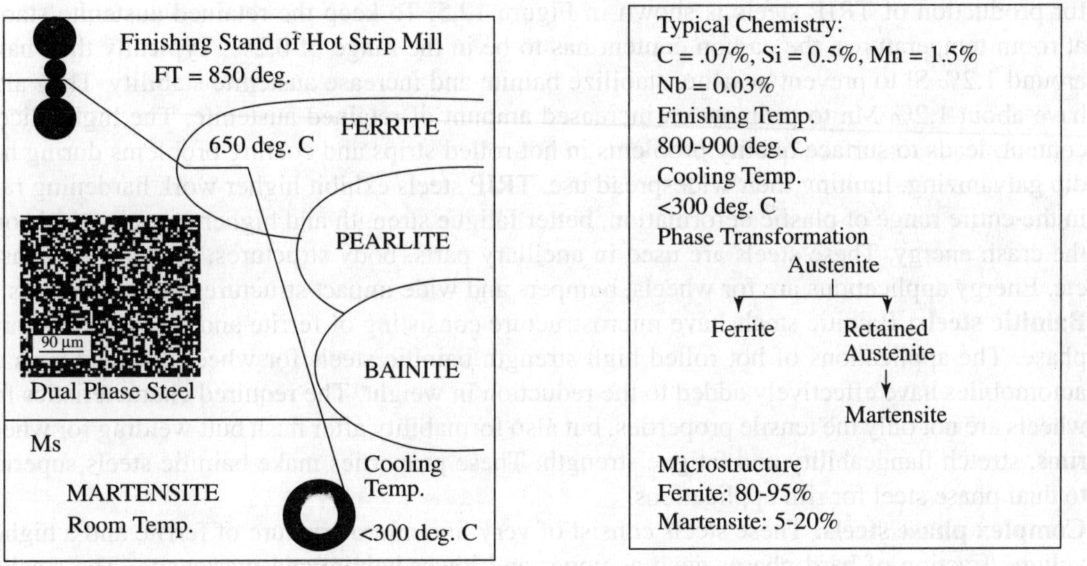

Fig. 12.4. Thermomechanical path for the production, microstructure, typical chemistry finishing and coiling temperatures of dual phase steels strip.

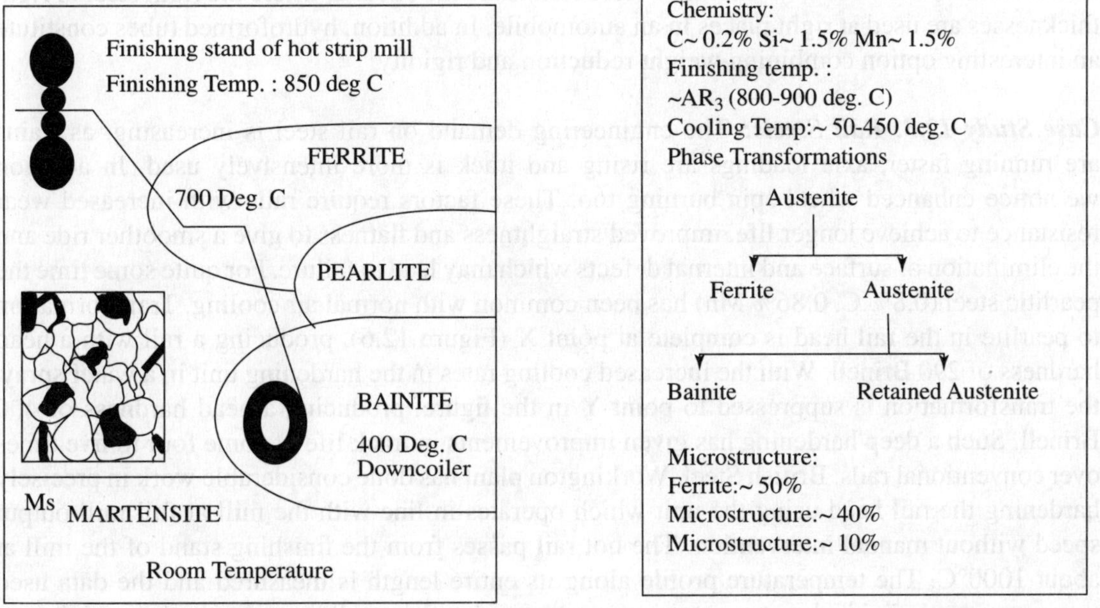

Fig. 12.5. Thermomechanical path for the production, microstructure, typical chemistry, finishing and coiling temperatures of TRIP steels strip.

for production of TRIP steels is shown in Figure 12.5. To keep the retained austenite stable at room temperatures, the carbon content has to be in the range of 0.2%. Typically they have around 1.2% Si to prevent pearlite, stabilize bainite and increase austenite stability. They also have about 1.2% Mn to promote an increased amount of retained austenite. The high silicon content, leads to surface quality problems in hot rolled strips and coating problems during hot dip galvanizing, limiting their widespread use. TRIP steels exhibit higher work hardening rate in the entire range of plastic deformation, better fatigue strength and higher capacity to absorb the crash energy. These steels are used in ancillary parts, body structures, suspension, chasis etc. Energy applications are for wheels, bumpers and wide impact structures of automobiles.

Bainitic steels: Bainitic steels have microstructure consisting of ferrite and 10 to 30% bainite phase. The applications of hot rolled high strength bainitic steels for wheel disc and rims of automobiles have effectively added to the reduction in weight. The required characteristics for wheels are not only the tensile properties, but also formability after flash butt welding for wheel rims, stretch flangeability and fatigue strength. These properties make bainitic steels superior to dual phase steel for rim applications.

Complex phase steels: These steels consist of very fine microstructure of ferrite and a higher volume fraction of hard phases such as upper and lower bainite and martensite. The steel is further strengthened by fine precipitates of microalloying elements, such as niobium, vanadium and titanium. Such steels provide tensile strengths of 800 MPa and higher and are largely hot rolled products. Such steels are useful in applications requiring very high strengths such as structural members of suspension, chasis and underbody parts of automobiles.

Other developments: Multi-thickness steel blanks are introduced, where the right steel of right thicknesses are used at right places in an automobile. In addition, hydroformed tubes constitute an interesting option combining weight reduction and rigidity.

Case Study 12.3: Rail Steels: The engineering demand on rail steel is increasing, as trains are running faster, axle loadings are rising and track is more intensively used. In addition we notice enhanced wheel spin burning too. These factors require rails with increased wear resistance to achieve longer life, improved straightness and flatness to give a smoother ride and the elimination of surface and internal defects which may lead to failure. For quite some time the pearlitic steel (0.8% C, 0.86% Mn) has been common with normal air cooling. Transformation to pearlite in the rail head is complete at point X (Figure 12.6), producing a rail with a head hardness of 290 Brinell. With the increased cooling rates in the hardening unit in a water spray, the transformation is suppressed to point Y in the figure, producing a head hardness of 400 Brinell. Such a deep hardening has given improvements in track life of some four to five times over conventional rails. British Steel, Workington plant has done considerable work in precisely hardening the rail head using the unit which operates in line with the mill at full mill output speed without manual intervention. The hot rail passes from the finishing stand of the mill at about 1000°C. The temperature profile along its entire length is measured and the data used to programme individual water spray in a cooling train where rollers maintain the straightness of the rail, as it is cooled. The rail leaves the unit at a uniform temperature and transformation from austenite to pearlite takes place in still air. The straight rail then passes down the cooling bank for finishing in the normal manner.

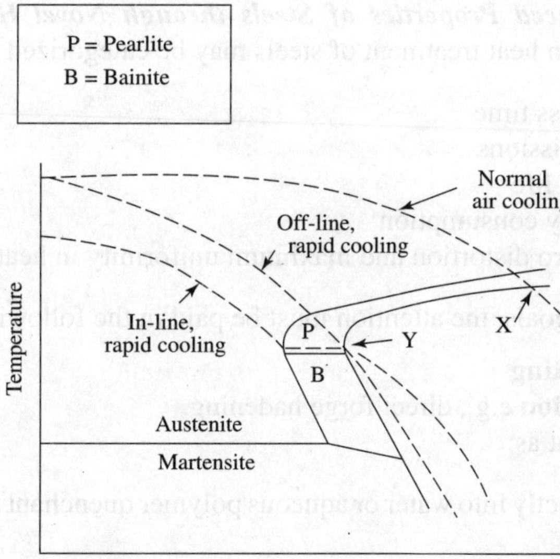

Fig. 12.6. Continuous cooling transformation curves for plain carbon rail steels (Reproduced with permission from 'The Hadfield Memorial Lectures, Vol. 1, Ed. P. Beeley, Maney Publishing on behalf of the Institute of Materials,Minerals and Mining London, 2000, p. 272).

Regarding hardening of the rail head, there have been two practices. In one, there is through heating of the rail followed by spray quenching of the rail head only. In the other, induction heating of only the rail head is carried out, followed by spray quenching. In either case, there is some distortion of the head. In the first case it is of convex type, while in the second case it is of a concave type.

Presently for a still better performance, bainitic steel rails are used; which are low-to-medium-carbon steels. British Steel has shown that the carbide free bainite developed in C-Mn-Si steel can make a tough, wear resistant rail steel. A typical composition is 0.23 C, 1.3 Mn, 0.4 Si, which has mixed microstructure of bainitic ferrite lathes and retained austenite at the lath boundaries. The advantage of silicon addition is to impede the carbide formation at the lath boundaries even with 0.4% C. The rails are auto-tempered during exit of the head hardening unit in which the interior heat facilitates the temperature of the outer head surface. The characteristics of the medium carbon lath martensitic rails are less wear resistance, and higher fracture toughness (110 MPa$\sqrt{m}$) as compared to fully pearlitic head-hardened rail.

We have mentioned earlier about 'wheel spin-burning'. To prevent this, a stable microstructure is essential. Otherwise brittle martensite can form when the rail gets heated transiently above the austenitizing temperature and rapidly cooled by heat absorption into the bulk wheel mass. Steels with lower carbon content shall be preferred, because of their low hardenability and high M_s temperature. Carbide free molybdenum-boron low carbon bainitic steels (0.25 C, 1.8 Si, 2.0 Mn, 0.45 Mo, 1.95 Cr, 0.35 Ti, 0.003 B) are useful in railway crossings and turnouts where impact wear and fatigue of the crossing nose and turnouts pose severe wear problems. From the cost point of view, the high alloying costs of low-C bainitic steels would still be not advantageous in comparison with in-line hardened pearlitic steels.

Case Study 12.4: Enhanced Properties of Steels through Novel Heat Treatments: The direction of development in heat treatment of steels may be categorized as follows:

- Reduction in process time
- Achieving zero emissions
- Increase in furnace life
- Reduction in energy consumption
- Achievement of zero distortion and maximum uniformity in heat treated parts.

To achieve some of these goals, the attention must be paid in the following:

(a) **Accelerated heating**

(b) **Process integration** e.g., direct forge hadening.
This is carried out as:

- Quenching directly into water or aqueous polymer quenchant solutions (or oils) after hot forging.
- Direct cooling after forging in forced air in order to obtain the desired micro-structure e.g., ferrite and pearlite/bainite.

Another example of process integration is that of casting and heat treatment processes. Aluminium castings can be directly transferred to heat treating furnaces, where three processes decoring and sand removal, thermal sand reclamation and heat treatment can be combined. This results in great energy savings compared to reheating of the castings after solidification. Similar processes have been developed using fluidized bed furnaces.

Another example of process integration is sinter hardening of steel parts.

(c) **High temperature carburizing:** High temperature carburizing conducted at temperatures higher than 1010°C provides reduction in carburizing cycle times. But in this case new class of steels is to be designed, to resist grain coarsening.

(d) **Intensive or shell quenching:** It is defined as a very rapid; but uniform cooling, usually in a water/polymer bath with a high agitation rate. A high pressure spray or water-mist quench, with very high evaporative cooling rates also qualifies as 'intense quench', but the spray or mist must be uniform over all part surfaces. Figure 12.7 shows a steel part with a varying thickness. During conventional quenching, the martensite forms first in the thinner section of the part, since this section cools faster. This gives rise to residual stress, which promotes distortion and possible part cracking. In case of intense quenching, the same steel part is cooled very rapidly; such that the martensite forms simultaneously over the entire part surface creating a hardened shell. Such a quenching offers following benefits:

- Elimination of cracking,
- Minimum distortion,
- High compressive surface stresses (similar to those developed in mechanical shot peening),

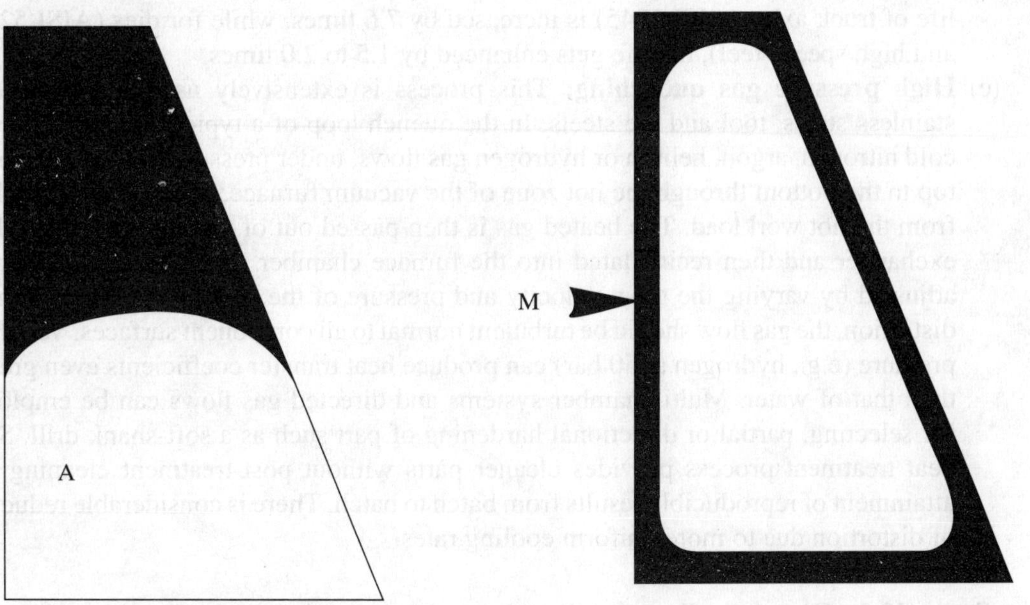

(a) Conventional quenching (b) Intensive quenching

Fig. 12.7. Schematic of martensite formation during (a) conventional quenching and (b) intensive quenching. A: austenite, M: martensite.

- Improved mechanical properties (yield and ultimate strength, wear resistance, depth of hardness, etc.),
- Longer part life,
- Usage of less alloy steels,
- Elimination of hazardous oil and hot salt quenchants with environmentally-friendly water or polymer/water solutions.

Table 12.6 shows the comparative picture of the properties of a steel bolt quenched under conventional and intensive manner.

Table 12.6. Comparison of Properties of a Steel Bolt Conventionally and Intensively Quenched.

Specimen size		*Hardness, HRC Mean*
10 mm dia, 62 mm long rod	Conventionally quenched	42.7
	Intensively quenched	54.8
5 mm dia, 110 mm long rod		Distrortion, mm mean
	Conventionally quenched	0.21
	Intensively quenched	0.18

By such quenching great improvement in part life has been noticed. For example the life of truck axles (AISI 1045) is increased by 7.6 times, while for dies (AISI 52100 and high speed steel), the life gets enhanced by 1.5 to 2.0 times.

(e) **High pressure gas quenching:** This process is extensively used for hardenable stainless steels, tool and die steels. In the quench loop of a typical vacuum furnace, cold nitrogen, argon, helium or hydrogen gas flows, under pressure (>5 bar), from the top to the bottom through the hot zone of the vacuum furnace, where it picks up heat from the hot workload. The heated gas is then passed out of the hot zone into a heat exchanger and then recirculated into the furnace chamber. The cooling rate can be adjusted by varying the type, velocity and pressure of the gas. In order to minimize distortion, the gas flow should be turbulent normal to all component surfaces. Very high pressure (e.g., hydrogen at 50 bar) can produce heat transfer coefficients even greater than that of water. Multi-chamber systems and directed gas flows can be employed for selecting, partial or directional hardening of part such as a soft-shank drill. Such heat treatment process provides cleaner parts without post-treatment cleaning and attainment of reproducible results from batch to batch. There is considerable reduction of distortion due to more uniform cooling rates.

Case Study 12.5: Cryogenic Storage Container: A cryogenic storage container is to be designed to withstand forces resulting from the internal pressure, the weight of the content and bending stresses. In addition, the material of construction must be compatible with low temperatures.

In Chapter 5, we have described how materials get embrittled at low temperatures. Table 12.7 gives the classes of materials suited for such applications. Figure 12.8 shows the variation of ductility of various materials with temperature in the low temperature range. Practically all plastic or elastomer become brittle at low temperature, except PTFE. The relatively high tensile elongation of copper even increases further at low temperatures. It is also noticeable that the increase in strain rate, e.g., impact loading, shifts the ductility/temperature plot to the right (Figure 12.9).

Table 12.7. Embrittlement of Structural Materials at Low Temperatures.

Materials that remain ductile at low temperatures, if ductile at room temperature	*Materials that become brittle at low temperatures*
Copper	Iron
Nickel	Carbon and low alloy steels
All copper-nickel alloys	Molybdenum
Aluminium and all its alloys	Niobium
Austenitic stainless steels containing more than 7% nickel	Tungsten
Zirconium	Zinc
Titanium	Most plastics
Polytetrafluoroethylene	

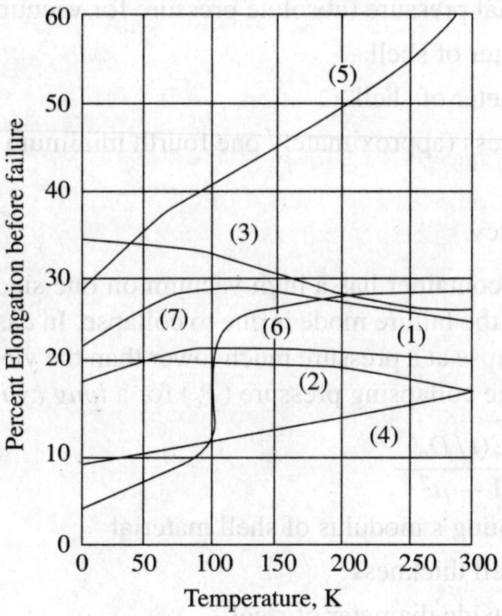

Fig. 12.8. Ductility vs. temperature relations for several structural materials: (1) 2024 T4 aluminum alloy, (2) beryllium copper, (3) K. Monel, (4) titanium, (5) 304 austenitic stainless steel, (6) C 1020 carbon steel and (7) 9% Ni-steel (Reproduced with permission from Cryogenic Engineering, T.M. Flynn, Marcel Dekker, Inc., New York, 1997, p. 190).

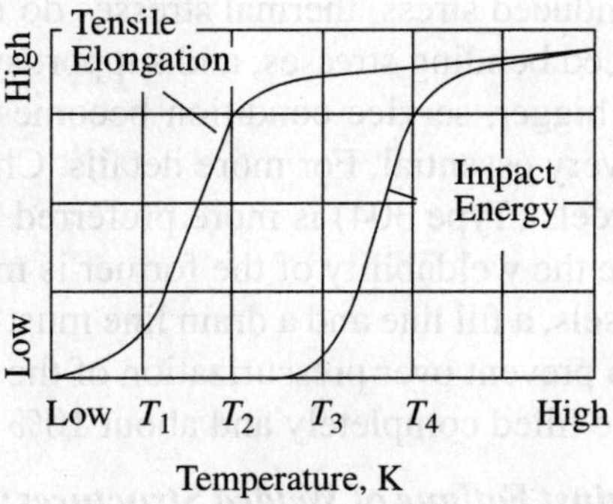

Fig. 12.9. Effect of strain rate on ductility.

Ti-5 Al-2.5 Sn extra low interstitial alloy is used for liquid hydrogen turbo-pumps.

From the viewpoint of materials economy, it is evident that ordinary carbon steel is not useful for cryogenic container. Since other alloys are expensive, designer attempts to make the inner vessel as thin as possible. This constraint also reduces cool down time and the amount of cryogenic liquid required for cooling down.

There are various design codes established by ASME. According to it, the minimum thickness of the inner shell for a cylindrical container is determined by:

$$T = \frac{PD}{2S_a.e_w} \quad or \quad \frac{PD_o}{ws_a e_w + 0.8P} \tag{12.8}$$

where P = design internal pressure (absolute pressure for vacuum jacketed containers)

D = inside diameter of shell

D_o = outside diameter of shell

s_a = allowable stress (approximately one fourth minimum ultimate strength of material)

e_w = weld efficiency.

The outer shell of the container has a high vacuum on one side and atmospheric pressure on the other and therefore, the failure mode is due to collapse. In case the wall thickness is low, the external shell may collapse at a pressure much lower than the yield strength for the material. One typical equation for the collapsing pressure (P_c) for a *long cylinder* is

$$P_c = \frac{2E(t/D_o)^3}{1-\mu^2} \tag{12.9}$$

where E = Young's modulus of shell material

t = shell thickness

D_o = outside diameter of steel

μ = Poisson's ratio for shell material.

The collapsing pressures for different shaped bodies e.g., hemispherical, elliptical or torispherical are expressed differently and can be obtained from the code book.

In addition to pressure-induced stress, thermal stresses do impose additional contribution. Other features are load induced bending stresses, and support systems.

In case the container is bigger, service condition becomes more demanding. The proper welding of thicker plates is very essential. For more details, Chapter 10 may be referred. Low carbon austenitic stainless steels (Type 304) is more preferred than the high carbon austenitic steel (say Type 302), because the weldability of the former is much much better.

For larger cryogenic vessels, a fill line and a drain line must be incorporated. A vapour vent line must also be provided to prevent over pressurization of the vessel. Cryogenic fluid storage vessels are not designed to be filled completely and about 10% ullage volume is common.

Case Study 12.6: Design against Fatigue of Welded Structures: A bridge is to be assembled by an army made out of welded steel structures. How would you design it against fatigue? In case the steel bridge is to be replaced by a welded aluminium alloy structure, what difference would it make in your evaluation? The temperature variation in the geographical location ranges from 20°C to sub-zero. In case the structures are situated near the seashore, what further problems will it pose?

The case study is an interesting one, as it involves interaction of material processing/environment and stress condition. We know that during fatigue, cracks can form and then grow under repeated or fluctuating loading.

In case of welded structures, the problem of fatigue is greater, since the fatigue strength can be much lower than those of unwelded components. Welded steel joints can fail by fatigue at stress range below 30 N/mm^2. Moreover, fatigue failures can also occur in welded joints under

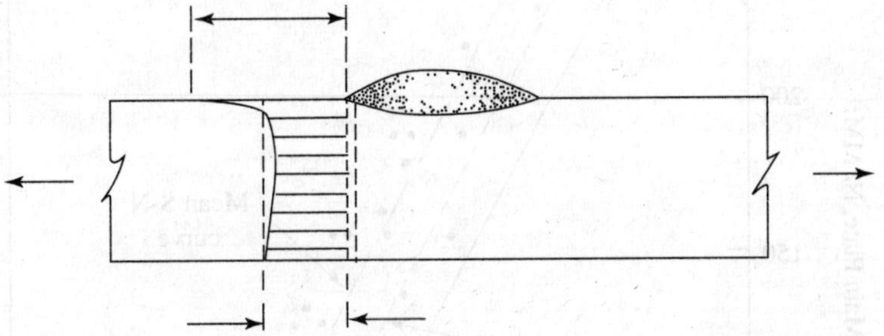

Fig. 12.10. Stress concentration at toe of weld bead.

nominally compressive stresses. A weld bead made across a piece of plate would produce a change in shape and hence, if loaded in the direction shown in Figure 12.10, a stress concentration would arise at the toe. In case the change in section is abrupt (more likely under normal welding conditions), the stress concentration will be high. The stress concentration conditions at weld toes can be further compounded by the existence of the crack like intrusions within a region of concentrated stress. The effect of weld toe is generally incorporated in to the S-N data for such a joint.

Another factor is the material strength. The fatigue strength of unwelded components normally increases with the material tensile strength. However, this is not the case with welded joints in which fatigue failure is by crack propagation. In the welded member, the crack growth is relatively insensitive to microstructure.

Selection of material is another factor. For example, the fatigue strength of a fillet weld in aluminium alloy is approximately 1/3 of that for the same joint in steel. This means that, the fatigue strengths are in the ratio of their elastic modulii.

The design rules are based on a statistical analysis of fatigue test data obtained from tests on welded joints, to produce the best-fit mean S-N curve by the method of least squares. A straight line relationship between log S and log N was assumed, where $S =$ stress range and $N =$ fatigue life. The standard deviation of log $N\sigma$, which is a measure of the scatter in fatigue lives obtained, is also calculated. An S-N curve two standard deviations below the mean (M-2SD), which, assuming a log-normal distribution of fatigue life, has the lower 95.4% confidence limit and represents approximately 2.3% probability of failure or a 97.7% survival limit, is taken as a reasonable basis for design.

Figure 12.11 exhibits the relationship between stress range and fatigue life for M, M-1SD and M-2SD conditions.

The next operation for the purpose of fatigue design is to designate each part of the welded joint into its class. There are nine such classes (For detail refer S.J. Maddox, Fatigue Strength of Welded Structures, 2nd ed., Abington Publishing, Cambridge, 1991, p. 82-98). Each class exhibits similar fatigue strengths. It must not be confused that it is the site for fatigue cracking which determines the classification and not the joint type. In general, three possible sites for fatigue cracking should be considered: the toes of the weld in each of the members joined,

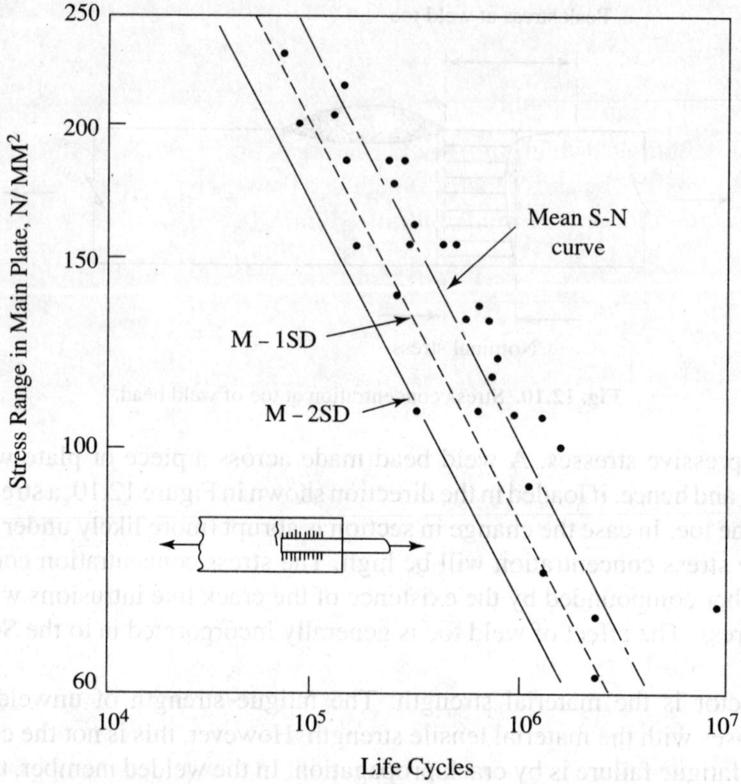

Fig. 12.11. Typical set of fatigue data for welded joint in steel.

giving two sites and the weld itself. The classification mentioned above refers to the fatigue performance of welded joints and should not be confused with the weld quality. The designer should ensure that conditions relating to welding material, weld quality or inspection details are adequately described to the fabricator.

For each class of joint, the general equation relating applied stress range, S and fatigue life, N, is

$$S^m N = a' \Delta^d \tag{12.10}$$

where m is the slope, a' is a constant which refers to the mean S-N curve, d is the number of standard deviations of log N below the mean, and Δ depends on the standard deviation of log N, σ, such that $\sigma = \log(1/\Delta)$. For a particular design S-N curve, the equation may be written

$$S^m N = A \tag{12.11}$$

where $A = a'\Delta^d$. For the mean S-N curve, $A = a'$, for the M-1SD, $A = a'\Delta$; for the M-2SD, $A = a'\Delta^2$, etc. Values of m and Δ together with values of A for the mean (also equal to a), M-1, M-2SD and S-N curves can be readily obtained in a welding Handbook (for as-welded joints). Equations for other confidence limits can be calculated easily by putting the required value of d into equation (12.10) above.

Welded joint dimensions: The effect of welded joint dimensions for fatigue behaviour is to be considered now. The effect of plate thickness, is introduced as a factor by which the permissible design stress range must be multiplied namely $\left(\frac{22}{B}\right)^{0.25}$, where B = plate thickness in millimeters. Weld details to which this thickness correction must be applied may be referred from welding Handbook. The above factor is deduced from fatigue strength results for members containing transverse fillet or butt welded joints (Figure 12.12).

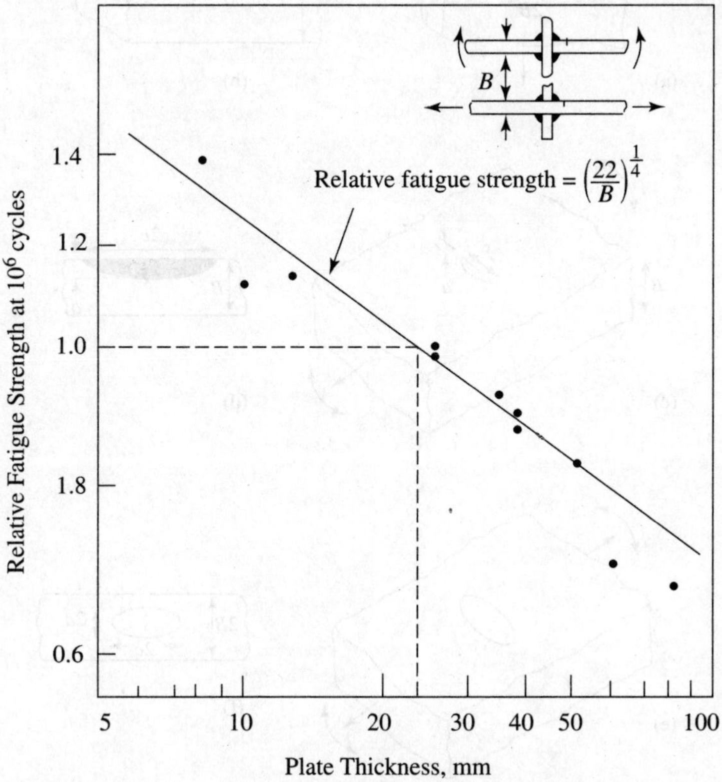

Fig. 12.12. Effect of plate thickness on fatigue strength ($R \sim 0$).

Fatigue crack propagation: Before going further, it is necessary to define 'a' correctly. It is the depth of a surface or edge, but only half of the depth or length of a buried or through-thickness crack (Figure 12.13). The basis of the application of fracture mechanics to fatigue is the relationship between rate of crack propagation $\frac{da}{dN}$, and stress intensity factor range, ΔK, based on applied cyclic stress range, which for many practical cases has the form (Refer Section 5.13.3)

$$\frac{da}{dN} = A(\Delta K)^m \tag{12.12}$$

where C and m are material constants which are found experimentally by measuring the rate of growth of a crack under known loading in a simple test specimen.

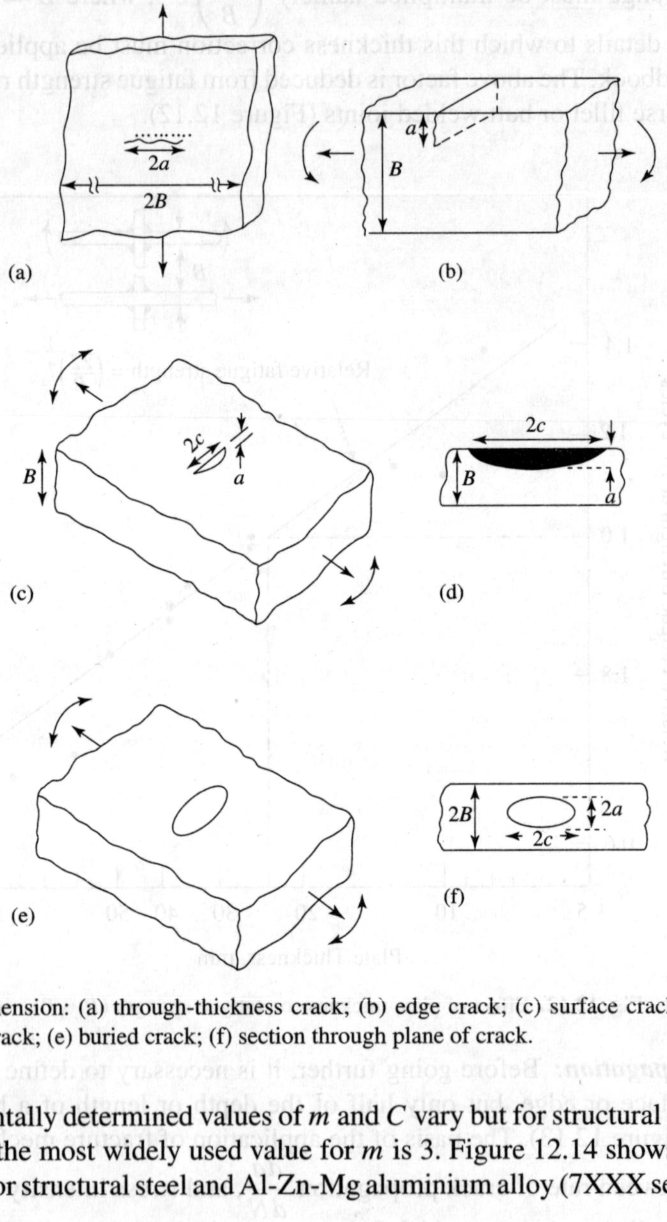

Fig. 12.13. Crack dimension: (a) through-thickness crack; (b) edge crack; (c) surface crack; (d) section through plane of crack; (e) buried crack; (f) section through plane of crack.

The experimentally determined values of m and C vary but for structural steel and weldable aluminium alloys the most widely used value for m is 3. Figure 12.14 shows the fatigue crack propagation data for structural steel and Al-Zn-Mg aluminium alloy (7XXX series). Substituting for ΔK, we get

$$\frac{da}{dN} = A(Y^{\Delta\sigma}\sqrt{\pi a})^m \tag{12.13}$$

The above relationship can be rearranged as

$$\int_{a_{init}}^{a_f} \frac{da}{(Y\sqrt{\pi a})^m} = A\Delta\sigma^m \int_O^N dN \tag{12.14}$$

This equation allows the fatigue process to be considered at any stage in the fatigue life, in contrast to the S-N curve which relates to one event i.e., final failure.

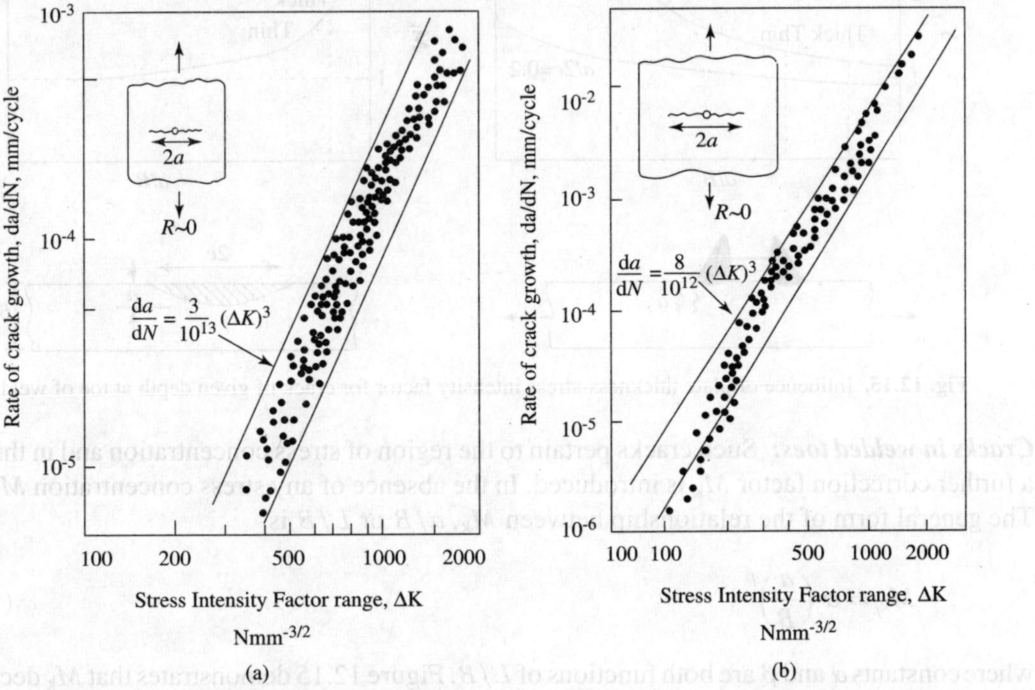

Fig. 12.14. Fatigue cracks propagation data for: (a) structural steels; (b) Al-Zn-Mg alloy.

In practice, it is often convenient to express equation (12.14), in terms of a/B, where B is the section size, especially as Y is a function of a/B. Then

$$\int_{a_{init}\,1B}^{a_f\,1B} \frac{d(a/B)}{\left(Y\sqrt{\dfrac{\pi a}{B}}\right)^m} = A^m \Delta\sigma B^{\left(\frac{m}{2}-1\right)} N \tag{12.15}$$

However, it must be realized that the above expression does not relate to variable range loading.

In the real system, the effect of the changes in the weld joint geometry, by varying B and/or L (attachment length) is to be incorporated.

Cracks in welded joints: Fatigue cracks which occur in welded joints are of the surface or embedded type, so that the crack front has an elliptical shape. In such case, the Y must have a correction factor for crack front shape $a/2c$ as well as a/B. In general, Y increases with increase in a/B and decrease in $a/2c$. In other words, the most severe shape of crack is one with a straight crack front $a/2c = 0$ (Figure 12.15)

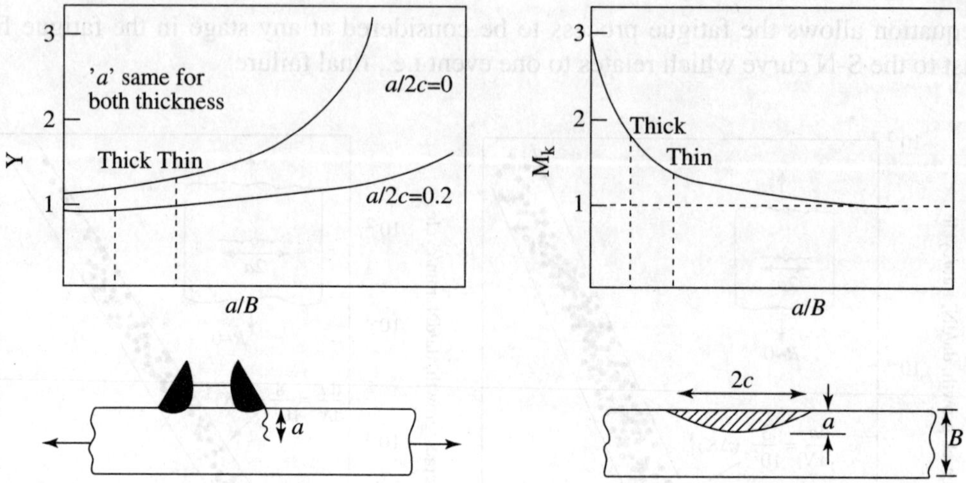

Fig. 12.15. Influence of plate thickness-stress intensity factor for crack of given depth at toe of weld.

Cracks in welded toes: Such cracks pertain to the region of stress concentration and in this case a further correction factor M_k is introduced. In the absence of any stress concentration $M_k = 1$. The general form of the relationship between M_k, a/B or L/B is

$$M_k = \alpha\left(\frac{a}{B}\right)^{\beta}, \tag{12.16}$$

where constants α and β are both functions of L/B. Figure 12.15 demonstrates that M_k decreases with increase in a/B, as the crack tip moves further away from the region of stress concentration, until it reaches unity.

In the absence of any stress concentration (i.e., $M_k = 1$), for a given crack depth, Y is less in thick plate (Figure 12.15) than in thin plate, with the result that fatigue strength will increase with thickness.

In conclusion it can be summarized that the design of welded structure against fatigue can be taken care from one of the following ways:

- Estimation of tolerable flaw size for specified fatigue strength (a_{init} unknown)
- Calculation of remaining life of a fatigue crack detected in service (N unknown)
- Calculation of inspection periods during which a critical crack will not develop (N unknown)
- Making allowance for material fracture toughness in fatigue performance calculations (the S-N curve unknown).

Selection of Al-alloy: Near sea shore the 6XXX series Al-alloy will be suited as it is immune to stress corrosion cracking (Refer Section 7.3). The disadvantage with 6XXX alloy is that its ultimate tensile strength is far lower than for 7XXX. Although from strengthwise 2XXX serves reasonably good, but here too, stress corrosion cracking is high. A costly Al-alloy structure is favourable in places too difficult to access, like high altitude mountaineous region. The transport there shall be less problematic.

Service temperature: In Chapter 7 (Section 7.1.2), we have described the ductility of structural steels. For structural steel, hot-rolled low alloy steel would be selected, but care should be taken as with increase in plate thickness ductile-brittle transition temperature increases. In case of Al-alloy with fatigue cracks the loss of impact value with decrease in service temperature is not high as in case of notches. It may be emphasized that the range of transition temperatures can be obtained simply by changing the specimen thickness and/or the crack length of the test bars. For this reason, in case of full scale component, a range of correction factors is ought to be incorporated. For details refer 'Deformation and Fracture Mechanics of Engineering Materials' by R.W. Hertzberg, 3rd ed., John Wiley & Sons, New York, 1989.

Case Study 12.7: Sintered Camshaft Pulley: Based on engine tests, designer's experience and empirical rules, distribution pulleys for automobiles are designed. The following are the major highlights of the steps to be taken by designer for a sintered camshaft pulley:

1. Analytical method may be developed using static measurements to apply the right loading on the pulley's teeth.
2. Use the information of above mentioned stage to simulate the dynamic behaviour of the component. Different degrees of belt tensions must be tried. The measuring programme must consist of stabilized and transient regimes for at least two different levels of the belt tension and engine load.
3. Predict the fatigue life of the pulley for different designs. One may use post processor software which calculates a 'safety factor' for each given design.
4. Federal Mogul of France have developed two designs of sintered pulley: one having 4 strut and other with 5 struts (Figure 12.16). It was found that Design A had higher safety factor against the fatigue endurance life (3.96) as compared to Design B (3.56). The margin in 'B' a still large, but the uncertainty of some parameters as belt tension, material parameter (roughness) were not all known.
5. Further study is required to take into account the different regimes according to their frequency in the engine life, and to estimate as accurate as possible the capability of the components, so as reducing the acceptable safety factor.

(For further details refer: E. Ayax, G. Laurent, L. Vinai, P. Blanchard and O. Girand, Design Optimization of the Camshaft Pulley: Conception, Verification and Validation, in Advances in Powder Metallurgy and Particulate Materials 2002, Part 2, Automotive, compiled by V. Arnhold, C. Chu, W.F. Jandeska, Jr. and N.I. Sanderow, Metal Powder Industries Federation, Princeton, U.S.A., 2002, pp. 2.64-2.70.)

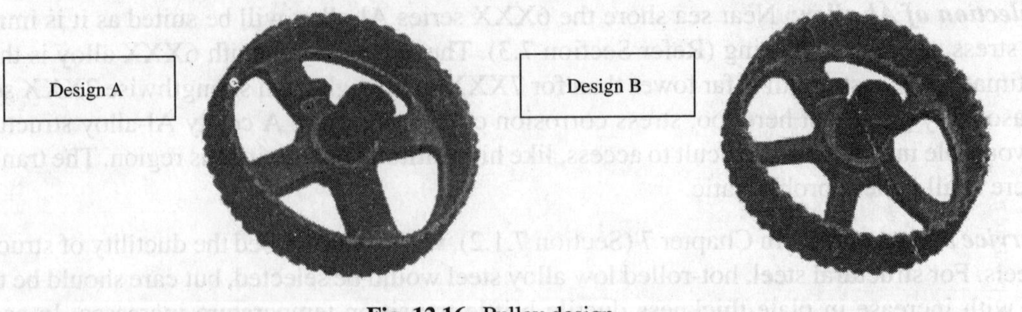

Fig. 12.16. Pulley design.

Case Study 12.8: Fatigue in Transmission Gear: The fatigue design of transmission gear requires special material properties. The data are more complex, because they must take into account interacting parameters of material/loadings/part geometry/manufacturing. Rolling contact of loading mode occurs on gear flanks. In such component, Hertzian pressure is felt, which causes a fatigue damage at or below the surface. The multiaxial stress state is transformed into an uniaxial one by the von Mises criteria (refer Chapter 8), because of the compressive nature of the stress state. The change is also dependent on the grade of sliding between the contact partners. The maximum sliding which can occur on gear flanks is ~24% and diminishes the fatigue strength in comparison with a loading without sliding (Figure 12.17). Figure 12.18 shows the test rig of rolling contact fatigue. However, the direct correlation of rolling contact fatigue data from test specimens to real part i.e., gear is not yet ideally achieved. Some of the parameters are enumerated in Figure 12.18. Figure 12.17 pertains to data for the steel made out of sintering route from a low alloy steel, followed by heat treatment to introduce surface hardening.

Case Study 12.9: Fatigue in Powder Metallurgy (P/M) Connecting Rod: Steel connecting rods for automobiles have been perfected through powder metallurgy route.

The connecting rod for an automobile offers a case of classical fatigue application, so the complexity is not that severe as in transmission gear. Fatigue testing of such part can be dealt with, as following:

(a) **Specimen shape:** The test piece specimens in sintered materials have to be looked from additional considerations. Almost all sintered parts have rectangular cross-section, edges and unmachined surfaces. Therefore, data obtained with machined, unnotched round specimens are difficult to be applied for fatigue design; especially when they are determined under fully reversed ($R = -1$) rotational bending.

(b) **Notched vs. unnotched specimen:** Notched specimens deliver data which are the closest to reality, because parts without notches do not exist. Round bars can be notched, but machining of P/M part influences the properties significantly. Central notched specimens are suitable for the assessment of bores (e.g., oil bore of connecting rods) only under axial loading. On the other hand, the double-edged notched specimens are suited for the assessment of fillet notches. Therefore, the specimens may be specified with different radii and consequently different loading mode dependent notch factors.

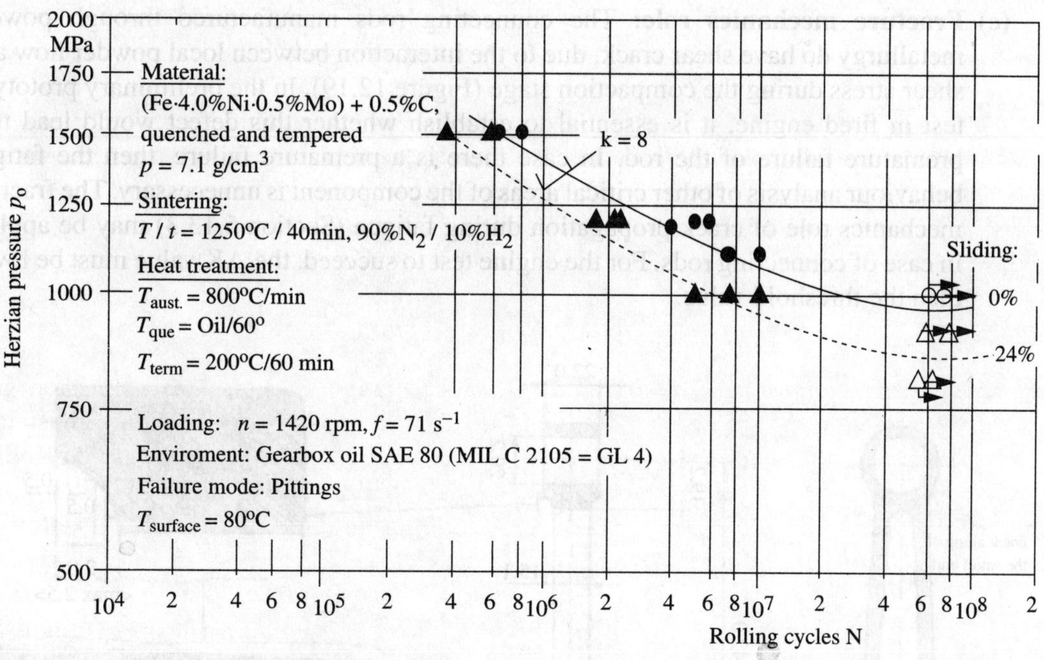

Fig. 12.17. Rolling contact fatigue curves for different sliding ratios (Reproduced with permission from Advances in P/M and Particulate Materials, 2002, Part 5–1, Metal Powder Industries Federation, Princeton, U.S.A.).

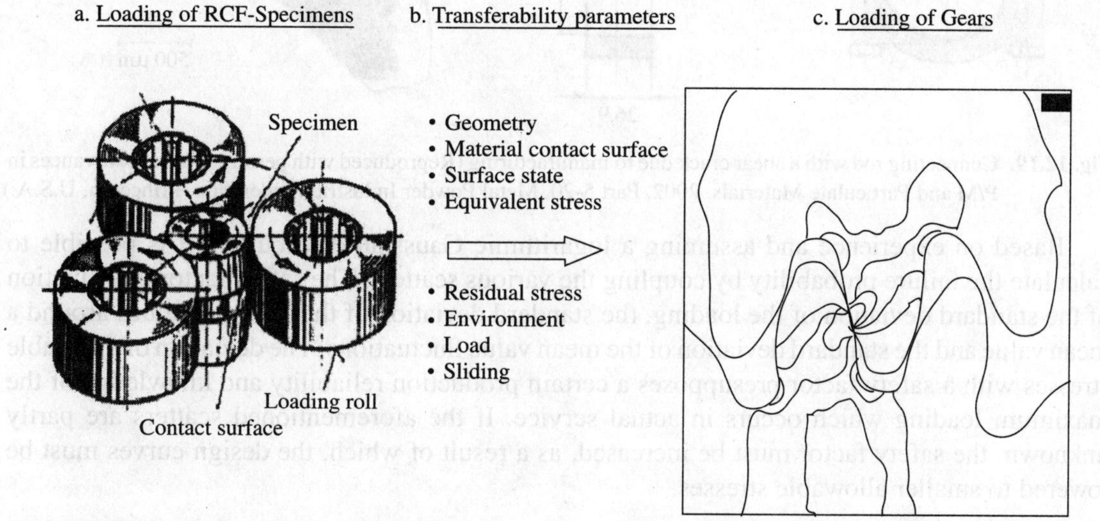

Fig. 12.18. Parameters for the transferability of rolling contact fatigue data for the design of gears.

(c) **Fracture mechanics role:** The connecting rods manufactured through powder metallurgy do have shear crack, due to the interaction between local powder flow and shear stress during the compaction stage (Figure 12.19). In the preliminary prototype test in fired engine, it is essential to establish whether this defect would lead to a premature failure of the rod. In case there is a premature failure, then the fatigue behaviour analysis of other critical areas of the component is unnecessary. The fracture mechanics role of crack propagation during fatigue (Section 5.13.3) may be applied in case of connecting rods. For the engine test to succeed, the ΔK value must be lower than the threshold value.

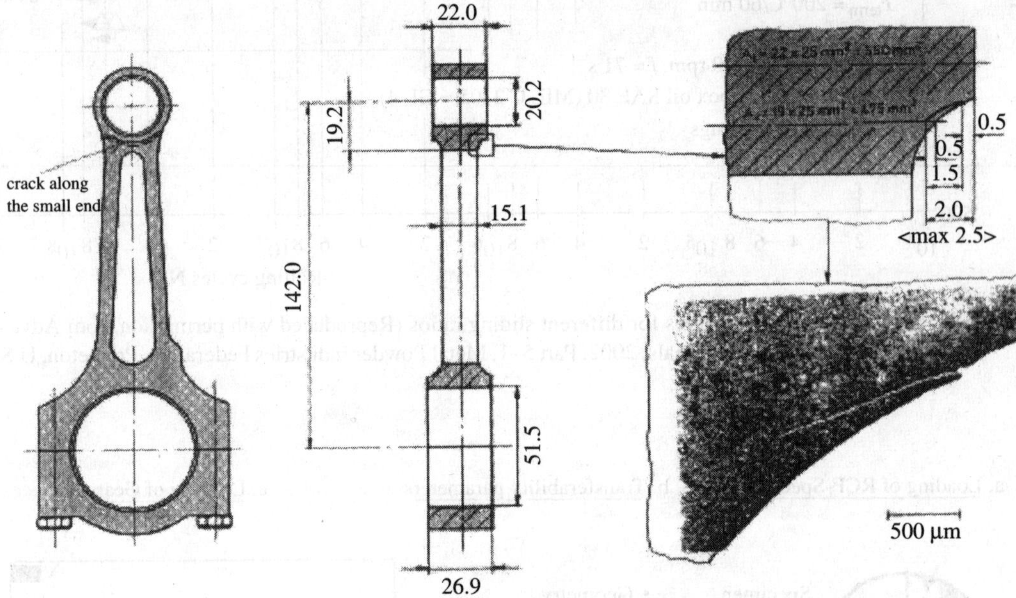

Fig. 12.19. Connecting rod with a shear crack due to manufacturing (Reproduced with permission from Advances in P/M and Particulate Materials, 2002, Part 5-20, Metal Powder Industries Federation, Princeton, U.S.A.).

Based on experience and assuming a logarithmic Gaussian distribution, it is possible to calculate the failure probability by coupling the various scatters. The safety factor is a function of the standard deviation of the loading, the standard deviation of the strength values around a mean value and the standard deviation of the mean value fluctuations. The deviation of allowable stresses with a safety factor presupposes a certain production reliability and knowledge of the maximum loading which occurs in actual service. If the aforementioned scatters are partly unknown, the safety factor must be increased, as a result of which, the design curves must be lowered to smaller allowable stresses.

Case Study 12.10: Austempered Ductile Iron (ADI): Austempered cast iron of following composition C 3.55, Si 2.37, Mn 0.41, P 0.034, S 0.008 and Mg 0.05 (all mass %) has as much as 25 vol. % retained austenite and an elongation of 4%. This ductile nodular iron has a microstructure of graphite nodules surrounded by a bainite structure. The presence of bainite is not suited for the improvement of the Charpy impact strength and ductility. For enhanced ductility it is desirable that the graphite nodules be surrounded by ferrite. This ferrite would prevent the propagation of cracks and improve ductility. The conventional austenitizing temperature is around 900°C and the austempering temperature 380°C. To precipitate ferrite suggested between these two temperatures, a furnace cooling of the cast iron be done followed by standard austempering treatment. It has been noticed that after this additional cooling to 700°C the ferrite thickness is 20μm, which increases to 40μm if the temperature is little elevated to 750°C. This treatment decreases the amount of retained austenite, being lower in case of 750°C treatment. This additional heat treatment has been found not to alter the brittle-ductile transition temperature (−63°C).

In case more strength is required, the ADI should be quenched to sub-zero after the austempering treatment. But this will be detriment to ductility.

The mechanisms of the bainitic reaction taking place during austempering of cast iron may be divided into three stages:

1. Growth of ferritic plates within an increasing amount of stabilized austenite
2. stabilization of the austenite
3. precipitation of carbides, growth of ferritic plates and transformation of the austenite phase.

In case of any cold working, the retained austenite transforms into plastically induced martensite. The wear resistance of ADI gear wheels can be attributed to this phenomenon. In order to further enhance wear resistance, surface hardening of ADI can be done. One of the typical method, is induction hardening or flame hardening.

In case one desires more content of martensite phase, it is suggested that the homogenization of nodular iron be done in nitrogen-rich atmosphere. Note nitrogen is an austenite stabilizer.

Secondary graphitization in quenched and tempered ductile cast iron results in small graphite precipitates throughout a ferrite matrix at temperatures as low as 430 or 540°C to a hardness level of less than HRC 31. This cannot be prevented by alloying element. It is, therefore, recommended to go to the austempering route followed by subsequent heat treatment, if any.

Case Study 12.11: Steel-Framed Building and Houses: Till 1980, the steel share of the multistorey building market was less than 35% relative to about 65% for concrete. For better durability and fire protection, now-a-days steel framed buildings are coming up. The load bearing components of such buildings are essentially a series of horizontal, vertical and diagonal roll-formed, galvanized steel 'U' channels. The frame is designed to withstand all the internal and external forces acting on the house since no contribution to strength or stiffness could be assumed from the inner plaster board lining or brick external skin. The combined action of vertical loading (dead weight, floor) and sideways loading (wind) are applied to the test frame plus an appropriate safety factor. In past the frames used to suffer due to lack of effective corrosion protection. The hot-dipped galvanizing steel solved this problem. The corrosion product films, which form on zinc act as barriers to prevent further attack. The need to prevent condensation on the frame or within the structures has led to the concept of 'warm frame'. Insulation on the outside of the steel frame prevents cold bridging and reduces the risk of surface condensation on the steel work. By careful selection of material, it is possible to reduce the risk of condensation by ensuring that dew point conditions only occurred outside the main frame.

Case Study 12.12: Superplastic Metallic Alloys: Zn-22 Al-0.5 Cu-0.5 Mg (ZAM), Al-6 Cu-0.4 Zr (Supral 2004) and Ti-6 Al-4 V alloys are amenable to superplastic forming. Some data are as follows:

Alloy	m (Max. value)	% Elongation	Superplastic temp., K
ZAM	0.65	1000	293 – 523
Supral 2004	0.5	1000	750
Ti-6 Al-4 V	0.85	1000	1023–

Describe critically the merits and demerits of each alloy under following categories:

(a) Composition
(b) '*m*' values
(c) applications
(d) cost.

The first alloy is zinc based binary eutectoid of Zn-22 Al composition containing minor additions of Cu and Mg. As Zn-22 Al is a low melting point material (~400°C). The creep resistance is too low. It creeps even at room temperature. To avoid this problem, Cu and Mg additions are made, making the alloy fairly satisfactory for creep properties at 100°C. 'm' value is fairly high. The application of this alloy may be for car body panel. The cost is reasonably low. The mechanism for forming is through the fine eutectoid microstructure. In order to obtain fine equiaxed microstructure cross rolling is done.

The latter two alloys pertain to the class where the superplasticity is caused by the presence of the fine second phase. In case of Supral 2004, the precipitates are of $ZrAl_3$ particles. The processing of this alloy is expensive, as compared to ZAM. The conventional aluminium alloy is cast at 700°C, but the superplastic one is cast (DC cast) from high temperature, over 900°C. This is because of the fact that at high temperature zirconium can go into solution. For this alloy too, in order to get equiaxed structure cross rolling is done. In order to precipitate fine dispersion of $ZrAl_3$, unlike conventional alloy, intermediate hold is necessary (say for 4 hrs). Grain growth is largely inhibited by the fine $ZrAl_3$ particles during the hot rolling stage. For superplastic grade alloy annealing is never done on the final cold worked structure, so that on recrystallization a fine grain results. The recrystallization results during the superplastic forming of the artifact. 'm' value is not that high as compared to the remaining two alloys. The material cost is twice to that of the conventional aluminium alloy. The cost of Zn-base superplastic metal is about half to that of Supral. The superplastic forming cycle time in the latter is larger (15 min.) as compared to former alloy (8 min.). The overall economy of superplastic forming over conventional pressing can be achieved in case the number of components to be produced is large. Supral is far better creep resistant than ZAM. The application of Supral is in light aircraft's fuselage panels.

The third candidate alloy i.e., Ti-6 Al-4 V is an excellent superplastic two phase ($\alpha + \beta$) alloy with a higher 'm' value than that of Supral 2004. This alloy is better creep resistant. The problem is the reactivity of titanium with atmosphere, which makes the processing tedious and costly. The advantage of Ti base alloy is that they are light (density 4.42 g cm^{-3}) with a good tensile strength (1000 MN m^{-2}). High performance and high cost mean that this material would be preferred for aerospace applications. The material is used mainly for hot forgings and machining from plate. As with other two alloys, to produce uniform equiaxed grains, cross rolling is done, which is an expensive and labour-intensive process. As compared to conventionally formed, the superplastically formed Ti-6 Al-4 V alloy has a little lower proof stress (810 Mn m^{-2} as compared to 860 MN m^{-2}). This is caused by grain growth during the forming cycle at elevated temperature (825–950°C). However, this drawback is not a serious one. In case the part is big, the tool is ought to be big, thus causing some temperature gradient and hence a gradient in microstructure and mechanical properties. The biggest advantage of superplastically formed Ti-6 Al-4 V alloy over the Supral is the complete absence of any cavitation problem. This is due to the fact that at the forming temperature the diffusion rate in the β-phase is pretty high. The optimum pressure cycle needed to form Ti-alloy is subject to the same constraints as for Supral, but titanium is far more forgiving material. Because of its higher 'm' value as compared to that for Supral, it is much less susceptible to drastic reductions in wall thickness during processing. Another advantage of Ti-6 Al-4 V alloy over Supral is that the former is amenable to diffusion bonding, making it possible to make the panel for aerospace applications in a cellular shape. This replaces all of the welding operations. The panel of the former alloy is less than half the weight of the aluminium design and much simpler. This gives rise a cost saving of 30% overall.

The main disadvantage of titanium is the ambient atmosphere, since the metal is very reactive, particularly at elevated temperature. Air quenching of the part allows the tooling to be reused. Although this entails some contamination of the surface layer, which can be etched away in an acid bath.

(For further details refer A. Demain, A History of Superplastic Metals, in Case Studies in Manufacturing with Advanced Materials, Vol. 1, J.H.W. deWit, A. Demaid and M. Orillon (eds.), Elsevier Science, Amsterdam, 1992, p. 35-71).

Case Study 12.13: Cost Effective Substitute for Bronze Bearings: Interest in Zn-Al alloy for bearing started as back as in 1940's, but the real impact was felt in 1980's in U.S.A., when Zn-Al-Cu alloys were developed:

- These alloys are best suited for lower speed bearing and bushing applications, at high loads.
- Cast zinc alloys show higher hardness range compared to sand cast bronze.
- Continuous casting results in increased material utilization in machining, increase in tensile strength, yield strength and hardness.
- Material cost for a given part can be decreased by 50% and finished cost by 20 to 40%.

The composition of some typical alloys are:

	Al	Cu	Mg	Sn	Fe	Pb	Cd	Zn
$ZA-12$	10.5 – 11.5	0.5 – 1.25	0.15 – 0.30	0.002max	0.075max	0.004max	0.003	maxBal.
$ZA-27$	25.28	2 – 2.5	0.01 – 0.02	0.002max	0.10max	0.004max	0.003	maxBal.

The maximum solid solubility of zinc is 18 mass % Al and 2.8 mass % Cu at 325°C and 0.8 mass % Al and 1.7 mass % Cu at 275°C. Figure 12.20 shows the isothermal section of diagram at 350°C. At 377°C a ternary eutectic reaction L $\rightarrow$(Al) $+\varepsilon+$ (Zn) takes place. ε phase has composition (81.4% Zn, 15.33% Cu and 3.3% Al). The liquid phase L has composition 80.9% Zn 15.4% Al, 3.7% Cu. ε phase is an intermetallic compound.

If one looks at the binary Zn-Al phase diagram, it is evident that at 78% zinc eutectic composition exists, where η (zinc rich solid solution) and L (Al-rich solid solution) phases

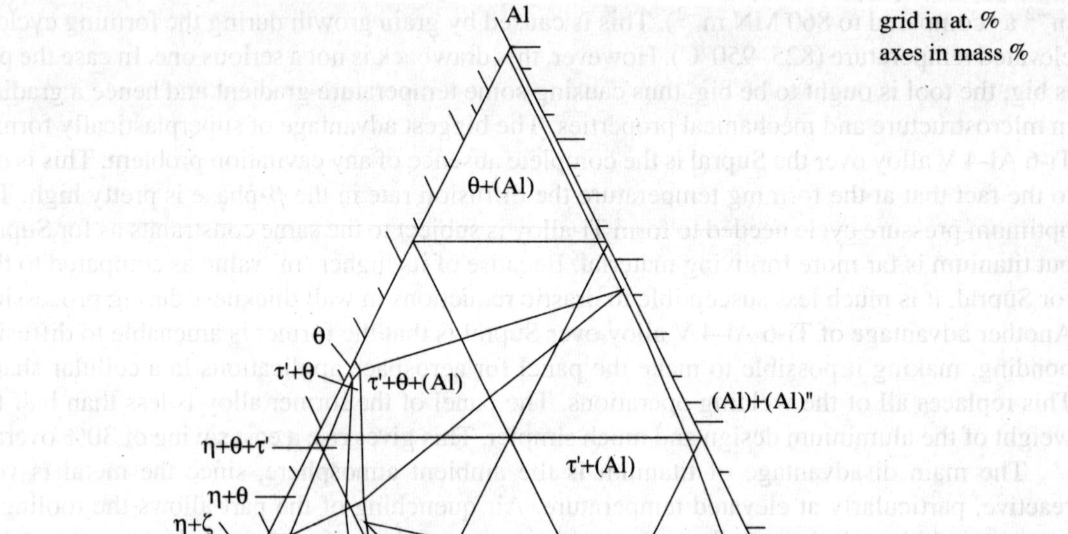

Fig. 12.20. Isothermal section of Al-Cu-Zn phase diagram at 350 °C (Reproduced with permission from Ternary Alloys, ed. G. Petzow and G. Effenberg, Vol. 5, Wiley-VCH , Weinheim, 1992, p. 109).

emerge. This means that in case of alloy ZA-27, the amount of eutectoid mixture $\eta + \alpha$ shall be more.

Commercial zinc-based bearing alloys are based on the eutectoid in the Zn-Al system and rely on a substantial copper content of the order of 5 wt % to obtain their good wear resistance. The cast microstructure reveals primary dendrites, eutectoid ($\alpha + \eta$) and ε phases. The ε phase is the hardest among all these phases. Both hardness and tensile strength of these alloys increase rapidly with increasing copper content. It has been shown that the growth which occurs on ageing these materials is due to the gradual conversion of metastable copper containing phases to those closer to equilibrium. Thus, for dimensional consistency either a stabilizing heat treatment must be applied or the copper content must be reduced to about 1 wt %. Heat treatment at 150°C for 10 days is necessary to stabilize the alloys fully but the coarsened structures which accompanied the conversion of the unstable copper-containing phases into a stable form produce a serious reduction in mechanical strength. Also, the smaller copper content reduces the growth on ageing; the reduced wear resistance may preclude its use in some applications. Figure 12.21 gives values of recommended bearing stresses at sliding speeds up to 0.3 m/s. For the concast alloys using hard journal strength (AISI 1144 steel, Rc 40–45), Alloy ZA-27 is capable of supporting higher stresses than is alloy ZA-12 at any given sliding speed.

Zn-Al alloys containing silicon are not as strong as copper containing alloys. However, the dimensional stability of Zn-Al alloys can be increased by the addition of silicon. The wear resistance of near eutectoid alloys with 3 and 5 wt % Si is superior to that of a similar alloy containing 3 wt.% Cu and SAE 660 bronze. The wear resistance increases with the amount of silicon in the alloy. The dimensional changes on ageing of Zn-Al-Si alloys are very small and there are no metastable phases involving silicon. The excellent wear behaviour of Zn-Al-Si alloys is attributed to the presence of discrete silicon particles which act as load carrying phase in the Zn-Al matrix. The distribution mode of silicon is found to be as important as the amount of silicon in influencing the wear resistance. The distribution is related to the silicon content; a lower silicon content gives a more uniform distribution.

Comparison with bronze:

(a) Cast ZA-12 has higher hardness range (92–96 HB) compared to 60–70 HB for sand cast SAE 660 bronze.
(b) Zn-alloys have lower cost than bronze. In addition they possess lower density.
(c) Zn-27 has high load capacity and comparable wear resistance over the entire speed range. It possesses low coefficient of friction at high load ranges.
(d) Zn-12 also possesses favourable properties, but not as good as Zn-27.
(e) Zn-based alloys do not compete with bronze at the highest operating temperature (>120°C).

Figure 12.22 illustrates the comparative wear rate of Zn-base and bronze alloys produced by different casting processes. The wear resistance of die cast and gravity cast alloys falls between that for sand cast and concast alloys. All the zinc base alloys are stabilized by heat treating for 12 hrs at 250°C followed by furnace cooling.

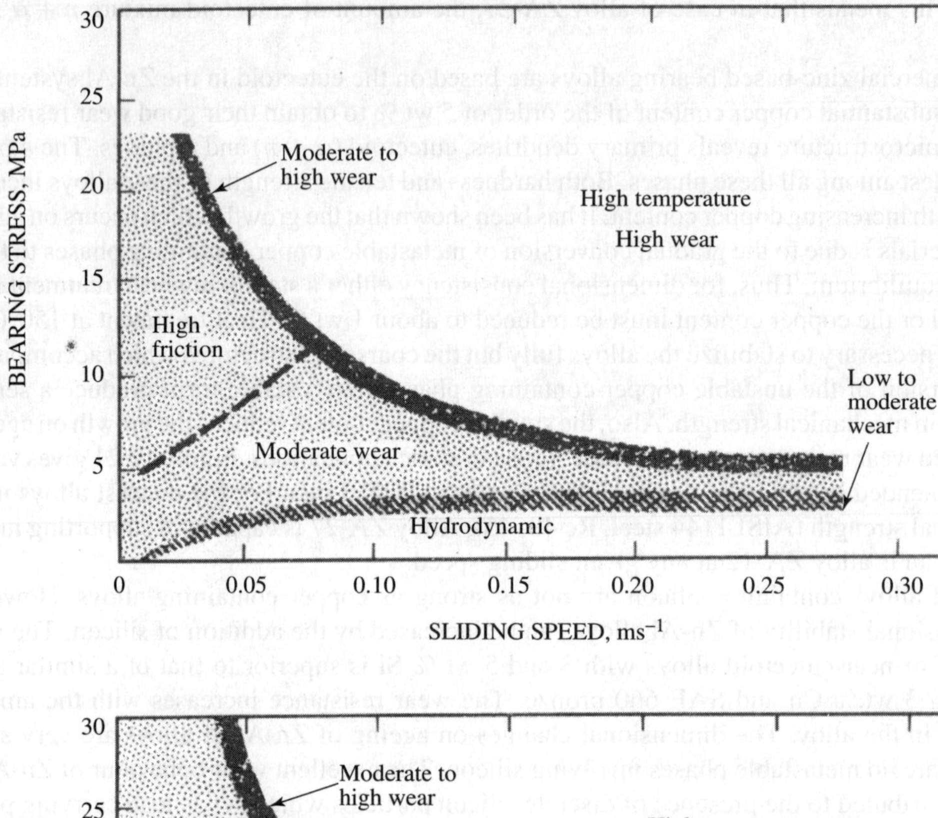

Fig. 12.21. Load-velocity (PV) curves for (top) alloy 12 and (bottom) alloy 27 (Reproduced with permission from Zinc Handbook, F. Porter, Marcel Dekker, Inc., N.Y., 1991, p. 487).

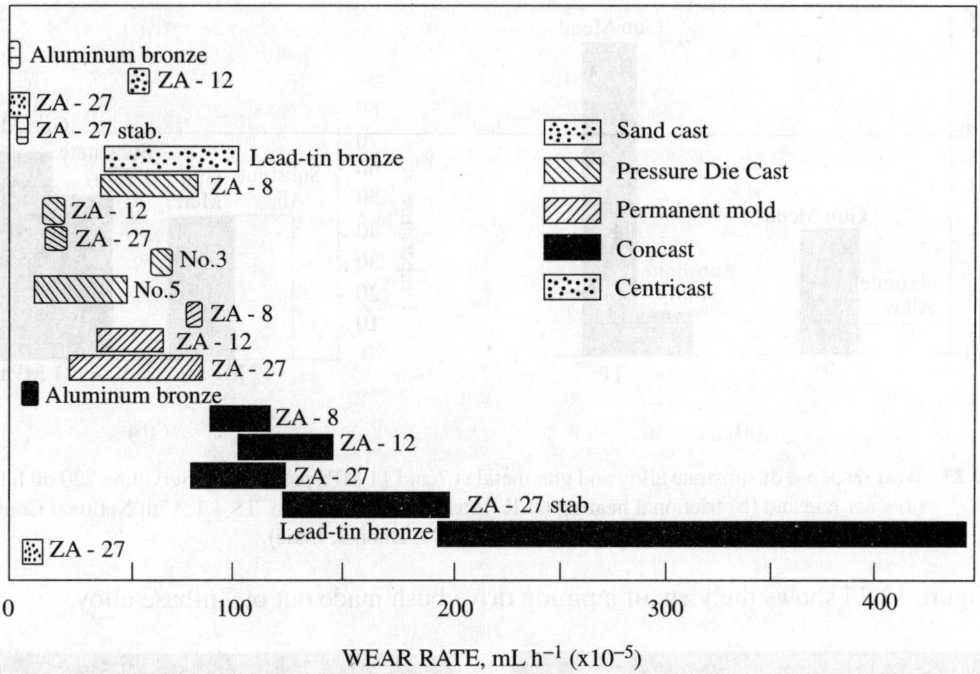

Fig. 12.22. Comparative wear rate of zinc and bronze alloys produced by different methods. All heat treated for 12 h at 250 °C and furnace cooled (Reproduced with permission from Zinc Handbook, F. Porter, Marcel Dekker, Inc., N.Y., 1991, p. 509).

The comparative properties of the two types of bearing materials are given in Table 12.8.

Table 12.8. Comparison of Properties of Zn-base Alloy and Gun Metal Alloy Bearings.

Property	*ZA-27 Bearing*	*Conventional Gun Metal Bearing*
Density, g/cm^3	5.02	8.90
Hardness, HV	130	74
Tensile strength, MPa (sand cast)	440-440	225
Yield strength, MPa (sand cast)	365-372	110
Tensile elongation, % (sand cast)	3-6	11.6
Abrasive wear rate, $\times 10^{-11}$ m^3/m	4.34	0.82
Specific heat, J/kg/K	525	376
Coefficient of thermal expansion 30-100 °C $\times 10^{-6}$/C	27	18
Surface finish	~1.0μm (Ra)	~1.0 μm (Ra)

Figure 12.23 shows the sliding wear behaviour of gun metal and Zn-base alloy. Improved wear response of the substitute alloy in terms of lower wear rate and less friction heating as compared to gun metal is noticeable.

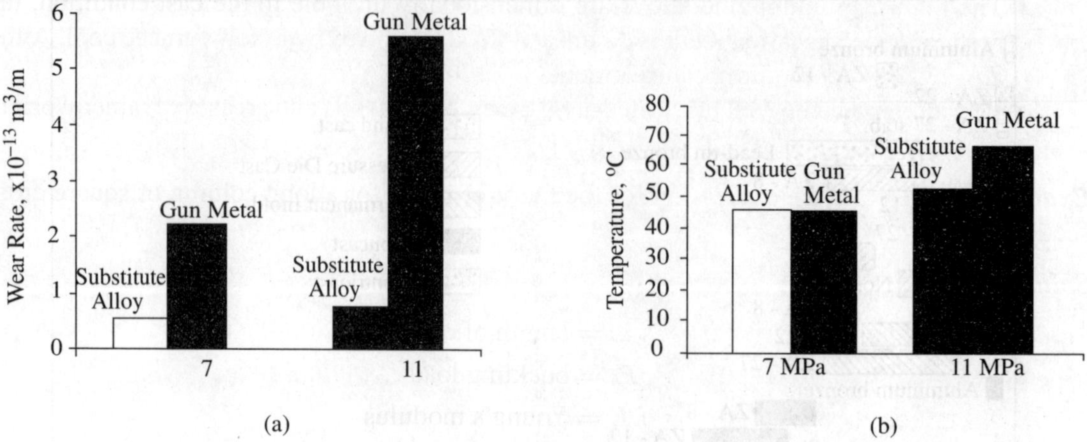

Fig. 12.23. Wear response of substrate alloy and gun metal at 7 and 11 MPa pressures in Servoline 220 oil lubricant: (a) wear rate and (b) frictional heating (B.K. Prasad et al., Paper No. TS 4.1, 17th National Convention of Metallurgists and Materials Engineers, Oct., 2003, Bhopal, India).

Figure 12.24 shows the view of a pinion drive bush made out of Zn-base alloy.

Fig. 12.24. Pictures of a pinion drive bush after machining made out of zinc base alloy.

Design guidelines for zinc-based bearings:

- Bearing length should not exceed 1.5 times the shaft diameter.
- Minimum wall thickness of bearings should be 0.045 D + 0.5 mm where D = shaft dia.
- ZA-27 bearings should have approximately 30% greater assembly clearance than bronze bearings, whereas ZA-12 bearings perform well with the same clearances conventionally used with bronze bearings.
- Hardness of shaft should not be less than 180 BHN and preferably should be harder. The difference between shaft and bearing hardness should be atleast 100 BHN.

- The high-aluminium zinc alloys are dimensionally unstable in the cast condition, the recommended heat treatment is 12 hrs at 480 °C, followed by a slow furnace cool. Alloy ZA-12 does not require heat treatment.
- Bearings should be well lubricated at all times. Neutral oils and greases of mineral origin are preferred.

Example 12.1 Derive the expression for the merit of index for a long column of square cross section.

Solution Let

L = length of column
P_b = buckling load
E = Young's modulus
I = second moment of area of the cross section
w = side of the cross section

and ρ is the density.

Buckling load $$P_b = \frac{kEI}{L^2} \quad (1)$$

where k = constant.

From the knowledge of the 'strength of solids',

$$I = \frac{w^4}{12}$$

Also $A = w^2$

Substituting in eqn. (1)

$$P = \frac{kEA^2}{12L^2} \quad (2)$$

Since lightness is inverse of mass (m), $\frac{1}{m} = \frac{1}{\rho LA}$, substituting the value for A; we have

$$\frac{1}{m} = \left(\frac{kE}{12L^2P}\right)^{1/2} \times \frac{1}{\rho L}$$

$$= \frac{\sqrt{E}}{\rho} \times \left(\frac{k}{12L^4P}\right)^{1/2}$$

Thus merit of index is $\sqrt{E}/\rho$.

Example 12.2 In an old automobile the front part made out of mild steel (yield stress 160 MPa, failure stress 250 MPa and failure strain 45%) is to be replaced by a thinner gauge high strength low alloy steel with yield stress 220 MPa, failure stress 355 MPa and failure strain 37.5%. If the mild steel of 1.8 mm thickness was used, what would be the thickness of the HSLA steel?

Solution Let b is the width and t the thickness of the sheet. Let θ be the angle of bending during impact loading. The bending moment M can be expressed as (refer any Mechanics of Solids textbook)

$$M = \frac{bt^2\sigma_y}{4} \qquad (1)$$

So the plastic work done during bending of the steel sheet is

$$E = \frac{bt^2\sigma_y}{4}\theta \qquad (2)$$

After replacing the mild steel by high strength steel sheet of thickness t' and yield stress σ'_y , , we get

$$E = \frac{bt'^2\sigma'_y}{4}.\theta \qquad (3)$$

In order to keep the plastic work done to be the same, we get

$$t' = \frac{t\sqrt{\sigma_y}}{\sqrt{\sigma'_y}} \qquad (4)$$

The above equation assumes a rigid plastic model, but in real material we can multiply the above expression (right hand side) by a coefficient α. If in present case we assume α to be unity, then the thickness of new steel, t', comes out to be 1.53 mm.

Example 12.3 How will you design pipelines against any fracture? How crack arrestor ring helps in preventing catastrophic failure?

Answer The leak detection is much easier to perform than the internal flaw detection. During service the crack grows both in length and depth until it cuts through the wall. At this point, the contents can leak through the crack to give a warning. The designer has to select the right material and wall thickness to promote leaking before fracture. The designer should ensure that the critical crack depth is greater than the wall thickness. In case the reverse is there, possibility of fracture before leakage is present.

The use of crack arrestors extends the maximum permissible crack length from the safety viewpoint. The spacing between the rings can be taken as the maximum permissible crack length.

Example 12.4 Some inclusions are invariably present in commercial cast steels, which adversely affect machining and fatigue behaviour of engineering steels. Which type of inclusions can be tolerated and what special needs eventually arise?

Answer Generally Al_2O_3 and TiCN rich inclusions are avoided, as they are very brittle. Plastic inclusion like SiO_2-rich ones (Si-Mg-Ca-Al-O) can be tolerated. Any reoxidation should be avoided. For this, it is essential to eliminate oxygen inputs after the vacuum period and casting in the RH degassing process. A total deslagging after tapping is indispensable for suitable inclusion control. In addition the use of large form (such as ingot or bloom) is preferable to smaller forms (such as billets), because their greater reduction ratio leads to smaller size inclusions. During steel making a close slag steel contact is desirable to promote suitable chemical reactions. Ferroalloy additions with a very low content of Al and Ti (Ti goal < 0.001 wt % in the steel) should be made. Clean and preferably new ladles with a MgO refractory should be used in order to avoid impurities of any type. With such inclusion engineering, the fatigue life of steel in certain cases is extended from $10 - 20 \times 10^6$ to 100×10^7 cycles.

Example 12.5 Suggest the solutions to achieve the following goals, while developing a microalloyed steel? (a) Higher strength, (b) improved toughness, (c) improved formability and Z-direction properties, (d) better weldability, (e) improved sour service performance, and (f) lower cost.

Answer

(a) Precipitation and grain refining
(b) Controlled rolling
(c) Deoxidation and cleanliness of steel
(d) Lower carbon
(e) Ca, Cu, rare earth, control of S, P and segregation
(f) Alloy redesign; plate mill cooling.

Example 12.6 Why niobium is very common in microalloyed steels? What are the major advantages?

Solution The reason why niobium is added, because it forms carbonitride precipitates in low-carbon steels. The precipitates are either completely or partially dissolved during the reheating of slabs prior to rolling thus the reprecipitation sequence and kinetics control to a large extent the benefits that are achieved.

Some of the important benefits are listed below and are shown schematically in Figure Ex. 12.6:

(i) Grain refinement of high temperature austenite when reheating prior to rolling, during the rolling process or when reheating for normalizing or for quenching treatment.
(ii) Retardation of the recovery and recrystallization processes which take place during rolling (particularly during low temperature thermomechanical treatments).
(iii) Modification of the hot-rolled crystallographite texture via selection of appropriate rolling regimes and steel compositions.
(iv) Niobium in solid solution lowers the austenite-to-ferrite transformation temperature giving a tendency toward the production of high dislocation density acicular ferrite or bainitic microstructures. Low carbon variants of these structures are useful for improving the strength and toughness of hot rolled plate and weld metal.
(v) Precipitation of fine niobium carbonitride particles strengthens ferrite formed in air-cooled plate.
(vi) The carbonitrides can be used to facilitate development of favourable crystallographic textures upon subsequent cold rolling and annealing. If the precipitation of interstitial carbon and nitrogen is substantially complete, the cold rolled and annealed steels have exceptional deep drawability.

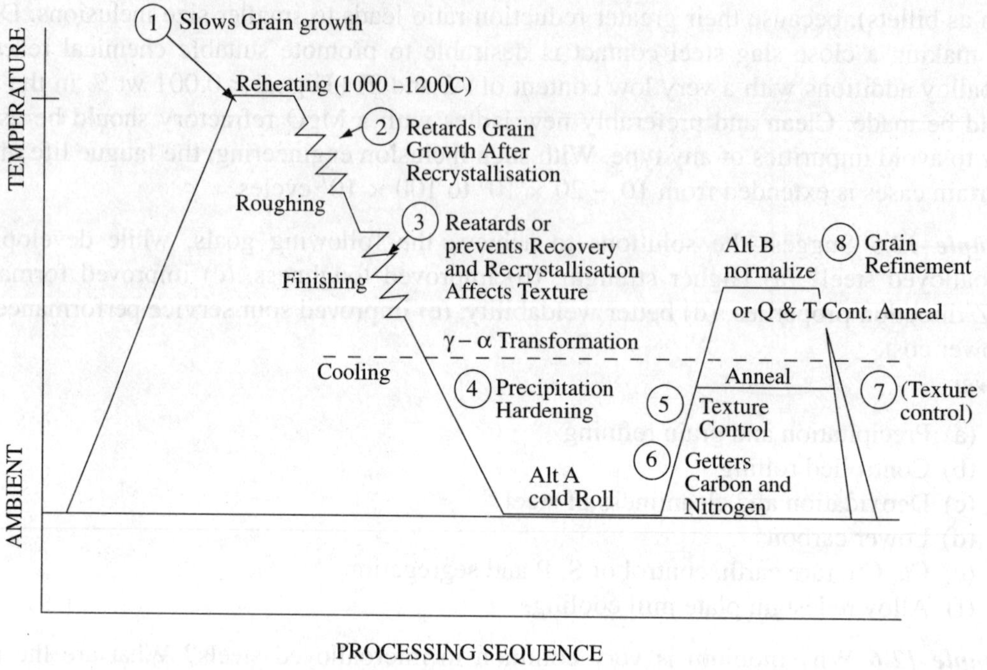

Figure Ex. 12.6

Now all of the above benefits can be achieved simultaneously and the mechanisms involved can become complex particularly in cold rolled steels that are processed by continuous annealing.

Example 12.7 For a successful performance of an engineering component, its evaluation for mechanical properties becomes more important than that of mere material. Show schematically what mechanical tests on a sintered synchronizing ring for an automobile transmission system will be recommended?

Answer Figure Ex. 12.7.

Even such a test of component in a laboratory, is not enough and an actual performance in field becomes necessary.

Example 12.8 A 45 mm diameter shaft is required to sustain a maximum torsional stress of 170 MPa and a maximum bending stress of 550 MPa. A company uses 4140 H steel for other parts' production and wishes to decide whether 4140 H grade would suffice for such an application. The company has an agitated oil quenching facility (60 m/min). From other similar parts, it is known, that at least 80% martensite structure should be present at the 1/4-radius position in the shaft.

Solution Since the maximum torsional (shear) stress is one-half the tensile bending strength (see Chapter 5), the maximum torsional stress is really 275 MPa and exceeds the stated required 170 MPa. Since the shaft is subjected to bending fatigue, we can assume 550 MPa as the fatigue limit. Referring the fatigue limit/hardness of through hardened steel (Figure Ex. 12.8a), we see that the shaft must have minimum quenched and tempered hardness of $35R_c$. To obtain this

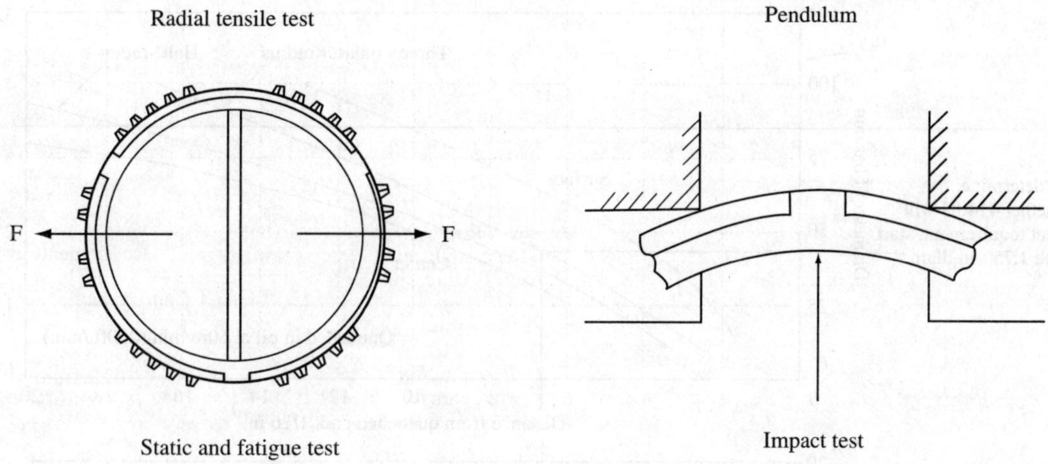

Figure Ex. 12.7

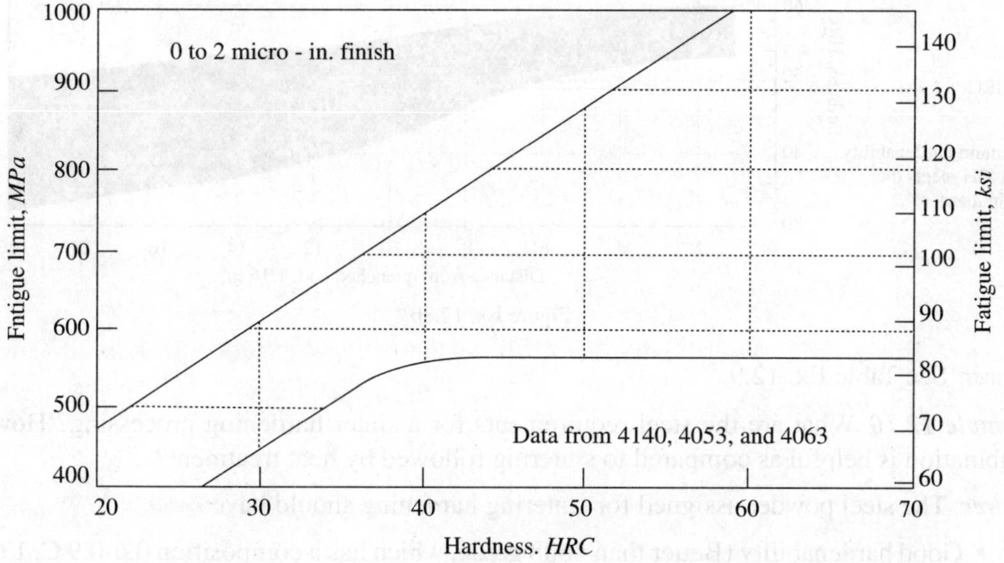

Figure Ex. 12.8a

tempered hardness, the quenched hardness would be around 45 R_C with 80% martensite ie 0.37%c (Refer Metals Handbook). Figure Ex. 12.8b shows the conversion of round to Jominy distance for that particular quenching oil for 4140 H steel. When we read the hardness corresponding to $6.5 \times 1/16$ inch, the value is 49 R_c. This confirms that the steel is more than adequate for the application.

Example 12.9 In a Table show how various surface heat treatments of steels and cast irons effect wear like abrasive, adhesive and fatigue. Also indicate their effect of any distortion. Incorporate various applications from engineering industries.

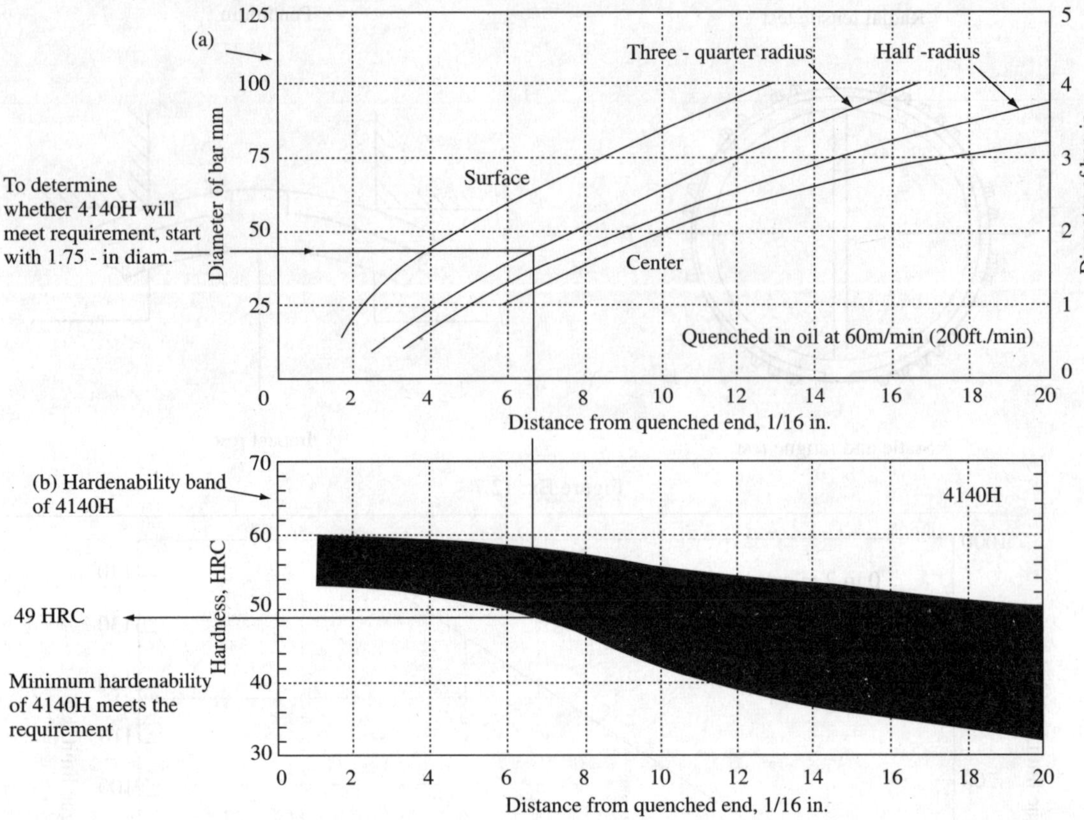

Figure Ex. 12.8b

Answer See Table Ex. 12.9.

Example 12.10 What are the steel requirements for a sinter hardening processing? How the combination is helpful as compared to sintering followed by heat treatment?

Answer The steel powder assigned for sintering hardening should have:

- Good hardenability (Better than 4608 grade, which has a composition 0.6-0.9 C, 1.6-2.0 Ni, 0.39-1.1 Mo, 1.0-3.0 Cu, balance Fe).
- Good compressability better than 6.8 g cm^{-3} at 550 MPa.
- The system should be sinterable up to 1120°C, i.e., in a continuous mesh belt furnace.

Advantage are:

- Reduced distortion of parts due to less severe quenching. This leads to better dimensional control.
- The tempering of sinter-hardened parts in air is easier than for quench-hardened parts. Sintered steel parts, quenched in oil bath, retain significant amount of oil in their pores. In that case the oil has to be first burned off at around 200°C, followed by tempering, which is generally carried out at a temperature >200°C.

Table Ex. 12.9.

Required Characteristics → *Process and Materials ↓*		*Abrasive Wear*	*Adhesive Wear*	*Fatigue Wear*	*Distortion*		Typical *Appications*
					Size	*Shape*	
Carburizing	Carbon Steels	H	M	H	M	M/H	Assorted Parts
	Alloy Steels	H	M	VH	M	M	Gears,Shafts,Machine components
Carbonitriding	Carbon Steels	H	M	M	L	M	Cams,Tappets, Textile Machinery
(Shallow < 0.3mm)	Alloy Steels	H	M	H	L	M	Gears,Ball Joints
Carbonitriding	Carbon Steels	VH	M	H	L/M	M	Shafts, Bearings, Bushess, Tappets
(Shallow > 0.3mm)	Alloy Steels	NA	NA	NA	NA	NA	Low Applicability
Nitriding	Carbon Steels	NA	NA	NA	NA	NA	Low Applicability
	Alloy Steels	H	H	H	VL	VL/L	Crankshafts,Gears,Bearings
Ferritic Nitrocarburising	Carbon Steels	L	VH	M/H	VL	VL	Crankshafts,Oil Pump Gears
	Alloy Steels	L/M	VH	H	VL	VL	Crankshafts,Tappets,Camshafts
	Iron Castings	L/M	VH	H	L	VL	
Deep F.Nitrocarburizing	Alloy Steels	M	VH	H	VL/L	L	Shafts,Molds,Dies,Gears
Austenitic Nitrocarburizing (720 ° C)		M	VH	M/H	L	L	Thinwall washers,Seives,Screens
Austenitic Nitrocarburizing (760 °C)		M	VH	M/H	L	L	Couplings
Iron Nitriding	Steels and ADI	M	VH	H	VL	VL	Gears,Wheels,Seed Feeders
Ferritic Nitrocarb+Post Oxidation	Carbon Steels	L	VH	M/H	VL	VL	Valves
	Alloy Steels	L/M	VH	H/VH	VL	VL	Valves,Bushes,Tools
Nitrocarburizing+Induction hardeining		H/VH	VH	H/VH	M	L/M	Synchronizers,Tappets
Carburizing+Nitrocarburizing		M	VH	H/VH		M	Differential Spiders,Gears etc
Induction Hardening	Carbon and Alloy Steels	H/VH	M	H	M	*	Shafts, RockerArms, Gears,
	ADI	VH	M	H	M	*	Sockets, etc
Boriding		VH	VH	M/H	M	M	Gears
Carburizing+Boriding		VH	VH	H	M	H	Gears, Drills
PVD(TiN)		L	VH	NA	VL	VL	Tools, Dies, Valves, Piston Rings
Modes of failure		Abrasion, Attrition, Lapping	Scuffing	Pitting, Spalling	Volume and dimensional change	Curving, Bending Twisting	
Aim of the surface Heat treatments		Harden surface	Produce case incompatible	Harden surface and Core	Minimize Volume change	Minimize Warpage	

References: (VH)-Very High, (H)-High, (M)-Medium, (VL)-Very Low, (*)-Depends on the Components, (NA)-Not Applicable ADI: Austempered ductile iron.

Example 12.11 In a Table distinguish between the behaviour of austenitic, ferritic and martensitic stainless steels under following categories: (a) Mechanical properties, (b) formability, (c) weldability, (d) oxidation resistance, (e) general corrosion and (f) stress corrosion.

Answer Refer Table Ex. 12.11.

Example 12.12 A cold storage firm is currently using an ammonia refrigeration system to maintain a food warehouse temperature of 255.6 K. Plans are being drawn to change to a spray liquid nitrogen system in which controlled amounts of liquid nitrogen are sprayed directly into the cold storage room. Because of this, it must be assumed that at times parts of the warehouse shelving will drop to 77.8 K.

Assume that the shelving was assembled at 300 K and that

Coefficient of linear expansion $= 1.656 \times 10^{-7}/°C$

Modulus of elasticity $= 1.86 \times 10^{13}$ Pa

Yield strength $= 2.76 \times 10^{8}$ Pa

Ultimate tensile strength $= 1.17 \times 10^{9}$ Pa

Working stress $= 1/4 \times$ ultimate tensile strength

(a) What is the material of shelve?
(b) Will the shelves rupture if exposed to liquid nitrogen?
(c) Will permanent set occur?
(d) From the working stress viewpoint, is the design satisfactory?

(Reproduced with permission from T.M. Flynn, Cryogenic Engineering, Marcel Dekker, New York, 1997)

Table Ex. 12.11.

	Grade Group		
Property	*Austenitic*	*Ferritic*	*Martensitic*
Strength	Low (cold rolled) → high	Low	Moderate (hot) → high
Ductility	High	Moderate	Moderate/Low
Toughness	High	Moderate(Low)	Moderate Low
Formability	Excellent	Good-drawing Poor-stretching	-
Weldability	Excellent	Satisfactory in thin section	Difficult
Oxidation resistance	To 1100 °C	To 800 °C	To 700 °C
General corrosion (and pitting)	Excellent	Good (Excellent)	Moderate
Stress corrosion	Poor	Excellent	

Solution

(a) Material of shelve is cold worked austenitic stainless steel (match the given data with Handbook).

(b) Modulus of elasticity E

$$E = \frac{S}{\varepsilon} = \frac{stress}{strain} = 1.86 \times 10^{13}\ Pa$$

Let β = Coefficient of linear expansion

$$\beta = \frac{\varepsilon}{\Delta T} \quad or\ \varepsilon = \beta \Delta T$$

Therefore;

$$S = \varepsilon E = \beta \Delta T E$$
$$= (1.656 \times 10^{-7})(255.6 - 77.8)(1.86 \times 10^{13})$$
$$= 5.48 \times 10^{8}\ \text{Pa}$$

Stress < ultimate strength of 1.17×10^9 Pa. Therefore material will not rupture.

(c) Stress > yield strength of 2.76×10^8 Pa. Therefore permanent set will occur. Not satisfactory.

(d) Working stress = $(1/4)(1.17 \times 10^9$ Pa$) = 2.92 \times 10^8$ Pa
Stress 5.48×10^8Pa > 2.92×10^8 Pa (working stress)
Stress exceeds working stress. Not satisfactory.

Example 12.13 Alloy 2205 is a common austenitic-ferritic duplex stainless steel (~40-60% austenite) with low carbon, nitrogen and chromium alloying. A typical composition is: 0.03 C, 2 Mn, 1 Si, 4.5-6.5 Ni, 21.0-23.0 Cr, 2.5-3.5 Mo and 0.08-0.2 N. The pitting index of the alloy is 34. Answer the following:

(a) General advantages of duplex stainless steels as a group.
(b) How the composition of 2205 s.s. was optimized against pitting corrosion?
(c) How the solution treatment parameters are established for this alloy?

Answer

(a) Refer Chapters 7 and 10.
(b) Figure Ex. 12.13a shows the optimization of the 2205 alloy composition, making it one of the preferred alloy in the group of duplex stainless steel. For better corrosion resistance, one can switch to super duplex stainless steels, but the problems of precipitate of secondary phases like sigma, chi are there.
(c) Such steels meet across the 475°C embrittlement. Hence the steel should not be used in process temperature above 300°C for long time. After hot working in the 650-1040°C range, the solution annealing must be carried out above 1050°C followed by a water quench. This annealing dissolves any detrimental intermetallic phase/s, that are formed during hot working. Figure 12.13b shows a TTT diagram of 2205 s.s., which gives the temperature range with time for various precipitates.

Example 12.14 Superferritic stainless steel products containing more than 30% Cr cannot be used in cast or wrought forms because of extremely brittle sigma phase. As certain property like corrosion resistance, particularly in chloride and reducing acid solutions, is excellent for such alloys, their need still exists. The only viable processing method seems to be powder metallurgy

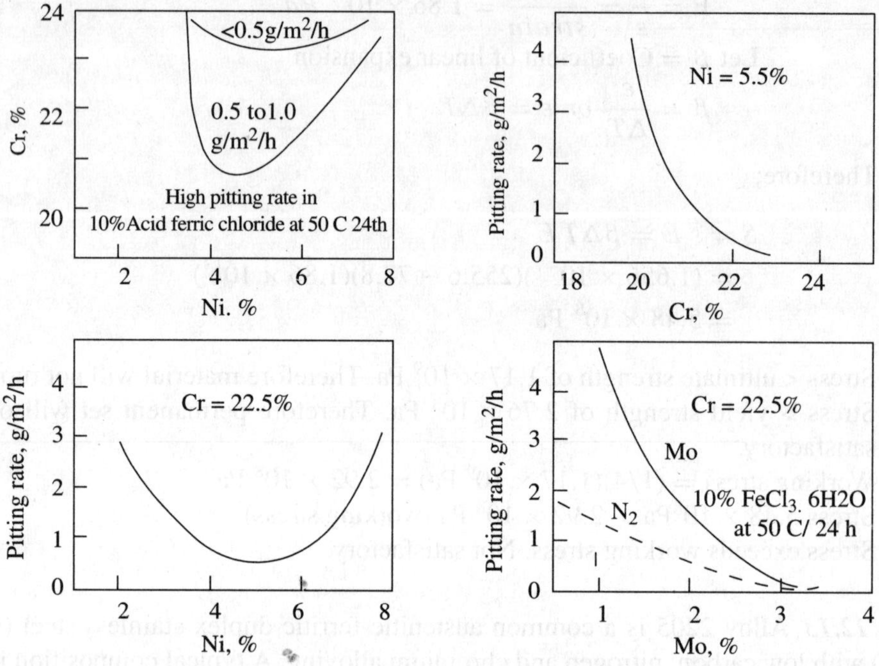

Figure Ex. 12.13a

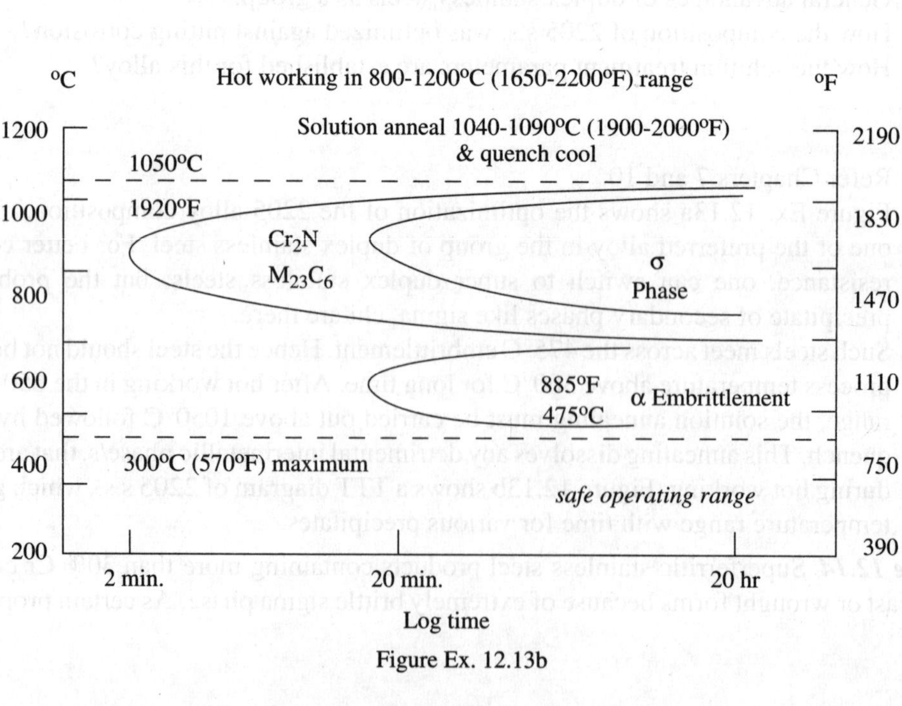

Figure Ex. 12.13b

processing of the milled sigma phase particles of such steels. Describe the processing strategy to be applied.

Answer If we look the Fe-Cr binary phase diagram, the equilibrium phase is sigma phase, which is an intermetallic (brittle). As the product must be in ferritic state, it means one has to add ferrite stabilizer. Out of many ferrite stabilizers, aluminium seems to be good, as its melting point is relatively low. Liquid phase sintering process could be therefore applied. Try with different amounts of aluminium. The main problem still remains in rather poor green strength of the part because of non-compressible sigma phase powder. For this, better lubricant is to be selected. Large volume fraction of melt phase is not necessarily good for the end properties. Therefore, optimization, is required. Too much aluminium, being soft, will give rise to galling problem during ejection of green part. MINTEK of South Africa has succeeded with 3 wt % Al containing material sintered at 1300°C for 1 hr.

Example 12.15 Stainless steel and nickel base alloys are used in pulp and paper mills. A design engineer suggests the use of fibre reinforced plastic (FRP) pipes in place of costly 316L stainless steel pipes. Do you think his suggestion is correct for long time service viewpoint?

Answer A careful analysis would infer that there is not much difference when the total installed cost is considered. No doubt 316L gives rise to some failure problems, but then other grade like 317L may be tried. When one considers requirements for special resins and resin-to-glass ratio in the FRP pipe for bleach applications, plus the possible requirements for safeguarding, joint design etc., the comparative cost favours the selection of metallic alloy.

12.3.2 Elevated Temperature Service

Case Study 12.14: Concrete-filled Hollow Steel Sections: A hollow-section steel filled with concrete provides an extremely compact and effective structural member particularly for building columns. It also provides fire resistance. The fire would have different effect depending on the applied load and the relative strength of the steel and concrete. While the concrete core has the capacity to absorb some of the heat, the effect is not substantial. Therefore, the temperature rise of the steel will not alter significantly with column size compared with the temperature rise experienced with an unfilled column. As steel temperature increases, the load originally resisted by both the steel and the concrete transfers more and more to the concrete; and the column fails when the combined strength of the steel and concrete falls below the level of the applied load. With further increase in temperature, the concrete alone needs to be capable of resisting the whole of the applied load. Concrete, too, begins to lose its strength as the temperature rises, but as its thermal conductivity is so low, this will occur much more slowly than in the steel. Improvement in the strength of concrete can be made by the addition of reinforcing bars or of chopped steel fibres to the concrete.

In the last two decades, HPC and UHPC (high and ultrahigh performance concrete) have been developed. The dual addition of silica fumes and superplasticisers enhances the strength. The ductility can be improved by the addition of steel, carbon or polypropylene fibres. The strength is enhanced further in RPC (reactive powder concrete) by using finer aggregates, adding steel fibres and admixtures, high temperature curing and multiaxial compaction. Self-compacting concrete was developed in Japan in 1988 and is now the material of choice in high rise buildings. Incomplete hydration of cement can be overcome by replacing 25% of the volume of the aggregate by pre-weighted lightweight aggregates.

As reinforcing bars are introduced in the concrete mass, the corrosion aspect must not be ignored. Accelerated corrosion results from the loss of alkalinity in the concrete or the penetration of chloride ions. Some of the methods to control corrosion are, cathodic protection, surface treatments of the rebars (epoxy coating, galvanizing, copper cladding, protective rust growth, surface oxidation and sand-blasting), the use of admixtures (organic and inorganic corrosion inhibitors, silica fume, fly ash, slag) and surface coating on concrete. In stead of using a corrosion-inhibiting admixture in the total volume of concrete, one may use the admixture to modify the cement slurry that is used as a coating on the steel rebar.

Case Study 12.15: Ceramic Liner for Piston Engines:

Although perfect adiabatic (no heat gained or lost) engine operation is not achievable in practice, the insulated or ceramic diesel concept is often generally referred to as the adiabatic diesel. A low conductivity TTZ ceramic has been favoured for the cylinder wall and piston material to minimize heat loss. An analysis shows that the major factor in reducing heat loss from the cylinder is not cylinder material conductivity, but elimination of the coolant. With no ceramic insulation, elimination of the coolant would, of course, raise the cast iron cylinder wall temperature and adversely effect engine life. This is shown in Figure 12.25 for a high power condition, where the time-averaged temperature of an all cast iron cylinder would be increased from approximately 190 °C to 630 °C and the heat loss would be reduced by over 80%. Changing the material from cast iron to TTZ would further reduce heat loss.

Two major approaches exist to reduce heat loss through elimination of the coolant. The first uses a low conductivity TTZ ceramic liner inside of a cast iron block. Here the main function of the ceramic liner is not to reduce heat loss, but to protect the cast iron interior surface from oxidation and high temperature. This arrangement is attractive because of the close match of thermal expansion between the iron and TTZ which permits a simple and effective ceramic-to-metal attachment.

The other more revolutionary approach is to use free-standing ceramic cylinders. In this case, since there is no metal containment, the ceramic cylinder becomes a structural member. The ceramic material selected would be based, not on its thermal conductivity, but on its contribution to system reliability. Material strength, Weibull modulus, elastic modulus and thermal expansion would be key factors.

Case Study 12.16: Ceramic Glow Plug:

Diesel engines have an advantage of better fuel economy over gasoline engine. But on the other hand, small diesel engines have difficulty in engine starting. At the begining the compressed air temperature is not raised easily in winter time and glow plugs are installed and are preheated to 800°C to promote fuel ignition. Diesel car drivers are supposed to wait for 4 to

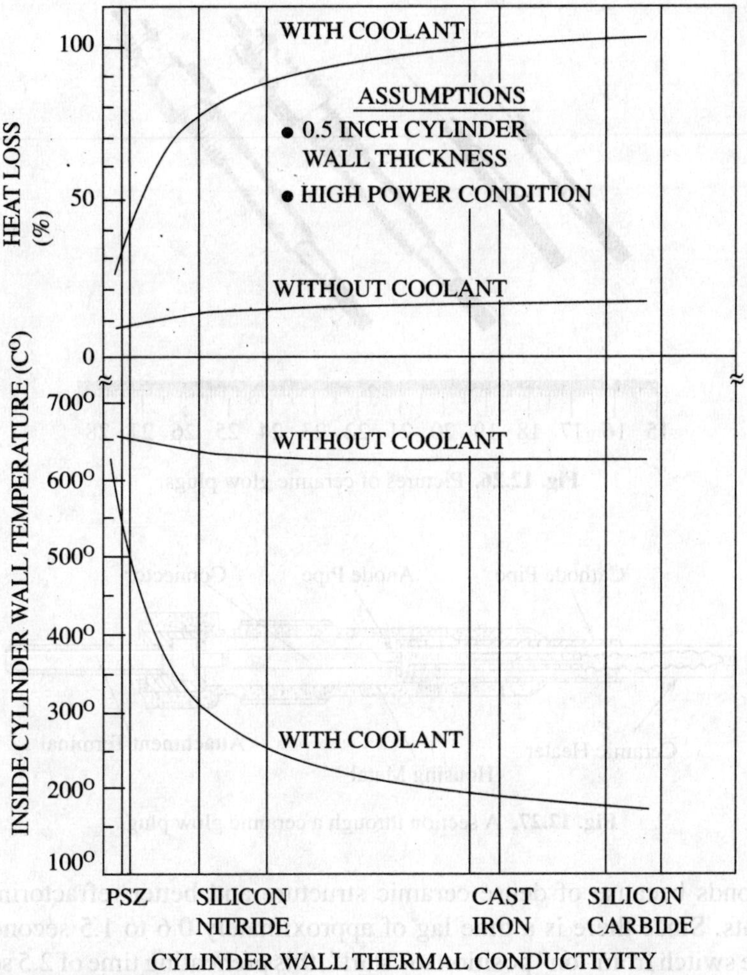

Fig. 12.25. Cylinder wall temperature variation with respect to ceramic insulated wall thermal conductivity of a diesel engine (Reproduced with permission from Fundamental Structural Ceramics, Ed. S. Somiya and R.C. Brandt, Terra Scientific Publishing Co., Tokyo, 1987, p. 457).

20 seconds, depending upon temperature controlling systems, prior to engine starting. Ceramic glow plugs have been developed to minimize such waiting time.

Conventional metallic glow plugs have nickel heating elements electrically insulated from steel sheath. The insulating ceramic powder also limits thermal flow and 4 to 20 seconds are necessary before the plug surface is heated to 800 °C. Of course, with higher power the time necessary can be shortened but the large temperature difference within a plug may destroy the plug.

Figures 12.26 and 12.27 show ceramic glow plugs which have tungsten heating elements embedded in dense silicon nitride by hot-pressing. Although the thermal conductivity of silicon nitride is lower than that of steel, these plugs have no porous insulating layer, and as a whole, they have less thermal flow resistance than conventional ones. The plug surface is heated to 900°C

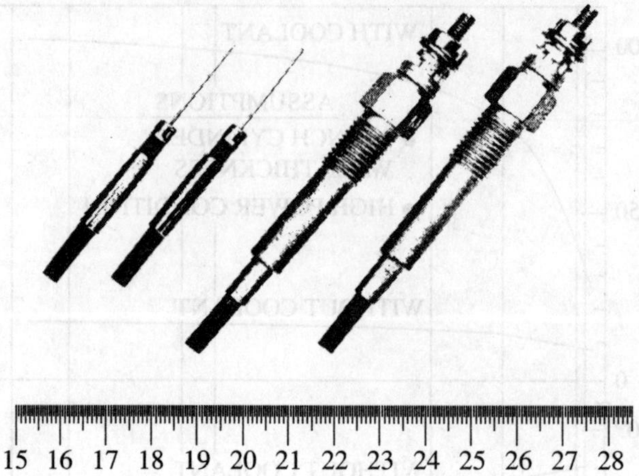

Fig. 12.26. Pictures of ceramic glow plugs.

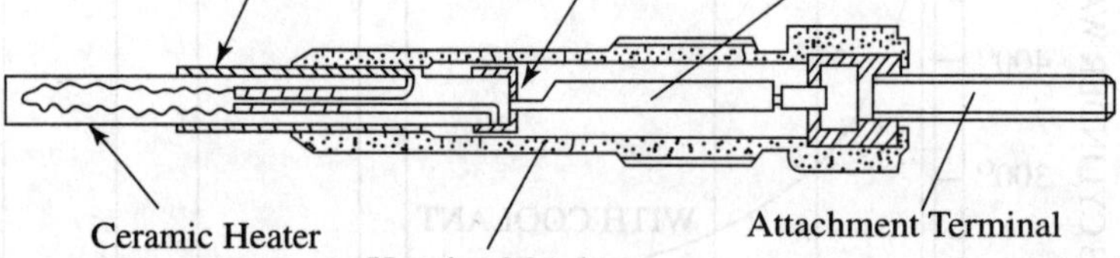

Fig. 12.27. A section through a ceramic glow plug.

within 2.5 seconds because of dense ceramic structure and better refractoriness of tungsten heating elements. Since there is a time lag of approximately 0.6 to 1.5 second when a driver turns the engine switch from "on" position to "start", this preheating time of 2.5 seconds virtually eliminates the waiting time and gives diesel car drivers the same starting action as gasoline car drivers.

Furthermore, with an introduction of turbo charging, the durability of metallic plugs is becoming serious problem because of metal sheath corrosion at higher combustion temperatures. Ceramic glow plugs show better durability (Figure 12.28). This characteristic is also effective in the use of after-glow operation after the engine has started, which shows advantages of lower idling noise and less white smoke emission immediately after starting.

Case Study 12.17: Microstructural Engineering of Silicon Nitride:

A preliminary description of silicon nitride has been made in Chapter 7. As solid state sintering of pure silicon nitride because of very strong covalent bond is very difficult, a liquid phase sintering is evolved with additions of sintering aids. Common sintering additives are mixtures of metal oxides (Li_2O, CaO, MgO, SrO, Al_2O_3, RE_2O_3, ZrO_2) or mixtures of oxides with non-oxides like AlN, ZrN and Mg_3N_2. Mostly, the amount and composition of the additive

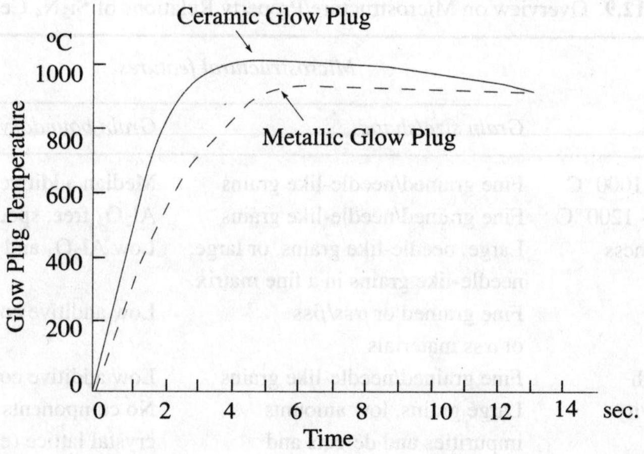

Fig. 12.28. Quick heating characteristics of ceramic glow plug.

mixtures have been developed empirically. For a desirable additive following features must be noticed:

(a) Under sintering condition Si_3N_4 should not convert into SiO_2.
(b) Additives must form eutectics with Si_3N_4 below the sintering temperature. Densification improves with lower eutectic temperature and viscosity of the eutectic melt. The densification is observed to decrease in the direction MgO/Al_2O_3 > MgO/RE_2O_3; MgO; RE_2O_3/Al_2O_3 >> RE_2O_3 (RE = Y, Sc, La and lanthanides).

Amount and composition of the additives, therefore influence phase relations and thus the sintering parameters e.g., temperature, pressure, time, atmosphere. A detailed discussion is beyond the scope of the present text. Table 12.9 gives an overview of microstructure/property relations of Si_3N_4 ceramics.

Case Study 12.18: Comparative Assessment of Silicon Nitride and Silicon Carbide: A comparison of the major characteristics of the dense forms of SiC and Si_3N_4 is summarized in Table 12.10. SiC is intermediate in strength to Si_3N_4. However, SiC experiences much less strength degradation by subcritical crack growth, and thus SiC is superior at temperatures greater than 1400°C. Similarly, SiC is more stable and oxidation-resistant at extremely high temperatures. It is, in general, a purer material and not as affected by impurities or secondary phases as Si_3N_4 is. SiC has extremely good creep strength, which coupled with its good oxidation resistance and microstructural stability, make SiC a likely candidate for very long time, high temperature applications. SiC has much higher thermal conductivity and thermal diffusivity compared to Si_3N_4. This means that SiC would be a good candidate for a combustor in a gas turbine engine, but would not be a good candidate as a cast iron diesel engine cylinder liner.

The major disadvantages of SiC when directly compared to Si_3N_4 are its lower fracture toughness and lower thermal shock resistance. The low toughness of SiC is due to its low critical stress intensity factor and low fracture surface energy. Poor thermal shock resistance is often cited as the most critical difference between silicon carbide and silicon nitride. The low thermal shock resistance of SiC is due to the combination of its higher thermal expansion and

Table 12.9. Overview on Microstructure/Property Relations of Si_3N_4 Ceramics.

Property	*Microstructural features*	
	Grain size/shape	*Grain boundary phase*
High strength up to 1000 °C	Fine grained/needle-like grains	Median additive content
High strength at T > 1200 °C	Fine grained/needle-like grains	Al_2O_3 free, special compositions
High fracture toughness	Large, needle-like grains, or large, needle-like grains in a fine matrix	Low Al_2O_3 and SiO_2 content
High hardness	Fine grained or αss/βss or αss materials	Low additive content
High fatigue strength	Fine grained/needle-like grains	Low additive content
High heat conductivity	Large grains, low amounts impurities and defects and solid solutions	No components solvable in the Si_3N_4 crystal lattice (e.g., Al, Be), no impurities
High creep resistance	Large, needle-like grains, composites with SiC, or refractory silicides	Al_2O_3 free with special compositions or no sintering additives
High oxidation resistance at T > 1200 °C	Large, needle-like grains, composites with SiC, or refractory silicides	Al_2O_3 free with special compositions or no sintering additives
High corrosion Resistance		Special compositions depending on the corrosive media
Good wear behaviour	Fine-grained microstructure	Homogeneous distribution

Table 12.10. A Comparison of Major Characteristics of Dense Forms of Silicon Carbide and Silicon Nitride.

Silicon carbide	*Silicon nitride*
• Intermediate strength	High strength
• High thermal expansion	• Low thermal expansion
• Poor thermal shock resistance	• Good thermal shock resistance
• Low fracture toughness	• Higher fracture toughness
• High temperature stability	• Strength degradation by slow crack growth
• High thermal conductivity	• Low thermal conductivity
• Good oxidation resistance	• Oxidation strong function of impurities
• High creep strength	• Low creep strength
• Low influence of impurities	• Second phases influence behaviour
• High elastic modulus	• Lower elastic modulus
• Transgranular fracture	• Intergranular fracture
• Little subcritical crack growth	• Nonlinear stress-strain

higher elastic modulus in comparison to Si_3N_4. These properties are more or less inherent, and cannot be modified to any meaningful extent by varying composition or processing method.

Case Study 12.19: Metal Matrix Composites and P/M Aluminium-SiC in particular: A brief introduction of metal matrix composites (MMC) has been made in Chapter 1. Also strengthening/weakening features of MMCs are described in Chapter 5.

Some of the special features of MMCs, which are unique and not found in other class of materials, are:

- The "composite" approach to metallurgical processing is the only pathway for the production of entire classes of metallic materials. Only in this way can aluminium, copper or magnesium be combined with significant volume fractions of carbide, oxide or nitride phases because, unlike iron, the solubility of carbon, nitrogen or oxygen in the molten metal is (with the exception of O in Cu) far too low.
- The approach facilitates significant alterations in the physical properties of metallic materials. Composites offer scope for exceeding the specific elastic modulus value of about 26 J kg^{-1}, which is exhibited by all the main engineering metals. Composites also offer the only pathway for producing materials with tailored physical property combinations: an example is that of low thermal expansivity combined with high thermal conductivity, a combination of importance for electronic packaging.
- MMCs offer significant improvements over their polymer matrix counterparts with regard to several properties, including tolerance of high temperature, transverse strength, chemical inertness, hardness and wear resistance, while significantly outperforming ceramic matrix composites in terms of toughness and ductility.

Table 12.11 summarizes comparisons of different classes of MMCs viz. continuous fibre, whisker and particulate reinforced ones.

Al-SiC Powder Metallurgy Particulate Composites **:** Silicon carbide particle reinforced aluminium alloys show promise as metal matrix composites for stiff and high strength light weight applications requiring adequate corrosion resistance. SiC also has advantages over other candidate reinforcing materials such as boron, graphite and alumina in Al-matrix alloys. These include the excellent thermal conductivity and corrosion resistance of the SiC, comparatively low particulate cost, high machinability and good workability of the Al-SiC composites. SiC is also chemically compatiable with aluminium and forms an adequate bond with the matrix without developing intermetallic phases.

The Al alloys used in such composites are not tailored alloys but are generally the commercial 2XXX, 7XXX, 5XXX and 6XXX compositions. The technique commonly employed in the preparation of such composites involves blending, cold compaction of the premix (Al-alloy and SiC particulates), followed by hot pressing of the resulting green body at a temperature above the solidus of the matrix. For preparing Al-alloy powder, atomization is the well established method. However the major impediment in the fabrication of aluminium powder is the oxide coating which reduces metal to metal contact and gives rise to poor end properties. Such a layer is undesirable as it creates safety hazard, when compacted at high temperature in the confined die due to rapid evolution of hydrogen. It is, therefore, necessary

Table 12.11. Comparative Study of Different Class of MMCs.

Type of composites / *Properties*	*Continuous fibre Reinforced*	*Short fibre/whisker reinforced*	*Particulate reinforced*
Reinforcement type, size and application	Long fibres of 5 to 100 mm dia, costliest among all composites, specialized use to defence and aerospace	Short fibres/whiskers of length 0.2 to 250 mm and L/D ratio 2 to 250. Whiskers are single crystal, where short fibres are polycrystalline. Cheaper than fibre composite, use is restricted to defence	Equiaxed particulate (dia 0.1 to 250 μm), polycrystalline in nature, cheapest among all composites, wide application in defence, automobiles and other household appliances
Role of matrix material	It holds the fibres together, align them in the desired stress direction. The transfer of the applied load to the fibre helps in improving fracture toughness. It protects the fibres from environmental corrosion and mechanical damage	If L/D is less than a critical value, matrix strength plays a dominant role in composite strengthening, else their role is similar to what observed in fibre composites	Matrix strength plays a critical role in composite strengthening. Reinforcement also helps in matrix strengthening by dispersion or precipitate strengthening mechanisms
Reinforcement volume % and isotropicity in mechanical properties	Very high volume fraction (~60%) of fibres can be introduced. Properties are highly directional (anisotropic) and by proper reorientation fibres quasi-isotropic properties can be achieved	Control of reinforcement distribution is difficult and beyond 40% agglomeration becomes severe, property is relatively less anisotropic than fibre composites	Control of reinforcement distribution is equally difficult and beyond 40% addition is avoided. Properties are nearly isotropic
Processing ease	Processing is most difficult and highly sophisticated. It is done in two stages of preform fabrication of fibres and metal infiltration	It is comparatively easier and require fewer steps (e.g., in P/M processing it involves blending, consolidation and secondary processing	Processing steps are same as what followed in whisker reinforced composites. All conventional route of secondary processing can be applied
Cost and applications	Cost is the main prohibitory factor of its growth, otherwise best properties are achieved in this case. Application is restricted to defence and aerospace	These are cheaper than fibre composites. Reinforcement cost is still high and there is consistant effort to reduce its cost, excellent prospect in non-defence application	There are the cheapest among all composites, and there are tremendous prospects for their applications in automobiles and other home appliances. Growth rate of this class of composites is the highest

to degas and preheat the aluminium powder and/or green compacts at initial stages. To achieve this, the powder should be degassed at a temperature equal to as greater than the temperature reached during processing, in order to get pore free and/or surface blister free components. Time and temperature of degassing could be economized by using vacuum heating of the powder. A temperature of 350°C and vacuum level of 3×10^{-4} torr has been found suitable for degassing the powder.

In order to obtain composites with aligned SiC particles, hot pressing is generally followed by hot extrusion of the hot pressed billet at temperatures close to the matrix solidus. In addition to possibly creating some internal matrix defects such as internal bursting, the extrusion step invariably causes additional particle breakage. The extrusion step can also result in bands of SiC-rich areas along the extrusion direction. These bands can be broken up and greater homogeneity achieved by hot rolling either cross-wise or parallel to the extrusion direction. The hot rolling can further break up the particles as well as possibly introduce matrix inhomogeneities such as fissures.

The final stage of the making Al-alloy composite involves heat treatment i.e., solution and ageing treatment as practiced for the matrix alloy. The prescribed temperature treatments of the matrix alloy may not be the most ideal one for the composite. For example, the kinetics of precipitation during ageing were, in one case, shown to be higher for the composite than for the matrix alone, presumably since the SiC particles can act as heterogeneous sites for nucleation during ageing. Other problems include either incipient solution of existing precipitates, which weaken the matrix grain boundaries.

Matrix/Reinforcement Interfaces: The successful production, applications and properties of MMCs are largely related to the properties of the interface between the matrix and the reinforcement. The critical aspects of the matrix/reinforcement interface region are: (i) the promotion of good wetting to aid production, (ii) the avoidance of detrimental reactions, leading to reinforcement degradation in production, fabrication and application, and (iii) the development and maintenance of sufficient bonding between matrix and reinforcement to permit the full potential properties of the composites to be realized.

The main features observed regarding wetting of reinforcements by metals in general are:

- Most matrix/reinforcement systems show a transition from wetting to non-wetting as temperature is increased.
- In many cases, the degree of wetting is found to be time dependent. This has been attributed to chemical reactions occurring at the interface and/or the necessity for metal to break through an oxide layer formed on the metal.
- The degree of wetting is often found to be dependent upon the oxygen partial pressure of the atmosphere indicating that the wettability is dictated by the presence of an oxide on the metal.
- Good wettability appears to be synonymous with reaction between the matrix and the reinforcement and/or removal of an oxide layer.

Three approaches designed to overcome the problem of non-wetting are reinforcement pre-treatment, alloying modification of the matrix and reinforcement coating.

Reinforcement pre-treatments seek to change the reinforcement surface energy by altering the chemical nature of the surface. This involves selective adsorption of additional species on the surface.

Alloying additions to the metal matrix takes two forms: firstly, those that promote reactions between the reinforcement and the matrix and thus increase the degree of wetting; secondly, those which modify the characteristics of the oxide layers. Examples of the former include additions of Li to Al for Al_2O_3 and SiC reinforcements and addition of Ti, Hf, Zr and Cr to Al for C reinforcement. An example of the latter is Mg addition which has been found to reduce the tenacity of the oxide layer formed on molten aluminium.

2124 Al-alloy-SiC Composites. Among various Al-alloy matrix, we dwell in this section, on a typical 2124 grade alloy (4% Cu-1% Mg-Rest Al). Our main concern is to show how the particulates in the matrix strengthens the composite at room temperature and also retains its strength at elevated temperature. Various strengthening mechanisms are already highlighted in Chapter 5 (Section 5.14). The particles selected are SiC and TiC (1.4 μm), the former having different sizes (0.25 μm, 1.9 μm and 17 μm). The volume fraction of particles are 10 and 20% respectively.

The P/M processing consisted of blending of matrix and electroless coated particulate ceramics/cold isostatic compaction/degassing and vacuum hot pressing/hot extrusion. (For full details refer Ph.D. Thesis of D. Kumar entitled Processing and Mechanical Behaviour of Particulate Reinforced 2124 Al Composite, October, 1993, Indian Institute of Technology, Kanpur). The final operation was the age hardening of the alloy.

As far as room temperature mechanical properties are concerned the main contributions to strengthening were matrix yield strength, grain size effect, dislocation density and load sharing by reinforcement particles. The main contribution is that of matrix yield strength. The other direct/indirect effects showed lesser contributions. Load sharing by reinforcement is influenced by the strength of the interface between matrix and the reinforcement which is largely dependent on processing parameters and chemical nature of the reinforcement. Lower yield strength in TiC composites than SiC one, when all other parameters are similar is due to this fact. Apart from matrix strength which is sensitive to heat treatment, particle size and its volume fraction have significant effects. With increasing volume fraction of the reinforcements, contribution to strengthening from the direct load sharing of the reinforcement and the indirect contribution of grains and dislocations, increases, thus improving the over all yield strength. Processing problem like tendency of agglomeration of fine particles with increasing volume fraction, hampers load transfer and may result in to micro-cavitation.

Figure 12.29 shows the results of yield strengths of the as fabricated composites based on theoretical calculations. Also the experimental data are incorporated. It is not necessary that the experimental results will correspond one-to-one with the calculated data, since there are many variables in the real systems. However, this comparison gives a fairly good guideline.

Coming to the elevated temperature compressive strengths of the composites, Figure 12.30 illustrates the variation of maximum possible temperature corresponding to a specific retained strength of 250 MPa for various composites. For the same strength level of 250 MPa, composites withstand up to higher temperature (~230°C) than control material (~180°C). SiCp

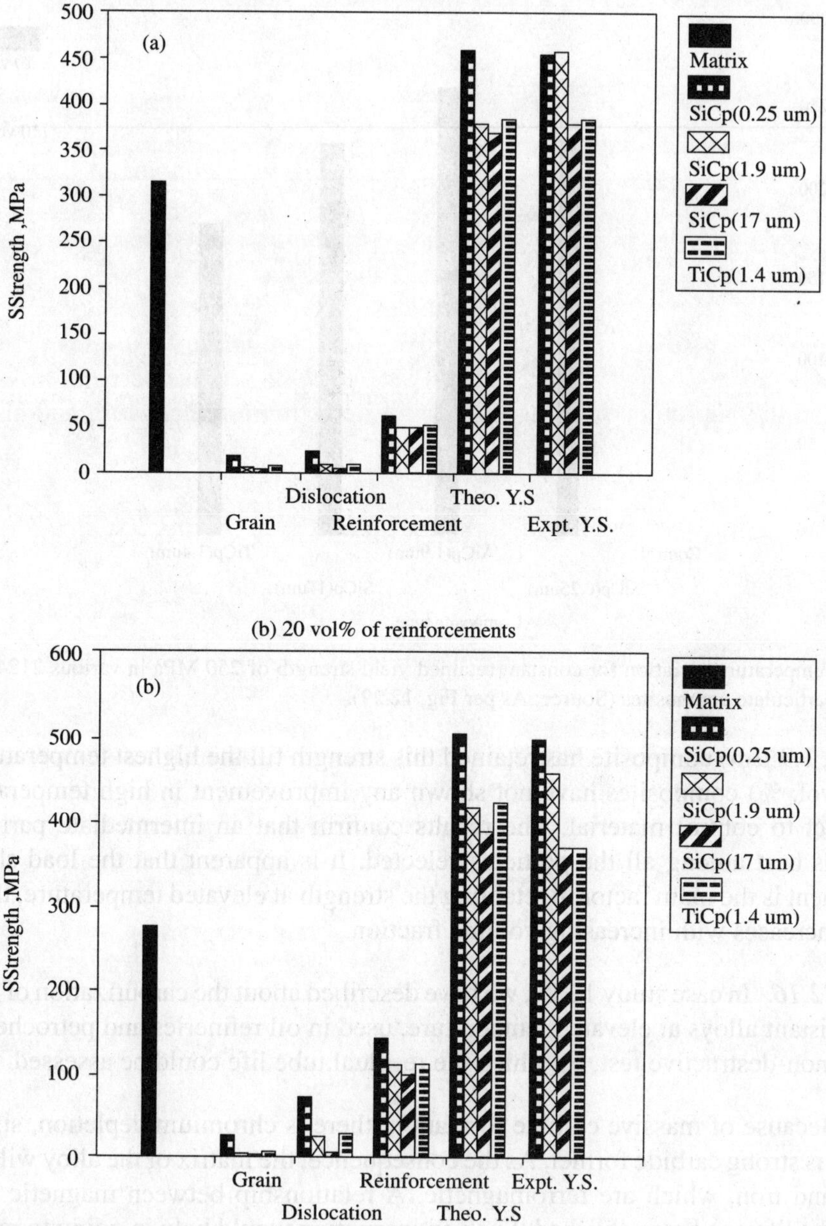

Fig. 12.29. Contribution to yield strength in 2124 Al-alloy matrix through various strengthening mechanisms: (a) 10 vol. % and (b) 20 vol. % ceramic particles (Ph.D. Thesis by D. Kumar, I.I.T., Kanpur, Oct. 1993).

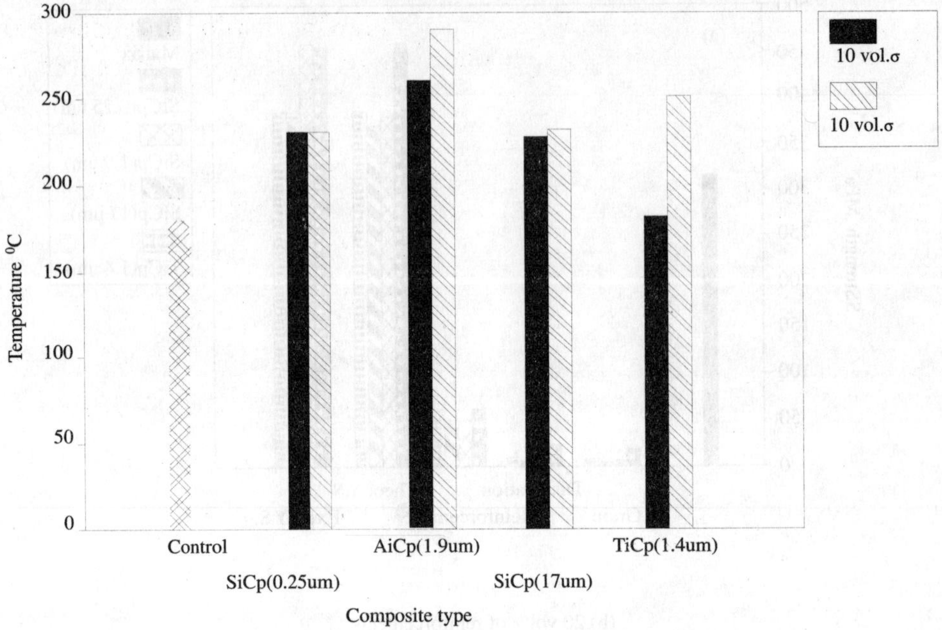

Fig. 12.30. Temperature variation for constant retained yield strength of 250 MPa in various 2124 Al-alloy based particulate composites (Source: As per Fig. 12.29).

(20 vol. %, 1.9 μm) composite has retained this strength till the highest temperature of 295°C. TiCp (10 vol. %) composites have not shown any improvement in high temperature strength with respect to control material. The results confirm that an intermediate particle size SiC (1.9 μm) is best among all the particles selected. It is apparent that the load sharing by the reinforcement is the main factor in retaining the strength at elevated temperature, the magnitude of which increases with increasing volume fraction.

Example 12.16. In case study 12.24, we have described about the carburization of furnace tubes of heat resistant alloys at elevated temperature, used in oil refineries and petrochemical plants. Suggest a non-destructive test, by which the residual tube life could be assessed.

Answer. Because of massive carbide formation, there is chromium depletion, since the metal concerned is strong carbide former. As the consequence, the matrix of the alloy will get enriched in nickel and iron, which are ferromagnetic. A relationship between magnetic response and residual tube life can be established. Such an inspection would help in going to repair welding. Repair welding after prolonged service would require solution annealing to increase ductility.

Example 12.17. The valve and valve seat assembly is partially responsible for the combustion chamber tightness in automobiles. The other parts are the piston rings and gaskets. The good wear resistance must be accomplished at high temperature and in chemically aggressive environment. The engine intake valve parts are subjected to less harsh conditions as compared to exhaust valve seats. Among various possible materials sintered Fe-Co alloy is used for exhaust valve seat. For economy reasons, if you wish to substitute it by cheaper ferrous material without any cobalt, what strategy would be adopted? Certain properties of Fe-Co alloy is as follows: sintered density 7.3 g cm^{-3}; hardness 25-30 HRC and radial crushing strength 280-300 Nm^{-2}.

Answer. Apart from high mechanical stresses, wear, corrosion and erosion at high temperature, proper machinability of the alloy is very essential. High speed steel sintered part can be tried, but it is still not that cheap. Another alternative may be the develop composite of iron-high speed steel material by powder metallurgy route using copper infiltration. In this case the sintering temperature shall be much lower than that for pure high speed steel. In case the mechanical strength is not adequate, refractory hard carbide secondary additive can be thought over. Care should be taken that the carbide distribution is fine and uniform.

The uniformity of microstructure i.e., ferrite-pearlite area, isolated copper rich areas, carbide and porosity should be maintained by proper processing. In case it is not so, then the machining of surface would not be uniform, giving rise to inefficient functioning of the part. The sintered properties of the new composition i.e., hardness and radial crushing strength should be compared with the original Fe-Co material and some fine tuning in material composition/processing may be adopted.

Example 12.18. What are the drawbacks of TiO_2 as a refractory material? How ZrO_2 is different from TiO_2?

Answer.

1. TiO_2 has no advantage over mullite refractory, while it is more costlier than mullite.
2. With SiO_2, TiO_2 forms eutectic at 1540°C (10% TiO_2).
3. Resistance of pure TiO_2 to molten slags and glasses is poor, because of the existence of low melting complex silicates.
4. The oxide is not very stable in reducing atmosphere and changes to $Ti_3O_5 \rightarrow Ti_2O_3 \rightarrow$ TiO. These changes are accompanied with volume changes.

ZrO_2 has no lower oxide like TiO_2. It is thermodynamically stable. ZrO_2 exhibits a greater range of oxygen deficiency than does TiO_2. It can therefore be either a p-type or an n-type semiconductor depending on the oxygen partial pressure. Its use as an oxygen sensor in hot combustion gases is related to its ionic mobility. It functions equally well in oxidizing and reducing (up to 1600°C) conditions.

Problem 12.19. Taking into consideration the chemistry and the processing of refractories rank the following refractors in the ascending order of their prices.

Tar bonded magnesite, bloating clay, 50-70% Al_2O_3, Tar-bonded dolomite, 80% Al_2O_3, MgO-20 graphite, direct bonded, 60% MgO.

Answer. Price in increasing order:

Bloating clay;	$50-70\%Al_2O_3$;
Tar-bonded dolomite;	$80\%Al_2O_3$;
Direct bonded;	60%MgO;
Tar-bonded magnesite;	MgO – 20% graphite.

The price of the last item i.e., MgO-20% graphite is roughly 12 times greater than that of clay.

Example 12.20. What are the features which can increase the thermal conductivity of a hot facing refractory brick in a furnace?

Answer. Either partial melting of the hot face or condensation of vapour can raise the thermal conductivity. Partially molten mass fills the pores of the outer surface of the brick, thus raising the conductivity. The change in temperature gradient is the function of thickness of the melt penetrated zone. This is not a serious problem in practice, since the fraction of the penetrated layer is far far less than the total brick thickness.

Example 12.21. Various types of processing methods used for Si_3N_4 ceramics are given below:

Sintered Si_3N_4	SSN
Gas-pressure sintered Si_3N_4	GPSN
Hot pressed Si_3N_4	HPSN
Presintered hot isostatically pressed Si_3N_4	HIP-SN (Sinter-HIP-SN)
Encapsulated hot isostatically pressed Si_3N_4	HIP-SN
Reaction bonded Si_3N_4	RBSN

Identify the processing methods for the following attributes:

(a) Lowest sintering cost
(b) Highest sintering cost
(c) Highest productivity
(d) Most time consuming
(e) Highest fracture toughness
(f) Products with no or very low additive
(g) Refractories
(h) Cutting tools/ball bearing production.

Answer.

(a) RBSN
(b) HIP-SN
(c) GPSN or HIP-SN
(d) RBSN
(e) GPSN/HIP-SN
(f) HIP-SN
(g) RBSN
(h) GPSN/HIP-SN.

Example 12.22. What are the major issues in the development of fibre reinforced ceramic matrix composites?

Answer. Ceramic matrix composites reinforced with fibres (CMCs) are devoted to *high-performance thermomechanical applications* without the brittle fracture behaviour typical of conventional ceramics. For the future, the main obstacle concerning these materials is costs of both, raw materials and manufacturing. In addition, several properties (such as *oxidation resistance, creep and fatigue*) have to be improved. Following are the main issues to be considered:

- *Reinforcement:* There is a need for cheaper fibres with improved creep resistance. The best SiC fibres are very expensive and exhibit extremely high creep and oxidation rates. Oxide fibres, like alumina, are limited because of very high creep rates. Carbon fibres need improved oxidation-protection solutions.
- *Protection against oxidation:* Development is needed of sophisticated multilayer matrices and multilayer *surface coatings* in order to promote the *matrix cracks healing* by formation of glasses in the whole temperature range.
- *Reliability of inhomogeneous, anisotropic and porous materials:* Development is needed of adapted nondestructive examination methods, *specific strength and quality criteria* and *new damage* indicators.

12.3.3 Multi-Constraint Service

Case Study 12.20: Wear Aspects in Petrol Engine: In running of a new engine, surface layers of metallic body is removed during running in. The process involves establishment of deformed sub-layer which will be hard enough to resist steady-state wear but surface peaks must deform and wear in order to offer conformity to the interacting couple. In this stage the higher fuel consumption is obviously due to extra mechanical work done in deformation and wear. The prediction of wear rate is complicated, because apart from load and surface speed, there are other factors too. The presence of abrasives is not ruled out, which will further promote wear. A high surface speed is advantageous, as it gives a higher interfacial temperature which facilitates the formation of an oxide film. In case of petrol engine, about same volume of water is produced as the volume of the petrol consumption. The combustion products contain oxides of carbon, sulphur and small amounts of bromine and chlorine compounds. If the engine wall temperature is low, the water and the combustion products form acids, which corrodes the cylinder wall and piston rings. Figure 12.31 illustrates this feature.

In brief, the whole concept of wear is complicated. In case the system operates in a corrosive environment, the problem gets further accentuated. The design must be done in keeping the service conditions as realistically as possible.

Case Study 12.21: Boiler Tubing Steels for Fossil Fuel Power Plants: A boiler is a steel pressure vessel designed to transfer heat produced by combustion to a fluid. It is used to produce superheated steam for a power plant. Fossil fuel fired boilers are of two types: fire tube boiler and water tube boiler. In the former the hot gases flow inside tubes that are submerged within a shell. In water tube boiler, water flows through tubes that are surrounded by hot combustion gases in a shell. These are used for high pressure steam and, therefore, require more instrumentation and control. Figure 12.32 shows a schematic of a typical oil-fired boiler. The upper drum is the

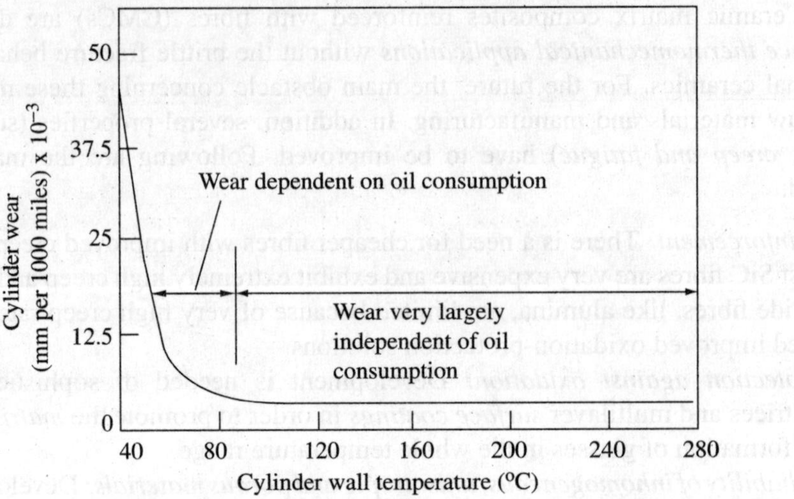

Fig. 12.31. Wear of an engine as a function of cylinder wall temperature.

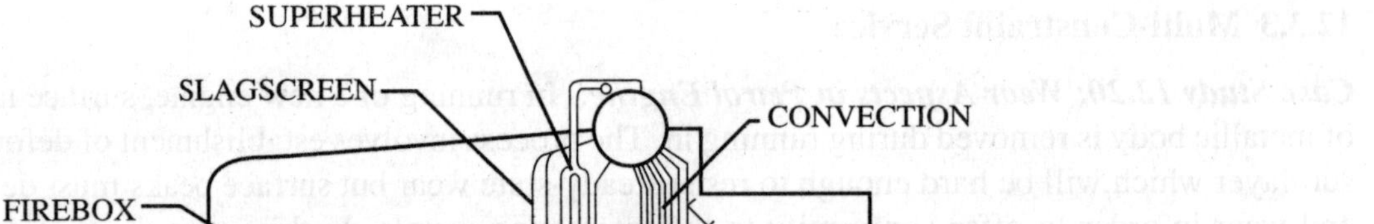

Fig. 12.32. Typical oil-fired boiler. The auxiliary equipment (water treatment, oil storage etc.) is not shown (Source: Fundamentals of Boiler Efficiency, Exxon Co., U.S.A., 1976).

Table 12.12. Target Values for Boiler Water.

Boiler pressure at outlet psig	*Total solids ppm*	*Alkalinity (total) ppm*	*Suspended solids ppm*	*Silica ppm*	*Specific conductance (micromhos/cm)*
0-300	3500	700	300	125	7000
301-450	3000	600	250	90	6000
451-600	2500	500	150	50	5000
601-750	2000	400	100	35	4000
751-900	1500	300	60	20	3000
901-1000	1250	250	40	8	2000
1001-1500	1000	200	20	2.5	150
1501-2000	750	150	10	1.0	100

steam drum, while the bottom one is the water drum. The water used in the power plant is a treated one, which prevents the following:

- sludge and scale deposits (mostly calcium and magnesium salt)
- corrosion and pitting (dissolved gases and acids)
- embrittlement (high caustic presence)
- carryover (foaming).

The target values for boiler water are given in Table 12.12. It is clear that the limits vary with operating pressure. Even after treatment, the role of corrosion should not be minimized, which we shall look into further.

The choice of steel grade for boiler tubes is a function of the design pressure of the boiler. The problems in proper functioning of steel tube are manifold: corrosion, high temperature and fatigue (corrosion fatigue). The boilers can be classified into coal or oil fired, although the use of natural gas is increasing. The flue gas or environment resulting from these fossil fuels are different in respect of impurities in the fuel. These impurities have direct effect on the resultant corrosion of those components within the furnace and exposed to these gases. Corrosion attack of furnace tubes is due to high chlorine in the coals together with impingement of burning pulverized coal particles. There is also the formation of complex sulphate at these temperatures. In oil fired plants, the major impurities in the oil are vanadium and sulphur and under operating conditions complex vandates are formed. Another cause is the formation of complex nickel sulphide compounds due to the presence of vanadium pentoxide, which acts as a catalyst. There is a direct relationship between corrosion attack and nickel content of the steel, the higher the nickel content the greater the corrosion. In general, carbon and 1 Cr 0.5 Mo steels are used for furnace and primary super heating tubes for temperatures up to 420 and 450°C respectively; 2.25 Cr 1 Mo, 12 Cr Mo V(W), 316, for secondary superheater and reheater tubing in coal fired power stations. Austentic steels are not employed in oil fired units. Maximum temperatures for these steels can be 580°C for 2.25 Cr 1 Mo and 620°C for the austenitics.

Corrosion problems can be discussed under following categories:

(a) **Pitting:** Pitting is a localized form of corrosion attack and can, at times, penetrate the tube wall. Such an attack can be caused by gas bubbles (air or CO_2) clinging the tube wall causing local liquid/gas reactions. In a partial type of pitting, known as scab pitting, corrosion products build up and form a cap over the locally corroded area. Both 'hard' and 'soft' scabs can be formed, the former under conditions of high dissolved oxygen and low pH and is more common in the fire side of the tube. 'Soft' scabs formation results from water droplets sticking to the metal surface during shutdown or in partially drained boilers or pipe work. In this case, the oxide is muddy and contains a large amount of red iron oxide Fe_2O_3. In another type of pitting, known as acid pitting, corrosion occurs when the acid cleaning solution is not fully inhibited. An ingress of sea water also causes such a corrosion.

(b) **On load corrosion:** This occurs when the boiler pressure is increased from 4.13 MPa or so to greater than 6.89 MPa. Under such high pressures, the cause of corrosion is high oxygen and/or imported iron oxide from the feed system. The latter deposits on high heat transfer surfaces giving rise to 'wick' boiling.

(c) **Corrosion under stress:** Corrosion fatigue is a typical example, which is common in failures of flexing tubes and welded-on attachments. The high thermal stress can be due to poor or irregular water circulation and alternate wetting and drying. Caustic embrittlement is caused by high stress, presence of hydroxide ions and crevice to allow concentration of these ions. In riveted joints all the above may be present.

Case Study 12.22: Future Fossil Fuel Plant Materials: Future fossil fuel power plants are going to operate with steam temperatures as high as 750°C. Ferritic steels are useful only up to 650°C. Austenitic stainless steel may be used in thin walled compartment but they suffer from a high thermal expansion coefficient and a low thermal conductivity. Keeping these factors, nickel based superalloys can be useful. However these alloys contain expensive alloying elements such as Mo, Co, Ta, Nb, or even Hf, Re.

The engineering requirements demand a creep rupture life of 1000,00 h at 750°C under a stress of 100 MPa. In addition, the ultimate tensile strength to yield strength ratio should be as high as possible, and in any case in excess of 1.3 at room temperature. The alloy must be easily forgeable and workable i.e., it must be free of γ' at all temperatures within ~200°C of melting. Furthermore, the quantity of γ' must be less than ~25% by volume after heat treatment in order to ensure weldability and ductility. Another requirement shall be corrosion resistance.

The following alloying elements excluding the costly ones in the superalloy may be considered favourably:

(i) Cr: About 20 wt % to achieve a good high temperature corrosion resistance.

(ii) Al and Ti: To form γ' precipitates. Their relative amount must be adjusted to achieve a low γ/γ' lattice misfit.

(iii) W: It enhances the solid solution strengthening of both the γ matrix and γ' precipitates.

(iv) C: To precipitate grain boundary carbides in order to limit grain boundary sliding. Its content should not be more, so that the formation of grain boundary carbide films is avoided. The appropriate amount is 0.07 wt %.

(v) B: To segregate at grain boundary so as to limit the grain boundary sliding. The usual level is 0.005 wt %.

(vi) Fe: About 5 wt %, so as to reduce the cost. As pure chromium is very expensive, added as ferrochrome scrap.

(vii) Si: As a deoxidant. A typical level is 0.4 wt %.

A typical composition of the alloy proposed by Tancret et al (Materials Science and Technology, March, 2003, p. 283-362) is

Ni-20 Cr-3.5 W-2.3 Al-2.1 Ti-5 Fe-0.4 Si-0.07 C-0.005 B.

The heat treatment consists of 4 h at 1175°C, 4 h at 935°C and 24 h at 760°C, with air cooling in all cases.

Case Study 12.23: Alloy Cast Iron Brake Drums: The material for automobile brake and disc must have

- High heat capacity
- Good thermal conductivity
- High emissivity
- Adequate high temperature strength
- Resistance to thermal shock.

Low silicon chromium-molybdenum grey alloy cast iron is preferred for brake drum or disc applications. In general, decreasing silicon from 2.5 to 1.54 increases thermal conductivity by about 10%. Although most alloying elements decrease the thermal conductivity of grey iron, their effect is not as great as that of silicon. In addition, carbon equivalent must be kept relatively high to ensure solidification as grey iron rather than the white iron. Finally, alloying imparts more stable structure and better stress-rupture properties than the unalloyed cast irons. A suitable composition for the category of heat resistant grey irons may be selected depending on the economic factor.

Case Study 12.24: Heat Resistant Materials for Oil refineries and Petrochemical Plants: In oil refinery and petrochemical plants a number of speciality stainless steels, cast and wrought are used in the temperature range of 800-1150 °C. A selection of alloys commonly used in this range is shown in Table 12.13.

Some of the high temperature problems faced in ethylene pyrolysis, steam-methane reforming for hydrogen production and ammonia plants are:

- stress rupture failure
- fatigue
- carburization
- metal dusting.

Stress rupture failure. It has been observed that an increase in temperature of 55°C above design can shorten the tube life of HK-40 alloy to 1.4 yr. Similarly a 0.7 MPa increase in operating pressure can shorten the life to 3.2 yr. Other problems may be non-uniform temperature in the

Table 12.13. Heat Resisting Alloys for Service at 800-1200 °C.

Alloy Designations	*Cr/Ni, others*	*Stress to produce rupture in 100,000 h, psi*			
		1400 F (746 °C)	*1600 F (857 °C)*	*1800 F (968 °C)*	*2000 F (1079 °C)*
JK-40	25/20	4500	2250	875	-
IN-519	24/24 Nb	6550	3050	1160	-
HP Mod.	25/35 Nb	6550	3650	1450	360
HP Mod.	25/35 Nb	-	3500	1320	350
H110	30/33 W	-	2700	1300	570
800 H cast	20/32 Nb	5100	2200	680	-
800 H forged	29/32	4200	1600	490	-
310 SS	25/20	-	-	-	-
800	20/32	4000	1600	620	-
800 H	20/32	5200	2100	765	-
802	20/32	8300	3200	950	-
800 DS	18/37				

fire box (20°C deviations not uncommon), and faulty readings of thermocouples. This leads to operation of HK-40 at 20°C below designer utilization of higher strength cast alloys. Candidate alloys are 24/24 Nb and 25/35 Nb. The 24/24 Nb alloy can either provide the capability of operating at 30% more strength at 950°C or provide the capability of operating at 30°C higher temperature. Such replacement has given a life expectancy in excess of 20 yr. This 24/24 Nb alloy at approximately 50% cost above HK-40 is very effective in extending the operating range and provides increased operating reliability.

Other advantages of the new alloy are:

- Economy: As the wall thickness is reduced.
- Increased plant capacity: Because of thinner tubes, about 30-40% more catalyst space is available.
- Low temperature gradients: Because of better rate of heat transfer in the thinner tubes.
- Increased savings: At a constant pressure, the upgradation of alloy improves hydrogen purity and efficiency.

Fatigue. Due to stresses caused by thermal cycling, one out of four steam-methane reformer furnaces meets unanticipated failures of riser, manifolds and transfer headers every year. Thermal cycling failures are especially important in ethylene pyrolysis plant that require periodic coke burn off. The development of a cast version of alloy 800 H with 0.1% C and 0.5-1.5% Nb, provided a cast metal that behaved in a ductile manner. It has high strength below 900 °C and a low embrittlement tendency, and found wide acceptance by furnace constructors. The 20/30 Nb cast alloy is less expensive than 800 H and has a higher creep and rupture strength and permits the use of larger and fewer outlet headers. The centrifugal casting process offers flexibility in that many combinations of diameter and wall thickness of tube can easily be obtained. Its relatively low coefficients of expansion gives good performance under cyclic conditions. Figure 12.33 shows the comparative properties of HK 40 and 20/32 Nb alloys.

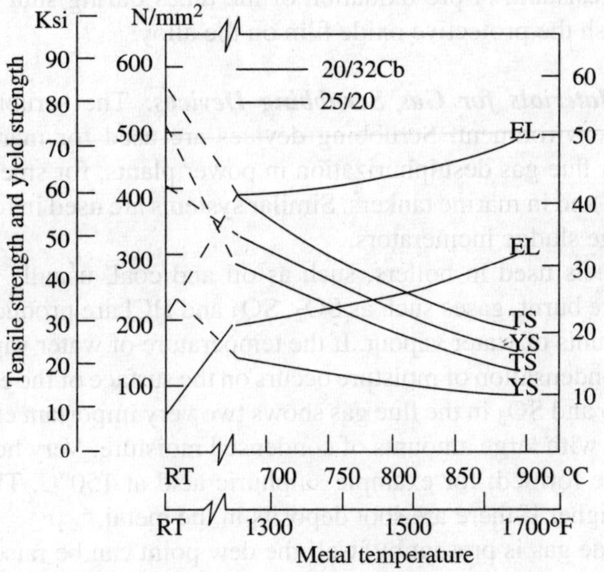

Fig. 12.33. Mechanical properties of HK-40 and cast alloy 800 H after exposure for 10,000 hours at 850 °C.

Carburization. At elevated temperature the furnace tube wall gets carburized and as much as 4.5% C level has been measured in the carburized inner surfaces of ethylene cracker tubes. The effects of carburization are:

- Increase in volume of metal, promotes internal stress
- Increase the coefficient of expansion promotes internal stress
- Reduction in creep strength and ductility
- Fissuring adjacent to the carburized layer.

The rate of carburization, as a general rule, is controlled by:

- Temperature – It roughly doubles for every 55°C increase.
- High nickel content and silicon – Alloys can be compared by using % Ni + 9 X % Si as the resistance factor.
- ID tube surface finish – Cast tubes should be bored and honed to a smooth finish of 125 rms.
- Protective and regenerative oxide films – Cr, Si and Al contents are beneficial.

Metal dusting. In the presence of hydrocarbons containing a high CO contact, there is strong tendency of carburization at elevated temperature. The degree of carburization is a function of temperature. At relatively lower temperatures such as 650-750°C, diffusivity of carbon is low and a very steep carbon gradient develops. This results in local mechanical fracturing of surface layers with resulting pitting or a break-up of protective scales in the form of filaments. This phenomenon is called *metal dusting*, and is common in 430 stainless steel and alloy 800. A CO/CO_2 ratio of 5 gives rise to a severe attack. An empirical formula is generally used for expressing the resistance towards metal dusting. The sum of the % Cr + 2 x % Si should exceed

24 for an alloy to be resistant. A pre-oxidation of the tubes during shut down of the plant is beneficial to re-establish the protective oxide film on the alloy.

Case Study 12.25: Materials for Gas Scrubbing Devices: The scrubbing of waste gases produces a corrosive environment. Scrubbing devices are used for many different types of chemical processes for flue gas desulphurization in power plants, for smelting of ores, in pulp and paper manufacture and in marine tankers. Similar systems are used in fertilizer plants, waste incinerators and sewage sludge incinerators.

As an example, fuels used in boilers, such as oil and coal, usually contain sulphur and chlorine. When they are burnt, gases such as SO_2, SO_3 and HCl are produced. These flue gases also contain large amounts of water vapour. If the temperature of water vapour falls below dew point of about 45°C, condensation of moisture occurs on the surface of the equipment containing it. The presence of SO_2 and SO_3 in the flue gas shows two very important effects. The dew point is raised to 100-150°C with large amounts of condensed moisture. Very hot, strong sulphurous and sulphuric acids are formed; for example sulphuric acid at 150°C. The dew point can be raised even 30-40°C higher if there are soot deposits in the metal.

If hydrogen chloride gas is present by itself the dew point can be raised to 60°C and up to 10% hydrochloric acid can be formed.

Scrubber systems to remove the corrosive gases have to cope with these dew point effects together with handling aggressive scrubbing liquor and often erosive particulate matter.

As the gas cools in the system the dew point may be reached and acid gases will begin to condense out to form sulphuric, sulphurous and possibly hydrochloric acid. Often some SO_3 will still remain in the gas stream after leaving the scrubber. This may then condense as sulphuric acid in the ducting, the fan or the stack.

If chlorides are present, together with the low pH caused by acid conditions, they produce an environment conducive to pitting and crevice corrosion.

Scrubber towers may be rubber-lined or made from a corrosion resisting alloy. The internals, i.e., gas inlet, baffle plates and demisters, are often metallic. Induced-draft fans are often of corrosion resisting alloys. Also, there are several examples of stacks being lined with highly alloyed materials.

Figure 12.34 shows the corrosion trends of several alloys exposed in various SO_2 scrubbing systems. The chart indicates pH and chloride content where field data have displayed no severe localized corrosion. These areas are indicated below the line for each alloy. The above figure clearly infers the beneficial effect of increasing molybdenum and chromium contents in an alloy for combating localized corrosion, especially in low pH, chloride containing environments. The two stainless steels (2-3% molybdenum, 17-19% chromium) show significant susceptibility to attack at less than 10 ppm Cl and pH 1). The increased molybdenum and chromium content of INCOLOY alloy 825 (3% molybdenum 21% chromium) yields significantly better corrosion resistance than Type 316 or 317 stainless steel. The highest degree of corrosion resistance is observed for INCONEL alloy 625. Its excellent corrosion resistance is due to its high molybdenum content.

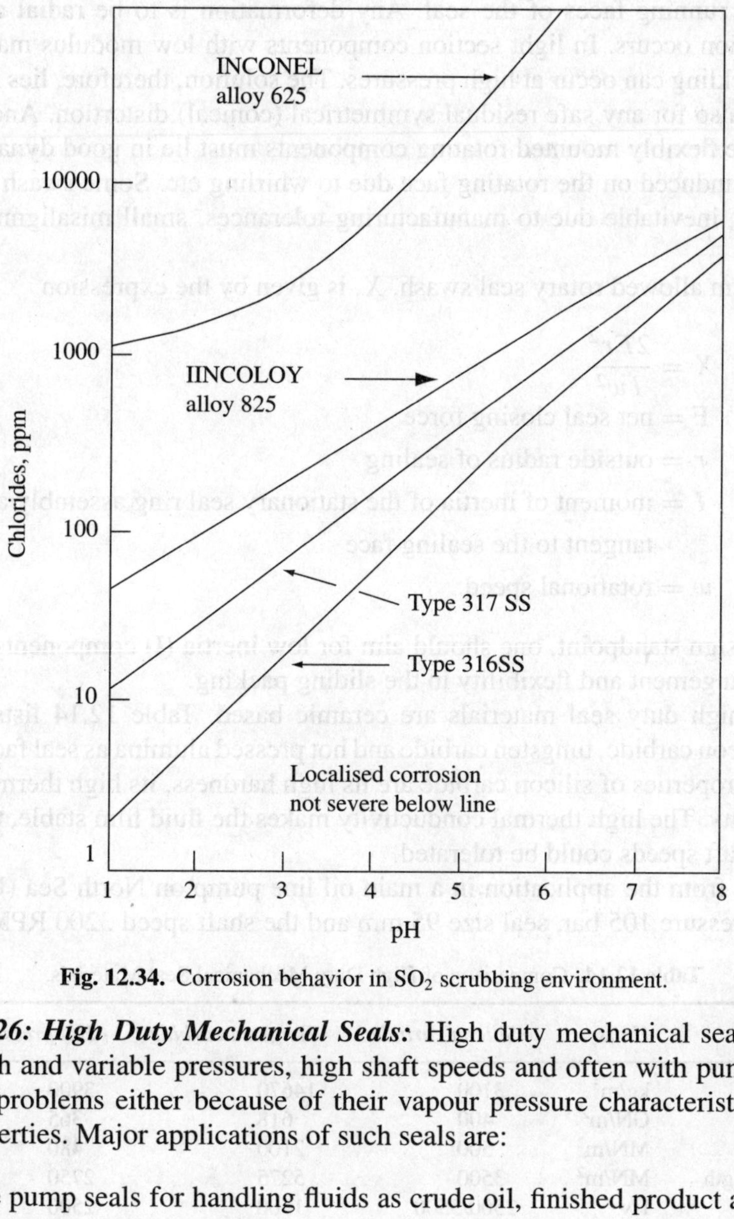

Fig. 12.34. Corrosion behavior in SO_2 scrubbing environment.

Case Study 12.26: High Duty Mechanical Seals: High duty mechanical seals are designed to deal with high and variable pressures, high shaft speeds and often with pump fluids which present unique problems either because of their vapour pressure characteristics and/or their lubricating properties. Major applications of such seals are:

- Pipe line pump seals for handling fluids as crude oil, finished product and fresh or sea water.
- Water injection pump seals.
- Boiler feed pump seals.

The running faces constitute the most important part of a mechanical seal. High pressure fluids create high loads at the running faces and thus high thermal loadings. This can induce distortion in the running faces of the seal. Any deformation is to be radial and no rotation of the cross-section occurs. In light section components with low modulus materials, such as carbon, axial buckling can occur at high pressures. The solution, therefore, lies in use of robust component and also for any safe residual symmetrical (conical) distortion. Another important feature is that the flexibly mounted rotating components must lie in good dynamic balance so that no swash is induced on the rotating face due to whirling etc. Some swash on the rotating face is, however, inevitable due to manufacturing tolerances, small misalignments and shaft vibration.

The maximum allowed rotary seal swash, X, is given by the expression

$$X = \frac{2Fr^2}{Iw^2} \tag{12.17}$$

where F = net seal closing force

r = outside radius of sealing

I = moment of inertia of the stationary seal ring assembly about a tangent to the sealing face

w = rotational speed.

From the design standpoint, one should aim for low inertia (I) components, a low friction anti-rotation arrangement and flexibility in the sliding packing.

The major high duty seal materials are ceramic based. Table 12.14 lists the important properties of silicon carbide, tungsten carbide and hot pressed alumina as seal face material. The most attractive properties of silicon carbide are its high hardness, its high thermal conductivity and its low density. The high thermal conductivity makes the fluid film stable, thus higher seal pressures and shaft speeds could be tolerated.

Typical data from the application in a main oil line pump on North Sea (U.K.) platforms are maximum pressure 105 bar, seal size 95 mm and the shaft speed 3200 RPM. The problem

Table 12.14. Comparison of High Duty Mechanical Seal Materials.

Property	*Units*	*Silicon carbide*	*Tungsten carbide*	*Hot pressed alumina 99.5%*
Density	kg/m^3	3100	14670	3900
Young's modulus	GN/m^2	400	618	365
Bending strength	MN/m^2	500	2100	480
Compressive strength	MN/m^2	3500	5275	2750
Hardness	Hv	2500/3500	1500	2500
Poissons ratio	-	0, 24	0, 26	0, 27
Weibull modulus	-	10	7, 5	10
Thermal conductivity				
at 20 °C	W/mK	210	85	32
at 500 °C	W/mK	84	66	8, 4
Expansion coefficient	μm/mK	4, 4	5, 5	9
Specific heat	J/kg K	900	218	1040
Heat shock factor	W/m	20800	21100	1200

there is complicated as the oil is light and gassy and, furthermore, it contains a varying amount of water. One major effect of this is that the addition of free water in oil lowers the bubble points of the mixture at all pressures. This, then, lowers the margin at which the fluid film between the faces becomes unstable. In case design is improper, vapourization of the face film takes place, with resultant leakage. Also the running becomes dry with consequent rapid face wear.

Most seal makers often sell the systems as complete packages, which provide the user with the highest possible level of safety in shaft sealing. Mechanical seal material developers must include real service tests apart from properties evaluation. At a fixed shaft speed, the rig must be allowed to stabilize and various readings like absorbed motor power, circulation flow and inlet and outlet temperatures are taken. The absorbed power includes all viscous and turbulent losses within the seal test module and is not face friction alone. The apparent coefficient friction is the total absorbed power compared with the total theoretical hydraulic and mechanical closing forces at the seal faces. Figure 12.35 shows the theoretical vapourization limits for silicon carbide and tungsten carbide mechanical seals in water. This confirms that silicon carbide is a better choice as compared to tungsten carbide.

Fig. 12.35. Comparison of theoretical vaporization limits for silicon carbide and tungsten carbide mechanical seals on water.

12.4 MATERIALS DESIGN FOR TOOLS:

Case Study 12.27: Conventional and Powder Metallurgy High Speed Steels: High speed steels (HSS) are basically a group of iron-based alloys containing some 20-30% of mainly carbide-forming alloying elements and can be hardened to a level of up to 65-70 HR_c and no appreciable softening takes place until temperatures in the region of 600°C are reached. The microstructure consists of a tough tempered martensite matrix with a dispersion of wear resistance, high-hardness alloy carbides.

HSS have a sufficient temperature gap between liquidus and solidus temperatures. In conventional steel making, this makes it prone to carbide segregation during solidification, which gives an embrittling effect. To avoid these problems, measures such as electroslag remelting, inoculation of the melt, casting relatively small ingots, etc. have been attempted with partial success.

Powder metallurgy processes are attractive in solving these problems. The main idea behind adopting P/M processing is that the segregation is limited within individual powder particle (maximum 250 μm dia.). Due to relatively high cooling rate achieved during atomization processes, segregation within every powder provide is minimized. Additionally, P/M HSS particle can contain a large quantity of small carbide particles distributed homogeneously in the matrix, which is not possible by conventional processing. Other advantages of P/M HSS are increased alloying flexibility, improved toughness, superior grindability, less distortion during heat treatment, isotropic mechanical properties, near net shape processing, good machinability and improved tool life.

The basic alloying elements of HSS include: carbon, chromium, tungsten, molybdenum, vanadium and sometimes cobalt. The other alloying elements which can also be added are niobium, tantalum, titanium, silicon, nitrogen and nickel.

Carbon can affect the carbide as well as the matrix. It produces the required amount and type of carbide, and also has an important role in the hardening operation. With increasing carbon content, the amount of retained austenite also increases, which results in a decrease in as-quenched hardness and coarser grain size. Chromium increases the fraction of retained austenite and leads to secondary hardening, after tempering. It also prevents scaling. Tungsten and molybdenum perform similar functions and are more or less interchangeable on an atomic basis: 1 mass % molybdenum is equivalent to ~1.6-2 mass % tungsten. Both the elements promote resistance to tempering. The austenite in molybdenum HSS is reported to be less stable than that of tungsten steels, giving some practical advantages during heat treatment. Vanadium acts as a grain refiner as it provides a very stable carbide which prevents grain growth during heat treatment. Cobalt enters into the matrix and raises the solidus temperature and thus provides an opportunity to increase the hardening temperature. This results in enhanced carbide dissolution, which in turn increases the amount of retained austenite on hardening.

There are two major classes of HSS: molybdenum high speed steels, called group 'M', and tungsten high speed steels, called group 'T'. The main advantage of group 'M' steels is lower initial cost due to lower atomic weight of molybdenum than tungsten. A typical specification of M2 steel is C (0.78-0.88), Mn (0.15-0.4), Si (0.2-0.45), Cr (3.75-4.50), Ni (0.3 Max), Mo (4.5-5.5), W (5.50-6.75) and V (1.75-2.20). On the other hand a typical specification of T15 grade is C (1.5-1.0), Mn (0.15-0.40), Si (0.15-0.40), Cr (3.75-5.00), Ni (0.3 Max), Mo (1.00 Max), W (11.75-13.00), V (4.50-5.25) and Co (4.75-5.25). It is evident that grade T15 contains an additional alloying element cobalt.

Powder metallurgy HSS is processed through supersolidus sintering, hot isostatic pressing, sinter-HIP or hot extrusion methods. For mass production of cutting tool inserts super-solidus sintering in vacuum is very common. A proper control of carbon is important, as it effects the critical solidus temperature (Figure 12.36).

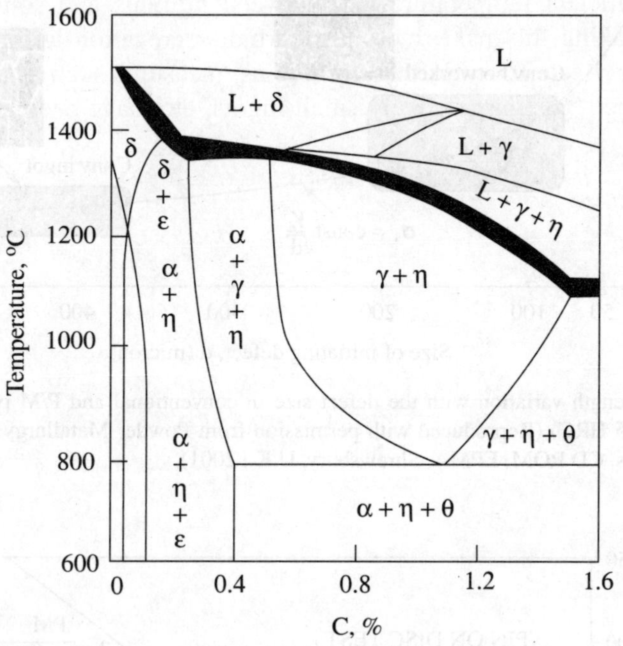

Fig. 12.36. Phase diagram of Fe-W-Cr-C system at W 18%, Cr 4%, showing the region with liquid phase and the supersolidus sintering temperature range.

As for mechanical properties are concerned a typical property i.e., bending strength variation is illustrated in Figure 12.37. It is evident that the strength of P/M HSS is far far greater than the conventional cast ingot material, as the defect size in the latter is much bigger. However, all properties are not negative in conventional high speed steel. For example the conventional HSS is much better in abrasive wear performance, when the particle size of the abrasive is greater than 120 μm (Figure 12.38).

The HSS cutting tool inserts can be coated by TiC or TiN in order to improve the tool life. For this physical vapour deposit (PVD) process is suited, as the operation is at lower temperature. Since chemical vapour deposition (CVD) process is an elevated temperature processing the properties and microstructural stability of high speed steel is lost. The details of these processes are already described in Chapter 11.

(For more details refer G.S. Upadhyaya, Sintered Metallic and Ceramic Materials, John Wiley & Sons, Ltd., Chichester, U.K., 1999, p. 143-172).

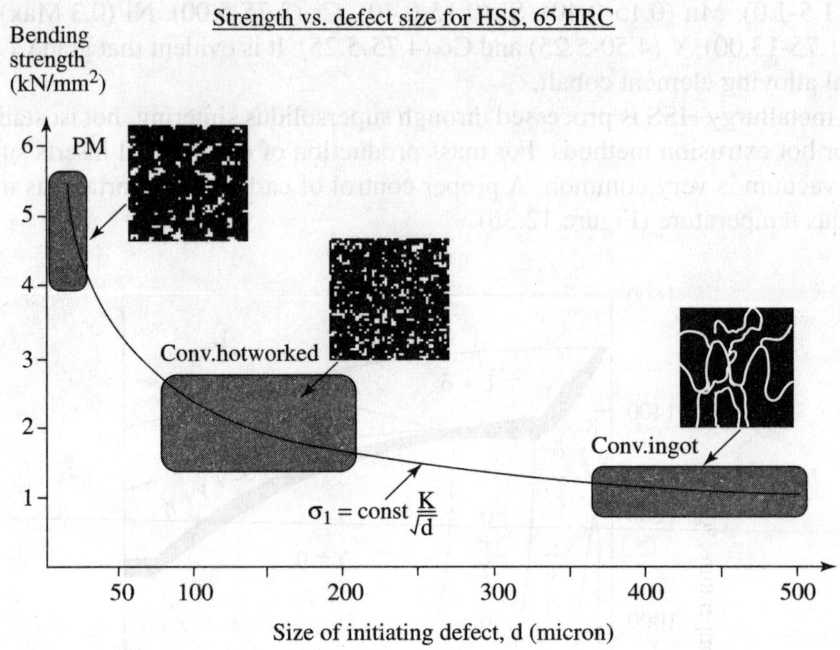

Fig. 12.37. Bending strength variation with the defect size in conventional and P/M processed high speed steel (hardness 65 HRC) (Reproduced with permission from Powder Metallurgy: Materials, Processes and Applications, CD ROM, EPMA, Shrewsbury, U.K., 2001).

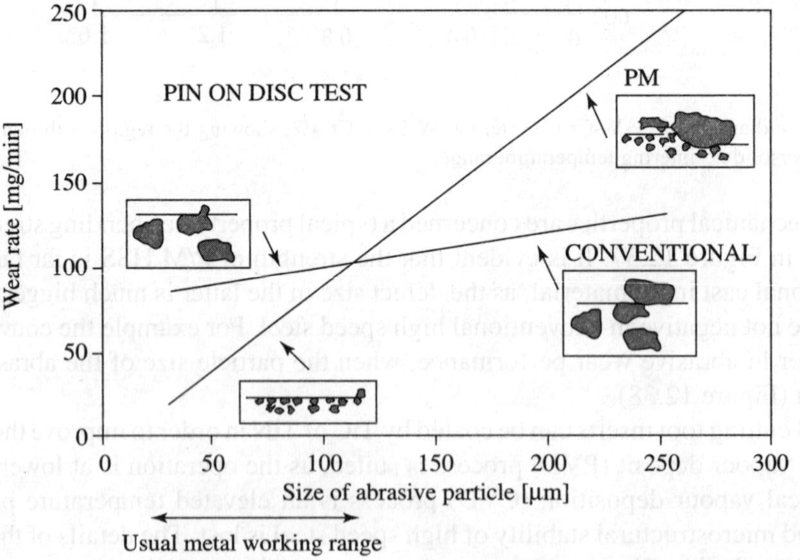

Fig. 12.38. Abrasive wear rate of conventional and P/M high speed steel with respect to abrasive particle size.

Case Study 12.28: Cermets Using Spinodal Transformation: In Chapter 3, a brief description of spinodal transformation is made. The principle has been applied successfully in developing cermet based on (Ti, Mo)(C, N) containing nickel binder.

The excellent wear resistance of TiC is well known. Nickel is a good compatible binder with TiC, but because of oxygen contamination the wettability is not ideal. To improve wettability molybdenum is either added in the binder or partial substitution of molybdenum for titanium is done. In order to enhance the properties of this alloy, it was proposed to add nitrogen in the lattice, so that the nitrogen and titanium rich phase would provide the necessary wear resistance, while the molybdenum-rich phase would be mainly responsible for providing adequate bonding with the binder phase. These aims are the driving force for developing spinodally transformed cermet systems. Figure 12.39 shows a partial isothermal section of the double-pseudo binary solid solution (Ti, Mo)(C, N) at approximately 1450°C, depicting the extent of miscibility gap and the location of the critical point at this temperature. With increasing temperatures, the critical point P_c shifts more towards the centre of the composition quadrangle, while lowering the temperature causes an expansion of the miscibility gap and a shift of P_c in the direction of the TiC corner. Because the coexisting

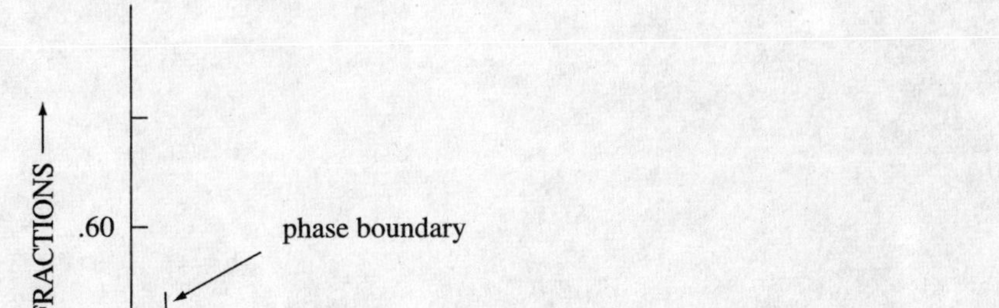

Fig. 12.39. Partial isothermal section of the (Ti, Mo)(C, N) system at 1450 °C.

compositions α' and α" have practically identical lattice dimensions, the coherency stresses at the α'/α" interface is small. The solubility of the hard phases in the binder metal is related to their stability and therefore preferential transport of the α" phase during liquid phase sintering is to be expected. For details of liquid phase sintering processing refer Chapter 8. The sintered properties of the spindonal carbonitride alloy will, thus, be mainly determined by those of the better-wetting α" phase. The electron micrograph clearly shows the contiguity of α" with nickel binder, isolating relatively poorly wetting α' phase from the binder (Figure 12.40). A typical composition of such alloy is $(Ti_{0.82}Mo_{0.17})(C_{0.83}N_{0.17})_{0.94}$, 13 wt % Ni, 3.3 wt % Mo.

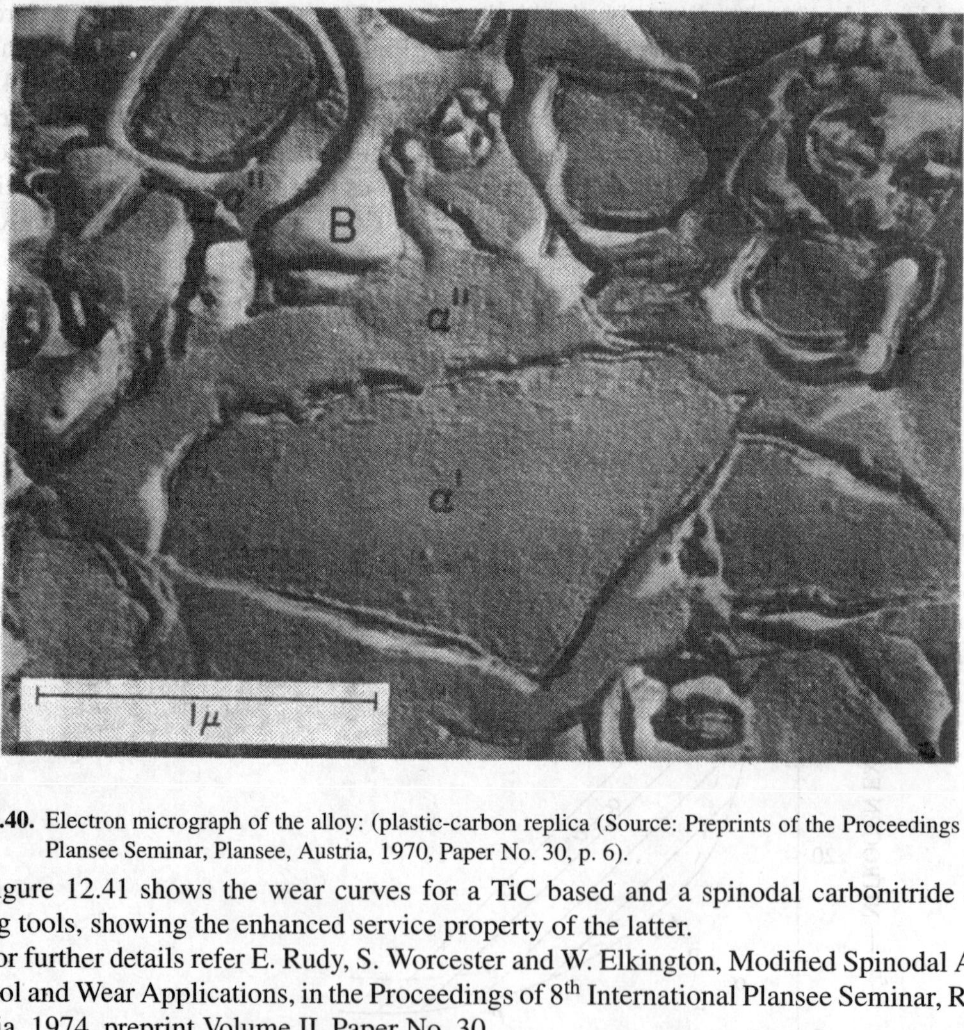

Fig. 12.40. Electron micrograph of the alloy: (plastic-carbon replica (Source: Preprints of the Proceedings 8th Int. Plansee Seminar, Plansee, Austria, 1970, Paper No. 30, p. 6).

Figure 12.41 shows the wear curves for a TiC based and a spinodal carbonitride based cutting tools, showing the enhanced service property of the latter.

For further details refer E. Rudy, S. Worcester and W. Elkington, Modified Spinodal Alloys for Tool and Wear Applications, in the Proceedings of 8th International Plansee Seminar, Reutte, Austria, 1974, preprint Volume II, Paper No. 30.

Case Study 12.29: Diamond Tools for Stone Sawing: The production of saw blades for stone cutting and construction applications is the biggest segment accounting for 50% of the overall consumption of industrial diamonds. Cobalt is the most common tough enduring matrix.

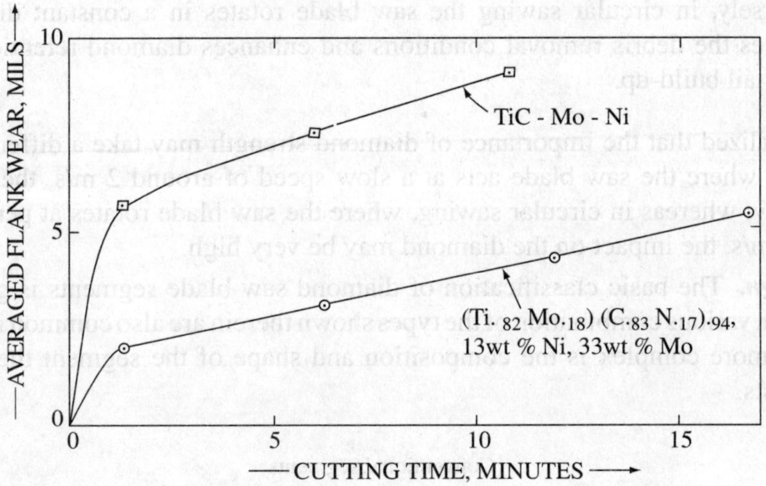

Fig. 12.41. Wear curves for a TiC-based finishing tool and a spinoidal alloy in machining medium-hard low alloy steel: Test condition:

Work metal	4340 steel, R_c 31-35
Tool geometry	SNG 433
Cutting speed	750 sfm
Depth of cut	0.060"
Feed rate	0.0010" per rev.
Coolant	None

(Source: same as for Fig. 12.40).

The fabrication of diamond impregnated saw blade segments is carried out by powder metallurgy route (Chapter 8), usually applying hot pressing method in the range 200-900°C.

From the engineering viewpoint following criteria are to be considered:

- segment manufacturing method and parameters,
- work piece properties,
- sawing conditions,
- cooling efficiency,
- quality of the segment-to-metal core joint,
- metal core design and tensioning,
- machine condition.

There are two main sawing methods:

- Circular sawing,
- Frame sawing.

The sawing method affects to a great extent the conditions of debris removal and retention of diamond grits.

- In frame sawing the reciprocating movement of the saw blade slows down the removal of stone detritus from the cut thus creating severe wear conditions for the matrix.

- Conversely, in circular sawing the saw blade rotates in a constant direction, which improves the debris removal conditions and enhances diamond retention through the matrix tail build-up.

It has to be realized that the importance of diamond strength may take a different priority. In frame sawing, where the saw blade acts at a slow speed of around 2 m/s, the impact on the diamond is low; whereas in circular sawing, where the saw blade rotates at peripheral speeds exceeding 25 m/s, the impact on the diamond may be very high.

Segment design. The basic classification of diamond saw blade segments is given in Figure 12.42, although various combination of the types shown therein are also common in the industrial practice. The more complex is the composition and shape of the segment the higher are its production costs.

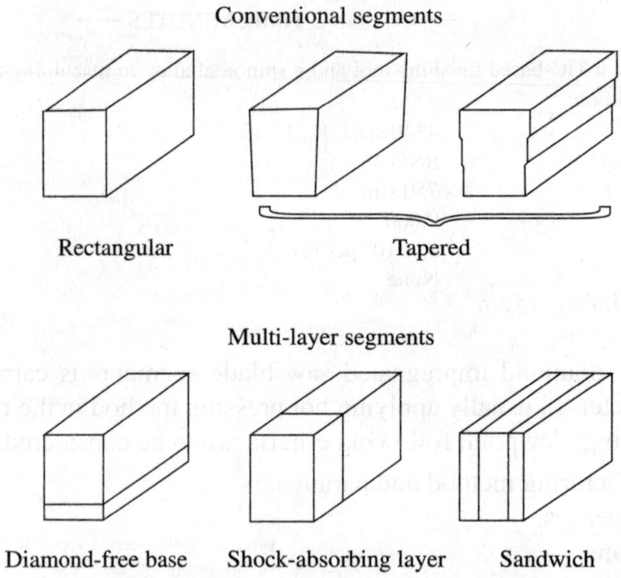

Fig. 12.42. Basic types of saw blade segments.

The application of tapered segments has the advantage of decreasing the power consumed for sawing through reducing the lateral friction of segments against the stone.

There are two reasons for using segments incorporating a diamond-free base. Firstly, they can hardly be worn down to the steel support and it is cost-effective to withdraw the diamond from the base section of a segment. And secondly, the diamond-free layer of suitable fusion characteristics is required if the laser welding technique is to be used.

The shock-absorbing layer, on the leading end, is produced with the intention to improve the segment's behaviour under high cutting rates.

The so-called sandwich is a three or more layer segment in which the outer layers differ from the inner one(s) in their susceptibility to wear. As illustrated in Figure 12.43, a desirable saddle-like wear pattern is developed at work, which imparts a self-guiding characteristic to the tool. This prevents the saw blade from deviating in the cut as is frequently observed with uniform segments.

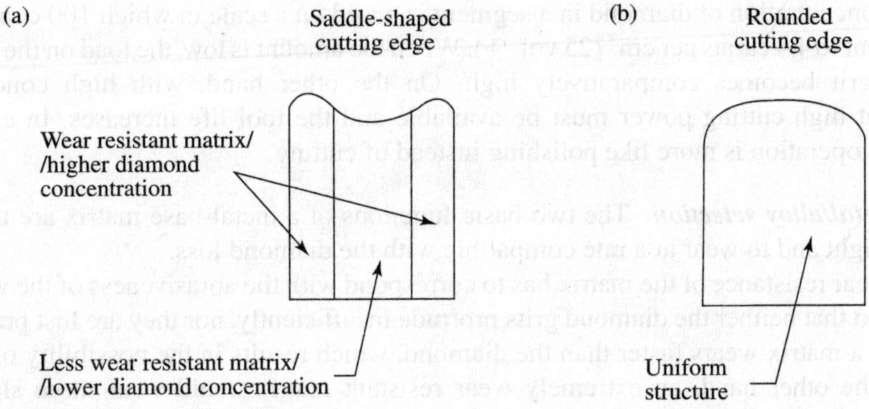

Fig. 12.43. Wear patterns of sandwich (a) and conventional (b) diamond impregnated segments.

Material Selection:

Diamond grit- The major factors in selection of grits are

- Diamond type
- Grit size
- Amount of grit.

There are a wide variety of natural and synthetic grits of varying properties. The major difference between the natural and synthetic diamond is the impurity content. Natural grit made by crushing mined diamond are free from metallic inclusions and hence show markedly higher thermal stability. They can retain their strength even up to 1400°C, whereas synthetic grits begin to loose strength beyond 800°C. From the matrix retention viewpoint crushed crystals possess excellent bonding characteristics, which results from the numerous re-entrant surfaces available. This makes the natural diamond advantageous for frame sawing the marble limestone and other less hard stones. For more rigorous applications rounded diamonds are used. However, in this case higher cutting forces are generated and, therefore, rigid and powerful machines are required.

The shape of the diamond grit affects the particle integrity and fracture characteristics. Diamonds, which have a regular, cubo-octahedral shape are stronger than irregular crystals with less defined edges and rough faces.

The size of the diamond grit determines its height of protrusion above the matrix surface and influences the clearance of swarf from the cutting zone. The recommended sizes for various applications are

50/60 Mesh - for very hard, fine ground materials (e.g., flint, very hard granite)

40/50 " - for hard, fine grained materials (e.g., granite, basalt)

30/40 " - for medium hard materials (e.g., marble)

20/30 " - for relatively soft materials (e.g., large grained limestone or sandstone).

The coarser grit offers the advantage of faster cutting rate.

The concentration of diamond in a segment is based on a scale in which 100 concentration is equivalent to 4.4 carats per cm^3 (25 vol. %). When the amount is low, the load on the individual diamond grit becomes comparatively high. On the other hand, with high concentration, a sufficient high cutting power must be available and the tool life increases. In case power is less, the operation is more like polishing instead of cutting.

Matrix metal/alloy selection. The two basic functions of a metal-base matrix are to hold the diamond tight and to wear at a rate compatible with the diamond loss.

The wear resistance of the matrix has to correspond with the abrasiveness of the work piece material, so that neither the diamond grits protrude insufficiently, nor they are lost prematurely. A too soft a matrix wears faster than the diamond, which results in the possibility of diamond loss. On the other hand, an extremely wear resistant matrix could wear more slowly than the breakdown of the diamond, causing the segment surface to polish. This phenomenon is commonly known as *glazing*.

The very wide range of cobalt powder grades, available in commercial quantities, offers great flexibility in planning the matrix microstructure and properties. The powders differ in particle size and size distribution, particle shape and chemical purity.

Modifications to the properties are possible through further alloying. The additive powders may alter to a great degree the matrix hardness and ductility thus matching perfectly its diamond retention and wear characteristics to the application requirements.

- Elements such as iron, manganese and nickel, when added in the form of either elemental iron powder or pre-alloyed manganese-nickel bronze, aid in stabilizing the fcc structure, which decreases hardness and mechanical strength of the matrix but markedly improves its ductility. Addition of 10 wt % iron exerts much stronger effect than the same amount of bronze, which, unlike iron, proves ineffective in a moderately oxidized environment.
- Admixtures of tungsten carbide or tin bronze to cobalt allow widening the hardness and mechanical strength range with marked interference to the ductility of the matrix. Larger amounts of fine tungsten carbide may raise the hardness, offset yield strength and Young's modulus, through particulate strengthening mechanism, far beyond the ranges offered by unalloyed cobalt.
- In the case of tin bronzes, larger amounts of either pre-mixed or pre-alloyed material are applicable shifting the alloy composition, in extreme cases, into the bronze-base field. Generally, by increasing the bronze content of the alloy it is possible to markedly decrease its hardness and mechanical strength.

Example 12.23. For communications and direct broadcast missions satellites are used, which have a life time of 10 years. The typical satellite spacebus 100 had a lift-off weight of 1200-1700 kg and a mass of 650-850 kg in geostationary orbit. Highlight the main elements of such satellite and select the structural aluminium alloy.

Solution. The main elements of the spacebus structure are: (1) Antennae, (2) Solar arrays, (3) Service module, (4) Communication module and (5) Panels. It has been observed that a reduction in 3 kg weight would amount to increasing the payload by 1000 additional simultaneous telephone communications. The aluminium alloy (Al-Li) can be replaced by stiffer composite material (Al-SiC). It is worth mentioning that in order to reduce the brittleness of binary Al/Li alloy, small amount of other elements such as Cu, Mg, Si must be incorporated. Although Be-alloy have highest specific Young's modulus, its biggest problem is handling because of its toxicity. Its application, therefore, is very limited.

Example 12.24. Collect the data on physical properties of cemented carbides and cermets in a tabular form. What are the major differences and how they affect technological properties and machining details?

Answer. Data are given in Table Ex. 12.24.

Table Ex. 12.24.

Cutting materials	*Cemented carbide* *P10*	*P20*	*Cermets* *Roughing Grade*	*Finishing Grade*
Density in g/cm^3	10.8	12.4	6.1	8.1
Hardness (Vickers)	1600	1500	1600	1600
Modulus of elasticity in kN/mm^2	540	550	500	500
Bending strength in N/mm^2	1700	2100	1500	1700
Compressive strength in N/mm^2	4400	4400	4200	4400
Coefficient of thermal Expansion in 10^{-6} K^{-1}	7.9	6.9	7.6	7.9
Thermal conductivity in W.m^{-1}.K^{-1}	27	38	12	10

The major difference is in thermal conductivity which proportionately affects the thermal shock resistance. The lower thermal shock resistance of cermets limits the application of coolants to finish turning and threading.

Cermets are capable of operating over a wide range of cutting speeds. The higher plastic deformation resistance and better chemical inertness of titanium carbonitrides in comparison to tungsten carbides permits higher cutting speeds and reduces edge build up at lower cutting speed.

Cermets achieve excellent surface finishes and close size control in turning, grooving, threading and milling applications. More expensive grinding operation can be replaced through machining with cermet cutting tools. Coolent is applicable, except in rough machining operations, and generally results in longer tool life.

Since the density of cermet material is lower than WC-Co, a relatively larger number of inserts per unit mass can be produced.

Problem 12.25. Grey cast iron is the most abundantly used material for high volume manufacture, such as automotive and earth-moving equipment components. Straight cemented carbide cutting tools has been used for machining. If your boss for better productivity in the plant asks your recommendation for tool replacement by either oxide or nitride ceramic tools, what would you suggest? Keep in view the economic factor as well. Using the fracture toughness and hardness data from the Handbook, plot a graph between abrasive wear resistance parameter (inverse of volume removed) and $K_{Ic}^{3/4}\,\mu^{1/2}$ and justify your answer.

Solution. Alumina tools can be operated at a speed of 300-500 m/min. To enhance its speed further, one can use Al_2O_3-ZrO_2 and Al_2O_3-TiC in the given order. However, the best solution shall be to use Si_3N_4 based cutting tool (speed 3000 m/min), but the price is pretty high. One can, therefore, make compromise with Al_2O_3-TiC tool. Another alternative is to braze silicon nitride tip on the substrate of more complex and less expensive cemented carbide.

Plot is shown in Figure Ex. 12.25.

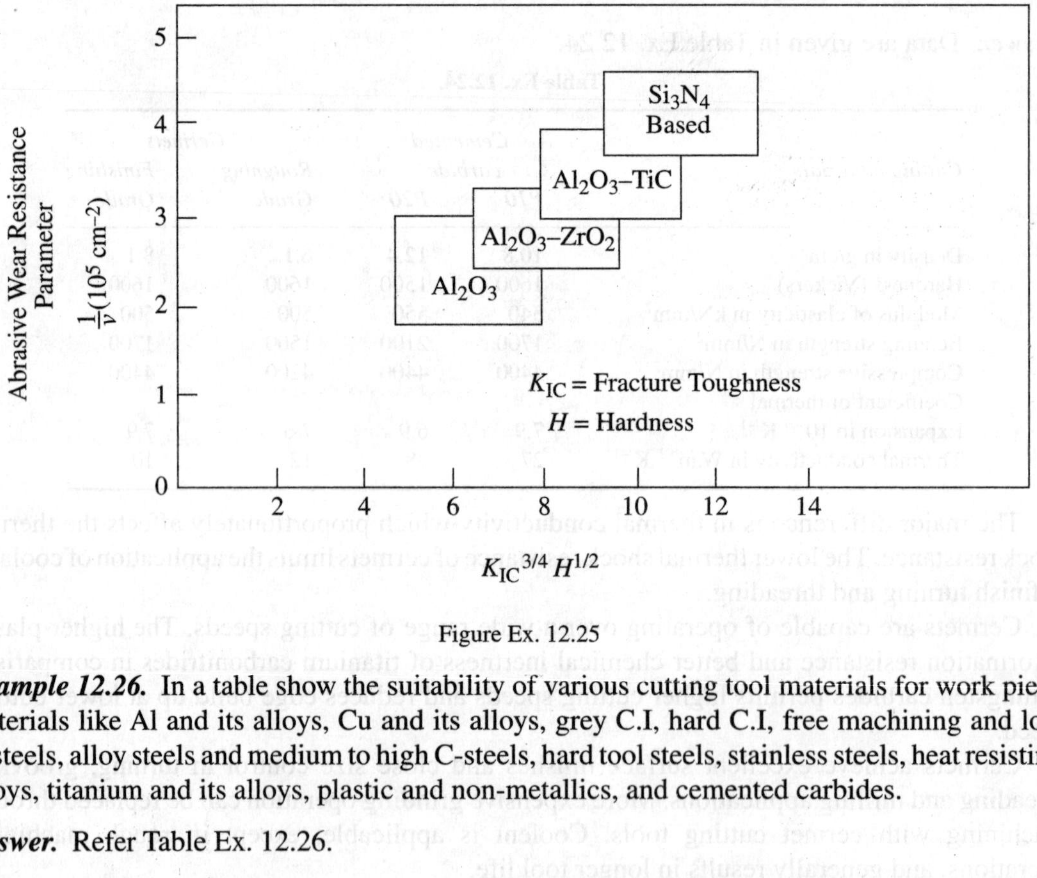

Figure Ex. 12.25

Example 12.26. In a table show the suitability of various cutting tool materials for work piece materials like Al and its alloys, Cu and its alloys, grey C.I, hard C.I, free machining and low C-steels, alloy steels and medium to high C-steels, hard tool steels, stainless steels, heat resisting alloys, titanium and its alloys, plastic and non-metallics, and cemented carbides.

Answer. Refer Table Ex. 12.26.

12.5 MATERIALS DESIGN FOR ELECTRICAL APPLICATIONS

Case Study 12.30: Conductors for Elevated Temperature: There has been a long felt need for high conductivity copper to be used at elevated temperature without undergoing any creep.

Table Ex. 12.26.

Workpiece material	*High speed Steel*	*Cemented Carbide*	*Al_2O_3 ceramic*	*Si-based ceramic*	*PCD*
Aluminium and Al alloys	Y	Y			Y
Copper, brass, bronze	Y	Y			Y
Gray cast iron		Y	Y	Y	N
Hard cast iron	Y	Y		N	N
Free-machining steels and low-carbon steels	Y	Y		N	N
Alloy steels and medium to high carbon steels		Y		N	N
Tool steels					
hard condition	N	Y	Y		N
Stainless steel		Y			N
Heat-resisting alloys		Y		Y	N
Titanium and Ti alloys		Y			
Plastic and non-metallics		Y			Y
Cemented carbide	N	N	N	N	Y

Y = Yes, N = No.

This problem was technologically solved by SCM Corporation of U.S.A. (now OMG America), which developed Cu-Al_2O_3 dispersion strengthened powder metallurgical wrought products.

The oxidant consists of a fine mixture of cuprous oxide and a carefully controlled amount of Al_2O_3. After reduction, the *in-situ* reduced copper oxide forms dispersion-strengthened copper due to the presence of the Al_2O during consolidation. These patented alloys contain normally 0.1, 0.2, 0.35 and 0.6 wt % Al_2O_3 and carry the designations Glidcop Al-10, 20, 35 and 60 respectively. Figure 12.44 shows the variation of electrical conductivity and mechanical properties of such ODS alloys. If the oxygen is present in free or reducible form (cuprous oxide), the alloys are prone to hydrogen embrittlement at high temperatures. The melting point, density, modulus of elasticity and coefficient of thermal expansion of these alloys are similar to those of pure copper. High electrical and thermal conductivities of these ODS coppers are the most distinctive properties. At room temperature, these range from 78% to 92% of those for pure copper. Such a combination of properties enables reduction of section sizes for component miniaturization without loss of structural strength or current and heat-carrying capabilities. Alumina particles in the alloy provide excellent thermal stability, and permit the retention of hardness, electrical conductivity and thermal properties even after exposure to relatively high temperatures.

Apart from alumina particles, titanium carbide particles can be used successfully for such application. Cu-5 Ti base alloy offers advantage of age hardening. These alloys are not yet produced commercially.

In case the temperature is relatively not high, Cu-Cr, Cu-Zr, Cu-Cr-Zr, Cu-Ni-P, Cu-Ni-Si, Cu-Co-Be heat treatable alloys can be used. These are susceptible to precipitation hardening. Table 12.15 gives the mechanical and electrical properties data of such alloys.

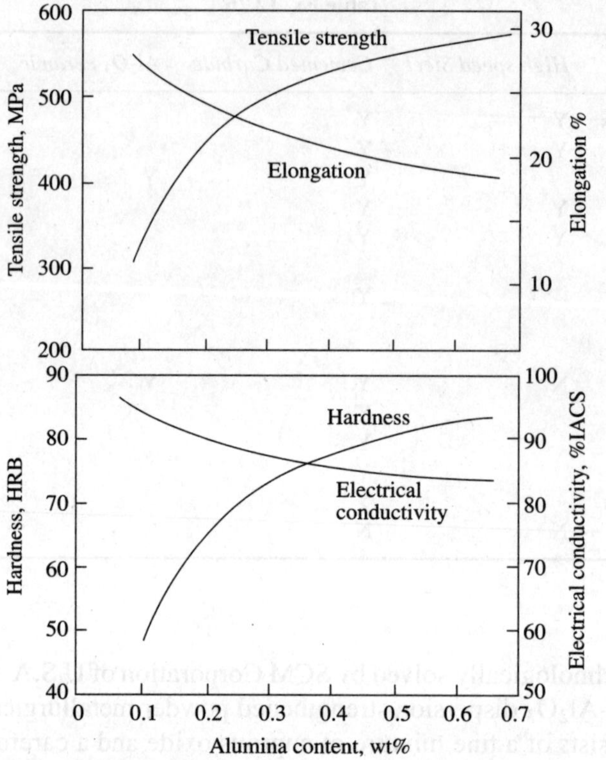

Fig. 12.44. Properties of alumina dispersion strengthened copper.

Case Study 12.31: Magnet Conductor Wires: The conductor wires used in transformers, chokes, relays etc. are known as 'magnet wires'. Some of the characteristics are:

- High dielectric strength to avoid voltage breakdown between adjacent turns -combined with good space factor
- good flexibility for winding
- compatibility with oils, impregnating varnishes, potty resins etc.
- stability under variations of temperature
- stability under winding pressure
- adequate resistance to withstand winding processes.

Most magnet wires are made of copper, but aluminium is also now-a-days used; particularly in strip form. Most common grade of copper used is electrolytic tough pitch copper, but special purpose wires e.g., oxygen free copper, silver bearing tough pitch and oxygen-free silver bearing grades are also used. Advantages of strip form conductors are in lower voltage gradient between turns, better use of winding space, and better heat transfer characteristics. Forces that tend to disrupt a winding on short circuit are a problem in wire wound coils, but are not so severe when strip is used. The strip must have edges contoured.

Copper clad aluminium wire is offered as a copper saving replacement. The copper percentage is in the range 10-20%. High strength copper conductors contain Cd, Zr, Cd-Cr.

Table 12.15. Typical Properties of Some Heat-treatable Conductivity Alloys.

Alloy	*Added Elements, %*	*0.1% Proof stress, MPa*		*Elongation, %*		*Hardness, HV*		*Conductivity, % IACS*	*Forms Available*
		Fully heat-treated	*Fully heat-treated and cold worked**	*Fully heat-treated*	*Fully heat-treated*	*Fully heat-treated and cold worked**	*Fully heat-treated and cold worked**		
Cu-Cr	0.4-1.0 Cr	258	350-425	25	25-15	125	130-160	80-85	Wire, rod, bar, strip
Cu-Zr	0.1-0.2 Cr	91	275-395	35	25-12	70	100-135	85-90	Wire, rod, bar, strip
Cu-Cr-Zr	0.4-1.0 Cr 0.1-0.2 Zr	258	350-425	25	25-15	125	130-160	80-85	Rod, bar
Cu-Ni-P	0.8-1.2 Ni 0.15-0.25P	334	350-411	25	25-20	140	150-175	55-60	Rod, bar, strip
Cu-Ni-Si	2.0-3.5 Ni 0.4-0.8 Si	471	532-639	15	15-10	170	180-210	40-45	Rod, bar, strip
Cu-Co-Be	2.0-3.0 Co 0.35-0.7Be	638	684-836	15	15-10	230	235-285	45-50	Wire, rod, bar, strip

• The range of figures indicates the effect of various amounts of cold work on material of moderate section.
Source: CDA, London, 1965.

For relatively high temperature operation there is difficulty in achieving magnet-wire installations that not only meet temperature requirements, but are also flexible enough to be satisfactorily wound. Although organic wire insulations exhibit adequate flexibility, they cannot function at the higher temperature levels. The converse, unfortunately, has been typical of many of the experimental inorganic insulations. However, an "all-ceramic" magnet wire insulation capable of withstanding 650°C has been commercially announced.

Table 12.16 is a chart of electrical insulating and conductor materials categorized by their suitability for operation in eight temperature regions. It will be noted that the organic insulations fall into the four lower regions, whereas the inorganics are in the upper four.

For temperatures higher than 200°C, copper requires protection against oxidation. In the temperature range from 250° to 400°, nickel-plating is effective. The plating can be as low as 4% by weight of the wire, but experience shows that heavier platings (10%) are more effective. However, in sizes finer than A.W.G. 22, 10% nickel-plated copper appears to suffer from oxygen penetration. To apply the finer sizes at temperatures over 400°C, it becomes necessary to use nickel-clad copper (that is, nickel tubing drawn over copper wire). Treating temperatures as high as 700°C can be used for short time periods.

Nickel has also been used to prevent oxidation of silver up to 650°C. Stainless-steel-clad copper has been used at 650°C with little oxidation effect. Above 750°C, it is necessary to use alloys such as Inconel.

The limitation of Cu-Ni assembly at elevated temperature (500°C or higher) must also be realized, as diffusion in solid state shall be effective. Small wire sizes, in particular, exhibit an increase in electrical resistance. The solution for this may be found in following alternatives:

- To select non-migrating cladding metal; example stainless steel
- To use a barrier layer to prevent mutual diffusion, example iron
- To find a substitute for copper to prevent diffusion with nickel altogether; example silver.

Case Study 12.32: Electric Contact Materials: In order to provide the electric current-carrying capacity in the contact material, the contact resistance should be low. The limitation in the amount of current that can be broken depends more upon the speed of break, the nature of the contact material, the inductance and voltage of the circuit being switched and the atmosphere surrounding the contact, than on the amount of contact material. Figure 12.45 shows the variation of control current with respect to voltage for different contact materials.

Another form of welding associated with contacts is that due to bounce. The arc struck during bounce causes liquid metal to sputter away the oxide film. The metal vapour prevents ingress of air, and when contacts finally make, they may weld.

Another feature while describing electric contacts is the contact noise, which is lowest with low resistivity contact materials such as rhodium, gold, silver and platinum and their alloys.

Most contact materials have a positive temperature coefficient of resistance, so that an increase in current through the constriction resistance still further increases the resistance. An associated effect is a softening of the contact area caused by heating, which reduces the resistance because thermal expansion or the closer approach of the contacts increases the spot area.

Dust plays its role, particularly in unsealed contacts. The small particles can be thermally decomposed in an arc and thereby produce contaminated contacts. Such feature may disconnect the circuit.

Table 12.16. Thermal Spectrum for Electrical Insulation and Conductors from 700 °C to 1500 °C.

Class of Materials	70 °C-90 °C	90 °C-120 °C	120 °C-170 °C	170 °C-250 °C	250 °C-400 °C	400 °C-650 °C	650 °C-1100 °C	1100 °C-1500 °C
Electrical Insulation	Untreated,Cotton Paper, Silk,etc	Oil-filled or varnished cotton, papersilk, enamels such as formvar, nylon etc.	Varnished glass and mica, mylar and other polyster films,enamels such as polyeurathane, Polyster,epoxy and combinations.	Polyimide silicon, TFE, fluorocarbon, Silicone-varnished glass fibers, mica, etc, resins + ceramic, polyimide + glass fiber	Ceramic -coated wires, glass-bonded fiberglass, glassenamel, mica and asbestos,etc	Glass -bonded fibers, glass + ceramics, Glassbonded synthetic	Glass + Refractories, crystallized glass, quartz,ceramic fiber + glass	Pure refractory oxides, Beryllia Magnesia,
Conductors	Copper aluminium	Copper aluminium	Copper aluminium	Nickelplated copper, aluminium-copper at 180 °C	Nickel-plated copper, nickel clad aluminium	Nickel clad copper, stainless-steel clad copper, Cufenic (nickel iron-clad copper), Nickel clad silver	Inconel barrier over D.S.copper, Inconel-clad silver	Platinum

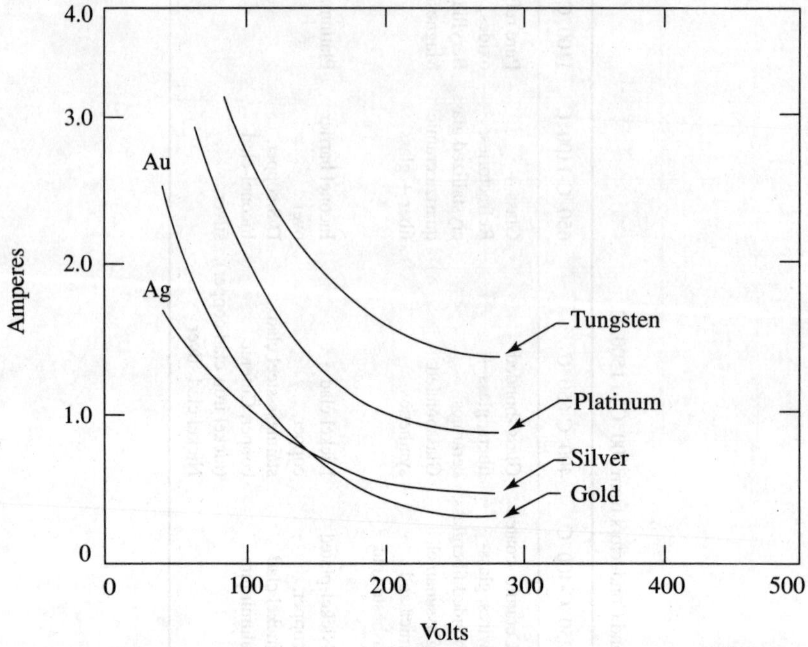

Fig. 12.45. Volt-ampere characteristics required to maintain a direct-current arc. When the critical values of current at a given voltage are exceeded, arcing occurs.

The contact materials are often classified according to their ability to resist tarnishing. Tarnishing characteristics are shown in Table 12.17. The metals/alloys in Group 1, 2 and 3 are used frequently in light and medium duty contacts, whereas Group 4 materials are used only for heavy current duty where resistance to arcing and mechanical wear are of great significance.

Table 12.17. Tarnishing Characteristics of Contact Metals.

Group 1 *Complete resistance to tarnishing*	*Group 2* *Resistant to sulphiding but oxidize at 350°C*	*Group 3* *Resist oxidation but sulphide*	*Group 4* *Oxide and sulphide*
Platinum	Palladium	Silver	Tungsten
Gold	Copper-palladium	Gold-silver	Molybdenum
Rhodium	Platinum-ruthenium	Palladium-silver	Silver-tungsten
Platinum-Iridium		Silver-copper	Silver-molybdenum
Platinum-gold-Silver		(sterling and Coil silver)	Silveer-nickel
			Silver-graphite Copper

Welding of a metal contact can be reduced by avoiding light contact pressures and by the elimination of bounce. Figure 12.46 shows the relative tendency to material transfer after 400,000 operations at 110 volts DC.

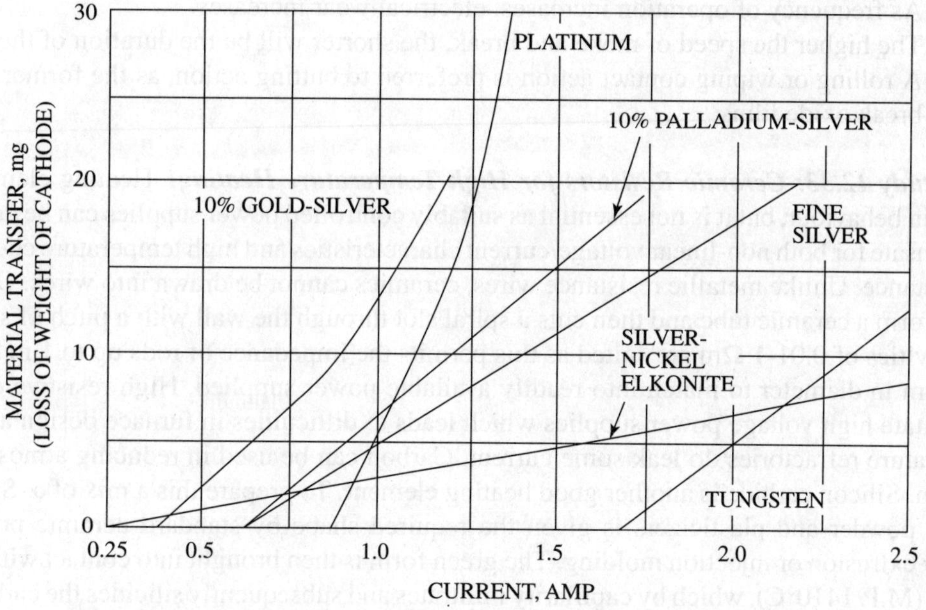

Fig. 12.46. Relative tendency to material transfer after 400,000 operations at 110 V d-c.

The current that can be switched, in a resistant load, without forming an arc depends upon the contact material, the circuit voltage, and to a lesser degree the shape of the contact, the ambient temperature and the atmosphere surrounding the contacts.

The criteria for successful operation of electric contacts can be summarized as:

- *Chemical Properties*

  Contact material must resist tarnishing and oxidation to preserve low contact resistance.
- *Physical Properties*

  High hardness contributes to low electrical wear (arc loss).
  High melting point, like hardness, also contributes to low electrical wear.
  High thermal conductivity is necessary to dissipate heat to prevent temperature rises.
  High electrical conductivity is required to obtain current-carrying capacity.
- *Electrical Conditions*

  High voltage at make and break favours electrical wear.
  Increasing current, to be carried and interrupted, favours electrical wear.
  Electrical wear on direct current is greater than on alternating current for a given voltage.
  Inductance of circuit: Interrupting an inductive circuit will cause more wear than interrupting a noninductive circuit on the same voltage.
  Arc suppression reduces electrical wear; condensers, resistors used singly or together, and, on large equipment, magnetic blowout and deionizing devices are helpful.
- *Mechanical Conditions*

  Size and form of contacts are related to the heating effect.
  Increasing contact pressure increases, within limits, the contact resistance, and greater currents can be carried.

As frequency of operation increases, electrical wear increases.
The higher the speed of make and break, the shorter will be the duration of the arc.
A rolling or wiping contact action is preferred to butting action, as the former tends to break oxide films.

Cast Study 12.33: Ceramic Resistors for High Temperature Heating: Heating elements are ohmic in behaviour, but it is not essential as suitably controlled power supplies can be devised to compensate for both non-linear voltage/current characteristics and high temperature coefficients of resistance. Unlike metallic resistance wires, ceramics cannot be drawn into wires. Often one has to form a ceramic tube and then cuts a spiral slot through the wall with a pitch of say 1 cm. Resistivities of 0.01-1 Ωm are suited as this permits the impedance of rods up to 1 m long and 0.5-2 cm in diameter to match into readily available power supplied. High resistive elements necessitate high voltage power supplies which leads to difficulties in furnace design as at high temperature refractories do leak some current. Carbon can be used in reducing atmosphere or vacuum. Silicon carbide is another good heating element. To prepare this a mix of α-SiC grain, carbon powder and plasticizers is given the required shape by standard ceramic processing usually extrusion or injection molding. The green form is then brought into contact with molten silicon (M.P. 1410°C), which by capillarity infiltrates and subsequently silicides the carbon, thus forming a secondary phase of SiC. This secondary SiC bonds to the primary SiC phase. The material is fired again in a recrystallization process at around 2500 °C in an inert atmosphere. This process transforms any β-SiC to the α-modification and partially removes residual silicon, permitting control over electrical resistivity. The final product has a low porosity and hence longer life. The resistance of the hot section of tubular element is adjusted to the required value by spiralling, in the green state. In case of rods, the low resistance 'cold ends' may be joined to high resistance hot-zone lengths by reaction bonding. Other methods of production are just pressureless sintering or 'recrystallization' process or in situ formation of SiC from carbon and SiO.

To prevent the oxidation during the use of SiC elements, they are invariably glazed by a silicate at around 1450°C. The ends of the heating elements are flame sprayed with aluminium in order to impart good electrical contact. It is worthwhile to mention that below approximately 800°C, the resistance of SiC varies from batch to batch, but above this temperature the behaviour is stable. The reason for the variability of resistance is due to the impurities.

Another important ceramic resister is $MoSi_2$, which is used in air above 1500°C. Its resistivity behaves as is expected for a metal, increasing from about 2.5×10^{-7} Ωm at room temperature to about 4×10^{-6} Ωm at 1800°C. The commercial $MoSi_2$ heating element is often known as 'Kanthal Super', which is a composite consisting of a mixture of particles bonded with an aluminosilicate glass phase (~20% by volume). Because of softening problem during high temperature service, such heating elements are commonly mounted vertically. This is the major difference in comparison to SiC elements, which are pretty rigid at operating elevated temperatures.

Lanthanum chromite $LaCrO_2$ is another ceramic material, but its wide usage as heating elements is now decreased. It is used mainly as interconnect in high temperature solid oxide fuel cells.

Zirconia or more properly stabilized zirconia is another candidate as high temperature furnace elements. However, they require preheating to reduce the resistance to a level at which Joule heating is effective.

Wrong selection of heating element, particularly with reference to furnace atmosphere, can cause very high furnace maintenance. A mere comparison in hours for the same resistor in commercial furnaces operated at the same temperature would not give a true picture of the action of different atmospheres. Other contributory factors for the life of a resistor material are watt-density loading, cross sectional area, operating temperature, frequency of switching the current on and off, support of the resistor in the furnace, physical shape and design of the resistor.

Case Study 12.34: Capacitors in Electronic Industry: The negative factor in the aluminium capacitor usage is the incompatibility of construction with the requirements of surface mount technology. However, the aluminium capacitor manufacturers are now providing a "surface mount" configuration. Perhaps more important is that the capacitor users are *modifying* their soldering techniques by lower temperatures and shorter times to enable the survival of aluminium electrolytic capacitors in surface mount applications.

Tantalum capacitors have better volumetric efficiency than aluminium capacitors while multilayer ceramics are better than single layer ceramics or films. A similar qualitative choice relates to electronic characteristics. The incompatibility of a chip form factor deletes single layer ceramics, films and to a lesser extent aluminium capacitors. A high temperature requirement dictates the choice of a tantalum or multilayer ceramic capacitor. Where cost is a driving factor, designers are driven towards single layer ceramics or aluminiums.

The relative high cost of tantalum capacitors is a clear detriment to their wider use. A very large deterrent for designs into new equipment is the instability of the cost of tantalum.

Single layer ceramic capacitors represent a mature technology and product. They remain the lowest cost style of capacitor and are widely used in consumer applications. However, their usage growth rate is declining as multilayer ceramics are substituted in newer designs. The multilayer ceramics have higher volumetric efficiency, are rapidly reducing in cost, and are ideal for the new labour and space saving surface mount technology. These capacitors are displacing not only single layer ceramics in new applications, but small tantalum capacitors and film capacitors. The growth in multilayer ceramic capacitors is also being accelerated by electronic application trends toward higher frequencies and digital techniques which reduce the electronic size of capacitance required in the application.

Table 12.18 shows some of the qualitative design factors used by equipment designers for choosing a dielectric and style to fill a capacitor function requirement.

Example 12.27. Enumerate the materials used as electrical conductors for following applications: (a) trolley wires, (b) bus conductors, (c) power conductors, (d) high voltage transmission.

Answer. Trolley wires: Hard-drawn copper wire or bronze alloy must have current-carrying and abrasion-resistance characteristics. Wire is often grooved to facilitate support.

Bus conductors: Copper and aluminium must have current-carrying capacity and requisite

Table 12.18. Capacitor Design Factors.

FACTOR	*Single Layer Ceramic*	*Multilayer Ceramic*	*Tantalum*	*Aluminium*	*Film*
Reliability	+	+	+	+	+
Volumetric Efficiency	–	+	+	–	–
Electronic Characteristics		+	+		
Available as "Chip"	–	+	+	–	–
High Temperature	–	+	+	–	–
Cost	+		–	+	

Source: Proceeding of International Symposium Ta and Nb, Nov. 1988, Tantalum-Niobium Study Center, Brussels, Belgium, p. 483.

mechanical strength.

Power conductors: Bare, solid or stranded copper or copper alloy. For specific applications, copperweld , aluminium and steel. For overhead lines, hard-drawn or medium-hard tempers. Insulated, soft or annealed grounding or bonding, weatherproof cables with copper or aluminium wires reinforced with steel wire.

High-voltage transmission: Hollow copper. Large outside diameter reduces corona loss. Current-carrying capacity greater when used for alternating current.

Example 12.28. A brass alloy is to be used in an application that will require a minimum ultimate strength of 275 MPa and an electrical resistivity of less than 50 ohm.nm. What percent zinc should the brass possess? Determine the "window" resulting from the two specifications. Data may be searched in Metals Handbook.

Solution.

$$\rho_{Cu} = 17 \text{ ohm.nm; therefore}$$

$$\sigma_{Cu} = 60 \times 10^6 \text{ ohm}^{-1}.\text{m}^{-1}$$

$$\rho_{Brass} < 50 \text{ ohm.nm}$$

and

$$\sigma_{Brass} > 20 \times 10^6 \text{ ohm}^{-1}.\text{m}^{-1} \quad \text{and} \quad > 0.33\sigma_{Cu}$$

Therefore, for ρ: Max. = 23%Zn For strength points of view, after referring data, the minimum zinc content must be 15%. Use an 80 Cu – 20 Zn brass.

Example 12.29. A conductor rail has a cross-section of 15 cm^2 and a specific resistance of 7.6 μ ohm.cm at 0°C. If the temperature coefficient of the material 0.005 per °C, estimate its resistance in ohms per kilometer when the temperature is 50°C.

Solution. The resistivity at 50°C is related to the resistivity at 0°C by the relation

$$\rho_{50} = \rho_0(1 + 50\alpha) \qquad (1)$$

where α = temperature coefficient of resistance

Here $\rho_0 = 7.6\mu\Omega.\text{cm} = 7.6 \times 10^{-6}\Omega.\text{cm}$

and $\alpha = 0.005$ per °C

Substituting the value of α and ρ_0 in (1), we get

$$\rho_{50} = 7.6 \times 10^{-6}(1 + 50 \times 0.005)$$
$$= 7.6 \times 10^{-6} \times 1.250$$
$$= 9.5 \times 10^{-6} \text{ ohm.cm}$$

Now, $R_{50} = \rho_{50} \times \dfrac{\ell}{a}$ (2)

Here, $\ell = 1$ km

$= 100,000$ cm

and $a = 15\ \text{cm}^2$

Substituting the values of ρ_{50}, ℓ and a in (2), we get

$$R_{50} = \frac{9.5 \times 10^{-6} \times 100,000}{15}$$
$$= 0.06334 \text{ ohms.}$$

Example 12.30. From the Table Ex. 12.30, it is obvious that Cu-25 Cr alloy possesses better electrical conductivity as compared to Cu-50 Cr alloy. This alloy is proposed to be used as contact material. Research carried out revealed that optimization of properties like electrical conductivity, hardness and breakdown voltage can be achieved by proper additives (X,Y) in Cu-25 Cr alloy. Suggest what these additives are. Justify as best you can. What about the processing routes?

Table Ex. 12.30.

Materials	*Average grain size of Cr phase /μm*	*Relative density /%*	*Conductivity /(MS.m^{-1})*	*Oxygen content /10^{-6}*	*Nitrogen content /10^{-6}*	*Brinell hardness /HB*	*Breakdown field /10^8 V.m^{-1}*
CuCr25	20 - 100	99.01	25	490	60	80	2.62
CuCr25	10 - 25	99.22	23.1	550	35	86	3.78
CuCr25	5 - 15	99.27	22.4	460	25	91	4.05
CuCr50*	70 - 150	99.12	18.7	430	60	97	2.52

* –Conventional CuCr50 prepared by sintering and infiltrating methods.

Solution. Alloy Cu-50 Cr is processed through sintering and infiltration route, as ingot metallurgy route would be difficult for this hard alloy. In case, one lowers the quantity of Cr, the electrical conductivity rises and hardness drops down. Here one can use induction melting/casting/working route reasonably well. Regarding additive, one desires electrical conductivity, but still having high breakdown voltage. For this inert but compatible particles may be added, e.g., graphite and tungsten. Processing needs to be done in absence of air, otherwise loss of carbon and oxidation of tungsten are present. Additive 'X' appears to be carbon 'Y' tungsten. The exact quantity (low in amount) needs to be optimized after experimentation. Since hardness of alloy containing Y is high (91 BHN), the answer is justified.

Example 12.31. You are given the task of developing a film (glaze) resistor for precision potentiometer based on palladium? What alloying addition in palladium would be required? Describe the steps involved in such a development.

Answer. For glaze film we do require a glassy phase. The major advantage of having palladium-glass is that resistivity changes uniformly over long time. As regards selection of the alloying addition, one has to select a system with minimum temperature coefficient of resistivity. Literature suggests that it is silver. Figure Ex. 12.31 shows the variation of resistivity and its temperature coefficient for Pd-Ag binary system. From this it is evident that Pd-40% Ag alloy shall be ideal.

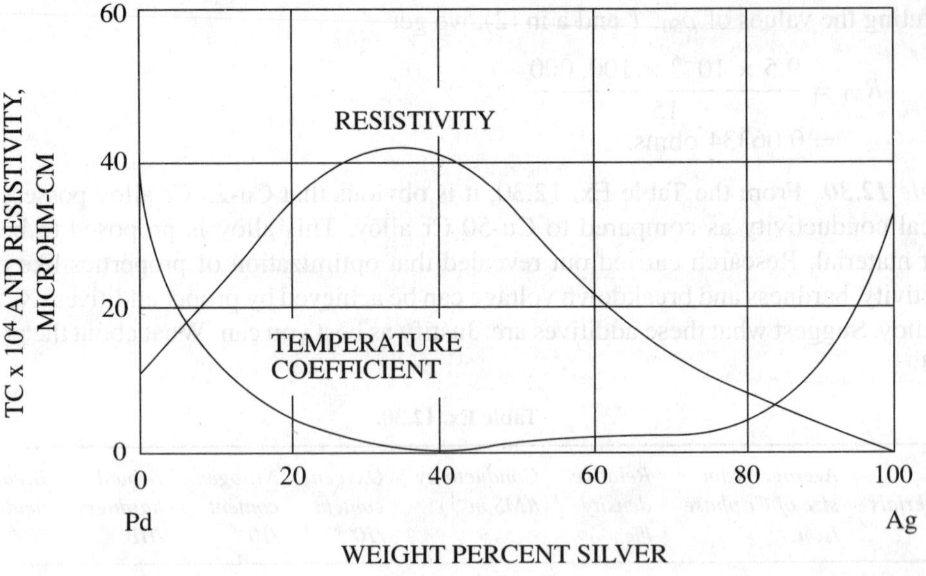

Figure Ex. 12.31

Films are produced for a mixture of finely ground powder of Pd, Ag and glass carried in an organic medium. It is applied to a ceramic substrate, which is then fired in the temperature range 750-850°C. This results in the formation of palladium oxide, thus giving the mixture a high resistivity. By careful control of the firing cycle and thus the amount of palladium oxide formed, the resistivity is determined. The glass provides the bond of the system to the substrate and gives moisture protection. If the glass content is less than about 40%, the resistor becomes more sensitive to moisture.

Example 12.32. Discuss the design requirement for insulated connecting wires under following properties (a) mechanical, (b) dielectric, (c) physical and (d) chemical.

Answer. Design requirements affecting the choice of wires may be separated into four categories: mechanical, dielectric, physical and chemical properties. These may be subdivided as follows:

1. Desirable mechanical properties:

   - Good flexibility without cracking or breaking over the temperature limits required
   - Good abrasion resistance for rough handling
   - In covered wires, insulation must be reasonably adherent to the wire but capable of being easily stripped
   - Homogeneity in composition
   - Good vibration and shock resistance
   - No mechanical change with time under relevant conditions of use.

2. Desirable dielectric properties:

   - High dielectric strength
   - Low power factor
   - Good insulation resistance (particularly after exposure to climatic or radiation conditions)
   - High surface-leakage resistance
   - Low dielectric constant.

3. Desirable physical properties:

   - Wide operating temperature range without softening or becoming brittle (may be from -175° to +250°C)
   - Noncombustible
   - Adequate life under operating conditions
   - Resistant to all forms of radiation
   - Little weight loss due to condensation or evaporation.

4. Desirable chemical properties:

   - Must be chemically inert and impervious to oils, acids, gasoline and the like, both in liquid and gaseous form
   - Noncorrosive
   - Not affected by pure oxygen in space vehicle environments
   - Colouring must remain stable under all operating conditions
   - Resistant to fungus and vermin attack
   - Must not outgas or generate toxic fumes in space environments.

Example 12.33. How for the effective design of insulating systems, a knowledge of the corona extinction level (CEV) of electrical insulating system and the conductor is a must? How ageing response is vital in predicting reliability?

Answer. To answer this question, we shall refer Figure Ex. 12.33a. At low voltage levels, the conduction current increases. At higher voltage, the concentration of charge carriers begins to increase at a greater rate. The portion of current-voltage curve is called the ionization region. As the voltage is increased beyond the ionization region, an increasing number of charge carriers attain sufficient energy to multiply: current increases rapidly and corona occurs.

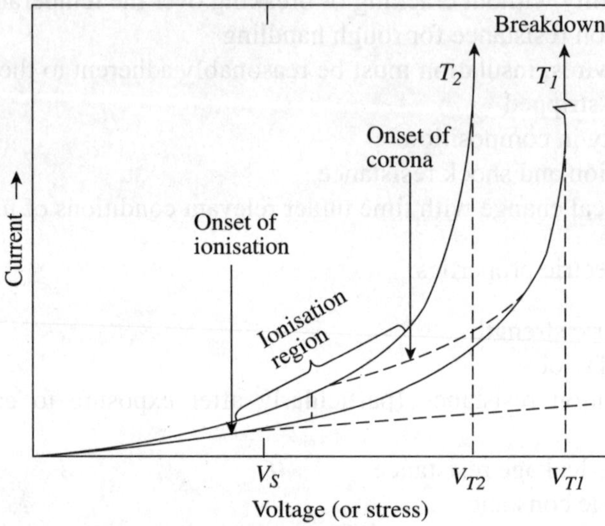

Figure Ex. 12.33a

Internal heating also increases, causing a higher corona level: this generates more heat, so that eventually thermal or corona runaway occurs and the dielectric breaks down. The corona region is sensitive to the rate of voltage increase. At slow rates of voltage increase, there is more time for heating of the material and consequently, a lower breakdown voltage.

In carefully controlled laboratory tests, the magnitude of the temperature rise can be estimated. With a voltage decrease, the conduction current decreases (but at a slower rate than that at which it increased), intersecting higher temperature curves at each point. This intersection with a higher temperature curve occurs because of the heating of the dielectric by the internal corona and ionisation. The point at which corona current is no longer detectable is known as the *corona extinction voltage* (CEV). If sufficient time has elapsed for cooling to take place, a second increase in voltage causes the current to retrace the original curve, leaving an open-loop area.

The usual voltage-breakdown test provides data which are influenced to varying extents by the environment, the dissipation factor, the temperature coefficient, the thermal conductivity, and the geometry (dimensions and design of the electrodes) of the dielectric. In general, the dielectric strength decreases with increasing temperature (Figure 4.9) and as an exponential function with increasing thickness of the test specimen. High-frequency dielectric strength is lower than that at power frequencies, an effect usually attributed to increased dielectric heating of the material and high corona levels.

The effect of ageing on breakdown voltage is shown in Figure Example 12.33b.

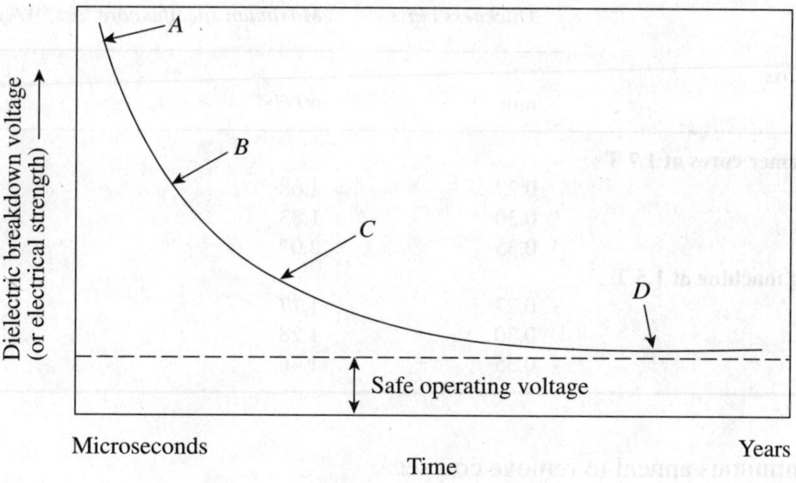

Figure Ex. 12.33b

12.6 MATERIALS DESIGN FOR MAGNETIC APPLICATIONS

Case Study 12.35: Oriented Fe-Si Alloy Sheets for Soft Magnetic Applications and Other Related Materials: Many times these alloys are referred in text books as Si-steels, which is a misnomer, for they are high purity alloys of iron and silicon. In Chapter 4, a brief description on the magnetic properties has been made. Silicon is added in iron because, it is the most effective element of increasing the electrical resistivity of iron, thereby, reducing eddy current losses in alternating current magnetization. The nonoriented grades of Si-steel are low carbon iron alloys containing from 0.8 to 3.5% silicon and aluminium. Such grades are used in small motors for household applications. Oriented grades are made at silicon content 3.0 to 3.3%. The limit is set by the necessity for severe cold works in the production process. Such alloys are used mainly for transformer cores and rotating electrical machinery and are sold in only three thickness 0.27, 0.3 and 0.35 mm. The core loss data are given in Table 12.19.
The processing schedule, a complex one, is as follows:

1. Reheat slab to 1315-1370°C
2. Hot roll, coil
3. Pickle
4. Cold roll about 40%, or
5. Continuous anneal (to recrystallize)
6. Cold roll about 40%
7. Continuous anneal to decarburize
8. Coat with separating medium
9. Anneal at about 1150°C in dry H_2 in coils

Table 12.19. Core Loss Types and Thickness.

ASTM core loss Type	Thickness class mm	Maximum specific core loss, W/kg 60 Hz	50 Hz
For transformer cores at 1.7 T			
27H076	0.27	1.68	1.27
30H083	0.30	1.83	1.39
35H094	0.35	2.07	1.57
For rotating machine at 1.5 T			
27G053	0.27	1.17	0.89
30G058	0.30	1.28	0.97
35G066	0.35	1.46	1.11

10. Continuous anneal to remove coil set
11. Apply insulating coating
12. Epstein tests to classify.

(One can omit Steps No. 5 and 6, if the steel is cold rolled ~85% in Step No. 4).

The slab heating temperature is very high. This is dictated by the Fe-Si binary phase diagram, which forms closed gamma loop (Figure 12.47). The presence of carbon greatly enlarges the $\alpha + \gamma$ field, with the amount of austenite sharply increasing with carbon content. This confirms the need for as little carbon possible in the alloy as impurity. Fortunately the formation of austenite during cooling the alloy is quite sluggish, so that hot rolling can be coupled in the single phase ferrite region. Any amount of austenite after hot rolling would affect the texture and the final magnetic properties are deteriorated. The other reason for the high slab heating temperature is to compatibly solutionize the second phase particles so that these can be precipitated later during the processing and serve as grain boundary pinning agents. This pinning leads to the secondary recrystallization that develops the desired $\{110\} < 001 >$ texture. This texture is also known as *Goss texture*, on the name of the metallurgist, who developed this in 1934. The common pinning agents are AlN, MnS, BN, TiC, Si_3N_4, VN and SiO_2.

The control of variables in the two cold rolling and two continuous annealing is mostly a proprietory information and each producer has its own schedule depending on the material chemistry. The separating medium is put on the sheet before coiling for the high temperatures anneal, to prevent sticking of the wraps of the coil. The high temperature anneal in dry hydrogen has two purposes; (a) to develop a strong cube on edge $\{110\} < 001 >$ texture and (b) to improve magnetic properties by removing C, N, O and S.

The significance of secondary recrystallization is to develop texture. The temperature range for such treatment is 815-870°C. The whole idea of such recrystallization is to allow grain boundaries to break away from the pinning points. The final grain size of such alloys are large, with each grain extending through the thickness of the sheet. Figure 12.48 shows a typical microstructure of cube-on-edge oriented 3.2 Si-Fe sheet.

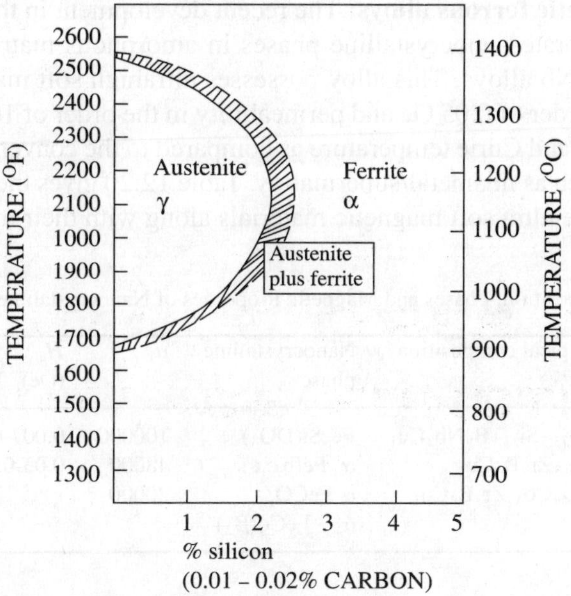

Fig. 12.47. Section of the ternary Fe-Si-C diagram at 0.01-0.02% C.

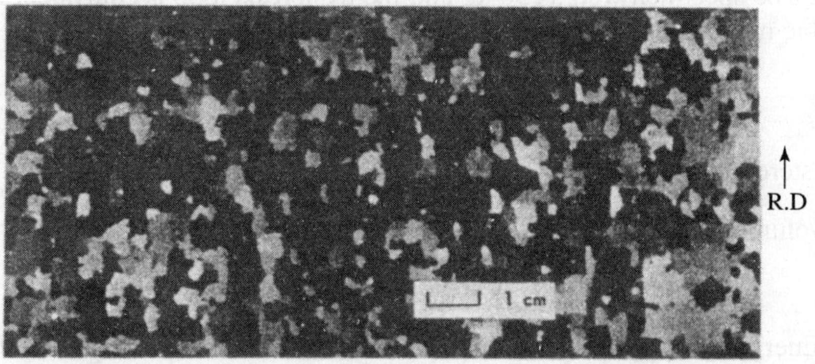

Fig. 12.48. Typical macrostructure of cube-on-edge grain orientation 3.2% silicon-iron sheet (0.3 mm thick).

Before final Epstein test, an insulating coating is applied over the sheet. In this test the specific core loss is reported in watts per unit mass, as given in the expression

$$\frac{W}{kg} = \left(W - \frac{E^2}{R} \right) \frac{\ell}{0.0235m} \tag{12.18}$$

where m = total specimen mass in kg

ℓ = actual strip length

W = watts read on a wattmeter

E = rms volts

R = parallel resistance of wattmeter potential circuit.

Advanced soft magnetic ferrous alloys: The recent development in the area of soft magnetic materials is to incorporate nanocrystalline phases in amorphous matrix. Alloy FINEMET is based on Fe-Si-B-Cu-Nb alloys. This alloy possesses ultrahigh soft magnetic properties (very low coercivity in the order of 0.05 Oe and permeability in the order of 10^5) coupled with higher saturation flux density and Curie temperature as compared to the conventional nickel based soft magnetic materials such as mismetal/supermalloy. Table 12.20 gives the typical composition of various nanocrystalline ultra soft magnetic materials along with their magnetic properties.

Table 12.20. Composition, Phases and Magnetic Properties of Nanocrystalline Ultra Soft Magnets.

Alloy name	Typical composition (at%)	Nanocrystalline phase	μ_i	H_c (Oe)	B_S (kG)	T_C (°C)
FINEMET	$Fe_{73.5}Si_{13.5}B_9Nb_3Cu_1$	$Fe_3Si(DO_3)$	100000	0.007-0.05	10-12	550
NANOPERM	$Fe_{88}Zr_7B_4Cu_1$	α -Fe(b.c.c)	48000	0.03-0.10	15-18	770
HITPERM	$Fe_{44}Co_{44}Zr_7B_4Cu_1$	α-FeCO, α'– $FeCo(B_2)$	20000	-	16-21	965

Example 12.34. Calculate the energy lost per hour in a specimen of iron subjected to magnetization at 50 c/s. The specimen weighs 50 kg and the hysteresis loop is equivalent in area to 250 joules per cubic metre. Density of iron is 7500 kg/m^3.

Solution.

$$\text{Hysteresis loss per m}^3 = 250 \text{ joules/cycle}$$

$$\text{Volume of specimen} = \frac{50,000}{7500} = \frac{20}{3}\text{ m}^3$$

$$\text{Energy loss} = \text{Hysteresis loss per m}^3 \times \text{Vol. of specimen} \times \text{No. of cycles}$$

$$\text{Energy loss per sec.} = \frac{20}{3} \times 250 \times 50$$

$$= \frac{250,000}{3}\text{ joules or watt-s.}$$

$$\text{Energy loss per hour} = \frac{250,000}{3}\text{ watt.hr} = \frac{250}{3}\text{ kWh}$$

$$= 83.33\text{kWh.}$$

Case Study 12.36: Nd-Fe-B Permanent Magnets: RCO_5 permanent magnets are well known. Interest arose to develop rare earth-iron permanent magnets, since iron is far cheaper than cobalt. The progress has been difficult because of following reasons:

- Fe forms much fewer intermetallic compounds with rare earths than cobalt
- Stable compounds of the RFe_5 composition are absent
- Compounds which are stable e.g., R_2Fe_{17} have low Curie temperature and planar preference anisotropy.

Binary Fe-Rare earth phases do not appear to be promising for potential permanent magnet materials. Two alternative approaches may be worthwhile to consider:

- extending the study to metastable/non-equilibrium phases instead of being limited to equilibrium phase.
- extending the study to ternary/quaternary systems to stable phases.

In the first category melt-spinning method has been applied to obtain amorphous materials. In the second category the development of Nd-Fe-B intermetallic $Nd_2Fe_{14}B$ was achieved, which is carried out by powder metallurgical process. The high reactivity of the rare-earth and their alloys and the critical dependence of the magnetic properties on the chemical composition, made it essential to avoid any contamination during the alloy preparation and the subsequent P/M processing. The oxidation of the rare-earth components by O_2 or H_2O must be kept to a minimum throughout fine powder handling and heat treatment stages.

The electron metallographic study of sintered permanent magnet reveals the following (Figure 12.49):

- The hard magnetic major tetragonal phase 27 wt % Nb (A)
- A grain boundary phase - 38 wt % Nd (B)
- A boron-rich phase - 38 wt % Nd (C)
- Numerous oxide inclusions.

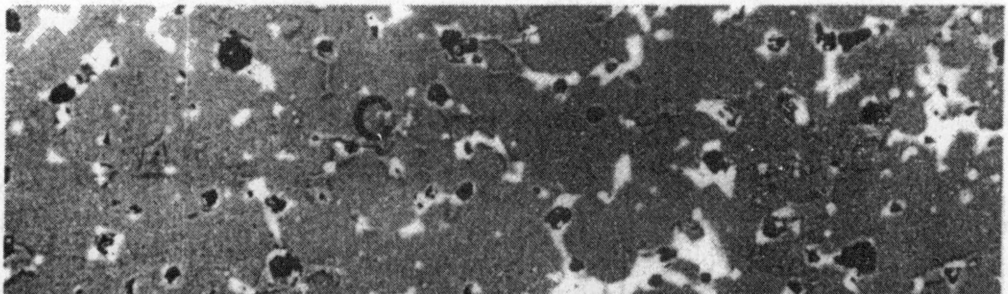

Fig. 12.49. Back scattered electron image of an as-sintered Nd-Fe-B magnet (from I.V.Mitchell (ed),Nd-Fe Permanent Magnets:Their Present and Future Applications,Kluwer Academic Publishers, 1985, p.88; with kind permission from Springer Science and Business Media, Dordrecht.)

The starting alloy composition is generally hyper-stoichiometric of the compound $Nd_2Fe_{14}B$ with respect to Nd and B in order to compensate for oxidation during processing. The Nd-rich grain boundary phase is eutectic phase, which facilitates densification through liquid phase sintering.

The major disadvantage of this permanent magnet is relatively low Curie temperature, which causes a high reversible temperature coefficient of remanance. Increase in Curie temperature can be achieved by substituting some Fe by Co, e.g., a substitute of 10% of the iron by cobalt produces an increase in Curie temperature from 585 K to 671 K, with a reduction in αBr to 0.08 K

. One of the biggest application of Nd-Fe-B magnets is in stepper motors. Figure 12.50 shows a magnet made out of Alnico 550 (left). The length of the magnet depends on the coercive field strength. The RCo 190 magnet, shown in the middle, has nearly the same remanance, but a much higher coercivity so that the magnet can be much shorter. The system at the right pertains to Nd-Fe-B; which is of the same size, has the same stability but 50% more flux. The torque of the motor can increase accordingly. (For more details refer "Nd-Fe Permanent Magnets: Their Present and Future Application", ed. I.V. Mitchell, Elsevier Applied Sc. Publisher, London, 1985).

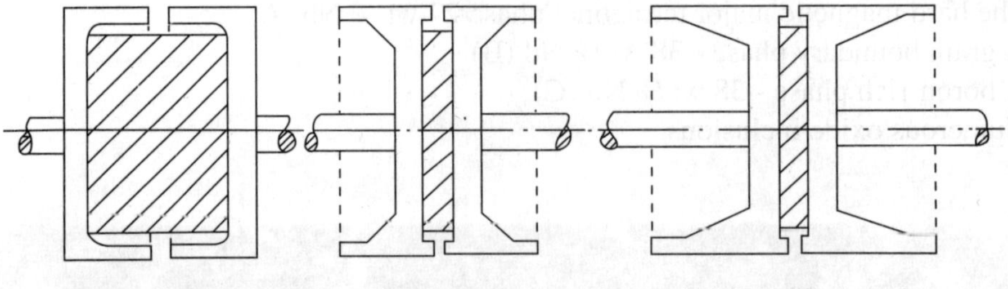

Fig. 12.50. Rotors for stepper motors with different permanent magnets.

Case Study 12.37: Magnetic Oxide Powders for Tapes and Discs: Magnetic oxide powders are prepared in order to offer tailored properties. In case of tapes, these are deployed in a passive binder on a flexible polymer sheet, e.g., PET. In case of discs, the deployment is on a rigid aluminium disc. By opting single domain particles, we can have control over remanance and coercivity. To promote high coercivity we need to have certain magnetization direction preferable to others, that is to say that we need same type of anisotropy. Some sources of anisotropy are composition and shape. We must attempt to have remanance as large as possible in order to have signal strength. Another important feature is the coating thickness and its uniformity, which must be controlled tightly. No doubt more flux shall be available for reading when the coating is thick, but there is no point in using coatings thicker than the desired bit size. In general practice, the volume fraction of magnetic particle coatings are 20% and 40% in rigid discs and flexible media respectively. A very thin films can be deposited by chemical or physical vapour deposition methods.

As far powder oxides are concerned, γ-iron oxide (Fe_2O_3) is very common. This is a metastable phase reverting to the α-form on heating above 250°C in air. This means that its normal Curie temperature (590°C) is never reached (Figure 12.51).

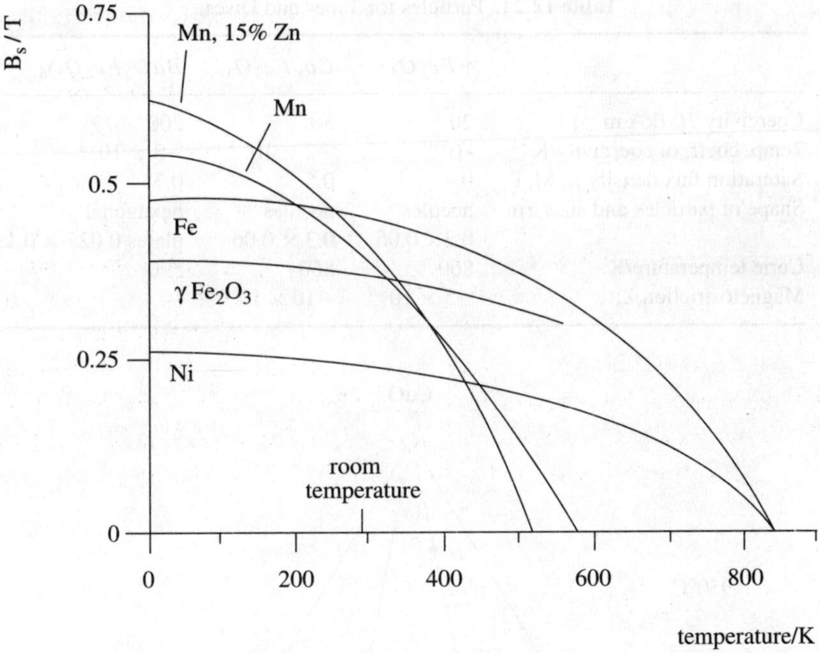

Fig. 12.51. Saturation flux density vs. temperature for some ferrites.

As a candidate material for coating on tapes, magnetite Fe_3O_4 is not suited to form single domain particles, as it further oxidizes to higher oxide Fe_2O_3. α-Fe_2O_3 is non-magnetic as Fe^{3+} ions pair up. γ-Fe_2O_3 is magnetic and its powder is synthesized to produce needle shaped powder. Such a morphology offers higher degree of shape anisotropy. An Iron II hydroxide slurry is oxidized to particles of α-$Fe_2O_3.H_2O$. Dehydration of this form gives γ-Fe_2O_3, which is reddish in colour, followed by Fe_3O_4. Finally the ferrite particles are oxidized at about 200°C in presence of moisture to γ-Fe_2O_3. This is brown in colour and possesses the inverse spinel structure.

Further development has been made by coating the fine particles of the oxide with cobalt hydroxide $Co(OH)_2$, before final oxidation. This does have an increased coercivity through the modification of the crystalline anisotropy. This material is known as cobalt modified $Co\gamma Fe_2O_3$, where cobalt content is around one per cent or so. Table 12.21 compares some of the properties of γ-Fe_2O_3 with other ferrite.

Case Study 12.38: Processing of Y-Ba-Cu-O Based High Tc Superconductors: In Chapter 4, we have mentioned about the high T_c superconductors. Figure 12.52 shows the pseudo-ternary Y_2O_3-BaO-CuO system. The point marked 1 2 3 represent $YBa_2Cu_3O_{7-x}$ phase composition. This system undergoes a number of isothermal reactions involving liquid phase.

Table 12.22 summarizes the liquidus temperature of YBCO system in argon, air and O_2. Sintering with liquid formation produces dense components but because melting is incongruent, phase separation occurs and the superconducting grains become coated with non-superconducting phases causing poor electrical and mechanical properties.

Table 12.21. Particles for Tapes and Discs.

	γFe_2O_3	$Co_\gamma Fe_2O_3$	$BaO(Fe_2O_3)_6$
Coercivity H_c (kA m^{-1})	20	50	200
Temp. coeff. of coercivity/K^{-1}	-10^{-1}	-2×10^{-3}	-3×10^{-3}
Saturation flux density $\mu_0 M_s T$	0.4	0.5	0.3
Shape of particles and size/μm	needles 0.3× 0.06	needles 0.3 × 0.06	hexagonal plates 0.025 × 0.1
Curie temperature/K	860	860	590
Magnetostriction λ_s	-5×10^{-6}	-10×10^{-6}.	–

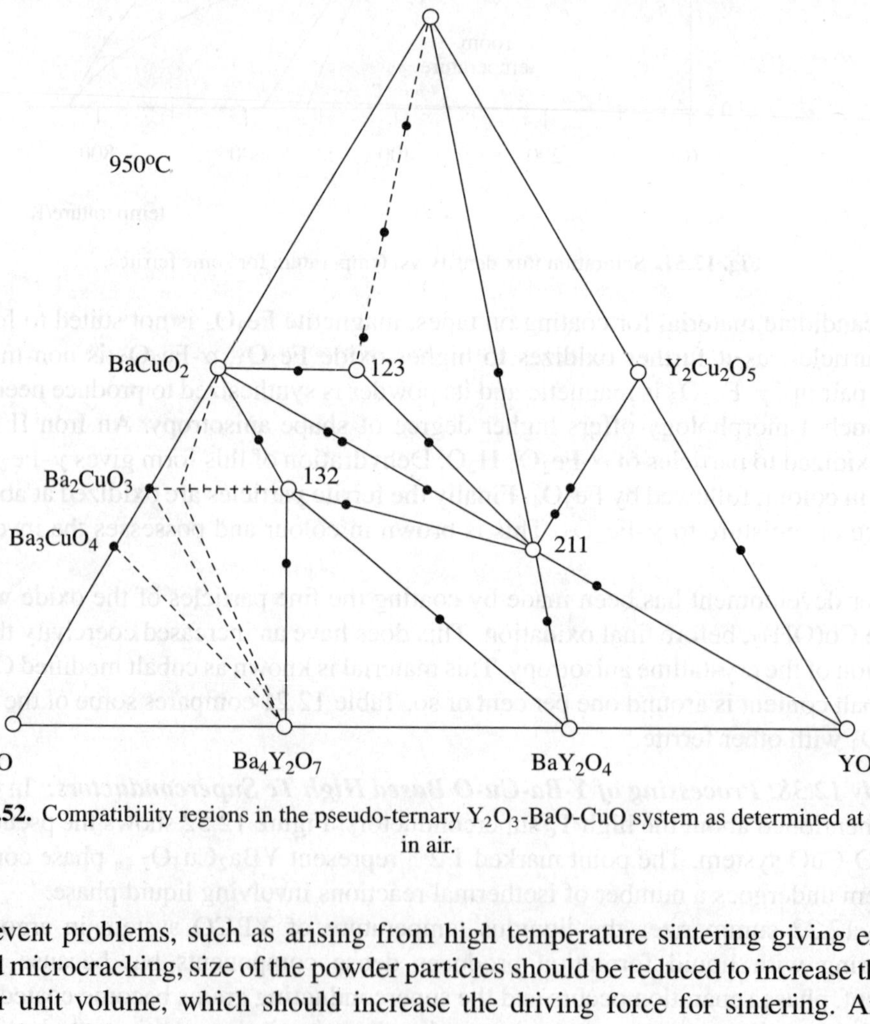

Fig. 12.52. Compatibility regions in the pseudo-ternary Y_2O_3-BaO-CuO system as determined at 950 °C in air.

To prevent problems, such as arising from high temperature sintering giving extraneous phases and microcracking, size of the powder particles should be reduced to increase the surface energy per unit volume, which should increase the driving force for sintering. Anomalous grain growth, which can trap the original interparticle pores within grains and prevent full

Table 12.22. Liquidus Temperatures in the Y-Ba-Cu-O System.

*Invariant Point**	*Reaction⁺*	*Liquidus Temperature* (°C) *Argon*	*Air*	O_2
e^1	$123 + 011 + 001 \leftrightarrow L$		898	925
e^2	$011 + 001 \leftrightarrow L$	816	902	938
p^1	$123 + 001 \leftrightarrow 211 + L$	862	933	962
p^2	$211 + 001 \leftrightarrow 202 + L$	897	956	990
p^3	$123 + 011 \leftrightarrow 211 + L$			1,010
e^3	$211 + 011 \leftrightarrow L$			1,015
m1	$123 \leftrightarrow 221 + L$		999	1,028

The letter "e" is used for eutectics, "p" represents pseudo-peritectic or type II four- phase equilibrium, and "m" stands for peritectic decomposition. ⁺The numbers in this column indicate the x, y and z, respectively, in the general formula $R_xBa_yCu_zO_?$.L represents a liquid phase.

densification, can be suppressed by having a uniform size of powder particles. Heating rate to the sintering temperature should be as fast as possible since surface, rather than bulk diffusion dominates at lower temperatures.

Though sintering above the eutectic temperature allows more rapid sintering, an accompanying detrimental effect on the critical current density, $J_{c,t}$ has usually been attributed to residual insulating eutectic phase at grain boundaries. However, investigations of the grain boundary areas and $J_{c,t}$ reveal the possible major importance of microcracks as a cause of poor current carrying ability rather than the impurity segregation to grain boundaries.

It was suggested that low oxygen partial pressure increased the sintering kinetics possibly due to the increased oxygen vacancy concentration and or to the removal of trapped gases (mostly nitrogen) from the pores.

Figure 12.53 gives the plots of bulk density and J_c of the YBCO samples sintered at different temperatures for 8 and 24 hours. With increased sintering temperature and/or time, the bulk density increases, but J_c does not increase monotonically. Samples sintered at 975°C for 8 hrs give rise to the formation of $BaCuO_2$ and CuO as secondary phases along the grain boundaries, which inhibits the transport currents. Formation of high density microcracks along grain boundaries is not unusual, which is attributed to the reduced cohesive strength due to the grain boundary precipitates.

Ceramic wire preparation: Wire can be fabricated either by wire drawing or by extrusion. Wire drawing is a more economical and commercially desirable process because, it can be easily made into a continuous operation. On the other hand, hydrostatic extrusion appears to provide a better wire product. It assures higher compressive stress on the powder during the deformation, leading to better compaction of the ceramic powder. In extrusion, the solids content of slip is much higher than that available for tape casting hence the resulting plastic mass must be forced at high pressure through an assembly of dies. Extrusion can be used to produce various useful configurations of intricate complexities.

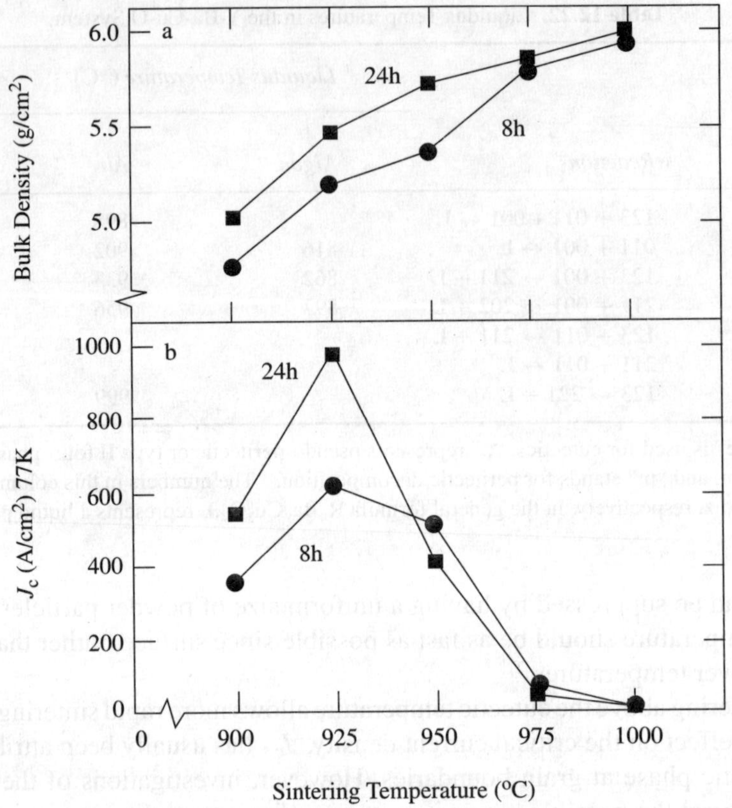

Fig. 12.53. Plots of bulk density vs. sintering temperature (a), and Jc (77 K) vs. sintering temperature (b) for the YBCO samples sintered for 8 hours and 24 hours respectively. (Reproduced with permission from Powder Processing of High T_c Oxide Superconductors and Their Properties, A.C. Vajpei and G.S. Upadhyaya, Trans. Tech. Publications, Aedermannsdorf, Switzerland, 1992, p. 156).

The various approaches to the fabrication of ceramic wire can be basically subdivided into three geometric categories: (a) bare wire, (b) metal core composite wire and (c) metal-clad composite wire (powder-in-tube).

In metal core composite wire configuration, a metal wire, e.g., silver is coated with a superconductor-slurry-mixture of YBCO powder and solvent-diluted organic binder. The coated composite wire is wound into a coil shape and then heat treated by slow heating (to burn off the binder) to ~930°C/16-40 hours and furnace cooled in oxygen.

In powder-in-tube configuration a silver sheath provides a barrier to protect the superconductor from damage and environmental degradation and also provides a path to shunt current across any defects. The powder is loaded in the silver tube and the assembly is compacted by swaging, drawing or rolling. Heat treatment near the melting point followed by cooling gives dense oriented microstructures with large J_c values of 1.5×10^4 A/cm² at a magnetic fields of 26 T. The choice of a proper cladding metal is critical in fabrication of the composite superconductor wire. Investigations into the suitability of various metals in contact with the YBCO reveal that most metals except Ag and Au interact with the compound at moderate high temperature and degrade the superconducting properties.

In sheathed YBCO wires, the reduction of the oxygen content is observed to cause loss in superconducting properties. For coping with this problem a two-step heat treatment is necessary:

1) recovery heat treatment at T > 900°C for reordering the metallic lattice and
2) a prolonged oxidation heat treatment at <500°C in order to compensate the oxygen losses during the recovery anneal.

The sheath material should have following requirements:

- ductility
- thermal-expansion-coefficient compatibility with the ceramic superconductor
- mechanical strength
- melting temperature (high enough to withstand the recovery heat treatment)
- little or no reaction with the ceramic superconductor
- compatibility with the oxygen content requirements of the superconductor
- inertness to oxygen reaction
- costs.

Table 12.23 shows the comparative view of the sheathed YBCO wires and pressed and sintered YBCO samples. The J_c of the Ag sheathed samples is more than five times larger than that of the sintered sample whereas, the superconducting volume fraction (as estimated from magnetic moment measurements) is larger for the sintered sample. In Ag sheathed samples, the average size of each grain is usually far smaller than that in the sintered sample and they form clusters, with cracks and voids around each cluster.

12.7 MATERIALS DESIGN FOR DEVICES

Case Study 12.39: Semiconductor Device and its Metallization: Before describing metallization, it is desirable that we know the general fabrication sequence for integrated circuits (Figure 12.54). Silicon is advantageous as its oxide (SiO_2) is an excellent insulator and can be used for both isolation and passivation purposes. However silicon has some limitations, because of which compound semiconductors were introduced. In Chapter 4, we have already described these aspects.

Table 12.23. Basic Parameters of Ag Sheathed and Sintered YBCO.

Parameters	*Ag-sheathed*	*Sintered*
J_c at 77.3 K (A/cm^3)	1370	195
Density (g/cm^3)	5.8	5.65
T_c onset, K	92	92
T_c zero, K	90.5	91.3
ρ at T_c ($\mu\Omega$ cm)	550	300

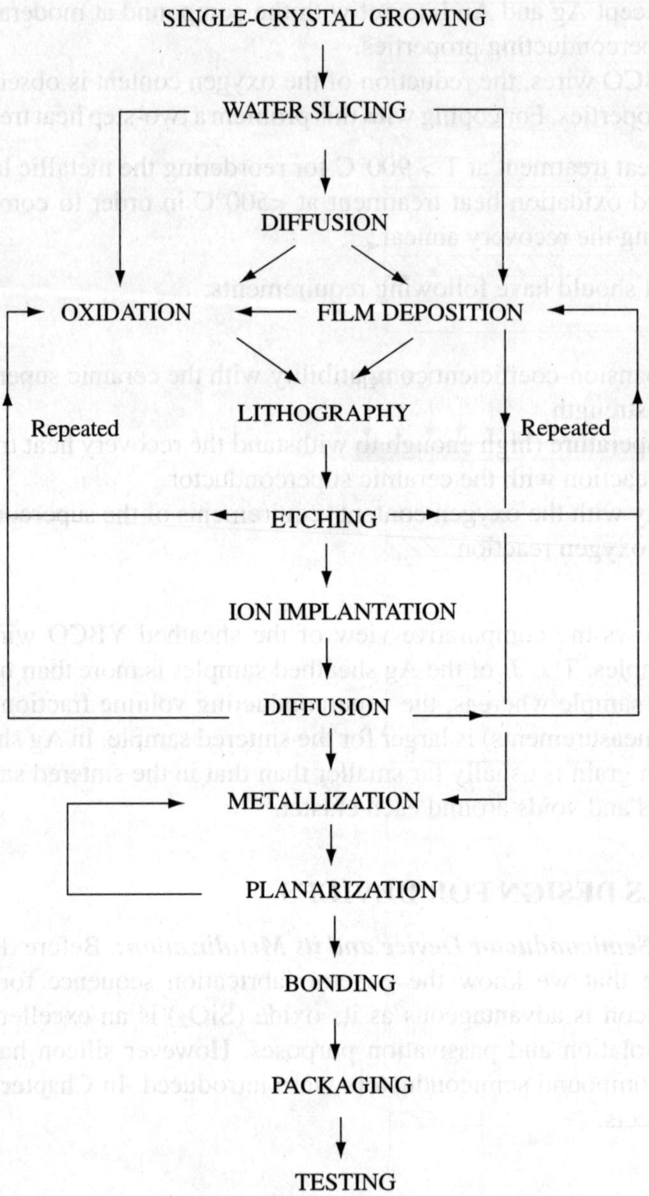

Fig. 12.54. General fabrication sequence for integrated circuits.

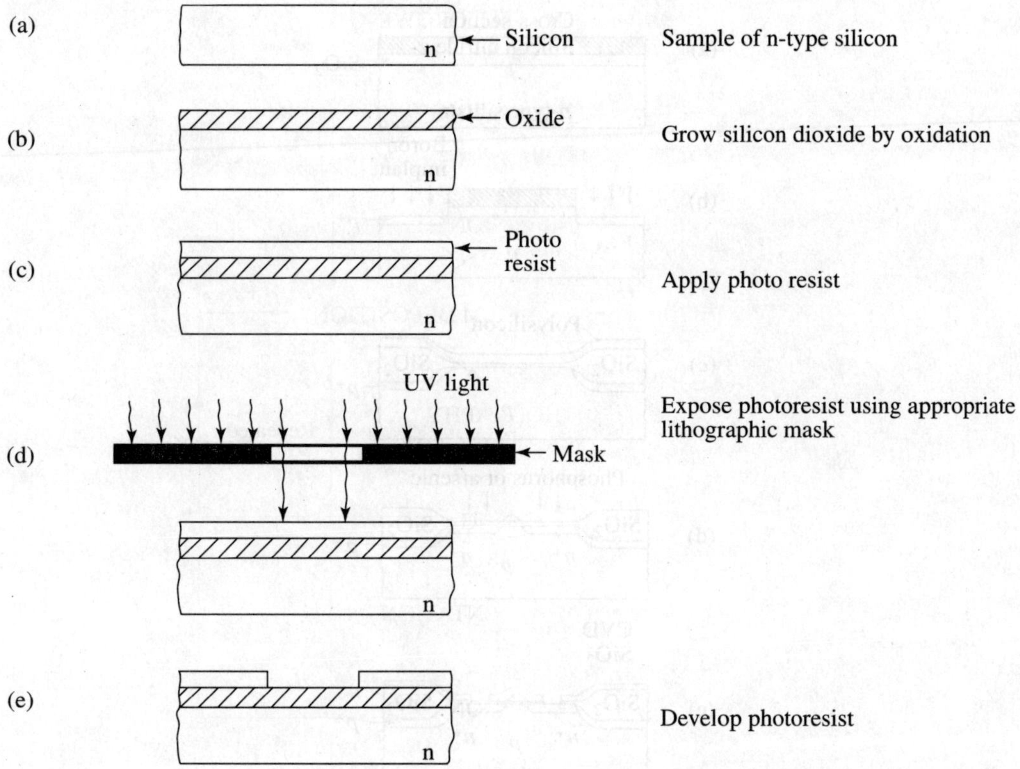

Fig. 12.55. Processing sequence for a p-type region in n-type silicon.

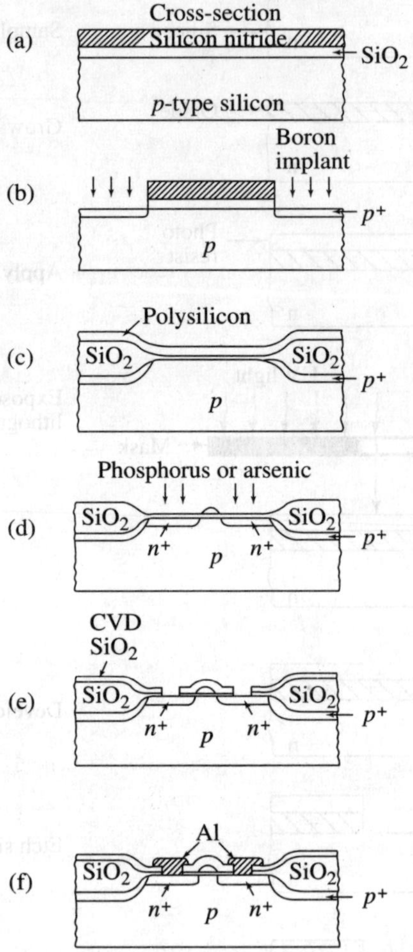

Fig. 12.56. Cross-sectional views of the fabrication of a MOS transistor.

A more detailed flowsheet to prepare a p-n junction diode is given in Figure 12.55. Figure 12.56 shows the details of a metal-oxide-semiconductor field effect transistor, an abbreviated form known as MOSFET. In some ways the operation of a MOSFET is very similar to that of a junction transistor. The primary difference is that the gate current is very very small in comparison to the base current of a junction transistor. Another difference is that although majority carriers dominate in the functioning of MOSFETs, minority carriers do play a role with junction transistors.

Metallization: To serve as effective interconnect metallization for silicon, the metal chosen must meet or exceed all the following requirements:

- The metal must make a low resistance electrical contact with the silicon.
- The metal must have a limited reactivity with silicon, in order to produce a stable contact.
- The metal must have high electrical conductivity, so that current is easily carried without any voltage drops.

- The metal must adhere well to the underlying silicon, silicon dioxide, or other dielectric used.
- A pattern must be easily definable in the layer using photolithographic techniques.
- The deposition method must be compatible with already existing structures.
- The metallization must not exhibit "electromigration" (migration of atoms in the metallization caused by the flow of current).
- The metallization must not corrode under normal operating conditions.
- It must be possible to bond easily to the metallization, in order to allow for external connection.
- The metallization must be economically competitive.

No one metal perfectly meets all of these requirements. However aluminium does meet them satisfactorily. The performance of aluminium is improved by the introduction of small amounts of other elements. For example; the tendency of aluminium to react with silicon is stopped by the introduction of a small percentage of silicon during deposition. Electromigration resistance of aluminium is greatly increased by including a small percentage of copper in aluminium. However, for advanced electronic applications, aluminium and its alloys cannot be used since they are vulnerable to electromigration and/or have rather high resistivity. Only silver, copper and gold have resistivities lower than aluminium. Gold has high resistance to electromigration but only shows a small improvement in resistivity. Furthermore, diffusion of gold into silicon will create deep levels in the band gap and thus dramatically affect electronic properties of devices. Silver with the lowest resistivity, also creates deep levels in the silicon band gap and is a very fast diffuser in SiO_2. Making diffusion barrier for silver has been proven difficult. In addition, silver has low resistance to electromigration due to its low melting point. Copper with close to half the resistivity (1.7 $\mu\Omega$.cm compared to 3.0$\mu\Omega$.cm for Al-0.5% Cu alloys) and with electromigration on the order of ten times better appears to be the most appropriate material for interconnect. Although, copper also creates deep levels in the silicon band gap, several materials, that can act as diffusion barrier for copper have been identified. Among those, tantalum, tantalum nitride and titanium nitride are the most well known materials.

Case Study 12.40: Glass-to-Metal Seals and Heat Sinks: Glass-to-metal seals are of great significance in hermetic package for electronic devices. For optimum thermal and mechanical integrity of the seal it is necessary that expansion of the glass and metal be substantially the same over the temperature range in which both materials behave elastically. It has been estimated that if differential contraction is less than 100 ppm, the integrity of an assembly can be assured. Glass is an insulator, while metal is conductor. Thus, the combination of both these must provide a hermetic environment within its capsule.

Among glasses, soft glasses have thermal expansion coefficient of about 9×10^{-8} °C, whereas hard glasses have lower values of $4 - 6 \times 10^{-8}$/°C. As the latter has better thermal shock resistance, it is required for such sealings. Now our problem gets narrowed down to find a suitable metal/alloy. Although Fe-Ni and Fe-Ni-Cr alloys have low values of thermal expansion coefficient, none of these have coefficients matching that of a novel glass up to the strain point of the glass. Molybdenum and tungsten are suited, but their utility is questioned due to fabrication problems. An alloy known as 'Kovar' (55 Fe, 29 Ni, 17 Co) is ideal for the seal.

In reality, the problem is more complex. Only single consideration of thermal expansion is not enough. Metallurgical problems like nonmetallic inclusion, residual gas content, transformation products or excess carbon content on either sealability or processing must also be considered. Another important metallurgical feature is oxidation. Both bond strength and adherence are enhanced by accomplishing the seal through an intermediate oxide, which is primarily a mixture of iron oxides. The oxide is both adherent to the alloy and soluble in glass while its expansion coefficient does not offer any mismatch to either.

It is well known that, semiconductors and ceramic substrates used in electronics are brittle materials with coefficient of thermal expansion (CTE) in the range 3-7 ppm/K. Semi-conductor and ceramic substrates are typically attached to supporting component, such as packages, printed circuit boards and heat sinks with solder or an adhesive. Figure 12.57 shows thermal conductivity as a function of CTE for various metals used in electronic packaging. It shows that aluminium, copper and silver have relatively high thermal conductivity, but have CTEs much greater than desirable for most electron packaging applications. This offers an idea for developing composites. The figure shows the data for a number of composites e.g., C/Cu, C/Al, C/C, Cu-Mo/W, C/Ep (Ep = epoxy resin), SiCp/Al, BeOp/Be, diamond p/Cu and Eglass/Ep. With the exception of Eglass/Ep, C/Ep and C/C, all other composites have CTEs in the desired range. The advantages of composites are even greater than those of conventional packaging metals when weight is concerned. When we consider specific thermal conductivity, the improvement in performance is an order-of-magnitude.

Case Study 12.41: Piezoelectric Transducers: A brief account of piezoelectricity is given in Chapter 4 (Section 4.2.5). During piezoelectric effect, direct interconversion of mechanical to electrical energy or vice-versa takes place. When the material changes its state of polarization under stress, the effect is known as '*generator effect*'. On the other hand, when the metal changes its shape then its ambient electric field changes and the effect is called '*motor effect*'. The coefficients of motor effect (m) and 'generator effect' (g) may be expressed as:

$$m = \left(\frac{\partial s}{\partial \varepsilon}\right)_{\sigma} \tag{12.19}$$

and
$$g = \left(\frac{\partial P}{\partial s}\right)_{\varepsilon} \tag{12.20}$$

where s = strain, ε = electric field and P = polarization.

Piezoelectric behaviour is usually described by a linear relationship between polarization and strain and between strain and electric field. The coefficients which describe the relationship are called the piezoelectric coefficients. A mathematical framework using Cartesian tensor, is used to describe this behaviour.

A transducer is a device which transforms nonelectrical energy into electrical energy or vice-versa. Electromechanical transducers are very common in which piezoelectric materials are used. Common piezoelectric material include quartz, barium titanate ($BeTiO_3$), lead zirconate titanate (PZT), and polyvinylidene fluoride (PVDF). Piezoelectric transducer have been used to measure almost any thing: strain, pressure, force, acceleration, light modulation etc. The semicrystalline material (e.g., PVDF) is not piezoelectric in its natural form, but can be

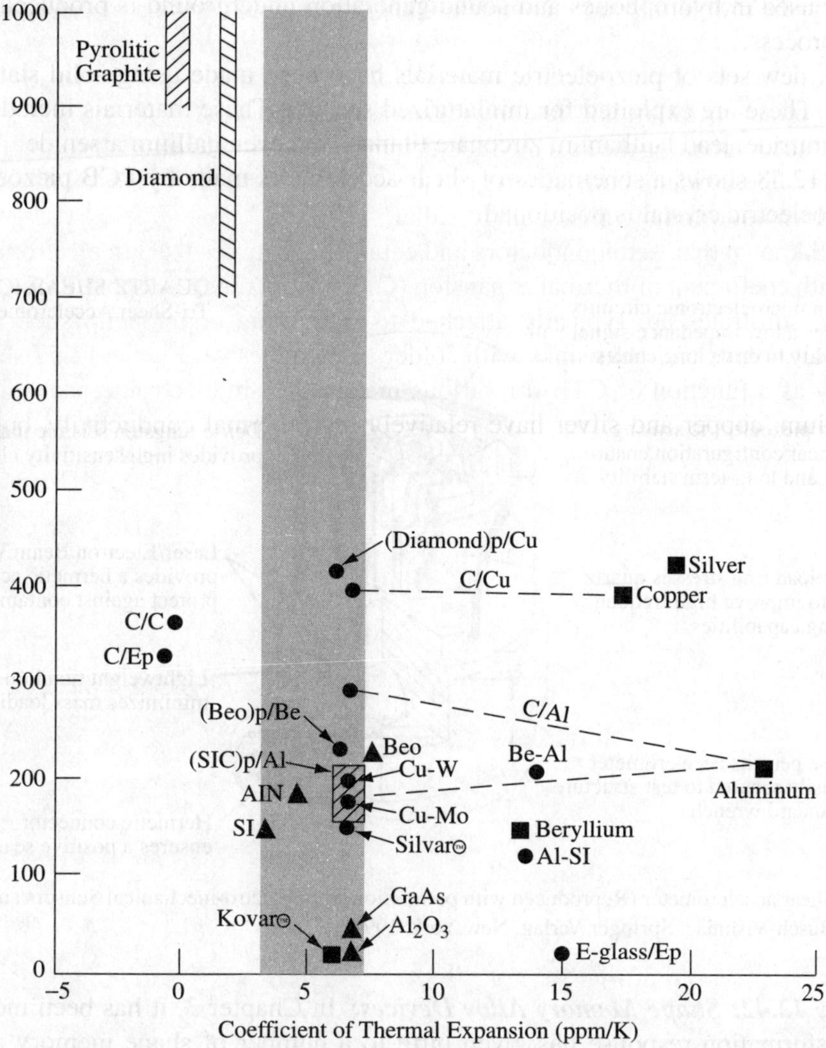

Fig. 12.57. Thermal conductivity as a function of coefficient of thermal expansion for selected monolithic materials and composites used in electronic packaging (Reproduced with permission from Handbook of Materials Selection, Ed. M. Kutz, John Wiley & Sons, Inc., Hoboken, NJ, 2002, p. 389).

made to exhibit piezoelectric behaviour through polishing and stretching procedures. Polymer composites have been developed using small pieces of ceramic piezoelectric materials suspended in a polymer matrix. The advantage of this approach is that the material can be designed so that its mechanical properties are primarily determined by the polymer matrix, while the transduction behaviour is dominated by the material suspended in the matrix.

Quartz has significant piezoelectric coupling in shear. On the other hand barium titanate and lead zirconate titanate offer greater coupling in compression than in shear. The Curie temperature of barium titanate is about 115°C, while for PZT it varies from about 300-365°C. PZT-polymer composites used in hydrophones and sound generation underground is produced by injection moulding process.

Of late, new sets of piezoelectric materials have been made using solid state fabrication techniques. These are exploited for miniaturized sensors. These materials include zinc oxide, aluminium nitride, lead lanthanum zirconate titanate and even gallium arsenide.

Figure 12.58 shows a schematics of shear accelerators made by PCB piezoelectronics in which piezoelectric crystal is positioned.

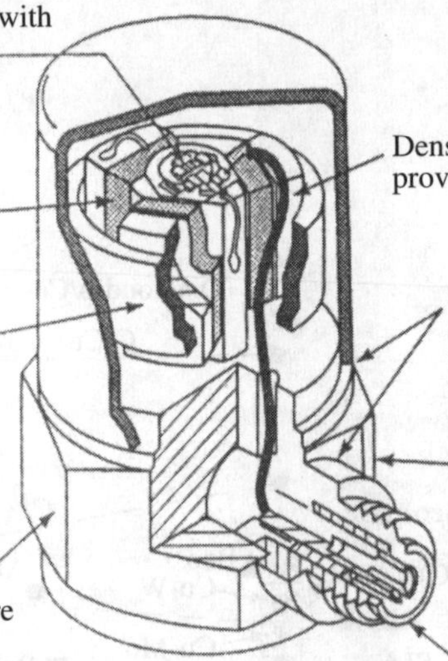

Fig. 12.58. Shear accelerometer (Reproduced with permission from Electro-mechanical Sensors and Actuators, I.J. Busch-Vishniac, Springer Verlag, New York, 1999, p. 166).

Case Study 12.42: Shape Memory Alloy Devices: In Chapter 3, it has been mentioned how phase transformation response has given birth to a number of shape memory alloys, which possess large heat recoverable strains (upto 8% in case of NiTi alloy). In addition to the normal shape memory effect, where strain is recoverable on heating the superelastic effect can also be used where, the material goes round a transformation cycle with no external temperature change, simply by loading and unloading just above A_f (austenite formation temperature). Many recent devices use superelasticity. It is worth mentioning that there are shape memory ceramics and polymers too.

In 1971, the large scale application of shape memory alloys began as reliable coupling to connect titanium hydraulic tubing in the Grumman F-14 aircraft. NiTi coupling which shrunk on removal from liquid nitrogen was known as cryofit ® couplings. Other applications are in medical applications, which moneywise provides largest market place.

Since NiTi is a high cost alloy, attempts have been made on Cu-based and more recently Fe-based couplings. Because of the lower recovery force of these alloys they result in a lower performance system. NiTi alloy is light weight, and easy to install with proven reliability. The couplings are made by machining a cylinder, usually with circumferential sealing lands on the inner diameter (ID), while the alloy is in the austenitic condition, cooling to form martensite or at least to where martensite can be stress induced, and expanding by forcing a tapered mandrel through. If tubing is inserted whose outer diameter (OD) is intermediate between the as-machined and as-expanded ID of the coupling, then when the coupling is heated through it's A_s-A_f transition it will recover, compress the tube and create an excellent joint.

Another shape memory alloy device is electrical connectors. Unlike most tube or pipe couplings, electrical connectors must be able to sustain several connect/disconnect cycles. This is done by utilizing the strength differential between the austenite and martensite phases, usually in combination with a compliance such as a mechanical spring. Connectors with highest forces have been deviced, which survive high inertial forces such as in missile control systems, where shock and vibration resistance are important prerequisites.

Case Study 12.43: Solid Oxide Fuel Cells (SOFC): A brief mention of fuel cells has been made in Chapter 6. There are two types of fuel cells: low temperature and high temperature cells. SOFC operate at high temperatures in the range 650-1000°C. The electrolyte is solid oxide, unlike melt in case of carbonate fuel cells, which conducts oxygen ions. Both the electrodes and the electrolyte form an all solid-state-system, while the reacting species (fuel and oxygen) are in the gas phase. The purpose of operation of these cells is illustrated in Figure 12.59. The electrolyte is typically ZrO_2 with 8-10 mol % Y_2O_3. The requirements for an ideal electrolyte are

- High oxygen ion conducting and minimum electronic conductivity.
- Good chemical stability with respect to the electrodes and inlet gases.
- A high density to inhibit fuel crossover.
- Compatibility of thermal expansion with other components.

In order to minimize the voltage losses in the electrolyte, the electrolyte layer should be as thin as possible. The thickness of the electrolyte depends on the cell design and the mechanical stability of the electrolyte. In case of a flat electrolyte supporting the electrodes, a minimum thickness of 100-150 μm is required for ZrO_2-8 mol % Y_2O_3 ceramic. The mechanical stability of this ceramic can be raised by lowering the yttria content or by addition of alumina. However, both of these additions lower the ionic conductivity. Thinner electrolytes can be tenable, only when it is supported by another component of the fuel cell (e.g., an electrode), or by a special shape.

The anodes requirements are: effective oxidation catalyst, high electronic conductivity, stability in the reducing environment; thermal expansion compatibility with the electrolyte and other fuel-cell components; a physical structure offering low fuel transport resistance; chemical and mechanical stability; and tolerance to sulphur contaminants. The anode reaction is

$$H_2 + O^{2-} \rightarrow H_2O + 2e^-$$

In the reducing fuel environment, metals are stable. Nickel is suited because of its high electrochemical activity for the hydrogen oxidation reaction, low costs and chemical compatibility reasons. However, to match the thermal expansion coefficient of nickel to that of electrolytes, the former is to be alloyed with oxide to form a cermet. Ni-yttria stabilized zirconia (40% Ni) is

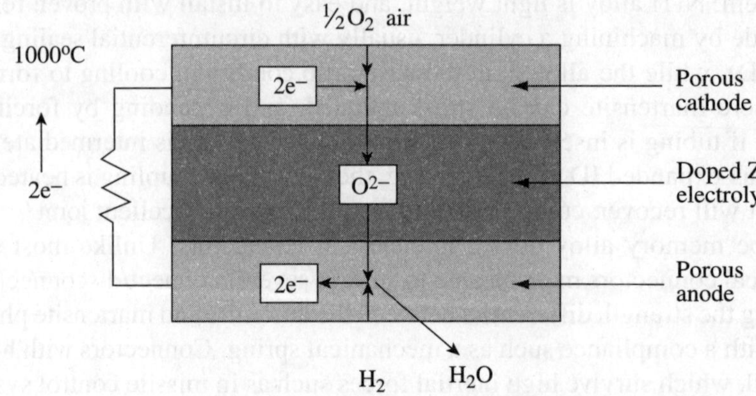

Cathode: $\frac{1}{2}O_2 + 2e^- \rightarrow O^{2-}$

Anode: $H_2 + O \rightarrow H_2O + 2e^-$

Cell: $H_2 + \frac{1}{2}O_2 \rightarrow H_2O$

$(CO + \frac{1}{2}O_2 \rightarrow CO_2)$

Fig. 12.59. The principles of operation of a high temperature solid oxide electrolyte fuel cell.

used as anode. No doubt this material is good for hydrogen, but numerous problems arise, when operating the cell in pure or reformed hydrocarbon fuels. Nickel has a high activity for coke formation, which decreases the electro-catalytic activity. The solution for this lies to introduce water in the fuel, which reduces the equilibrium potential and thus the maximum voltage of the cell, leading to a reduced power output. Sulphur in fossil fuels leads to a drastic fall in the anode performance due to catalytic deactivation. The alternate anode materials are Gd-doped ceria, Ti and Y-doped zirconia and lanthanum chromite.

At the cathode of the fuel cell, the gaseous oxygen is reduced and the resulting oxygen ions are incorporated into the electrolyte

$$O_2 + 4e^- + 2V_0'' \rightarrow 2O_0^x \qquad (12.21)$$

As the cathode environment is highly oxygen, the use of metals as electrode material is out of question. The common cathode material is electronically conducting perovskite ceramic lanthanum magnetite $LaMnO_3$, in which La is partially replaced by Sr. This material is known as strontium doped lanthanum magnetite $(La,Sr)MnO_3$ and abbreviated as LSM. These perovskites are p-type electronic conductors, where the electrical properties depend on La/Sr ratio. Like the anode, this material is also prepared in the form of composite with YSZ as the second phase ceramic. In order to obtain optimized current collection in the cathode, multiple layer electrodes are used. The cost of the cathode material can be reduced by replacing mixture of lanthanides for pure lanthanum in the ceramic.

The fabrication of SOFC components is similar to the ceramic fabrication techniques, described in Chapter 8. The wet ceramic processing or gas phase preparation routes are applied.

In the latter case, a thin layer below 1 μm is possible. As the electrodes are to be porous, sintering route is the only successful processing method.

SOFC assembly: For an engineer, a brief description of SOFC assembly, is worth while. There are other parts to be tackled. These are interconnection and sealing. In order to connect multiple cells, an electronic conductor is required, which are exposed to both highly reducing and oxidizing conditions at the electrodes. Another requirement is the dimensional stability and matching of the thermal expansion coefficient with other components i.e., the electrodes and the electrolyte. The selection of the interconnection material depends on the cell operating temperature. At temperatures above 850°C, Cr-alloys or $LaCrO_3$ ceramic interconnects are used. Both of these have limitations. The chromium alloy gives rise to CrO_3 which can be prevented by $La(Sr)MnO_3$ or $La(Sr)CoO_3$ coating at the air side. The ceramic interconnect, on the other hand, undergoes expansion after reduction, which leads to bending and loss of electrical contact to the electrodes. At a relatively lower operating temperature (up to 800°C), ferritic steel can be used as interconnects. But in this case too, under oxidizing condition, the need of protective layer at the cathode side arises.

The sealing material's choice is dictated by the operating temperature of the fuel cell and the area that has to be sealed. The sealing must be gas tight, electrically insulator and thermally and chemically compatible with the adjacent components. Glasses are used, but it should be seen that they do not undergo recrystallization at elevated temperature. Otherwise embrittlement takes place. Trace elements like Na, K, Ca and Ba in glass increases the propencity of its recrystallization.

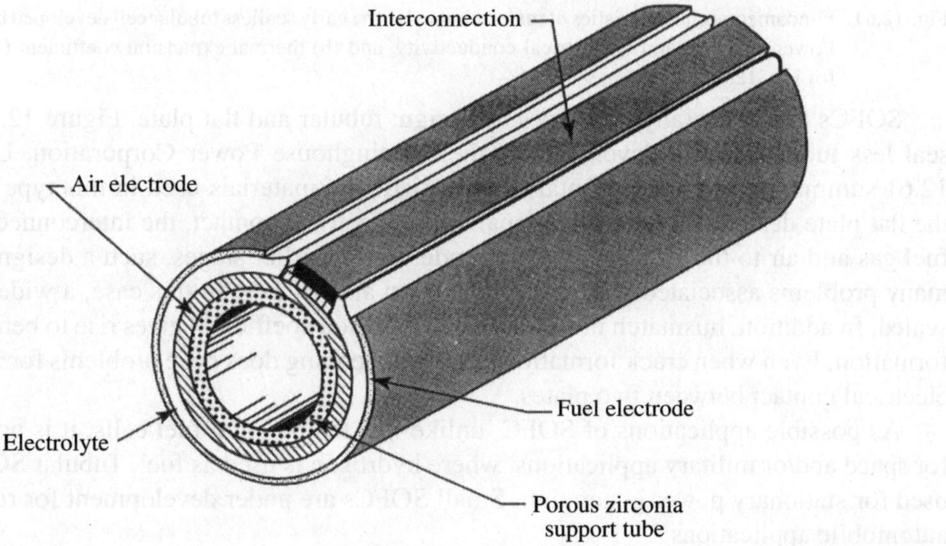

Fig. 12.60. Sealless tubular design developed by the Westinghouse Power Corporation (Reproduced with permission from Handbook of Fuel Cells, Vol. 1, Ed. W. Vielstich, A. Lamn and H.A. Gasteigen, John Wiley & Sons Ltd, Bognor Regis, West Sussex, U.K., 2002, p. 227).

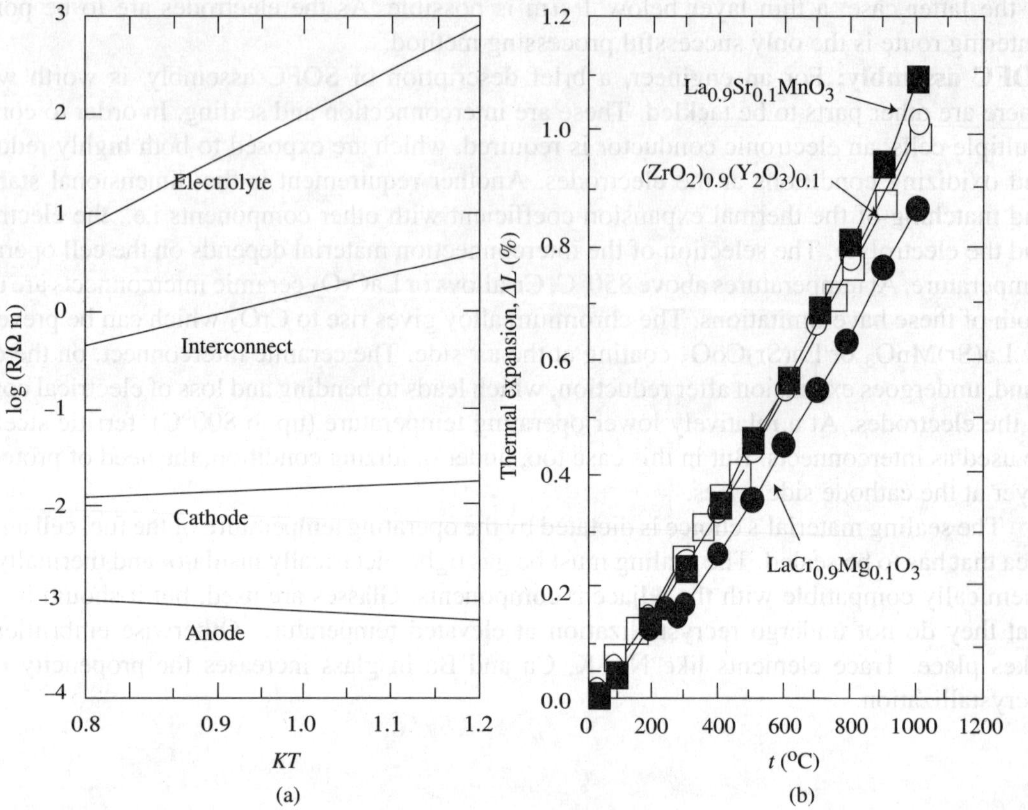

Fig. 12.61. Fundamental characteristics of materials used in the early sealless tubular cell developed by Westinghouse Power Corporation (a) electrical conductivity, and (b) thermal expansion coefficient (Source: same as for Fig. 12.60).

SOFCs have generally two types of design: tubular and flat plate. Figure 12.60 shows the seal less tubular design developed by the Westinghouse Power Corporation, U.S.A. Figure 12.61 summarizes the fundamental characteristics of materials used in this type of design. In the flat plate design (Figure 12.62), apart from electrical contact, the interconnection supplies fuel gas and air to the anode and the cathode. In the earlier stages, such a design encountered many problems associated with crack formation and sealing. In this case, a wide area is to be sealed. In addition, mismatch in the thermal expansion coefficient gives rise to bending or crack formation. Even when crack formation is solved, bending does pose problems for side seals and electrical contact between two plates.

As possible applications of SOFC, unlike low temperature fuel cells, it is not appropriate for space and/or military applications, where hydrogen is used as fuel. Tubular SOFC has been used for stationary power generation. Small SOFCs are under development for residential and automobile applications.

Case Study 12.44: Fiber-Optic Cables: In the Section 4.51, we have already described the optical aspects of such fibres. Both glass and polymer has been used for such applications.

Glass optical fibres are nearly always vitreous silica doped with small amounts of germanium oxide (GeO_2), phosphorus oxide (P_2O_5), or fluorine, but can be mixtures of oxides, sulfides,

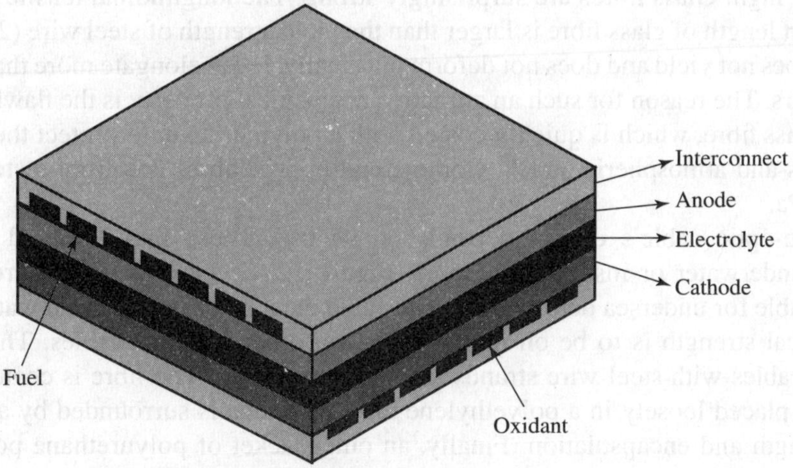

Fig. 12.62. Concept of the flat plate design for planar cells.

and selenides that create a random covalent network. The largest class of optical fibres is made from oxide glasses, in particular silica (SiO_2), with a refractive index of 1.458 at 850 nm. The refractive index of a material may be changed by controllably adding impurities or dopants. Such dopants include fluorine and metal oxides. Inclusion of GeO_2 or P_2O_5 increases the index, whereas the inclusion of fluorine or B_2O_3 decreases the refractive index.

Silica's properties make it a formidable candidate for optical fibre due to its low thermal expansion, high transparency, and high resistance to thermal shock. However, the high melting point needed for its fabrication from the molten form gives rise to the need to explore alternative low melting point silicates. These materials include soda-lime silicates, borosilicates and germanosilicates. Soda-lime silicate glasses are combinations of silica, alkaline oxides (such as Na_2O or LiO_2) and other oxides (MgO, ZnO or BaO). As a result, these materials are easy to melt and fabricate into fibre.

In some cases it is necessary to boost the signal periodically due to attenuation and losses. The signal can be amplified by doping the fibre with erbium. The EDFA (erbium-doped fibre amplifier) has proven to be an enabling technology for optical communication. The promise of a low-cost alternative to the EDFA is the EDWA (erbium-doped waveguide amplifier). An advantage of EDWAs is the ability to be integrated with other optical components due to the fabrication of the waveguide on a silicon optical bench.

For short data transmission ranges (several hundred meters) where high losses are tolerable, plastic-clad fibres are an inexpensive alternative.

The fibres are composed of silica cores with lower refracting plastic cladding made of silicone resin (with an index of 1.405 at 850 nm) or perfluoronated ethylene propylene (Teflon) (with a refractive index of 1.338). Only used for step indexed fibres, the core diameters are larger (150-600 μm) than conventional all-glass cores and cladding systems and permit low cost interfaces with light sources and detectors.

For short distance (maximum 100 m) an all plastic multimode step-index alternative may be used. As an alternative for short haul application plastic optic-fibre has been developed, but costs are still high. Glass fibres are surprisingly strong. The longitudinal tensile strengths of 5 GPa in a short length of glass fibre is larger than the yield strength of steel wire (2 GPa). Unlike steel, glass does not yield and does not deform plastically. It can elongate more than 10% before fracture occurs. The reason for such an attractive mechanical property is the flawless surface of the drawn glass fibre, which is quickly coated with a polymer layer to protect the surface from any scratches and atmospheric attack. Commercial optical fibres are proof tested at levels of 0.09-1.38 GPa.

The fibre-optic cable's design depends on its use. These may be used underground, submerged underwater or installed outdoors. Figure 12.63 shows a typical cross section of optic-fibre cable for undersea deployment. The cable must be protected from water vapour and the mechanical strength is to be on higher side than other types of cables. This is achieved by forming cables with steel wire strands and polymer jacket. The fibre is coated with buffer material and placed loosely in a polyethylene tube. The tube is surrounded by aramid yarn to provide strength and encapsulation. Finally, an outer jacket of polyurethane polyethylene or nylon binds the encapsulated fibres together.

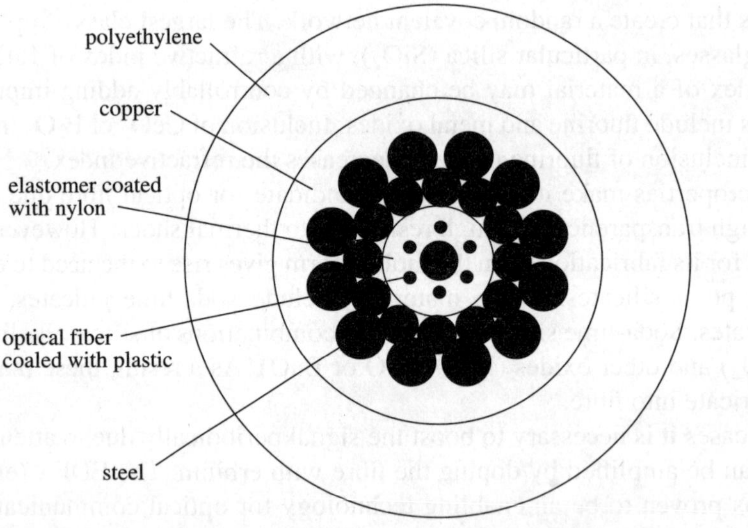

Fig. 12.63. Cross-section of optical fiber cable for undersea deployment.

Example 12.35. If two semiconductor A and B are mixed ideally via an appropriate growth techniques, describe the properties under (a) crystal structure and (b) band structure.

Answer.

(a) Crystal structure: Generally the two semiconductors have similar crystal structure and therefore the lattice constant of the alloy obeys Vegard's law i.e.,

$$a_{\text{alloy}} = xa_{\text{A}} + (1-x)a_{\text{B}}$$

(b) Band gap: The band structure of alloy is difficult to calculate in principle, since alloys are not perfect crystals even if they have a perfect lattice. A simple approach used for this called the virtual crystal approximation according to which the band structure of the alloy A_xB_{1-y} is simply the weighted band structure of the individual band structures of A and B. Thus the band structure is given by

$$E_{\text{alloy}}(k) = xE_{\text{A}}(k) + (1 - x)E_{\text{B}}(k)$$

The above two relations hold true if the alloy is a random alloy. It is essential that an alloy not be phase separated, otherwise the material is not very useful for devices and cannot be used in any reliable device process.

Example 12.36. Could you think of prospective developments in semiconductor materials for next decade?

Answer. In the next ten years, wide-gap semiconductors are intended to be developed, for example the III-V nitrides, GaN, AlN for applications in optoelectronics (blue-violet laser), the wide-gap II-VI (and to a lesser extent ZnSe), and "very large gap" materials and materials with exceptional thermal properties such as BN, diamond C, and SiC. Ternary and quaternary compounds are likely to possess interesting properties, such as the effects of the variation of the magnitude of the forbidden band, which in a given system can go through a minimum as a function of the composition, the so-called "bowing" effect (e.g., GaN-GaAs).

Example 12.37. What are the major features of materials to encapsulate the IC package? Grade the following classes of polymers in a Tabular form against the following properties: (a) dielectric strength, (b) elastic modulus, (c) tensile strength, (d) precursor viscosity, (e) adhesion to package and (f) moisture diffusion rate: (i) epoxy resins, (ii) silicones, (ii) polyurethane, (iv) polysulphides.

Answer. The material used to encapsulate the package should:

- Be electrically insulating
- Be easily molded to the desired shape around the chip die and its wire leads
- Be highly impervious to the penetration of moisture and contaminants
- Be able to form strong adhesive bonds with the chip surface, wires and other leadframe components
- Exhibit mechanical and chemical stability for the expected lifetime of the package
- Not require exposure to excessively high temperatures during installation
- Have a coefficient of thermal expansion similar to those of other package components to avoid thermal stresses capable of fracturing the wire leads.

The various groups of polymers are characterized as in Table Ex. 12.37.

Example 12.38. A spark ignitor has piezoelectric ceramic block (3 mm square x 1 mm thick) in the device. The electrodes are on its square faces. If a load of 20 N is applied to the face by

Table Ex. 12.37.

	Epoxy Resins	*Silicones*	*Polyurethanes*	*Polysulphides*
Dielectric strength	Good	Good	Good	Good
Elastic modulus	High	Low	Wide range	Low
Tensile strength	High	Low	Wide range	-
Precursor viscosity	Low	Low	Low	High
Adhesion to package	Excellent	Poor (to ceramics)	Good	Good
Moisture diffusion rate	High	High	Low	Very low

thumb pressure, establish whether the block can produce enough voltage to make a 1 mm spark in air. Given following data for ceramic:

$$\text{Break down voltage of ceramic} = 30 \text{ kV cm}^{-1}$$
$$\text{Young's modulus} = 70 \text{ GN m}^{-2}$$
$$\text{Generator coefficient} = 30 \text{ C m}^{-2}$$
$$\varepsilon_r = 70$$
$$\text{Permittivity in air } \varepsilon_0 = 9 \times 10^{-12} \text{ F m}^{-1}$$

Solution. The generator coefficient 'g' is given by

$$g = \frac{EA\Delta P}{F}$$

$$\text{or} \quad \frac{gF}{AE} = \Delta P$$

where A = Area

E = Young's modulus

ΔP = Change in polarization

F = Force

The change in polarization is related to the voltage (V) generated by the capacitance (C). Therefore

$$A\Delta P = C\Delta V$$

$$\text{By equation (4.18)} \quad = \frac{\varepsilon_r \varepsilon_0 A}{\ell}\Delta V$$

(ℓ = separation between electrodes).

$$\text{Therefore} \quad \frac{gF}{AE} = \frac{\varepsilon_r \varepsilon_0}{\ell}\Delta V$$

$$\text{Here} \quad \Delta V = \frac{gF\ell}{\varepsilon_1 \varepsilon_0 EA}$$

$$= \frac{30 \times 20 \times 10^{-3}}{70\varepsilon_0 \times 70 \times 10^9 \times 9 \times 10^{-6}} \text{ V}$$

$$= 1.5 \text{ kV}$$

This is enough for 1 mm spark since the break down voltage for this ceramic is 30 kV cm^{-1}.

Example 12.39. Compare and contrast between metallic shape memory alloy and ceramic anti-ferroelectric shape memory devices under the following heads (i) Driving power, (ii) strain range, (iii) generative force, (iv) response speed and (v) durability. Which type of material has more merits?

Table Ex. 12.39.

Properties	*Antiferroelectric*	*Shape memory alloy*
Driving power	Voltage (mW ~W)	Heat (W ~kW)
Strain ((ΔL/L)	10^{-3} ~10^{-2}	10^{-2} ~10^{-1}
Generative force	100 MPa	1000 MPa
Response speed	msec	sec ~min
Durability	> 10^6 cycles	10^4 cycles

Answer. Table Ex. 12.39 gives the comparison. Outstanding merits of the ceramics over the alloys are:

- quick response,
- good controllability by electric field to memorize and recover the shape without generating heat,
- low energy consumption as low as 1/100 of the alloy and
- wide space is not required to obtain the initial shape deformation.

Example 12.40. Propose a complex set up for smart structure involving various types of functional materials.

Answer. The smart materials and structural systems are based on the concept to incorporate sensors, actuators, etc. in the materials themselves and to integrate materials with the structures, thereby imparting such functions as to (1) self-evaluate the occurrence and deterioration of strains and damages, (2) self-control vibrations and noises, (3) self-change their own shape depending on conditions and (4) self-control the propagation of damages and repair them as the need arises. Figure Ex. 12.40 is a conceptual diagram of the smart materials and structural systems. While the mother structure supports strength, the optical fibre sensor and piezoelectric ceramic detect anomalies (such as defects) and the shape-memory alloy and piezoelectric actuator repair the anomalies. The electronic control circuit judges the signals detected by the sensor, and issues instructions.

Example 12.41. Ni-Ti and Cu-Al-Zn alloys used for commercial SMA's (Shape Memory Alloy) have temperature limitation of up to ~100°C. Suggest the development strategy for high temperature use SMA for jet engines, for example to switch or modulate fluid films of various sorts as the temperature changes.

Answer. Shape memory alloys use displacive transformation. Platinum noble metal is known to have excellent oxidation resistance, thus making it useful as SMAs. Now the question is to screen its based alloys, which undergo martensitic transformation. A cursory look suggests Pt-Fe but, it undergoes transformation below room temperature Pt-Cu, Pt-Mn show twins, but results are not significantly reproducible. Pt-Cr and Pt-V both show twinning, but the alloy

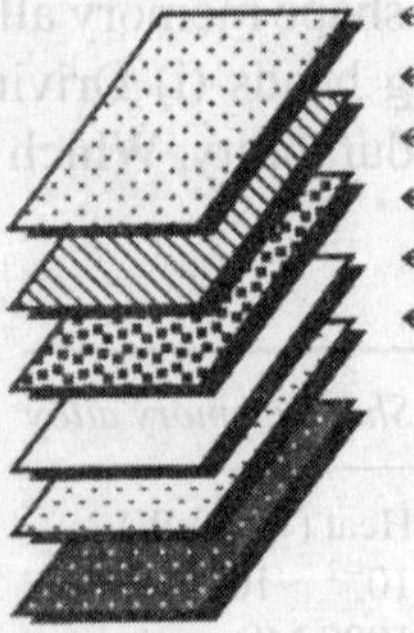

Figure Ex. 12.40

cracks even at the initial stage of rolling. This leaves Pt-Al and Pt-Ti as candidate systems. Pt-Al system at around 75 at. % Pt has a phase transformation which occurs from 132-337°C, depending on composition. This system is corrosion resistant and capable of withstanding elevated temperature. It has excellent oxidation resistance. However the transformation is exceedingly sensitive to compositional changes which makes it commercially unattractive. Pt-Ti system at around 50 at. % Pt has a displacive transformation from BCC to orthorhombic at ~1000, which varies slightly with composition. This alloy is cold and hot workable.

The ternary addition of nickel and ruthenium decreases the transformation temperature to ~600°C and 700°C respectively. The addition also increases the hardness. There is however still one shortcoming to Pt-Ti based alloys and this is tendency for oxygen contamination and associated internal oxidation. More work is being done in this direction.

(For more details refer 'Platinum Alloys for Shape Memory Applications', T. Biggs, M.B. Cortie, M.J. Witcomb and L.A. Cornish, Platinum Metals Review, Vol. 47, No. 4, 2003, p. 142-156).

12.8 MATERIAL PROCESSING SELECTION

Material processing involves forming, joining and surface engineering. These have been described in earlier chapters. Cost estimation can be done in either ways i.e., activity based costing or resource based costing. In the first case the cost of each unit operation involved is taken into account. This method is satisfactory only when the processing method is well established. In the second method the resources of materials, energy, capital, time and information associated with manufacture of the component are taken into consideration. Ashby has done elaborate study in this aspect (refer Materials and Design, Vol. 24, No. 8, 2003, 605-616).

Figure 12.64 shows the cost per unit with respect to the production volume of a particular component for three forming methods. At small batch sizes the unit cost is determined by the 'fixed' cost of tooling. As the batch size increases, the contribution of this to the unit cost falls, provided the tooling has a life greater than the batch production. The curves get flattened and is dominated by the 'variable' cost of material, labour and other overheads. Competing processes make the curves to intersect. Economic batch size for a particular process is one, where it is definitely less expensive than competing processes. As an illustration, thermoforming at low batch sizes, is much less expensive than injection moulding, but at batch size of approximately 1000 injection moulding becomes the cheaper process.

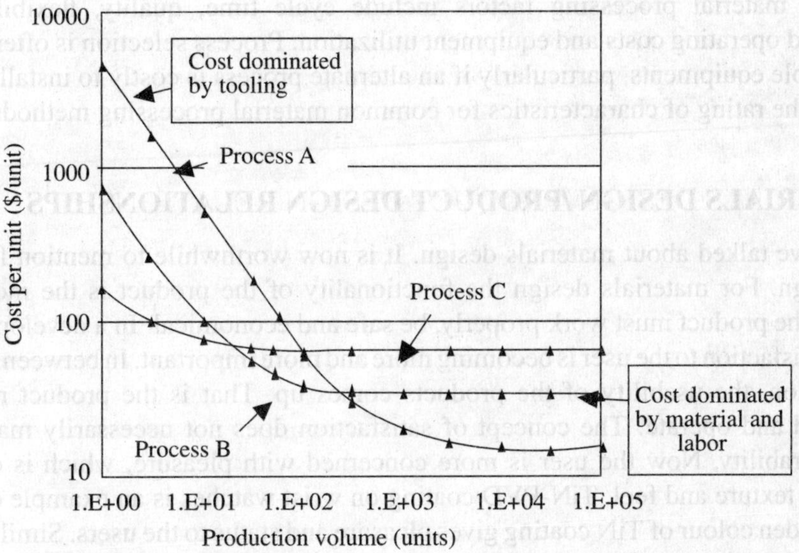

Fig. 12.64. A schematic showing the dependence of unit cost on the number of units produced, for three processes A, B and C (Reproduced with permission A.M.K. Esawi and M.F. Ashby, Materials and Design, Vol. 24, No. 8, 2003, p. 608, Elsevier, Oxford).

The competition between processing method is extremely active. There are only short-term equilibrium as technologies continually develop and improve in the manufacturing sector. The introduction of new processing methods and rejection of old methods is brisk and occurs even more rapidly for advanced technologies such as electronics. The choice of material for a function is never constant. Cast iron connecting rods have been replaced by forged steel and then back to cast ductile iron and back to forged steel and now on to sinter-forged powder metallurgy products. Door handles have changed from cast metal to filled injection moulded polymer and back to cast metal in the same model year of the car. One time zinc phosphate coated steel was used for hydraulic fittings. A drop in the copper price led in a change to a brass alloy. Similarly, the competition in investment casting and metal-injection moulding (MIM) is worth watching. MIM has certain advantages for the smallest parts and parts made from costly materials. The cost of investment costing feed stock is still much larger than MIM grade powder. Methods to increase the material efficiency for investment casting have been developed and are in use. Improvements in casting practice have resulted in lower material losses and high quality castings. The competition for some applications is very close and minor changes from within either MIM or investment casting can change the economics enough to cause the customer to change source.

Another feature to be taken into account in the economy of a particular society. The translation of processing method followed by a developed country is not necessarily going to be fruitful for a developing country at a particular moment. Lower labour cost in the third world nations would make the adoption of robotics and automation uncompetitive.

It must not be forgotten that the selection of processing method cannot be done in isolation, but has to be interlinked with the chemistry and structure of the material.

The key material processing factors include cycle time, quality, flexibility, material utilization and operating costs and equipment utilization. Process selection is often constrained by the available equipments, particularly if an alternate process is costly to install. Table 12.24 summarizes the rating of characteristics for common material processing methods.

12.9 MATERIALS DESIGN /PRODUCT DESIGN RELATIONSHIPS

So far we have talked about materials design. It is now worthwhile to mention few issues on product design. For materials design the functionality of the product is the most important feature, i.e., the product must work properly, be safe and economical. In a developed economy, the idea of satisfaction to the user is becoming more and more important. In between functionality and satisfaction, the usability of the products comes up. That is the product must be easy to understand and operate. The concept of satisfaction does not necessarily matter with the consumer durability. Now the user is more concerned with pleasure, which is derived from form, colour, texture and feel. TiN-PVD coating on wrist watches is an example of aesthetics, where the golden colour of TiN coating gives pleasure and status to the users. Similarly different items in household uses are like lampshades, plastics giving the feeling of timber. Elegant frames of bike, where untidy visualization of welded joints is absent is another example. However, a caution is needed. Any arbitrary product configuration is to be justified with adequate materials design. A configuration may appear pleasing, but may be inconvenient or cumbersome in use. One shape of cup of a specific material may be more convenient to handle than the other. Much use of adhesives, should be correlated with its durability. In a developing society, the robustness (not over design), of a product is more attractive than its personality. In the recent book Professor Ashby has done in depth study on the relationship between technical and industrial design (M.F. Ashby and K. Johnson, Materials and Design - The Art and Science of Materials Selection in Product Design, Butterworth Heinemann, Oxford, 2002). Figure 12.65 adequately summarizes the philosophy built in their treatment.

12.10 CAD/CAM SYSTEMS

It is important to understand few rudimentary facts on CAD/CAM systems, which has extensive relevance in material design. CAD and CAM are computer-aided design and computer-aided manufacturing. The stages in the design process and the possible uses of a computer are illustrated in Table 12.25. It is worth noting that during the early stages of design, the creativity of the designer is more important, than the use of computer. CAD includes all aspects of the design process including drafting work. Some of the main types of software packages available for CAD systems are

- Interactive computer graphics
- Engineering analysis of a design problem
- Software to handle, store, retrieve and present the large amount of data involved in design
- For final design drawings and instructions for numerically controlled (NC) machines.

Table 12.24. Rating of Characteristics for Common Manufacturing Processes.

Process	*Shape*	*Cycle time*	*Flexibility*	*Material utilization*	*Quality*	*Equipment Loading costs*
Casting						
Sand casting	3D	2	5	2	2	1
Evaporative form	3D	1	5	2	2	4
Investment casting	3D	2	4	4	2	3
Permanent mold casting	3D	4	2	2	3	2
Pessure die casting	3D solid	5	1	4	2	1
Squeeze casting	3D	3	1	5	4	1
Centrifugal casting	3D hollow	2	3	5	3	3
Injection molding	3D	4	1	4	3	1
Reaction injection molding (RIM)	3D	3	2	4	2	2
Compression molding	3D	3	4	4	2	3
Rotational molding	3D hollow	2	4	5	2	4
Monomer casting, contact molding	3D	1	4	4	2	4
Forming						
Forging, open die	3D solid	2	4	3	2	2
Forging, hot closed die	3D solid	4	1	3	3	2
Sheet metal forming	3D	3	1	3	4	1
Rolling	2D	5	3	4	3	2
Extrusion	2D	5	3	4	3	2
Superplastic forming	3D	1	1	5	4	1
Thermoforming	3D	3	2	3	2	3
Blow molding	3D hollow	4	2	4	4	2
Pressing and sintering	3D solid	2	2	5	2	2
Isostatic pressing	3D	1	3	5	2	1
Slip casting	3D	1	5	5	2	4
Machining						
Single-point cutting	3D	2	5	1	5	5
Multiple-point cutting	3D	3	5	1	5	4
Grinding	3D	2	5	1	5	4
Electrical discharge machining	3D	1	4	1	5	1
Joining						
Fusion welding	All	2	5	5	2	4
Brazing/soldering	All	2	5	5	3	4
Adhesive bonding	All	2	5	5	3	5
Fasteners	3D	4	5	4	4	5
Surface treatment						
Shot peening	All	2	5	5	4	5
Surface hardening	All	2	4	5	4	4
CVD/PVD	All	1	5	5	4	3

Rating scheme: 1 – poorest; 5 – best ratings.

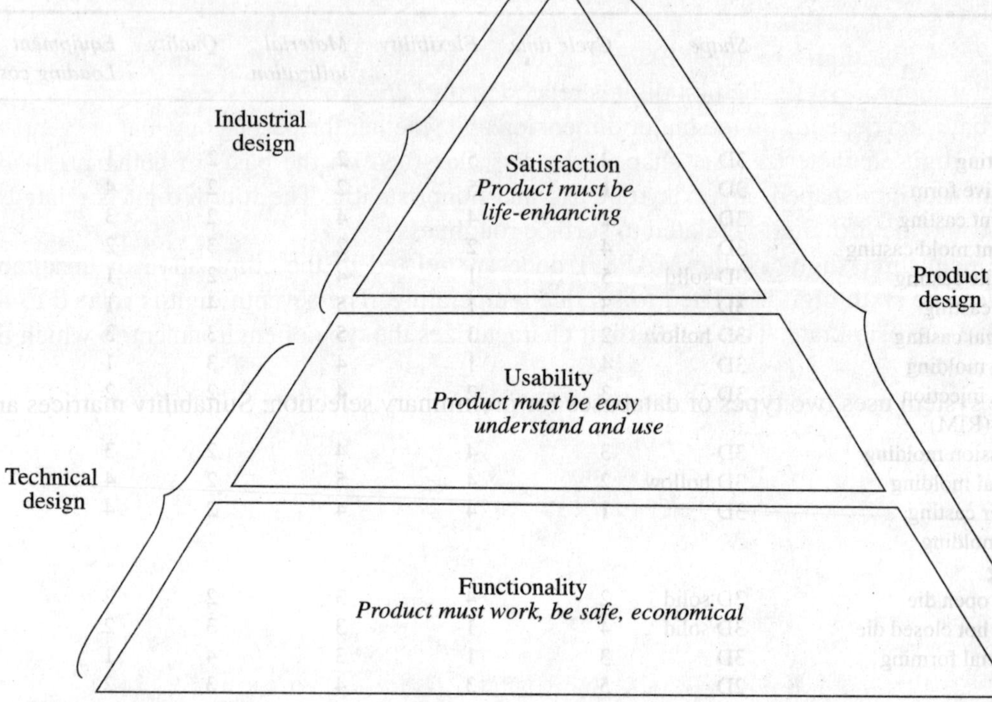

Fig. 12.65. The pyramid structure for product design (Reproduced with permission M.F. Ashby and K. Johnson, Materials Today, Vol.6,No 12,. 2003, pp. 25-35, Elsevier , Oxford).

Table 12.25. Potential for Computer Utilization in a Design Study.

Design stage	*Computer potential*
Recognition of a need and problem definition	Desirable, but few applications
Feasibility study	Limited application
Preliminary design	Certain specific applications and use of specially prepared programs or packages
Testing, evaluating and improving the design	Yes, preparation of standard specification and drafting
Final design and documentation	Yes, preparation of standard specification and drafting
Communicating the design	Computer graphics
Manufacture	CAM and NC
Feedback	Storage on files of subsequent results and Report

APT (automatically programmed tools) was very early language developed at MIT, U.S.A. Another version ADAPT (Air Material Command Developed APT) is a simplified version of the APT language and can run on a much smaller computer. UNIAPT was specifically developed to be run on minicomputers.

CAD is very useful for initial screening of materials and processes. Dargie proposed MAPS 1 system. It is worthwhile to give a brief description of the system.

The first five digits of the MAPS 1 code are related to the elimination of unsuitable manufacturing processes. The first digit is related to the batch size. The second digit characterizes the bulk and depends on the major dimension and whether the part is long, flat or compact. The third digit characterizes the shape, which is classified on the basis of being prismatic, axisymmetric, cup shaped, nonaxisymmetric and nonprismatic. The fourth digit is related to tolerance and the fifth digit is related to surface roughness.

The next three digits of the MAPS 1 code are related to the elimination of unsuitable materials. The sixth digit is related to service temperature. The seventh digit is related to the acceptable corrosion rate. The eighth digit characterizes the type of environment to which the part is exposed.

The system uses two types of databases for preliminary selection: Suitability matrices and compatibility matrix. The suitability matrices deal with the suitability of processes and materials for the part under consideration. Each of the code digits has a matrix. The columns of the matrix correspond to the value of the digit and the rows correspond to the processes and materials in the database. The elements of the matrix are either 0 indicating unsuitability or 2 indicating suitability.

The compatibility matrix expresses the compatibility of the different combinations of processes and materials. The columns of the matrix correspond to the materials while the rows correspond to the processes. The elements of the matrix are either 0 for incompatible combinations, 1 for difficult or unusual combinations, or 2 for combinations used in usual practice.

Based on the part code, the program generates a list of candidate combinations of materials and processes to produce it. This list helps the designer to identify possible alternatives early in the design process and to design for ease of manufacture.

The present trend is to have CAD/CAM system, which establish a direct link, through electronic channeling between the product design and manufacturing departments. This eliminates the duplication of efforts by designing and manufacturing personnel. CAD/CAM systems are particularly advantageous, when the shop includes different types of NC machine tools.

Presently the CAM has a much wider role to play. Some are:

- Computerized cost estimation
- Computer aided process planning
- Computerized machinability data systems
- Computer-aided monitoring and control of manufacturing process.

In the last variant, the real-time optimization of a performance index is obtained through an adaptive control system. As an example, the adaptive control system in case of machining operation, performs its function by detecting any variability in the condition of the work piece being machined and adjusts the feed and the cutting speed to account for that variability and to maximize the performance index. The variability can take different forms, such as hard spots, which require reduction in feed and cutting speeds, or the presence of air gaps in the work piece, where the feed should be increased to minimize the idle time during which the tool travels across the air gap. Adaptive control has found application in chipless manufacturing processes too like sheet metal working or welding.

A truly integrated CAD/CAM system is also known as CIM (computer-integrated manufacturing). It is a close-looped feedback system, whose prime inputs are product requirements and product concepts and whose prime outputs are finished products. CIM comprises a combination of software and hardware e.g., product design, production planning, production control, production equipment and production processes.

After CAD, the important job of an engineer is to generate a first full-scale model of a product known as prototype. Prototypes are of two types: physical and virtual. Some aspects of physical prototypes are already described in Chapter 8. Virtual prototype is totally software form of prototyping, which uses advanced graphics and virtual-reality environments to allow designers to examine a part. 3D-graphics has allowed views to change the view of a part on a computer screen. One of the examples of such type of prototyping is in case of Boeing 777 aircraft, where mechanical fits and interfaces were successfully evaluated virtually. However, a caution is made. One must not neglect the physical prototype, as an evaluation, since it is more fault free than in case of virtual prototyping.

Expert Systems **:** Expert systems also called knowledge-based systems, are computer programs that simulate the reasoning of a human expert in a given field of knowledge. Expert systems are different from conventional computer programs, in that they use heuristics, or rules of thumb, to extract information from a large knowledge base. Expert systems typically consist of three major components i.e., knowledge base, inference engine and user interface. MATSELEX is a rule-based expert system for materials and process selection at the conceptual design stage. The materials selection process starts with a whole range of materials in the database. A range of sift criteria are then used to define the requirements and to direct the search. The number of materials that survive the shifting process depends on the severity of the criteria used.

QUESTIONS AND PROBLEMS

Structural Applications

12.1 Engineers who previously designed riveted ships paid little attention to transition temperatures, unlike those who currently design welded ships. Why?

12.2 Examine the following materials for beverage containers under the following headings (A) Limitations; (B) Functional adaptations: (i) steel, (ii) paper/cardboard, (iii) plastics, (iv) glass, (v) aluminium.

12.3 Nuclear transportation flasks can be made from a steel either by welding of a rolled plate or one piece cube forging. Can the steel grade be same for either? Justify your answer.

12.4 What do you understand by tangent modulus and secant modulus in a structural material?

12.5 Give a brief account of failure of a structure due to instability.

12.6 Aluminium clad sheets of 2024-T3 and 7075-T6 aluminium alloys are very common in aircrafts. Describe the role of primary and secondary modulus of elasticity in a clad aluminium alloy on fatigue response.

12.7 Cast iron pipes are used extensively for water supply. High corrosion rate causes leakage of water. In addition high down time for maintenance is incurred involving additional cost. If a 13 mm thick C.I. pipe is to be replaced by stainless steel (high strength to weight ratio) pipe, what thickness of the pipe would you design? Select the grade of stainless steel.

12.8 LPG cylinders of 26.5 litre capacity for cooking purposes are made out of 2.7 mm thick carbon steel plates. To make the cylinders lighter as well as dutiful, the P.C. steel is to be substituted by stainless steel flats. Estimate the thickness of the stainless steel and the weight saving achieved thereby. Which grade of stainless steel would you select?

12.9 A gear had a hardness of 50 R_c at its centre when it was made of an SAE 4068 steel. What hardness would you expect the same gear to have if it were made out of an SAE 4140 steel?

12.10 One evolution in the design of automobile engines has been the change from in-line long-stroke engines to compact four- and six-cylinder engines. As a result, the crankshaft material has been changed from quenched and tempered steel forgings to cast crankshafts made from pearlitic malleable cast iron or nodular iron. Discuss this change in materials and processing in terms of the service performance and the properties of the materials.

12.11 Why high nickel alloys are used in the chemical and allied industries?

12.12 Automobile valve springs permit intake and exhaust valves to alternately open and close as the engine is in operation. The temperature of the spring during engine operation is around 80°C. The spring is made of low alloy steel. Could you identify the steel composition? What types of failure may be possible? Is the stress cycle completely reversed? Discuss its remedy from the point of compositional change and of redesign.

12.13 The automobile industries are using bake hardening (BH) steels where hardness levels are achieved during the paint baking or curing process to offset the reduction in formability that is associated with alloy additions and the processing of steels. The yield strength requirement is in the range 210-255 N/mm^2. Suggest the grades of such steels. Will there be much difference in welding response of BH and non-BH grades of steels?

12.14 Mild steel twisted bars are common reinforcement in building industry. As low alloy steels have greater strength than the mild steel interest arose to use it as reinforcement since the cross-section used shall be lower. Corrosion resistance of alloy steel was found to be better than the mild steel. When an over enthusiastic builder wanted to use the alloy steel bar in cold twisted condition, he met disappointment. Why was it so?

12.15 Name some engineering components from the automotive and aerospace industries where the following fatigue conditions are applicable:
(a) classical fatigue
(b) wear, rolling contact and classical fatigue
(c) impact and fatigue
(d) creep and fatigue.

12.16 To avoid corrosion problems aluminium alloys are generally claded with pure aluminium. Can you think of similar thing in case of steel claded by stainless steel? What are the merits and demerits of the latter combination?

12.17 The following three materials are being considered for making the outer body of a freshwater valve in a power generation plant:

Material	Al 770 Aluminium Alloy	Bronze ASTM B22	AISI 302 stainless steel
Yield strength (MPa)	173	168	280
Wear resistance	2	5	5
Fatigue strength (MPa)	150	120	315
Corrosion resistance	2	4	4
Processability	3	5	3
Cost relative to aluminium	1	2	3

Select the optimum material for use in making the valve body.

12.18 Copper has excellent electrical conductivity and ductility, but rather lower strength. Purposely we do not prefer to strengthen it by alloying in order to maintain the conductivity of pure copper. Describe the strategy you will employ to enhance its strength manifold, without the loss of ductility. Such a material will be useful in microelectromechanical systems (MEMs).

12.19 Write an essay on the theme of material selection which combines the technical 'best' at the lowest cost, with the minimum of problems due to outside influences. Select the development of a lead based alloy for applications as radiation shields in nuclear industry. Lead does pose environmental problems. What alternate material will you select to avoid this problem?

12.20 Copper possesses excellent electrical conductivity, but considered relatively expensive and scarce material. Obtain statistical data for the last 10 or 20 years in your country, showing the fluctuations in the price, production and usage of 'pure copper'. Discuss why these variations have occurred and what trends can be predicted for the next five years.

12.21 Nickel is an important constituent in stainless steel. Because of its scarcity, attempts in past had been made in substitution by manganese and nitrogen. This attempt was discarded due to some technical problems. If your national planners are urging you to relook their old decision, describe the steps to be taken in making a team work to solve the problems.

12.22 You are asked to develop a fully 'green' polymer matrix composite, which is fully biodegradable by combining natural fibres with biodegradable resins. How shall you begin the selection of candidate materials and how shall you tailor the properties?

12.23 Aluminium killed drawing quality steel sheets cost 5-10% more than rimming steels. One stamping plant saved millions of dollars annually by lowering the specification on certain parts from aluminium killed to rimmed steel:

(a) If you were in a stamping plant, how would you select the parts for similar lowering of specifications?
(b) Would substituting of rimmed for killed steel result in an inferior product? Explain.

12.24 In developed country motor cycles are usually used for racing and sporting purposes, while in developing country they provide the means of personal transportation. From materials engineer's viewpoint suggest the selection of materials in both the categories. Justify your answer.

12.25 For light weight cars, aluminium alloy sheets are common. If we wish to develop a still stiffer structure with built in insulation against noise and vibration, describe the schematic approach for the material selection and its processing. The main goal must be the best combination of properties.

12.26 The frames of bicycles are usually made of Cr-Mo 4340 steel tubes of 25 mm diameter and 1.25 mm wall thickness. There was a need to increase the stiffness and also making the bicycle lighter. You are free to do any change in the dimension of the frame tube. Could you replace the steel frame by a 7075 Al alloy frame and if so how much gain in stiffness is obtained? Given

	σ_c(MPa)	Density (kg/m^3)	E (GPa)	G(GPa)
7075Al	500	2700	70	27
4340 steel	1350	7800	210	83

12.27 Show schematically how the curvature of the mild steel surface being carburized influences the case depth.

12.28 How will you start a design for selecting material for pressure vessels, which hold liquid nitrogen?

12.29 What are the properties needed for steel ball bearings? How they are distinctive for steel race in which the bearings roll?

12.30 Among the following factors affecting the fatigue life of bearings, mention the parameter/s involved for each of them:

(a) Load
(b) Speed
(c) Contact geometry
(d) Seal design
(e) Lubricant
(f) Temperature
(g) Alignment.

12.31 In case of porous sintered metals, the bending strength is higher than the tensile strength. Give reasons.

12.32 How does the elastic modulus of porous sintered material vary with applied stress?

12.33 Why are oil impregnated sintered bearings not appropriate at higher sliding speeds and higher loads?

12.34 Substitution of a set of material by another many times gets necessitated. Give an account how would you proceed in following substitutions: lead for tin in soldering and coatings, coated steel for copper in cartridge casings and molybdenum for tungsten as an alloying addition in steels.

12.35 Which material is better to use for a mixer shaft, Ti-6 Al-4 V or 1020 steel? Explain.

12.36 You wish to make a new design for an Allen wrench. Specify the material and heat treatment.

12.37 The fatigue trend in automotive vehicles is toward hybrid vehicles, where diesel, fuel cells, electric and gasoline will be used in one form or the other. Presently about 6 kg of powder metallurgy parts are used per car. What you think shall be the potentiality of P/M parts in hybrid vehicles? List the names of various new parts and their possible processing through P/M route for such high power density vehicles.

12.38 In order to enhance the fatigue strength of Al-11 wt % Si castings for automobile pistons, copper is a valuable addition. Describe a road map of such a development from the point of view of following:

a. intermetallic formations and their role

b. wear resistance

c. creep-fatigue interaction.

Despite attractive strength to weight ratios, why both metal matrix composites and magnesium alloy pistons are still not in vogue?

12.39 The mild steel exhaust system in automobile shows about 8% of failure due to corrosion. For this ferritic stainless steels were used. However, it was noticed that the fatigue now became the predominant mode of failure. Is the problem because of material or something else?

12.40 A cylindrical steel pressure vessel with yield strength of 360 MPa is subjected to a maximum internal pressure of 1.5 MPa. The critical stress intensity factor of the steel is, $K_{IC} = 20$ MPa m$^{1/2}$. If the radius of the vessel is 1 m, estimate the minimum thickness of the vessel wall to avoid fast fracture. Assume, $\alpha = 0.72$ for semi-elliptical surface cracks parallel to cylinder axis. Suppose a residual tensile stress of 80 MPa is developed in the steel walls during the fabrication of the pressure vessel. What would be its effect? Consider the above pressure vessel be subjected to cyclic changes in internal pressure varying from 0 to 1.5 MPa. The vessel is routinely inspected to detect flaws by non-destructive techniques that can detect cracks larger than 0.1 mm. Estimate the design life of the vessel. Assume that the crack growth rate can be given by: $da/dN = 2.1 \times 10^{-7}\ (\Delta k)^{2.25}$.

12.41 Non-destructive test of the steel tie rod in an extrusion press revealed a surface crack of 4.5 mm long. The diameter of each tie rod (4 in numbers) is 300 mm and a total force of 1850 tons is exerted on the press. If the number of press cycle per month is approximately 9600 and the critical crack depth is about 60 mm, how long will it be possible to put off replacing the tie rod. The following alternatives are available to the plant engineer:

(a) 24 months, when the tie rod was due to be written off

(b) 14 months, when the next scheduled maintenance period is due

(c) 2.5 months, which is earliest a new rod of identical specification could be delivered

(d) 2 weeks, which is earliest a makeshift tie rod could be installed.

(e) Press operation is already unsafe and must be suspended until the tie rod is replaced.

Crack propagation during fatigue for the tie rod steel was experimentally established by relation

$$\frac{da}{dN} = 1.7 \times 10^{-1}\ (\Delta K)^{3.5}\ \text{mm/cycle}$$

where ΔK is measured in N.m$^{-3/2}$.

Given for $\lambda = 0.03$, $Y = 1, 88$
and for $\lambda = 0.4$, $Y = 1.01$; where$\lambda = a/R$
a = crack depth and R= radius of the tie rod.

12.42 An AISI grade 310S stainless steel muffle for heat treatment furnace is successfully used up to 18 months at 900°C in Monogas (93.5% N_2, 2.5 H_2, 4% CO) environment. Explain what alloy design would you envisage either for a longer service at 900°C or a still higher temperature say, 1100°C service.

12.43 In a boiler, the maximum metal temperature in the superheater and reheater section is limited because of the risk of tube failures caused by corrosion. A good solution is the use of metallurgically bonded composite tubes made of a creep resistant ferritic steel covered with a high temperature corrosion resistant austenitic steel. Describe the following:

(a) Production of such tubes by coextrusion
(b) Butt welding aspects of such tubes
(c) radial stress distribution across the total tube thickness and its nature.

12.44 Among the following operations of a 500 MW generator's rotor made out of 1% Cr MoV steel, mention in which case (i) fracture toughness is most vital and (ii) creep control is crucial
(a) Cold start with 30 min hold at 1500 rev/s, bored rotor
(b) cold start with pre-warming, bored rotor
(c) cold start with 30 min hold at 1500 rev/s, solid rotor
(d) warm start after 3 h shut down, bored rotor.
In which case the stress peak is minimum?

12.45 A typical steel used for rivets in brake discs of a fighter aircraft has a nominal composition of 0.35 C, 3% Cr, 0.4 Mo. A mechanical engineer in absence of such composition replaced the rivets by a steel of composition 0.3% C, 1% Si, 0.8-1.1% Mn, with the stipulation that the carbon level and strength of the latter was similar to those of former. However, in service, this substitution led to failures. Could you as a materials engineer comment on this?

12.46 For aircraft structures, light Be, magnesium, titanium and aluminium based alloys are candidate materials. Collect the data of their following properties and present them in a tabular form: (a) melting point, (b) fracture toughness, (c) coefficient of thermal expansion, (d) thermal conductivity, (e) machinability, (f) forming temperature and (g) corrosion resistance.
Justify, which one would be given top priority. In case economic factor is to be taken, will this priority change?

12.47 From creep fatigue perspective the requirements for fuselage material in the future supersonic aircrafts are
(i) stress 130-170 MPc
(ii) in service time 80,000 flight hours (20,000 cycles)
(iii) temperature 90-100°C with a maximum service temperature of 130°C
(iv) deformation <0.1% total plastic strain within 60000 h at 130°C and 150 MPa
(v) improvement in damage tolerance and fatigue with static properties 20% greater than for subsonic aircraft.
What composition developments in the existing alloys would be likely?

12.48 There is a general lack of high temperature aluminium alloys with good fracture toughness. Discuss the critical aspects which you shall keep in mind before developing such alloys.

12.49 The turbine blades in jet engines are based on Ni-base superalloys, which has room temperature proof stress of about 250 MPa. If each blade weighs 0.2 kg and has a cross sectional area of 10 cm^2, calculate the stress level generated in the blade. Assume a rotational speed of 10000 rpm and mean vane radius as 0.5 m. Are you confident that these blades would suffice for actual service? If not, what more data do you require?

12.50 A380 super jumbo jet of Airbus company has a capacity of 800 passengers. For this more and more uses of fibre reinforced polymers shall be in demand. The resins are cured at 177°C in mould, which must have matching coefficient of thermal expansion. What you suggest the composition of the metallic mould material?

12.51 For inner tank vessel of water heaters which material of the following would you recommend: (a) stainless steel, (b) copper, (b) galvanized steel.
Justify your answer under the following heads: (i) tensile strength, (ii) thermal conductivity, (iii) surface finish, (iv) corrosion resistance, (v) weight and recyclability.

12.52 A glass container after heating in an oven at 350°C is dipped in to boiling water. Assuming that the surface stress should not exceed the tensile stress (70 MPa) of the glass, select which of the following glasses (α values given) would safely work. Assume Young's modulus of glass as an average 70000 MPa: Fused silica (5.5×10^{-7}/°C), Borosilicate (33), Aluminosilicate (42), (iv) soda lime (92) and potash-lead (102).

12.53 In case you intend to replace the ceramic tiles in a space shuttle by a refractory metal system, what design changes would arise? Would active cooling be required? What might be some of the design challenges?

12.54 Figure P.12.54 shows a ceramic heat exchanger panel from a solar receiver, originally designed by Black and Veatch for EPRI. The use of ceramics is limited to the straight tubes directly exposed to the sunlight. The headers and U-tubes connected to the hot and cold legs are made of stainless steel. The outlet header is carrying 1040 °C air and is internally insulated. The most critical design feature is the joint between the ceramic tubes and the hot header. Discuss some of the important parameters as best as you can.

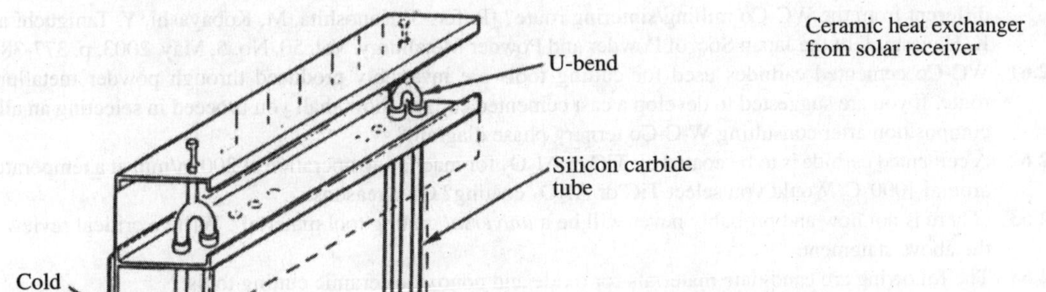

Figure P. 12.54

12.55 Discuss what types of compositions and microstructures you will develop for hard surface and a tough core of a Si_3N_4 ceramic to be used for wear applications.

12.56 In silicon nitride both active and passive oxidation takes place. During active oxidation, the oxidation products are immediately removed from the surface, causing weight loss, whereas during passive oxidation a weight gain takes place, because of oxide layers. Describe the factors affecting the active to passive oxidation transition.

12.57 TiB_2-graphite is a good electrode for aluminium extraction from Hall-Heroult process as compared to straight graphite. The requirement is the cell running at ≥10,000 amperes for an extended period. Describe the scheme for development of this composite and its evaluation.

12.58 Carbon refractory is common for iron blast furnace hearths. In order to prevent the degradation of the refractory, addition of titanium either as titaniferrous iron ore or ilmenite sand in the sinter charge is common. Explain the mechanisms of prevention both at pig iron and slag levels. [Hint: Refer CaO-TiO_2 phase diagram].

Tool Materials

12.59 What are the most important factors influencing the properties of hardmetal (cemented carbides)?

12.60 Japanese material scientists have researched on sintered cemented carbides (WC-Co) by direct milling of W, graphite and cobalt powders followed by sintering. They have claimed to produce highly oriented plate like WC grains. Could you analyze how the sequence of consolidation and microstructure evolution is different from the WC-Co milling/sintering route? [Refer: S. Konoshita, M. Kobayashi, Y. Taniguchi and K. Hayashi, J. of the Japan Soc. of Powder and Powder Metallurgy, Vol. 50, No. 5, May, 2003, p. 377-384].

12.61 WC-Co cemented carbides used for cutting tools are invariably produced through powder metallurgy route. If you are suggested to develop a cast cemented carbide, how shall you proceed in selecting an alloy composition after consulting W-C-Co ternary phase diagram?

12.62 A cemented carbide is to be coated by TiC or Al_2O_3 for machining operation at 200 m/min at a temperature around 1000°C. Would you select TiC or Al_2O_3 coating? Give reasons.

12.63 'There is not now and probably never will be a *universal* cutting tool material'. Write a critical review on the above statement.

12.64 The following are candidate materials for oxide and nonoxide ceramic cutting tools:
Tetragonal zirconia polycrystalline (TZP), PSZ (partially stabilized zirconia), Zirconia toughened alumina (ZTA), SiC and Si_3N_4. From the basic knowledge of bonding (Chapter 2), rank them from the viewpoint of application temperature. Also arrange them from the stand point of stress bearing capability during machining operation.

12.65 Describe the microstructural control, one would take into consideration for a successful ceramic cutting tool material.

12.66 In ceramic cutting tools, what considerations one has to take so as to achieve high fracture toughness and thermal shock resistance?

12.67 Why the final processing operation for manufacturing diamond saw consists of tensioning of the centre section?

12.68 If cobalt matrix used for diamond saw making is alloyed by two materials: (a) tungsten carbide and (b) 90 Cu-10 Sn bronze, what you think shall be trend of the bending strength with respect to the amount of alloying addition? Which additive shall have better effect on the strength of cobalt.

12.69 In case of diamond stone saw the cobalt matrix (hardness HV 229) is suggested to be substituted by Co-10 wt % Fe matrix (hardness HV 171). Is the hardness data enough to accept or reject the proposal? Why?

12.70 What are the effects of oxygen and sulphur in cobalt powder produced by hydrometallurgical process to be used for diamond saw making on the overall properties? Are their effects similar? Are their presence advantageous? (Refer: Co-S binary phase diagram).

Electrical Conductors

12.71 In a household appliance, what type of conducting wire do you use and why?

12.72 Anaconda Co. of U.S.A. under the Air Force contract developed a conductor 'Cufenic'. What are the advantages of this conductor?

12.73 The word 'bus' is used to denote a conductor, or a group of conductors, that serves as a common connection for two or more circuits. The sections are solid bars (round and rectangular) and tubes (round and square). Referring any Electrical Handbook, search to the application for the following
Rectangular cold drawn bars; Laminated bus structures; Tubes.

12.74 The induction-motor end rings are frequently made by forming from strip and butt joining the ends together by brazing or welding. For this purpose, what type of copper conductor alloy shall be selected and why?

12.75 A comparative picture of two copper-base conductor alloys is as follows:

Alloy	Tensile strength MPa	% elongation	Conductivity % IACS
Cu-Cd-Cr	340-490	20-7	85
Cu-Zr	280-460	20-4	90

Which alloy is suited for air-craft electrical equipments? Justify.

12.76 A general complaint in use of enamelled copper wire has been distortion in coils due to springiness in the winding wire. Suggest some reasons from the physical metallurgy viewpoint and the possible remedy.

12.77 Trolley wires for high speed electric trains are generally of CuCd0.7 composition. An increase in the train speed and environmental cost of cadmium alloying necessitates a substitution.Cu Mg 0.1 alloy was found to be of choice, since it has relatively good ductility of 3% (in contrast to 2% for CuCd) and tensile strength of 510 N/mm^2 (in contrast to 430 N/mm^2 of CuCd).With electrolytic tough pitch copper as the feed stock , 8.00 mm diameter conductor bars were produced by conventional method.After redrawing,the conductor wires failed mainly (90%) due to impurity inclusions.The inclusions were of two types : ferrous inclusions (50% of failure) and refractory inclusions (30% of failure).In order to avoid such problems,what decisions would you take for the following:

(a) Type of feed stock
(b) Type of heating for the melting furnace
(c) Type of refractory for the crucibles
(d) Factors to minimize iron pick up during working.[Hint:This emanates mainly from rod mill rolls and guide.]
(e) Type of casting process.Does it affect the item (d) above.?

[Data courtesy Rautomead International Ltd., Dundee,Scotland,U.K.]

12.78 What are the properties of copper which are crucial to its use as the host matrix in Nb-Ti alloy superconducting wire? Can we use some copper based alloy?

12.79 In electronic circuits, what types of metallic conductor useful?

12.80 For better applications, we wish to replace conventionally sintered Ag-SnO_2 (17%) composite used for electric contacts by one processed for obtaining nanocomposite by reactive milling Ag_2O and Ag_3Sn powder mixture. The sintered product revealed following problems: (a) ex-diffusion of silver from Ag-SnO_2 agglomerates and (b) a drastic swelling of sintered pellets. Could you suggest what may be the possible reasons? How will you minimize the problem?

12.81 In printed circuits copper tracks between components are used. However in integrated circuits we use aluminium tracks. Why?

12.82 What you can infer from Figure 12.46 from the material selection viewpoint? The materials are Ag, Pt, Cu-Pd, Ag-CdO.

12.83 What can be done against the tendency of sintered tungsten to weld in the arc formed during switching?

12.84 Calculate the maximum rate of internal heat generation per unit length in a silicon carbide heating element as a function of its radius. Assume that such heating elements are used to heat a laboratory furnace measuring $90 \times 90 \times 90\text{cm}^3$ with an inner hot zone of $60 \times 60 \times 60$ cm^3. Assume that the walls of the furnace are insulated with a fire-clay refractory. The furnace needs to be operated at 1000°C. How many of these silicon carbide heating elements of a given size are needed to maintain this temperature at steady-state operation? For a first estimate you can ignore any effects at the corners. Would you ever use only one or two elements? If not, why not? Why would a larger number of elements be preferred?

12.85 The resistance of a tungsten bulb is 144 ohms under normal working conditions. Find its temperature assuming

(i) cold resistance of the bulb at 20°C is 10 ohms, and
(ii) average temperature coefficient of tungsten wire at 20°C is 5×10^{-3} per °C.

Insulators Semiconductors and Magnetic Materials

12.86 A current of 1 μA flows through 1 meter of round copper wire conductor insulated by polymer. If the insulation has a thickness of 100 μm with an electrical resistivity of $10^{16}\Omega$m, what fraction of current is leaked if the over-voltage is 200 V?

12.87 PZT is synthesized from original oxide powder mixtures. PbO has high vapour pressure at elevated temperature and therefore is added a bit excess than required. Fine grain size microstructure gives rise to better quality piezoelectric devices and, therefore, fast sintering cycle was applied. It is found that this method is beneficial in case of specimens without excess PbO, but in case of excess PbO the end properties were inadequate. Give reasons why it is so?

12.88 The usefulness of a dielectric material as an electrical insulator or in a capacitor very often is limited by electrical conduction on its surface
(a) Qualitatively explain why surface conductivity might be greater than bulk conductivity.
(b) What environmental conditions will tend to increase the surface conductivity of a dielectric? Why?

12.89 Describe how the design of a practical piezoelectric device must take into account of materials properties such as strength, fatigue resistance, thermal properties, and of product features such as electrical and mechanical impedance.

12.90 PP polymer serves at temperature above its T_g. Why this does not affect the dielectric constant. What is the effect of ultraviolet radiation and heat?

12.91 How do high capacitance and high working voltage have conflicting design implications in an electronic circuit?

12.92 Using Vegard's law for the lattice constant of an alloy (i.e., lattice constant is the weighted average), find the bandgaps of alloys made in InAs, InP, GaAs, GaP which can be lattice matched to InP.

12.93 How one can exploit magnetostriction effect?

12.94 Refer Table 4.9 for the properties of Alnico, Ba ferrite, $SmCo_5$ and $Nd_2Fe_{14}B$ permanent magnets and comment on the suitability of each one for motor applications.

Electrical and Other Devices

12.95 Briefly describe the features in which a grown silicon dioxide layer in a semiconductor device differs from one deposited from a chemical vapour.

12.96 A transistor operates between 10 and 120 mV across the emitter. At the lower voltage, the collector current is 3 mA; at the higher voltage, it is 300 mA. Estimate the current when the voltage is 30 mV.

12.97 AlN ceramic is replacing BeO advantageously for heat-sink application in electronic devices since the latter is toxic. Y_2O_3 addition is done as a sintering aid; but still the sintering temperature is pretty high even for liquid phase sintering. It has been observed that molten Al_2O_3-Y_2O_3 phase has good wettability with AlN, which increases with increase in Al_2O_3 content. Referring Y_2O_3-Al_2O_3 phase diagram, suggest the composition of the additive mixture and the possible sintering temperature. Note, a relatively low sintering temperature is contributing in process economy.

12.98 Why ion implantation of the dopant atoms in a semi-conductor silicon is followed by an annealing treatment? Is there any special precaution at the annealing stage?

12.99 What is the primary difference in a junction transistor and a MOSFET?

12.100 Zinc sulphide (energy gap 3.7 eV) is used in thermistor, a device to detect any change in temperature. Such devices are important in microcalorimetric studies involving chemical or biological reactions. Determine to what fraction sensitivity, the semiconductor's resistivity be measured to detect 0.001°C change at 20°C.

12.101 Suggest a material suitable for the magnetic strip of electronic banking card.

12.102 Give at least three desirable properties, required for the building of microcircuits that include superconducting material.

12.103 Referring Figure P.12.103, suggest the pressure required at ambient temperature to obtain an explosively driven power supply from PZT 95/-2 Nb.

12.104 The absorption coefficient near the band-edges of GaAs and Si are $\sim 10^4$ cm^{-1} and 10^3 cm^{-1}, respectively. What is the minimum thickness of a sample in each case which can absorb 90% of the incident light?

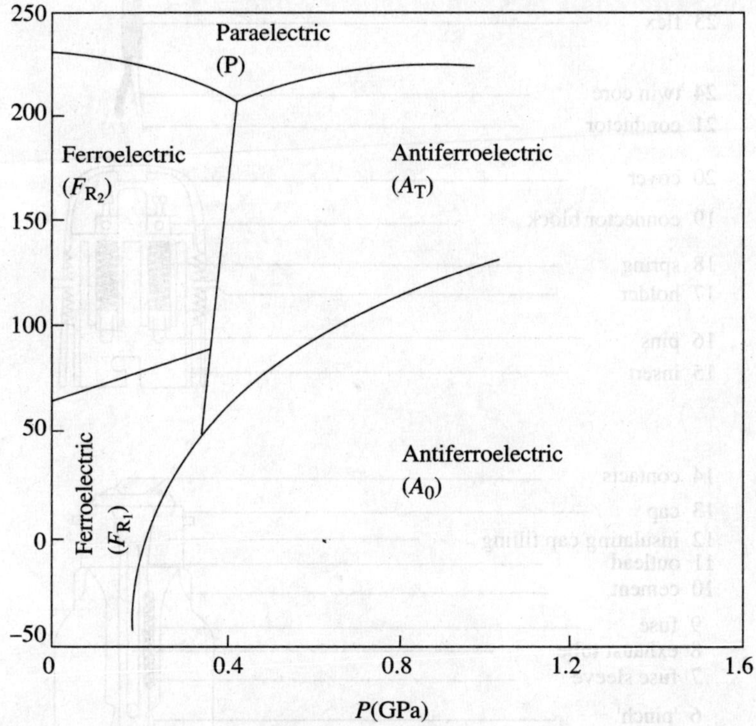

Figure P. 12.103

12.105 For long-haul optical communication, the optical transmission losses in a fibre dictate that the optical beam must have a wavelength of either 1.3 μm or 1.55 μm. Which alloy combinations lattice matched to InP have a band-gap corresponding to these wavelengths?

12.106 Could you think of replacing Kovar alloy and hard glass by say Fe-Ni alloys and solder glass (or plastic) without much loss in hermeticity and integrity? Discuss.

12.107 Discuss the role of magnetostriction in developing KOVAR alloy (55 Fe, 29 Ni, 17 Co) used for glass-to-metal seal in a piezometric electronic package.

12.108 Describe how with careful use of materials, the power supply device for a microprocessor can take the similar small size of an integrated circuit.

12.109 Why a coil made of shape memory alloy is better than the straight wire of same length from the shape recovery viewpoint?

12.110 What is the additional feature in a spring made out of shape memory alloy over a normal metallic spring?

12.111 Figure P.12.111 shows the details of the parts of an incandescent bulb. Describe the material for construction for each part along with the main manufacturing process. Also comment on material selection.

12.112 In development of SOFC the functionally graded cathodes right from YSZ to LSM to LSCo shall be ideal. Justify this statement.

12.113 Discuss the engineering problems one might face in mass production of fuel cells for automobile.

REFERENCES

Engineering Design, 3rd ed., G.E. Dieter, McGraw Hill Book Co., New York, 2000.
Elements of Engineering Design, M.S. Ray, Prentice Hall, N.J. 1985.

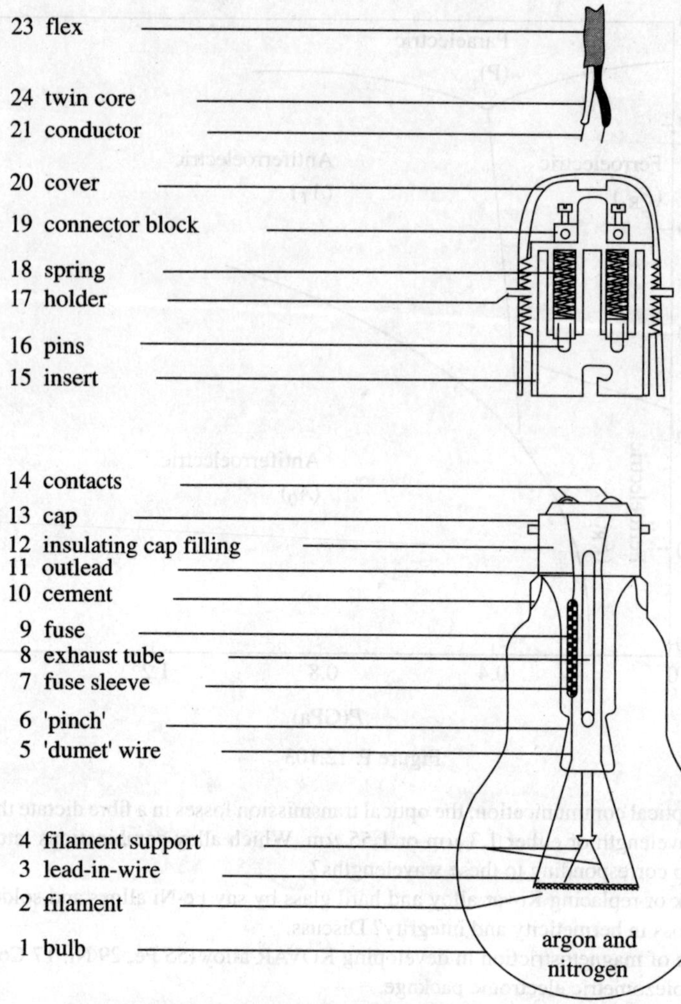

Figure P. 12.111

Engineering Materials: Selection and Value Analysis, H.J. Sharp, Heywood Books, London, 1960.
Selection of Engineering Materials, G. Lewis, Prentice Hall, N.J., 1990.
Handbook of Materials Selection for Engineering Applications, Ed. G.T. Murray, Marcel Dekker, Inc., New York, 1997.
Materials and the Design, E.H. Cornish, Cambridge University Press, Cambridge, 1987.
Steel Selection, R.F. Kern and M.E. Suess, John Wiley & Sons, New York, 1979.
An Introduction to Ceramic Engineering Design, D.E. Clark, D.G. Folz and T.D. McGee, American Ceramic Society, Westerville, Ohio, 2002.
Fatigue Strength of Welded Structures, 2nd ed., S.J. Maddox, Abington Publishing, Cambridge, 1991.
The Technology and Application of Engineering Materials, M.S. Ray, Prentice Hall Int., N.J., 1987.
Materials Consideration in Design, A.M. Ruskin, Prentice Hall, N.J., 1967.
Materials Selection in Mechanical Design, 2nd ed., M.F. Ashby, Butterworth-Hinemann, Oxford, 1999.
Materials in Electronics, C.E. Jowett, Business Books Ltd., London, 1971.

Materials Technology for Electrical Appliances, G.N. Dubinin and Yu S. Avraamov, Oxford Press, N. Delhi, 1985.
Design with Non-ductile Materials, W.E.C. Creyke, i.e.,J. Sainsbury and R. Morrell, Applied Science Publishers, London, 1982.
Plastic Component Design, P.D.Q. Campbell, Industrial Press Inc., New York, 1996.
Fatigue Design of Welded Joints and Components, A. Hobbacher, Abington Publishing, Cambridge, 1996.
Sintered Metallic and Ceramic Materials, G.S. Upadhyaya, John Wiley & Sons, Chichester, U.K., 1999.
Handbook of Electromagnetic Materials, P.S.Neelkanta, CRC Press, Boca Raton , Fl.,1995.
Handbook of Materials Selection, M.Kutz (ed.), John Wiley & Sons,Inc,Hoboken, NJ,2002.
Materials Selection for Process Plants, R.E.Gackenbach, Reinhold Publishing Corporation, New York,1960.
Handbook of Materials for Product Design,3rd ed. Ed. G.A.Harper, MacGraw Hill, New York, 2001.
Principles of Polymer Engineering, 2nd ed.,M.G.McCrum, C.P.Buckley, C.B.Bucknall, Oxford University Press, Oxford, 1997.
Materials for High Temperature Engineering Applications, G.W.Meetham and M.H.Van de Voorde, Springer, Berlin, 2000.
Semiconductor Materials, B.G.Yacobi, Kluwer Academic, New York, 2003.
Powder Metallurgy Diamond Tools, J. Konstanty, Elseviev, Amsterdam, 2005.

ANSWERS TO PROBLEMS

CHAPTER 2

2.3 Body centered orthorhombic
2.4 15.8 °
2.7 0.4049 nm
2.11 60.8°
2.12 No
2.18 −3.3%
2.19 $NiAl_3$
2.21 0.667; 2.26 g cm^{-3}
2.23 4518 kg m^{-3}
2.29 6 μm; decreases
2.30 46%
2.32 2.33×10^{-15}
2.33 226 K

CHAPTER 3

3.2 0.3509 nm; 8.93 g cm-3
3.9 Either 0.62% C or 1.96% C
3.10 2.6 w/o
3.17 7.4 h
3.18 ≈ 100 min.
3.19 400 hrs.
3.23 $\square/O^{2-} = 0.15/1.85$
3.43 40 hrs.
3.44 2490 ergs cm^{-2}
3.52 −18.5°C

CHAPTER 4

4.4 4.0×10^{-3} m^2 v^{-1} s^{-1}
4.5 0.0125 Ω
4.8 1×10^{-8} ohm.m
4.9 6.47×10^{-8} Ωm
4.13 4.93×10^{22} m^{-3}
4.17 (a) 0.23 m^2/v.s; (b) Any Group III element
4.20 4.8×10^{-6} m
4.21 2.29
4.22 (a) 33 V
(b) 113 V
(c) 6×10^{-12} F; 1.76×10^{-12} F
(d) 2.25×10^{-7} C/m^2
(e) 1.82×10^{-7} C/m^2
4.30 2.21
4.31 0.57
4.34 0.32 T
4.38 3.13 K
4.41 113.98 MPa, Compressive;
48.85 MPa, Tensile
4.44 >3.27 eV
4.45 5.9 K
4.47 15 at % In.

CHAPTER 5

5.3 9.692×10^2 kN m^{-2}
5.5 Glissile, 2.55 Å
5.6 1.64×10^7 J m^{-3}
5.22 0.27 kN
5.24 1 kN.m
5.25 2.8×10^9 N m^{-2}; 1.4×10^{-4} rad.
5.29 198 HV
5.30 57 HB
5.36 $K = 1.356 \times 10^{33}$;
$a = 11.35$; $N_f = 262.0 \times 10^9$ cycles
5.38 (a) & (b) does not fail;
but (c) is near the limit
5.39 55 h
5.41 17.7 h
5.49 6.6×10^8 N m^{-2}
5.50 {111}
5.54 ~11 μm
5.57 3.4 mm
5.63 4135 Mpa
5.70 (a) 95%; (b) 30 MPa
5.71 (a) Copper; (b) 540 N; (c) 130×10^3 MPa
5.77 −2.8 k bars
5.78 Volume fraction of precipitate 0.04;
Interparticle spacing = 23 nm; particle size = 0.7 nm

CHAPTER 6

6.1 1.564 V
6.2 $[H^+] = 4 \times 10^{-9}$ M;
$[OH^-] = 3.5 \times 10^{-6}$ M
6.5 (a) B; (b) A; (c) B; (d) B; (e) B
6.9 3.68 μm
6.11 −0.3 V
6.18 Yes
6.23 −600 kJ
6.25 Parabolic
6.31 $> 10^{-9}$ atm

CHAPTER 7

7.16 600°C
7.21 7.2 vol. % graphite
7.28 Cu 0.28; Pb 0.62
7.44 Cristobalite fraction 0.35;
Mullite fraction 0.65
7.48 2:1
7.49 0.38
7.50 37 kg
7.51 47 kg sulphur; 32% sulphur by weight
7.54 72.3% Polysterene;
27.3% Polybutadien
7.62 238 g/mole; 588 mers

CHAPTER 8

8.14 3 passes
8.27 64.66 kN;
37.5 mm
8.40 10.12 mm
8.44 0.59; 7.9%
8.45 −0.8

CHAPTER 9

9.5 (a) 0.625; (b) 34°; (c) 0.375 m/s
9.7 2728 A; 4.17 min.

CHAPTER 10

10.10 0.48
10.16 5
10.18 18 Ni equivalent; 35 Cr equivalent
10.19 3

CHAPTER 11

11.6 0.044 A/cm^2
11.7 4600 s
11.8 215 s

CHAPTER 12

12.7 3 mm
12.8 1.5 mm
12.9 R_c 41
12.26 Two times stiffness if tube diameter 50 mm
12.40 Minimum thickness 9.2 mm; 316 cycles
12.41 Can be safely used upto the scheduled maintenance
12.44 (i) a, (ii) d, (iii) c
12.52 Only used silica and borosilicate suit
12.76 Cu-Cd-Cr
12.85 2700°C
12.86 6.6×10^{-13} A
12.96 6.9 mA
12.100 0.00025
12.103 0.3 GPa
12.104 Ga As 2.3 μm; Si 23 μm.

INDEX